13th International Conference on Civil and Architecture Engineering (ICCAE-13)

IOP Conference Series: Materials Science and Engineering
Volume 974

Cairo, Egypt
7 - 9 April 2020

ISBN: 978-1-7138-3296-6
ISSN: 1757-8981

Printed from e-media with permission by:

Curran Associates, Inc.
57 Morehouse Lane
Red Hook, NY 12571

Some format issues inherent in the e-media version may also appear in this print version.

Institute of Physics
Dirac House, Temple Back
Bristol BS1 6BE UK

Phone: 44 1 17 929 7481
Fax: 44 1 17 920 0979

techtracking@iop.org

Additional copies of this publication are available from:

Curran Associates, Inc.
57 Morehouse Lane
Red Hook, NY 12571 USA
Phone: 845-758-0400
Fax: 845-758-2633
Email: curran@proceedings.com
Web: www.proceedings.com

TABLE OF CONTENTS

IOP Conf. Series: Materials Science and Engineering **974** (2020) 011001 doi:10.1088/1757-899X/974/1/011001

10th International Scientific Conference of The Military Technical College 7-9 July 2020

المؤتمر الدولي العلمي العاشر للكلية الفنية العسكرية
٧-٩ يوليو ٢٠٢٠

The 13th International Conference on Civil and Architecture Engineering

المؤتمر الدولي الثالث عشر للهندسة المدنية والمعمارية

ICCAE – 13

Conference Program and Abstracts

IOP Conf. Series: Materials Science and Engineering **974** (2020) 011001 doi:10.1088/1757-899X/974/1/011001

Contents

IOP Conf. Series: Materials Science and Engineering **974** (2020) 011001 doi:10.1088/1757-899X/974/1/011001

Preface

The Military Technical College is pleased to organize the Eight International Conference on Civil and Architectural Engineering. The conference is a chance to bring together all scientists and engineers from Egyptian armed forces and their counterparts in the universities, research center, academic and technical institutions from Egypt and other countries. It is our hope that the conference will provide a good opportunity to exchange knowledge, and to find appropriate solutions to present problems in Civil and Architectural Engineering fields. The conference has the aim to provide an environment of submitting and discussion of modern advances and experience on various Civil and Architectural Engineering activities that cover the following fields:

1- Concrete Structures
2- Geotechnical Engineering
3- Material Quality and Control
4- Metallic Structures
5- Project Management
6- Sanitary Engineering
7- Structural Analysis and Design
8- Structure Dynamics
9- Surveying and Photogrammetry
10- Transportation Engineering
11- Water Resources Engineering
12- Architectural Engineering
13- Building Technology
14- Environmental Engineering
15- Urban Planning

Forty-six manuscripts were submitted after receiving our call for papers. Professors from M.T.C., Egyptian universities and other research organization refereed these papers. From the Forty-six papers, only forty-one papers were accepted for presentation in the conference. These papers will be published in the conference proceeding after their discussion in the conference sessions.

The conference committee would like to acknowledge the valuable response received from all contributors. We are also grateful to the members of scientific committee for their great help in assessing all the received papers.

We hope that the conference is successful in solving some problems in the armed forces and different civil engineering sectors.

IOP Conf. Series: Materials Science and Engineering **974** (2020) 011001 doi:10.1088/1757-899X/974/1/011001

Conference High Committee

Maj.Gen Prof. Dr.Ismail Kamal	Chairman
Maj. Gen.Prof .Dr.Yehia Mehasseb	Vice- Chairman
Maj.Gen.R. Prof.Dr. Mohamed. T. Sallam	General rapporteur
Brig. Assoc.Prof.Dr. Mohamed Misbah Elkhatib	General rapporteur Assistant
Brig. Gen. Prof. Osama M. Moussa	Conference rapporteur

Conference Steering Committee

Maj. Gen. Dr. Ismael M. Kamal
Maj. Gen. R. Dr. Nabil H. Abdel-Metaal
Brig. Gen. R. Dr. Osama M. Moussa
Maj. Gen. R. Dr. Hassan M. Farag
Maj. Gen. R. Dr. Mohamed A. Barakat
Maj. Gen. R. Dr. Ehab H. Mahmoud
Brig. Gen. Assoc. Prof. Dr. Mohamed F. Abd El aleem
Brig. Gen. Assoc. Prof. Dr. Hossam EL Hanafy
Coll. Dr. Ashraf M. A. Osman
L. Coll. Dr. Shereif A. Ibrahim
L. Col. Dr. Mohamed S. Zahran

Conference Organizing Committee

Brig. Gen. R. Dr. Osama M. Moussa
Brig. Gen. Assoc. Prof. Dr. Mohamed F. Abd El aleem
Coll. Dr. Ashraf M. A. Osman
L. Coll. Dr. Shereif A. Ibrahim

IOP Conf. Series: Materials Science and Engineering **974** (2020) 011001 doi:10.1088/1757-899X/974/1/011001

Conference Scientific Committee

Name	Faculty
Prof. Abd El-Reheam Khalel	Ain Shams
Prof. Ahmed Abdelkhalea	Future Univ.
Prof. Adel H. Salem	Ain Shams
Prof. Ahmed A. A. Hassan	Ain Shams
Brig.Gen. Prof. Ahmed El-Desoky	Shobra
Maj. Dr. Ahmed M. Elshesheny	M.T.C.
Prof. Ashraf A. El-Ashaal	Irrigation Ministry
Prof. Ayman Hassan	Cairo Univ.
Assist. Prof. Beter H. Sobhy	Ain Shams
Prof. Ebrahim Abd EL-Rashed	Ain Shams
Maj. Gen. Prof. Ebrahim M. Goda	M.Accademy
Maj. Eng. Ebrahim M. Elway	M.T.C.
Prof. Ebrahim M. Mahdy	FUE
Maj. Gen. Prof. Ehab H. Mahmoud	M.T.C.
Col. Dr. Hany M. M. Abd El- Khalek	M.T.C.
Assoc. Prof. Hatem M. Fathy	EJU
Col. Dr. Hesham a. El-Kady	M.T.C.
Assistant Prof. Maged Gorge	CIC
Maj. Gen. Prof. Mohamed A. Barakat	M.T.C.
Prof. Mohamed A. El-Aghory	Ain Shams
Lt. Col. Dr. Mohamed A. El-Sayed	M.T.C.
Brig. Gen. Assoc. Prof. M. F. Abd El-Aleem	M.T.C.
Col. Dr. Mohamed M. Mahdy	M.T.C.
Prof. Mohamed O. R. El-Hariry	Shoubra
Assoc. Prof. Mona A. E. Hagras	Ain Shams
Prof. Morad A. Abd El-Mohsen	Ain Shams
Prof. Mostafa K. Zedan	Ain Shams
Maj. Gen. Dr. Nabil H. Abd El-Metaal	M.T.C.
Brig. Gen. Prof. Nabil M. Nagy	M.T.C.
Prof. Sherif S. Ezz El-Din	Alazhar
Assoc. Prof. Waael A. M. Sheta	Alazhar
Prof. Youssry B. Shahin	Monofia

IOP Conf. Series: Materials Science and Engineering **974** (2020) 011001 doi:10.1088/1757-899X/974/1/011001

SCIENTIFIC SESSIONS

IOP Conf. Series: Materials Science and Engineering **974** (2020) 011001 doi:10.1088/1757-899X/974/1/011001

Tuesday

7 July 2020

IOP Conf. Series: Materials Science and Engineering **974** (2020) 011001 doi:10.1088/1757-899X/974/1/011001

Military Technical College	***13th International Conf. on Civil & Arch. Eng.*** **7-9 July 2020**

Session No. 1	
Title	**Water Resources & Transportation Engineering (WR&TE)**
Date	**Tuesday 7-7-2020**
Time	**9:00 - 10:45**
Place	**Room B**
Session Chairmen	**Maj. Gen. R. Dr. Nabil H. Abd El Metaal** **Brig. Gen. R. Prof. Dr. Osama M. Moussa** **Col. Dr. Ashraf M. Osman**

Comparison of gridded datasets for the simulation of stream flow in Africa ***Mostafa Tarek, François P. Brissette and Richard Arsenault***	**WR3**
Using Remote Sensing Techniques to Determine New Route for the Nile Valley ***Osama M. Moussa, Ahmed S. Khalil, Hasan E. Elhefnawy and Mohamed M. Kamal***	**WR5**
Studying effects of asphalt modification by Nanosilica and Nanoclay on asphalt binder and hot mix asphalt properties ***Ahmed W. Oda, Ahmed El-Desouky, Hassan Mahdy, Osama M. Moussa***	**TE1**

Military Technical College
Kobry El-Kobbah,
Cairo, Egypt

13th International Conference on Civil and Architecture Engineering ICCAE-13-2020

WR-3

Comparison of gridded data sets for the simulation of stream flow in Africa

Mostafa Tarek[1,2], François P. Brissette[1] and Richard Arsenault[1]
[1]École de technologie supérieure, Université du Québec, 1100 Notre-Dame West, Montréal, Québec, Canada, H3C 1K3
[2]Department of Civil Engineering, Military Technical College, Egypt.
Correspondence to: Mostafa Tarek (mostafa_tarek45@ mtc.edu.eg)

Abstract: In recent decades, many parts of the African continent have experienced high precipitation variability with periodic drought and flood events. However, the network of streamflow gauges is too sparse in most countries to adequately capture these variations. In addition, no observed reference climatological dataset exists to adequately represent precipitation and temperature changes within all topographic and climatic zones. Consequently, the use of global gridded datasets needs to be considered.

This paper aims to use the different available gridded datasets as inputs to a hydrological model to evaluate dataset performance. Eleven precipitation and two temperature gridded datasets are used to this effect. The precipitation datasets include two gauged-only products, four satellite products corrected using ground-based observations, four reanalysis products and one merged product of gauge, satellite, and reanalysis. The two temperature datasets include one gauged- only and one reanalysis product. The ten precipitation and two temperature datasets were combined in their 20 possible arrangements for analysis purposes. Each combination was used to force the HMETS lumped hydrological model. The model parameters were calibrated individually for each combination against the streamflow records 850 African catchments. The Kling-Gupta Efficiency (KGE) was used to evaluate the semi performance. Results show that both temperature datasets performed equally well. were however observed between precipitation datasets. The MSWEP merged performing precipitation dataset, followed by two satellites (TRMM and CHIRPS) and one reanalysis (ERA-5) product. The performance of both gauged- only datasets (CPC and GPCC) was inferior, outlining the limitations of extrapolating information in data-sparse regions.

IOP Conf. Series: Materials Science and Engineering **974** (2020) 011001 doi:10.1088/1757-899X/974/1/011001

Military Technical College
Kobry El-Kobbah,
Cairo, Egypt

13th International Conference on Civil and Architecture Engineering ICCAE-13-2020

WR-5

Using Remote Sensing Techniques to Determine New Route for the Nile Valley

Osama M. Moussa[1], Ahmed S. Khalil[1], Hasan E. Elhefnawy[2] and Mohamed M. Kamal[2]

[1]Military Technical College

[2]Egyptian Armed Defense

Abstract: The collapse of dam structure leads to serious disasters for surrounding areas along the waterway. The objective of this research is to determine a new route for the waterway by using remote sensing techniques.

Drainage pattern is determined from Digital Elevation Model (DEM) of Shuttle Topographic Radar Mission (STRM) by using the ARC-GIS software. The most appropriate route for the new valley is determined by using DEM and assisted by topographic maps of scale 1:500000 starting from Toshka lake until Qattara downrange and passing through the depressions of El-Dakhla, El-Farafra and El-Baharia Oases.

The route selection is matched with the expected new valley that proposed by Prof.Dr. Farouk El-Baz in his project " The Development Project / Corridor of Development Project [1].

Keyword: DEM , ARC-GIS, Remote sensing, Drainage pattern.

IOP Conf. Series: Materials Science and Engineering **974** (2020) 011001 doi:10.1088/1757-899X/974/1/011001

Military Technical College
Kobry El-Kobbah,
Cairo, Egypt

13th International Conference on Civil and Architecture Engineering ICCAE-13-2020

TE-1

Effects of asphalt modification by Nanosilica and Nanoclay on asphalt binder and hot mix asphalt properties

Ahmed W. Oda[1], Ahmed El-Desouky [2], Hassan Mahdy[3], Osama M. Moussa[1]

[1] *Department of Civil Engineering, Military Technical College, Cairo, Egypt.*
[2] *Canadian International College, Cairo, Egypt.*
[3] *Ain Shams University, Cairo, Egypt.*

Abstract. In the Hot mix asphalt (HMA) is widely used in asphalt paving in most of countries. The goal of this paper is to study the effect of adding nanomaterials on the properties of asphalt binder and HMA. An experimental program has been conducted to study how to evaluate the performance of the modified asphalt binder and the modified HMA. Nanosilica (NS) and Nanoclay (NC) were used as modifiers. NS and NC were added at 3%, 5% and 7% by bitumen weight. Tests were performed to evaluate enhancement occurred in Penetration, softening point (SP), rotational viscosity (RV). Marshall stability tests, loss of stability and indirect tensile strength were performed on modified asphalt mixtures. Results were compared with control mix. Results shows significant enhancement in most of properties of asphalt binder and HMA. Penetration decreased by 24.07% at 3% NC and 20.37% at 5% NS. 3% NC and 5% NS were chosen to be optimum nanomaterial content (ONMC). Modified bitumens by 3% NC and 5% NS were used to produce modified asphalt mixtures. The modified mixtures showed a significant enhancement in stability and indirect tensile strength (IDT), but they, in contrast, had a harmful effect on the loss of stability (LOS).

Keywords: Modified Asphalt; Asphalt; Hot Mix Asphalt; Nanomaterial; Nanosilica; Nanoclay; Modified Bitumen; Loss of stability

IOP Conf. Series: Materials Science and Engineering **974** (2020) 011001 doi:10.1088/1757-899X/974/1/011001

Military Technical College	***13th International Conf. on Civil & Arch. Eng.*** **7-9 July 2020**

Session No. 2

Title	**Geotechnical Engineering (GE)**
Date	**Tuesday 7-7-2020**
Time	**11:00-12:45**
Place	**Room B**
Session Chairmen	**Brig. Gen. Prof. Dr. Nabil M. Nagy** **Dr. Ahmed M. El Shesheny** **Brig. Gen. R. Prof. Dr. Osama M. Moussa** **Col. Dr. Ashraf M. Osman**

Finite Element Analysis of Reinforced Sand Under Circular and Ring Footing ***A. Abd El Rahman, Nabil M. Nag and Ismail M. Kamal***	**GE2**
Displacement assessment of rock socketed shafts: a numerical approach ***A. M. H. Mahmoud and A.M. Samieh***	**GE3**
Geological, Geotechnical and Geophysical Aspects of Zafarana Wind Farms Sites and Their Expansion at Gabel El Zeit Sites Egypt ***H. Kamal, G. Saudi and Abdel Aziz Abd Aal***	**GE4**

Military Technical College
Kobry El-Kobbah,
Cairo, Egypt

13th International Conference
on Civil and Architecture
Engineering
ICCAE-13-2020

GE-2

Finite element analysis of reinforced sand under circular footing

A. Abd El Rahman[1], Nabil M. Nagy[2], Ismail M. Kamal[2], Mostafa A. Hassan[3]
[1] *Msc. Candidate, Military Technical College, Cairo, Egypt.*
[2] Professor, Military Technical College, Cairo, Egypt.
[3] PhD, Military Technical College, Cairo, Egypt.

Abstract: This paper presents the results from numerical investigations carried out on circular footing rests on Geo-grid reinforced sand. The finite element program (Plaxis-2D) was used to validate experimental results carried out by [1]. This experimental work studied the effect of variation of Geo–grid reinforced layers (N= 0,1,2, and 3) under circular footing. The validation with experimental work was considerably agreed with an average percentage of 93%. Then a parametric study has been conducted to evaluate the soil performance of circular footing, under the same loading conditions. Several parameters have been investigated to identify the best soil behaviour that achieves maximum bearing capacity and minimum settlement under the footing. The parameters include number of Geo-grid layers (N), the Geo-grid layer width (L), the spacing between the footing base and topmost Geo-grid layer (U), the spacing between Geo-grid layers (h) and the Geo-grid stiffness (K). The results indicated that by increasing the number and the width of the Geo-grid reinforced layers the bearing capacity ratio increases up to a limited extent. Moreover, the optimum number of Geo-grid layer equals four layers for circular footing.

Keywords: bearing capacity, footing shape, reinforced soil, Geo-grid layer.

IOP Conf. Series: Materials Science and Engineering **974** (2020) 011001 doi:10.1088/1757-899X/974/1/011001

MilitaryTechnicalCollege
Kobry El-Kobbah,
Cairo, Egypt

13th International Conference on Civil and Architecture Engineering ICCAE-13-2020

GE-3

Displacement assessment of rock socketed shafts: a numerical approach

A. M. H. Mahmoud[1] and A.M. Samieh[2]

[1]Hot Engineering & Construction Company, Free Trade Zone Plot E65, Kuwait.
[2]Department of Civil Engineering, Helwan University, Ibrahim Abd El-Razik Street, El-Naam, Mataria, Cairo, Egypt.
eng_asmaa_mah@yahoo.com, abcan98@yahoo.com

Abstract. Assessment of shallow/deep foundations settlement, under the working load range, has to be assessed to complete the design of any foundation system. The settlement of shafts resting on or socketed into sound rock is generally negligible. Settlement may be significant for shafts socketed into weak rock. Numerical approaches are among the promising tools that can be adopted to assess the settlement of piles socketed into rock. This study presents the results of a proposed numerical modeling scheme to predict the vertical displacement of socketed shafts. A comparison between the predicted and the monitored responses is presented to evaluate the viability of the proposed numerical scheme to assess the socketed shafts vertical displacement. A good comparison is exhibited between the results of the proposed model and those of the monitored test piles. In addition, the results of a sensitivity study are presented to highlight the influence of some numerical analysis controlling parameters such as the rock-shaft interface strength, the shear strength characteristics and the deformation modulus of the rock formation on the predicted numerical response.

IOP Conf. Series: Materials Science and Engineering **974** (2020) 011001 doi:10.1088/1757-899X/974/1/011001

Military Technical College
Kobry El-Kobbah,
Cairo, Egypt

13th International Conference on Civil and Architecture Engineering ICCAE-13-2020

GE-4

Geological, Geotechnical and Geophysical Aspects of Zafarana Wind Farms Sites and Their Expansion at Gabel El Zeit Sites Egypt

H Kamal[1], G Saudi[2] and Abdel Aziz Abd Aal[3]

[1] Soil Mechanics and Geotechnical Engineering Department, Housing and Building National Research Center (HBRC), Cairo, Egypt
[2] Steel and Structural Engineering Department, Housing and Building National Research Center (HBRC), Cairo, Egypt
[3] National Research Institute of Astronomy and Geophysics (NRIAG), Cairo, Egypt

Abstract. The growing need for sustainable power generation has led to increasing number of power plants, of fossil-fuel type and of non-conventional type. The foundation types of wind turbine towers in power plant structures are mainly depend on the geotechnical and structural considerations and also and other geological and geophysical considerations. The safety and economic conditions also play a vital role in the design of foundations. Such foundations must effectively be designed taken into consideration large overturning moments and dynamic loading due to extreme wind and earthquake. The present research is focused on studying the geotechnical, geological and geophysical aspects and characterizes of soils in Zafarana Wind Farms sites and their expansion at Gabel El Zeit sites on the Suez Gulf of Egypt. The study first address and study the geological aspects of both sites Zafarana Wind Farms sites and Gabel El Zeit sites. Second the geotechnical and geophysical characteristics are studied based on establishing data base from 698 borings and 8 seismic refraction profiles across Zafarana wind Farm sites and 10 borings and 2 seismic refraction profiles at Gabel El Zeit site. The geotechnical testing and geophysical profiles were compiled to determine the variation of the soil profile as well as the characteristics of the soil layers within the study sites. The results showed construction of 8 representative boring logs at Zafarana Wind Farm sites and 2 representative boring logs Gabel El Zeit sites to represent the soil the characteristics of the soil layers and site statgerphy. Also, the study shows the result of P wave velocity and shear wave velocity across both sites and within the studied soil layers up to 30 m. All these results from this research will help to study the foundation and the structure of wind turbine tower , where these types of tall structures are similar to chimney structures in design and other engineering considerations.

IOP Conf. Series: Materials Science and Engineering **974** (2020) 011001 doi:10.1088/1757-899X/974/1/011001

Military Technical College	***13th International Conf. on Civil & Arch. Eng.*** **7-9 July 2020**

Session No. 3	
Title	**Structural Analysis & Project Management (SA & PM I)**
Date	**Tuesday 7-7-2020**
Time	**13:00 -14:45**
Place	**Room B**
Session Chairmen	
	Maj. Gen. Dr. Hasan M. Farag
	Brig. Gen. R. Prof. Dr. Osama M. Moussa
	Col. Dr. Ashraf M. Osman

Numerical Modeling of New Connection Pontoon under Static and Dynamic Loads ***Ahmed S Elaresh, Mostafa M Abdel Wahab, Shreif A Mazek and Ashraf M Osman***	**SA1**
New Model to Calculate Blast Loading on Above and Underground Structures ***Hassan M. Farag***	**SA2**
Simulation and Optimization of a Metallic Ferry under MLC70Loading ***Mohamed N. Lotfy, Yasser A. Khalifa, Abdelrahim k. Dessouki and Elsayed Fathallah***	**SA3**
Multi-objective optimization of sandwich composite pressure hull for decreasing weight and drag force and increasing buckling load capacity ***Mohamed Helal and Elsayed Fathallah***	**SA4**

Military Technical College
Kobry El-Kobbah,
Cairo, Egypt

13th International Conference on Civil and Architecture Engineering
ICCAE-13-2020

SA-1

Numerical Modeling of New Connection Pontoon under Static and Dynamic Loads

Ahmed S Elareshy[1], Mostafa M Abdel Wahab[2], Shreif A Mazek[3], and Ashraf M Osman[4]

[1] M.Sc. Student, Department of Civil Engineering, Military Technical College, Egypt.
[2] Professor at Department of Civil Engineering, Military Technical College, Egypt.
[3] Professor at Department of Civil Engineering, Military Technical College, Egypt.
[4] Asst. Professor at Department of Civil Engineering, Military Technical College, Egypt.

Abstract. The development of floating bridges is a major engineering challenge to make the existing bridges durable and easy to construct. The connection pontoon unit between the heavy communication bridge (HCB) and the assault bridge (AB-PMM71) is discussed as developed by Mazek et al. [1]. The new connection pontoon (NCP) should offer fast assembly time to operate the combat floating bridge.

The floating bridge system including the HCB, the AB-PMM71, and the NCP is modeled. The 3-D numerical model of the floating bridge system is proposed by the nonlinear finite element analysis under static and dynamic Military load Class MLC70. The field test for the new floating bridge system was conducted by [1].

The proposed finite element model of the floating bridge system is verified based on the practical field test developed by [1]. The numerical results have good agreement with those obtained by the reviewed field test.

Keywords: Floating Bridges, Dynamic Analysis, Military Load Class 70 ton (MLC70).

IOP Conf. Series: Materials Science and Engineering **974** (2020) 011001 doi:10.1088/1757-899X/974/1/011001

Military Technical College
Kobry El-Kobbah,
Cairo, Egypt

13th International Conference on Civil and Architecture Engineering
ICCAE-13-2020

SA-2

New Model to Calculate Blast Loading on Above and Underground Structures

Hassan M. Farag (MTC)

Abstract: Blast loading is a type of extraordinary dynamic load to which structures or any object may be exposed to. The load resulted on structures or any targets from explosion is a complicated task to do. This paper presents an accurate and so simple model and program "BLAST" to predict the blast or shock wave resulted from the explosion that termed "Over Pressure" which comes from conventional and non-conventional weapons (such as Nuclear weapons). The pressure from the mass of air flow at and behind the shock front that termed the "Dynamic Pressure" is also predicted. The loads on any structure or target from the explosion are combination of overpressure Ps(t) and dynamic pressure Pd(t), this dynamic load on different faces of a structure either above or under the ground is predicted and presented in this paper. The over pressures resulted from different charges or weapon yields that are calculated in this study are compared to those obtained from "ConWep". The load on structures predicted by the present study is also compared to that obtained from "AUTODYN" software. A study is performed to investigate the effect of depth of burial on the load affecting the different faces of the buried structure.

KEYWORDS: Explosions, Nuclear, Overpressure, Buried structures, Dynamic loading.

**Military Technical College
Kobry El-Kobbah,
Cairo, Egypt**

**13th International Conference
on Civil and Architecture
Engineering
ICCAE-13-2020**

SA-3

Simulation and Optimization of a Metallic Ferry under MLC70Loading

Mohamed N. Lotfy[1], Yasser A. Khalifa[1], Abdelrahim k. Dessouki[2]and Elsayed Fathallah[1]

[1]Department of Civil Engineering, Military Technical College, Cairo, Egypt.
[2]Ain Shams University, Cairo, Egypt.

Abstract. Floating ferries are used for both civilian and military purposes. This study concerned with a ferry composed of sixteen connected floating pontoons. This ferry is simulated and optimized to carry Military Load Capacity MLC70 (Tank load).Consequently to the increasing demand of evolution and cost optimization, the design optimization is performed in this paper to obtain the optimum minimum weight which minimize both the cost and the buoyancy factor. The simulation of the ferry is performed using the finite element program ANSYS software. Furthermore, different grades of the structural steel and aluminium alloy are incorporated in this study. The simulation is verified with both practical and mathematical results. The performance of the ferry is investigated, In addition to the design parameters, constraints and objective functions are determined. The optimum weight of the ferry is obtained, followed by a reduction in the buoyancy factor; accordingly the capacity of the ferry can be increased. Comparison between the behaviour of the different ferries using different materials is operated considering stresses, deformations and weight. Conclusions and recommendations are then stated.

Keywords: Floating; Pontoon; Optimization; Ferry; Steel; Aluminium; Modelling; ANSYS

IOP Conf. Series: Materials Science and Engineering **974** (2020) 011001 doi:10.1088/1757-899X/974/1/011001

Military Technical College
Kobry El-Kobbah,
Cairo, Egypt

13th International Conference
on Civil and Architecture
Engineering
ICCAE-13-2020

SA-4

Finite Element Modelling and Multi-Objective of Sandwich Composite Hull for Decreasing (weight and Drag Force) and Increasing Buckling Load Capacity

Mohamed Helal[1,2] and Elsayed Fathallah[3]
[1,2]Taif University, Saudia Arabia and Mansoura University, Egypt
[3]M.T.C., Egypt

Abstract: In under water applications, space vehicles, and aircrafts the weight becomes important factor. Additionally, the design of composites structures greatly depends on number of layers and the fiber orientation angle. Therefore, this work presents the optimization of sandwich composite pressure hull in order to minimize (weight and drag force) and maximize buckling load capacity using ANSYS Parametric Design Language (APDL). Tsai-Wu and maximum stress failure criteria were incorporated for predicting the first-ply failure. The major and minor radius of the pressure hull, the ring and long beams dimensions, the fiber orientation angle and layer thickness are taken as design variables. The results illustrated that, core thickness (T_{core}) has a great effect to resist the shell buckling. While, it has a little effect on both Tsai-Wu and maximum stress failure index.

IOP Conf. Series: Materials Science and Engineering **974** (2020) 011001 doi:10.1088/1757-899X/974/1/011001

Military Technical College	*13th International Conf. on Civil & Arch. Eng.* **7-9 July 2020**
Session No. 4	
Title	**Structural Analysis & Project Management (SA & PM II)**
Date	**Tuesday 7-7-2020**
Time	**15:00 -16:45**
Place	**Room B**
Session Chairmen	**Maj. Gen. Dr. Hasan M. Farag** **Brig. Gen. R. Prof. Dr. Osama M. Moussa** **Col. Dr. Ashraf M. Osman**

Optimization of the Panels Used in Free-Form Buildings and Its Impact on Building Cost ***Sara M. Elkabany, Ahmed M. Elkordyand Hesham A. Sobh***	**SA5**
Optimization of Load-Bearing concrete Wall Using Genetic Algorithm To achieve Mechanically Integrated Behavior ***Sara M. Elkabany, Ahmed M. Elkordyand Hesham A. Sobh***	**SA6**
Deflection of Thin-Walled Panels Loaded in Shear with Different Types of End Stiffeners ***E. A. Mousa, M. H. El-Boghdadi and N. M. Yossef***	**SA7**
Elements of Safety Management System in the Construction Industry and Measuring Safety Performance ***Mostafa Elsebaei,Omar Elnawawy, Ayman Othman and Mohamed Badawy***	**PM1**

IOP Conf. Series: Materials Science and Engineering **974** (2020) 011001 doi:10.1088/1757-899X/974/1/011001

Military Technical College
Kobry El-Kobbah
Cairo, Egypt

13th International Conference on Civil and Architecture Engineering ICCAE-13-2020

SA-5

Optimization of the Panels Used in Free-Form Buildings and Its Impact on Building Cost

Sara M. Elkabany [1], Ahmed M. Elkordy [2] and Hesham A. Sobh[3]

1 Assistant lecturer at Architecture Department, Faculty of Engineering, Al-Azhar University, Cairo, Egypt
2 Professor & Head of Architecture Department, Faculty of Engineering, Al-Azhar University, Cairo, Egypt
3 Professor of Architecture & Vice Dean of Faculty of Engineering, Al-Azhar University, Cairo, Egypt

Sara.alkabany@azhar.edu.eg

Abstract: The outside surfaces of the free form buildings include panels with very complicated curvatures in their design. As the curvature of the panels becomes higher, the more costly the building costs, a free-form building designer needs to perform the procedure to minimize curvature of the panels at the early design stage while at the same time calculating the cost of structure panel optimization without compromising the features of its free-form facade. Nevertheless, in the early design stages, it is very difficult to estimate costs for design options and to find a good approach to the design intent and budget cap. Many designers lack panel optimization knowledge and understanding technology. This research proposes how Grasshopper can be used to optimize the panels for the free-form facade, a commonly used program which can be easily approached by a building designer and how to measure building costs in accordance with the process of optimization. The optimization panel is mainly done by linking the continuity analysis function Grasshopper to the Function of continuity analysis. The analysis of free-form ventures is conducted to check the validity of the results.

Keywords: Free-Forms; Cost; Optimization; Panel And Algorithm.

Military Technical College
Kobry El-Kobbah
Cairo, Egypt

13th International Conference on Civil and Architecture Engineering ICCAE-13-2020

SA-6

Optimization of Load-Bearing concrete Wall Using Genetic Algorithm To achieve Mechanically Integrated Behavior

Sara M. Elkabany[1], Ahmed M. Elkordy[2] and Hesham A. Sobh[3]
[1]Assistant lecturer at Architecture Department, Faculty of Engineering, Al-Azhar University, Cairo, Egypt
[2]Professor & Head of Architecture Department, Faculty of Engineering, Al-Azhar University, Cairo, Egypt
[3]Professor of Architecture & Vice Dean of Faculty of Engineering, Al-Azhar University, Cairo, Egypt

Sara.alkabany@azhar.edu.eg

Abstract: This paper proposes mechanical and growth characteristics of nature inspire shapes to be applied to load-bearing walls. In this study, this is performed by analyzing the case studies of O-14 Building, Victoria Gate and The Broad Museum. A model was created through the SolidThinking Inspire program using the morphogenesis tool. The model design is based on the algorithm for topology optimization and will be compared in terms of structural and material performance with the selected case studies. Optimizations results are found to be effective and designed with maximize stiffness in order to prevent deformation which is clearly reflected in the cost estimate in the initial design stages.

Keywords: Genetic Algorithm; Optimization; Load Bearing Walls and Free-forms.

Military Technical College
Kobry El-Kobbah,
Cairo, Egypt

13th International Conference on Civil and Architecture Engineering
ICCAE-13-2020

SA-7

Deflection of Thin-Walled Panels Loaded in Shear with Different Types of End Stiffeners

E. A. Mousa*, M. H. El-Boghdadi*, N. M. Yossef*1,

[*]*Department of Structural Engineering, Faculty of Engineering, Tanta University, Tanta, Egypt*
[*1]Corresponding author (Mobile: +201225151616; Fax: *+20403315860*; Email: *nashwa.yossef@gmail.com*, *nashwa_abdeltawab@f-eng.tanta.edu.eg.*

ABSTRACT
Numerical analysis for the load/deflection behavior of thin-web plate girders will be studied in this paper using a finite element program. An analytical approach presented in literature was discussed, and hence, it was used to calculate the load/deflection values in pre-bucking and post buckling stages. Conforming finite element load-deflection curves within the theoretical approach is the main goal of this study. The effect of different parameters on load-deflection curves is proposed. Stiffener thickness (t_s), end post type (no end post (NEP), with rigid end post (REP), and with non-rigid end post (NREP)) and end distance (e) are the main parameters considered in this paper. The analytical approach estimates well the load/deflection behaviors REP and NREP plate girders with ($a/h_w < 1.4$). On the other hand, the analytical approach can't estimate either for the shear strength or the deflection of NEP plate girders.

KEYWORDS
Plate girders; Shear behaviour; In-plan deflection; End stiffeners; Finite element mode

IOP Conf. Series: Materials Science and Engineering **974** (2020) 011001 doi:10.1088/1757-899X/974/1/011001

Military Technical College
Kobry El-Kobbah,
Cairo, Egypt

13th International Conference on Civil and Architecture Engineering ICCAE-13-2020

PM-1

Elements of Safety Management System in the Construction Industry and Measuring Safety Performance

Mostafa Elsebaei[1]*, Omar Elnawawy[2], Ayman Othman [3], Mohamed Badawy[4]

[1] Teaching Assistant at The British University in Egypt, Post Graduate Student at Ain Shams University; Mostafa.adel@bue.edu.eg
[2] Professor, Structural Engineering Department, Ain Shams University; nawawyomar@hotmail.com
[3] Professor, Head of Architecture Department, The British University in Egypt; Ayman.othman@bue.edu.eg
[4] Ph.D., Structural Engineering Department, Ain; Mohamed.badawy@eng.asu.edu.eg

* Correspondence: Mostafa.adel@bue.edu.eg

Abstract. Construction industry is considered to be one of the most hazardous industries in the world. The reason could be attributed to its hazardous nature as it is an accident-prone industry. Thus, a need for better understanding of safety management system is essential for improving safety performance in this sector. This paper discusses briefly the elements of safety management by presenting different systems (such as Oregon OSHA Occupational Health and Safety Administration, and OTAR Overseas Territories Aviation Circle) and elaborating their elements. It also discusses two types of measuring safety performance the first is the lagging indicators and the second is the leading indicator. In addition, a field study was conducted to explore contractors' perception on safety management. A questionnaire was distributed to construction firms. 200 responses were collected and analyzed. All of the results showed positive answers which indicate that safety in performance in Egypt is slightly above average as all means were close to average.

IOP Conf. Series: Materials Science and Engineering **974** (2020) 011001 doi:10.1088/1757-899X/974/1/011001

Wednesday
8 July 2020

IOP Conf. Series: Materials Science and Engineering **974** (2020) 011001 doi:10.1088/1757-899X/974/1/011001

Military Technical College	***13th International Conf. on Civil & Arch. Eng.*** **7-9 July 2020**

Session No. 5	
Title	**Architecture Engineering (AE)**
Date	**Wednesday 8-7-2020**
Time	**9:00 -10:45**
Place	**Room B**
Session Chairmen	
	Maj. Gen. Prof. Dr. Mohamed A. Barakat
	Brig. Gen. Assoc. Prof. Mohamed F. Abd El Aleem
	Col. Dr. Mohamed M. Mahdy

Cairo Airport Between Origination And Future Vision ***Mostafa Mahmoud Saber , Alaa El Din Abdel Rahman Ibrahim and Salah Zaki Saeid***	**AE1**
The Reflective Impact of Different Teaching Strategies in Teaching Architectureand Interior Design Practical Courses ***Marwa Abdelalim, Noura Aji, Sherif Mahmoud and Mohammad Fahmy***	**AE2**
Appropriate Housing Projects Management of a Random Sample in the Greater Cairo Region ***Maged Moneer Gad, Ahmed Galal El Din Aly***	**AE3**
The history of monasteries in Egypt as self-sustained settlements ***Gena Romel, L Sherif, and S Ashour***	**AE4**
Flexible Design: An Innovative Approach for Achieving Sustainability in Primary Public Schools in Egypt ***Yara A Fahmy and Ayman A E Othman***	**AE5**

IOP Conf. Series: Materials Science and Engineering **974** (2020) 011001 doi:10.1088/1757-899X/974/1/011001

Military Technical College
Kobry El-Kobbah,
Cairo, Egypt

13th International Conference on Civil and Architecture Engineering ICCAE-13-2020

AE-1

Cairo Airport Between Origination And Future Vision

Mostafa Mahmoud Saber[1] , Alaa El Din Abdel Rahman Ibrahim[2] and Salah Zaki Saeid[2]

[1]Department of Architecture - College of Engineering - Al-Azhar University
[2]Prof. Department of Architecture, Faculty of Engineering, Al-Azhar University

mostafa2ms@yahoo.com

Abstract: Airports are the latest means of transportation that reflect the progress of human civilization and reflect the distinctive identity of each country. The passenger building is the main component of the airport, which in turn contains many dynamic and constantly evolving facilities depending on the increase in passenger traffic and the continued entry of new aircraft and modern operating and service systems that are constantly heading to try to facilitate and accelerate the travel procedures, as the development of airports in recent years is a remarkable development that promotes the level of booms as a result of increased air traffic in the world. The airports have become one of the most important international projects. In this research, Cairo Airport has been chosen because it is considered the largest airport used in Egypt, and the interface that arrives from the outside world to Cairo. It accommodates all foreign air movements for all foreign airlines operating in Egypt. In this research, a detailed explanation about Cairo Airport will be provided by studying its components from the ground side, passenger buildings and the air side, and studying the area of Cairo Airport and how to benefit from unexploited places at the airport by studying the elements of the airport city and how to benefit from the application of the components of the airport city at Cairo International Airport.

Keywords: Airport - Airport City – landside – Terminal Building - Airside.

IOP Conf. Series: Materials Science and Engineering **974** (2020) 011001 doi:10.1088/1757-899X/974/1/011001

Military Technical College
Kobry El-Kobbah,
Cairo, Egypt

13th International Conference on Civil and Architecture Engineering
ICCAE-13-2020

AE-2

The Reflective Impact of Different Teaching Strategies in Teaching Architecture and Interior Design Practical Courses

Marwa Abdelalim[1], Noura Aji[2] and Mohammad Fahmy[3]

[1]Assistant Professor, Prince Sultan University, Architectural Engineering Department, Riyadh, KSA
[2]Lecturer, Prince Sultan University, Interior Design Department, Riyadh, KSA
naji@psu.edu.sa
[3]Associate Professor, Military Technical College, Head of Civil & Architecture Engineering Branch, Cairo, EGYPT

Corresponding email address: mabdelalim@psu.edu.sa

Abstract. Achieving learning outcomes in a creative manner, actually considers a vision for all instructors whether theoretical courses or practical courses, in a reflective way where such reflection interfere to boost students' self-confidence while enhancing their knowledge. The research investigates some effective teaching strategies that attempt to bridge the gap between theories and real application trying to close the loop of getting both interior/architectural design projects, and construction technology integrated.

The research adopts an ontological assumption of applying questionnaire and surveys to recognize this silent war, is it in the students' mind as just fear of *MAKE NO MISTAKE* or fear from the future career. According to *Tony Robbins. He said if you want a better result then ask yourself a better question.* The main research problem is thinking about the rational beyond the reflection of effective teaching strategies as well as solving the problems that fear junior architects and interior designers, to know exactly where is the battlefield? A whole mechanism of some teaching and learning strategies as well as direct assessment tools, in accordance with well-defined rubrics were applied.

The researcher presents a considerable hypothesis that assures that believing in the architecture and interior design students' abilities and the instructor's variable teaching strategies can bring out examples of best practices in teaching philosophies.

Keywords: Flipped classroom strategy, One to one basis, circle talk, assessment tools

IOP Conf. Series: Materials Science and Engineering **974** (2020) 011001 doi:10.1088/1757-899X/974/1/011001

Military Technical College
Kobry El-Kobbah,
Cairo, Egypt

13th International Conference on Civil and Architecture Engineering ICCAE-13-2020

AE-3

Appropriate Housing Projects Management of a Random Sample in the Greater Cairo Region

[1]Maged Moneer Gad and [2]Ahmed Galal El Din Aly

[1] Doctor Engineer, Architecture Department in Obour high Institute for Engineering and Technology, PhD (Wuppertal University, Germany).

[2]Assistant Professor Civil Engineering Department, Modern University for Technology and Informations, Cairo, Egypt.
Corresponding Author: Dr.Eng./ Maged Moneer Gad, gadmaged@yahoo.de

Abstract: The housing industry is an essential part of the urban development that requires full cooperation of different parties including public sector institutions, individuals, and private sectors institutions and groups. This is to achieve economic, social, and urban developments. Everyone's right of having a decent home is a basic demand makes it an essential requirement to offers different appropriate housing solutions. After the industrial revolution and the development of mechanization with the technology development, also the economy and social changes, the human concepts through the years changed and evolved the housing patterns, methods forms and uses, as the human needs concepts and cultures changed, from the need of only a shelter above the head to protect from weather into home that provides not only protection but also physiological and psychological comfort, recreational means to practice indoor various activities, all that resulted into the development of houses from two or three stories to include a family into the current form of residential multistory towers that contain large number of residents families.
This research paper is based on the multistory buildings residents' opinion, their point of view, their requirements for designing a suitable residential house that consider natural elements that affect the urban planning and the shape of the city, therefore a questionnaire was made to collect the opinion for some samples Of the population. Accordingly, the researcher used the descriptive analytical method by designing a survey questionnaire to evaluate and collect information about a suitable housing for the residents and to derive (the requirements) for a comfortable housing that provides the necessary needs for the population and that was distributed to a sample of (500) individuals. Income, social, and educational levels were taken into consideration, and valid survey lists were emptied and analyzed.
The study ended to several results and recommendations, the most important of which are:1) The residential building should be residential only without the presence of other usage in it; 2) The residential buildings' height should not exceed 5 floors; 3) The residential building should not contain more than 10 apartment units; and 4) There should be some special services for residential buildings, as a suitable number of parking spaces, as well as the elevator in the buildings and a terrace

Key words: residents, residential needs, populations, multistory buildings, residential buildings, office buildings, commercial buildings.

IOP Conf. Series: Materials Science and Engineering **974** (2020) 011001 doi:10.1088/1757-899X/974/1/011001

MilitaryTechnicalCollege
Kobry El-Kobbah,
Cairo, Egypt

13th International Conference
on Civil and Architecture
Engineering
ICCAE-13-2020

AE-4

The history of monasteries in Egypt as self-sustained settlements

G Romel, L Sherif, and S Ashour

Department of Architecture, Faculty of Engineering, Arab Academy for Science, Technology & Maritime Transport, Cairo, Egypt

gina_romel@aast.edu

Lobna.sherif@aast.edu

shaimaa.ashour@aast.edu

Abstract. Monasticism initiated in Egypt and has spread worldwide. It introduced a new self-sustained architectural innovation called “monasteries”. Monasteries can be considered as homogeneous self-sustained settlements through history. This paper aims to document the origins of monasticism, leading to the development of self-sustained monasteries: conceptually, economically, and architecturally. It relies on observations, maps, and historical references. The main challenges in collecting data for this paper were the lack of published references about origins of monasticism. Most of the resources are rare documents stored in the Coptic Clerical and Theological College of Cairo, and libraries of monasteries. Thus, the paper analysis present monasteries as self-sufficient, self-sustained settlements that survived through history since the fourth century to the current state through: (zoning, architectural elements, building materials, and building techniques). The architectural analysis shows that the architectural solutions of monasteries evolved in respect to the geographical location, and context. Monasteries architecture relies on trials and errors, until it reached an applicable architectural model.

Keywords: History of Monasteries, Architecture, self-sustained settlements, Egypt.

IOP Conf. Series: Materials Science and Engineering **974** (2020) 011001 doi:10.1088/1757-899X/974/1/011001

MilitaryTechnicalCollege
Kobry El-Kobbah,
Cairo, Egypt

13th International Conference on Civil and Architecture Engineering ICCAE-13-2020

AE-5

Flexible Design: An Innovative Approach for Achieving Sustainability in Primary Public Schools in Egypt

Yara A Fahmy [1] And Ayman A E Othman [2]
[1]Senior Student, Architecture Engineering Department, Faculty of Engineering, The British University in Egypt, Cairo-Suez Desert Road, El-Shorouk, Cairo, Egypt.
[2]Professor of construction and project management, head of architectural engineering department, faculty of engineering, The British University in Egypt, Cairo-Suez Desert Road, El- Shorouk, Cairo, Egypt.

Yara141478@bue.edu.eg

Abstract: Public schools in Egypt are considered a valuable asset that should be preserved and sustained, as they play a paramount role for the development of a sustainable community. Whereas, currently, the schools in Egypt are not sustainable from social, economic and environmental perspective as they are generally heavy, fixed and normally irreversible once construction has been completed. This is due to the traditional construction method used in the construction of schools which consumes time and generates a vast amount of waste. In addition, the changing demands of the occupants may confront the need for future expansion or complete changeover. With the increase in labour costs, lack of skilled labours for different trades, demand for faster construction, increased considerations for safety and environment and above all meeting these demands at an economic and reasonable price requires us to change the approach used in construction. Hence, these faults could be sufficed with the introduction of green prefabrication methods as a flexible design approach to enhance sustainability for primary public schools in Egypt. Therefore, this paper aims to investigate the role of flexible design for achieving sustainability in primary public schools in Egypt through Green Prefabrication. A research methodology consists of literature review to investigate the topics of primary public schools, traditional construction, flexible design, Green prefabrication and sustainability.

Keywords: Green prefabrication, Sustainability; Primary Public Schools; Egypt; Flexible Design; Matrix

IOP Conf. Series: Materials Science and Engineering **974** (2020) 011001 doi:10.1088/1757-899X/974/1/011001

Military Technical College	***13th International Conf. on Civil & Arch. Eng.*** **7-9 July 2020**

Session No. 6	
Title	**Building Technology (BT)**
Date	**Wednesday 8-7-2020**
Time	**11:00-12:45**
Place	**Room B**
Session Chairmen	
	Maj. Gen. Prof. Dr. Mohamed A. Barakat **Brig. Gen. Assoc. Prof. Mohamed F. Abd El Aleem** **Col. Dr. Mohamed M. Mahdy**

Paths Towards Energy Efficient Building Using Nano Architecture ***Nader Gharib and Basant Ahmed***	**BT1**
PV Grid-Connected Systems: Performance and Architectural Integration Aspects ***Hala Alaa and Nader Gharib***	**BT2**
Comparison of Egyptian Innovation in Affordable Housing with Global Models ***Hebatalrahman Ahmed***	**BT3**

Military Technical College
Kobry El-Kobbah
Cairo, Egypt

13th International Conference on Civil and Architecture Engineering
ICCAE-13-2020

BT-1

Paths Towards Energy Efficient Building Using Nano Architecture

Nader Ghariband Basant Ahmed[1, 2,*]

[1] Head of Architectural Engineering and Environmental Design Department, Engineering & Technology College, Arab Academy of Science, Technology and Maritime Transport, Port Said 42511, Egypt.
[2]Architectural Engineering and Environmental Design Department, Engineering & Technology College, Arab Academy of Science, Technology and Maritime Transport, Alexandria 21913 P.O. Box 1029, Egypt.

*Arch_Nader@gmail.com

Abstract. Nanotechnology has been in a massive use in the last two decades in many fields specially the architecture and construction field. Applying nano technology in architecture is taking place and has become the biggest challenge in our time. The challenge of nano architecture is to improve the material properties, adding new properties or production new materials which used in architecture design of the building or in construction phase. The study aims to reduce the energy usage of the building by using nano architecture and to achieve one of the environmental sustainability factors.The research adopts the descriptive theoretical approach of nano technology and nano material in field of architecture and its impact on the energy use of the building. This followed by the second analytical part which examines the nano materials and nano devices through examples of global sustainable projects and nano architecture applications which leads us to a huge difference in the architecture design and reduces the total energy usage of the building. In conclusion, the research shows the effect of using nano material and devices on the building energy use which can help to save the environment by reducing the energy consumption.

Keywords: Nano architecture – Nano material – energy efficient building – environmental sustainability

IOP Conf. Series: Materials Science and Engineering **974** (2020) 011001 doi:10.1088/1757-899X/974/1/011001

Military Technical College
Kobry El-Kobbah,
Cairo, Egypt

13th International Conference on Civil and Architecture Engineering ICCAE-13-2020

BT-2

PV Grid-Connected Systems: Performance and Architectural Integration Aspects

Hala Alaa[1,*] and Nader Gharib[1,2]

[1] Architectural Engineering and Environmental Design Department, Engineering & Technology College, Arab Academy of Science, Technology and Maritime Transport, Alexandria 21913 P.O. Box 1029, Egypt.

[2] Head of Architectural Engineering and Environmental Design Department, Engineering & Technology College, Arab Academy of Science, Technology and Maritime Transport, Port Said 42511, Egypt.

* eng.hala.alaa1@gmail.com

Abstract. Considering the high global rates of carbon dioxide emissions, there is a great concern about renewable energy especially solar energy in buildings. Despite its growth, there has been limited acceptance of architects and stakeholders concerning the use of solar energy technologies in buildings, and subsequently, insufficient architectural integration. Here, the research aims to develop better understanding of architects for the design stages, considerations, and possibilities regarding solar PV grid-connected systems in buildings using inductive methodology, thus analysing the technology specifications and good integration practices. The research determines the factors that influence the early design and conception phase of the building. It develops a framework for architects concerning the technical and physical aspects affecting the design process, through building needs, cell types, system performance, site conditions, space and environmental requirements, and BIPV integration possibilities. In conclusion, this study will help in the implementation of PV grid-connected powered buildings in new projects by addressing previous design challenges and providing data for architects and researchers on the system design and performance.

IOP Conf. Series: Materials Science and Engineering **974** (2020) 011001 doi:10.1088/1757-899X/974/1/011001

Military Technical College
Kobry El-Kobbah
Cairo, Egypt

13th International Conference on Civil and Architecture Engineering ICCAE-13-2020

BT-3
Comparison of Egyptian Innovation in Affordable Housing with Global Models

Hebatalrahman Ahmed
Consultant in Materials Science and Applications
hebatalrahman11@yahoo.com

Abstract: Affordable housing is a set of building models which is considered as type of low-income housing primarily by providing residential products that are within their reach and difficult living conditions. The models of affordable housing are often linked to rapid and constructive housing in the face of disasters.

In these work Survey about different types of affordable housing was done with the advantages and disadvantages of each type. The combtability of each type with sustainable in construction according to national and international standards was considered. The risk assessment related to this type of construction was evaluated.

The Egyptian innovation in the affordable housing which is registered as patent in the Egyptian patent office is studied. The new building has one floor with spherical external shape for energy conservation, it is consisting of multilayers movable walls moving on the streams. The metallic structure is installed in the soil surface by fixation elements, it is equipped with sensors which is controlled automatically to detect climate changes.

The movement of successive layers can be controlled manual or automatic according to climate changes. The movement of insulated layers is done by sliding on dedicated channels, it is equipped with upper and lower ducts located on several axes positioned at the top and bottom of the ball so that the upper and lower positions are located on one axis. The unit has conical top contain the layers folded to the top and prevent the accumulation of dust and water. The layers can be stored inside the conical top. The Egyptian patent model is compared with the common global models of affordable housing to explain its advantages and disadvantages.

Keywords: Affordable Housing, Risk, Sustainable Building, Innovation.

IOP Conf. Series: Materials Science and Engineering **974** (2020) 011001 doi:10.1088/1757-899X/974/1/011001

Military Technical College	***13th International Conf. on Civil & Arch. Eng.*** **7-9 July 2020**

Session No. 7	
Title	**Environmental Engineering (EE I)**
Date	**Wednesday 8-7-2020**
Time	**13:00-14:45**
Place	**Room B**
Session Chairmen	
	Maj. Gen. Prof. Dr. Ehab H. Mahmoud **Brig. Gen. Assoc. Prof. Mohamed F. Abd El Aleem** **Col. Dr. Hesham A. El Kady**

Design out waste as an approach for achieving sustainability in Egyptian housing projects ***Laila A. Elsawaf and Ayman A. E. Othman***	**EE1**
Passive Solar Technique Using Trombe Wall: Simulation of Energy Performance in Hot Climates ***Ayah Mohamed Ramadan.***	**EE2**
The philosophy of visual perception of the external space with a vision of the biophilic design that achieves quality of life ***Enjy Ibrahim Youssef, Khaled Mustafa Khorshed, Sherif Sabri***	**EE3**
Environmental Assessment of Coastal Resorts: Recreational Spaces as a Case Study ***Mohamed Sobhi Hassan, Wael Ahmed Sheta , Ahmed Mohamed El Kordy***	**EE4**
Performative Design Optimization for Shading Screens ***Ahmed Mohamed ELkordy, Hesham Ahmed Sobh and Mohga Ali Abdelslam Youssef***	**EE5**

Military Technical College
Kobry El-Kobbah,
Cairo, Egypt

13th International Conference
on Civil and Architecture
Engineering
ICCAE-13-2020

EE-1

Design out waste as an approach for achieving sustainability in Egyptian housing projects

Laila A. Elsawaf and Ayman A. E. Othman

Senior Student, Architectural Engineering Department, Faculty of Engineering, The British University in Egypt, Cairo-Suez Desert Road, El-Shorouk, Cairo, Egypt. E- mail[1]: Laila141068@bue.edu.eg

Professor of Construction and Project Management, Head of Architectural Engineering Department,
Faculty of Engineering, The British University in Egypt, Cairo-Suez Desert Road, El- Shorouk, Cairo, Egypt. E-mail[2]: ayman.othman@bue.edu.eg

Abstract: Housing projects have been increasing rapidly during the previous years to fulfil community needs and achieve the national and international sustainable development objectives. Due to their nature, housing projects generate a significant amount of waste that accounts for approximately 40% of the solid waste that the construction industry generates annually. Many of these wastes are a result of inappropriate decisions adopted during the design process such as using non-sustainable materials and inefficient use of natural recourses. Meanwhile, research has revealed that achieving sustainability within housing projects is a crucial challenge, yet an essential one. This called to think differently and adopting creative strategies for reducing construction waste during the design process. Therefore, this research aims to develop matrix that correlates the aspects of sustainability with design out waste strategy as an approach of achieving sustainability in housing projects. A research methodology consists of literature review and case study was designed to accomplish three objectives. Firstly, literature review was used to investigate the topics of sustainability, design out waste strategy and housing projects. Secondly, a case study was presented and analysed to identifies the design out waste strategy for reducing waste as an approach to achieve sustainability. Finally, developing a matrix that correlates sustainability and design out waste strategy.

Keywords: Waste Reduction, Design out waste strategy, Sustainability, housing projects, Waste.

IOP Conf. Series: Materials Science and Engineering **974** (2020) 011001 doi:10.1088/1757-899X/974/1/011001

Military Technical College
Kobry El-Kobbah,
Cairo, Egypt

13th International Conference on Civil and Architecture Engineering ICCAE-13-2020

EE-2

Passive Solar Technique Using Trombe Wall: Simulation of Energy Performance in Hot Climates

Ayah Mohamed Ramadan[1].

[1]Lecturer at Faculty of Engineering, Modern Academy for Engineering and Technology, Cairo, Egypt.
ayahezzat54@gmail.com

Abstract. The research aimed to evaluate a Trombe wall, a solar wall, is an important component of passive solar strategies. It is a very compact wall and can be south-facing or north-facing, which is painted black and made of exclusive material that absorbs a lot of heat. The main advantage of Trombe wall is that they are often easily built from locally obtainable materials, very dependable and having less repair, maintenance and operation costs. The analysis of gathered data concludes that Trombe wall can provide great advantage to indoor environment thermal comfort and enhancing energy performance compared to single skin facades. The research has shown that Trombe wall has possibility of providing acceptable internal thermal comfort through "Mixed-Mode" ventilation strategy in hot climate, this is important in determining the possibility of Trombe wall in incorporating "Mixed-Mode" ventilation reducing energy usage in cooling demands and is proven by applying a simulation using Design-Builder software of "Energy plus "simulation engine on a case-study of a model of office building with two façades alternative base case. The conclusion of this research that Trombe walls can reduce the cooling load by 75% compared to the base case with annual energy savings.

Keywords: Trombe wall; Passive solar heating and cooling; Energy efficiency; Thermal comfort.

IOP Conf. Series: Materials Science and Engineering **974** (2020) 011001 doi:10.1088/1757-899X/974/1/011001

Military Technical College
Kobry El-Kobbah,
Cairo, Egypt

13th International Conference on Civil and Architecture Engineering ICCAE-13-2020

EE-3

The philosophy of visual perception of the external space with a vision of the biophilic design that achieves quality of life

Enjy Ibrahim Youssef[1], Khaled Mustafa Khorshed[2], Sherif Sabri[3]

[1]Ph.D. researcher at Al-Azhar University e-mail:
[2]Professor, Department of Engineering, Al-Azhar University
[3]Professor, Department of Urban Planning, Al-Azhar University

enjyyusuf123@gmail.com

Abstract.-The research deals with the study of the integrated relationship between the elements of visual perceptions of the void and its interaction with the vital elements of it through the study and address the concept and philosophy of visual perceptions of the void and its development through the ages and the effect of these visual perceptions on the perception of life in the vacuum at its levels the three vitals It is the first level: belief in the existence of a life of visual formations and the sense of awareness of users in a dynamic interaction, and comes the second level: with regard to the mental and mental path of the user aware of these visual perceptions as physical elements with vital visions to convey the perception to the existence of real life and not Imagine it, level 3: The presence of the user in the scope of visual perceptions interactively with the characteristics of the formations that he realized between his mental perceptions and his inner feelings and his impression of the place and his human interaction towards it in the conscious and unconscious mind of his vital qualities surrounding the pulsating quality Life has vivid images that interact with the pulse of natural ingredients in the void The research has found that the interaction between human and visual perceptions in the void has exceeded the limits of time and space and has become more oriented towards biophilic design that enhances communication between nature and human, the architect has an important role in applying the concept of biophilic and expressing it to design elements space.

Key words: - Quality of Life - biophilic Design - External space - Visual Perception

Military Technical College
Kobry El-Kobbah,
Cairo, Egypt

13th International Conference on Civil and Architecture Engineering
ICCAE-13-2020

EE-4

Environmental Assessment of Coastal Resorts: Recreational Spaces as a Case Study

Mohamed Sobhi Hassan[1], Wael Ahmed Sheta[2], Ahmed Mohamed El Kordy[3]
[1] lecturer Assistant of Architecture
Department of Architecture, Faculty of Engineering, Al-Azhar University- Cairo.
sobhieg@hotmail.com
[2]Assistant Professor of Architecture,
Department of Architecture, Faculty of Engineering, Al-Azhar University- Cairo
wael.sheta@azhar.edu.eg
[3]Professor of Architecture
Department of Architecture, Faculty of Engineering, Al-Azhar University-
Cairoahmed.elkordy@ymail.com

Abstract. Tourism is one of the most important resources of national income in Egypt. Tourism as a vital role in bringing economic benefits to local communities, helping with conservation efforts or placing value on aspects of cultural heritage. Egypt is one of the world's most attractive countries in terms of its natural and historical characteristics, which makes it eligible to play an important role in the international tourism market. The vision of Egypt 2030 urges all sectors of the state to adhere to the axes of sustainability. Thus, the tourism sector is studying the current situation in the hotel industry and exploring the possibility of applying sustainability to hotels and resorts. This paper studies some of the sustainable practices in outdoor recreational spaces (especially in landscape) affects positively on conservation and the reduction of operation cost. The research method applies the seven principals of Xeriscape, which is considered as one of the sustainable benchmarks to evaluate landscape practices. The research aiming at assessing the coastal resorts in the Red Sea, Egypt from an environmental standpoint, analysing the operation practices in the recreational landscape which affects positively resources, conservation and the reduction of operation cost.

Keywords—Environmental Assessment, Recreational Spaces, Sustainability, Tourism Development.

IOP Conf. Series: Materials Science and Engineering **974** (2020) 011001 doi:10.1088/1757-899X/974/1/011001

Military Technical College
Kobry El-Kobbah,
Cairo, Egypt

13th International Conference on Civil and Architecture Engineering ICCAE-13-2020

EE-5

Performative Design Optimization for Shading Screens

Ahmed Mohamed ELkordy[1], Hesham Ahmed Sobh[2]
[1,2]Professor - Department of Architectural Engineering – Al -Azhar University
Mohga Ali Abdelslam Youssef Assistant lecture - Department of Architectural Engineering – Al - Azhar University

E.mail: Mohga.ali@azhar.edu.eg

Abstract: The need for energy- conscious design starting from early design phases has become a vital issue. It has become necessary to take immediate action to avoid dangerous consequences for future generations. Among strategies and solutions used in discussing problem of a buildings' energy consumption, the most important methods of saving energy in building are by carefully designing its facade. Building façade plays a significant role in architecture; it is not only a mean to express the design concepts but it is the main moderator between exterior environment and internal spaces. Ecological facades are defined as the ability to response and adapt to the changes of the environmental conditions. The strategies mainly focus on the functions and performances of Ecological facades in the context of indoor daylight quality.

The geometry of the proposed screen is not a focal point of this paper, as the aim is to find a method for designing and evaluation non-conventional solar screens that improve indoor daylight quality, while the specified shading screen is installed in front of a transparent full floor to ceiling glass facade. It is assumed that by using the shading screen, the required lighting levels are being maintained, while the heat gain is being reduced. In addition, the designed patterns play a role as an architectural feature of the building. Ultimately, the presented shading screens are based on geometric ornamental patterns and arranged according to the lighting requirements. Results are indicating a multi-disciplinary approach in the design of the shading screens, which can be employed in creating similar prototypes with different climatic and loading requirements.

IOP Conf. Series: Materials Science and Engineering **974** (2020) 011001 doi:10.1088/1757-899X/974/1/011001

Military Technical College	***13th International Conf. on Civil & Arch. Eng.*** **7-9 July 2020**

Session No. 8
Title **Environmental Engineering (EE II)**
Date **Wednesday 8-7-2020**
Time **15:00-16:45**
Place **Room B**
Session Chairmen
Maj. Gen. Prof. Dr. Ehab H. Mahmoud
Brig. Gen. Assoc. Prof. Mohamed F. Abd El Aleem
Col. Dr. Hesham A. El Kady

Comparison of occupant thermal comfort with and without passive design for an educational building: a case study in Cairo. *Mohammad Fahmy, Sherif A. Mahmoud, Ibrahim M. Olwy and Marwa Abdelalim*	**EE-7**
Life cycle assessment of a residential building in Egypt: A case study *D. M. Morsi, W. S. Ismael, Mohamed Elmahdy and A. E. Abd El Hamed*	**EE-8**
The impact of an external obstacle daylight in deep office spaces at hot arid zone *Doha M. Saeed*	**EE-9**

IOP Conf. Series: Materials Science and Engineering **974** (2020) 011001 doi:10.1088/1757-899X/974/1/011001

Military Technical College
Kobry El-Kobbah,
Cairo, Egypt

13th International Conference
on Civil and Architecture
Engineering
ICCAE-13-2020

EE-7

Comparison of occupant thermal comfort with and without passive design for an educational building: a case study in Cairo.

Mohammad Fahmy[1], Sherif A. Mahmoud[2], Ibrahim M. Olwy[3]

[1]Associate Professor, Architecture Department, Military Technical College, Cairo, Egypt.
[2]Senior Lecturer, Architecture Department, Military Technical College, Cairo, Egypt.
[3]Assistant Lecturer, Architecture Department, Military Technical College, Cairo, Egypt.

md.fahmy@live.com

Abstract: Occupant thermal comfort in educational buildings is important for students as it support the educational process itself. The passive design of naturally ventilated buildings plays a crucial role in achieving occupant thermal comfort but the passive design itself might be of higher cost to convince the owner even with proofs from the pre and while design simulation outputs. In this paper, the as built of a 1000 students' educational building is simulated using Design Builder and compared with the same thermal zones of the as built but with passive cooling applications. Two wind catchers, shading devices, low e-glass windows, double skin façade, and double roof were applied. The status of student comfort was improved noticeably, results show that air temperature, air speed and thermal comfort achieved …… for the as built and recorded …….. for the passively modified version of the building.

Keywords: passive cooling, educational buildings, occupant thermal comfort, design builder

IOP Conf. Series: Materials Science and Engineering **974** (2020) 011001 doi:10.1088/1757-899X/974/1/011001

Military Technical College
Kobry El-Kobbah,
Cairo, Egypt

13th International Conference on Civil and Architecture Engineering
ICCAE-13-2020

EE-8

Life cycle assessment of a residential building in Egypt: A case study

D M Morsi[1], W S Ismael[2], Mohamed Elmahdy[3]and AE ABD EL-HAMED[4]

[1] Teaching assistant at the higher institution of engineering, Sherouk Academy, Civil Engineering Department, Egypt
[2]Lecturer at the British University in Egypt, Architectural Engineering Department El Sherouk City, Suez Desert Road
[3] Deptartment of Structural Engineering, Cairo University, Giza, 12613, Egypt
[4] Department of Civil Engineering, Higher Institution of Engineering, Sherouk Academy.

Abstract. The present study quantified the significant environmental impacts of a residential building located in New Cairo, Egypt, covered all life cycle steps from cradle to cradle with a projected 60-year life span: (i) an inventory of all the construction materials were analysed, covering the building structure as well as the energy consumption; (ii) Three types of functional units were defined; (iii) the six top building materials were examined, and a sensitivity analysis was conducted to investigate the impact associated with the choice of building materials. The result shows that two life cycle phases, product and operation, were more significant in all of the impact categories, and two building assemblies, the structure and the floor most of the environmental loads. In terms of the sensitivity analysis, the structural concrete had the largest impact, dominating four of the seven selected impact categories, then hot-rolled structural steel took the second material of the largest impact after concrete. Finally, limitations and challenges are discussed to explore better design decisions in future studies.

IOP Conf. Series: Materials Science and Engineering **974** (2020) 011001 doi:10.1088/1757-899X/974/1/011001

Military Technical College
Kobry El-Kobbah,
Cairo, Egypt

13th International Conference on Civil and Architecture Engineering
ICCAE-13-2020

EE-9

The impact of an external obstacle daylight in deep office spaces at hot arid zone

Doha M. Saeed

Teacher of Architecture, Bader University in Cairo,

eng.doha.m@gmail.com,Egypt.

Abstract: The use of effective daylighting systems is one of the main factors that influences energy consumption in buildings especially in deep office spaces. Therefore, the need for innovative daylighting systems with optimum overall daylighting performance - especially in term of saving energy performance - is an essential design element for such spaces. Light shelf systems are the most major potential lighting systems that can be used to provide such innovative daylighting systems.
The parametric approach was implemented where all possible combinations of the various design parameters for the main two design elements of the light shelf system (light shelf and room specifications) were computed through Diva-for-Rhino (Grasshopper). The study of optimized light shelf systems in deep office spaces has been carried out in case of an external obstacle (neighbor) and without it – according to the Egyptian building codes - in order to identify the impact of such design elements on the performance of the optimized light shelf system in the main four directions in hot arid zone climate.

Keyword: Light shelf systems – obstacle - deep office spaces - Diva-for-Rhino (Grasshopper)

IOP Conf. Series: Materials Science and Engineering **974** (2020) 011001 doi:10.1088/1757-899X/974/1/011001

Thursday
9 July 2020

IOP Conf. Series: Materials Science and Engineering **974** (2020) 011001 doi:10.1088/1757-899X/974/1/011001

Military Technical College	***13th International Conf. on Civil & Arch. Eng.*** **7-9 July 2020**

Session No. 9	
Title	**Urban Planning (UP I)**
Date	**Thursday 9-7-2020**
Time	**9:00-10:45**
Place	**Room B**
Session Chairmen	**Maj. Gen. Prof. Dr. Ehab H. Mahmoud** **Col Dr. Hany Abd El Khalek** **Lt Col. Dr. Sherief Ibrahim**

Transition governance as a model for city transformation toward sustainable future ***Sherif Sabri, Maged Elmahdy, Abd Alrahim Kasem, Mohamad farrag Fayad.***	**UP3**
Major national projects as development poles and their impact on urban development. A case study of the new city of El Alamein in the north western axis ***Asmaa Gabr, Kamal Khalaf and Gamal Helewa***	**UP4**

IOP Conf. Series: Materials Science and Engineering **974** (2020) 011001 doi:10.1088/1757-899X/974/1/011001

Military Technical College
Kobry El-Kobbah,
Cairo, Egypt

13th International Conference
on Civil and Architecture
Engineering
ICCAE-13-2020

UP-3

Transition governance as a model for city transformation Toward sustainable future

Sherif Sabri, Maged Elmahdy, Abd Alrahim Kasem, Mohamad farrag Fayad.

Urban Planning Department, Faculty of Engineering, Al-Azhar University, Cairo, Egypt

engmohamad.farag@gmail.com

Abstract In this paper will brief the audience about the origins and status of the transition governance approach which was developed over a decade ago in the Netherlands and has since been adopted worldwide. Transitions are fundamental changes in the structures and practices of (parts of) our cities and societies. Transition governance has been developed based on empirical and theoretical insights and an understanding of the dynamics and mechanisms of transitions. It provides a governance philosophy and concrete instruments and methods aimed at influencing the direction and pace of urban change dynamics towards sustainability. The approach is generic for complex societal issues and persistent unsustainability. It is used to address transition challenges on the scale of regions, cities and neighborhoods, as well as to initiate transformations in socio-technological systems such as energy, water and mobility. Transition governance in practice creates a new interface between policy makers and other societal actors. Local governments can be (but are not necessarily) in the driving seat with this approach and, together with (often) external transition experts, co-develop pioneering frontrunner networks called transition arenas. These transition arenas form the basis for developing society based transition agendas, strategies and experiments. Through the specific process design and approach to structure the transition agenda a collective ambition and perspective is developed which fuels all sorts of innovative projects and experiments. To local governments and other actors (such as citizens, businesses, institutions and other organizations) , transition governance is a learning process during which opportunities are created to build upon each other's transformative capacity and play into local dynamics.

Key words: Transition, governance, transformation, city, Sustainability.

IOP Conf. Series: Materials Science and Engineering **974** (2020) 011001 doi:10.1088/1757-899X/974/1/011001

Military Technical College
Kobry El-Kobbah,
Cairo, Egypt

13th International Conference on Civil and Architecture Engineering ICCAE-13-2020

UP-4

Major national projects as development poles and their impact on urban development: A case study of the new city of El Alamein in the northwestern axis

Asmaa Gabr, Kamal Khalaf, Gamal Helewa
Al-Azhar University, Faculty of Engineering, Urban Planning Department, Egypt
Email: Assmaa.gabr2015@yahoo.com

Abstract- The major national projects are considered Egypt's vision in investing its various resources and one of the most important methods used to create urban development on all development sectors, starting from the local level through the regional level to the national level, which occurs a great development in its urban environment, especially the scope of those projects, and its role in terms of being development poles It helps in spreading urban development with its pivotal concepts, By measuring the extent to which these projects are able to physically attract urban investment to evaluate the investment orientation of these projects, by identifying the factors affecting the national projects 'capacity for urban polarization, and then defining the form of the relationship that links these factors to urban polarization through a statistical mathematical scale On comparative theoretical studies, And, through this scale, it is possible to measure the urban polarization capacity of any existing national project or any urban gathering in the light of the proposed national project, and predict the potential population of the project, as it is possible to determine the priorities for development and investment, which in turn reflects on the spread of urban development in all parts of the Republic. The research aims to identify national projects of various levels of planning, And to identify the major projects, including those with development poles, and to identify the elements of its success and determine its development returns by measuring its urban polarization capacity.

IOP Conf. Series: Materials Science and Engineering **974** (2020) 011001 doi:10.1088/1757-899X/974/1/011001

Military Technical College	***13th International Conf. on Civil & Arch. Eng.*** **7-9 July 2020**

Session No. 10	
Title	**Urban Planning (UP II)**
Date	**Thursday 9-7-2020**
Time	**11:00-12:45**
Place	**Room B**
Session Chairmen	**Maj. Gen. Prof. Dr. Ehab H. Mahmoud** **Col Dr. Hany Abd El Khalek** **Lt Col. Dr. Sherief Ibrahim**

Architectural Design Firms' Social Role to develop Sustainable Heritage Communities in Egypt ***Dalia A Elweshahy and Ayman A E Othman***	**UP5**
Evaluation of Slum Upgrading Interventions - Literature Review and Methodological Approaches ***Hesham El-Maradny, Mohamed M. Mahdy and Hesham El-Kady***	**UP6**
Managing of City Transition Toward Sustainable Eco. Urbanism (Proposed paradigm) ***Sherif Sabri, Maged Elmahdy, A. Alrahim Kasem and Mohamad F. Fayad***	**UP-8**

IOP Conf. Series: Materials Science and Engineering **974** (2020) 011001 doi:10.1088/1757-899X/974/1/011001

Military Technical College
Kobry El-Kobbah,
Cairo, Egypt

13th International Conference on Civil and Architecture Engineering
ICCAE-13-2020

UP-5

Architectural Design Firms' Social Role to develop Sustainable Heritage Communities in Egypt

Dalia A Elweshahy [1] And Ayman A E Othman [2]

[1]Senior Student, Architecture Engineering Department, Faculty of Engineering, The British University in Egypt, Cairo-Suez Desert Road, El-Shorouk, Cairo, Egypt
[2]Professor od construction and project management, head of architectural engineering department, faculty of engineering, The British University in Egypt, Cairo-Suez Desert Road, El-Shorouk, Cairo, Egypt.
Corresponding Author Email: Dalia145555@bue.edu.eg

Abstract: Cultural heritage buildings are a crucial element for future generations, as they preserve the cultural identity unique to where they are located. Moreover, they create one of the most dominant and memorable architectural achievement that have significant historical and cultural values. However, heritage buildings in Egypt suffer from a surge in demolition due to the deteriorated conditions of their surroundings areas, as well as, the maintenance of only the physical structures without considering the surrounding communities. Egypt constantly encounters many challenges such as poor coordination from the government due to shortage of skilled people, poor heritage awareness of local communities and lack of finance needed for achieving sustainable development. On the other hand, Architecture Design Firms (ADFs) act as a supporting tool for the government as they have a vital role in the local community development; which further act as the cornerstone of developing the historical regions. Hence, in order for the ADFs to achieve such development, they need to activate their Corporate Social Responsibility (CSR) role which is a driver for developing the heritage community. Consequently, this research aims to develop a matrix that correlates CSR with ADFs as an approach to achieve sustainable heritage communities in Egypt. A research methodology consists of literature review to investigate the topics of CSR, Heritage community development and ADFs. Finally, developing a matrix the relation between the architect's social responsibility and the indicators of sustainable development of local communities.

Keywords: *Corporate Social Responsibility, Heritage Community Development, Sustainable Development, Architecture Design Firms, Heritage Buildings*

.

IOP Conf. Series: Materials Science and Engineering **974** (2020) 011001 doi:10.1088/1757-899X/974/1/011001

Military Technical College
Kobry El-Kobbah,
Cairo, Egypt

13th International Conference on Civil and Architecture Engineering ICCAE-13-2020

UP-6

Evaluation of Slum Upgrading Interventions - Literature Review and Methodological Approaches.

Hesham El-Maradny [a] **Mohamed M. Mahdy** [a] **Hesham El-Kady** [a]

[a] Architecture Engineering Department, Military Technical College, Cairo, Egypt

helkady@mtc.edu.eg

Abstract. This work addresses slums definition according to UN-Habitat, and provides some objectives of impact evaluation of slum upgrading projects. The main objective is to clarify the importance behind the slum upgrading projects (interventions), through over viewing the slum upgrading interventions and their main characteristics then summarize methodology of evaluation studies. This paper gives an overview of counterfactual question, qualitative and quantitative methodologies discusses main strategies to control bias in Quasi-experimental approaches, and provides a summary about common past evaluation challenges and recommendations which can be applied to future work.

MilitaryTechnicalCollege
Kobry El-Kobbah,
Cairo, Egypt

13th International Conference on Civil and Architecture Engineering ICCAE-13-2020

UP-8

Managing of City Transition Toward Sustainable Eco. Urbanism (Proposed paradigm)

Sherif Sabri, Maged Elmahdy, A. Alrahim Kasem, Mohamad farrag Fayad.

Engmohamad.frag@gmail.com

Abstract: The Current research clarify the main basics of Mutation's strategy towards sustainable eco urbanism which they are (Climate , renewable energy, zero waste, water, landscape, sustainable transport ,local and sustainable materials, retrofitting of existing districts, green buildings ,livability, local food ,cultural heritage, urban governance, education and knowledge, strategies for urban areas) , and what spinoff the reliable Universal Criteria in addition to concept the managing of sustainability transition in cities.

The aim of the research was to present and elaborate a proposed model for managing the sustainable transformation of Egyptian cities, so that they are suitable to deal with and apply to most existing and new urban communities, with a view to transforming them into sustainable environmental cities and communities, by working to develop a future vision for the sustainable image that the city will reach, dependent on Scientific foundations and an understanding of the problems these regions suffer from, to explain sustainability strategies in light of the general data for these areas and the current and future needs of the population and the available capabilities.

The proposed model consists of ten consecutive stages of operations and is related to each other, where the subsequent stage is based on the data of the previous stage, and the success of the previous stage has a positive impact on the success of the subsequent stage and vice versa.

The main findings of the study are: preparing and implementing the management of Egyptian cities transformation into sustainable cities: include the flowing

Building an institutional framework to manage urban sustainability transitions at the national level

Prepare a strategy for existing and new cities

Provide the necessary financial resources for implementation

Information integration and capacity support

Institutional support and capacity building - Support capabilities

Key words: Managing, transition transformation, eco urbanism, sustainable city. Sustainability

IOP Conf. Series: Materials Science and Engineering **974** (2020) 011001 doi:10.1088/1757-899X/974/1/011001

Military Technical College	***13th International Conf. on Civil & Arch. Eng.*** **7-9 July 2020**

Session No. 11	
Title	**Material Quality and Control (MQ &C)**
Date	**Thursday 9-7-2020**
Time	**13:00 -14:45**
Place	**Room B**
Session Chairmen	**Maj. Gen. Prof Dr. Essam M. El-Tehawy** **Col. Dr. Ashraf M. Osman** **Lt. Col. Mohammed Aly Al Sayed**

Static and Dynamic Performance of Two Stage Concrete ***Mohamed A.E.M. Ali***	**MQC2**
Thermal Conductivity for Foamed Paste made of Cement and Cement kiln dust ***Rafik K. Abdel Wahab***	**MQC3**
Investigating the experimental process for partial replacement of cement with sugarcane bagasse in the construction industry ***Nadin Dawoud, Amany Michealb, Rania Rushdy Moussa***	**MQC4**
Structural Behavior of Light Weight Ferrocement Walls ***Yousry B. I. Shaheen, Mohsen Mousa, and Eman Gamal***	**MQC6**
Effect of crumbed rubber Treatment method on the Flexural Behavior of High Strength Rubberized Concrete Slabs ***Fatma M. Eid, A. A. Bashandy , M. A. Mazyad and M. A. Arab***	**MQC7**

Military Technical College
Kobry El-Kobbah,
Cairo, Egypt

13th International Conference
on Civil and Architecture
Engineering
ICCAE-13-2020

MQ&C-2
Static and Dynamic Performance of Two Stage Concrete

Mohamed A.E.M. Ali*

**Department of Civil Engineering, Military Technical College, Cairo, Egypt*

Abstract. Concrete is arguably the most important construction material in the world. In the last decades, it has been reported that infrastructures faced extreme static and dynamic loads which caused structural catastrophic disasters [1]. The increase usage of conventional concrete in engineering industries led to high consumption of mineral resources which caused several environmental damages related to reduction in the rocket layers [2], besides, the economical damages due to the high demand of natural resources materials which rises the cost of construction [3]. Therefore, the utilization of recycled waste rubber (steel wires) in concrete can be a sustainable solution for waste construction developments [4]. Recently, two stage concrete (TSC) (non-traditional concrete) was introduced. Placing of coarse aggregate followed by injecting the cementitious materials among the aggregate layers in the formwork are the basic steps of producing TSC. Evaluating the static and dynamic performance of TSC were the objectives of this study. Cement, aggregates, water, recycled steel wires of rubber tires were utilized in manufacturing TSC specimens. The compressive characteristic and impact resistance of TSC test specimens were evaluated. Results showed that steel wires improved the dynamic behaviour of TSC, while a slight to no effect was observed on its compressive characteristics. This study demonstrates the potential to engineer sustainable TSC mixtures.

Keywords: Two-stage concrete; impact; sustainability

Military Technical College
Kobry El-Kobbah,
Cairo, Egypt

13th International Conference on Civil and Architecture Engineering
ICCAE-13-2020

MQ&C-3

Thermal Conductivity for Foamed Paste made of Cement and Cement kiln dust

Rafik K. Abdel Wahab[1,2]

A. Prof .Civil Engineering Department, Faculty of Engineering, Qena Branch- Al-Azhar University -Egypt
Vice Dean of High Institute for Engineering and Technology In Obour City –Qalyobia -Egypt

Dr.rkhairy@gmail.com

Abstract As with most large manufacturing industries, by-product materials are generated. Cement kiln dust is a by - product material of cement manufacturing process. The aim of this paper is to investigate the properties of foamed paste made of CKD in addition of different percentage of cement. Compressive strength and thermal conductivity had been determinate. The results were encouraged to use the Foamed (Cement kiln dust – cement) Paste (FCKDC) as an insulation material. The results indicate a comparable value for thermal conductivity which vary from **(0.074- 0.3)** W/mk.

Keywords: Cement, Cement kiln dust (CKD),Thermal conductivity, Compressive strength, Foamed Paste

IOP Conf. Series: Materials Science and Engineering **974** (2020) 011001 doi:10.1088/1757-899X/974/1/011001

Military Technical College
Kobry El-Kobbah,
Cairo, Egypt

13th International Conference on Civil and Architecture Engineering ICCAE-13-2020

MQ&C-4

Investigating the experimental process for partial replacement of cement with sugarcane bagasse in the construction industry

Nadin Dawoud[a], Amany Micheal[b], Rania Rushdy Moussa[c]

a, b & c The British university in Egypt (BUE), Suez Desert Road
El Sherouk City, 33811, Cairo Egypt

Nadin150210@bue.edu.eg

Abstract: In the last few decades there has been speedily increasing in the agriculture and industrial wastes. This causes many environmental issues and raises the potential to contaminate the natural resources of living such as water, air and soil. Recently, the amount of organic waste produced daily has been rising, while it is poorly managed. It is either burned or disposed improperly, which effect negatively the environment and public health. On the other hand, during the cement production process many wastes, and pollutants are generated which have major negative impacts on the environment. Cement is considered as a substantial constituent of building materials in the construction industry. Many research's intended to explore the potential of utilizing organic waste products in the construction industry by partially replacing cement with organic wastes such as sugarcane bagasse to create an eco-friendly brick with appropriate structure properties. Previous research's used different treatment techniques to treat the organic waste and prepare it for construction industry. It was noticed that the treatment method used in previous research's affected the structural properties of the new bricks with organic waste partially replacing. This research intends to study and analyse the process and techniques used by other researcher's experimental work to treat and replace cement with sugarcane bagasse.

This research will present the best procedures for partially replacing cement with sugarcane bagasse in cement bricks without compensating the structural properties of regular cement bricks. This study will analyse and compare previous researcher's experimental work to obtain the best experimental program and to utilize sugarcane bagasse in brick industry.

KEYWORDS: sugarcane bagasse, cement, eco-friendly block, partial replacement, thermal conductivity, strength

IOP Conf. Series: Materials Science and Engineering **974** (2020) 011001 doi:10.1088/1757-899X/974/1/011001

Military Technical College
Kobry El-Kobbah,
Cairo, Egypt

13th International Conference on Civil and Architecture Engineering ICCAE-13-2020

MQ&C-6

Structural Behavior of Light Weight Ferrocement Walls

Yousry B. I. Shaheen[1], Mohsen Mousa[2], and Eman Gamal[3]

[1,2] Department of Civil Engineering, Faculty of Engineering, Menoufia University, EGYPT.
[3]Civil Engineer and Postgraduate Fellow

Abstract: This paper presents the experimental results of research to develop ferrocement sandwich panels for use as wall bearing units. The proposed panels are lighter in weight relative to the conventional reinforced concrete panels. The sandwich panels consisted of two thin ferrocement layers reinforced with one or two layers of closely spaced galvanized welded wire mesh and expanded steel mesh. The core of the panel was made of light weight brick. Steel wires were used to tie the steel meshes of the two skin layers together and to act as shear connectors to transfer shear between the two ferrocement skin layers. These steel wires were embedded in the mortar joints of the brick. The thickness of the ferrocement skin layer was 10mm when single layer of wire mesh was used and 15mm when two layers of mesh reinforcement were used. Experimental investigation was conducted on the proposed panels. A total of 10 sandwich panels Having the dimensions of 600mm width and 700mm height were tested under compressive loadings until failure. The deformation characteristics and cracking behavior were measured and observed for each wall panel. The results showed that high ultimate and serviceability loads, crack resistance control, high ductility, and good energy absorption properties could be achieved by using the proposed panels. This could be of great construction advantages for both developed and developing counties.

Keywords: Ferrocement; Sandwich Panels; Deformation characteristics; Serviceability load; Ductility and Energy absorption properties; Cracking pattern; Ultimate loadings.

IOP Conf. Series: Materials Science and Engineering **974** (2020) 011001 doi:10.1088/1757-899X/974/1/011001

Military Technical College
Kobry El-Kobbah,
Cairo, Egypt

13th International Conference on Civil and Architecture Engineering ICCAE-13-2020

MQ&C-7

Effect of crumbed rubber Treatment method on the Flexural Behavior of High Strength Rubberized Concrete Slabs

Fatma M. Eid[1]*, A. A. Bashandy [2], M. A. Mazyad [3] and M. A. Arab [4]

[1]Fatma M. Eid, Associated prof., Civil Engineering Department, Minoufiya University, Egypt (E-mail: dr.fatmaelzahraa@gmail.com)
[2]Alaa Ali Bashandy, Associated prof., Civil Engineering Department, Minoufiya University, Egypt (E-mail: eng_alb@yahoo.com)
[3]Mohammed Ahmed Mazyad, Assistant lecturer, Civil Engineering Department, Thebes academy, Egypt (E-mail: mohammedmazayd79@gmail.com)
[4]Mohammed Abd El-Salam Arab, Assistant prof., Civil Engineering Department, Beni-Suef University, Egypt (Corresponding Author; E-mail: emarab@eng.bsu.edu.eg)
*Correspondence: dr.fatmaelzahraa@gmail.com , fatma_elzahraa2002@yahoo.com

Abstract. The most important research trends that the world is currently interested in are those that have a positive impact on the environment. The burning of the waste tyres is one of the most used means to get rid of these tyres, which causes a lot of gases to the air. The use of crumbed tyre rubber in concrete is considered one of the methods with a positive environmental impact. Using crumbed rubber as a partial replacement material of concrete aggregate is found to possess many engineering applications and holds promise within the future. Replacing 5% of the sand volume by crumbed tyre rubber was used. Flexural behavior of treated and untreated rubberized concrete slabs was investigated. Experimental program shows that the behavior of the high strength rubberized concrete slabs can be improved by the used treatment methods. Beside that the coarser of the rubber particles, the more negative effect on the behavior of the high strength rubberized concrete slabs.

IOP Conf. Series: Materials Science and Engineering **974** (2020) 011002 doi:10.1088/1757-899X/974/1/011002

Peer review declaration

- **Types of peer review : double-blind**
- **Conference submission management system : Online**
- **Number of submissions received : 52**
- **Number of submissions sent for review: 52**
- **Number of submissions accepted: 40**
- **Acceptance rate : 76.9 %**
- **Average number of reviews per paper : 2**
- **Total number of reviewers involved : 30**
- **Any additional info on review process : No**
- **Contact person for queries:**

- Prof. Osama M. Moussa eramo1987@gmail.com

- Dr. Mohamed Elkhatib melkhatib@mtc.edu.eg
elkhatib.m@gmail.com

Military Technical College Kobry El-Kobbah, Cairo, Egypt

13th International Conference on Civil and Architecture Engineering ICCAE-13-2020

Comparison of gridded datasets for the simulation of streamflow in Africa

Mostafa Tarek[1,2], François P. Brissette[1] and Richard Arsenault[1]

[1] École de technologie supérieure, Université du Québec, 1100 Notre-Dame West, Montréal, Québec, Canada, H3C 1K3

[2] Department of Civil Engineering, Military Technical College, Egypt.

Correspondence to: Mostafa Tarek (mostafa_tarek45@mtc.edu.eg)

Abstract. In recent decades, many parts of the African continent have experienced high precipitation variability with periodic drought and flood events. However, the network of streamflow gauges is too sparse in most countries to adequately capture these variations. In addition, no observed reference climatological dataset exists to adequately represent precipitation and temperature changes within all topographic and climatic zones. Consequently, the use of global gridded datasets needs to be considered. This paper aims to use the different available gridded datasets as inputs to a hydrological model to evaluate dataset performance. Nine precipitation and two temperature gridded datasets are used to this effect. The precipitation datasets include two gauged-only products, two satellite products corrected using ground-based observations, four reanalysis products and one merged product of gauge, satellite, and reanalysis.The two temperature datasets include one gauged-only and one reanalysis product. The ten precipitation and two temperature datasets were combined in their 18 possible arrangements for analysis purposes. Each combination was used to force the HMETS lumped hydrological model. The model parameters were calibrated individually for each combination against the streamflow records of 850 African catchments. The Kling-Gupta Efficiency (KGE) was used to evaluate the simulation performance. Results show thatboth temperature datasets performed equally well. Large differences were however observed between precipitation datasets. The MSWEP merged-product was the best-performing precipitation dataset, followed by CHIRPS satellites and ERA5 reanalysis products, respectively. The performance of both gauged-only datasets (CPC and GPCC) was inferior, outlining the limitations of extrapolating information in data-sparse regions.

Keywords:precipitation datasets, gridded datasets, reanalysis products, streamflow simulation, hydrological modeling, African catchments.

1. Introduction

Ground meteorological stations are consideredthe most accurate source of climate data, as they offer physical record of data in a specified area. However, stations may suffer from many limitations such as missing measurements or short temporal coverage[1]. In recent decades, many regions have experienced high variability in precipitation with periodic drought and flood events[2-5]. However, the spatial coverage of station networks is not sufficient to adequately represent these changes within all topographic and climatic

IOP Conf. Series: Materials Science and Engineering **974** (2020) 012001 doi:10.1088/1757-899X/974/1/012001

zones [4]. In addition, a gradual but steady decrease in the number of weather stations with long record listed in the Global Historical Climatology Network (GHCN) has started in the early 1990. To resolveall these problems, a large effort has been put into producing global gridded meteorological datasets. Such datasets providecontinuous spatial and temporal coverage and, typically, with no missing data.

Over recent decades, several precipitation products have been producedwith different spatial and temporal characteristics. These datasets differ in terms of data sources (gauge, radar, satellite, reanalysis or combinations thereof), spatial resolution (0.05° to 2.5°), spatial coverage (continental to global), temporal scale (30 minutes to annual) and temporal coverage (from 1 to several years).Several studies addressed the importance of evaluating these datasets to stand on their advantages and limitations. Most studies quantified the accuracy of these products through a direct comparison against data from weather stations[6-11], while others assessed the performance indirectly using a hydrological model to compare against observed streamflow[12-16].

2. Study Area

In this study, the African continent was chosen as the main research area. Africa is considered the second largest continent with an area of 30.3 million km^2 covering about 20% of the global land area[17]. Africa is considered to be the hottest continent on Earth. The northern half is mostly covered bydrylands and desert, while the central and southern parts contain savanna and rainforests [18]. Based on the combination of temperature, precipitation and evapotranspiration, Africa canbe divided into four main climatic zones; 1) arid and semi-arid, 2) tropical, 3) equatorial, and 4) temperate [19].

3. Data and Methods

3.1 Data

For many African regions, observed meteorological data are not easily available, either due to the lack of weather stations or the high fees to access the data. Most studies in Africa therefore depend on using satellite-derived data as a reference dataset [20-23]. Hence, this paper aims to evaluate the performance of other several types of gridded datasets over a large set of hydrologically heterogeneous watersheds. The dataset performance is assessed through their ability as generating accurate streamflow through the use ofa hydrological model.

3.1.1 Precipitation gridded datasets

Gridded datasets can be classified as a function of their data source. Gauge-basedgridded datasets are obtained by interpolating the information measured ata small scale(typically a point measurement at a weather station) and mapped ontoa predefined spatial and temporal resolutions grid. However, variation in gauge types or instrument replacements affect error characteristics on the long-term records. In addition, observations areaffected by systematic biases from evaporation and wind effect or due to, for example, elevation placement of gauges in mountainous regions[24].

A different approach to measure precipitation is using ground weather radars, as it partially addresses the issue of rain gauge coverage. Moreover, it provides much larger spatial coverage to measure precipitation than the point measurements provided by gauges. However, radar coverage is limited to developed regions that have a high population. In addition, they sense the real rainfall rate at a certain observational level above ground. Therefore,the presence of weather stations is required for the calibration and correction processes[25].

Nowadays, satellite products are available at the global scale and can cover large areas athigh spatial and temporal resolutions and near real time coverage. They are mainly suitable for rainfall estimation in the tropics and data sparse regions. However, satellites are relatively insensitive and generally miss a significant quantity of light precipitation and tend to fail over snow and ice-covered surfaces [26]. Some studies evaluated the uncertainties of these datasets and showed that high resolution satellite products perform better when bias correctedusing gauge observations [27, 28].

Retrospective-analysis / reanalysis systems are vital sources of data in weather and climate studies. A typical reanalysis system consists of two main components, the forecast model and the data assimilation system. The role of the data assimilation system is to integrateobserved databases of many sources of observations with the numerical weather forecast models to produce consistent gridded datasets [29]. Although reanalysis are not

Table 1. The selected global gridded datasets

	Short name	Data Source	Spatial Resolution	Spatial Coverage	Temporal Resolution	Temporal Coverage
1- Precipitation datasets						
1	CPC Unified	Gauge	0.50°	Global	Daily	1979 - Present
2	GPCC	Gauge	1.0°	Global	Daily	1979 - 2016
3	PERSIANN-CDR	Gauge, Satellite	0.25°	60°N - 60°S	6 hours	1983 - 2012
4	CHIRPS	Gauge, Satellite	0.05°	50°N - 50°S	Daily	1981 - Present
5	NCEP-CFSR	Reanalysis	0. 50°	Global	6 hours	1979 – 2010
6	ERA-Interim	Reanalysis	0.75°	Global	3 hours	1979 – 8/2019
7	ERA5	Reanalysis	0.25°	Global	hourly	1979 – 2017
8	JRA-55	Reanalysis	0.5625°	Global	3 hours	1959 - Present
9	MSWEP V1.0	Gauge, Satellite, Reanalysis	0.25°	Global	3 hours	1979 - 2015
2- Temperature datasets						
1	CPC Unified	Gauge	0.50°	Global	Daily	1979 - Present
2	ERA5	Reanalysis	32 Km.	Global	hourly	1979 - 2017

direct observations, they provide analyzed variables in areas where stations are minimal [30]. Overall, no single precipitation product could be considered ideal for measuring precipitation. In fact, all precipitation products tend to miss a significant volume of rainfall [12].

As discussed earlier, there is now a rather large number of gridded datasets from stations, satellites, reanalysis or a combination thereof. However, not all those datasets can be used for climate change impact studies. Appropriate datasets would have the following desirable characteristics: 1) spatial resolution (between 0.05° to 1°); 2) daily scale or finer temporal resolution, 3) long temporal coverage (~ 30 years), and 4) all datasets should cover approximately the same time interval. Based on those criteria, ten precipitation and two temperature gridded datasets were chosen in this study as shown in 'table 1'. The precipitation datasets include two gauged-only products (GPCC and CPC),two satellite products corrected using ground-based observations (CHIRPS and PERSIANN), four reanalysis products (JRA55, NCEP-CFSR, ERA-Interim and ERA5) and one merged product of gauge, satellite, and reanalysis (MSWEP).

3.1.2 Temperature gridded datasets

Land surface temperature is a key variable for meteorological monitoring and forecasting services [31]. It is also a key variable for climate and hydrological studies. In hydrological modelling, the air temperature is the

IOP Conf. Series: Materials Science and Engineering **974** (2020) 012001 doi:10.1088/1757-899X/974/1/012001

key driving variable for the evapotranspiration and snowmelt processes. Hence, accurate temperature data is a vital issue. However, the lack of adequate gauge network can result in improper estimates of temperature. Therefore, temperature gridded datasets are also crucial in many fields.Temperature products are generally thought to be less complex than precipitation datasets due to its much smaller spatial and temporal variability. Therefore, much fewer studies have compared and evaluated the uncertainty of using different temperature datasets. On this study, two temperature datasets have been included: the gauge-based CPC dataset, and the ERA5 reanalysis.

3.1.3 Observed streamflow data

Streamflow records from the Global Runoff Data Centre (GRDC) were used to calibrate the hydrological models and evaluate the hydrological modelling performance. The GRDC database has streamflow data from 1150 African stations. In this study, 850 stations were chosen based on two criteria. First, stations should havedata during the1983-2012 study period. Second, stations that have less than three years of consecutive data during this period were excluded. The spatial distribution of these stations is shown in 'figure 1'.

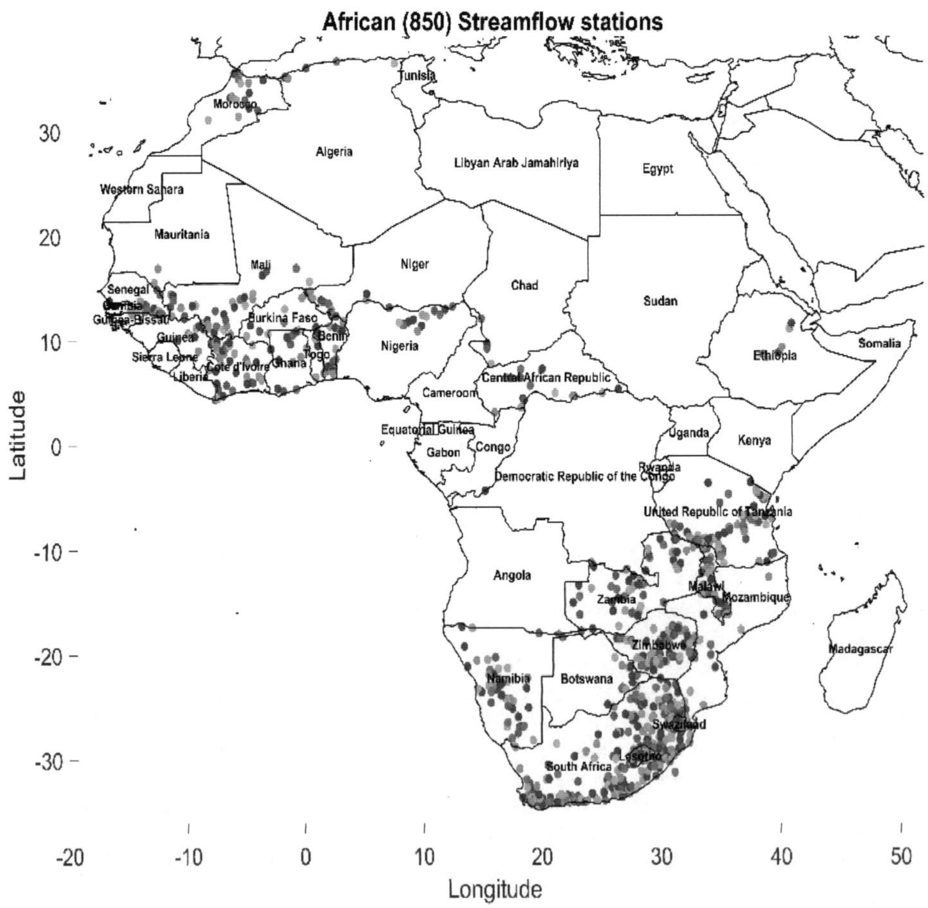

Figure 1. Spatial distribution of the African 850 streamflow stations

3.2 Hydrological model
In this study, the useof a distributed model was discarded due the scale of the study. The lumped hydrological modelHMETS[32]was used to evaluate the performance of the various climate datasets. This modelhas shown an overall good performance in a wide range of climates and hydrological studies[32-34].The model requires daily precipitation, temperature and potential evapotranspiration (PET) as inputs. The Oudin'stemperature-based formula [35] was used to calculate PET as it has shown an overall good performance and robustness on large-scale hydrological studies[36].

3.3 Hydrological model calibration
As will be detailed in the following section, the nineprecipitation and two temperature datasets were combined in their 18 possible arrangements for analysis purposes and the hydrological model parameters were calibrated for each catchment and each dataset combination.The15300 calibrations to be performed (9 precipitation datasets x 2 temperature datasets x 850 catchments) required the application of an automatic model parameter calibration method. For this study, the CMAES algorithm was applied because of its flexibility[37]. Moreover, it is considered as one of the best auto-calibration algorithms for hydrological modelling [38].

The Kling-Gupta Efficiency (KGE) calibration objective function was used to evaluate the simulation performance. KGE is a modified version of the Nash-Sutcliffe Efficiency (NSE) metric that was introduced by Gupta[39]and modified by Kling[40]. It is defined as a combination of three elements;correlation, bias and variability as shown in 'equation 1'.Pearson's correlation coefficient used to represent the correlation component (r), the ratio of estimated and observed means used to calculate the bias component (β) and the ratio of the estimated and observed coefficients of variation represent the variability component (γ).

$$KGE = 1 - \left((r-1)^2 + (\beta-1)^2 + (\gamma-1)^2\right)^{1/2} \quad (1)$$

The theoretical value for KGE to be equal 1 means that there is a perfect fit between the observed and simulated flows. Generally, KGE values above 0.6 are considered good.

4. Results and discussion

4.1 Analysis of precipitation and temperature
'Figure 2' presents mean annual temperature over the 1983-2012 period for the two selected temperature datasets.

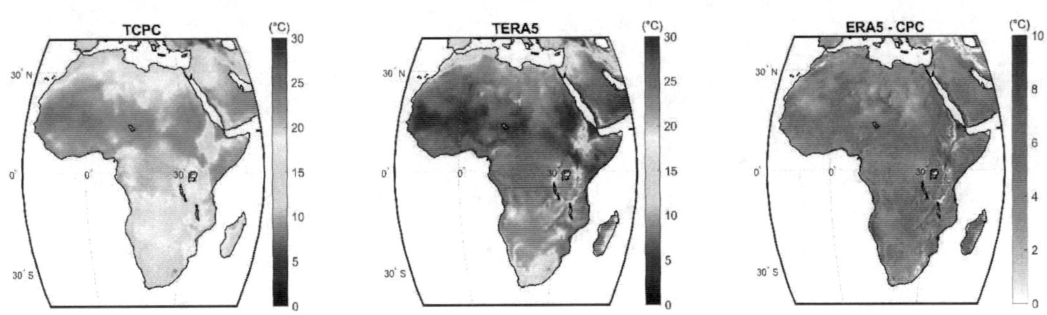

Figure2. Mean annual temperature for the two datasets and the bias between them

Both datasets display the same temporal patterns. ERA5 is however significantly warmer than CPC with a typical warm bias of 5-6 degrees over most of Africa. This difference is very large and canpotentially affect evapotranspiration.However, the specific calibration of the hydrological model to each dataset has the potential to take this into account.

To better present the differences between the datasets, the bias in the mean annual precipitation was calculated between each individual dataset and the average of all datasets. Results are shown in 'Figure 3'. The average here is considered as the reference bench mark. Since all the gridded datasets have different spatial resolution, the datasets were first interpolated to the finest grid scale. A red color indicates that the dataset is wetter than the average, while the blue color indicates it is dryer.Results show important differencesbetween the different precipitation datasets. All the datasets are generally similar in the desert and semi-desert regions butlarge differences are obvious in the tropical western and central regions.

Overall, the reanalysis (middle row) are wetter over the intertropical zone, with ERA5 being much closer than the other three considered reanalysis. The CPC gauge-based dataset is much drier than all other datasets. The large differences between both gauge-based datasets (CPC and GPCC) outline the complexity of interpolating in data-spare regions. Differences between the other datasets are comparatively smaller. In the absence of any reliable reference datasets, it is difficult to interpret the differences observed here. While an outlier dataset (e.g. CPC) may lead to suspicion, the limitations associated with each dataset does not allow for any firm conclusion. This is why hydrological modeling is used as an indirect validation method in this study. Even though streamflow gauges records do contain errors[41], in the context of this study, they are considered as the most reliable source for validation of the precipitation and temperature datasets.

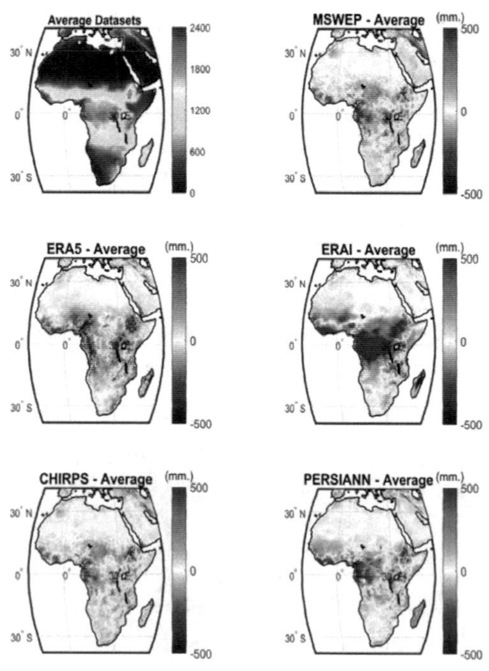

Figure3. Mean annual precipitation for the average of the all datasets (top left) and the bias from the average

4.2 Hydrological model simulations

This section presentsthe results obtained from the hydrological modelling simulations.'figure4' shows the distribution of KGE scores for each of the 18 combinations of precipitation (9 sets) and temperature (2 sets). Each boxplot in 'figure 4' contains the KGE scores of all of the catchments in this study.

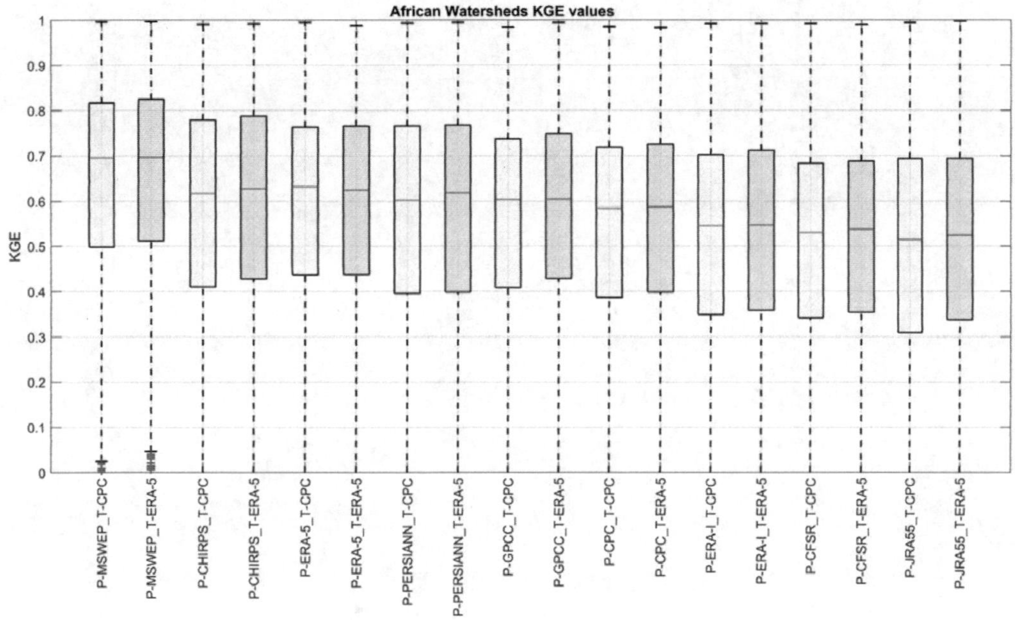

Figure4.KGE boxplots for nine precipitation and two temperature datasets

Many conclusions can be drawn from 'figure 4'. Both temperature datasetsperform very similarly across all precipitation datasets, although ERA5 gives very small but consistently better results. Most of the differences observed in 'figure 4' therefore originate from the precipitation datasets.

All precipitation datasets result in acceptable KGE median value larger than 0.5, showing they can all be used for hydrological modeling. There are however large differences with some datasets clearly outperforming others. The CPC and GPCC gauge-based datasets are outperformed by fivedatasets. The MSWEP merged-product isquite clearly the best-performing precipitation dataset, followed by theCHIRPSsatelliteand theERA-5 reanalysis datasets. The ERA-I, CFSR and JRA reanalysisare the least-performing datasets in this study.

In order to study the impact of spatial variability, 'figure 5' present the spatial distribution of KGE values for all nine precipitation datasets used in conjunction with ERA5 temperature.

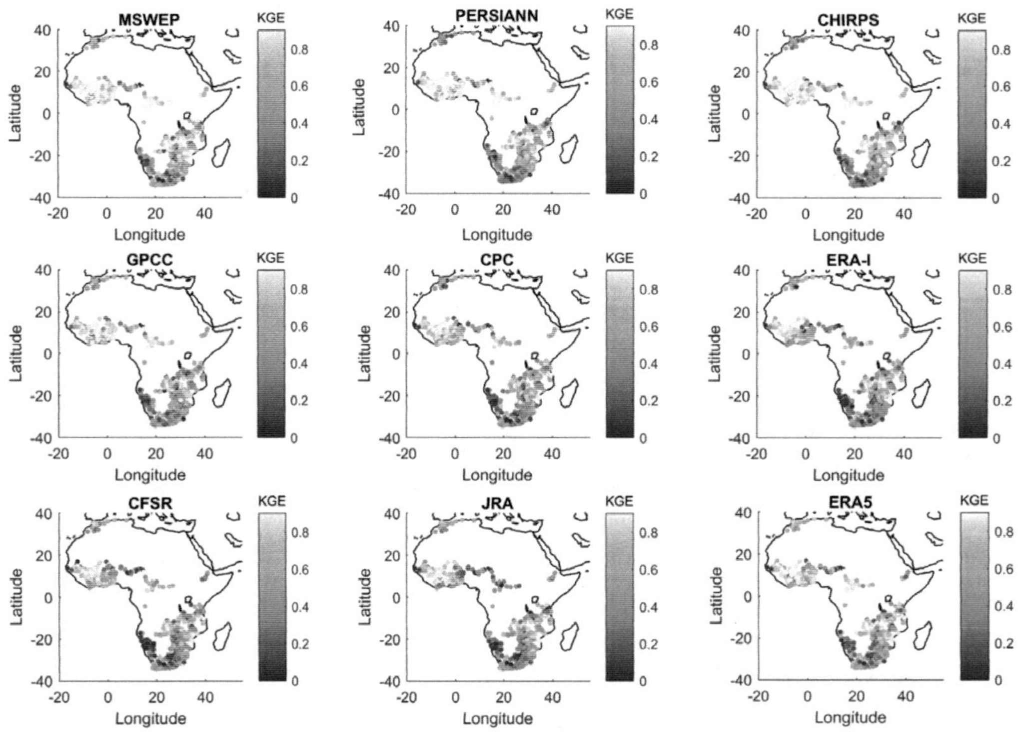

Figure5. Spatial distribution of Kling-Gupta efficiency metrics for nine precipitation datasets and ERA5 temperature datasets

The spatial patterns are consistent for all precipitation datasets. Hydrological modelling performance is general quite good everywhere with the exception of South Africa. This could either be due to less reliable streamflow records in this region or more likely to the hydrological model difficulties in dealing with the arid climate of south Africa. Rainfall-runoff models have long been known to have difficulties in such climates [42].

IOP Conf. Series: Materials Science and Engineering **974** (2020) 012001 doi:10.1088/1757-899X/974/1/012001

5. Conclusion

The main objective of this study was to evaluate the performance of nineprecipitation and twotemperature datasets to simulate streamflows of 850 African catchments over the 1983-2012period.The MSWEP merged-product dataset was clearly the best performing one, followed by CHIRPS and ERA5 products, respectively. The performance of both gauged-only datasets (CPC and GPCC) was inferior, outlining the limitations of extrapolating point-based measurement in data-sparse regions. Both temperature datasets performed similarly.

6. Acknowledgements and data access

This study was partly funded by the National Science and Engineering Research Council of Canada and the Egyptian Armed Forces (Ministry of Defense).

The CPC, GPCC and NCEP datasets can be downloaded from the Earth System Research Laboratory (ESRL), available here:https://www.esrl.noaa.gov/psd/data/gridded/tables/precipitation.html

ERA-Interim, ERA5 and JRA55 dataset are available on the Research Data Archive: https://rda.ucar.edu/datasets/ds628.0/

The GRDC streamflow data can be downloaded from the Global Runoff Data Centre, available here: https://www.bafg.de/GRDC/EN/Home/homepage_node.html;jsessionid=814972125050CA97A6F1CE67230E5CE6.live21303.

MSWEP data are available through the PCA servers at: https://platform.princetonclimate.com/PCA_Platform/mswepRetro.html.

CHIRPS satellite dataset can be downloaded from the Climate Hazards Center: https://www.chc.ucsb.edu/data/chirps

Finally, the HMETS hydrological model is available on the MATLAB File Exchange: https://www.mathworks.com/matlabcentral/fileexchange/48069-hmets-hydrological-model.

7. References

[1] F. J. Tapiador *et al.*, "Global precipitation measurement: Methods, datasets and applications," *Atmospheric Research,* vol. 104, pp. 70-97, 2012.

[2] P. Tschakert, R. Sagoe, G. Ofori-Darko, and S. N. Codjoe, "Floods in the Sahel: an analysis of anomalies, memory, and anticipatory learning," *Climatic Change,* vol. 103, no. 3-4, pp. 471-502, 2010.

[3] O. Rojas, A. Vrieling, and F. Rembold, "Assessing drought probability for agricultural areas in Africa with coarse resolution remote sensing imagery," *Remote sensing of Environment,* vol. 115, no. 2, pp. 343-352, 2011.

[4] S. E. Nicholson, "The West African Sahel: A review of recent studies on the rainfall regime and its interannual variability," *ISRN Meteorology,* vol. 2013, 2013.

[5] P. A. o. Omondi *et al.*, "Changes in temperature and precipitation extremes over the Greater Horn of Africa region from 1961 to 2010," *International Journal of Climatology,* vol. 34, no. 4, pp. 1262-1277, 2014.

[6] D. A. Vila, L. G. G. De Goncalves, D. L. Toll, and J. R. Rozante, "Statistical evaluation of combined daily gauge observations and rainfall satellite estimates over continental South America," *Journal of Hydrometeorology,* vol. 10, no. 2, pp. 533-543, 2009.

[7] C. Andermann, S. Bonnet, and R. Gloaguen, "Evaluation of precipitation data sets along the Himalayan front," *Geochemistry, Geophysics, Geosystems,* vol. 12, no. 7, 2011.

[8] S. Jiang *et al.*, "Comprehensive evaluation of multi-satellite precipitation products with a dense rain gauge network and optimally merging their simulated hydrological flows using the Bayesian model averaging method," *Journal of Hydrology,* vol. 452, pp. 213-225, 2012.

[9] S. Prakash, M. R. Kumar, S. Mathew, and R. Venkatesan, "How accurate are satellite estimates of precipitation over the north Indian Ocean?," *Theoretical and Applied Climatology,* vol. 134, no. 1-2, pp. 467-475, 2018.

[10] T. G. Romilly and M. Gebremichael, "Evaluation of satellite rainfall estimates over Ethiopian river basins," *Hydrology and Earth System Sciences,* vol. 15, no. 5, p. 1505, 2011.

[11] S. Chen *et al.*, "Evaluation of high-resolution precipitation estimates from satellites during July 2012 Beijing flood event using dense rain gauge observations," *PloS one,* vol. 9, no. 4, 2014.

[12] A. Behrangi, B. Khakbaz, T. C. Jaw, A. AghaKouchak, K. Hsu, and S. Sorooshian, "Hydrologic evaluation of satellite precipitation products over a mid-size basin," *Journal of Hydrology,* vol. 397, no. 3-4, pp. 225-237, 2011.

[13] M. Tarek, F. P. Brissette, and R. Arsenault, "Evaluation of the ERA5 reanalysis as a potential reference dataset for hydrological modeling over North-America," *Hydrology and Earth System Sciences Discussions,* pp. 1-35, 2019.

[14] Z. Wu *et al.*, "Hydrologic evaluation of multi-source satellite precipitation products for the upper Huaihe River Basin, China," *Remote Sensing,* vol. 10, no. 6, p. 840, 2018.

[15] Z. Duan *et al.*, "Hydrological evaluation of open-access precipitation and air temperature datasets using SWAT in a poorly gauged basin in Ethiopia," *Journal of hydrology,* vol. 569, pp. 612-626, 2019.

[16] H. Zhu, Y. Li, Y. Huang, Y. Li, C. Hou, and X. Shi, "Evaluation and hydrological application of satellite-based precipitation datasets in driving hydrological models over the Huifa river basin in Northeast China," *Atmospheric Research,* vol. 207, pp. 28-41, 2018.

[17] A. P. Sayre, *Africa*. Twenty-First Century Books, 1999.

[18] M. Hulme, R. Doherty, T. Ngara, M. New, and D. Lister, "African climate change: 1900-2100," *Climate research,* vol. 17, no. 2, pp. 145-168, 2001.

[19] J. K. W. Ngaira, "Impact of climate change on agriculture in Africa by," *Scientific Research and Essays,* vol. 2, no. 7, pp. 238-243, 2007.

[20] C. J. Skinner, T. J. Bellerby, H. Greatrex, and D. I. Grimes, "Hydrological modelling using ensemble satellite rainfall estimates in a sparsely gauged river basin: The need for whole-ensemble calibration," *Journal of hydrology,* vol. 522, pp. 110-122, 2015.

[21] K. A. Adjei, L. Ren, E. K. Appiah-Adjei, and S. N. Odai, "Application of satellite-derived rainfall for hydrological modelling in the data-scarce Black Volta trans-boundary basin," *Hydrology Research,* vol. 46, no. 5, pp. 777-791, 2015.

[22] S. A. Koriche and T. H. Rientjes, "Application of satellite products and hydrological modelling for flood early warning," *Physics and Chemistry of the Earth, Parts A/B/C,* vol. 93, pp. 12-23, 2016.

[23] K. M. Bâ *et al.*, "Spatially distributed hydrological modelling of a Western Africa basin," *EPiC Ser. Eng,* vol. 3, pp. 343-350, 2018.

[24] F. A. Isotta *et al.*, "The climate of daily precipitation in the Alps: development and analysis of a high-resolution grid dataset from pan-Alpine rain-gauge data," *International Journal of Climatology,* vol. 34, no. 5, pp. 1657-1675, 2014.

[25] B. Martens, P. Cabus, I. De Jongh, and N. Verhoest, "Merging weather radar observations with ground-based measurements of rainfall using an adaptive multiquadric surface fitting algorithm," *Journal of hydrology,* vol. 500, pp. 84-96, 2013.

[26] Y. Tian *et al.*, "Component analysis of errors in satellite-based precipitation estimates," *Journal of Geophysical Research: Atmospheres,* vol. 114, no. D24, 2009.

[27] P. Xie *et al.*, "A gauge-based analysis of daily precipitation over East Asia," *Journal of Hydrometeorology,* vol. 8, no. 3, pp. 607-626, 2007.

[28] J. Awange, V. Ferreira, E. Forootan, S. Andam-Akorful, N. Agutu, and X. He, "Uncertainties in remotely sensed precipitation data over Africa," *International Journal of Climatology,* vol. 36, no. 1, pp. 303-323, 2016.

[29] M. Di Luzio, G. L. Johnson, C. Daly, J. K. Eischeid, and J. G. Arnold, "Constructing retrospective gridded daily precipitation and temperature datasets for the conterminous United States," *Journal of Applied Meteorology and Climatology,* vol. 47, no. 2, pp. 475-497, 2008.

[30] M. G. Bosilovich, "Regional climate and variability of NASA MERRA and recent reanalyses: US summertime precipitation and temperature," *Journal of applied meteorology and climatology,* vol. 52, no. 8, pp. 1939-1951, 2013.

[31] H. Nieto, I. Sandholt, I. Aguado, E. Chuvieco, and S. Stisen, "Air temperature estimation with MSG-SEVIRI data: Calibration and validation of the TVX algorithm for the Iberian Peninsula," *Remote Sensing of Environment,* vol. 115, no. 1, pp. 107-116, 2011.

[32] J.-L. Martel, K. Demeester, F. P. Brissette, R. Arsenault, and A. Poulin, "HMET: a simple and efficient hydrology model forteaching hydrological modelling, flow forecasting and climate change impacts," *The International journal of engineering education,* vol. 33, no. 4, pp. 1307-1316, 2017.

[33] C. Perrin, C. Michel, and V. Andréassian, "Improvement of a parsimonious model for streamflow simulation," *Journal of hydrology,* vol. 279, no. 1-4, pp. 275-289, 2003.

[34] R. Arsenault, F. Brissette, and J.-L. Martel, "The hazards of split-sample validation in hydrological model calibration," *Journal of hydrology,* vol. 566, pp. 346-362, 2018.

[35] L. Oudin *et al.*, "Which potential evapotranspiration input for a lumped rainfall–runoff model?: Part 2—Towards a simple and efficient potential evapotranspiration model for rainfall–runoff modelling," *Journal of hydrology,* vol. 303, no. 1-4, pp. 290-306, 2005.

[36] P. Baguis, E. Roulin, P. Willems, and V. Ntegeka, "Climate change scenarios for precipitation and potential evapotranspiration over central Belgium," *Theoretical and applied climatology,* vol. 99, no. 3-4, p. 273, 2010.

[37] N. Hansen, S. D. Müller, and P. Koumoutsakos, "Reducing the time complexity of the derandomized evolution strategy with covariance matrix adaptation (CMA-ES)," *Evolutionary computation,* vol. 11, no. 1, pp. 1-18, 2003.

[38] R. Arsenault, A. Poulin, P. Côté, and F. Brissette, "Comparison of stochastic optimization algorithms in hydrological model calibration," *Journal of Hydrologic Engineering,* vol. 19, no. 7, pp. 1374-1384, 2014.

[39] H. V. Gupta, H. Kling, K. K. Yilmaz, and G. F. Martinez, "Decomposition of the mean squared error and NSE performance criteria: Implications for improving hydrological modelling," *Journal of hydrology,* vol. 377, no. 1-2, pp. 80-91, 2009.

[40] H. Kling, M. Fuchs, and M. Paulin, "Runoff conditions in the upper Danube basin under an ensemble of climate change scenarios," *Journal of Hydrology,* vol. 424, pp. 264-277, 2012.

[41] G. Di Baldassarre and A. Montanari, "Uncertainty in river discharge observations: a quantitative analysis," *Hydrology & Earth System Sciences,* vol. 13, no. 6, 2009.

IOP Conf. Series: Materials Science and Engineering **974** (2020) 012002 doi:10.1088/1757-899X/974/1/012002

Military Technical College
Kobry El-Kobbah,
Cairo, Egypt

13th International Conference on Civil and Architecture Engineering ICCAE-13-2020

WR-5

Using Remote Sensing Techniques to Determine New Route for the Nile Valley

Osama M. Moussa[1], Ahmed S. Khalil[1], Hasan E. Elhefnawy[2] and Mohamed M. Kamal[2]

[1]Military Technical College,[2]Egyptian Armed Defense

Abstract: The collapse of dam structure leads to serious disasters for surrounding areas along the waterway. The objective of this research is to determine a new route for the waterway by using remote sensing techniques.
Drainage patterns were determined from Digital Elevation Model (DEM) of Shuttle Topographic Radar Mission (STRM) by using the ARC-GIS software. The most appropriate route for the new valley is determined by using DEM and assisted by topographic maps of scale 1:500000 starting from Toshka lake until Qattara downrange and passing through the depressions of El Kharga oasis, El-Dakhla, El-Farafra and El-Baharia Oases.
The route selection is matched with the expected new valley that proposed by Prof. Dr. Farouk El-Baz in his project " The Development Project / Corridor of Development Project [1].

Keyword: DEM , ARC-GIS, Remote sensing, Drainage pattern, Route selection.

1. Introduction

Most dams have a spillway or weir over through which water flows, either discontinuously or continuously and some dams have hydroelectric power generation systems installed.
Dam failures can cause enormous damage and loss of souls when they occur. The most common reason behind the failure of dam is overtopping. Overtopping could be a major failure because of significant floods.
Drainage patterns were used to delineate the appropriate route selection for the flooded water that yielded from dam failure. These patterns were determined from Digital Elevation Model (DEM) of Shuttle Topographic Radar Mission (STRM) by the aid of ARC-GIS software.
Topographic maps of scale 1:500000 for the Western desert starting from Toshka lake until Qattara downrange were used to clarify and assist the recommended route that been selected by drainage patterns inspection.
To transfer water from the Toshka spillway to the Qattara downrange, some hydrological constructions are needed. In order to justify these constructions, the mentioned topographic maps were used. The route selection is matched with the expected new valley that proposed by Prof.

IOP Conf. Series: Materials Science and Engineering **974** (2020) 012002 doi:10.1088/1757-899X/974/1/012002

Dr. Farouk El-Baz in his project "The Development Project / Corridor of Development Project [1].

2. Reasons and causes of dam failures

There are many reasons of dam failure that face most countries all over the world and well known to the hydrologists, engineers and scientists who works in this field. The most common reason behind the failure of dam is overtopping which caused by water spilling over the top of a dam and usually a precursor of dam failure. Foundation issues and defects, including slope instability, is the root cause behind about 30% of all dam failures. Moreover, cracking caused by movements like the natural settling of a dam is one of important reason of dam failure. On the other hand, inadequate maintenance and upkeep is of the same effect of dam failure. Another reason of dam failure is piping, when seepage through a dam is not properly filtered and soil particles continue to progress and form sink holes in the dam, see Figure (1).

There are many causes of dam's failure. These causes can be summarized as follows: (I) Human error, error in design of the spillway and poor maintenance; (II) Natural disasters, earthquakes, extreme inflow, earth's instability etc.; (III) construction materials & techniques used.

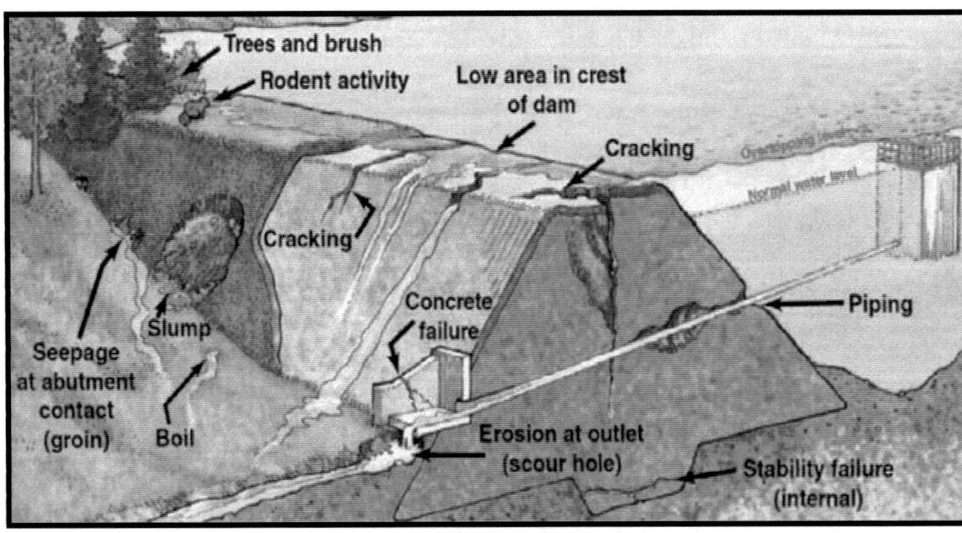

Figure (1) Various Reasons of Earth dam's Failure [8].

3. Study Area

The study area extended from the Aswan high dam (AHD) reservoir in upper Egypt and ended by El- Qattara downrange, north part of Egypt (Close to the Mediterranean Sea).

The study area includes the western dessert starting from the Toshka Lake and passing through El Kharga, El-Dakhla, El-Farafra, El-Baharia and ended by the El-Qattara downrange.

3.1 Aswan High Dam

The Aswan High Dam is an embankment dam that was built across the Nile in Aswan, Egypt between 1960 and 1970, see Figure (2). The dam, completed in 1968 at a distance of 7 km south of Aswan City, is a rock fill dam made of granite rocks, sands, and provided with a vertical cutoff wall consisting of impermeable clay [2]. The Aswan High Dam is 3,830 Meter (12,570 ft) long, 40 m (130 ft) wide at the crest, 980 m (3,220 ft) wide at the base and 111m (364ft) tall. It contains 43,000,000 cubic Meter (56,000,000 cu yd) of material. At maximum, 11,000 cubic Meter per second (390,000 cu ft/s) of water can pass through the dam [2].

Figure (2) satellite image of Aswan High Dam [2].

3.2 Aswan High Dam Reservoir (AHDR).

Aswan High Dam Reservoir, see Figure (3) extends for 500 km along the Nile River and covers an area of 6,000 km^2, it holds 132 km^3 of water of which northern two-thirds (known as Lake Nasser) is in Egypt and tierce (called Lake Nubia) in Sudan. [3]
The reservoir has 100 side arms called khors. The total capacity of the reservoir (163.3 km^3) consists of the dead storage of 31.6 km^3 (85 - 147 m) of lake water amount, the active saved storage of 90.7 km^3 (147 - 174 m) and the emergency saved storage for flood protection of 41 km^3 (175 - 182 m). [4]

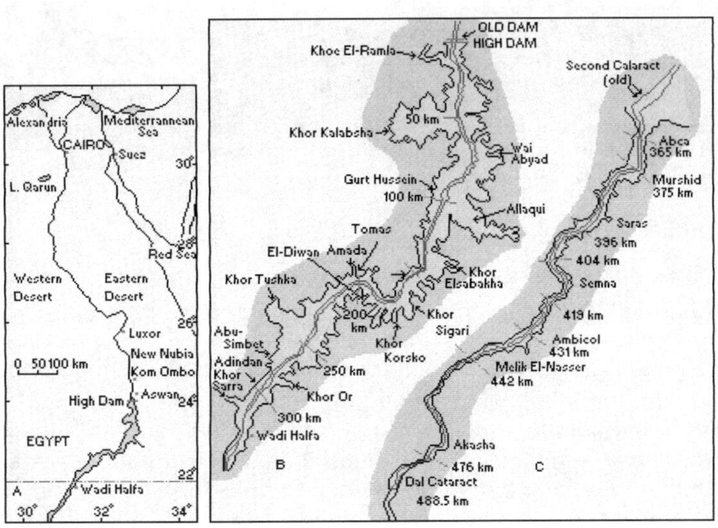

Figure (3) Aswan High Dam Reseviour [5]

IOP Conf. Series: Materials Science and Engineering **974** (2020) 012002 doi:10.1088/1757-899X/974/1/012002

3.3 Toshka Depression

NASA satellite image (October2000) of Lake Nasser and the Western Desert of Egypt. Shows the four lakes, known as the Toshka Lakes which lie west of Lake Nasser. In 1978, the Egyptians constructed a canal known as the Toshka Spillway or Sadat Canal from Lake Nasser into a low area to the west to allow spill over of Lake Nasser water if the reservoir filled above 178m. Despite the fact that the spillway was created in 1978, It was not until the late1990 that Lake Nasser actually filled up to the level of the spillway canal and the overflow lakes began to form. In November of 1998, US astronauts orbiting within the space craft detected the lakes filling for the primary time, and that they have had water in them ever since. Figure (4) close-up views of the Toshka Depression in 2004.

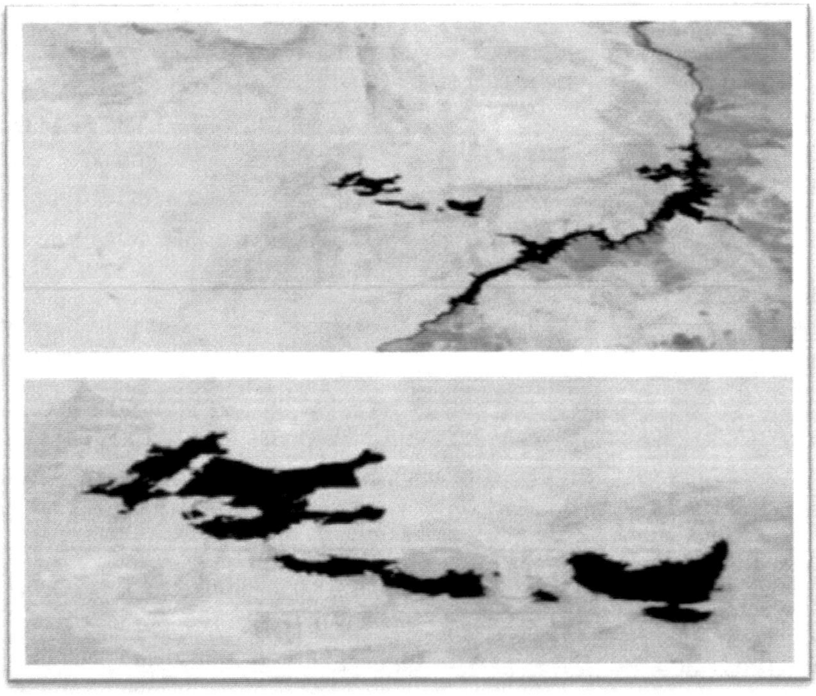

Figure (4) Toshka Lakes in 2004 [6]

3.4 Western Desert.

The area is characterized by a hot and dry summer with rare winter rainfall and bright sunshine through the year. The average annual temperature is 26.4 °C, while the average of evaporation is 7.76 mm.

El Kharga oasis is the nearest oasis to Luxor city and it is the capital of the New Valley. The Kharga depression is hemmed in by 300 m high cliffs, with belts of dunes advancing across the oasis. Many desert routes converged on the oasis, noted by the Forty Days Road.

IOP Conf. Series: Materials Science and Engineering **974** (2020) 012002 doi:10.1088/1757-899X/974/1/012002

El Dakhla oasis lies in the Nile Valley governorate, 350 Km from the Nile and lie between El Farafra and El Kharga oases. It is extended about 80 Km from East to West and 25 Km from North to South and covers an area of 2000 Km^2.
EL Farafra oasis lies at a distance of 170 Km South el Bahria oases. It belongs to the New Valley governorate and lies at level of 76 m above mean sea level. EL Farafra Formation tops with clear slopes are made out of independent bowls, which differ in size. Al Gunnah playa, which is situated at the upper east foot incline of Al Quss Abu Said level, covers region of more than 100 km^2 and It is one of the most established agrarian and settlement zones in the Farafra.
El Bahariya oasis is approximately 370 Km away from Cairo. The roughly oval valley extends from North East to South West, has a length of 94 Km with maximum width of 42 Km and covers an area of 2000 Km^2; with an average width of 21.28 Km. It is located in El Giza governorate.

3.5 Qattara Depression.

The Qattara Depression is the largest underground depression below the sea level in the world. This depression extends about 300 km from the Oasis of Mugra, south of El Alamein, with a distance of about 53 km to Siwa Oasis in the west. While its maximum width is about 137 km. The area of Qattara depression is about 19516 km^2. Several sub depressions were attached to this depression with a total area of about 1572 km^2. Thus, the total area of the main depression and its attachments from the sub depressions is 21088 km^2. An area of 20695 km^2 is under sea level with a maximum of -139.0 meters below sea level and average of 50m.

4. Methodology

DEM was constructed from shuttle Topographic Rader Mission (STRM), then using Geographic Information System (Arc-GIS) version 10.3 to construct watershed for each depression along the route of the study area.
The available data is 30m Digital Elevation Model (DEM) for the study area, extracted from Shuttle Topographic Rader Mission (STRM).
The SRTM (Shuttle Radar Topography Mission) database is a remarkable worldwide digital elevation data set for about 80% of the Earth's land surface. It took 10 days to collect all of the data. The data set covers all land areas between 60°N and 56°S latitude, higher latitudes were excluded because of the inclination of the Shuttle orbit which was 56°S. The resolution of SRTM knowledge is thirty meter per pixel for the best resolution knowledge.
Digital Elevation Model (DEM) represents ground surface and permit computer software direct extraction of hydrological options, thus it gives advantages in terms of processing efficiency, cost effectiveness and accuracy assessment, compared with traditional methods based on topographic maps, field surveys or photographic interpretations [7]. For the last few decades, DEMs are widely used for resource management, urban planning, transportation planning, earth sciences, and environmental assessments and Geographic Information System (GIS) applications. In this study the raster DEM is used as a base map layer to derive the topographic phenomenon of the watershed such as area, slope and field slope length. Resolution of pixel represents the horizontal accuracy of DEM. The topography of the hydrologic modeling is based on spatial resolution of the model that perform optimally.

5. Expected Route Selection using ARC – GIS

In order to avoid flow disaster that's caused from dam failure, it's obligatory to select a new route of water far away from the old river valley. Primary inspection for route selection was carried out by reviewing the topographic maps of scale 1:500,000 and satellite images which cover the western desert of Egypt.

IOP Conf. Series: Materials Science and Engineering **974** (2020) 012002 doi:10.1088/1757-899X/974/1/012002

The expected route is started from the western side of Lake Nasser, the long arm of Lake Nasser that focuses at the Toshka Lakes as shown in the satellite image, see Figure (4). It extends over the Toshka Depression, and broadens north along the old Drab el-Arba'in through the desert springs of Baris and El-Kharga. This route extends through El-Farafra Oasis, passing by El-Baharia Oasis and ends to the Qattara downrange.

DEM for western desert from Toshka spillway to el-Qattara downrange is obtained from Shuttle Topographic Rader Mission STRM as shown in Figure (5.1). Then Geographic Information System (Arc-GIS) was utilized for watershed display activities.

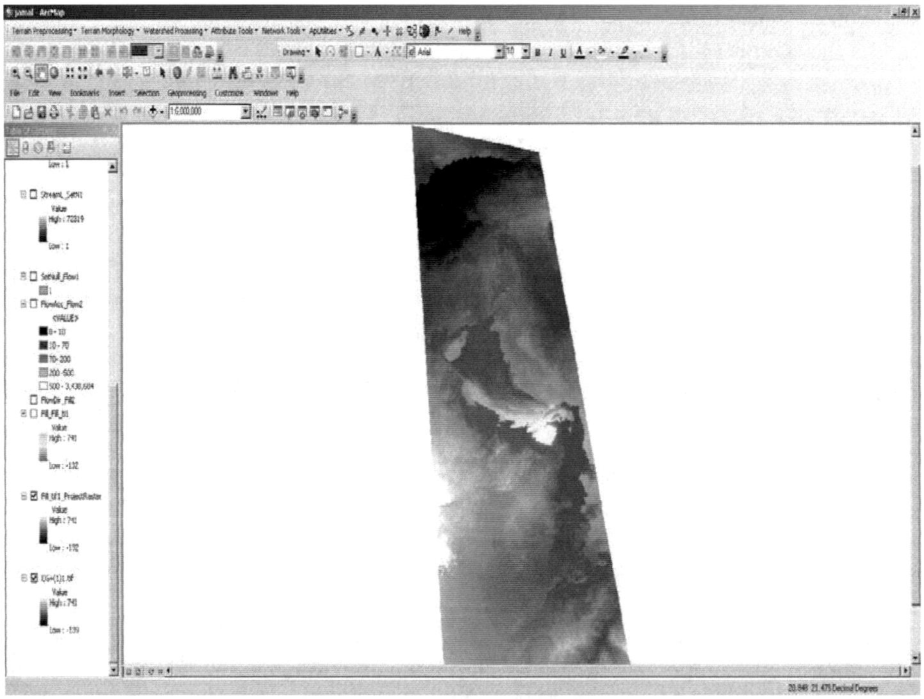

Figure 5.1 DEM of western desert

Image of western desert was embedded in Arc-GIS model and the dataset were transformed from (Latitude, longitude) projection to WGS_1984_UTM projection framework (N, E) Coordinates.

Hydrology selected from spatial analysis tool. This subroutine consists of the following steps, subtitles **5.1** through **5.8**, until the drainage pattern of each basin clarified.

5.1 Fill (Spatial Analyst):

Fills and sinks in a surface raster is used to remove small imperfections of the data as shown in Figure (5.2).

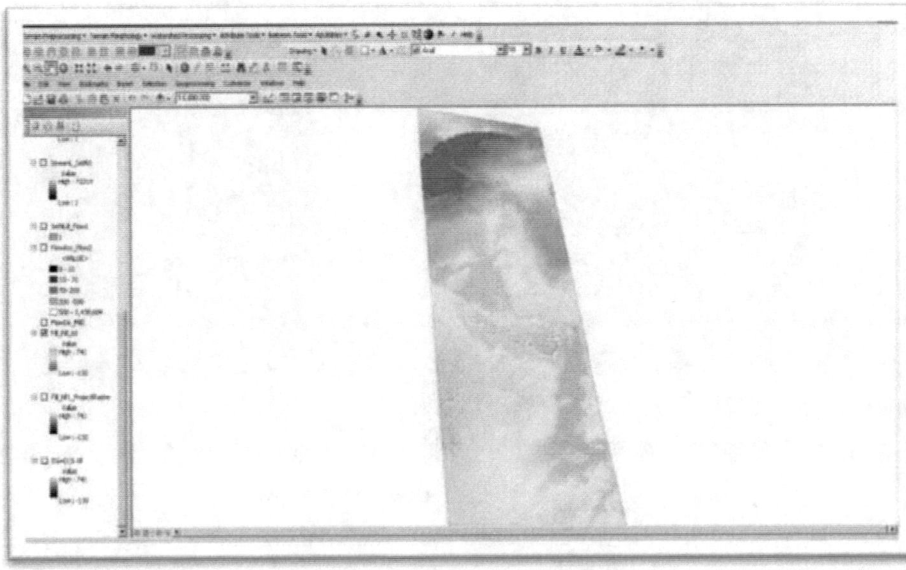

Figure 5.2 fill (Spatial Analyst)

5.2 Flow direction:

This creates a raster of flow direction from each cell to the steepest down slope neighbor as shown in Figure (5.3).

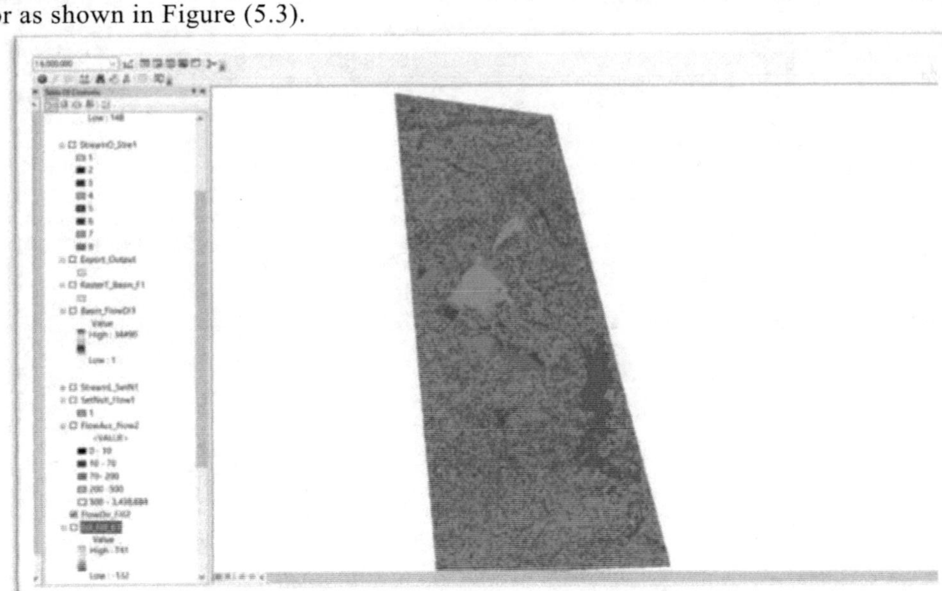

Figure 5.3 flow direction

5.3 Flow Accumulation:

This creates a raster of accumulated flow into each cell. A weight factor can optionally be applied see Figure (5.4).

IOP Conf. Series: Materials Science and Engineering **974** (2020) 012002 doi:10.1088/1757-899X/974/1/012002

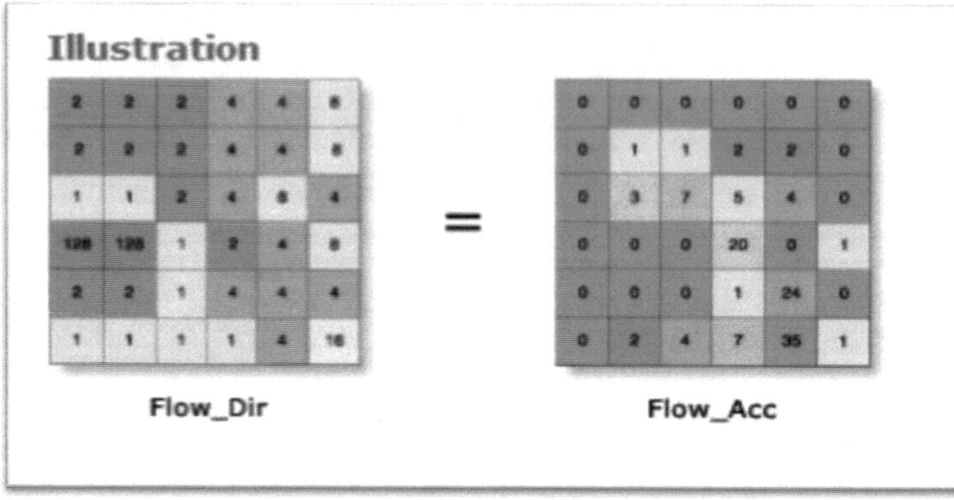

Figure 5.4 flow dir. & flow acc.

5.4 Stream link:
This step assigns unique values to sections of a raster linear network between intersections.

5.5 Stream Order:
This assigns a numeric order to segments of a raster representing branches of a linear network as shown in Figures (5.5 & 5.6).

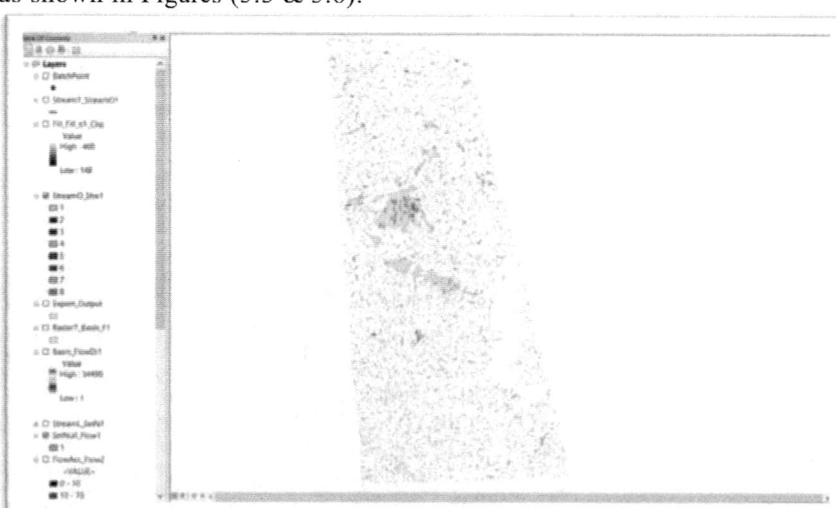

Figure 5.5 Stream Order

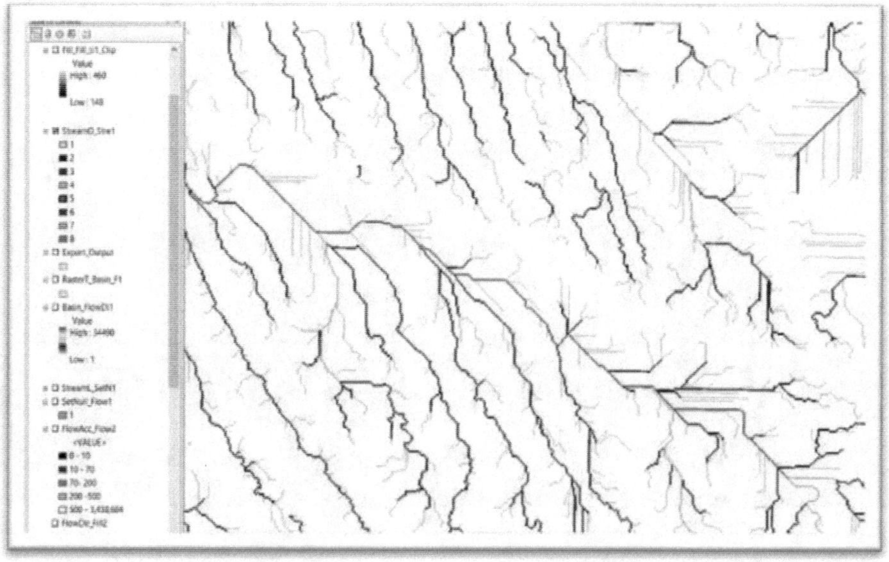

Figure 5.6 Stream Order (drainage pattern)

Stream order of the drainage pattern is divided into eight colored elevation levels from high to low order which can follow the direction of water related to the slope of drainage pattern orders as shown in Figure (5.6), then we can follow the red one which describe the slope in the study area.

5.6 Stream to feature:

This step converts a raster of linear network to features representing the linear network, see Figure (5.7).

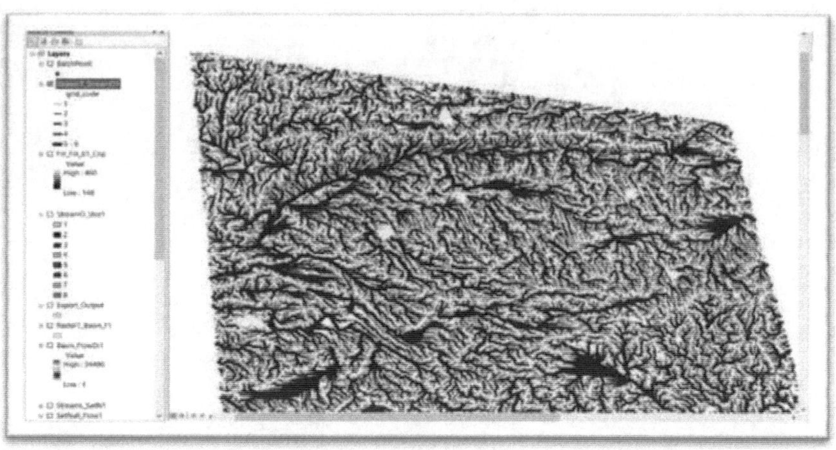

Figure 5.7 Stream to future

IOP Conf. Series: Materials Science and Engineering **974** (2020) 012002 doi:10.1088/1757-899X/974/1/012002

5.7 Define Basin:

Which Creates a raster delineating all drainage basins, see Figure (5.8).

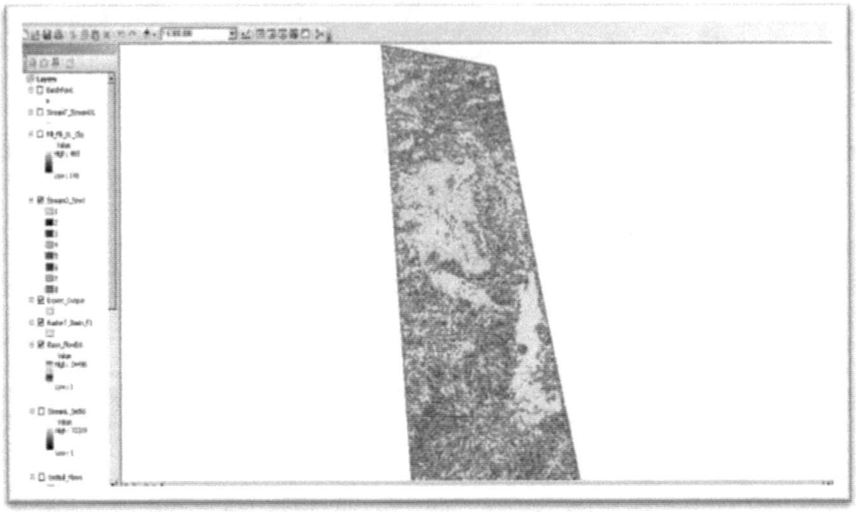

Figure 5.8 Define basin

5.8 Basin flow direction:

The drainage basins are delineated within the analysis window by identifying ridgelines between basins. The flow direction raster is analyzed to find all sets of connected cells that belong to the same drainage basin. The drainage basins are created by locating the pour points at the edges of the analysis window (where water would pour out of the raster), as well as sinks, then identifying the contributing area above each pour point. This results in a raster of drainage basins, see Figure (5.9).

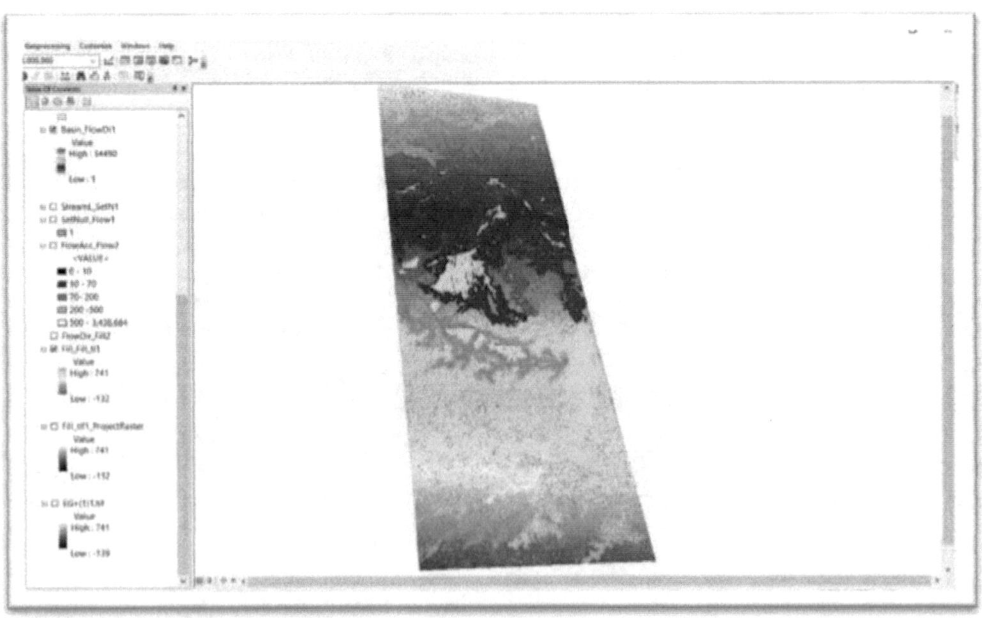

Figure 5.9 Basin Flow directions

IOP Conf. Series: Materials Science and Engineering **974** (2020) 012002 doi:10.1088/1757-899X/974/1/012002

Drainage pattern from Toshka to El-Qattaea downrange:

The Arc-GIS model is delineating the boundary of the watershed to collect the water flooded from Aswan reservoir.

a- From Toshka to El_Dakhla:

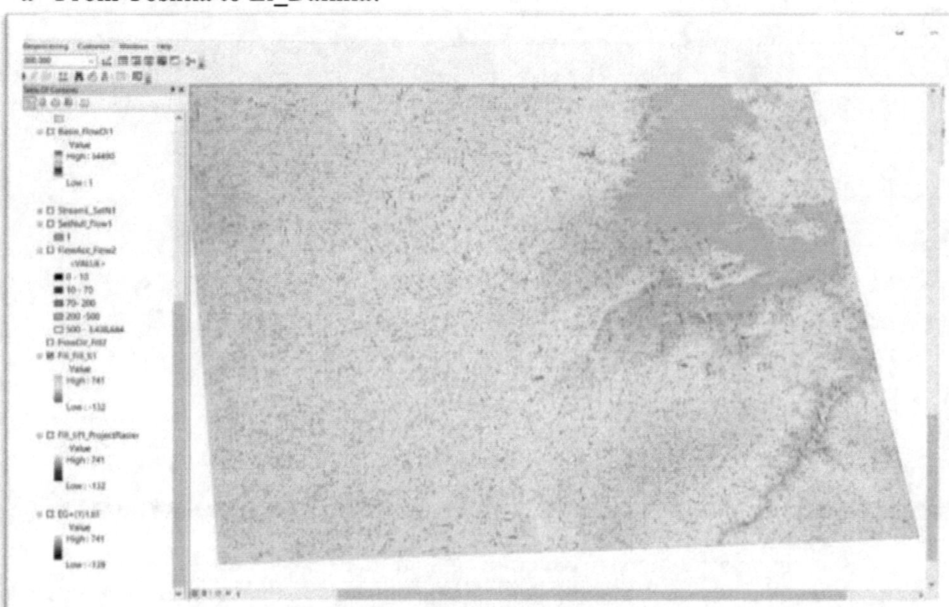

Figure 5.10 drainage pattern from Toshka to El_Dakhla

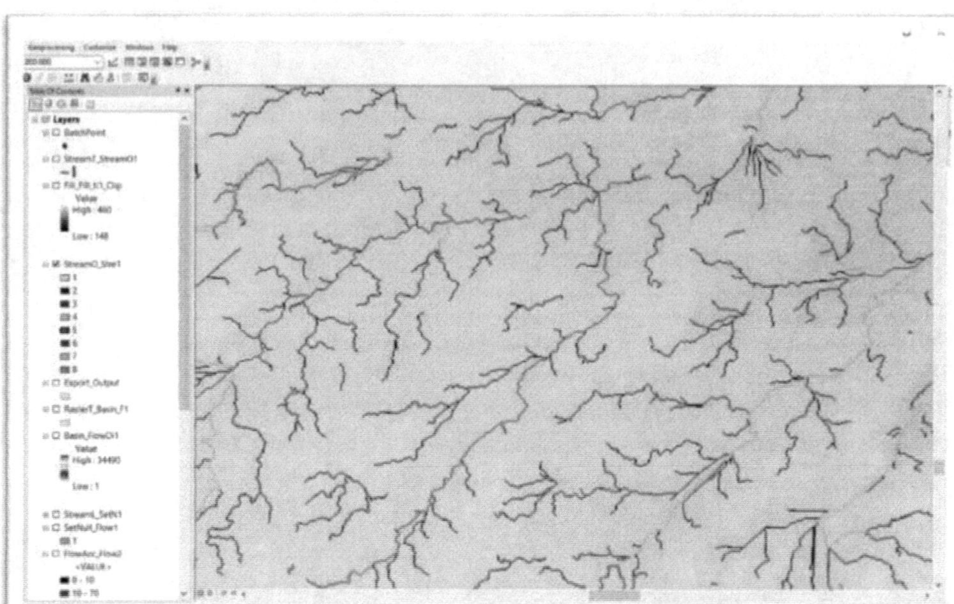

Figure 5.11 drainage pattern from Toshka to El_Dakhla.

b- From Farafra to Elbaharea:

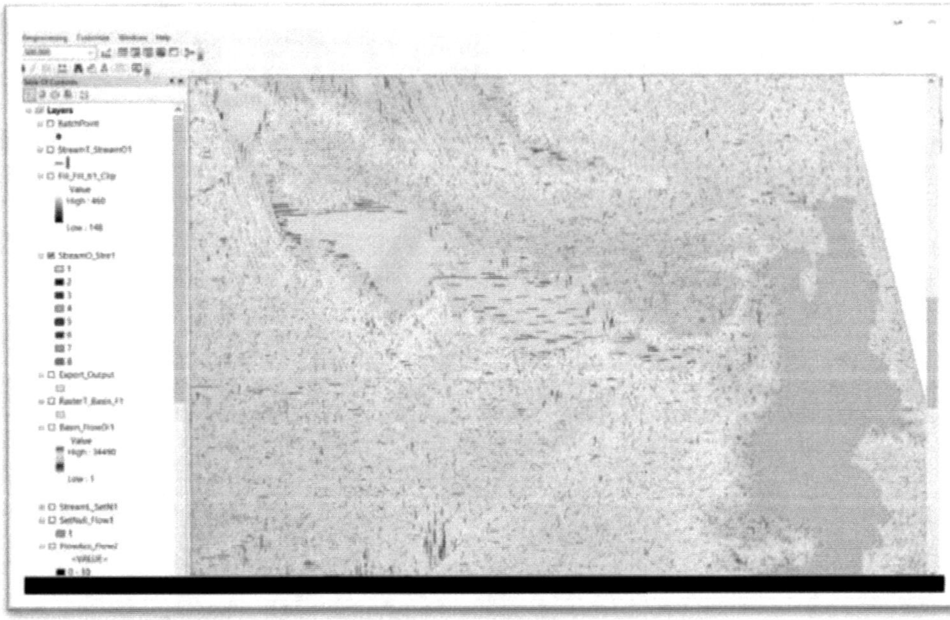

Figure 5.12 drainage pattern from Farafra to Elbaharea

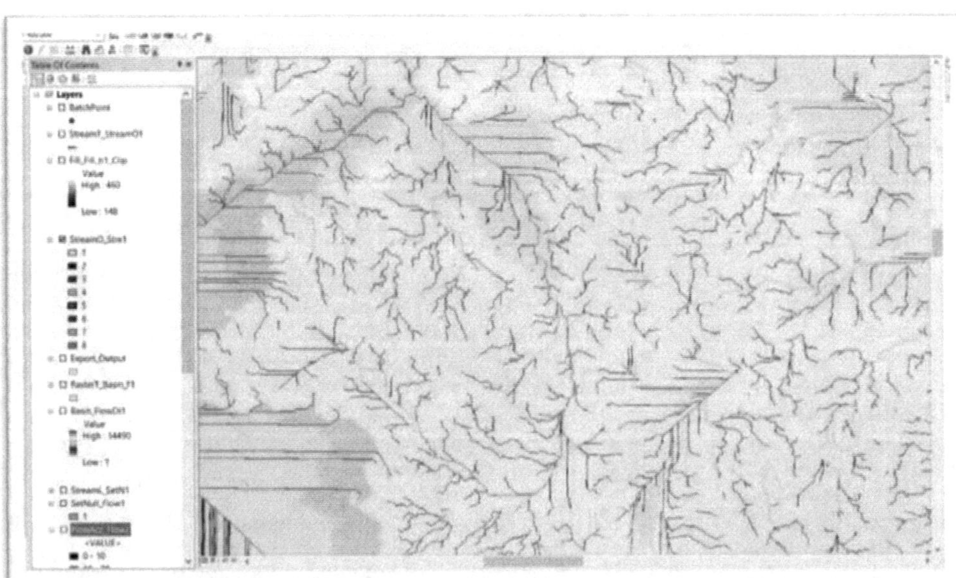

Figure 5.13 drainage pattern from Farafra to Elbaharea.

c- From Elbaharea to El-Qattara downrange:

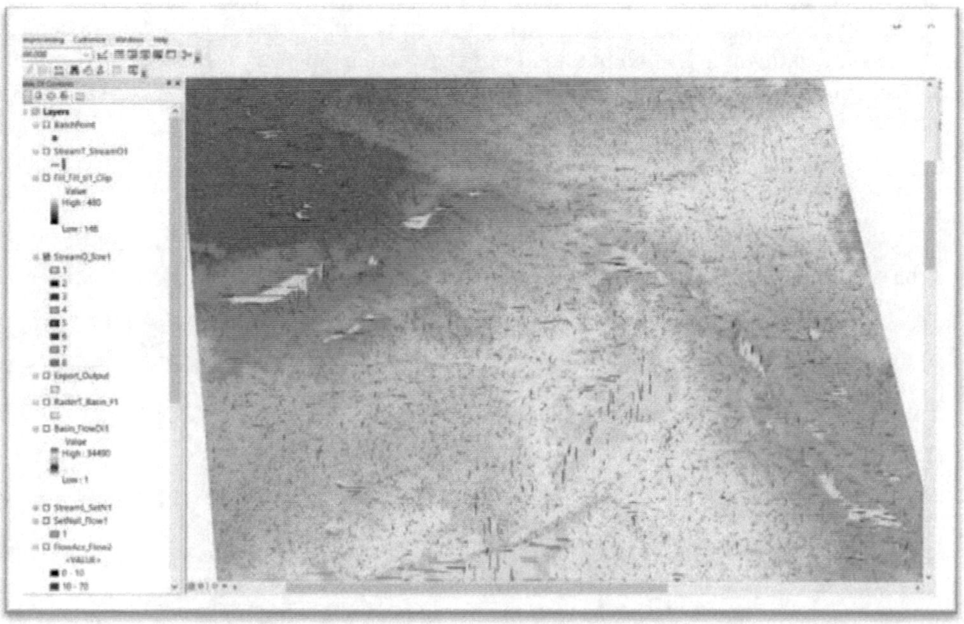

Figure 5.14 drainage pattern from Elbaharea to El-Qattara downrange

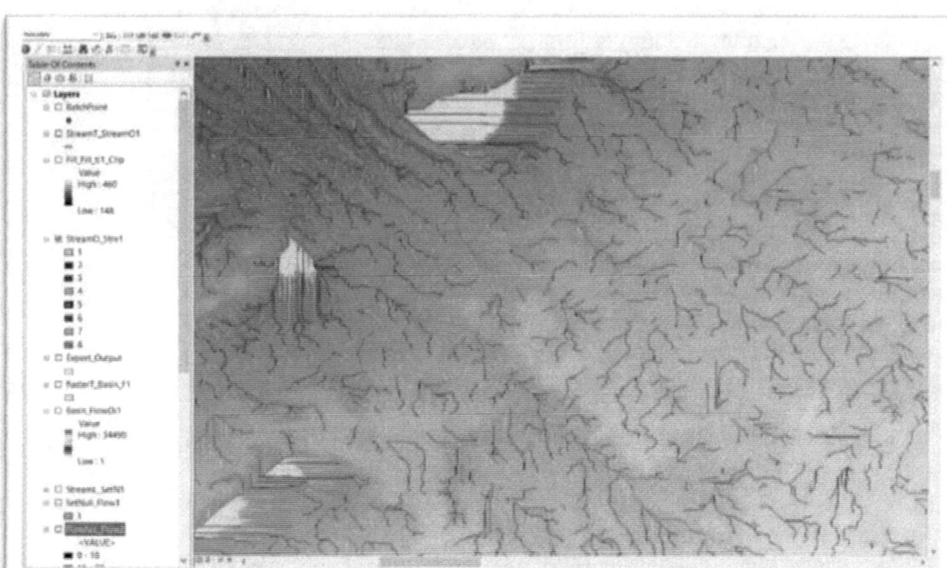

Figure 5.15 drainage pattern from Elbaharea to El-Qattara downrange

IOP Conf. Series: Materials Science and Engineering **974** (2020) 012002 doi:10.1088/1757-899X/974/1/012002

6. Topographic Maps usage for route construction projects.

New Valley route selection can be determined from hydrological study of the region in order to choose the most appropriate path for the construction to transfer water from the Toshka spillway to the Qattara downrange and to reduce the budget of industrial works to be carried out at that distance, estimated at 800 km, topographic maps of the scale (1: 500000) were used to justify these works.

1- Luxor map
   Constructing a tunnel corridor at coordinate 30° 38′ E & 24° 27′ N through the Darb al-Arba'een road, see Figure (6.1)

Figure 6.1 Construction works in Luxor

2- Dakhla map
   a-Construct bridge at coordinate 28° 58′ E & 25° 28′ N or made tunnel in the same valley path and pass down the road.
   b- Construct tunnel at coordinate 29° 38′ E & 25° 15′ N for a distance 1km then another tunnel at coordinate 29° 30′ E & 25° 20′ N for a distance 2km, see Figure (6.2).

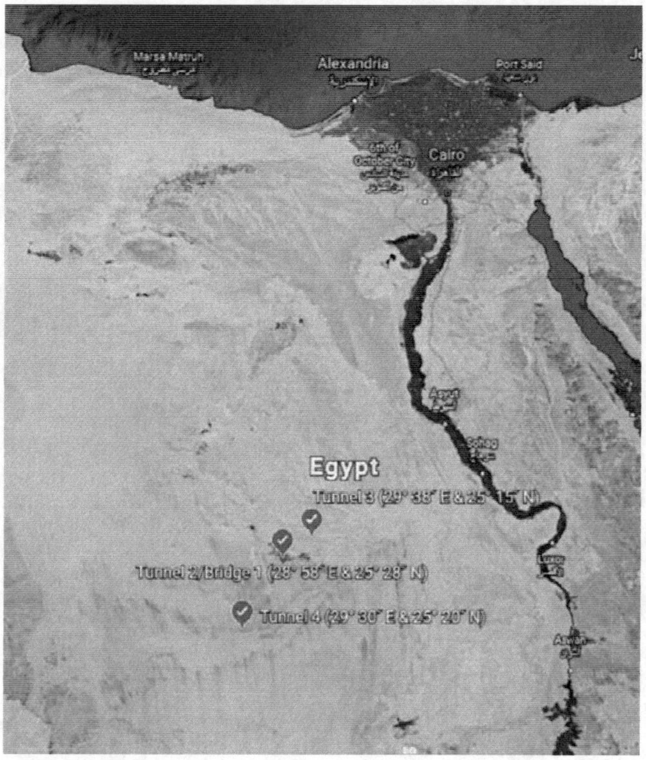

Figure 6.2 Construction works in Dakhla

3- Farafra map
   a-Construct industrial work (bridge or tunnel) at coordinate 27° 50′ E & 26° 41′ N.
   b- Construct industrial work (bridge or tunnel) at coordinate 28° 12′ E & 27° 28′ N.
   c-Digging a distance of about 17.5km from point 27° 31′ E & 26° 33′N to point 27° 42′ E & 26° 38′ N, see Figure (6.3).

IOP Conf. Series: Materials Science and Engineering **974** (2020) 012002 doi:10.1088/1757-899X/974/1/012002

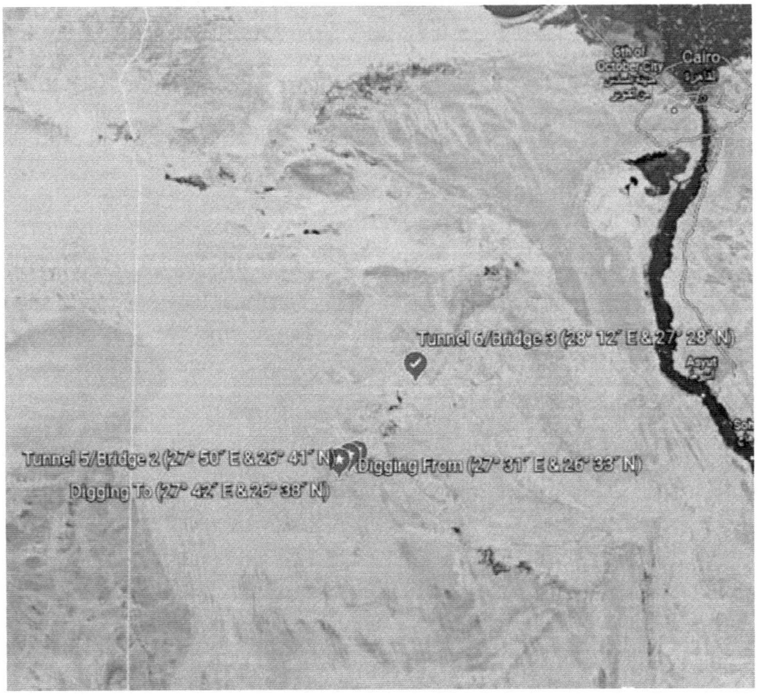

Figure 6.3 Construction works in Farafra

4- Baharia Map
Construct industrial work (bridge or tunnel) at coordinate 27° 40´ E & 28° 41´ N, see Figure (6.4).

IOP Conf. Series: Materials Science and Engineering **974** (2020) 012002 doi:10.1088/1757-899X/974/1/012002

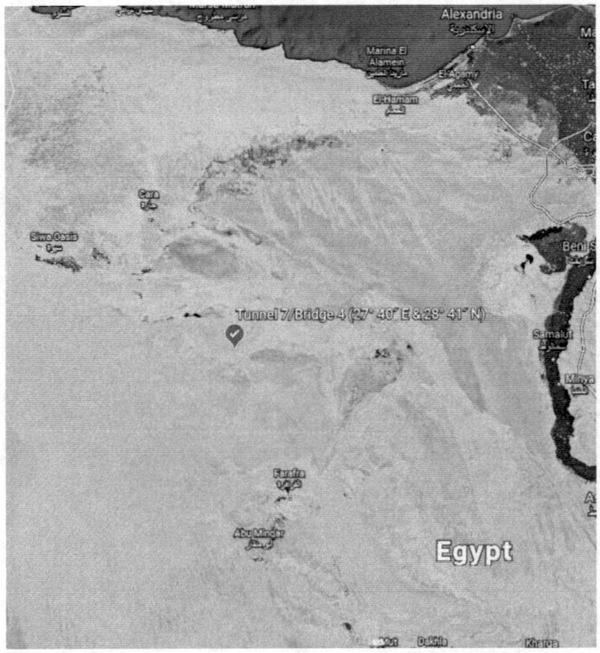

Figure 6.4 Construction works in Baharia

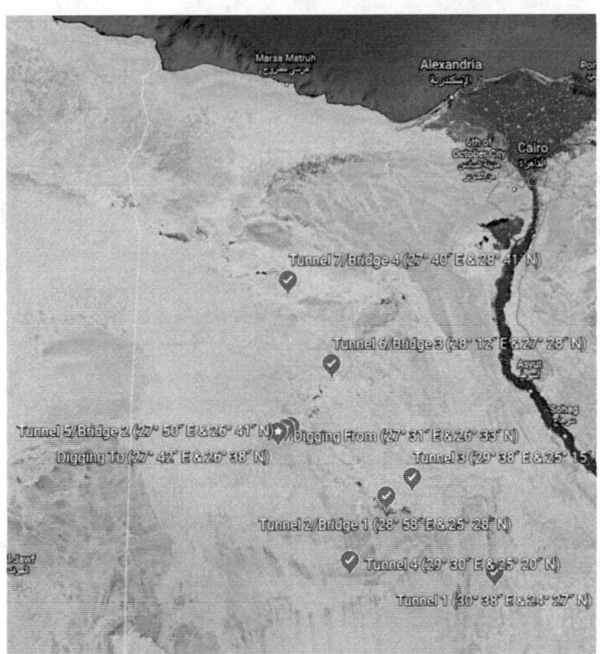

Figure 6.5 Construction works from Luxor to El Baharia Oasis

IOP Conf. Series: Materials Science and Engineering **974** (2020) 012002 doi:10.1088/1757-899X/974/1/012002

The route selection shown in Figure (6.6) is matched with the expected new valley that Dr. Farouk El Baz Proposed in his project (The Development Project/Corridor of Development Project) as shown in Figure (6.7).

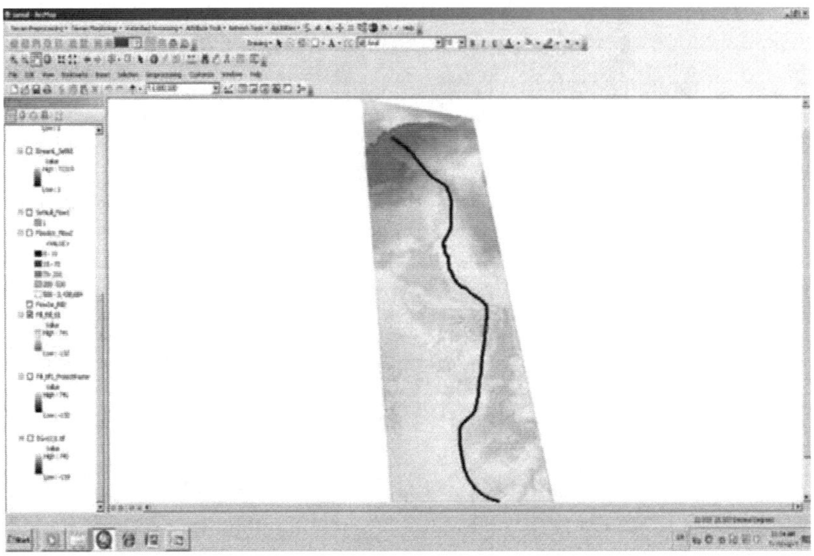

Figure 6.6 Final route selection

Figure 6.7 Final route selection [8]

IOP Conf. Series: Materials Science and Engineering **974** (2020) 012002 doi:10.1088/1757-899X/974/1/012002

7. Conclusion & Recommendations for future work.

The following statements are concluded:

1- The DEM is an effective way to represent the surface of the earth and extract hydrological characteristics.
2- Route selection for a new valley in western desert is carried out by using ARC-GIS to protect many cities from tremendous flood.
3- The volume of El-Qattara downrange (1054.4 km^3) is about 6.5 times the water volume of AHDR (131.7km^3), without any losses that contain evaporation, infiltration, etc.
4- Outlet stations of different watersheds or basins in each oasis delineate the path of the new route of the Nile River.

It is recommended to determine the new route by matching the outlet station of each neighbor watershed or basin consequently for each oasis in order to obtain the best route for the new valley. It is also recommended to determine the volume of construction works to be carried out (digging - pipelines - tunnels - bridges) on the new valley.

References

1- Farouk El-Baz.,2011. " Desert Development Corridor" Plan Accepted by Egypt's Government

2- S.M Rashad, M.A Esmail., 2000. "Environmental-impact assessment of hydro-power in Egypt", Applied Energy.

3- Stock, Jill Kamil; photographs by Michael (1993). Aswan and Abu Simbel: history and guide. Cairo: American University in Cairo Press.

4- Abu Zeid, M.A., 1989. "Environmental impacts of the High Dam". Water Resources Development.

5- Abdel Samie Elewa. "Effect of flood water on the salt content of Aswan High Dam Reservoir", Hydrobiologia, 09/1985.

6- El-Shabrawy, Gamal M; Henri J. Dumont (2009). The Toshka Lakes. In the Nile. Monographiae Biologicae. 89.

7- Vaze, J., 2010. "Impact of DEM accuracy and resolution on topographic indices",

Environmental Modelling and Software.

8-.https// the constructor.org, Causes of Failures of Earthfill Dams.

Military Technical College
Kobry El-Kobbah,
Cairo, Egypt

13th International Conference on Civil and Architecture Engineering ICCAE-13-2020

Effects of asphalt modification by Nanosilica and Nanoclay on asphalt binder and hot mix asphalt properties

Ahmed W. Oda[1], Ahmed El-Desouky [2], Hassan Mahdy[3], Osama M. Moussa[1]

[1] *Department of Civil Engineering, Military Technical College, Cairo, Egypt.*
[2] *Canadian International College, Cairo, Egypt.*
[3] *Ain Shams University, Cairo, Egypt.*

Abstract. In the Hot mix asphalt (HMA) is widely used in asphalt paving in most of countries. The goal of this paper is to study the effect of adding nanomaterials on the properties of asphalt binder and HMA. An experimental program has been conducted to study how to evaluate the performance of the modified asphalt binder and the modified HMA. Nanosilica (NS) and Nanoclay (NC) were used as modifiers. NS and NC were added at 3%, 5% and 7% by bitumen weight. Tests were performed to evaluate enhancement occurred in Penetration, softening point (SP), rotational viscosity (RV). Marshall stability tests, loss of stability and indirect tensile strength were performed on modified asphalt mixtures. Results were compared with control mix. Results shows significant enhancement in most of properties of asphalt binder and HMA. Penetration decreased by 24.07% at 3% NC and 20.37% at 5% NS. 3% NC and 5% NS were chosen to be optimum nanomaterial content (ONMC). Modified bitumens by 3% NC and 5% NS were used to produce modified asphalt mixtures. The modified mixtures showed a significant enhancement in stability and indirect tensile strength (IDT), but they, in contrast, had a harmful effect on the loss of stability (LOS).

Keywords: Modified Asphalt; Asphalt; Hot Mix Asphalt; Nanomaterial; Nanosilica; Nanoclay; Modified Bitumen; Loss of stability

1. Introduction

Nowadays, there is a growing concern in highway community on improving quality of asphalt mixes and asphalt binders to enhance design life of flexible pavements. Bitumen is also known as the most popular asphalt binder. It is widely used in asphalt pavements. It mainly consists of different chemical compounds as carbon, hydrogen, and nitrogen compounds[1]. Virgin bitumen (not modified by additives or others) has multiple properties and characteristics that have major effects on asphalt pavement performance as consistency, adhesion, visco-elasticity and temperature susceptibility[2]. These properties could be called physical and rheological properties[3, 4, 5].

Due to the high traffic loads and weather effects, many problems and distresses occurs in the asphalt pavement, such as cracks and various distresses. So, Asphalt modification with different modifiers has become a trend for many researchers to improve the asphalt binders and asphalt pavement performance [6 , 7]. There are several requirements and conditions when choosing the appropriate modifier type as maintaining asphalt distinctive characteristics, physical and chemical stability and achieving suitable viscosity at normal application temperatures[2]. Through research and

development, researchers identified several good asphalt modifiers such as: crumbed rubber, styrene butadiene styrene (SBS), polyethylene, ethylene vinyl acetate, and nanomaterials [6, 8].
As is agreed upon, the term nanomaterial is the material which particles reach the nano size (10^{-9}) m [9]. Because of nanomaterials large surface area and small size, they represent better characteristics compared to other materials[10]. From research experience and history, it was observed that most of modified asphalt samples with nanomaterials had lower values of penetration and higher degrees of temperature of softening point than virgin asphalt. From this there is an indication of improvements of the flexibility and stiffness of modified asphalt binders[11].
Khattak et.al [12] studied the behaviour of HMA designed according to Superpave™ method. A comparison between properties of control mix and modified HMA was performed to evaluate the rate of enhancement in asphalt properties. Carbon Nano Fibers (CNF) was used as bitumen modifier. It was found that the complex shear modulus (E*) increased by 24% - 46% and phase angle ( ) decreased by 6% - 25%. That means higher resistance to permanent deformation for CNF modified asphalt mixtures.
Yao Zhang et.al[13] studied the effect of asphalt modification by SBS. Four types of SBS were used in HMA mixtures preparation. An empirical vertically loaded wheel test was created to measure rutting performance of modified HMA mixtures. From the analysis of results, they calculated the compressive dynamic modulus that indicates the rutting development. It was concluded that SBS enhanced the permanent rutting performance of asphalt mixtures.
Shu Wei Goh et.al [14] studied the effect of adding Nanoclay on superpave HMA mixtures. Asphalt binder used is a PG58-28 bitumen. Modified specimens were classified into two groups. The first group was exposed to water in freeze-thaw conditions. The second remained in dry condition. It was observed that the first group gave higher IDT with the increasing in NC content. On the contrast the second group gave opposite results.
A. E. A. Mostafa [7] studied the effect of bitumen modification by adding Nanosilica (NS) and Nano-Carbon (N-C). NS was added at percentages of 1%, 3%, 5%, 7% AND 9% by weight of bitumen. N-C was added at the percentages 0.01%, 0.1%, 0.5% and 1% of weight of bitumen. It was observed that with the increasing in NS and N-C penetration values decreased but values of softening point and viscosity increased.

2. Materials

The bitumen used in this research was produced by El-Nasr Petroleum Company. It was brought from one of highway projects in the New Administrative Capital established by Orascom Company for constructions.
Aggregate (course and fine) was completely produced by crushers, supplied from Attaqa crushers. Validation tests were performed to evaluate the quality of aggregate particles.
Nanomaterials used in this experimental work were NS and NC. They were manufactured locally by the laboratories of the Faculty of Science, Beni Suef University.
A large part of the experimental tests were conducted in the roads laboratory of the Faculty of Engineering, Ain Shams University.

3. Experimental work

An experimental program was established to test the behaviour of modified bitumen compared with virgin bitumen (bitumen without any additives). The program is mainly based on evaluation of physical and rheological properties of bitumen. Tests performed were Penetration, SP and RV tests.
Conventional tests of bitumen, aggregate and HMA were performed. Aggregate validation tests were conducted to ensure aggregate suitability for use in the production of HMA. Tests were LAA test, flat & elongated particles, plastic and liquid limits.
HMA control specimens and modified were tested as shown below, to evaluate the effects of asphalt modification.
Modifiers used in this research were NS and NC. Both Nanomaterials were characterized described and characterized using the SEM, XRD and TEM techniques.

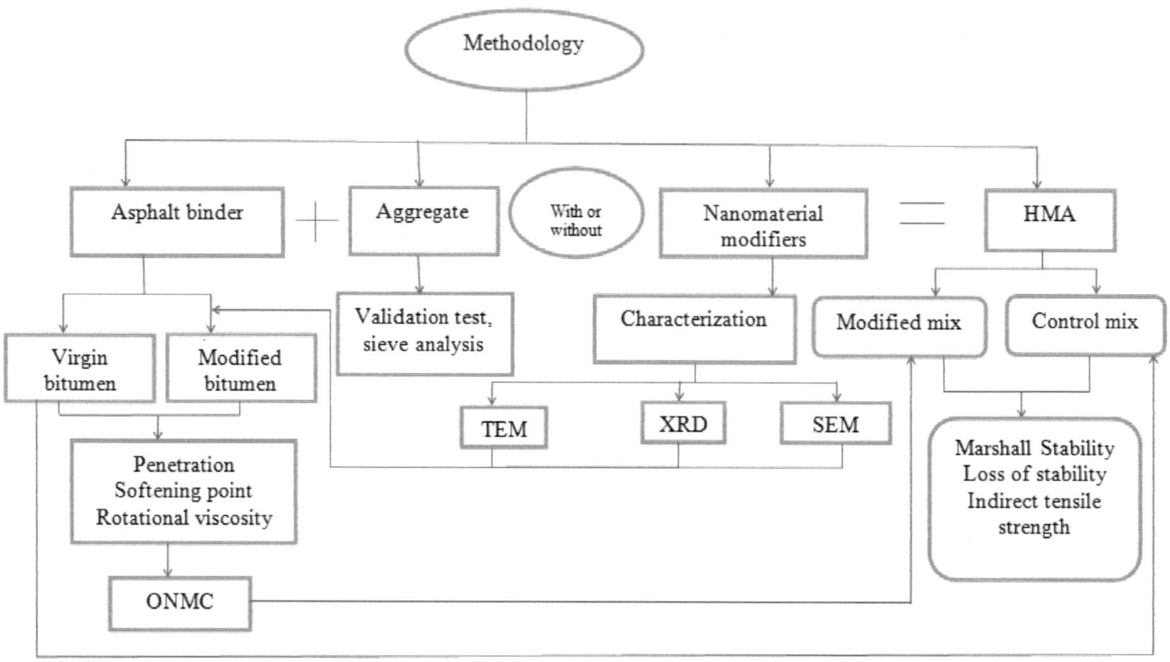

3.1 Bitumen

Bitumen is divided into small samples to be ready for testing. Tests were performed according to the procedures of the American Society of Testing Materials (ASTM) and American Association of State Highway and Transportation Officials (AASHTO). Results were compared with limits of Egyptian Code of Practice (ECP, 2008), [2,16,17, 13].

- Penetration test[15]: (ASTM D5/ 5M) Penetration testing measures standard needle penetration values for a solid and semi-solid bitumen surface. The needle is left to penetrate the surface of bitumen -in standard container- freely under its weight. The unit of penetration is 0.1 mm. the penetration value expresses the consistency of bitumen.
- SP test[16]: (ASTM D36/ 36M) This test measures softening point of bitumen by ring and ball apparatus. In this method two rings with standard dimensions are filled with bitumen. Two steel spheres with standard weights are put over the surface of bitumen in the rings. All of these components are put in a water bath. Under the influence of heat, the two spheres fall with the bitumen to touch a small horizontal plate under the rings with a standard distance. The temperature at which contact occurs is considered softening point. Average readings are taken. SP is an indication that expresses the tendency of bitumen to flow at high temperatures, encountered in service.
- RV test: (ASTM D4402/ 4402M) Brookfield viscometer is used to determine RV. A spindle of weight 27 gram is let to rotate at speed 20 rpm under various temperatures. It can measure viscosity at different units as Pa.s or CP.

IOP Conf. Series: Materials Science and Engineering **974** (2020) 012003 doi:10.1088/1757-899X/974/1/012003

3.2 Aggregates

Tests were performed on coarse and fine aggregate that were completely resulting from crusher millers. Aggregates used are:

- Aggregate A (Agg. A): The aggregates grain size ranges from 12 to 24 mm.
- Aggregate B (Agg. B): The aggregates grain size ranges from 9 to 12 mm.
- Aggregate C (Agg. C): The aggregates grain size ranges from 5 to 9 mm.
- Crushed sand (C.S): The aggregates grain size < 5mm.
- Mineral filler (M.F): Limestone powder.

Aggregates were tested according to ASTM and AASHTO specifications and compared with ECP, 2008 limits. Tests were[17]:

- Los Angeles Abrasion test (LAA)[18]: Test was performed according the procedures of ASTM C131.
- Flat and elongated particles: Test was conducted according the procedures of ASTM D 693.
- Water absorption: Test was conducted according the procedures of ASTM C 127.
- Liquid and plastic limits (LL, PL)[19]: Test was performed according the procedures ASTM D 4318.

3.3 Modifiers

In this research two different nanomaterials are used as asphalt modifiers. Nanomaterials were produced by the laboratories of the Faculty of Science, Beni-Suef University. The materials used in this research as modifiers are Nanosilica (NS) and nanoclay (NC). Various methods of scanning were used to show particle size distribution of the two nanomaterials. NS and NC were characterized using various techniques. They were characterized using X-ray diffraction (XRD), and Transmission Electronic Microscope (TEM). Scanning Electron Microscope (SEM).

For NS, The TEM images of the synthetic silica nanoparticles reflected their formation as agglomerated grain in the nano-scale. The grains appear in spherical form with size less than 10 nm (Fig.1). The XRD pattern shows broad peak around 2Theta 22.5° which characterize silica particle of non-crystallize properties (amorphous). NS mainly consists of SiO_2, where SiO_2 represents about 99.8% of the chemical composition of NS.

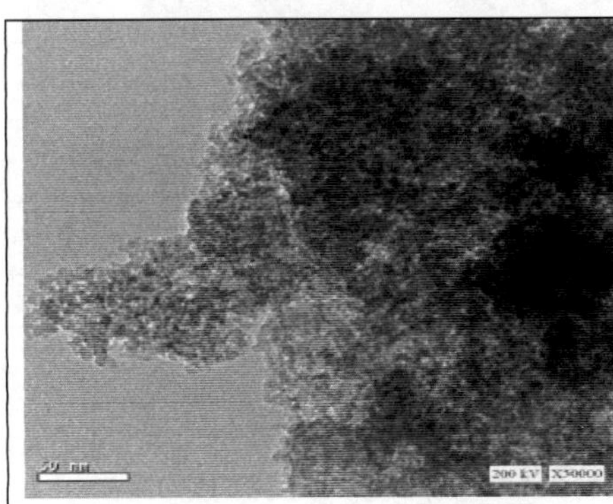

Fig. 1.TEM image of silica nanoparticles

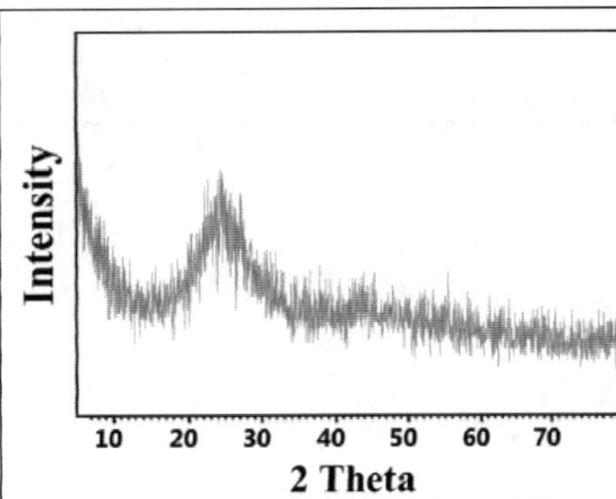

Fig. 2. XRD pattern of silica nanoparticles

For NC, The chemical analysis of the sample reflected its composition from MgO (40.2%), SiO_2 (46.3%) and Al_2O_3 (5.5%) in addition to other traces for other elements that related to the impurities of the using natural raw materials. The SEM image reflected the formation of the clay sample as aggregates of nano nudes agglomerated to each other in some samples.

Table 1: Major Oxides of NC

Major oxides of NC	Percentage %

IOP Conf. Series: Materials Science and Engineering **974** (2020) 012003 doi:10.1088/1757-899X/974/1/012003

MgO	40.2
Al_2O_3	5.48
SiO_2	46.3
TiO_2	0.41
K_2O	0.34
Na_2O	0.46
Cr_2O_3	1.5
Fe_2O_3	4.2
CaO	1.13

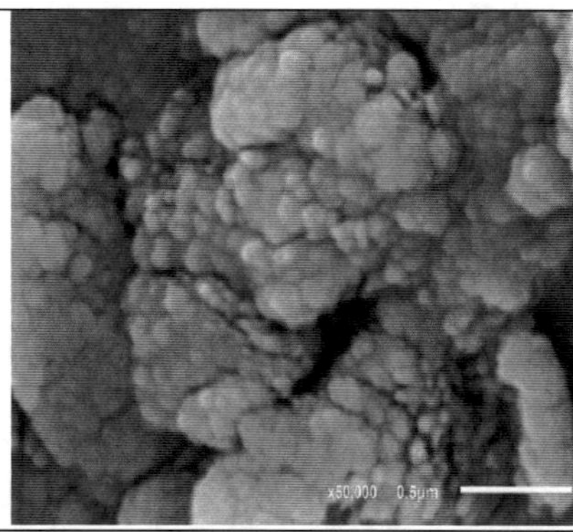

Fig. 3. SEM of the synthetic NC

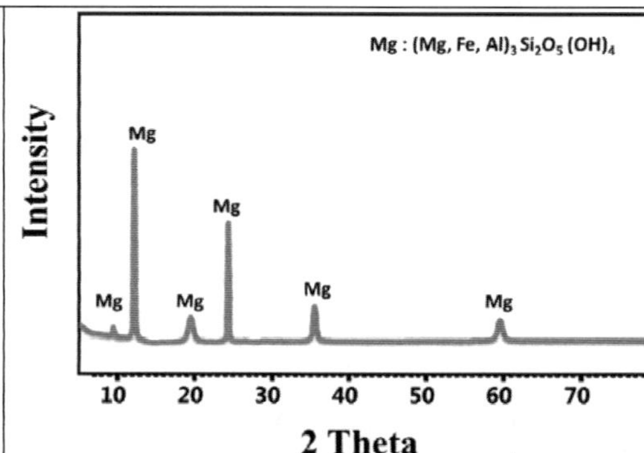

Fig. 4. XRD pattern of clay nanoparticles

3.4 Control mix design

Sieve analysis tests [21 , 20]were conducted Aggregate 1, Aggregate 2, Aggregate3, crushed sand, and limestone powder. The gradations were as showed in table 5. Gradation chosen is gradation 4C. Bitumen was added with percentages 4%, 4.5%, 5%, 5.5% and 6% of the weight of aggregates.

Table 2 job mix formula and ECP, 2008 limits

sieve	JMF	Specifications	
		L.L	U.L
1"	100	100	100
¾ "	98.7	80	100
3/8"	74.6	60	80
#4	52.0	48	65
#8	41.6	35	50
#30	22.6	19	30
#50	16.5	13	23
#100	9.7	7	15
#200	5.2	3	8

3.5 modified mixtures

NS and NC was added to virgin bitumen at percentages 3%, 5% and 7%. Modified mixes were prepared by using modified bitumen by NS and NC. They were tested and evaluated by Marshall Stability, loss of stability and IDT

4. Results and discussion

4.1 Bitumen:

The results of tests performed on bitumen were as shown in table 3. The table shows a numerical comparison between the results of tests performed. The comparison was used to evaluate the effects of bitumen modification.

Table 3 Properties of asphalt cement (Bitumen)

Test	specifications	Virgin bitumen	3% NS	5% NS	7% NS	3% NC	5% NC	7% NC
Penetration(0.1mm)	ASTM D5	54	52	43	55	41	57	56
		Changes (%)	-3.7	-20.37	+1.85	-24.07	+5.56	+3.7
Softening point(°C)	ASTM D36	42.5	44.25	46	45.75	47.5	45	40
		Changes (%)	+4.12	+8.24	+7.65	+11.76	+5.88	-5.88
Viscosity at 135°C(CP)	ASTM D4402	345.67	429	447	496	495.67	404	433
		Changes (%)	+24.11	+29.31	+43.5	+43.2	+16.87	+25.26
Viscosity at 165°C(CP)	ASTM D4402	75	91	95.67	125.2	120.67	87	91.33
		Changes (%)	+21.33	+27.56	+66.67	+60.89	+16	+21.77

The results were recorded in the previous table of tests conducted on virgin bitumen and modified bitumen (Table 3). The following charts represent comparisons between the penetration values, the softening point, and the rotational viscosity of the different NS and NC percentages.

Figure 5 represents the relationship between softening point and the amount of modifiers. In case of NS, the chart represents a direct relationship between increasing the amount of NS and increasing the value of the softening point. The relationship remains increased until it reaches 5% NS, after which the softening point value when increasing the ratio decreased to 7%. The softening point has decreased in value very slightly. In the case of NC, the value of SP increased sharply at the amount of 3% NC, and then decreased, decreasing thereafter to be an inverse relationship with the ratio of the modifier.

Figure 6 represents the relationship between penetration in unit of 0.1 mm, and the amount of modifiers. In the case of the NC, the curve decreased dramatically with adding NC by ratio 3%. After that ratio, the penetration value increases to the highest value at 5% then returns to decrease slightly. In the case of the NS, the value of penetration began to decrease gradually to the value of 3% NS, and then the curve took a greater rate of decline to reach the lowest value at 5% NS, Then the relationship turns to be a direct relationship after the 5% ratio.

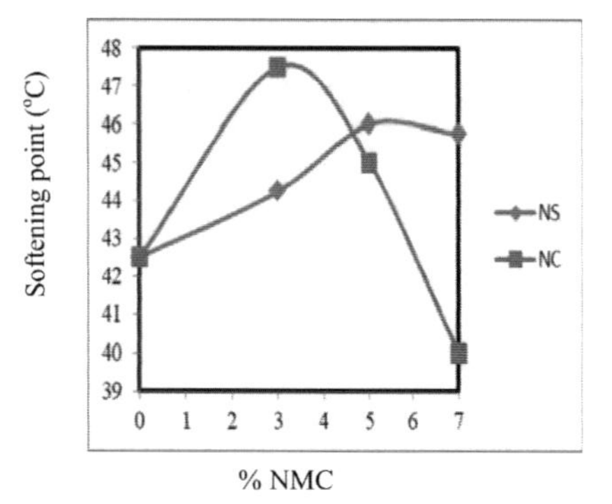

Fig. 5 softening point values versus Nanomaterial content (NMC)

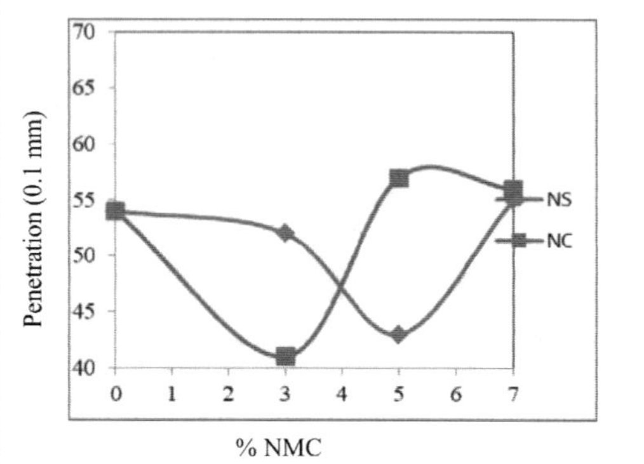

Fig. 6 Penetration value versus NMC

The rotational viscosity of the bitumen was measured at two different temperatures. The temperatures at which viscosity is measured are 135ºC and 165ºC. At temperature of 135ºC, from figure 7, and in the case of NC, it was found that the relationship between the amount of modifier and RV behaved as follows, RV increased in accelerated rate to reach 495.67 CP increasing by 43.2%. After reaching the peak value at 3% NC, the manner of the RV reacts diversely. RV decreases from the highest point to reach 404 CP at 5% NC. In the case of NS, In the case of the NS, it was found that the relationship was constantly increasing viscosity with increasing quantity of NS. Figure 8 shows the relationship between the quantity of modifier and RV. It was found that the proportion between the quantity of the modifier and the RV takes almost the same behavior as their behavior at 135 ° C with different values of the RV.

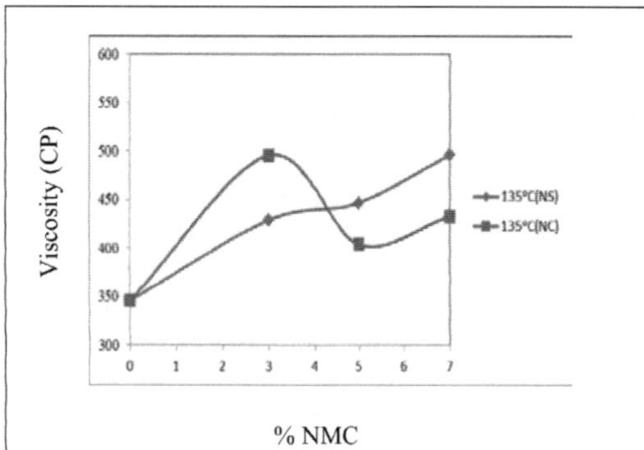

Fig. 7 Rotational viscosity at 135°C

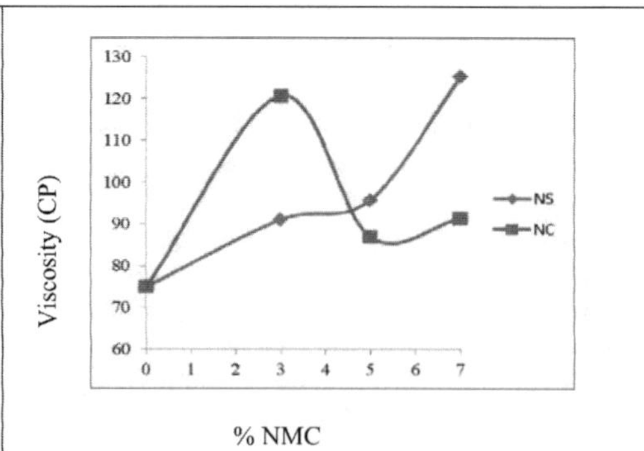

Fig. 8 Rotational viscosity at 165°C versus NMC

To determine the optimum quantities of modifiers (ONMC), those are the ratios that achieve the best performance of the asphalt bond for different properties. The results of the previous tests for the bitumen reached were analyzed. The amount of 3% of the NC has the highest softening point values, the lowest penetration values and the viscosity achieves its highest values, either at 135 or 165

temperatures. The quantity of 5% of the NS has the highest softening point values, the lowest penetration values and the viscosity achieves suitable values of viscosity at 135°C and 165°C
The results indicate noticeable improvements in the properties of the modified bitumen. From analyzing the results and identifying the improvement rates for the different bitumen properties, the 3% NC and 5% NS ratios were found to be the ONMC.

4.2 Aggregates:
The aggregate was tested according to the following specifications [20,21]. LAA test is the resistance of coarse aggregate to degradation under the conditions of Los Angeles machine. Flat and elongated particles represent a weak point for the aggregate if it exceeds the permissible limits in the *ECP, 2008*, where inertia is weak in one or more directions, which may expose the particles to breakage and fragmentation during handling, use or operation of the pavement. Table 2 shows results of the tested aggregates.

Table 4 Aggregates properties and specifications

Test	Test spec.	Agg. A	Agg. B	Agg. C	Crushed Sand
LA abrasion	ASTM C131	23.8	24.1	25.4	--
Flat particles	ASTM D 693	3	3	2	--
Elongated particles	ASTM D693	3	4	3	--
Water absorption	ASTM C 127	2	3	2	1.1
LL, PL	ASTM D 4318	--	--	NP	NP

4.3 Asphalt mixture:
A sieve analysis was performed for all types of aggregate involved in the production of asphalt mixtures. Asphalt mixture for surface layer is designed according to gradation 4C. A blend of aggregates was collected in varying proportions of each type of aggregate for the design of Job Mix Formula (JMF). Figure 9 represents the gradation of aggregate types Agg. A, Agg. B, Agg. C, Crushed sand and Mineral filler (limestone powder). Figure 10 shows that he JMF gradation is located between upper and lower limits of *ECP, 2008.*

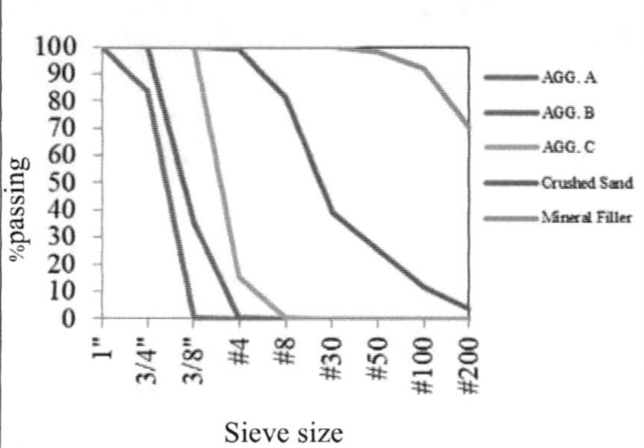

Fig. 9 Aggregate Particle Size Distribution

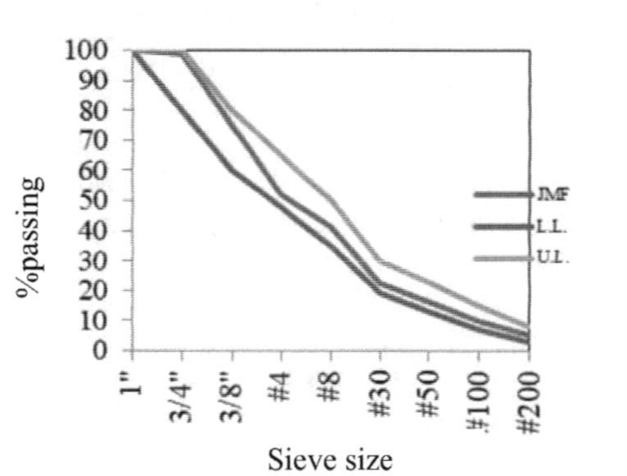

Fig. 10 Job mix formula according to gradation 4C

IOP Conf. Series: Materials Science and Engineering **974** (2020) 012003 doi:10.1088/1757-899X/974/1/012003

The control mix was designed by Marshall Mix design method and results were as follows:
Aggregates are graded by sieve analysis test. Gradation 4C was selected to be the gradation of HMA. Bitumen was added with 4, 4.5, 5, 5.5 and 6% of the weight of aggregates. Mixes were prepared according to procedures of Marshall Mix design method.
Figure 11 shows the relationship between stability of specimens prepared and asphalt content. The highest stability was at asphalt content 5%. The limits of air voids in asphalt mix used in surface layer should be not less than 3% and not more than 5%. At 4% air voids, the asphalt content was 5.5% as shown in figure 12

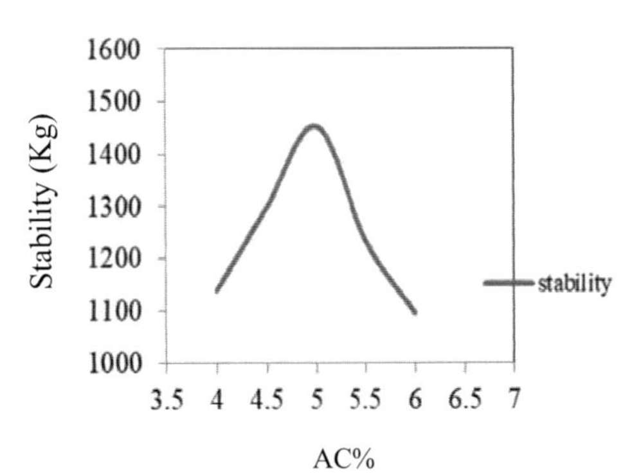

Fig. 11 Asphalt content versus stability

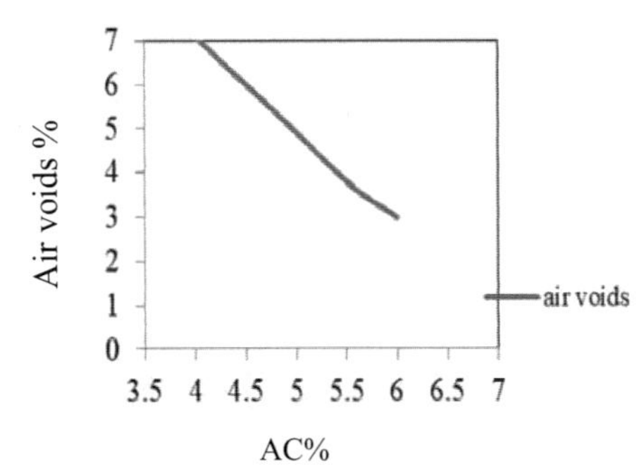

Fig. 12 Asphalt content versus air voids

In figure 13 the highest value of unit weight was at 5.1% asphalt content. The optimum asphalt content (OAC) is the average of the previous three values of asphalt content. OAC is 5.2%. The percentage 5.2% achieved the limits of ECP, 2008 in the case of flow as shown in figure 14 and VMA as shown in figure 15.

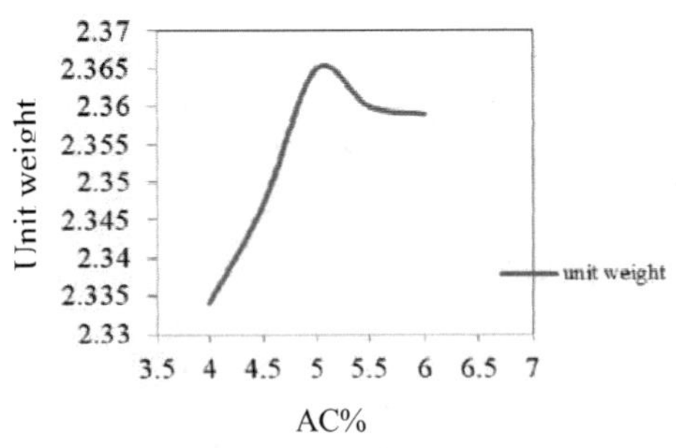

Fig. 13 Asphalt content versus unit weight

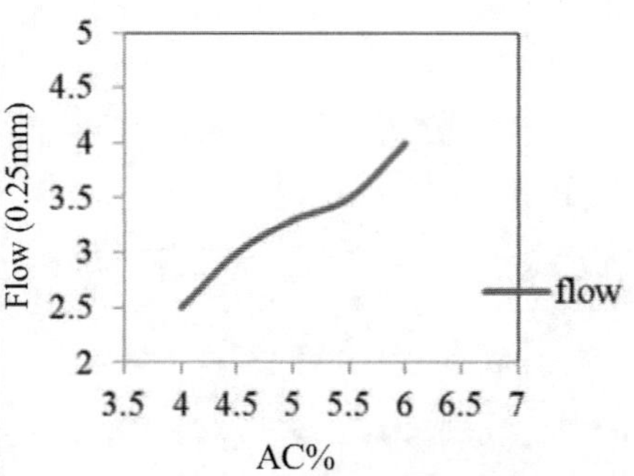

Fig. 14 Asphalt content versus flow.

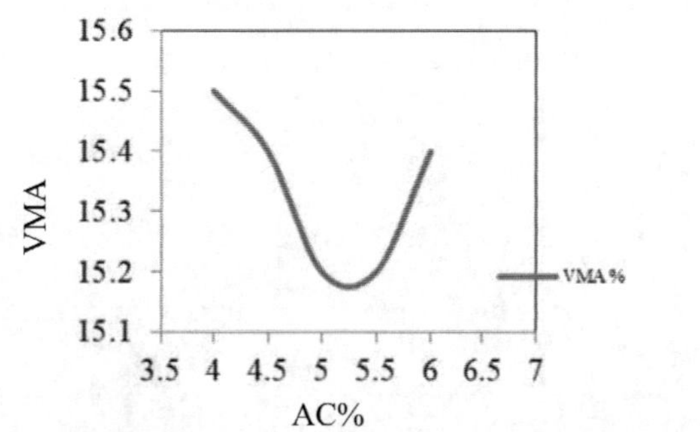

Fig. 15 Asphalt content versus voids in mineral aggregates (VMA)

Modified bitumens with 3% NC and 5% NS were used to produce modified HMA mixtures. Modified specimens were tested by Marshall Stability, flow, and Indirect Tensile Strength (IDT). The results of the mentioned tested specimens were compared with control mix results.

Table 5 results of testing HMA mixtures

Specimen	Stability	%change in stability	Flow (0.1 in)	Density (gm/cm^3)	Marshall Quotient	%change in MQ	Stability after 24 hrs in waterbath	(LOS)%	Change in LOS%	IDT	%change in IDT
Control	2134	--	13.65	2.29	615.54	--	1879	11.95	--	2.36	--
Modified 3%NC	2499	+17.1%	15.49	2.35	635.34	3.22	2046	18.13	+9.06	4.17	+76.77
Modified 5%NS	3027	+41.85 %	13.91	2.35	856.69	39.18	2109	30.33	+12.42	6.63	+92.28

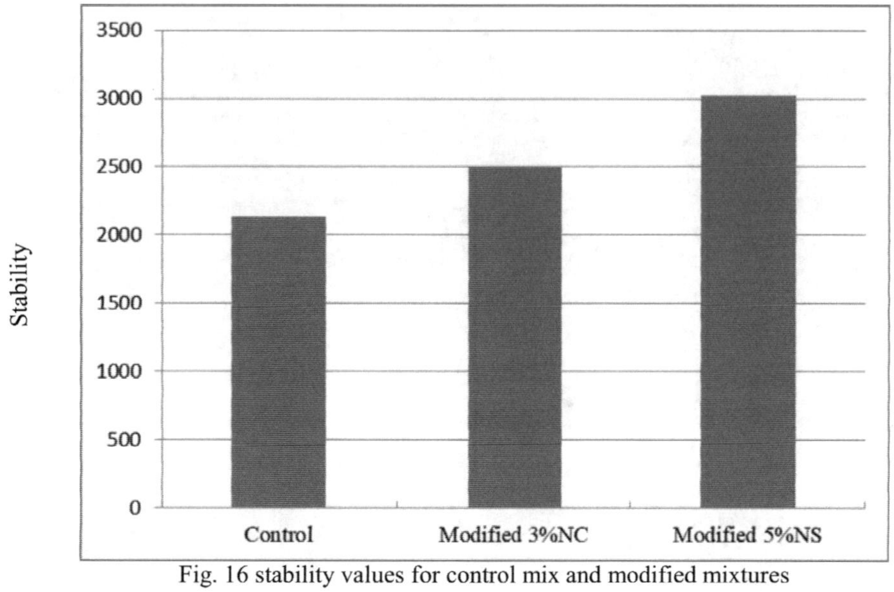

Fig. 16 stability values for control mix and modified mixtures

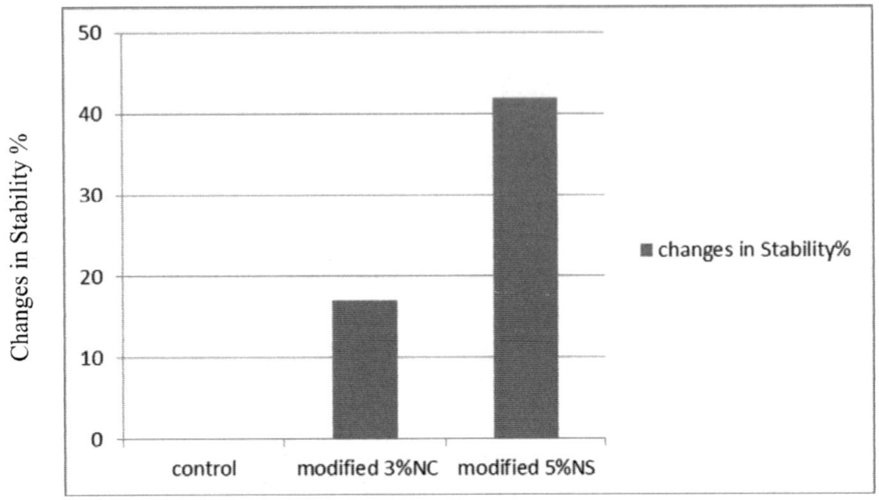

Fig. 17 Changes in Stability %

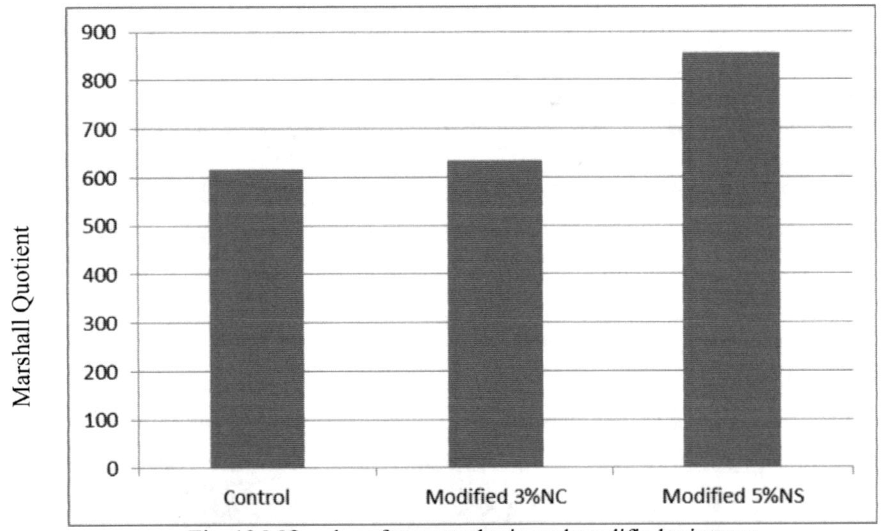

Fig. 18 MQ values for control mix and modified mixtures

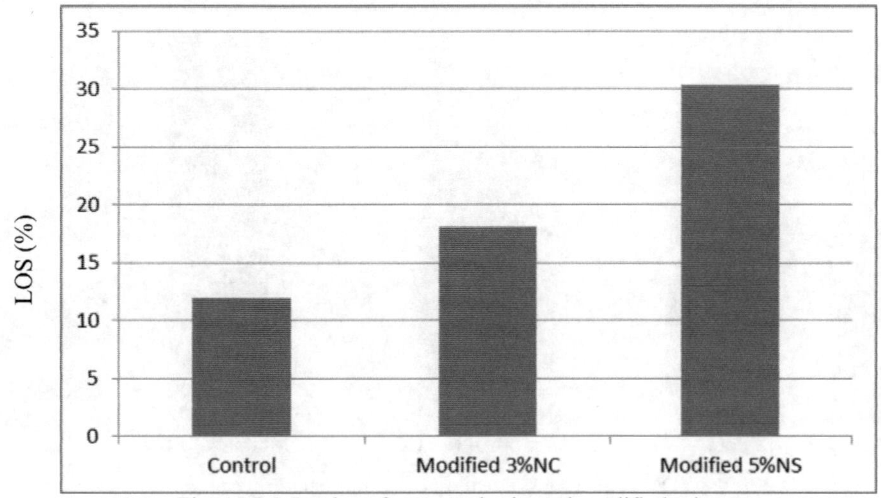

Fig. 19 LOS values for control mix and modified mixtures

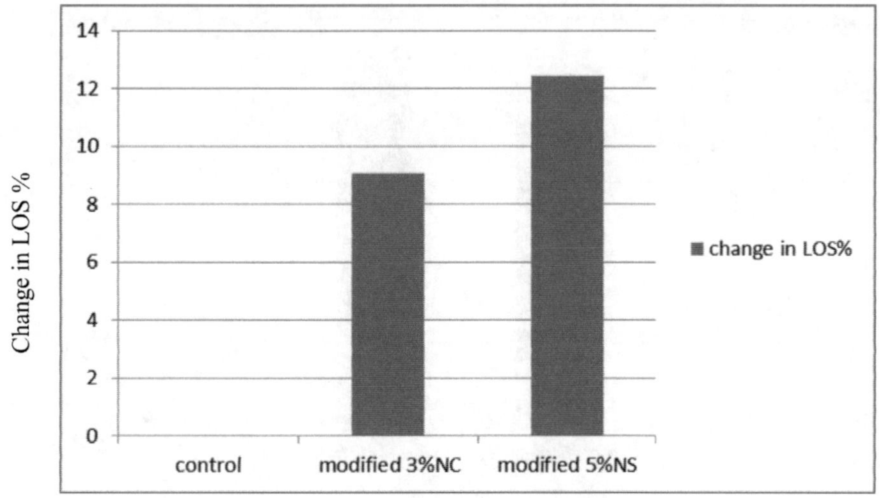

Fig. 20 change in LOS %

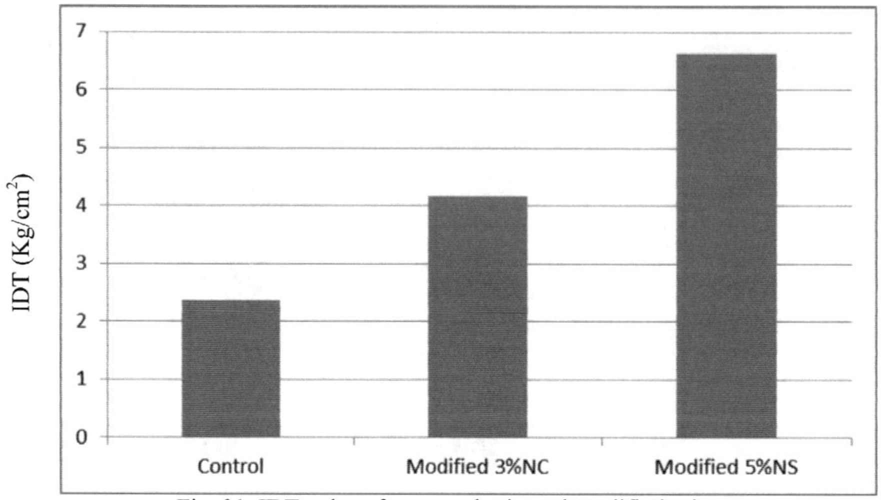

Fig. 21 IDT values for control mix and modified mixtures

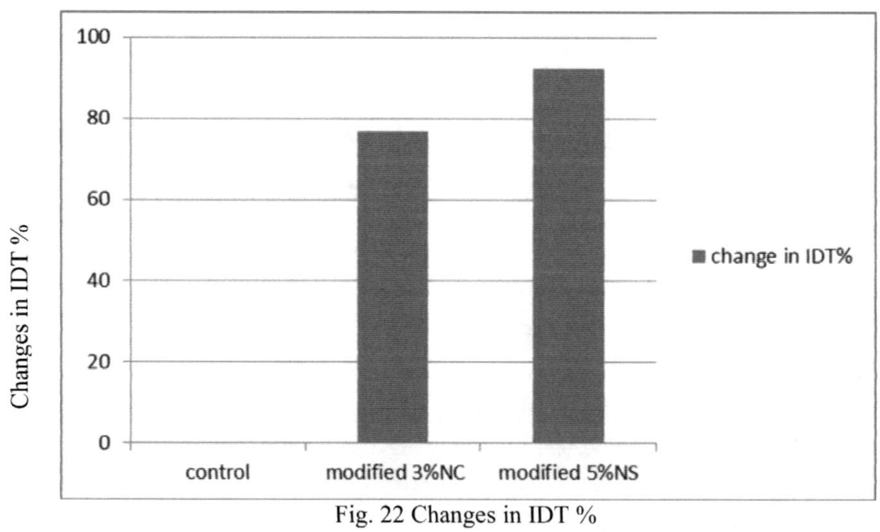

Fig. 22 Changes in IDT %

As shown in table 5 and figures 16 to 22, there is a clear trend to increase different values of HMA properties. Figure 16 and figure 17 are comparisons between the values of stability of control specimen and modified specimens. The two types of modified specimens show significant improvement in stability. Modified specimens with 3% NC showed an increase in the stability value of the modified samples by 17.1%. Modified specimens with 5% NS showed an increase in the stability value of the modified samples by 41.85%.

Figure 18 shows a comparison between the values of MQ for control and modified specimens. In the case of modified specimens with 3% NC, MQ increased by 3.22%. In the case of 5% NS modified specimens, the MQ dramatically increased to reach an increase by 39.18%.

Figure 19 and figure 20 show also a comparison between loss of stability for control and modified specimens. Control specimen lost 11.95% of stability after one cycle of submerging in a water bath. 3% NC modified specimens lost 18.78% of stability under the same previous conditions. 5% NS

IOP Conf. Series: Materials Science and Engineering **974** (2020) 012003 doi:10.1088/1757-899X/974/1/012003

modified specimens lost about 30.33% after the same period under the same previous conditions. The change occurred on modified specimens compared with control specimens that LOS increased by 9.06% in case of modified specimens with 3% NC and LOS increased by 12.42% in the case of modified specimens with 5% NS.

Figure 21 and figure 22 show the effect of adding the two nanomaterials on IDT. The two modifiers caused very large increase in the IDT. In the case of 3% NC, IDT increased very significantly by 76.77%. The same behavior occurred in the case of 5% NS. IDT increased dramatically to reach 92.28%.

5. Summary and conclusion

In this study, an experimental investigation was conducted to evaluate the effect of asphalt modification by using two different nanomaterials NC and NS. The study included the behaviours of bitumen and HMA mixtures. The conclusion of the study is as follows:

- NS and NC have great effects on enhancing the properties of asphalt binders and HMA.
- Penetration values, which are an effective indication of the consistency of asphalt binder, were strongly influenced by the modification of asphalt with nanomaterials. Lower values of penetration mean more viscous consistency and harder binder. Penetration decreased with increasing NS content in bitumen until reaching 5%, after that penetration value returned to increase.
- NS increased rotational viscosity of bitumen at different temperatures as 135°C and 165°C significantly.
- For HMA mixtures, the results of tests performed on control mix and modified mixtures points to noticeable enhancement in stability, MQ and IDT values for the modified mixtures. But in contrast, loss of stability in modified mixtures is higher than of control mixture. That means faster aging in asphalt mixtures.
- It is recommended to study the performance of bitumen when modified with both nanomaterials. It is also recommended to evaluate the resistance to permanent deformation by wheel tracker in case of modified mixtures with NS and NC. Skid resistance is also recommended to be evaluated to know the effect of NS and NC on surface roughness of asphalt mixtures.

6. References

[1] M. El-Shafie, I. M. Ibrahim, and A. M. M. Abd El Rahman, "The addition effects of macro and nano clay on the performance of asphalt binder," *Egypt. J. Pet.*, vol. 21, no. 2, pp. 149–154, 2012.

[2] D. W. John Read, *The Shell Bitumen Handbook*, Fifth Edit., no. 7. London: Thomas Telford Publishing, 2003.

[3] NATHANIEL TETTEH, "Rheological Evaluation of Field-Aged Asphalt Cements," Queen's University, 2018.

[4] I. Amin, S. M. El-Badawy, T. Breakah, and M. H. Z. Ibrahim, "Laboratory evaluation of asphalt binder modified with carbon nanotubes for Egyptian climate," *Constr. Build. Mater.*, vol. 121, pp. 361–372, 2016.

[5] S. Massoud, A. E.- Badawy, M. Henry, and Z. Ibrahim, "Laboratory Evaluation of Asphalt Mixtures Modified With Nanomaterials," Mansoura University, 2016.

[6] C. Fang, R. Yu, S. Liu, and Y. Li, "Nanomaterials applied in asphalt modification: A review," *J. Mater. Sci. Technol.*, vol. 29, no. 7, pp. 589–594, 2013.

[7] A. E. a Mostafa, "Examining the Performance of Hot Mix Asphalt Using Nano- Materials," vol. 06, no. 02, pp. 25–34, 2016.

[8] M. H. M.N. Partl , R. Gubler, "NANO-SCIENCE AND -TECHNOLOGY FOR ASPHALT PAVEMENTS," in *EMPA Swiss Federal Laboratories for Materials Testing and Research*, 2007, pp. 2–4.

[9] R. A. Sultan and M. M. Mourad, "THE EFFECT OF USING ENHANCED MATERIALS WITH NANO TECHNOLOGY ON INDOOR AIR QUALITY IN THE EGYPTIAN," in *nano-technology in construction (green and sustainable construction)*, 2019, pp. 1–21.

[10] R. Li, F. Xiao, S. Amirkhanian, Z. You, and J. Huang, "Developments of nano materials and technologies on asphalt materials – A review," *Constr. Build. Mater.*, vol. 143, pp. 633–648, 2017.

[11] S. I. Sarsam, "Effect of Nano Materials on Asphalt Cement Properties," *Int. J. Sci. Res. Knowl.*, vol. 1, no. 10, pp. 422–426, 2013.

[12] M. Jamal Khattak, A. Khattab, and H. R. Rizvi, "Characterization of carbon nano-fiber modified hot mix asphalt mixtures," *Constr. Build. Mater.*, vol. 40, pp. 738–745, 2013.

[13] Y. Zhang, X. Luo, Y. Deng, S. Hou, X. Shi, and R. L. Lytton, "Evaluation of rutting potential of flexible pavement structures using energy-based pseudo variables," *Constr. Build. Mater.*, vol. 247, p. 118391, 2020.

[14] S. W. Goh, M. Akin, Z. You, and X. Shi, "Effect of deicing solutions on the tensile strength of micro- or nano-modified asphalt mixture," *Constr. Build. Mater.*, vol. 25, no. 1, pp. 195–200, 2011.

[15] S. K. P. SAMY, "Effects of warm mix additives and dispersants on rheological aging and failure properties of asphalt cements," Queen's University, Ontario, 2013.

[16] ASTM D 36 – 95, *Standard Test Method for Softening Point of Bitumen (Ring-and-Ball Apparatus)1*, vol. 04.04, no. d. West Conshohocken: ASTM Committee, 2006.

[17] B. Shirini and R. Imaninasab, "Performance evaluation of rubberized and SBS modified porous asphalt mixtures," *Constr. Build. Mater.*, vol. 107, pp. 165–171, 2016.

[18] American Society for Testing and Materials, "Standard Test Method for Resistance to Degradation of Small-Size Coarse Aggregate by Abrasion and Impact in the Los Angeles Machine," vol. i, pp. 1–5, 2014.

[19] A. D4318, *Standard Test Methods for Liquid Limit , Plastic Limit , and Plasticity Index of Soils*, vol. 04. Annual Book ofASTM Standards, Vol 14.02, pp. 1–14.

[20] I. Kett, *Dry Sieve Analysis of Fine and Coarse Aggregates (C 136)*, vol. 04.02. WestConshohocken, 1996.

[21] *Standard Test Method for Sieve Analysis of Fine and Coarse Aggregates(C 136-01)*, vol. 04.03. WestConshohocken, 2001.

IOP Conf. Series: Materials Science and Engineering **974** (2020) 012004 doi:10.1088/1757-899X/974/1/012004

MilitaryTechnicalCollege
Kobry El-Kobbah,
Cairo,Egypt

13th International Conference on Civil and Architecture Engineering ICCAE-13-2020

Finite element analysis of reinforced sand under circular footing

A.Abd El Rahman, Nabil M. Nagy, Ismail M. Kamal, Mostafa A. Hassan

Corresponding author:hamany9742@gmail.com

This paper presents the results from numerical investigations carried out on circular footing rests on Geo-grid reinforced sand. The finite element program (Plaxis-2D) was used to study experimental results carried out by [1]. This experimental work was used to study the effect of variation of Geo–grid reinforced layers number (N = 0, 1, 2, and 3) under circular footing. The numerical validation with experimental work was considerably agreed with an average percentage of 93%. Then a parametric study was conducted to evaluate the soil performance under circular footing, while applying the same loading conditions. Several parameters have been investigated to identify the optimumsoil behaviour that achieves optimum bearing capacity and minimum settlement under the footing. The parameters include number of Geo-grid layers (N), the Geo-grid layer width (L), the spacing between the footing base and topmost Geo-grid layer (U), the spacing between Geo-grid layers (h) and the Geo-grid stiffness (K). The results indicated that by increasing the number and the width of the Geo-grid reinforced layers the bearing capacity ratio increases up to a limited extent. Moreover, the optimum number of Geo-grid layer equals four layers for circular footing.

Keywords: bearing capacity, footing shape, reinforced soil, Geo-grid layer.

IOP Conf. Series: Materials Science and Engineering **974** (2020) 012004 doi:10.1088/1757-899X/974/1/012004

1. Introduction

Construction of superstructures on loose soil requires enhancement to the soil performance to increase the bearing capacity of the soil and decrease the settlement. Soil reinforcement using Geo-grid products has been widely expanded in the last three decades. The optimum Geo-grid soil reinforcement behaviour could be achieved by investigating several reinforcement parameters such as Geo-grid type, number of layers (N), width of reinforced layers (L), spacing between the reinforced layers (h), distance between the footing and the top reinforced layer (U), footing width (B), and the depth of footing (Df). As shown by [2], [3] footing geometry rests on reinforced soil significantly affect the soil performance. These parameters are investigated to achieve the optimum soil behaviour in the study. However, some soil properties which affect the behaviour of the reinforced soil are not considered in this paper such as relative density (Dr), cohesion(C), and friction angle (φ)

Ultimate bearing capacity of reinforced soil could be calculated using several methods, as showed by [4]. In this paper, the tangent method was adopted to determine the ultimate bearing capacity for unreinforced and reinforced soil. Bearing capacity ratio (BCR) is a suitable technique to determine the degree of enhancement in the bearing capacity of any type of soil. It is defined as the ratio between the ultimate bearing capacity of reinforced soil to the ultimate bearing capacity of unreinforced soil, as shown in eq (1)

$$\mathrm{BCR} = \frac{q_{u\ \mathrm{reinforced}}}{q_{u\ \mathrm{unreinforced}}} \quad \text{............ eq (1)}$$

1.1 Effect of Geo-grid layers number.

[1], [3], [5] and [6] showed experimentally that the soil performance could be improved by increasing the number of Geo-grid layers to three layers, that improvement could be translated as increasing in the bearing capacity and decreasing of the settlement. [7] showed that The optimum number of reinforcing layers (N) ranges between four to six–layers. [8],[9] and [10] proved numerically and experimentally that the best soil behaviour occurs when the number of reinforced layers is four. Also, it was found that the relationship between the number of reinforcing layers and the bearing capacity ratio (BCR), is directly proportional also, it is inversely proportional between the footing width and the BCR. Best BCR value could be achieved when the number of layers is three.

1.2 Effect of Geo-grid width.

[11] showed that the optimum width of the reinforced layers to footing diameter (L/D) equals 4.9 for circular footing. The width of reinforcing layers is about four times the width of the square footing as showed by [12] and [13]. [6], [14] and [15] showed experimentally that the best results for ultimate bearing capacity of Geo-grid reinforced soil could be achieved when the Geo-grid width is greater than or equal five times the footing width (L/B = 5). [8] showed numerically that the optimum width of the reinforced layers (L/B) equal to 7.5B. From the experimental and numerical study carried out by [10] the optimum ratio between the width of Geo-grid and width of foundation (L/B) is 4.5.

1.3 Effect of the spacing between footing to topmost Geo-grid layer.

[14] showed experimentally that the best results on strip footing could be achieved when the topmost Geo-grid reinforcement layer is placed at 0.35B from the base of the footing. [7] showed that BCR is better when U/B is less than 0.67. [8] proved that it is more effective to use (U/B) equals to 0.25. In another experimental study [13], it was found that the bearing capacity under shallow foundation increases by decreasing the vertical spacing between Geo-grid layers and footing until that the top Geo-grid layer lies in the range of 0.2 - 0.3 the footing width. From the experimental and numerical study carried out by [10], it was concluded that the ratio (U/B) = 0.25.

1.4 Effect of the spacing between Geo-grid reinforcing layers.

Experimental and numerical studies carried out by [12] on square footing resting on sand found that, the optimum spacing between the reinforced layers (h) is about 0.4 times the width of the footing. [14] showed experimentally

IOP Conf. Series: Materials Science and Engineering **974** (2020) 012004 doi:10.1088/1757-899X/974/1/012004

that the second layer should be placed at 0.25 B from the top layer. [8] showed that optimum (h/B) equals 0.75 under a strip footing. From the numerical study carried out by [16] the layers placed at a distance 0.4375B gives the best results for the bearing capacity. From the experimental and numerical study carried out in [10] the ratio between the spacings to width of foundation (h/B) is 0.2.

1.5 The effect of footing shape

According to [17], the shape of footing also control in BCR .the performance of the Geo-grid reinforced soil supporting circular footing is better than the performance of reinforced soil supporting square footing.

$$BCR = \frac{q_{u\ reinforced}}{q_{u\ unreinforced}}$$

2 Objectives

The aim of this numerical study is to improve the reinforced soil performance to increase its failure limit, which occurs at a settlement of 10 % of footing width [11]&[18]. Plaxis-2D was used to validate experimental results carried out by [1]. After validation with experimental work, a parametric study has been conducted to evaluate the soil performance under circular footing, keeping the same loading conditions (loading cell capacity). The parametric study was performed to investigate the influence of varying the following parameters (Figureure (1)):

1. The Geo-grid layers number (N).
2. The Geo-grid width (L) relative to footing width.
3. The spacing between the footing and top-most Geo-grid layer (U).
4. The spacing between the Geo-grid layers (h).
5. The Geo-grid stiffness (K).

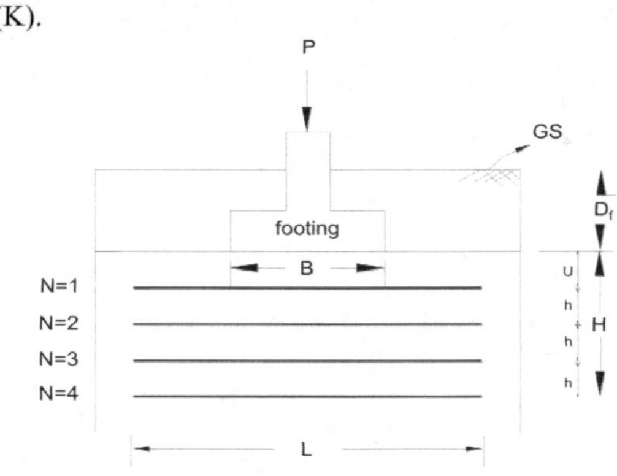

Figure 1: The Geometric parameters of Geo-grid reinforced sand

3 Experimental work

In the experimental work, [1], a cubic testing tank of side length 0.8 m was used. To avoid the boundary conditions effect, the ratio between the circular footing diameter and tank width is 6.5. A circular steel plate with thickness 15 mm and diameter 120 mm was used to represent the footing. Incremental loading was applied using a hydraulic jack. The settlement was measured by displacement reader as shown in Figure (2).The soil used is poorly graded sand with unit weight of 14.4 KN/m^3, coefficient of uniformity (C_u) 2.89, coefficient of curvature (C_c) 1.05, effective size (D_{10}) 0.27, internal friction angle of 39°, specific gravity was 2.65 with maximum and minimum dry densities 1.64 (g/cm^3), 1.44 (g/cm^3) and maximum and minimum void ratios of the sand were found to be 0.89 and 0.65, respectively. High-density polyethylene (HDPE) geo-textile layers with a tensile strength of 7.68 KN/m were

IOP Conf. Series: Materials Science and Engineering **974** (2020) 012004 doi:10.1088/1757-899X/974/1/012004

used. A square formed reinforcement layer, 4.5 times the diameter of circular footing (b /D = 4.5), was used, where the layer's configuration was U/D = h/D = 0.42.

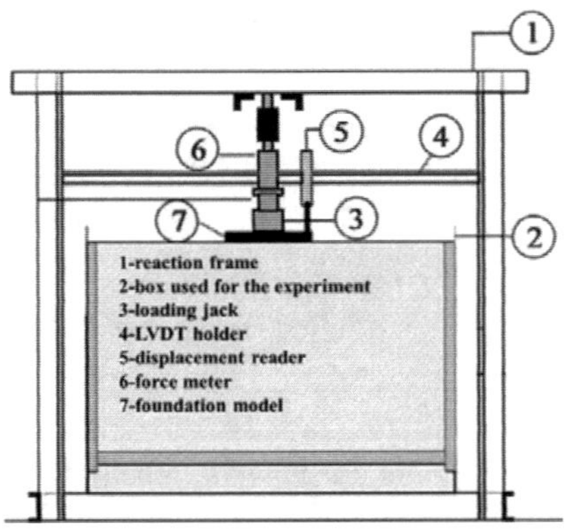

Figure 2 : Schematic view of the experimental equipment [1]

4 Finite Element model

a. Introduction

Plaxis-2D was used to simulate and analyze the Geo-grid reinforced soil. Two different models could be generated using Plaxis-2D, an axisymmetric and plain strain model. These models could be used to solve and simulate different types of foundation. Generally, the axisymmetric model is used to solve circular footing while plain strain model could be used for simulating strip footing. Meshing size could be varied from very coarse, coarse, medium, fine, and very fine. Solution accuracy and computational time effort depends on mesh size. Mohr-coulomb and hardening soil model are the most common models used for modeling the reinforced sandy soil [19].

The finite element model was created to verify:

1- Assuring model quality by validating its results with those of experimental results.
2- Studying the results of changing the parameters as shown in chart (1).

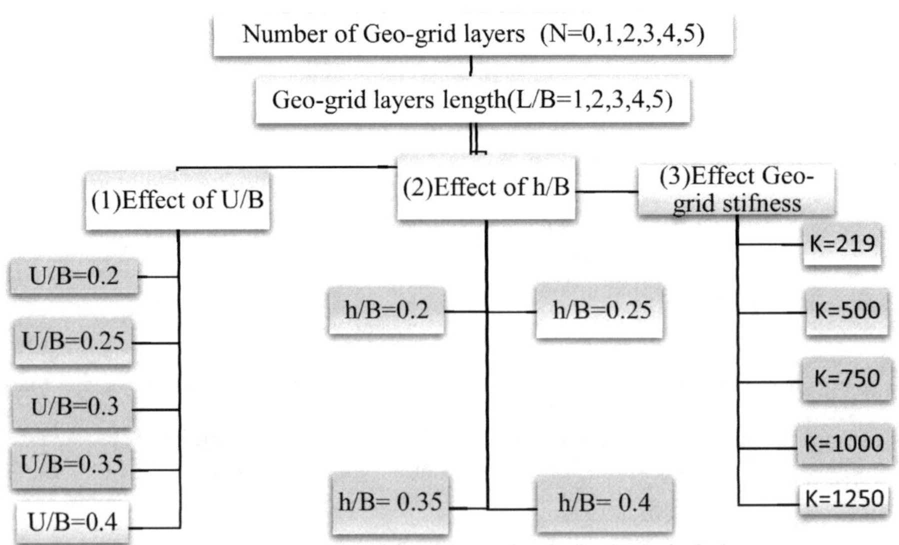

chart 1: Studying cases of the Geo-grid reinforced soil under circular footing

b. Numerical Model discretization

In this study, the same bin dimensions, as in experimental work, were used (0.8 x 0.8 m). Sandy soil with unit weight 14.4 KN/m^3,Young's modulus 40000 KN/m^2, Poisson's ratio 0.3, cohesion 1 KN/m^2, friction angle 39° and dilatancy angle 9°[20] was represented byhardening soil model. The axial stiffness (EA) of the Geo-grid reinforcing layer was 219 KN/m[21], [15], [7], and [11]. The same plan area of the steel plate was used to simulate the circular footing which equal (0.0113 m^2). Very fine meshing with 15-Noded element was used to increase the accuracy of the results. Refining mesh was performed in Geo-grid reinforcing layers to increase the simulation accuracy as shown in the Figureure 3. Axisymmetric model was used to reduce the computational time. The boundary conditions were fixed at the bottom of the bin and roller on the sides to allow the sand to move freely in the vertical direction. The load applied to the steel plate was distributed and applied gradually.

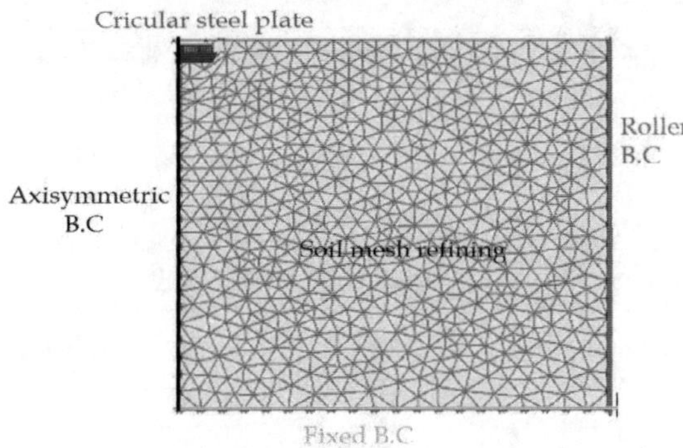

Figure3 : Boundary conditions for the numerical model

i. **Validation between the numerical model and experimental work**

The FE model was carried out to simulate the circular footing over unreinforced and reinforced soil. The soil was reinforced using 1, 2, and 3 Geo-grid layers. The Finite element model results were in a good agreement with the experimental model. It worth mentioning that the un-reinforced soil results were more accurate due to the absence of the interaction definition between the Geo-grid layers and the soil. However, all the results were acceptable, with a maximum tolerance of 10%.

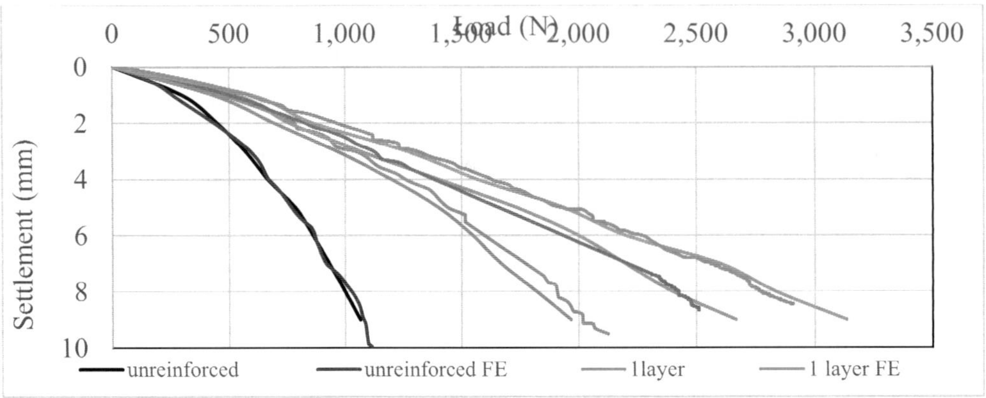

Figure 4 : The load settlement of validation with the experimental work

c. Parametric study

This parametric study was performed for the evaluation of the effect of each parameter separately, as shown in chart 1, while fixing the others. The individual parameter under the investigation will be varied and the values of the others are as follows: spacing between topmost layer and the footing is U/B=0.42, spacing between the layers is h/B = 0.42 which equals about five cm as shown in [1], the tensile stiffness for Geo-grid (EA)=220 KN/m, the number of geo-grid layers is four, and the width of geo-grid layers is equal 5B.

i.The effect of layers number (N)

The optimum number of Geo-grid reinforced layer was investigated. Figureure 5 showed numerically a comparison of Load-Settlement curve of a different number of reinforced layers N=0, 1, 2, 3, 4, and 5 under circular footing. Figureure5 showed that the inclusion of one reinforcing layer had a significant effect on the load capacity of the system compared with the unreinforced soil. This effect started to decrease by increasing the number of reinforcing layers. Data illustrated that insignificant enhancement in the load capacity was achieved after increasing the reinforcing layers number more than three layers. Consequently, the optimum number of layers is three layers.

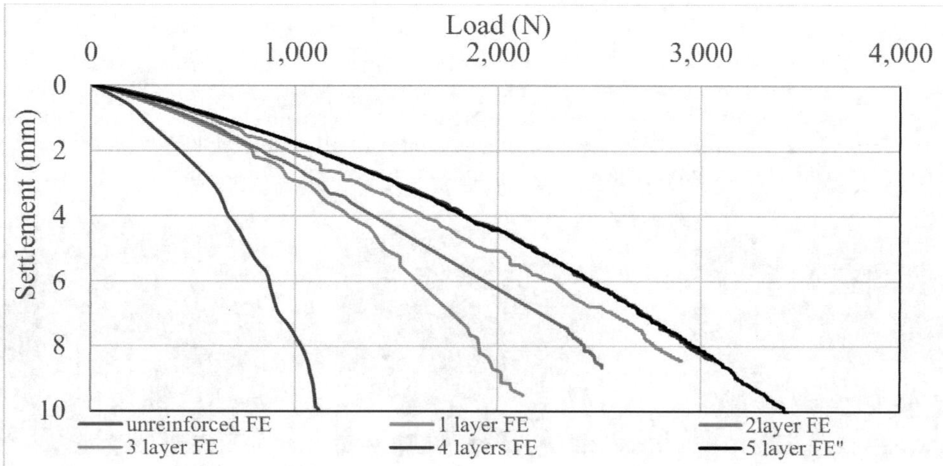

Figure5 : Numerical load settlement curve of reinforced layers N=0, 1, 2, 3, 4, and 5

Figure (6) showed that the stress distribution due to the inclusion of the reinforcing layers. Generated load transfer mechanisms due to the inclusion of the reinforcing layers resulted in redistribution process of stresses in the soil. Reinforcing layers and the trapped soil in between its apertures forms a stiff layer, at which stress is redistributed on a wider area, hence; a lower value of stress was transferred underneath it.

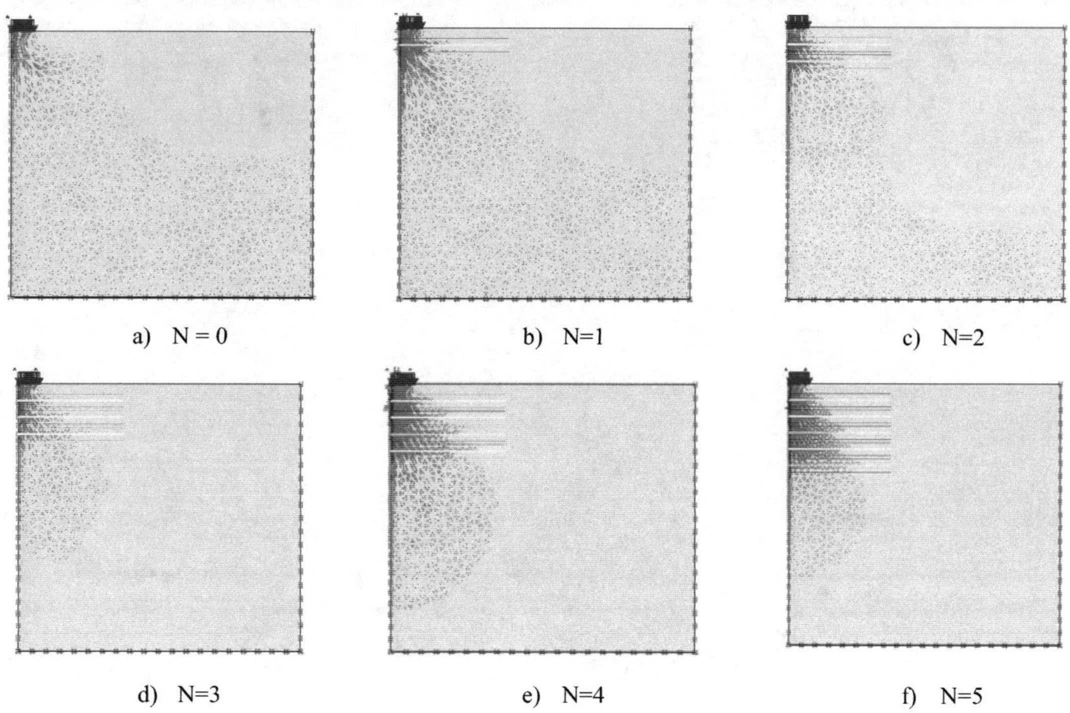

Figure 6 : plastic points under circular footing

ii. **The effect of layers width (L)**

The width of the Geo-grid layers to the footing diameter (L/D = 1,2,3,4, and 5) was studied as shown in Figureure (7). The Geo-grid layers should cover all soil failure zones assumed by Terzaghi (Figureure 8). Zone I is the elastic zone (Red zone), Zones II is the radial shear zone (Blue zone) and zone III is the Ranking passive zone (Yellow zone). Hence, the reinforced Geo-grid width depends mainly on the depth of Geo-grid layers and soil angle of friction.

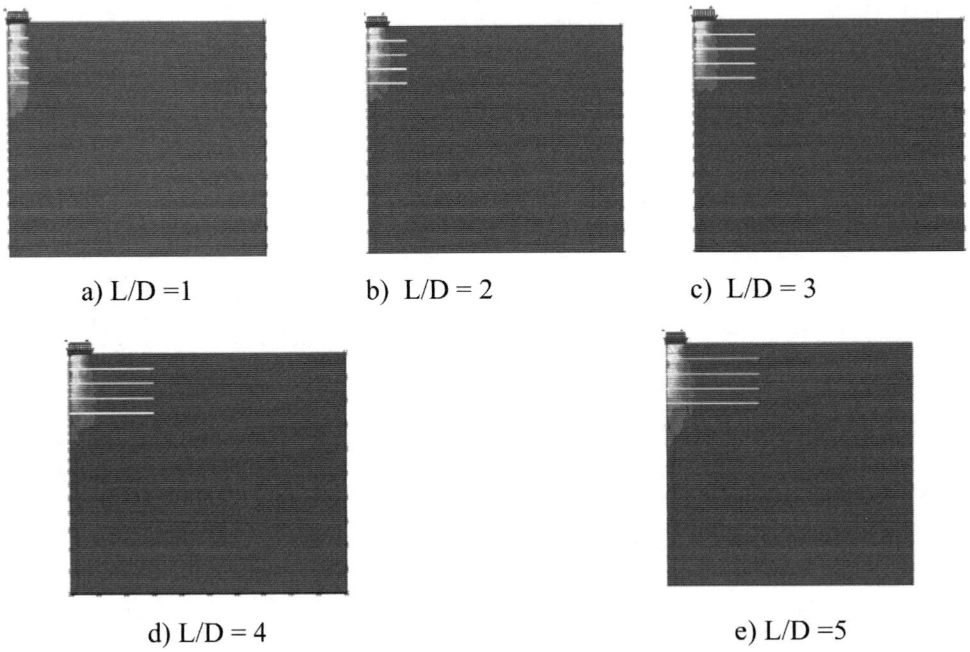

Figure 7 : The different geo-grid width under circular footing

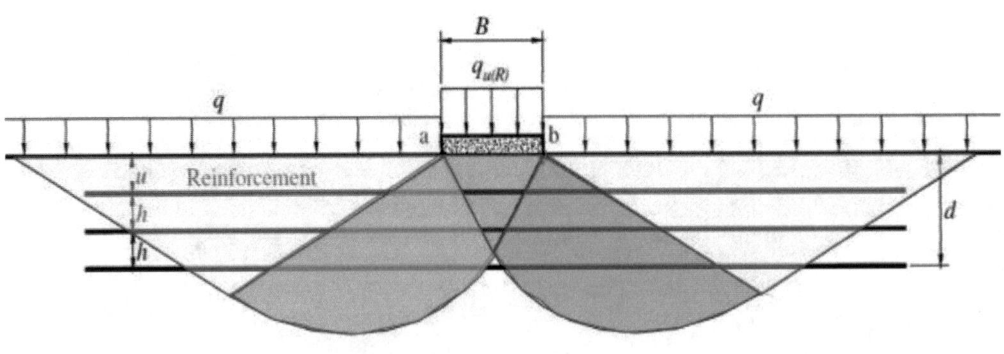

Figure 8: Failure zones assumed by Terzaghi.

Figure (9) shows the load settlement curve of the mentioned Geo-grid widths on reinforced soil performance under the circular footing. it is concluded from Figureure (9) that the bearing capacity ratio (BCR) increases by increasing

IOP Conf. Series: Materials Science and Engineering **974** (2020) 012004 doi:10.1088/1757-899X/974/1/012004

Geo-grid width. however, the more increasing in Geo-grid width the more increase in the bearing capacity ratio. Significant improvement achieved using first case (L/D =1) and gradually the improvement starts to decrease, until (L/D= 4) then the IR begins to be insignificant at L/D= 5. Finally, we can conclude that the optimum width of Geo-grid reinforced layers to footing diameter is L/D = 4.

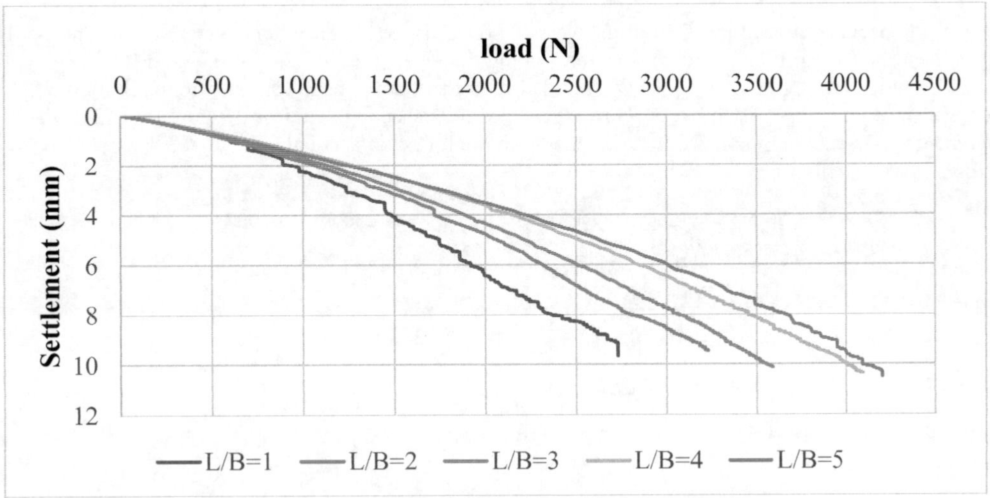

Figure (9) Load-settlement of reinforced soil with a different Geo-grid width

iii. The effect of distance between the foundation and top Geo-grid layer(U)

The optimum spacing between the topmost Geo-grid layer and the footing on soil performance was investigated (U/D = 0.2,0.25,0.3,0.35, and 0.4). Figureure (10) represents the load settlement curve using the mentioned ratios under the circular footing. Very small improvement was occurred by increasing U/D. it could be concluded from Figureure (10) that the soil behavior using U/D=0.4 has the highest bearing capacity ratio.

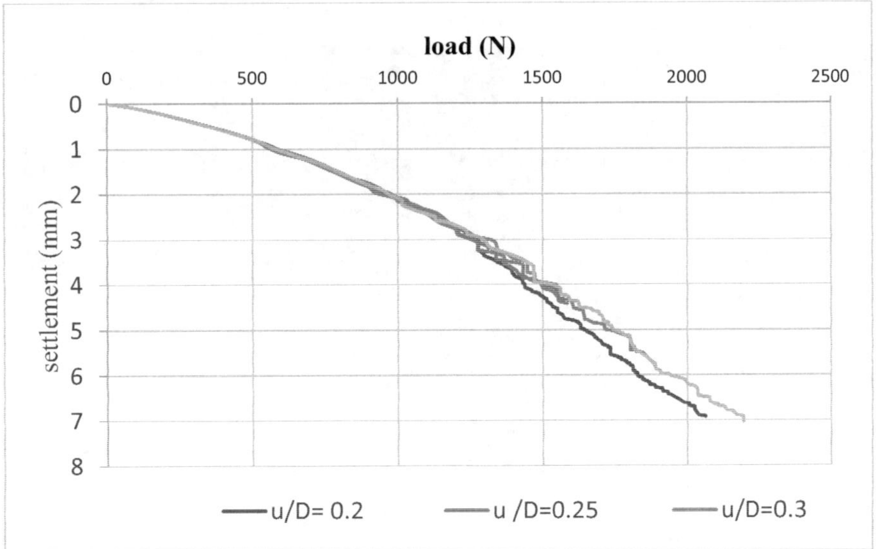

Figure 10 : Load settlement curve with different spacing between the top Geo-grid layer and the footing base

iv. The effect of distance between the Geo-grid layers (h)

The optimum spacing between the Geo-grid layers (h) on soil performance was performed. In this study, different ratios between the Geo-grid layers relative to the footing diameter (h/D = 0.2, 0.3, 0.4, and 0.5) and its effect on the BCR was investigated. Figureure (11) shows the load settlement curve using the mentioned h/D ratios. A very small difference was noticed by varying h/D. However, it could be concluded from Figureure (12), which shows the improvement ration under circular footing, that the optimum h/D ratio is equal 0.3.

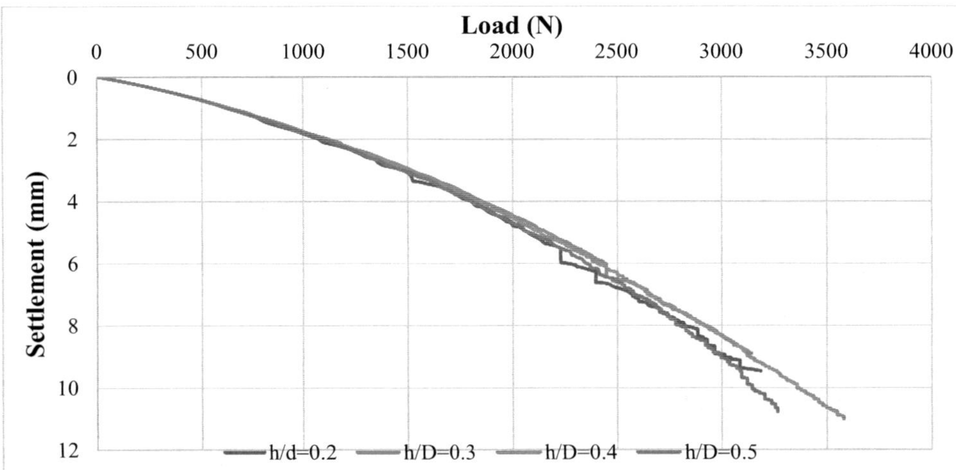

Figure11 : The load-displacement curve with different spacing between the Geo-grid layers.

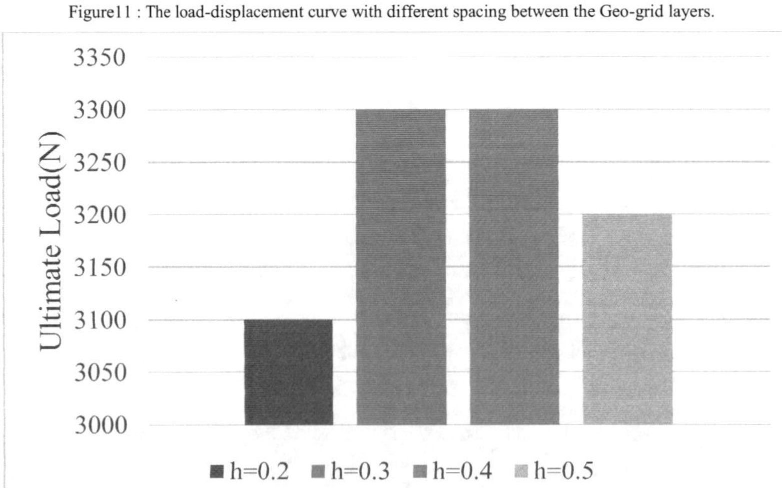

Figure12 : The improving ratio of the spacing between the Geo-grid layers

v. **The effect of varying the Geo-grid layers stiffness (K)**

Validation of Geo-grid reinforced layer stiffness was studied under the circular footing. Figureure (13) shows the effect of varying the stiffness of the Geo-grid layers (K) on reinforced soil performance. The Geo-grid layer was represented in the finite element program as an Elastoplastic material or elastic material.

Elastic material requires the Axial/Normal Stiffness (EA). While elastoplastic material requires ultimate tensile strength of the Geo-grid (Np). The values of the stiffness were varied as reported in Table (2)

Table 2 The properties of Geo-grid layers reinforced(KN/m)

	Case (1)	Case (2)	Case (3)	Case (4)	Case (5)
Ultimate tensile strength of the Geo-grid (Np) KN/m	Elastic material	45	50	55	60
Axial/Normal Stiffness (EA)KN/m	219	500	750	1000	1250

Figure (13) shows the load settlement curve using the aforementioned geo-grid stiffness. The higher the stiffness the higher bearing capacity and the improvement ration as well. However, a significant improvement was occurred by increasing the Axial/Normal Stiffness (EA) from 219 to 750 KN/m then the improvement starts to decrease above the (EA) equal 750KN/m .The improvement ratio or bearing capacity ratio decreases from 4% to 1.5% by changing the normal stiffness from EA =750 KN/m to EA = 1000 KN/m and the same reduction ratio by changing the normal stiffness from EA = 1000 KN/m to EA = 1250 KN/m. It could be concluded that an optimum Axial/Normal Stiffness (EA) equal 750KN/m with the ultimate tensile strength of the Geo-grid (Np) 50 KN/m. The key to determine the optimum Geo-grid layers stiffness (K) is that the cost of the high stiffness geo-grid is high comparatively with the improvement ratio.

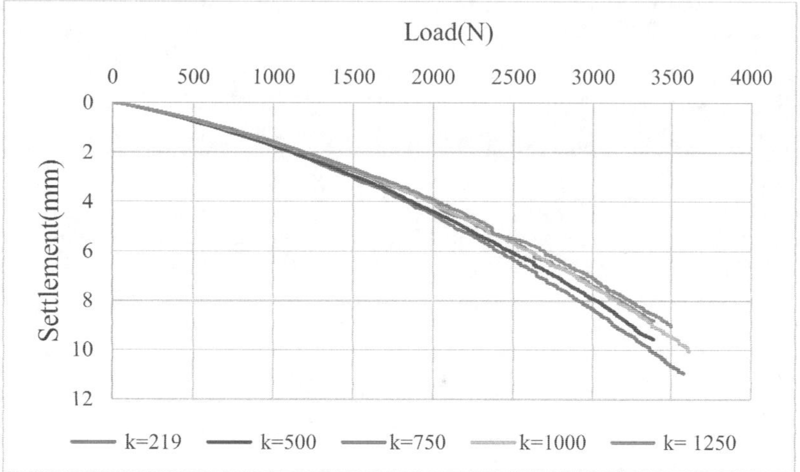

Figure 13: The effect of varied the Geo-grid layers stiffness (K)

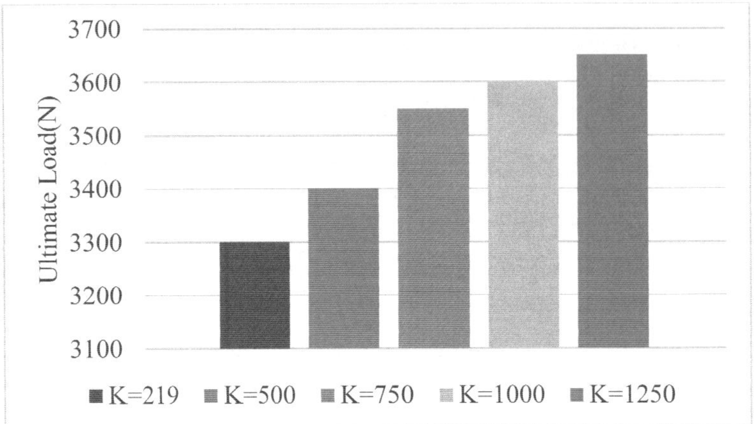

Figure14 : The improving ratio of the Geo-grid layers stiffness (K)

5 Conclusion:

In this paper, the reinforced medium sand behavior under the circular footing was studied numerically after a validation with a previous experimental work. A numerical model was used to simulate the bearing capacity ratio by varying five parameters. The five parameters that have been analyzed in this study are: Number of Geo-grid layers (N), geo-grid layers width (L), distance between the foundation and topmost Geo-grid layer (U), the distance between the Geo-grid layers (h), the Geo-grid layers stiffness (K) on soil performance under circular footing.

The conclusion could be drawn as the follows.

1. The numerical analysis of validation with the experimental results is agreeable by the ratio about 93%.
2. The optimum number of Geo- grid layer (N) is three layers for circular footing.
3. The more increasing of the Geo-grid width the more improvement in the bearing capacity results and the optimum ratio between the length of Geo-grid layer to footing diameter is (L/D =4).
4. The proper distance between the foundation and topmost Geo-grid layer (U/D) equal 0.4.
5. The optimum distance between the Geo-grid layers was (h/D) equals 0.3.
6. The optimum Axial/Normal Stiffness (EA) equal 750KN/m with ultimate tensile strength of the Geo-grid (Np) 50 KN/m for circular footing.

6 Reference:

[1] E. Badakhshan and A. Noorzad, "Effect of footing shape and load eccentricity on behavior of geosynthetic reinforced sand bed," *Geotext. Geomembranes*, vol. 45, no. 2, pp. 58–67, 2017.

[2] J. Agarwal and R. S. Jakka, "INVESTIGATIONS ON A GEOSYNTHETIC REINFORCED BEARING LAYER UNDER STATIC AND DYNAMIC LOADING," no. December, pp. 15–18, 2016.

[3] M. Touahmia, A. Rouili, M. Boukendakdji, and B. Achour, "Experimental and Numerical Analysis of Geogrid-Reinforced Soil Systems," *Arab. J. Sci. Eng.*, vol. 43, no. 10, pp. 5295–5303, 2018.

[4] J. H. Boushehrian and N. Hataf, "Experimental and numerical investigation of the bearing capacity of model circular and ring footings on reinforced sand," vol. 21, pp. 241–256, 2003.

[5] M. A. El Sawwaf, "Behavior of strip footing on geogrid-reinforced sand over a soft clay slope," *Geotext. Geomembranes*, vol. 25, no. 1, pp. 50–60, 2007.

[6] M. El Sawwaf and A. K. Nazir, "Behavior of repeatedly loaded rectangular footings resting on reinforced sand," *Alexandria Eng. J.*, vol. 49, no. 4, pp. 349–356, 2010.

[7] M. G. Arab, M. Omar, and A. Tahmaz, "Numerical analysis of shallow foundations on geogrid reinforced soil," vol. 06011, pp. 1–14, 2017.

[8] A. M. Gamal, A. M. Belal, and S. A. Elsoud, "Numerical Modeling of Geogrid Reinforced Soil Bed under Strip Footings using Finite Element Analysis," pp. 3–10, 2012.

[9] A. Belal, N. Nagy, and A. Elshesheny, "Numerical Evaluation of Bearing Capacity of Square Footing on Geosynthetic Reinforced Sand," *Proc. Int. Conf. Civil, Struct. Transp. Eng. Ottawa, Ontario, Canada*, no. 143, pp. 1–9, 2015.

[10] N. Hataf, A. H. Boushehrian, and A. Ghahramani, "Experimental and Numerical Behavior of Shallow Foundations on Sand Reinforced with Geogrid and Grid Anchor Under Cyclic Loading," vol. 17, no. 1, pp. 1–10, 2010.

[11] S. Sireesh, T. G. Sitharam, and S. K. Dash, "Bearing capacity of circular footing on geocell-sand mattress overlying clay bed with void," *Geotext. Geomembranes*, vol. 27, no. 2, pp. 89–98, 2009.

[12] G. M. Latha and A. Somwanshi, "Bearing capacity of square footings on geosynthetic reinforced sand," *Geotext. Geomembranes*, vol. 27, no. 4, pp. 281–294, 2009.

[13] A. Marto, "The Effect of Geogrid Reinforcement on Bearing Capacity Properties of Soil Under Static Load ; A Review," pp. 1881–1899.

[14] P. K. Haripal, R. N. Behera, and C. R. Patra, "Behavior of Surface Strip Footing on Geogrid Reinforced Sand Bed," pp. 1–6, 2008.

[15] A. B. Salahudeen, "NUMERICAL MODELLING OF SOIL REINFORCEMENT USING GEOGRIDS," no. March, 2016.

[16] T. Roychowdhury and J. N. Meegoda, "Numerical Modelling of Soil Improvement Using Geo-fabrics," vol. 5, no. 10, pp. 50–59, 2015.

[17] M. Makarchian, "Comparison of Bearing Capacity of Footings With Same Area Resting on Reinforced Sand," *Int. J. GEOMATE*, vol. 12, no. 29, pp. 9–16, 2017.

[18] S. K. Dash, K. Rajagopal, and N. R. Krishnaswamy, "Performance of different geosynthetic reinforcement materials in sand foundations," *Geosynth. Int.*, vol. 11, no. 1, pp. 35–42, 2004.

[19] H. Faheem and A. M. Hassan, "Journal of Engineering Sciences Faculty of Engineering 2D PLAXIS FINITE ELEMENT MODELING OF ASPHALT- CONCRETE PAVEMENT REINFORCED WITH GEOGRID," vol. 42, pp. 1336–1348, 2014.

[20] AHMED MOHAMED GAMAL EL-DIN MOHAMED MANSOUR, "Improving Sand Dunes Bearing Capacity Using Reinforced Soil Technique," ARAB ACADEMY FOR SCIENCE, TECHNOLOGY AND MARITIME TRANSPORT, 2013.

[21] D. O. F. Philosphy and F. O. F. Engineering, "MODELING AND EVALUATION OF STRIP FOOTINGS ON ANCHORED GEOSYNTHETIC REINFORCED SOIL OVERLAYING LOOSE SAND By Mohamed Kamal Hussein Mohamed MODELING AND EVALUATION OF STRIP FOOTINGS ON ANCHORED GEOSYNTHETIC REINFORCED SOIL OVERLAYING LOOSE SAND By Mohamed K," 2019.

Military Technical College
Kobry El-Kobbah,
Cairo, Egypt

13th International Conference on Civil and Architecture Engineering
ICCAE-13-2020

Geological, Geotechnical and Geophysical Aspects of Zafarana Wind Farms Sites and Their Expansion at Gabel El Zeit Sites Egypt

H Kamal[1] , G Saudi[2] and Abdel Aziz Abd Aal[3]

[1] *Soil Mechanics and Geotechnical Engineering Department, Housing and Building National Research Center (HBRC), Cairo, Egypt*
[2] *Steel and Structural Engineering Department, Housing and Building National Research Center (HBRC), Cairo, Egypt*
[3] *National Research Institute of Astronomy and Geophysics (NRIAG), Cairo, Egypt*

Abstract. The growing need for sustainable power generation has led to increasing number of power plants, of fossil-fuel type and of non-conventional type. The foundation types of wind turbine towers in power plant structures are mainly depend on the geotechnical and structural considerations and other geological and geophysical considerations. The safety and economic conditions also play a vital role in the design of foundations. Such foundations must effectively be designed taken into consideration large overturning moments and dynamic loading due to extreme wind and earthquakes. The present research is focused on studying the geotechnical, geological and geophysical aspects and characterizes of soils in Zafarana Wind Farms sites and their expansion at Gabel El Zeit sites on the Suez Gulf of Egypt. The study first address and study the geological aspects of both sites Zafarana Wind Farms sites and Gabel El Zeit sites. Secondly, the geotechnical and geophysical characteristics are studied based on establishing data base from 698 borings and 8 seismic refraction profiles across Zafarana wind Farm sites and 10 borings and 2 seismic refraction profiles at Gabel El Zeit site. The geotechnical testing and geophysical profiles were compiled to determine the variation of the soil profile as well as the characteristics of the soil layers within the study sites. The results showed construction of 8 representative boring logs at Zafarana Wind Farm sites and 2 representative boring logs Gabel El Zeit sites to represent the characteristics of the soil layers and site statgerphy. Also, the study shows the result of compression wave velocity and shear wave velocity across both sites and within the studied soil layers up to 30 m. All these results from this research will help to study the foundation and the structure of wind turbine tower , where these types of tall structures are similar to chimney structures in design and other engineering considerations.

Keywords: Zafarana Wind Farms, expansion at Gabel El Zeit, wind turbine tower, Geological aspect, geotechnical aspect, geophysical aspect.

1. Introduction

Wind turbine generator structures, as shown in Figure 1, are typically founded on massive shallow footings that are designed to transmit moderate vertical forces and large dynamic horizontal forces and overturning moments [1,2]. The lack of experimental data and numerical analysis of these types

of foundations may lead a designer to use conservative assumptions. These conservative assumptions can lead to overdesigned foundations, additional costs, and reduced life-cycle cost efficiency of wind farms.
Zafarana Wind Farms and their expansion at Gabel El Zeit are currently generating a capacity of 517MW, making it one of the largest onshore wind farms in the world. It is located in an active seismic zone along the west side of the Gulf of Suez. The main purpose here in this study is to obtain a proper understanding of the geological , geotechnical and geophysical aspects of these sites as well as the local subsurface conditions within the studied layers and their stategrapy. This has been performed by study the geological formations across both sites and establishing data base from 698 borings and 8 seismic refraction profiles across Zafarana wind Farm sites and 10 borings and 2 seismic refraction profiles at Gabel El Zeit site.

2. Zafarana Wind Farms

Zafarana Wind Farm site was analyzed as part of this research to investigate the geological, geotechnical and geophysical characterisation of this site and it consists of eight wind farms over an area round of 12 X 18 km as shown Figure 2. The expansion area in Gabel El Zeit was also analysed in this study which consist of ten wind farms under construction as shown in Figure 3. The sites of the eight wind farms in Zafarana and ten wind farms under construction in the expansion area in Gabel El Zeit were selected to investigate the geological, geotechnical and geophysical characteristics of these sites.

Fig .(1) A wind turbine generator at Zafarna Wind Farm

IOP Conf. Series: Materials Science and Engineering **974** (2020) 012005 doi:10.1088/1757-899X/974/1/012005

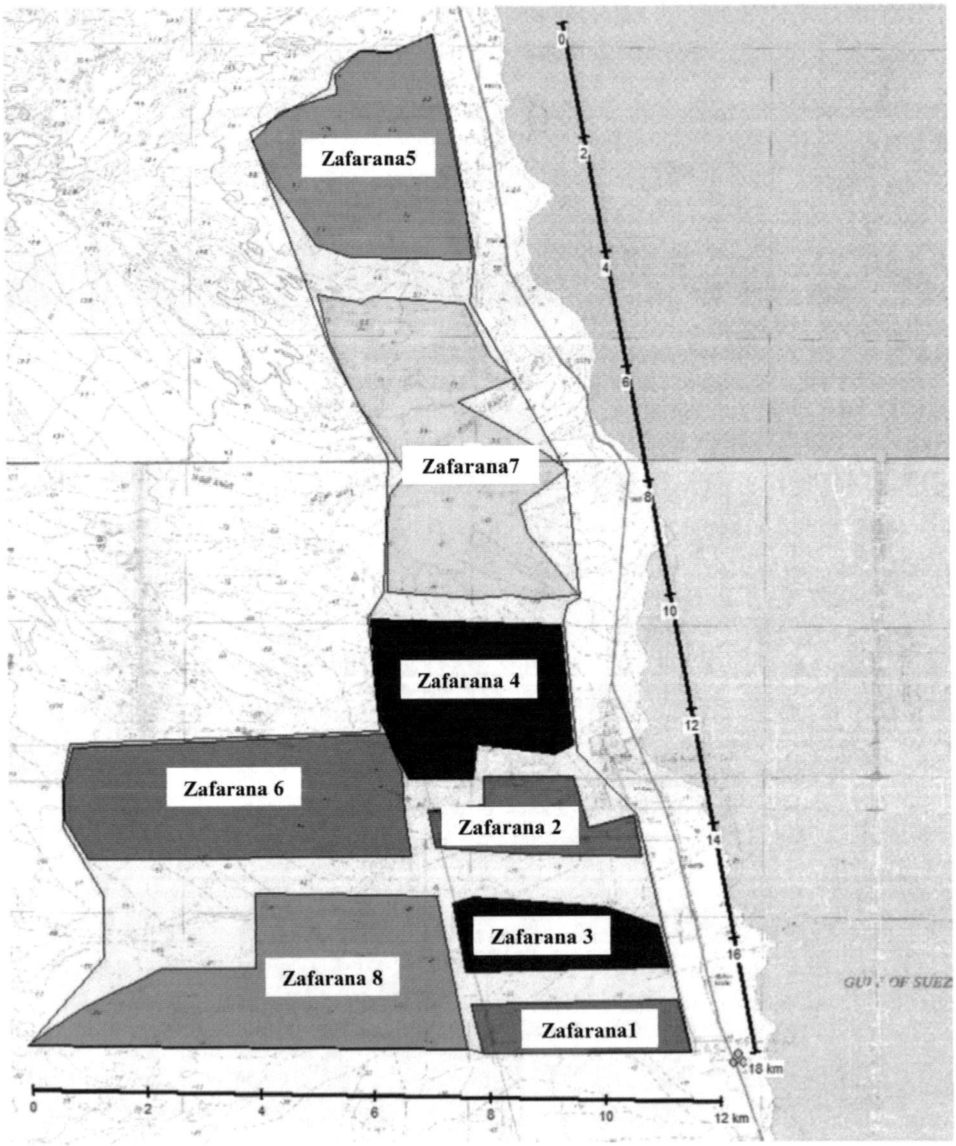

Fig. (2) Layout of Zafarana wind Farms

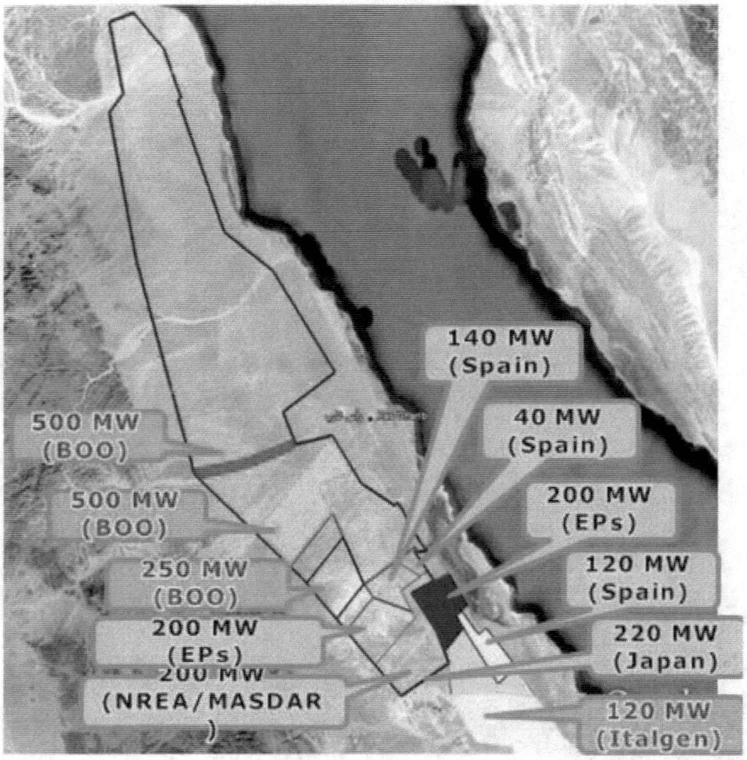

Fig. (3) Layout of the Expansion area of Gabal El Zeit wind Farms including 11 wind farms in this location from 140MW to 500 MW

3. Geological Investigation

The Red Sea Basin Province originated as an Oligocene continental rift impacted by left-lateral wrenching. Rift location and borders are defined by crustal weaknesses created more than 500 Million age , including the late Proterozoic to early Paleozoic cratonization of the Arabian-Nubian shield, its suturing to the African continent, and subsequent supercontinent breakup. Those events resulted in the juxtaposition of structurally and compositionally different basement terranes. Said [3], and El-Gezery and Marzouk [4] showed that the depth of the basement increases northwards towards the Mediterranean Sea. The Paleozoic rocks are characterized by continental clastic deposits. The marine episodes are minor in space and time. The Paleozoic period ends with the Upper Carboniferous Lower Permian marine deposition, that followed the Hercynian orogenic phase and the subsequent erosional period. Predominant continental deposition started again in the Mesozoic. The Cenozoic witnessed the transformation of the Tethys into the Mediterranean Sea during the Paleocene-Eocene. The Miocene was a period leading to the formation of present Red Sea coast sediments and sediments in the northwestern part of the Gulf of Suez. The Quaternary was determined by regression with minor transgression. Uplift and tectonic disturbances mark the Pliocene-Quaternary boundary in the Red Sea region [5]

IOP Conf. Series: Materials Science and Engineering **974** (2020) 012005 doi:10.1088/1757-899X/974/1/012005

The geological features indicate that the study area consists of coastal and wadi deposits. There are no intensive faults at Zafarana area. However, some faults, taking the NW-SE trend, appear at the area between Quseir and Ras Ghareb and also between Marsa Alam and Quseir. The oldest known formation in the study area is of Late Precambrian age, igneous and metamorphic rocks forming the northern edge of the African Shield [6] as shown in Figure 4 and Figure 5.

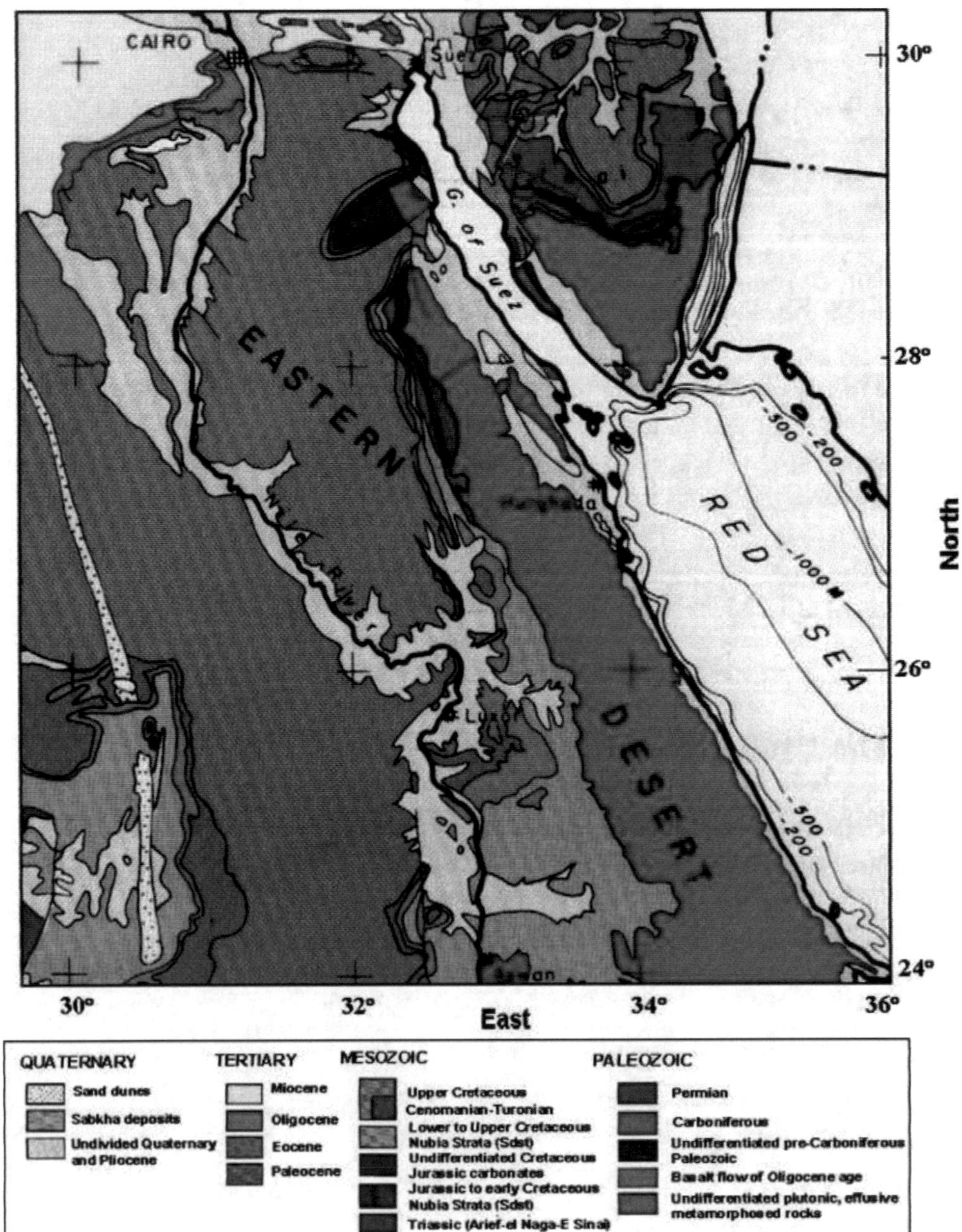

Fig. (4) Surface geological map of the Red Sea Governate [6]

IOP Conf. Series: Materials Science and Engineering **974** (2020) 012005 doi:10.1088/1757-899X/974/1/012005

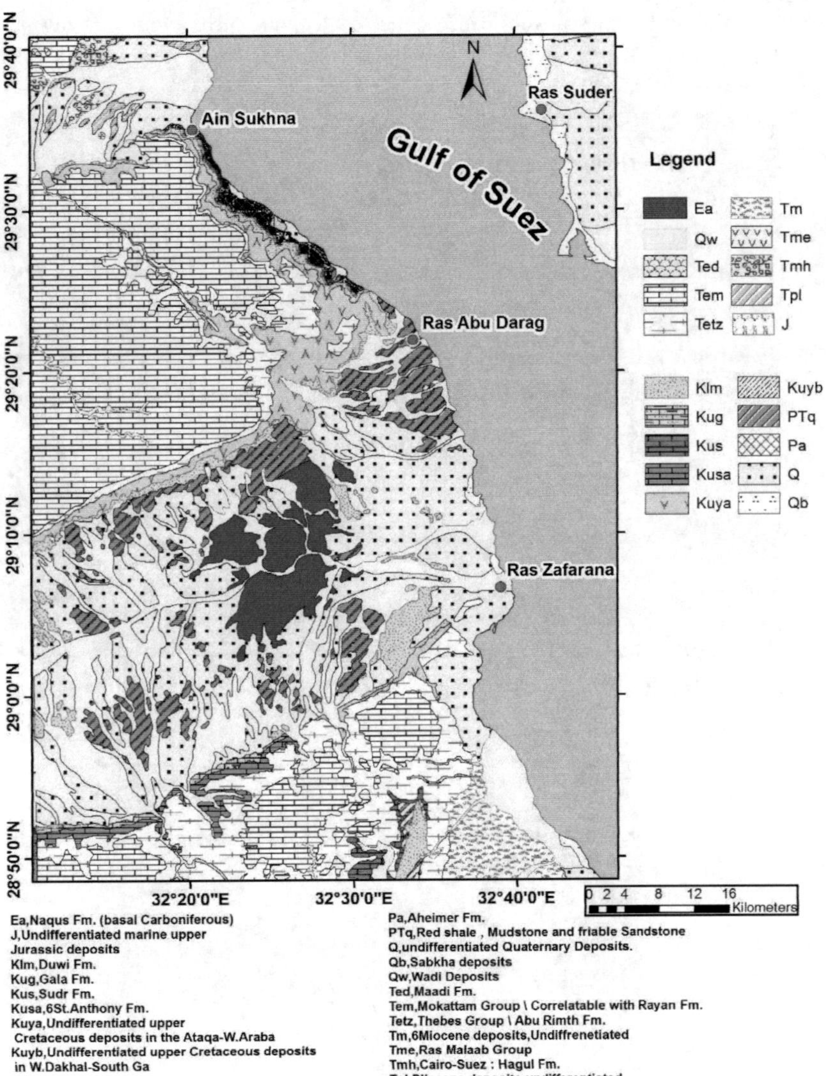

Fig. (5) Surface geological map of the Zafarana region [6]

4. Geotechnical Investigation

Geotechnical database was established for Zafarana sites and 698 borings were collected from previous geotechnical site investigations reports where the distribution of the 698 Borings have the same location of wind turbines towers cause under each wind turbine one boring was executed as shown in Figure 6. Also 10 borings were collected from previous preliminary geotechnical site investigations report for Gabel El Zeit sites as shown in Figure 7.

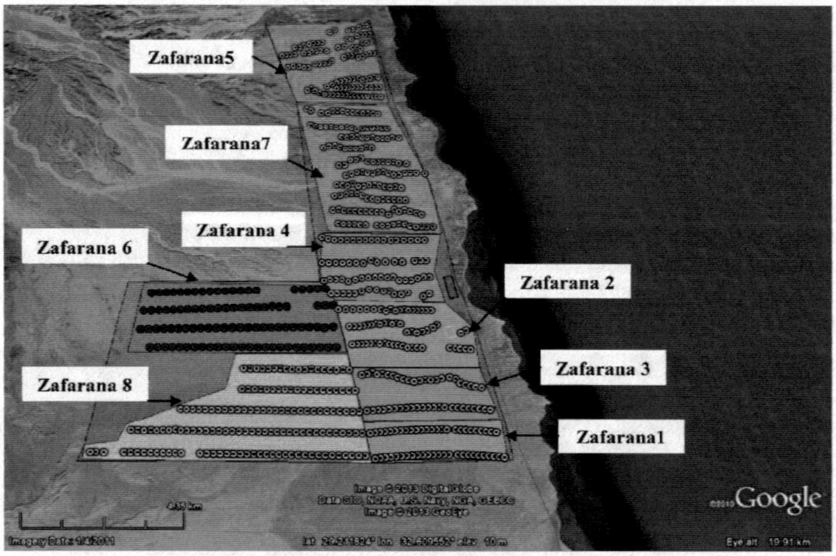

Fig. (6) The Distribution of 698 Borings at Zafarana wind farms

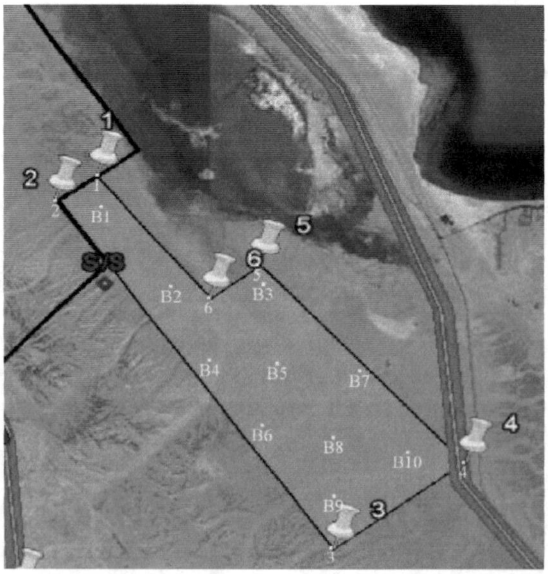

Fig. (7) The Distribution of 10 Borings at Gabal El Zeit wind farms for preliminary geotechnical of investigation.

4.1 Geotechnical Investigation of the Zafarana Sites

4.1.1 Subsurface formations across the Zafarana sites

Based on the results of the previous site investigation and laboratory testing the soil formations encountered in the boreholes show five main formations, which are Wadi deposits, Clay, Claystone, Sand and Sandstone. The subsurface formations at the site can be summarized as follows:

1. **Wadi Deposit:** Reddish brown to yellowish brown wadi deposit that consist of gravel, sand, silt, clay and iron oxides. The gravel is identified to be of dolomitic origin. This layer appears most boreholes from the ground surface to a depth varying from 0.25 m to 15.0 m from the ground surface.
2. **Clay:** Hard sandy laminated, lean or fat reddish yellow to greyish brown clay with interbedded band of claystone or sandstone pieces (5-15 cm) and traces of iron oxides, it appears at most boreholes from 3.7 m to 10 m.
3. **Claystone:** Reddish to yellowish brown very weak to medium hard sandy claystone interbedded by thin to thick band of sand or sandstone (5- 15 cm). It is stained by iron oxides, it becomes fragmented in some times. This layer appears at most boreholes from 7.5 m to 15 m .
4. **Sand:** Yellow to reddish brown sand with different percent of silt, could be cemented or clayey, it appears with interbedded of sandstone or clay stone pieces, calcanrous, and with traces of iron oxides. This layer appears most boreholes from 25 m to 6 m.
5. **Sandstone:** Yellow to reddish brown fine to medium grained sandstone, interbedded by thin to thick band of claystone (5- 15 cm). It is stained by iron oxides. This layer appears at most boreholes from 6 m to 15 m.

Groundwater is not encountered in any of the boreholes at the time of investigation.

4.1.2 Evaluation of soil Properties from field and laboratory testing

1. Wadi deposit

Reddish brown to yellowish brown wadi deposit that consist of gravel, sand, silt, clay and iron oxides. The gravel is identified to be of dolomitic origin. The SPT blow counts N values of this layer varies from 33 to 50. The blow counts 50 are counted after depth of penetration 6 to 13 cm. It has to be noted that the presence of gravel and/or cementation may contribute to the high values of blow count N values. Therefore, evaluation of soil parameters based on SPT results should be dealt conservatively. The following is an evaluation of the soil properties in order to deduce design parameters that can be used in the geotechnical analyses.

- **Bulk Unit Weight:** The bulk unit weight of the deposit is conservatively estimated to be about20 kN/m^3. This value is recommended to be used in the geotechnical analyses.
- **Friction Angle:** The friction angle is estimated based on SPT test results, taking into consideration that these SPT results may be affected by the presence of high gravel content and/or cementation, which will increase the SPT test results. Accordingly, a conservative value of 37°can be considered in the geotechnical analyses.
- **Deformation Modulus:** A conservative value of Young's modulus of 40 MPa can be considered in the geotechnical analyses.

2. Clay Layer

Hard sandy laminated, lean or fat reddish yellow to greyish brown clay with interbedded band of claystone or sandstone pieces (5-15 cm) and traces of iron oxides. The liquid limit, plastic limit and plasticity index are in the ranges of 25% to 63%, 14% to 27%, and 9% to 39%, respectively.

- **Bulk Unit Weight**: Bulk unit weight determinations were performed on samples selected from the clay layer. The results indicated a bulk unit weight ranging between 20.99 kN/m^3 and 23.84 kN/m3. An average value of 22.5 kN/m^3 can be considered in the geotechnical analysis.

- **Dry Unit Weight:** Dry unit weight determinations were performed on samples selected from the clay layer. The results indicated a dry unit weight ranging between 19.32 kN/m^3 and 21.89 kN/m^3. An average value of 20.6 kN/m^3 can be assumed for the clay layer.
- **Shear Strength:** The measured values of pocket penetrometer qu are of more than 500 kPa. The measured pocket penetrometer qu values indicate undrained shear strength of more than 250 kPa. Based on this interpretation, undrained shear strength of the clay layer of 100 kPa can be conservatively considered in the geotechnical analysis. Such value takes into consideration the fissured nature of the clay. Considering the plasticity index range of the clay, a drained shear friction angle in excess of about 25° can be conservatively used in geotechnical analysis. The drained cohesion intercept of 0 kPa can be conservatively assumed.
- **Swelling Potential:** One-Dimensional Swell tests were performed on samples selected from the clay layer at various depths from the ground surface. The results indicated swelling pressure of (0) MPa. When the samples were subjected to inundation, no sign of swelling was observed in the Oedometer. In fact a reduction in volume was observed. These values indicate that no expansiveness according to the classification by Snethen et al. [7] as shown in Table 1.

The free swell tests showed results of free swell less than 85% with the majority of the results showed less than 50% free swell.

Table 1. WES classification system based on Snethen et al. [7]

Classification of Potential Swell	Potential Swell %	Liquid Limit %	Plasticity Index %	Natural Soil Suction kPa
Low	<0.5	<50	<25	<150
Marginal	0.5-1.5	50-60	50-60	150-400
High	>1.5	>60	>50	>400

- **Deformation Modulus:** Based on the range of plasticity index (9-39%), a ratio of 300 between deformation modulus and undrained shear strength was used to estimate the defmmation modulus of the clay. Such ratio resulted in a value of deformation modulus of 30 MPa that can be considered in the geotechnical analyses.

3. Claystone Layer

Reddish to yellowish brown very weak to medium hard sandy claystone interbedded by thin to thick band of sand or sandstone (5- 15 cm). It is stained by iron oxides, it becomes fragmented in some times.

- **Bulk Unit Weight:** Bulk unit weight determinations were performed on samples selected from the claystone layer. The results indicated a bulk unit weight ranging between 21.48 kN/m^3 and 24.43 k:N/m^3. A value of 23 kN/m^3 can be considered in the geotechnical analysis.
- **Dry Unit Weight:** Dry unit weight determinations were performed on samples selected from the claystone layer. The results indicated a dry unit weight ranging between 19.33 kN/m^3 and 23.84 kN/m^3. A value of 21.5 kN/m^3 can be assumed for the claystone.
- **Unconfined Compression Test:** Unconfined compression tests were performed on selected from various depths throughout the layer. The results indicated unconfined compression strength ranging between 0.67 MPa and 35.39 MPa. An average value of 18 MPa can be assumed for intact samples. A reduced value of about 0.5 MPa can be conservatively considered for the rock mass in the geotechnical analysis.
- **Deformation Modulus:** The results of the unconfined compression tests were used to estimate the deformation modulus of the claystone samples. The average value of the estimated deformation modulus of the intact samples was further reduced by 10 to conservatively take into account the rock mass behaviour. The results indicated a very conservative reduced value of 90 MPa to be used in the geotechnical analysis.

4. Sand Layer

Yellow to reddish brown sand with different percent of silt, could be cemented or clayey, the layer is interbedded with sandstone or clay stone pieces, calcareous, and with traces of iron oxides.

- **Bulk Unit Weight:** Bulk unit weight determinations were performed on samples selected from the sand layer. The results indicated a bulk unit weight ranging between 17.66 kN/m^3 and 22.27 kN/m^3. A value of 18 kN/m^3 can be conservatively considered in the geotechnical analysis.
- **Dry Unit Weight:** Dry unit weight determinations were performed on samples selected from the sand layer. The results indicated a dry unit weight ranging between 15.61 kN/m^3 and 20.91 $k:N/m^3$. A value of 16 kN/m^3 can be assumed for the sand layer. The lower values are observed in the cemented sand samples.
- **Friction Angle:** The friction angle is estimated based on SPT test results, taking into consideration that these SPT results may be affected by the presence of high gravel content and/or cementation, which will increase the SPI test results. Accordingly, a conservative value of 37°can be considered in the geotechnical analyses.
- **Deformation Modulus:** A conservative value of Young's modulus of 40 MPa can be considered in the geotechnical analyses.

5. Sandstone Layer

Yellow to reddish brown fine to medium grained sandstone, interbedded by thin to thick band of claystone (5- 15 cm). It is stained by iron oxides.

- **Bulk Unit Weight:** Bulk unit weight determinations were performed on samples selected from the sandstone layer. The results indicated a bulk unit weight ranging between 20.99 kN/m^3 and 24.53 kN/m^3. An average value of 22 kN/m^3 can be considered in the geotechnical analysis.
- **Dry Unit Weight:** Dry unit weight determinations were performed on samples selected from the sandstone layer. 1he results indicated a dry unit weight ranging between 18.15 kN/m^3 and 23.5 kN/m^3. A value of 21 kN/m^3 can be assumed for the sandstone.
- **Unconfined Compression Test:** Unconfined compression tests were performed on samples selected from various depths throughout the layer. The results indicated unconfined compression strength ranging between 0.72 MPa and 23.71 MPa. An average value of 12 MPa can be assumed for intact samples. A reduced value of about 0.5 MPa can be conservatively considered for the rock mass in the geotechnical analysis.
- **Deformation Modulus:** The results of the unconfined compression tests were used to estimate the deformation modulus of the sandstone samples. The average value of the estimated deformation modulus of the intact samples was further reduced by the ratio of 10 to conservatively take into account the rock mass behaviour. The results indicated a very conservative reduced value of 75 MPa to be used in the geotechnical analysis.

5.2 Geotechnical Investigation of the Expansion area in Gabel El Zeit

4.2.1 Subsurface formations across the Expansion area in Gabel El Zeit

Based on the results of the previous preliminary site investigation and laboratory testing the soil formations encountered in the boreholes show granular soil where this layers consisting mainly of sand and gravel mixtures with varying percentages. The granular soil were encountered at the location of all boreholes. It extend from ground surface to the end of borehole. At some boreholes the surface layer contains percentage of silt. At few boreholes sand was appeared without gravel.
Groundwater is not encountered in any the boreholes at the time of investigation.

5.2.2 Evaluation of soil Properties from field testing

During the drilling of boreholes, the standard penetration tests (SPT) were performed at approximately 3.00 m – interval in order to obtain approximate relative densities of the soils. No. of blows are mentioned in borehole logs. S.P.T results indicate that the granular soil layers are very

IOP Conf. Series: Materials Science and Engineering **974** (2020) 012005 doi:10.1088/1757-899X/974/1/012005

dense, and giving high bearing capacity at level 2.0 m – 3.0 m. However, a lot of the S.P.T results were affected because of the high percentage presence of gravel.

Table 2. Results of Standard Penetration Test

No. of Blows (N)	**Relative Density**
0-4	Very Loose
4-10	Loose
10-30	Medium
30-50	Dense
Over 50	Very Dense

4.3 Representative Boring Logs

4.3.1 Representative Boring Logs of Zafarana Sites

The 698 collected borings from previous geotechnical site investigations were analyzed to produce eight representative boring logs for the eight Zafarana wind Farms as shown in Figure 8, where these boring logs were used for ground site response analysis .

4.3.2 Representative Boring Logs of the Expansion area in Gabel El Zeit

The 10 collected borings from previous preliminary geotechnical site investigations were analyzed to produce two representative boring logs for the under construction wind farms in the Expansion area in Gabel El Zeit as shown in Figure 9, where these boring logs were used for ground site response analysis .

IOP Conf. Series: Materials Science and Engineering **974** (2020) 012005 doi:10.1088/1757-899X/974/1/012005

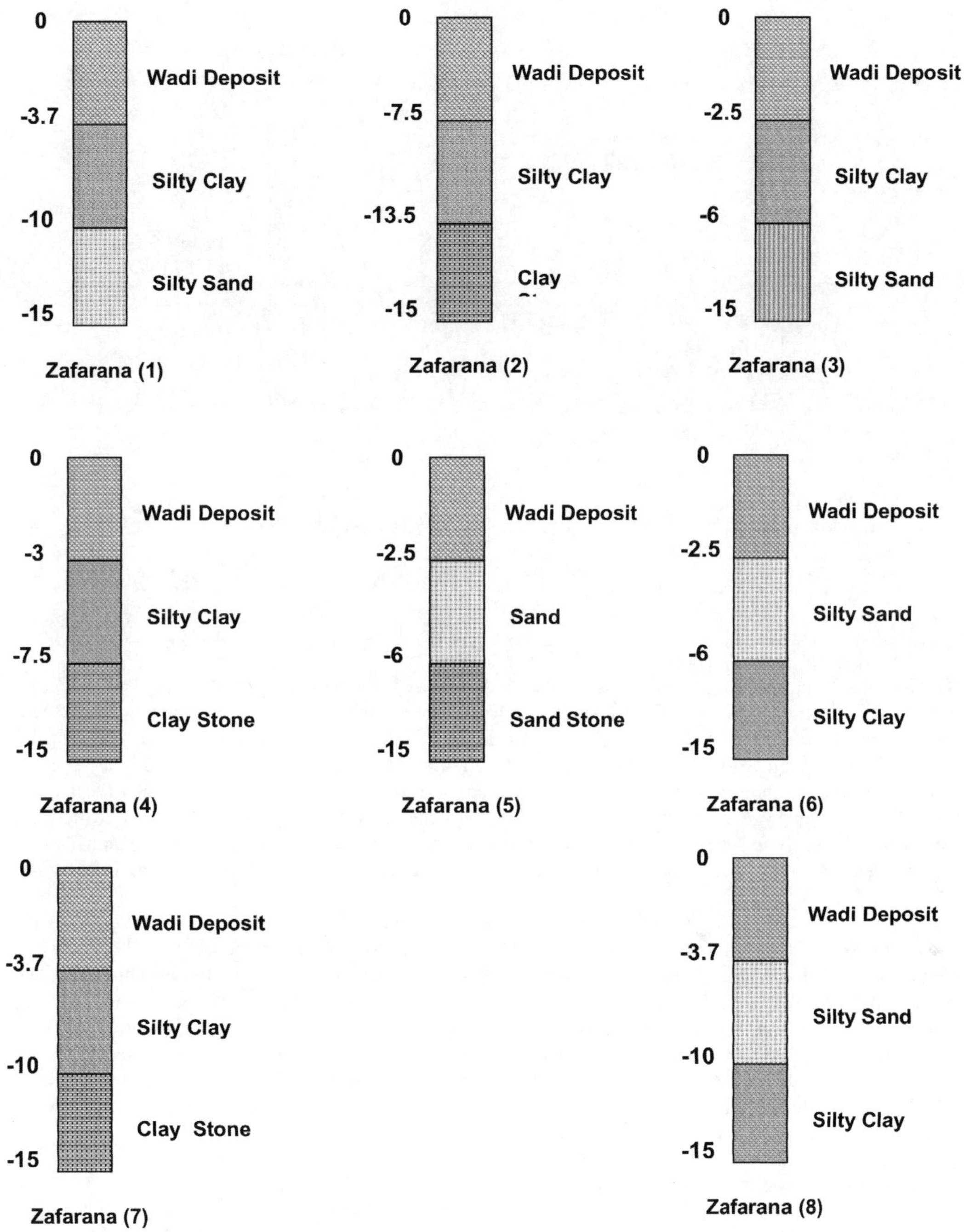

Fig. (8) Eight representative boring logs for the eight Zafarana farms where each farm has a one representative boring logs

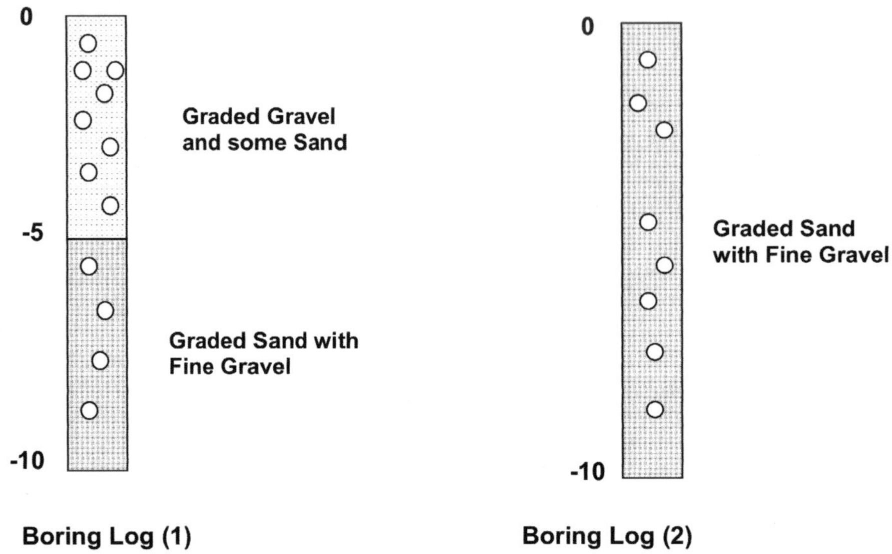

Fig. (9) Two representative boring logs for the expansion area in Gabel El Zeit

6. Geophysical Investigation

Geophysical exploration as shallow seismic refraction has been widely applied in civil engineering; i.e. road tunnelling, dam sites, quarries, hydroelectric power plants, subway constructions, nuclear power plants, cavity and fault detection and many other purposes. P- and S- wave velocity measurements obtained from shallow seismic refraction surveys are used to evaluate the bed rock in order to determine the elastic properties of the materials in the sites under assessment. SH (horizontal shear) waves can be used to study the vibration characteristics of the subsurface layers, which are important for earthquakes resistance structural design. Increasing source energy, stacking and filtering are considered as ways to acquire high quality refraction records, which contain sharp first breaks [8,9].

The refraction method consists of measuring (at known points along the surface of the ground) the travel times of compressional waves generated by an impulsive energy source [8]. The energy is detected, amplified and recorded by special equipment (Seismograph) designed for this purpose. The instant of explosion, or "zero-time," is recorded on the record of arriving pulses. The raw data, therefore, consist of travel times and distances, and these time-distances information is then manipulated to convert it into the format of velocity variations with depth. The subsurface structure is inferred from interpretation methods based on the laws of energy propagation. The propagation of seismic energy through subsurface layers is described essentially by the same rules that govern the propagation of light rays through transparent media [8,9].

The refraction that a seismic pulse undergoes when passing from one material to another depends upon the ratio of the transmission velocities of the two materials. The fundamental law that describes the refraction of seismic pulse is Snell's Law, and this together with the phenomenon of "critical incidence," is the physical foundation of any seismic refraction surveys. Snell's Law and critical incidence are illustrated in Figure 10, which shows a medium with a velocity V_1, underlain by a medium with a higher velocity V_2. Snell's Law is given in the following relation:

$$\sin \alpha / \sin \beta = V_1 / V_2 \quad (1)$$

IOP Conf. Series: Materials Science and Engineering **974** (2020) 012005 doi:10.1088/1757-899X/974/1/012005

Critical incidence occurs when $\beta = 90\ °$, at this case:

$$\sin \alpha = V_1 / V_2 \quad (2)$$

Until the critical angle of incidence is reached, almost all of the compressional energy is refracted into the higher velocity medium. When the critical angle is exceeded, the energy is almost totally reflected and no energy is refracted into the high-speed layer.

In shallow seismic refraction surveying of the current study, the basic instrument includes seismic source (a weight drop 150 kg is used), detectors (48 geophones with natural frequency of 40 Hz are used, as shown in Figure 11 for vertical and horizontal geophones), cable (geophone cable connected to the data logger with one meter interval and 48 takeouts), and data logger (Strata View 48 channels seismograph manufactured by Geometrics Company) is utilized with a menu involves an option for standard settings, 24 dB, single-stage, Instantaneous Floating Point (IFP) amplifier, followed by an A/D converter sampling at 1/4 microsecond intervals. The floating point number is fed to a 32-bit, floating-point, digital signal processor (DSP) which performs multi-sample processing to increase the effective resolution and dynamic range. A 24-bit number is generated from the A/D process. The digital signal is stored in a semiconductor memory, where it can be viewed on a display unit, plotted on a paper record, or saved on floppy disk Figure 12. A weight drop 150 kg is used Figure 13.

5.1 Seismic Data Acquisition

In order to classify the shallow foundation section into soil and bed rock and to define the competence scales and geotechnical characteristics, the seismic refraction was carried out through applying the forward, mid-point and reverse acquisition system vertically and horizontally to create the compressional waves (P-waves) and shear waves (S-waves). The object of such P-waves and S-waves is to conclude the variations of velocities (compressional, Vp, and shear, Vs,) with depth and correlate them with the subsurface geologic information, if present. The ground refraction field work executed in the Zafarana wind farm and the expansion in Gabel El Zeit sites is consisted of eight and two seismic profiles distributed regularly in the sites Figures 14 and Figure 15. Accordingly, for each seismic site, two seismic refraction profiles are acquired, one for the compressional waves and one for the shear waves.

The P-waves are acquired by generating seismic energy using energy source, sending the created seismic waves inside the earth. The direct (head) and refracted (diving) waves are detected through geophones, which are motion sensitive transducers, that convert the mechanical ground motion into an electrical signal, whose voltage amplitude is proportional to the received energy. The Eight and two profiles have 96 meters long spread, the total spread length should be three to five times the maximum depth anticipated [10]. The geophone spacing was fixed and equal to 2 m. The geophones must be firmly coupled to the ground.

The technique is to shot the profile (3 shots) 2 meters far from both ends, also from its mid-point. Figure 16 shows a representative example of the recorded P-wave seismograms at the P-waves seismic profiles.

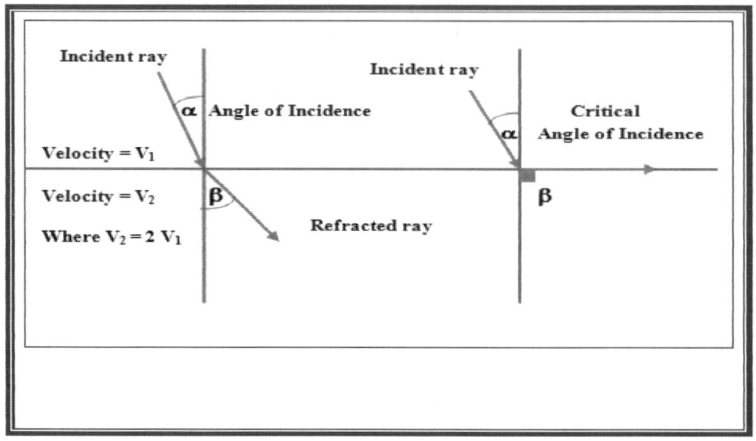

Fig. (10) Refraction of a ray transmitted across a boundary between two media with different velocities (V_2=2 V_1)

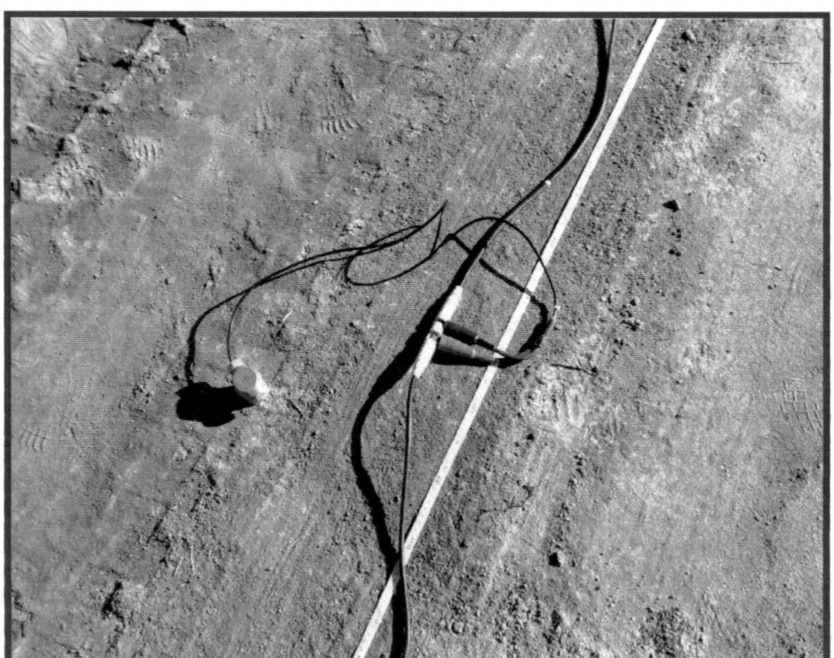

Fig. (11) The types of Vertical and horizontal geophones used in the data acquisition

Fig. (12) The StrataView data logger which used in the shallow seismic refraction survey

Fig. (13) A weight drop 150 kg is used

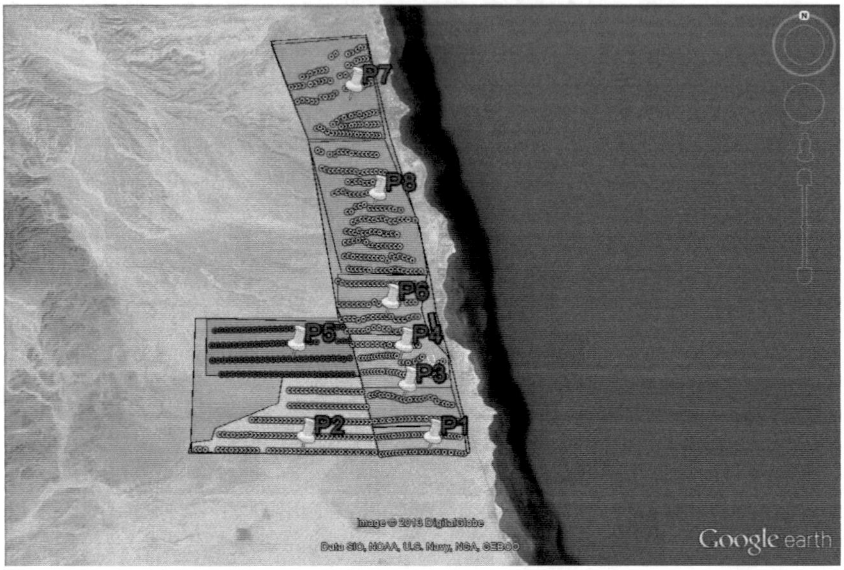

Fig. (14) Location of seismic profiles conducted at Zafarana wind farm site

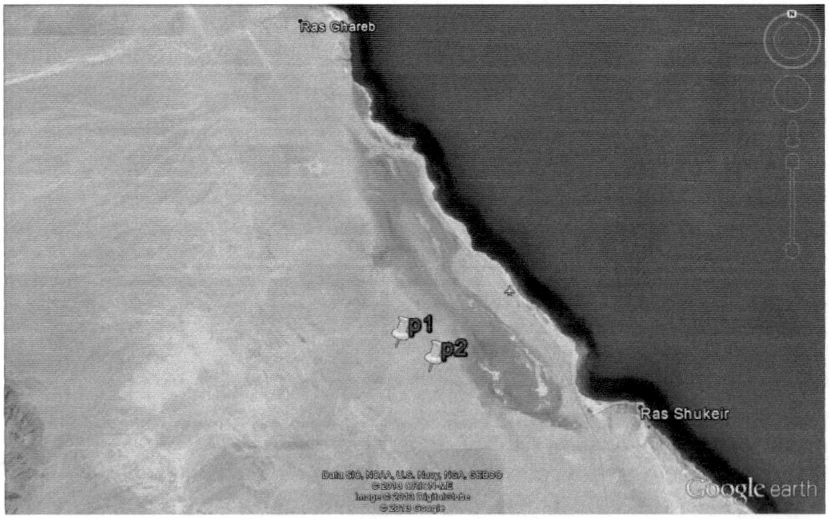

Fig. (15) Location of seismic profiles conducted at the expansion of Gabel El Zeit

IOP Conf. Series: Materials Science and Engineering **974** (2020) 012005 doi:10.1088/1757-899X/974/1/012005

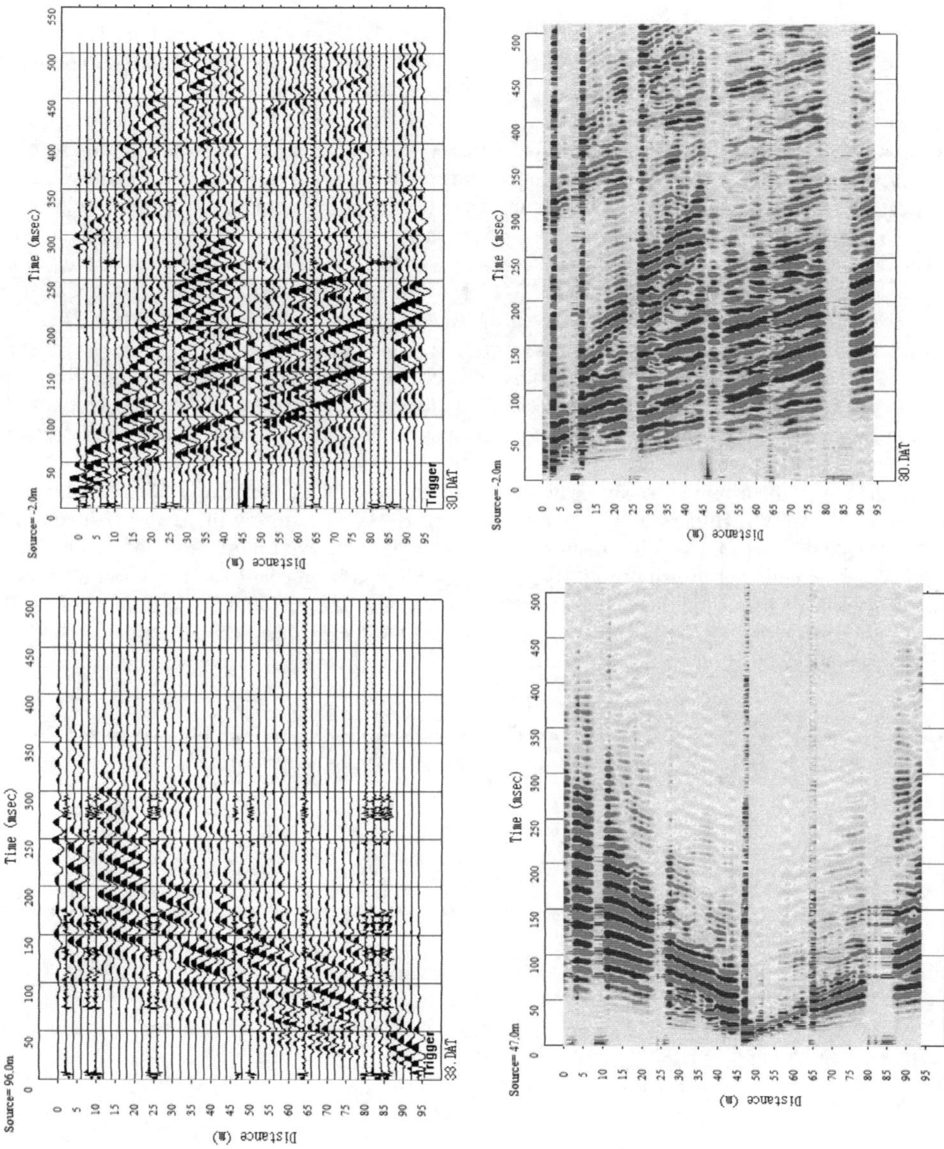

Fig. (16) Seismograms of P- wave data at Zafarana wind farm site

The generation of shear waves in contrast with the P-waves is more difficult. The shear sources create compressional waves simultaneously with the shear energy [11]. The generation of shear energy is discussed by many authors; such as: [12,13,14,15] .They conclude several steps to provide unambiguous identification of the horizontal shear waves (SH) arrival on the seismic waveform as follow: 1- The first step is to use a source, which produces the largest possible shear waves and the smallest possible waves of other types, 2- The second step is to make measurements at a location, where the source radiation pattern predicts the largest shear wave amplitudes with respect to other types. For a horizontal surface force, measurements along a perpendicular line would detect principally horizontal shear waves (SH), and 3- The third step is to orient the detector to take maximum advantage of the directionality of the shear waves. Transverse detectors would be used.
The thrust of shear wave source design has therefore taken the direction of suppressing other unwanted wave types. Particular emphasis has given to create pure horizontal shear wave (SH) type motion and suppressing P and SV by impact first in one direction and reverse the polarity mode in the seismograph then impact in the opposite direction. By this technique, the amplitude of the horizontal shear waves will enhance, while all the other signals will vanish.
Another technique to determine the shear wave velocities is used in this study. This is the most common type of MASW (Multi channel Analysis of Surface Waves) survey that can produce a 2-D Vs profile [16]. The overall setup is illustrated in Figure 17. The maximum depth of investigation (Zmax) that can be achieved is usually in 10-30 m range, but this can vary with sites and types of active sources used. Field procedures and data processing steps are briefly explained below. 8 MASW profiles are carried out at Zafarana wind farm while 2 MASW profiles were carried out at the expansion area in Gabel El-Zeit.

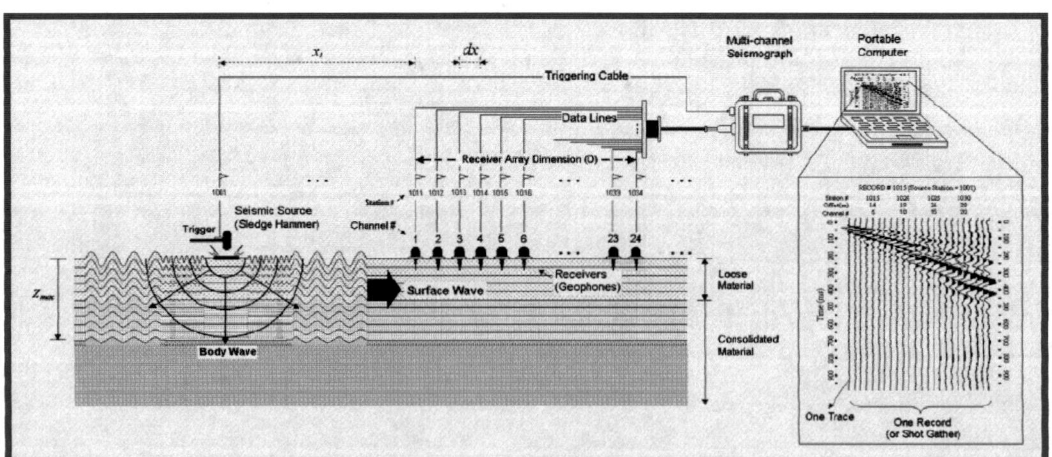

Fig. (17) The MASW (Multi channel Analysis of Surface Wave) data acquisition configuration for determining the 1-D shear wave velocities

A weight drop 150 kg was used as a source of active MASW. Vertical stacking with multiple impacts can suppress ambient noise significantly and is therefore always recommended, especially if the survey takes place in an urban area. Low-frequency of 4.5 Hz geophones is used. Length of the receiver spread (D) as shown in Figure 17 is directly related to the longest wavelength (λ) that can be analyzed, which in turn determines the maximum depth of investigation (Zmax): On the other hand, (minimum if uneven) receiver spacing (dx) is related to the shortest wavelength (λ) and therefore the shallowest resolvable depth of investigation (Zmin). A one millisecond of sampling interval is most common with a 2-sec total recording time (T=2 sec). Figure 18 shows example of MASW profile at Zafarana site.

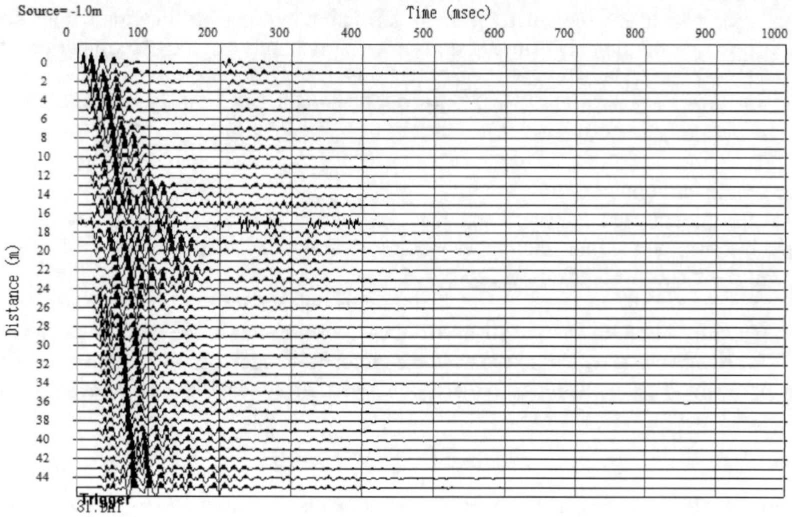

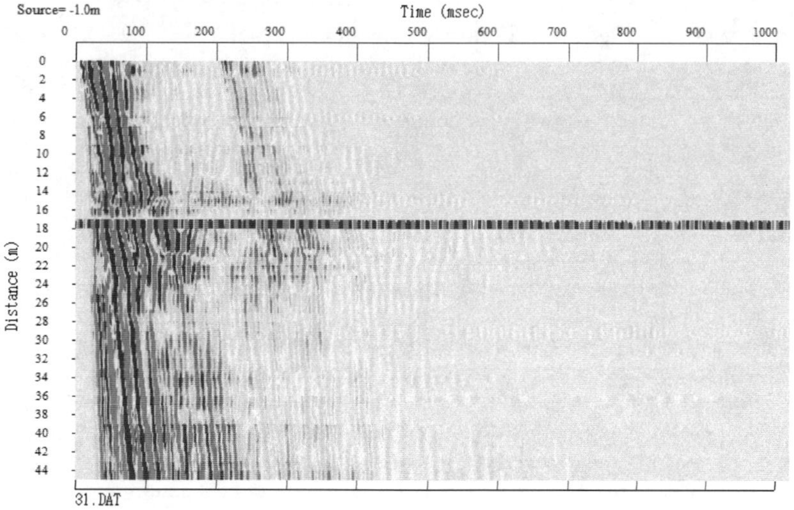

Fig. (18) Surface wave data examples at Zafarana wind farm site

5.2 Seismic Data Processing

It was pointed out that, the true refractor velocities can't be determined by shotting at only one end of a seismic line, but such velocities can be determined if the arrival times are recorded from both ends. Further, a depth computed from an intercept time actually represents the depth of the refracting surface projected back to the shot point. The reversed profile, however, offers a significant advantage in that, the true velocities and thicknesses of layers can be computed beneath each geophone to allow the mapping of irregular and dipping boundaries by using several methods. The delay time method was discussed by many authors, [17,18,19,20]. The Wave Front method elaborated by; [21,22,23,24]. Hagiwara's method was explained by Masuda, [25]. The Plus-Minus method was discussed by Hagedoorn [26] and the Generalized Reciprocal Method (GRM) is introduced by Palmer [27]. On the basis of the first arrival P-waves picking up, the wave forms are analysed by picking the first breaks and determining the travel time-distance (T-D) curves and depth models using SeisImager software package, which is a complete seismic refraction processing and modelling software. It is based in its processing on the delay time and ray tracing methods for the determination of the specific depth for each refractor beneath each geophone. The MASW profiles were analysed using the SeisImager software package in order to obtain the dispersion curve at each profile and then inverse the phase velocity into shear wave velocity.

5.3 Seismic Interpretation

According to the first arrival P-waves and S-waves picking up, the wave forms are analysed. The deduced time-distance curves and the corresponding 2-D depth model at each profile are obtained in order to interpret the subsurface features.

By applying the MASW technique, we obtain the dispersion curves (phase velocity versus frequencies) and by inversion of the dispersion curves, we obtain the shear wave velocity model. Figure 19 lower panels shows the shear wave velocity model of all conducted seismic lines at Zafarana wind farm. Figure 20 lower panels shows the shear wave velocity model of all conducted seismic lines at Gabel Zeit wind farm.

5.4 Seismic Results

The results reveal that, the obtained subsurface layers consist of four to five layers in all the seismic profiles up to a depth of 30 m at Zafarana site. The first layer with a thickness of (2.5-7) m is formed from gravels and sands and silty sand (S-wave velocity varies from 450 to 903 m/sec). The second layer with a thickness of (3.5 to 8) m is made up of sands and gravels and with less physical properties (S-wave velocity in the range 460 to 860 m/sec). The S-wave velocities of the third layer are ranges between 520-920 m/sec respectively which is composed of sandstone. While the S-wave velocities of the fourth layer are ranges between 540-975 m/sec respectively which it is composed of claystone. The fifth layer is exposed only in two sites (at profile 4 and 6 only) with the S-wave velocities ranges from 700 to 1000 m/sec. The obtained subsurface layers at the expansion area at Gabel El Zeit site consist of three layers in two seismic profiles up to a depth of 30 m. The first layer with a thickness of 5 m is formed from gravels and sands and silty sand (S-wave velocity varies from 299 to 368 m/sec). The second layer with a thickness of 10 m is made up of sands and gravels and with less physical properties (S-wave velocity in the range 360 to 427 m/sec). The S-wave velocities of the third layer are ranges between 419-466 m/sec respectively which is composed of claystone as shown in Tables 3 and 4.

The average shear wave velocity up to 30 m depth which is known as V_{S30} is obtained and was used to classify the site according to International Building Code [28]. Table 5 at Zafarana site, the location of profiles p1, 2, p5, p6, P7 and p8 are belong to Class C while the location of profiles p3, p4 are belong to Class B. While at the expansion area at Gabel Ziet site the location of seismic profiles are belong to Class C and D respectively.

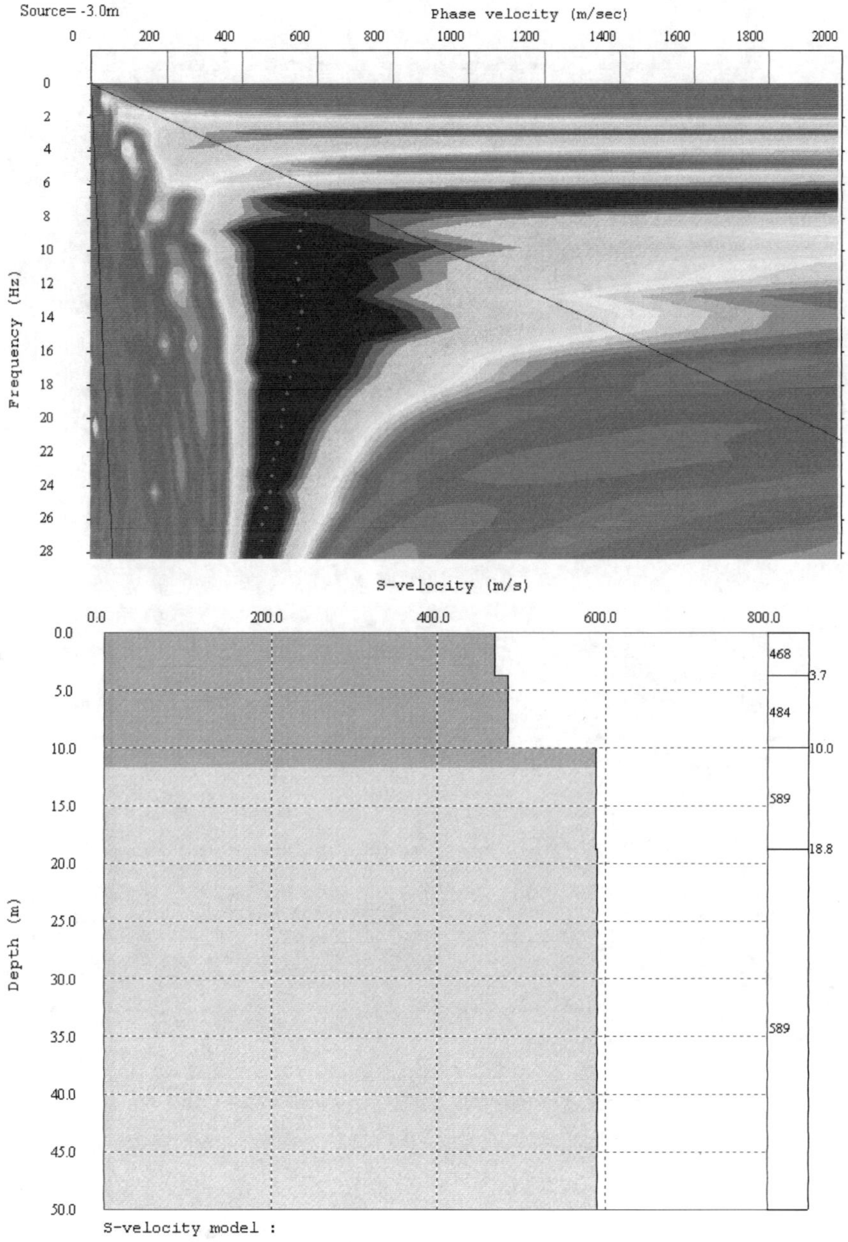

Fig. (19) The dispersion curve (upper panel) and S_wave velocity model (lower panel) for seismic line 1 at Zafarana wind farm

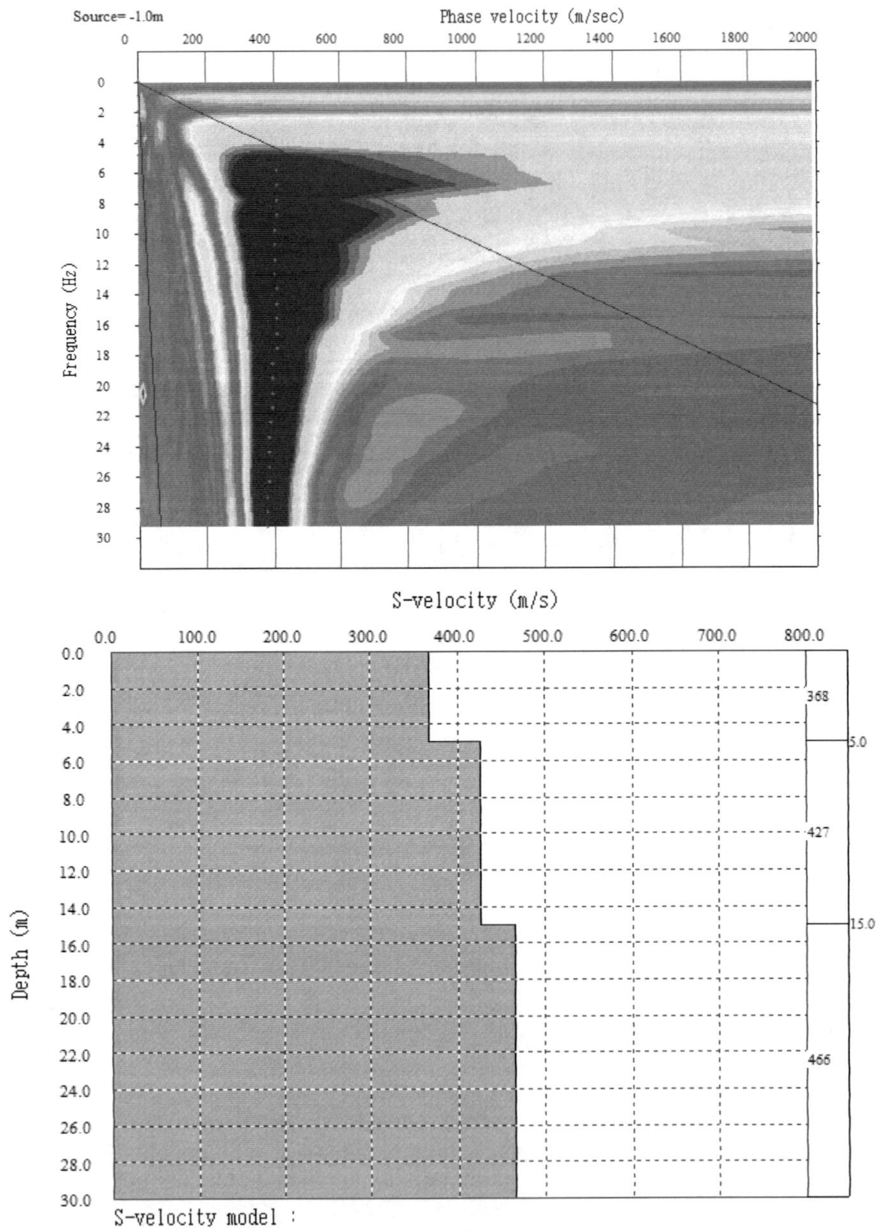

Fig. (20) The dispersion curve (upper panel) and S_wave velocity model (lower panel) for seismic line 1 at Gabel Zeit wind farm

IOP Conf. Series: Materials Science and Engineering **974** (2020) 012005 doi:10.1088/1757-899X/974/1/012005

Table 3. Seismic profiles location and parameters at Zafarana wind farm site

area	profile no	latitude	longitude	vs (m/s)	depth (m)	layer no
Zafarana (1)	1	29.15894	32.61821	468	0 - 3.7	1
(Class C)				484	3.7 - 10	2
				589	10 - 18.8	3
				589	18.8 - 30	4
Zafarana (8)	2	29.15895	32.56264	604	0 - 3.7	1
(Class C)				550	3.7 - 10	2
				685	10 - 18.8	3
				773	18.8 - 30	4
Zafarana (2)	3	29.17914	32.60731	903	0-7.5	1
(Class B)				864	7.5 - 13.5	2
				918	13.5 - 21	3
				970	21 -30	4
Zafarana (3)	4	29.19418	32.60572	894	0 - 2.5	1
(Class B)				861	2.5 - 6	2
				893	6 - 10.5	3
				983	10. 5 - 16	4
				993	16- 30	5
Zafarana (6)	5	29.19461	32.5585	572	0 - 3.7	1
(Class C)				495	3.7 - 10	2
				692	10 - 18.8	3
				756	18.8 - 30	4
Zafarana (4)	6	29.21088	32.59975	629	0 - 3	1
(Class C)				601	3 - 7.5	2
				636	7.5 - 13.5	3
				684	13.5 -21	4
				717	21 - 30	5
Zafarana (5)	7	29.29463	32.58329	448	0 - 2.5	1
(Class C)				470	2.5 - 6	2
				531	6 - 10.5	3
				545	10.5 - 30	4
Zafarana (7)	8	29.2529	32.59395	634	0 - 3.7	1
(Class C)				634	3.7 - 10	2
				638	10 - 18.8	3
				693	18.8 - 30	4

Table 4. Seismic profiles location and parameters at the expansion area at Gabel El Zeit site

area	profile no	latitude	longitude	vs (m/s)	depth (m)	layer no
(Class C)	1	28.16815	33.1395	368	0- 5	1
				427	5-15	2
				466	15 - 30	3
(Class D)	2	28.15585	33.1579	299	0- 5	1
				360	5-15	2
				419	15 - 30	*3*

Table 5. International building code (IBC 2006)

Site Class	Soil Profile Name	Average Properties in Top 100 Feet		
		Shear Wave Velocity Vs (ft/sec)	Standard Penetration Resistance N	Soil Undrained Shear Strength Su (psf)
A	Hard Rock	Vs> 5000	N/A	N/A
B	Rock	2500 < Vs <= 5000	N/A	N/A
C	Very Dense Soil and Soft Rock	1200< Vs <= 2500	N> 50	Su>= 2000
D	Stiff Soil Profile	600< Vs <= 1200	15<= N <= 50	1000<= Su <= 2000
E	Soft Soil Profile	Vs< 600	N< 15	Su> 1000
E	—	Any profile more than 10 feet of soil having the following characteristics: 1. Plasticity Index PI > 20 2. Moisture Content w>= 40% and 3. Undrained shear strength Su < 500 psf		
F	—	Any Profile containing soils having one or more of the following characteristics: 1. Soils vulnerable to potential failure or collapse under seismic loading such as liquefiable soils, quick and highly sensitive clays , collapsible weakly cemented soils. 2. Peats and / or highly sensitive clays (H > 10 feet pf peat and / or highly organic clay where H = thickness of soil) 3. Very high plasticity clays (H > 25 feet with plasticity index PI > 75) 4. Very thick soft/ medium stiff clay (H > 120 feet)		

6. CONCLUSIONS

This work presents one of the pioneer studies concerning the investigation of the geological, geotechnical and geophysical characterization and criteria of the zafarana wind farm sites and their expansion at Gabel Al Zeit sites Egypt. Both geotechnical design parameters and geophysical parameters are evaluated and presented. The conclusions from the current study can be summarized as below:-

- Geological investigation indicate that, the study area consists of coastal and wadi deposits. There are no intensive faults at Zafarana area.

IOP Conf. Series: Materials Science and Engineering **974** (2020) 012005 doi:10.1088/1757-899X/974/1/012005

- Geotechnical investigation at Zafarana wind farms shows that, the site consists of five main formations, which are Wadi deposits, Clay, Claystone, Sand and Sandstone.
- The geophysical analysis demonstrates that the subsurface section is composed of three to five layers at Zafarana wind farm, the second layer is characterized by its less physical properties (S-wave velocities). While the expansion area at Gabel El Zeit site the subsurface section composed of 3 layers up to 30 m depth.
- Geophysical investigation results reveal that, the obtained subsurface layers consist of four to five layers at Zafarana Wind Farm Sites. The first layer formed from gravels and sands and silty sand with average S-wave velocity varies from 500 m/sec. The second layer is made up of sands and gravels with average S-wave velocity 600 m/sec. The S-wave velocities of the third layer has an average value equal to 800 m/sec respectively which is composed of sandstone. While the S-wave velocities of the fourth layer with average 700 m/sec respectively which it is composed of claystone.
- Also Geophysical investigation results at Gabel El Zeit Wind Farm Sites shows that the subsurface soil conditions consists of three layers of gravely sands with average shear wave velocity 350 m/sec.
- The average shear wave velocity up to 30 m depth which is known as VS30 is obtained and was used to classify the site according to International Building Code (IBC-2006), Where at Zafarana sites, the location of profiles p1, 2, p5, p6, P7 and p8 are belong to Class C while the location of profiles p3, p4 are belong to Class B. While at the expansion area at Gabel Ziet sites the location of seismic profiles are belong to Class C and D respectively.

7. REFERENCES

[1] Tinjum, J.M., and Christensen, R.W. (2010). Site investigation, characterization and assessment for wind turbine design and construction. In Wind Energy Systems, edited by John D. Sorensen and Jens N. Sorensen, 26-45. Woodhead Publishing.

[2] Tinjum, J. M., and Lang, P. (2012). Wind energy geotechnics. Geo Strata, 2012: 18-26. Vestas V82-1.65 MW. Vestas Wind System A/S, Denmark. www.vestas.com.

[3] Said, R. (1962) The Geology of Egypt. Elsevier Publ. Co, Amsterdam and New York.

[4] El-Gezeery, M.V., Marsouk, I.M., (1974) Miocene rock stratigraphy of Egypt. Egypt. J. Geol. 18, 1-59.

[5] Said R. (1990) The Geology of Egypt, 2nd edition Balkema, Rotterdam, 439-449.

[6] Egyptian Geological Survey (1994) Geological map of Sinai, Arab Republic of Egypt, Scale 1:250 000.

[7] Snethen, D.R., Johnson, L.D. & Patrick, D.M. (1977) An evaluation of expedient methodology for identification of potentially expansive soils. Report No. FHWA-RD-77- 94, U.S. Army Engineer Waterways Experiment Station, USAEWES, Vicksburg, Miss., June 1977.

[8] Abd El-Aal, A.K., Mohamed, A.A. (2010) Near-surface seismic refraction applied to exploring subsurface clay layer at a new mining area in southeast Cairo, Egypt. (2010) Arabian Journal of Geosciences, 3 (2), pp. 105-112.doi: 10.1007/s12517-009-0036-2.

[9] Bruce B. Redpath (1973). Seismic Refraction Exploration for Engineering Site "Investigation, Explosive Excavation Research Laboratory Livermore, California, 63 pp.

[10] Redpath, B. B. (1973) Seismic refraction exploration for engineering site investigations. Technical report E-73-4 U.S. Army Engineer Waterways Experiment Station Explosive Excavation Research Laboratory Livermore, California.

[11] Lankestron, R. W. (1994) High-Resolution refraction seismic data acquisition and interpretation. Geotechnical and environmental geophysics, Vol. I, Society of exploration geophysicist, Tulsa, Oklahoma. Pp. 45-73.

[12] Edelmann, H.A.K. and Schmoll, J. (1983) Shear wave generation by vibration orthogonal to the polarization. Geophysical Prospecting 29, 409–442.

[13] Wright, C. and Johnson, P. (1982) On the generation of P- and S-waves energy in crystalline rocks. Geophysical Prospecting 30, 55–70.

[14] Helbig, K. and Mesdag, C. S. (1982) The potential of shear wave observations. Geophysical Prospecting 30, 413–431.
[15] Kaehler, S. and Meissner, R. (1983) Radiation and receiver pattern of shear and compressional waves. Geophysical Prospecting 31, 421–435.
[16] Park, C. B., Miller, R. D. and Xia, J. (1999) Multichannel analysis of surface waves (MASW); Geophysics, 64, 800-808.
[17] Gardner, L. W. (1949) Seismograph determination of salt-dome boundary using well detector deep on dome flank. Geotech. Eng. Div. 105, 715–726.
[18] Slotnick, M. M. (1950) A graphical method for the inerpretation of efraction profile data. Geophysics, Vol. 15, pp. 163-180.
[19] Tarrant, L. H. (1956) A rapid method of determining the form of a seismic refractor from line profile results. Geophysical prospecting, Vol. 4, pp. 131-139. Upper Saddle River, New Jersey, 653 pp.
[20] Wyrobek, S. M. (1956) Application of delay and intercept times in the interpretation of multi layer refraction time distance curves. Geophysical prospecting, Vol. 4, pp. 112-130.
[21] Gardner, G. H. F., Gardner, L. W. and Gregory, A. R. (1974) Formation velocity and density – The diagnostic basics for stratigraphic traps. Geophysics 39, 770–780.
[22] Hales, F. W. (1958) An accurate graphical method for interpreting seismic refraction lines, Geophysical prospecting, Vol. 6, pp. 285-294.
[23] Rockwell, D. W. (1967) A general wave front method. In A. W. Musgrave, (ed.). Seismic refraction prospecting, pp. 363-415. Tulsa, Soc. Of Exploration Geophysicists.
[24] Schenck, F. L. (1967) Refraction solutions and wave front targeting. In A. W. Musgrave, (ed.) seismic refraction prospecting, Soc. of Exploration Geophysicists, Tulsa, pp. 363-415.
[25] Masuda, H. (1975) Seismic refraction analysis for engineering study: OYO Technical Note, TN-10, Urawa Res., Japan.
[26] Hagedoorn, J. G. (1959) The plus-minus method of interpreting seismic refraction sections, Geophysical prospecting, Vol. 7, pp. 158-182.
[27] Palmer, D. (1980) The generalized reciprocal method of seismic refraction interpretation: Soc. Expl. Geophys.
[28] International Building Code IBC (2006) Site classification using the average shear wave velocity in the upper 30 m thickness.

ACKNOWLEDGMENTS
The authors are grateful to the Science & Technology Development Fund in Egypt (STDF) for funding this research project in the context of ongoing joint US– Egypt research "Seismic risk assessment of wind turbine towers in Zafarana wind farm Egypt", Project ID:4588. Results shown here are part of the work

Military Technical College
Kobry El-Kobbah, Cairo,
Egypt

13th International Conference
on Civil and Architecture
Engineering
ICCAE-13-2020

Numerical Modeling of New Connection Pontoon under Static and Dynamic Loads

Ahmed S Elareshy[1], Mostafa M Abdel Wahab[2], Shreif A Mazek[3], and Ashraf M Osman[4]

[1] M.Sc. Student, Department of Civil Engineering, Military Technical College, Egypt.
[2] Professor at Department of Civil Engineering, Military Technical College, Egypt.
[3] Professor at Department of Civil Engineering, Military Technical College, Egypt.
[4] Asst. Professor at Department of Civil Engineering, Military Technical College, Egypt.

Corresponding author: samy_mtc@yahoo.com

Abstract. The development of floating bridges is a major engineering challenge to make the existing bridges durable and easy to construct. The connection pontoon unit between the heavy communication bridge (HCB) and the assault bridge (AB-PMM71) is discussed as developed by Mazek et al. [1]. The new connection pontoon (NCP) should offer fast assembly time to operate the combat floating bridge.

The floating bridge system including the HCB, the AB-PMM71, and the NCP is modeled. The 3-D numerical model of the floating bridge system is proposed by the nonlinear finite element analysis under static and dynamic Military load Class MLC70. The field test for the new floating bridge system was conducted by [1].

The proposed finite element model of the floating bridge system is verified based on the practical field test developed by [1]. The numerical results have good agreement with those obtained by the reviewed field test.

Keywords: Floating Bridges, Dynamic Analysis, Military Load Class 70 ton (MLC70).

1. Introduction

The floating bridge is considered a very important structure in military operations. There is a great engineering interest to develop the floating bridges. One of the most important properties of the floating bridges is the assembly time to operate the heavy floating bridge which is a difficult process. The new connection pontoon (NCP) has been designed and validated experimentally by [1]. In this paper, a numerical study has been conducted using 3-D finite element analysis on the NCP. The numerical results have been compared with those obtained by the field tests conducted by[1].

The floating bridges offer an economic and efficient solution for crossing obstacles and barriers where the situation is unstable for traditional bridges [2]. The construction cost of conventional bridges is about four times more than the cost of floating bridges [3]. The structural system of floating bridges differs from one location to another. There are no defined rules to analyze and design floating bridges [4]. There is no code to find out the dynamic behavior due to the large variation in many factors such as water current, wave forces, depth of water, and the width of the water channel [4]. There is no engineering database for these floating structures. A direct design approach needs to be used for each bridge [5].

In recent years, engineering efforts have been conducted to increase the efficiency of floating bridges to be more lightweight, high-capacity, higher maneuverable and more competence [6]. Khalifa et al. [7] used the 3-D finite element method to study the best method to improve the assault floating bridge (AFB) performance from MLC60 to MLC70 for different five proposals. Khalifa et al. [8] presented the analysis of the old assault floating bridge under MLC60 and MLC70 and upgraded this bridge to carry MLC70. Viecili et al. [2] studied analytically and experimentally the dynamic behavior of ribbon pontoon floating bridges under two-axle vehicle loading. The maximum displacements of pontoon floating bridges at different vehicle speeds and weights were predicted and compared with experimental results.

Sun et al. [9] studied the hydrodynamic response of a newly designed pontoon floating bridge under regular waves by an experimental test. A numerical analysis was conducted to compare the results with the model test results. The bending moment along the entire floating bridge was obtained and analyzed to get the nonlinear hydroelastic effects. The results showed that the bending moment along the floating bridge model wasn't firmly linearly proportional to the wave heights, which might result from nonlinear connectors or nonlinear hydrostatic forces. It was concluded that nonlinear connections have an important influence on bending moment and should be considered in the design of floating bridges.

Zhang et al. [10] studied the dynamic responses with hydrodynamic influence coefficients for different water depths on two types of floating bridges; the continuous pontoon floating bridge and the separate pontoon floating bridge. Analytical models based on beam theory have been presented to study the dynamic responses of two types of floating bridges subjected to moving loads. The results showed that the displacement response curves happened due to a high-speed moving load is smoother than a low-speed one, and the water depth can be neglect during the floating bridge design.

Fu and Cui [11] studied the hydroelastic responses of a ribbon floating bridge consisting of several nonlinearly connected modules under moving loads from both experimental and numerical tests. The results showed that the downward dynamic response slightly increased with the moving load speed while the upwards dynamic response increased with the increase of the load moving speed and was more sensitive to the moving speed compared with the downwards response. The nonlinearity of the connectors had a great influence on the dynamic response of the floating bridge especially for the upward dynamic responses. Raftoyiannis et al. [12] studied the dynamic response of a continuous floating bridge under the action of a moving load with various constant velocities. This study was formed by an analytical method based on the Runge-Kutta method and numerical analysis to verify the analytical method. The analytical results showed that the velocity of the dynamic load had a great effect on the maximum oscillation amplitude of the bridge. The wave forces cause oscillations with significant amplitude in the cases of light-weighted pontoons, while for heavy-weighted pontoons this effect is significantly lower. The results showed that the difference between the analytical and numerical results was about 5%.

Shixiao et al. [13] studied the dynamic response of a nonlinear connected floating bridge taking into account the nonlinear properties of connectors and the vehicle's inertia effects. The equations of motions were proposed and solved with the Newton-Raphson iteration method. As a result, nonlinearities and the initial gap of the connectors must be considered for the nonlinear connected floating bridge since they have a noticeable effect on the static deflection, dynamic displacement responses and dynamic connection forces of the bridge. Qiu 2009 [14] analyzed numerically the dynamic response a free–free flexible beam floating in a boundless water field under the effect of moving loads. A finite element method was developed directly in a time domain and was applied to solve the 2-D problem. The results showed great effects of the traveling speed, the water depth and the beam length on the behavior of the floating elastic beam.

Xujun Chen et al. [15] investigated the hydrodynamic performances of a moored pontoon in finite depth water by numerical and experimental analyses. The dynamic behavior of the floating pontoon, as well as the chain forces in different wave heights and wave angles, were calculated with a wide range of wave periods by the numerical method. From the results, it can be concluded that both wave heights and wave angles have clear influences on the motions and the chain forces of the pontoon. The numerical results showed greet agreement with the model tests in majority of cases.

2. Problem statement

Many types of research on floating bridges focused on increasing the capacity of the bridge with decreasing own weight [1]. The decrease in the erection time is an important factor to operate the floating bridge during the war. For this reason, the bridge reoperation time has a great effect to operate the bridge after damage. A new steel floating structure called new connection pontoon (NCP) was used, designed and tested experimentally by [1]. The heavy communication bridge (HCB) takes a relatively long time in structure operations with respect to the assault floating bridge (AFB) [1]. Therefore, the NCP was designed to connect the HCB to the AFB PMM71 due to partial damage to operate again the bridge. The ANSYS 18 software is used.

In this paper, the static and dynamic responses of the NCP are investigated using the 3-D FEM under military load class of 70-ton (MLC70). The study introduces the techniques and procedures used in the ANSYS program to create a geometry model, to generate finite element meshes, to establish boundary conditions, and to select the type of analysis. The maximum bridge displacement is computed at different load speeds and cases of loads. The numerical results are compared to those obtained by the field test conducted by [1].

3. Model description

The bridge system model consists of three parts. The three parts are the HCB, the AFB, and the NCP, as shown in Figs. 1 and 2. The HCB consists of individual pontoons assembled together to form a rigid continuous beam rested on water modeled as elastic supports. The connection between the pontoons is modeled as a rigid connection type from both the top and bottom of the pontoon. The individual pontoon has dimensions of 2.4 m width, 5.25 m length, and 1.5 m height, as shown in Fig. 3. The HCB width is 9.9 m, as shown in Fig. 4. The HCB consists of four individual pontoons with 10 cm spacing between them. The bridge system model used the four individual pontoons for both the HCB and the AFB in the longitudinal direction.

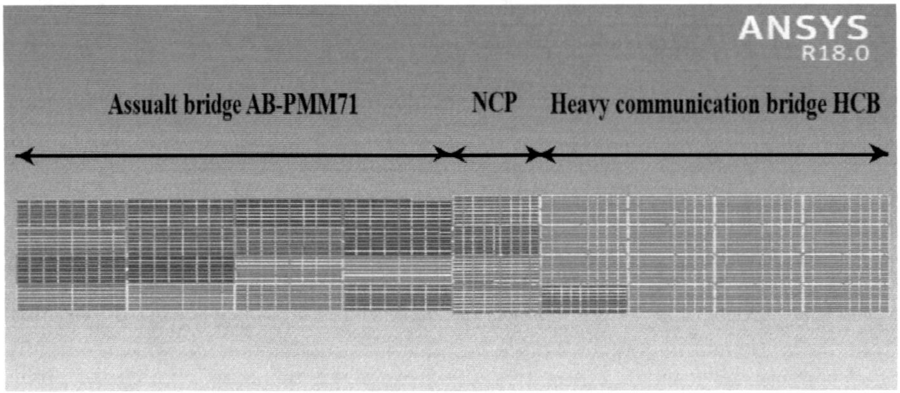

Figure 1: Plan of the bridge system model

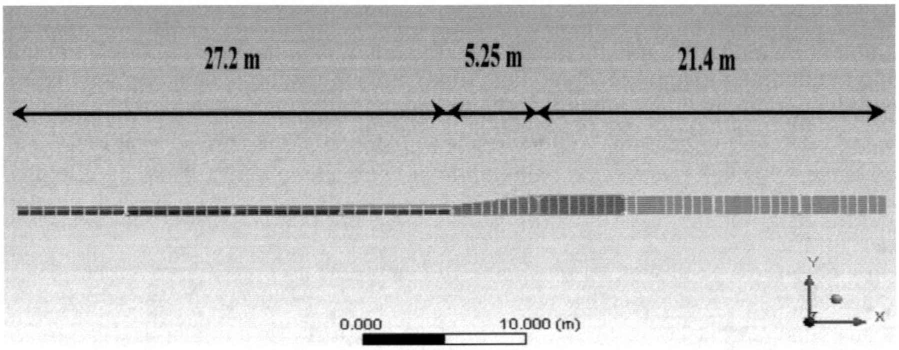

Figure 2: Side view of the bridge system model

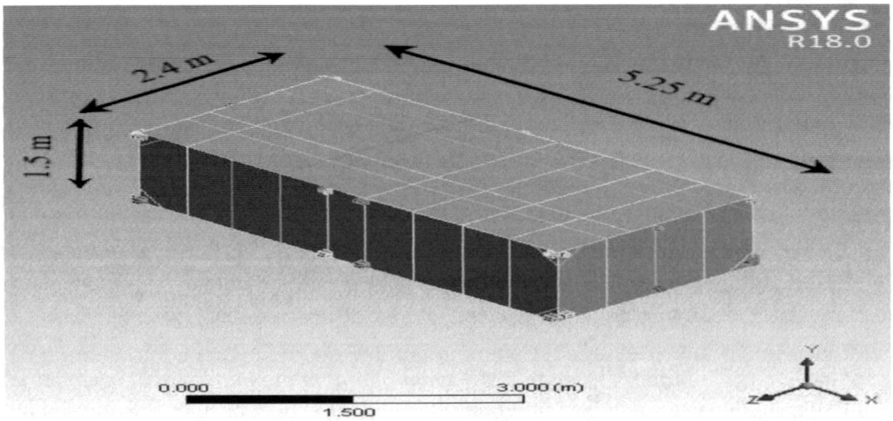

Figure 3: Single heavy communication pontoon

IOP Conf. Series: Materials Science and Engineering **974** (2020) 012006 doi:10.1088/1757-899X/974/1/012006

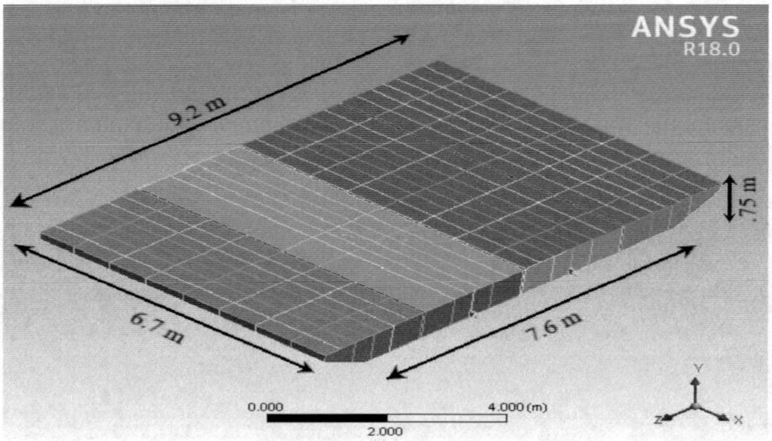

Figure 4: Single unit of the assault bridge AB-PMM7

The AFB consists of four pontoons already assembled together in W-shape. These pontoons were opened automatically as they were launched in water to take the straight shape, as shown in Fig. 4. The connections between the pontoons of the AFB are hinged joints to link the internal pontoons. The connections between the pontoons are set at the bottom without joints at the top to allow rotation about the lateral direction of the bridge. Two steel wires used at the top of the two external pontoons to limit this rotation. These connections resist bending moment beside the shear force. The dimensions of the AFB are 9.2 m width, 6.7 m length, and 0.75 m height. The bridge system model used the four individual pontoons for both the HCB and the AFB in the longitudinal direction [16].

The NCP is the floating structure connecting the two previous bridges together. The longitudinal section of the NCP is trapezoidal shape, as shown in Fig. 5. The height of the one end for the NCP is 1.5 m while the height of the other end is 0.75 m, as shown in Fig. 5. The boundary conditions of the NCP are a rigid connection at the deep side and a hinged connection at the other small side. The width of the NCP is 2.4 m and its internal structure is like the heavy pontoon bridge. The HCB, the AFB, and the NCP have rigid connections in the lateral direction.

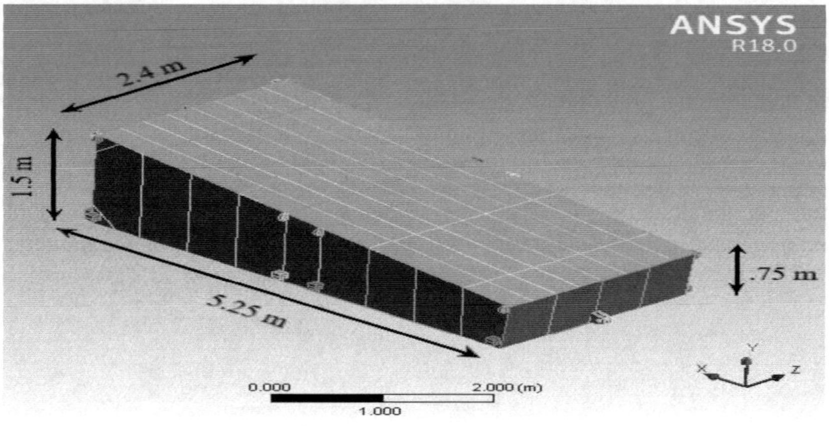

Figure 5: The new connection pontoon

IOP Conf. Series: Materials Science and Engineering **974** (2020) 012006 doi:10.1088/1757-899X/974/1/012006

The statical system of the NCP consists of internal structural members that include equal angle, UPN, and I beam sections, as shown in Fig. 6. The NCP is also covered by steel plates. Steel 44 is used in the HCB, the AFB, and the NCP. The water structure interaction is modeled as a beam resting on elastic supports. The stiffness of the elastic support is equal to the buoyancy force which acts upward and equal to the weight of water displaced by the submerged hull of the floating bridge system.

This numerical study was conducted by the 3-D FEA. 3-D beam elements and shell elements are used to model the internal structural members and the steel plates, respectively. The FEM of the floating bridge system consists of 124,093 nodes and 88,366 elements, as shown in Fig. 6.

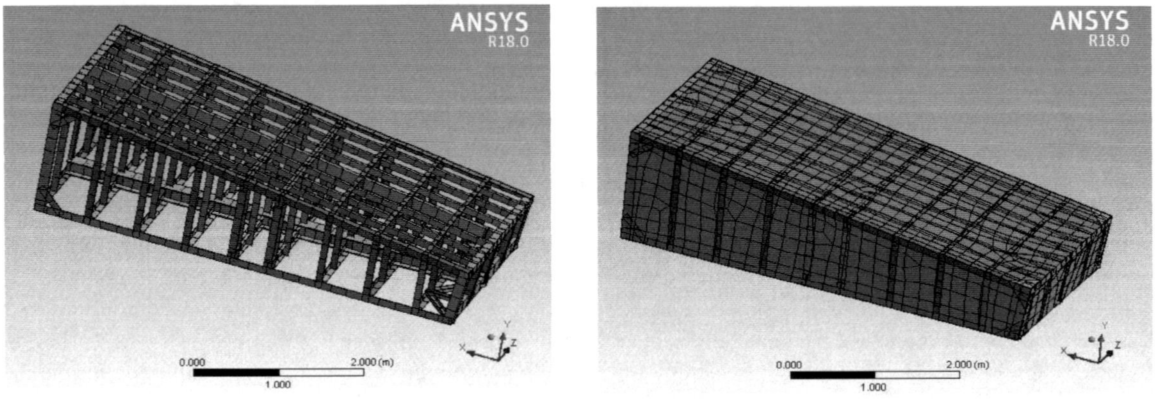

Figure 6: 3-D finite element model of the NCP

4. Floating stability of NCP

The trapezoidal shape of the NCP is considered an engineering challenge to study the stability of the pontoons in water. However, the center of gravity position was changed to be approximately above the center of buoyancy by increasing the weight of the NCP at the lighter side [1]. The draft depth of the HCB is 32 cm after the HCB is placed in water. Therefore, the draft depth of the NCP from the HCB side was the same as the draft depth of the HCB to facilitate the assembly process between the NCP and the HCB. Fig. 7 shows the floating stability and the draft depth of the NCP under its own weight only.

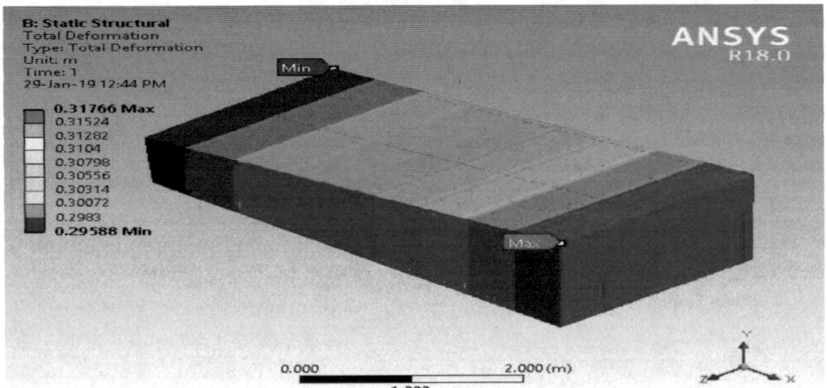

Figure 7: Draft depth of the NCP under its own weight only

IOP Conf. Series: Materials Science and Engineering **974** (2020) 012006 doi:10.1088/1757-899X/974/1/012006

5. Static analysis

A static load of 82 tons is used to compute the static deformation and stresses on the bridge using the 3-D finite element analysis (FEA). The static load has been applied in two positions on the bridge system as adopted in the field study [1]. The first case of the static load is placed only on the NCP, as shown in Fig. 8. The second case of the static load is placed on both the NCP and the AFB, as shown in Fig. 9. The static load acts on an area of 4.5 m length and 3.5 m width as this static load simulates the dimension of the tank.

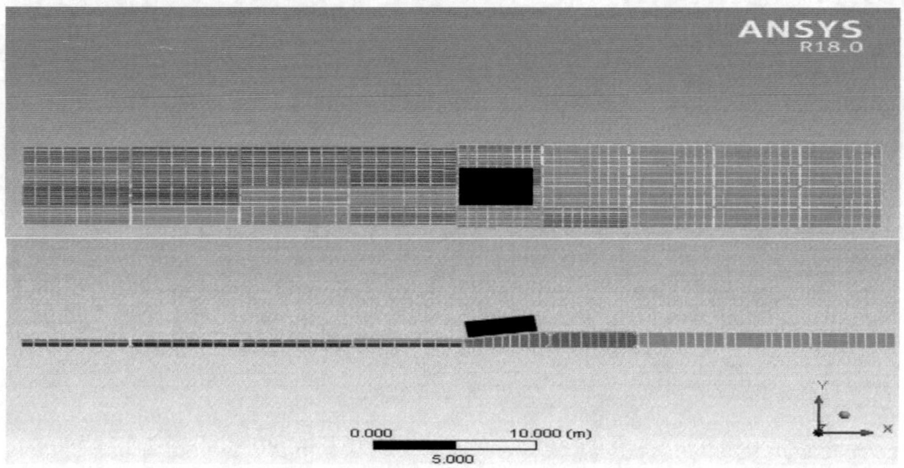

Figure 8: First case of static load

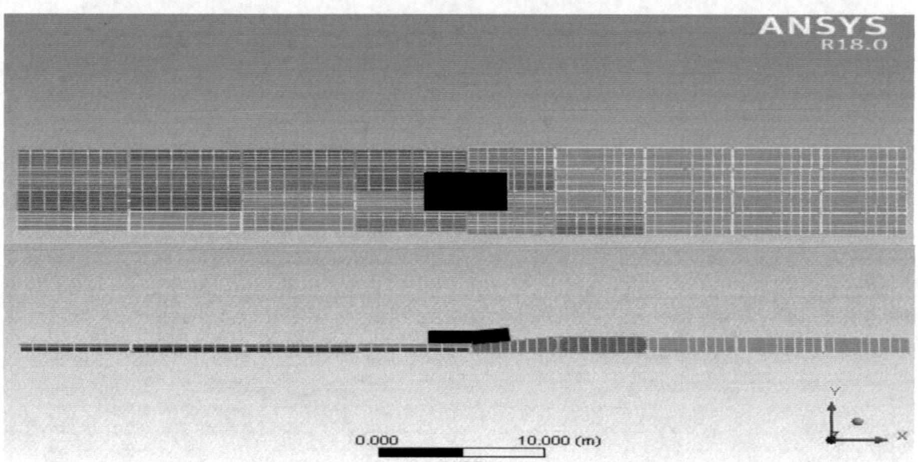

Figure 9: Second case of static load

The results of the first case of static load show that the maximum draft of the bridge is 0.733 m at the NCP edge. The maximum stress developed at the NCP is equal to 860 kg/cm^2. These results show good agreement with the results obtained by the field test [1]. The maximum measured draft depth is equal to .74 m. The discrepancy between the measured and the computed results is 2%.

IOP Conf. Series: Materials Science and Engineering **974** (2020) 012006 doi:10.1088/1757-899X/974/1/012006

The second case of static load shows that the maximum calculated draft depth of the NCP is 0.778 m. The maximum calculated stress of the bridge system is equal to 950 kg/cm^2. The second case of static load gives the maximum field draft depth of the NCP to be 0.79 m [1]. There is a good agreement between the calculated readings and the field readings for the draft depth of the NCP as shown in Figs. 10 to 15.

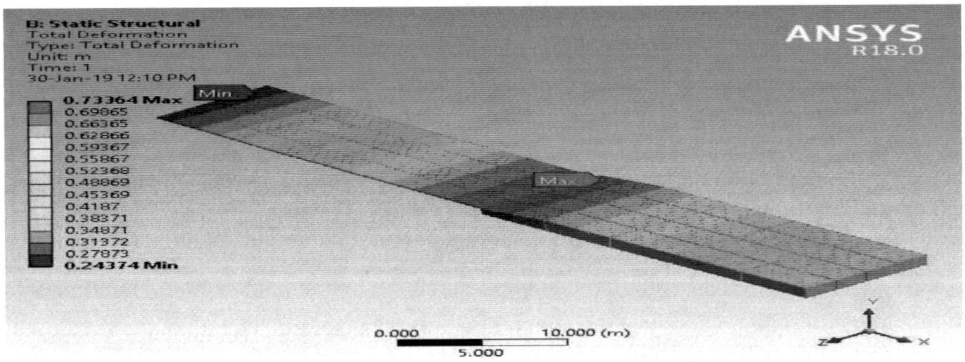

Figure 10: Calculated draft depth of the bridge system under the first static load case

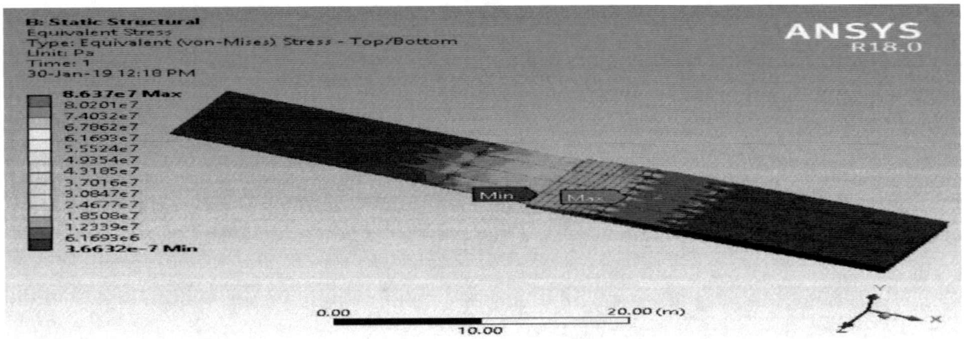

Figure 11: Equivalent stress of the bridge system under the first case of static load

Fig. 12: Field draft depth of the NCP under the first case of static load [1]

IOP Conf. Series: Materials Science and Engineering **974** (2020) 012006 doi:10.1088/1757-899X/974/1/012006

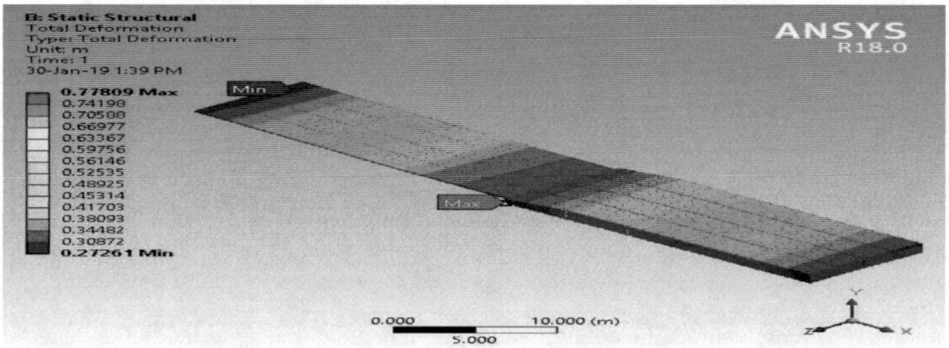

Figure 13: Calculated draft depth of the bridge under the second case of static load

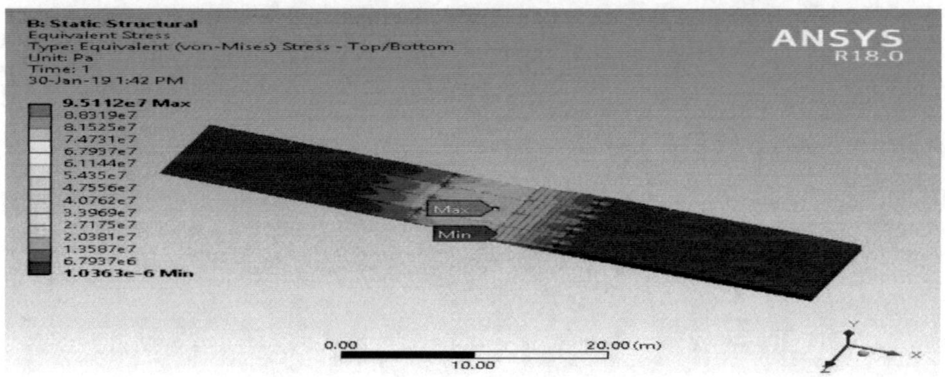

Figure 14: Equivalent stress of the bridge system under the second case of static load

Figure 15: Field draft depth of the bridge system under the second case of static load [1]

6. Dynamic analysis

The Trilateral Design and Test Code (TDTC, 2005) [17] was published to regulate an accepted design limits for different types of military bridges. A 70-ton tank was used in this study in which the 70-ton tank was called MLC70 according to TDTC. The 70-ton tank has two tracks. The footprint of the tank track has 4.57 m long and 0.6 m width. Two cases of operation speeds were used to study the performance of the floating bridge system. The operation speeds of the moving tanks are 8 km/hr. and 16 km/hr.

The results showed the draft depths of the bridge system as presented in Figs. 16 and 17 under the dynamic load MLC70 with speeds 8 km/hr. and 16 km/hr. respectively. The maximum draft depths of the bridge system were 0.756 m and 0.789 m, respectively. The maximum stresses of the bridge system are 1.58 t/cm^2 and 1.63 t/cm^2, respectively, as shown in Figs. 18 and 19.

The maximum draft depths of the bridge system were 0.71 m and 0.73 m at speeds of 8 km/hr. and 16 km/hr., respectively. These results showed good agreement between the numerical results obtained by the FEA and the results obtained by the field test [1].

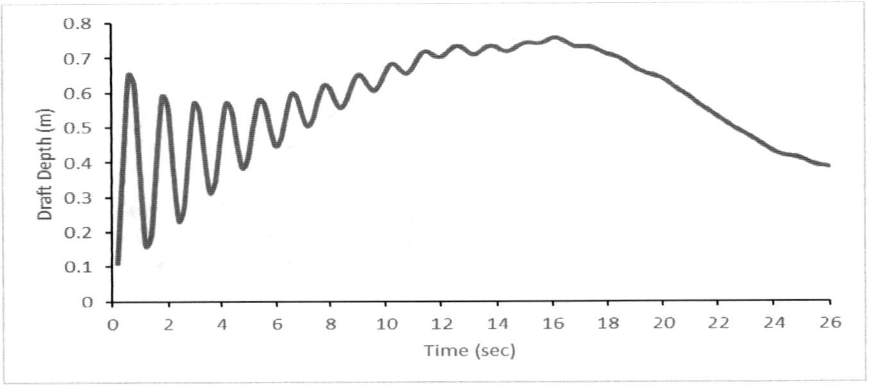

Figure 16: Draft depth of the bridge system under moving load with a speed of 8 km/hr

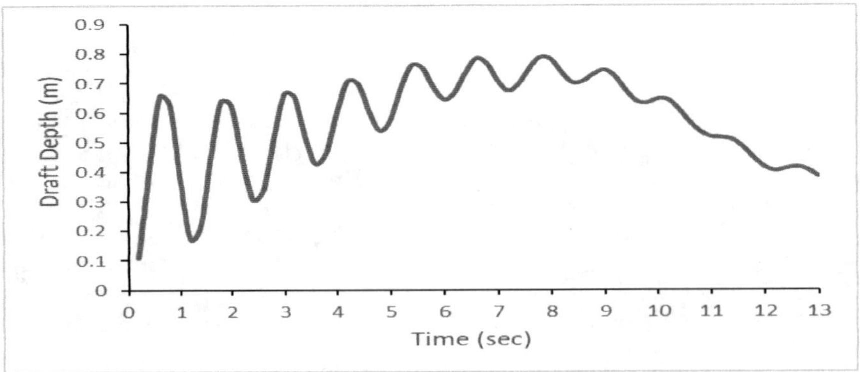

Figure 17: Draft depth of the bridge system under moving load with a speed of 16 km/hr

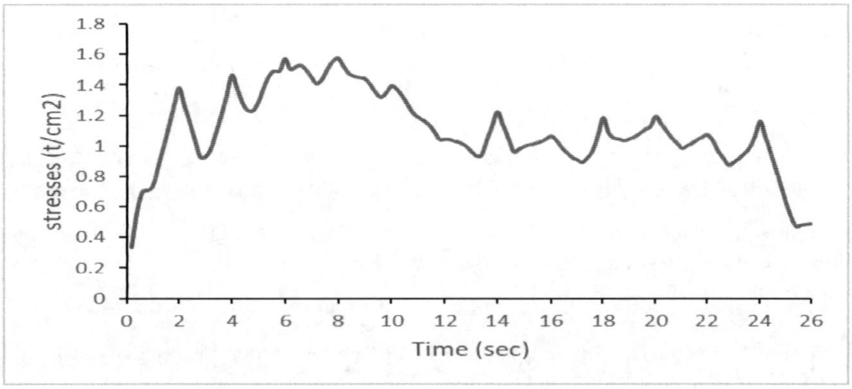

Figure 18: Equivalent stresses of bridge system under the dynamic load of speed 8 km/hr

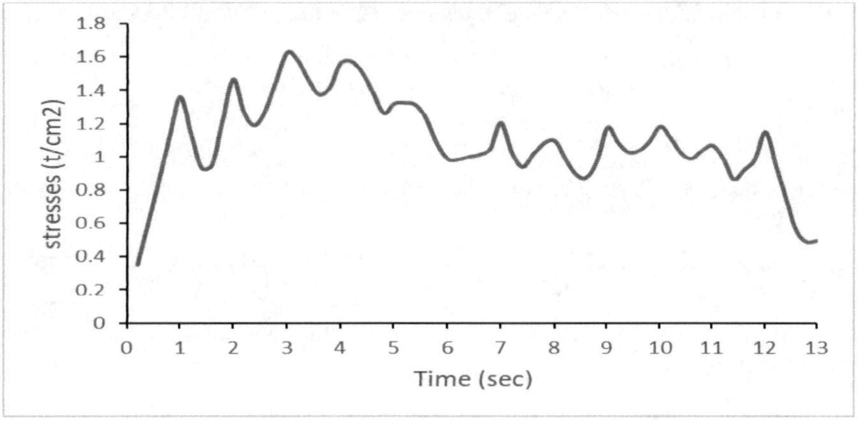

Figure 19: Equivalent stresses of bridge system under the dynamic load of speed 16 km/hr

7. Conclusions

- The 3-D finite element analysis is used to model the floating bridge system composed of the new connection pontoon (NCP), the heavy communication bridge (HCB), and the assault bridge AB-PMM71.
- The structure interaction between the floating bridge system and the water media is modeled based on the case study proposed by Mazek et al [1].
- The performance of the 3-D FEM for the floating bridge system is verified with the experimental results obtained from the field study presented by [1].
- The 3-D FEM of the bridge system is studied under static loads. However, the discrepancy between the numerical readings and the experimental readings is about 2 %.
- The 3-D FEM of the bridge system is studied under moving loads based on different tank speeds. However, the discrepancy between the numerical readings and the experimental readings reaches 6%.

8. References

[1] S. A. Mazek and his collegeas. "Design and Implementation of a New Connection Pontoon NCP between the Heavy Communication Bridge HCB and Assault Bridge AB-PMM71," Military Technical College, Cairo, Egypt, 2018.

[2] G. Viecili, A. E. H. O. A. El Halim, A. Braimah, and O. El-Desouky, "Transportation Optimization of Ribbon Floating Bridges: Analytical and Experimental Investigation," *Open Civ. Eng. J.*, vol. 8, no. 1, pp. 42–56, 2014.

[3] E. Watanabe, C. M. Wang, T. Utsunomiya, and T. Moan, "VERY LARGE FLOATING STRUCTURES : APPLICATIONS, ANALYSIS AND DESIGN," National University of Singapore, Singapore, 2004.

[4] O. E.-D. M. Ibrahim, "Dynamic Behaviour of Short-Term Floating Bridges," Ph.D. dissertation, Dep. of Civil and Environmental Eng., Carleton Univ., Carleton, 2011.

[5] A. Rahman, "Dynamic analysis of floating bridges," *Procedia Eng.*, vol. 194, pp. 44–50, Dec. 1997.

[6] D. Van-johnson, "Numerical and Experimental Investigation of Ribbon Floating Bridges," M.S. thesis, Civil Eng. Dep., Carleton University Ottawa, Ontario, 2018.

[7] Y. A. Khalifa, M. M. Abdelwahab, and A. H. Salem, "Structural Systems Effect on Stability of Ferries Using 3D FEM," in *the 9th Int. Conf. on Civil and Arch. Eng., Cairo, Egypt, May 29-31*, 2012.

[8] Y. A. Khalifa, M. M. Abdelwahab, and A. H. Salem, "Structural Analysis for Upgrading Metallic Rigidly Connected Floating Bridge Using 3D FEM," in *the 9th Int. Conf. on Civil and Arch. Eng., Cairo, Egypt, May 29-31*, 2012.

[9] J. Sun, P. Jiang, Y. Sun, C. Song, and D. Wang, "An experimental investigation on the nonlinear hydroelastic response of a pontoon-type floating bridge under regular wave action," *Ships Offshore Struct.*, vol. 13, no. 3, pp. 233–243, 2018.

[10] J. Zhang, G. ping MIAO, J. xun LIU, and W. jun SUN, "Analytical Models of Floating Bridges Subjected by Moving Loads for Different Water Depths," *J. Hydrodyn.*, vol. 20, no. 5, pp. 537–546, 2008.

[11] S. Fu and W. Cui, "Dynamic responses of a ribbon floating bridge under moving loads," *Mar. Struct.*, vol. 29, no. 1, pp. 246–256, 2012.

[12] I. G. Raftoyiannis, T. P. Avraam, and G. T. Michaltsos, "Analytical models of floating bridges under moving loads," *Eng. Struct.*, vol. 68, pp. 144–154, 2014.

[13] F. Shixiao, C. Weicheng, C. Xujun, and W. Cong, "Hydroelastic analysis of a nonlinearly connected floating bridge subjected to moving loads," *Mar. Struct.*, vol. 18, no. 1, pp. 85–107, 2005.

[14] L. Chao Qiu, "Modeling and simulation of transient responses of a flexible beam floating in finite depth water under moving loads," *Appl. Math. Model.*, vol. 33, no. 3, pp. 1620–1632, 2009.

[15] X. Chen, Y. Miao, X. Tang, and J. Liu, "Numerical and experimental analysis of a moored pontoon under regular wave in water of finite depth," *Ships Offshore Struct.*, vol. 12, no. 3, pp. 412–423, 2017.

[16] B. R. Russell and A. P. Thrall, "Portable and Rapidly Deployable Bridges: Historical Perspective and Recent Technology Developments," *J. Bridge. Eng.*, vol. 18, no. 10, pp. 1074–1085, 2013.

[17] *Trilateral Design and Test Code for Military Bridging and Gap-Crossing Equipments*. Department of USA Army, USA, 2005.

Military Technical Colle
Kobry El-Kobbah,
Cairo, Egypt

13th International Confere
on Civil and Architecture Enginee
ICCAE-13-2020

New Model to Calculate Blast Loading on Above and Underground Structures

Hassan M. Farag (MTC)

ABSTRACT

Blast loading is a type of extraordinary dynamic load to which structures or any object may be exposed to. The load resulted on structures or any targets from explosion is a complicated task to do. This paper presents an accurate and so simple model and program "***BLAST***" to predict the blast or shock wave resulted from the explosion that termed "Over Pressure" which comes from conventional and non-conventional weapons (such as Nuclear weapons). The pressure from the mass of air flow at and behind the shock front that termed the "Dynamic Pressure" is also predicted.

The loads on any structure or target from the explosion are combination of overpressure $P_s(t)$ and dynamic pressure $P_d(t)$, this dynamic load on different faces of a structure either above or under the ground is predicted and presented in this paper.

The over pressures resulted from different charges or weapon yields that are calculated in this study are compared to those obtained from "ConWep".

The load on structures predicted by the present study is also compared to that obtained from "AUTODYN" software. A study is performed to investigate the effect of depth of burial on the load affecting the different faces of the buried structure.

KEYWORDS: Explosions, Nuclear, Overpressure, Buried structures, Dynamic loading.

1. Introduction

Blast loading is a type of extraordinary dynamic load to which structures or any object may be exposed to. The load resulted on structures or any targets from explosion is a complicated task to do.

Explosion is always followed by a blast wave (a pulse of air in which the pressure increases sharply at the front) and is accompanied by blast winds. The destructive effects of a blast wave are generally characterised by the peak overpressure p_{os}. However, in many cases, the drag forces associated with the strong transient winds behind the shock front are responsible for the actual damage. The pressure from this mass of air flow at and behind the shock front is termed the "dynamic pressure". This pressure is a function of the density of the air through which the blast wave passes and the wind velocity behind

IOP Conf. Series: Materials Science and Engineering **974** (2020) 012007 doi:10.1088/1757-899X/974/1/012007

the shock front and it is responsible for what is called drag loading. These drag forces are a function of the shape and size of the structure and the transient dynamic pressure [1, 2, 3]. The blast or shock wave and the dynamic pressure are responsible for the load resulted on any object or structure exposed to explosion.

2.1 The peak overpressure p_{os}

The peak overpressure, p_{os} which is a function primarily of the weapon yield, the distance from the burst and the height of burst (in psi and $kft/kt^{1/3}$) will be: [1]

$$P_{os}(r,z)=\frac{10.47}{r^{a(z)}}+\frac{b(z)}{r^{c(z)}}+\frac{d(z).e(z)}{1.0+f(z)r^{g(z)}}+h(x,r,y) \tag{1}$$

Where

$$a(z)=1.22-\frac{3.908\,z^{2}}{1+810.2\,z^{5}}, \tag{2.a}$$

$$b(z)=2.321+\frac{6.195\,z^{18}}{1+1.113\,z^{18}}-\frac{.03831\,z^{17}}{1+.02415\,z^{17}}+\frac{.6692}{1+4164\,z^{8}}, \tag{2.b}$$

$$c(z)=4.153-\frac{1.149\,z^{118}}{1+1.641\,z^{18}}+\frac{1.1}{1+2.771\,z^{2.5}}, \tag{2.c}$$

$$d(z)=-4.166+\frac{25.76\,z^{1.75}}{1+1.382\,z^{18}}+\frac{8.257\,z}{1+3.219\,z}, \tag{2.d}$$

$$e(z)=1-\frac{.004642\,z^{18}}{1+.003886\,z^{18}}, \tag{2.e}$$

$$f(z)=.6096+\frac{2.879\,z^{9.25}}{1+2.359\,z^{14.5}}-\frac{17.15\,z^{2}}{1+71.66\,z^{3}}, \tag{2.f}$$

$$g(z)=1.83+\frac{5.361\,z^{2}}{1+.3139\,z^{6}}. \tag{2.g}$$

And

$$h(z,r,y)=\frac{-(64.67\,z^{5}+.2905)}{1+441.5\,z^{5}}-\frac{1.389z}{1+49.03\,z^{5}}+\frac{8.808\,z^{1.5}}{1+154.5\,z^{3.5}}+\frac{0.0014\,r^{2}}{(1-.158r+0.0486\,r^{1.5}+0.00128\,r^{2})(1+2y)} \tag{2.h}$$

$r=\sqrt{x^2+y^2}$ ($kft/kt^{1/3}$),

Where; x is the scaled ground rang, and y is the scaled height of burst.

2.2 Overpressure time history $P_s(t)$:

Overpressure time history $P_s(t)$ can be get from the expression: [1, 2]

$$p_s(x,y,t)=p_{os}(r,z)(1+a)(b.v+c) \quad psi \quad for\ x\geq xm\ ,\ y\leq 0.38 \tag{3.a}$$

$$p_s(x,y,t)=p_{os}(r,z)*b \quad psi \quad for\ x<xm\ or\ y>0.38 \tag{3.b}$$

In which, $p_{os}(r,z)$ = peak overpressure, psi as in (1)

$$xm = \frac{170\,y}{1+337\,y^{0.25}} + 0.914\,y^{2.5}, \tag{4.a}$$

$$a = (d-1).\left(1 - \frac{e^{20}}{1+e^{20}}\right), \tag{4.b}$$

$$e = \left|\frac{x - xm}{xe - xm}\right|, \qquad xe = \frac{3.039\,y}{1+6.7\,y}, \tag{4.c}$$

$$b = \left[f.\left(\frac{t_a}{t}\right)^g + (1-f).\left(\frac{t_a}{t}\right)^h\right]\left[1 - \frac{(t-t_a)}{t_o}\right], \tag{4.d}$$

$$f = \left(\frac{0.01477\,t_a^{0.75}}{1+.005836\,t_a} + \frac{7.402*10^{-5}\,t_a^{2.5}}{1+1.429*10^{-8}\,t_a^{4.75}} - 0.216\right).s + 0.7076 - \frac{3.077*10^{-5}\,t_a^3}{1+4.367*10^{-5}\,t_a^3}, \tag{4.e}$$

$$ga = \frac{t - t_a}{dt}, \qquad dt = 475.2\,y.(x - xm)^{1.25}, \tag{4.f}$$

2.3 Time of arrival t_a and positive phase duration t_o:

The time of arrival t_a is given by the following expression:

$$t_a = u \qquad for\; x < xm \tag{4.g}$$

$$t_a = u\frac{xm}{x} + w\left(1 - \frac{xm}{x}\right) \qquad for\; x \geq xm \tag{4.h}$$

$$u = \frac{(0.0543 - 21.8\,r + 386\,r^2 + 2383\,r^3)\,r^8}{[2.99*10^{-14} - 1.91*10^{-10}\,r^2 + 1.032*10^{-6}\,r^4 - 4.43*10^{-6}\,r^6 + (1.028 + 2.087r + 2.69\,r^2)\,r^8]} \tag{4.i}$$

$$w = \frac{(1.086 - 34.605\,r + 486.3\,r^2 + 2383\,r^3)\,r^8}{3.0137*10^{-13} - 1.2128*10^{-9}\,r^2 + 4.128*10^{-6}\,r^4 - 1.116*10^{-5}\,r^6 + (1.632 + 2.629\,r^2)\,r^8} \tag{4.j}$$

And the positive phase duration t_o is given by:

$$t_o = \left(\frac{1640700 + 24629\,t_a + 416.15\,t_a^2}{10880 + 619.76\,t_a + t_a^2}\right)\left[\left(0.4 + \frac{.001204\,t_a^{1.5}}{1+0.001559\,t_a^{1.5}}\right) + \left(0.0426 + \frac{0.5486\,t_a^{0.25}}{1+0.00357\,t_a^{1.5}}\right).s\right] \tag{4.k}$$

$$c = \left[\frac{\left(1.04 - \frac{240.9\,x^2}{1+231.7\,x^4}\right).(ga)^7}{(1+a).\left[1 + .923\,(ga)^{8.5}\right]}\right].\left[\frac{2.3*10^{13}\,y^9}{1+2.3*10^{13}\,y^9}\right]\left[1 - \left(\frac{(t-t_a)}{t_o}\right)^8\right], \tag{4.l}$$

$$v = 1 + \left(\frac{3.28*10^{11}\,y^6}{1+1.5*10^{12}\,y^{6.75}}\right).\left(\frac{ga^3}{6.13 + ga^3}\right).\left(\frac{1}{1+9.23.e^2}\right), \tag{4.m}$$

$$s = 1 - \frac{1.1*10^{10}y^7}{1+1.1*10^{10}y^7} - \left(\frac{2.441*10^{-8}y^2}{1+9*10^{10}y^7}\right)\left(\frac{1}{4.41*10^{-11}+x^{10}}\right), \tag{4.n}$$

$$h = 2.753 + \frac{0.05601\,t_a}{1+1.473*10^{-9}t_a^5} + \left(\frac{0.01769\,t_a}{1+3.207*10^{-10}t_a^{4.25}} - \frac{0.03209\,t_a^{1.25}}{1+9.914*10^{-8}t_a^4} - 1.6\right).s \tag{4.o}$$

$$g = 10.0 + \left(77.58 - \frac{64.99\,t_a^{0.125}}{1+0.04348\,t_a^{0.5}}\right).s\,, \tag{4.p}$$

2.4 Dynamic pressure time history $p_d(t)$:

Dynamic pressure time history $p_d(t)$ [3]

$$p_d(t) = p_{ds}(1-\omega)^2.\left[d.e^{(-\delta\omega)} + f.e^{(-\phi\omega)}\right] \tag{5}$$

Which is accurate within ± 2 % for $2 \le p_{os} \le 100{,}000$ psi.

Where

$$p_{ds} = \frac{5}{2}\frac{p_{os}^2}{7p_a + p_{os}}, \qquad for\ p_{os} < 300\ psi \tag{6.a}$$

$$p_{ds} = \frac{5}{2}\frac{p_{os}^2(1+0.0604\lambda+0.0345\lambda^2)}{7p_a+p_{os}-72.2\lambda^2+23.4\lambda^3}, for\ p_{os} > 300\ psi \tag{6.b}$$

Where p_a is the ambient pressure and $\lambda = p_{os}/1000$. This form is accurate within 6 % for $p_{os} \le 200{,}000$ psi.

$$\omega = (t - t_a)/t_d \quad for\ 0 \le \omega \le 1, \tag{6.c}$$

$$t_a = \left(0.004276 + \frac{112.51}{p_{os}^{0.80424}} + \frac{253.9}{p_{os}}\right) W_{Mt}^{1/3}\ \text{sec}, \tag{6.d}$$

$$f = 1 - d,\ \delta = 0.38\,p_{os}^{0.8605},\ \phi = 5.4\,p_{os}^{0.604}, \tag{6.e}$$

$$d = \frac{0.83\,p_{os}^{0.035}}{1+0.147\left(\frac{p_{os}}{100}\right)^3} + \frac{0.266\left(\frac{p_{os}}{500}\right)^3}{1+6\left(\frac{p_{os}}{500}\right)^{3.5}} \tag{6.f}$$

3. Program BLAST

The above relations were implemented in a program called ***BLAST***, which is also used to calculate the load resulted from the blast on structures above and below ground as will be seen. Although the above expressions were made for weapon yields in ranges of kilotons (Kt) and megatons (Mt), it can be used for small scale yields. The overpressure time histories for different weapon yields compared to that resulted from "***ConWep***" [6, 7], are shown in Fig (1) to Fig (6).

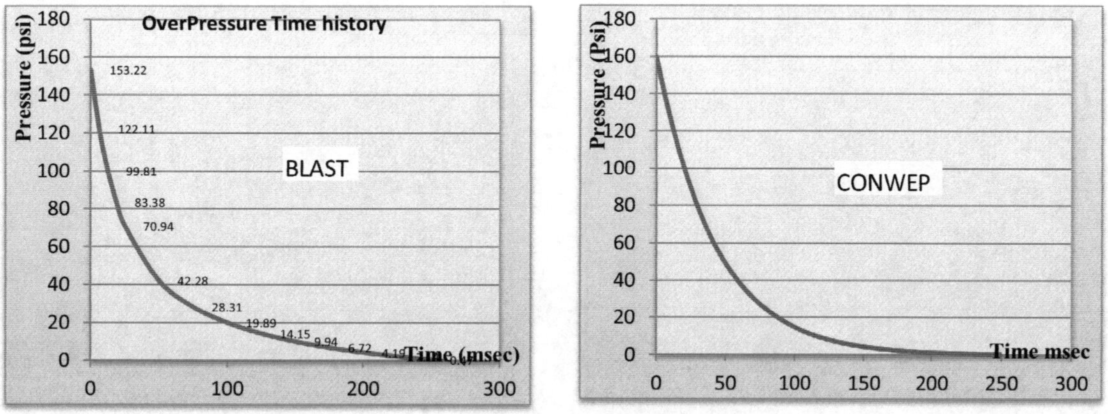

Fig (1) 10 kt at 200 m

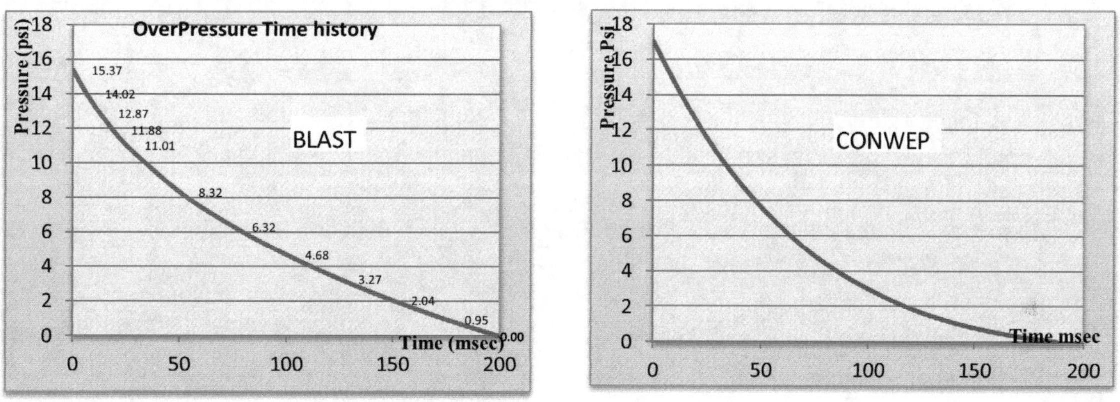

Fig (2) 0.5 kt at 200 m

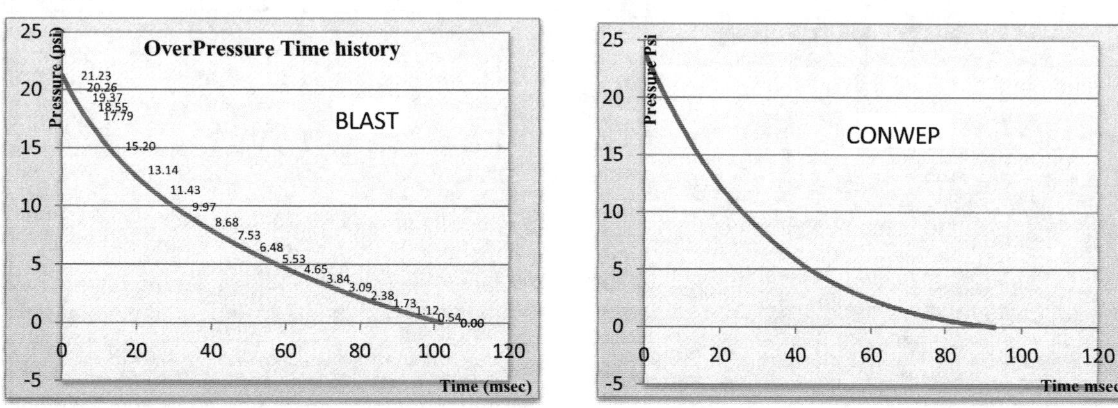

Fig (3) 0.1 kt at 100 m

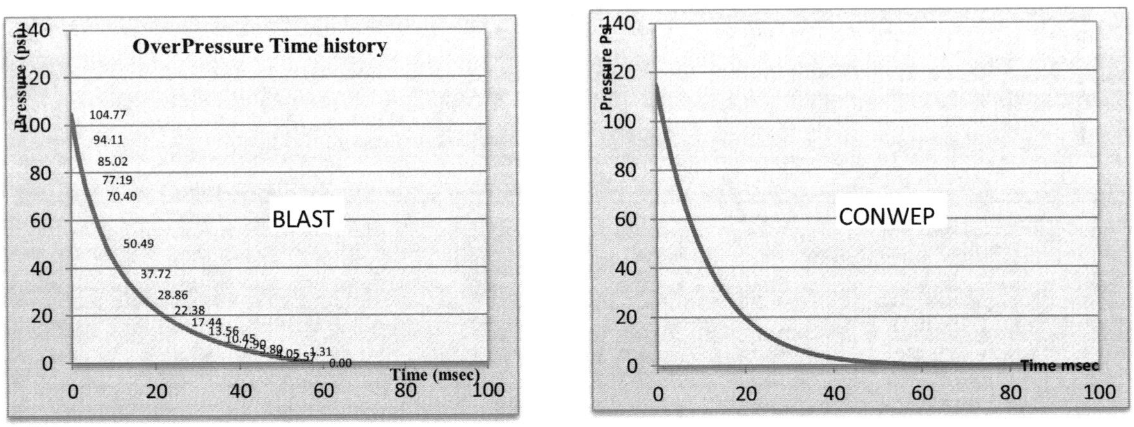

Fig (4) 0.1 kt at 50 m

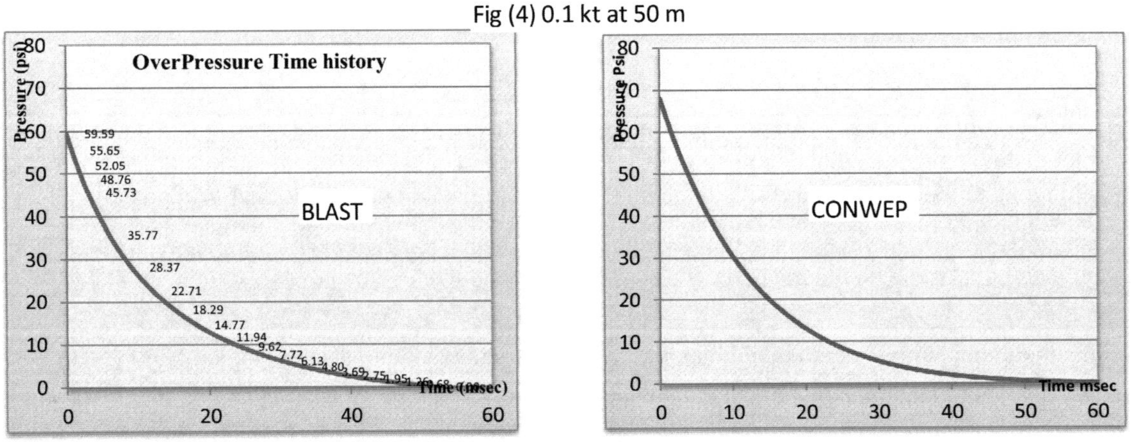

Fig (5) 50000 kg at 50 m

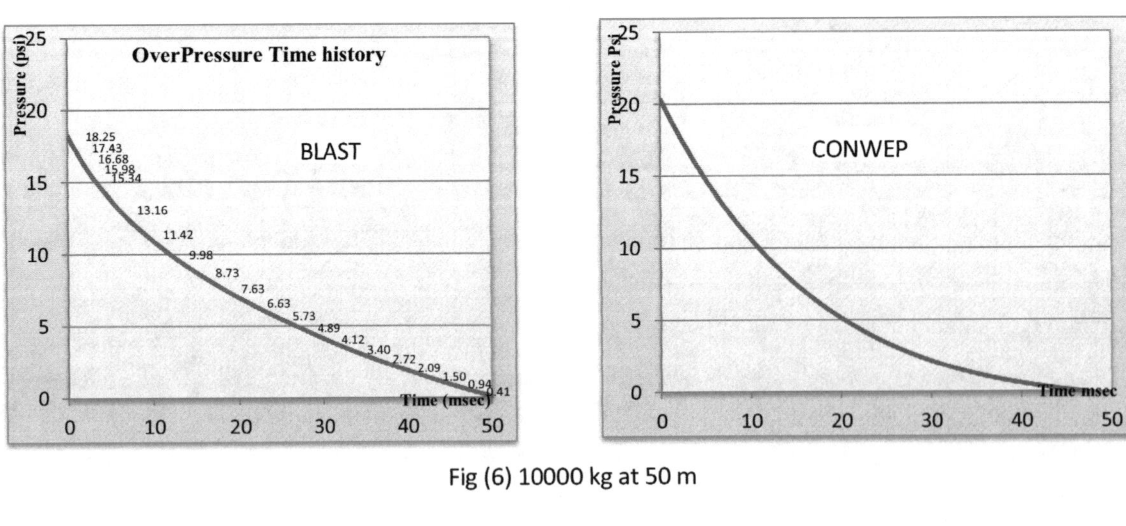

Fig (6) 10000 kg at 50 m

IOP Conf. Series: Materials Science and Engineering **974** (2020) 012007 doi:10.1088/1757-899X/974/1/012007

4.1 The blast loading on structures:

The air blast loading on exposed structure surfaces is a function of the free field overpressure (p_{so}) and the dynamic pressure (p_d), the size, the shape and the orientation of the structure. The total loading on a structure can be assumed to consist of three main parts [3, 10]

1) Effect of initial reflected overpressure

2) Effect of general overpressure

3) Drag loading "is related to dynamic pressure by the expression (drag pressure = $C_d * P_d$)"

Where: C_d is the drag coefficient and P_d .is the dynamic pressure

For front wall $C_d = 1$,

For roof, back and sides C_d = -0.2 for overpressure range 50 to 130 psi (34.5 to 89.6 N/cm^2)

= -0.3 for overpressure range 25 to 50 psi (17.2 to 34.5 N/cm^2)

= -0.4 for overpressure range 0 to 25 psi (0 to 17.2 N/cm^2)

The average loading on the front face of the structure P_{front} will consists of the reflected pressure P_r and decreasing linearly up to the clearing time t_c, where the load intensity at time t_c can be calculated as follows:-

$$P_c = P_s(t_c) + C_d P_d(t_c) \quad (7.a)$$

From time t_c to the end of duration of the blast wave (time t_o), the load intensity can be calculated as follows:

$$P(t) = P_s(t) + C_d P_d(t) \quad (7.b)$$

The clearing time (t_c) for a solid flat surface such as the wall of the building can be determinate from the following expression:-

$$t_c = 3S/U_r \quad (7.c)$$

Where S is the least of height (h) or half the length of the structure in the direction of propagation of the wave (L/2) and U_r is the velocity of sound in the region of the reflected overpressure. And it can be calculated from the expression (feet/s) [3]:-

$$U_r = 422\left(\frac{1.0088P_o + 70P_{so} + 720}{102.9 + 6P_{so}}\right)^{1/2} \quad (7.d)$$

Prediction of loads on buried structures is further complicated by the fact that stress wave propagating down ward reflect at structures surfaces that are in the path of the propagation [3].

The pressure acting on the roof of shallow buried rectangular structures (Depth of Burial "DoB" between 0.2L & 1.5L, L is the clear span of the structure) may be approximated as follows:

$$P(t) = P_{ff}(t)\left(2 - \frac{t}{t_d}\right) \quad \text{for } t \le t_d \quad (8.a)$$

$$P(t) = P_{ff}(t) \quad \text{for } t > t_d \quad (8.b)$$

Where $P(t)$ is the pressure acting on the roof at time t; and $P_{ff}(t)$ is the incident wave pressure in the free field at the location of the structure roof; t_d=12 D/C

Where D is the thickness of the roof and C is the compressional wave velocity of the structural material (2000-2500 m/sec), and t is taken as zero at the time the incident wave strikes the roof. It can be assumed that for a shallow buried structure the vertical free field pressure acting on the soil having the same magnitude as the ground surface overpressure the time t_d is probably in the range **1** to **5** milliseconds

An acceptable approximation for the pressure acting on the base of structure is:

$$P_b(t) = P_{ff}(t), \quad (8.c)$$

In which $P_b(t)$ is the pressure on the base of the structure; $P_{ff}(t)$ is the incident wave pressure in the free-field at base of the structure.

IOP Conf. Series: Materials Science and Engineering **974** (2020) 012007 doi:10.1088/1757-899X/974/1/012007

As an acceptable approximation, the pressure acting on the side of a structure can be taken as:

$$P_s(t) = K_o\ P_{ff}(t), \quad (8.d)$$

In which $P_s(t)$ is the uniform lateral pressure on the side of the structure; K_o is the coefficient of lateral earth pressure at rest; and $P_{ff}(t)$ is the incident wave pressure in the free-field at mid-height of the structures sidewall.

4.2 Blast Load using AUTODYN

In AUTODYN, air domain surrounds Reinforced Concrete panel and explosion zone and has boundary condition called FLOW OUT. The flow out permits to translate blast wave to hit the RC panel as in blast field test (i.e. RC panel responds similar to actual practical case). Fig (7) illustrates the boundary conditions of the panel, the air media, and the explosion sphere [8, 9]. In numerical model, air and TNT are simulated by Euler formulation.

The blast load levels are implemented through the application of 10 and 15 kg of TNT explosive at the same stand-off distance of 4 m. The pressure-time histories for both charges are illustrated in Fig (8) [8, 9].

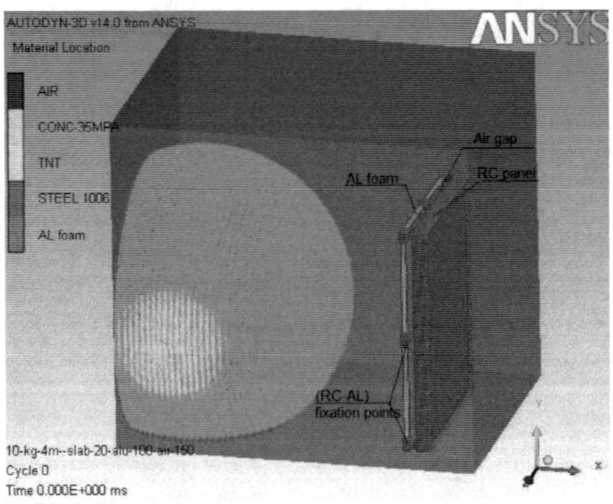

Fig (7) the boundary conditions, air medium and explosion sphere.

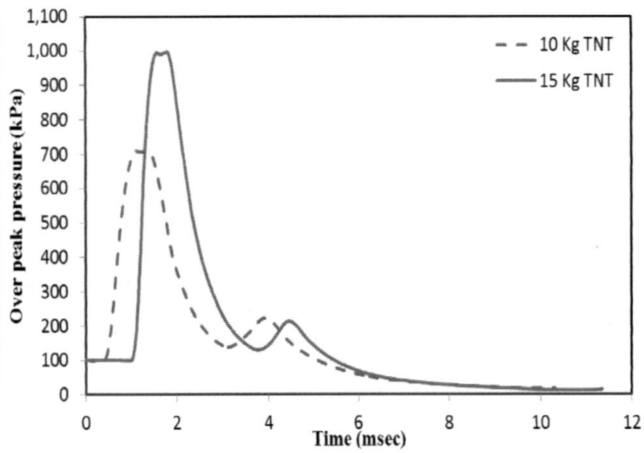

Fig (8) Loads on target from different charges at 4 m

4.3 Output results from Program *BLAST* for 15 kg TNT at 4 m range

P_{so} (peak side on pressure) = 39.4655 psi = 2.72105 bar = 2.7747 kg/cm^2 =272.105 kPa.

P_{do} (peak dynamic pressure) = 27.4182 psi = 1.89042 bar = 1.92769 kg/cm^2 =189.042 kPa.

t_a (time of arrival of the shock wave) = 4.93856 m sec.

t_o (the duration of the shock wave) = 4.12101m sec.

P_r (reflected pressure) = 144.573 psi = 5.65 P_{so} = 996.795 kPa.

U_r (velocity of sound in the region of the reflected over pressure) = 1354.07 ft/sec = 412.722m/sec.

U_o (the shock front velocity) = 2029.5 ft/sec =618.591 m/sec.

t_c (the clearance time) = 0.002908 sec = 2.908 msec.

t_s (the time of the stagnation pressure to take place) = 0.000647 sec = 0. 647 msec.

t_d (the time of the pressure to reach the back face) = 0.001293 sec =1.293 msec.

t_b (the built up time) = 0.002587 sec = 2.587 msec.

P_c (the pressure o the front face at the clearance time) = 3.41 psi =23.48057 kPa.

P_a (the average maximum pressure on the roof and sides) =21.1892 psi =146.0943 kPa.

P_{sb} (the over pressure of the blast wave at the time it arrives to back face) = 0.51876psi =3.576741kPa.

$P_{(back)max}$ (the maximum. pressure on the back face) = 0.23618 psi =1.628428 kPa.

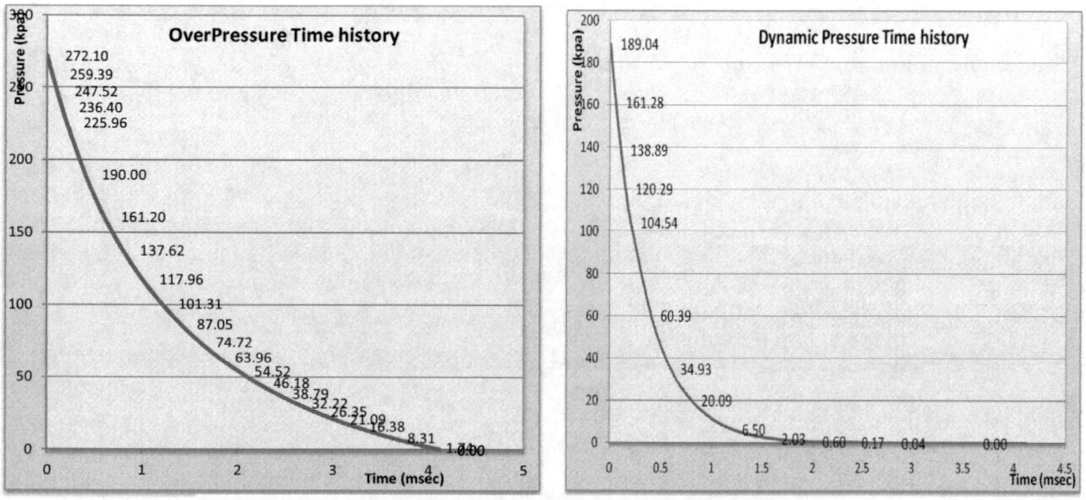

Fig (9) Over-pressure and Dynamic Pressure resulted from 15 kg at 4 m

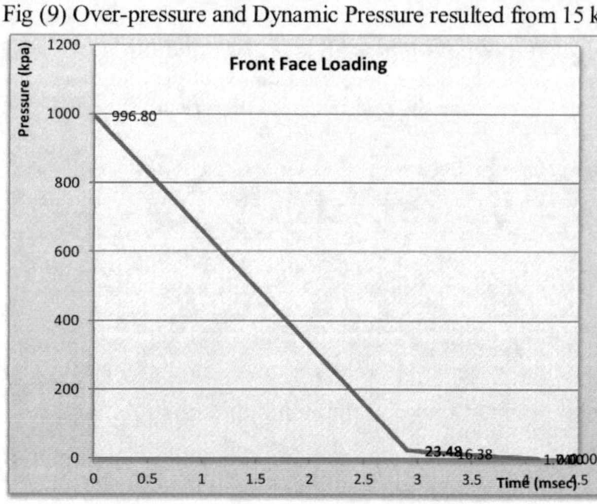

Fig (10) Load on target from 15kg at 4 m

The peak load on front face Fig (10) is 996.8 kPa with duration of 3 msec. which agrees well with that obtained from "AUTODYN" software Fig (8).

5. A case study:

If 200Kt weapon yield explodes at 600 m from the structure with the dimension as shown in Fig (11.a), the over-pressure and the dynamic pressure resulted from the explosion at the place of the structure will be as shown

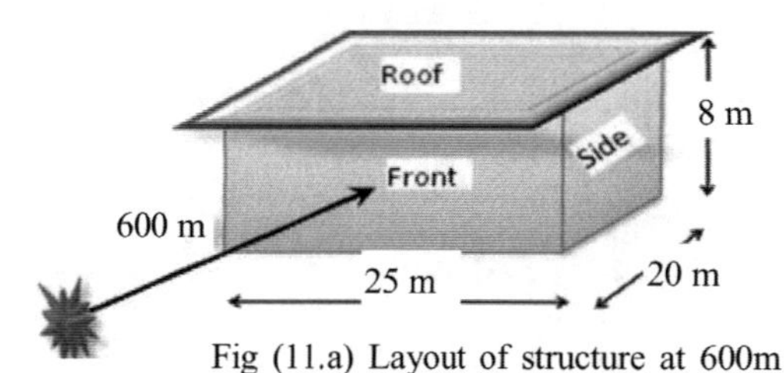

Fig (11.a) Layout of structure at 600m from 200Kt weapon yield

IOP Conf. Series: Materials Science and Engineering **974** (2020) 012007 doi:10.1088/1757-899X/974/1/012007

(Fig 11.b). The loads affecting the different sides of the structure will be as shown; if it is above ground (Figs 11.c-11.f) and if it is buried with DoB equal 10 m (Figs 12.a-12.c).

5.1 Sample output from program *"BLAST"*

P_{so} = 118.455 psi = 8.16721bar = 8.32823kg/cm^2 = 816.7 kPa
P_{do} = 158.725 psi = 10.9437 bar = 11.1595kg/cm^2 =1094.37 kPa
t_a = 410.09 m sec
t_o = 707.654 m sec
P_r (psi) = 617.249 = 5.21 P_{so} = 4255.78 kPa
U_r = 1405.6 ft/sec = 428.428 m/sec
U_o = 3140.94 ft/sec = 957.358 m/sec
t_c = 0.056019 sec = 56.0187 msec
t_s = 0.010445 sec = 10.4454 msec
t_d = 0.020891 sec = 20.8908 msec
t_b = 0.074692 sec = 74.6916 msec
P_c = 89.76 psi = 618.844 kpa
P_a = 88.1091 psi = 607.491 kPa
P_{sb} = 54.0968 psi = 372.985 kPa
$P_{(back)max}$ = 25.5908 psi = 176.442 kPa

Underground Structure:

t_d (time for load on roof to decrease to overpressure) = 12 msec.
σ_{ff} (the incident wave pressure in the free field at the location of the roof of structure)=111.364psi =767.829kPa
t_m (the time for the wave takes to reach the mid-height of the structure) = 10.4454 msec.

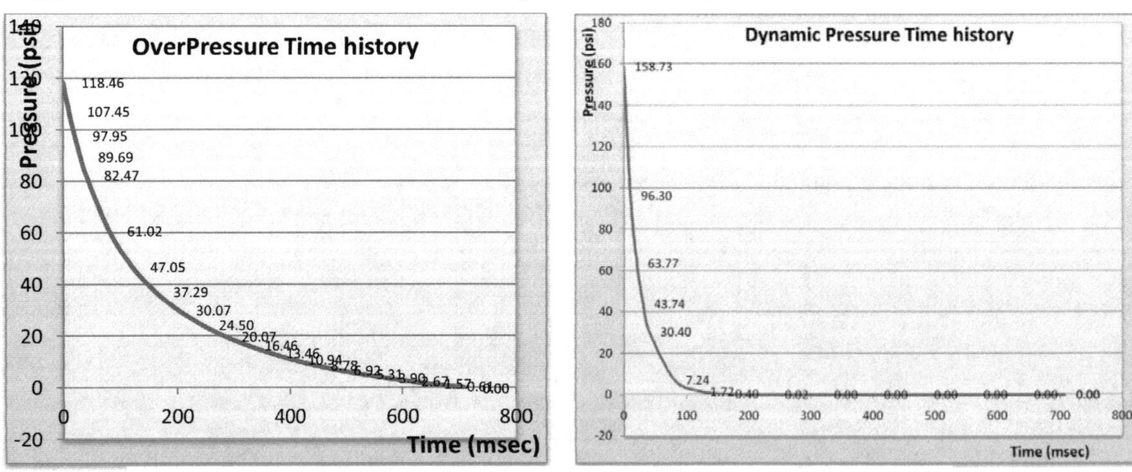

Fig (11.b) Over-pressure and Dynamic Pressure resulted from 200Kt at 600 m

5.2 Loads resulted from the explosion on the Above Ground Structure

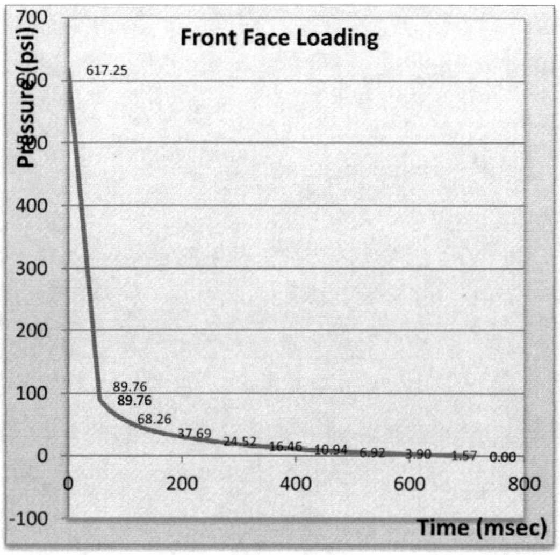

Fig (11.c) Front face loading

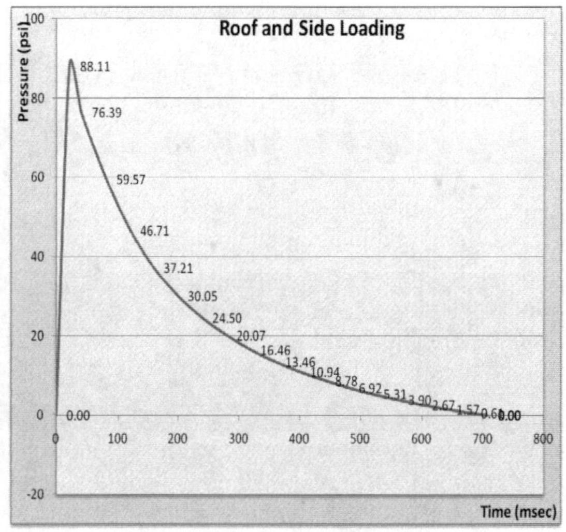

Fig (11.d) Loads on roof and sides

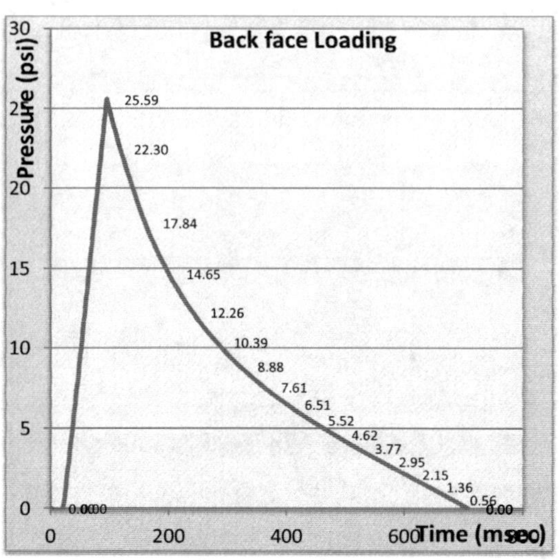

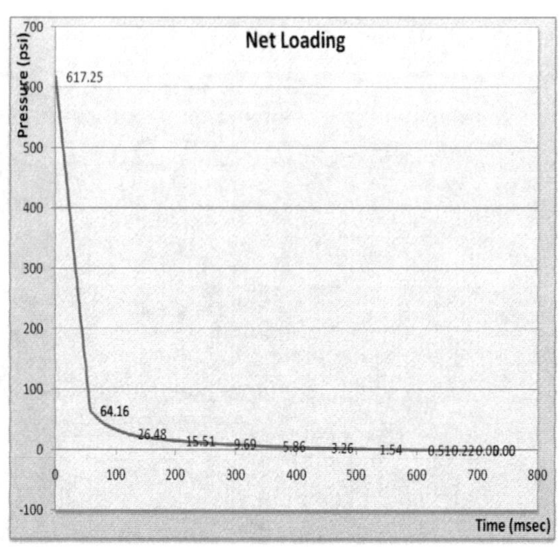

Fig (11.f) Net Loading on Structure

IOP Conf. Series: Materials Science and Engineering **974** (2020) 012007 doi:10.1088/1757-899X/974/1/012007

Fig (11.e) Loads on back face

5.3 Loads resulted from the explosion on the Under Ground Structure:

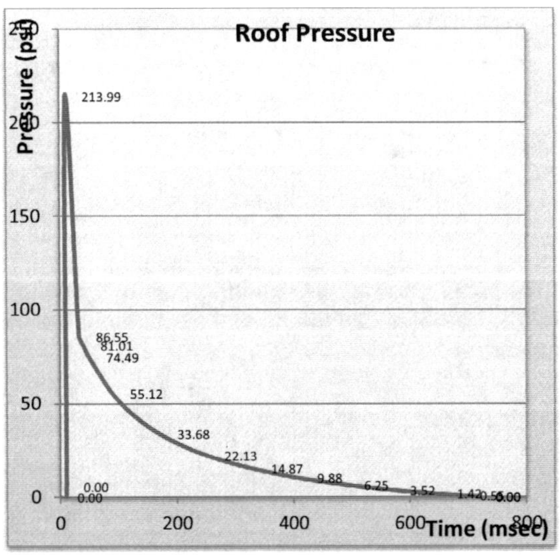

Fig (12.a) Loads on roof of buried structure

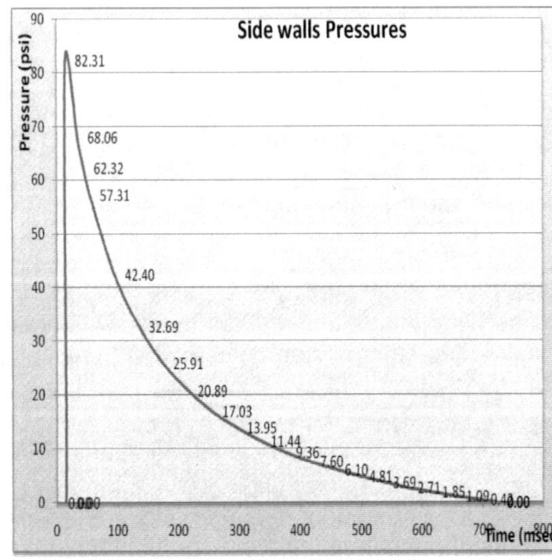

Fig (12.b) Loads on side walls of buried structure

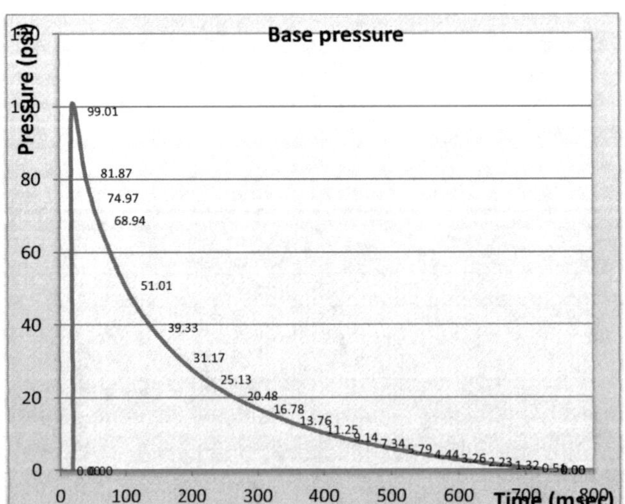

Fig (12.c) Loads on base of buried structure

6. Effect of Depth of Burial (DoB) on Peak loading on a buried Structure:

The effect of burial for the buried structure on the peak loads on roof, sides, and base of the structure were summarized in table (1). A normalized relation is shown in Fig (13). Using curve fitting, equation (9) gives the relation between normalized DoB and normalized peak pressure.

Table (1)

DoB (m)	Normalized DoB=(DoB/L)	Roof Peak Load (psi)	Normalized P_{Roof}/P_{so}	Side Peak Load (psi)	Normalized P_{Side}/P_{so}	Base Peak Load (psi)	Normalized P_{Base}/P_{so}
4	0.2	227.3069	1.91893	87.31631	0.737126	104.90807	0.885636486
6	0.3	222.7284	1.880278	85.59539	0.722598	102.88432	0.86855194
8	0.4	218.2908	1.842816	83.92646	0.708509	100.92054	0.851973661
10	0.5	213.9885	1.806496	82.30745	0.694842	99.014412	0.835882082
12	0.6	209.8161	1.771273	80.73643	0.681579	97.16371	0.82025841
14	0.7	205.7686	1.737104	79.21153	0.668706	95.36631	0.805084716
16	0.8	201.8411	1.703947	77.73097	0.656207	93.620185	0.790343886
18	0.9	198.0288	1.671764	76.29305	0.644068	91.923397	0.77601956
20	1.0	194.3274	1.640517	74.89615	0.632275	90.274094	0.762096104
22	1.1	190.7326	1.610169	73.53872	0.620816	88.670506	0.748558575
24	1.2	187.2404	1.580688	72.21928	0.609677	87.110941	0.735392689
26	1.3	183.8468	1.552039	70.93641	0.598847	85.593781	0.722584787
28	1.4	180.5482	1.524192	69.68875	0.588314	84.117476	0.710121785
30	1.5	177.341	1.497117	68.47503	0.578068	82.680546	0.697991187
32	1.6	174.2219	1.470785	67.293981	0.568097	81.281572	0.686181014
34	1.7	171.1876	1.44517	66.144437	0.558393	79.919196	0.674679802
36	1.8	168.235	1.420244	65.025258	0.548945	78.592119	0.663476586
38	1.9	165.3611	1.395982	63.935357	0.539744	77.299096	0.652560854
40	2.0	162.5631	1.372362	62.873695	0.530781	76.038932	0.641922519

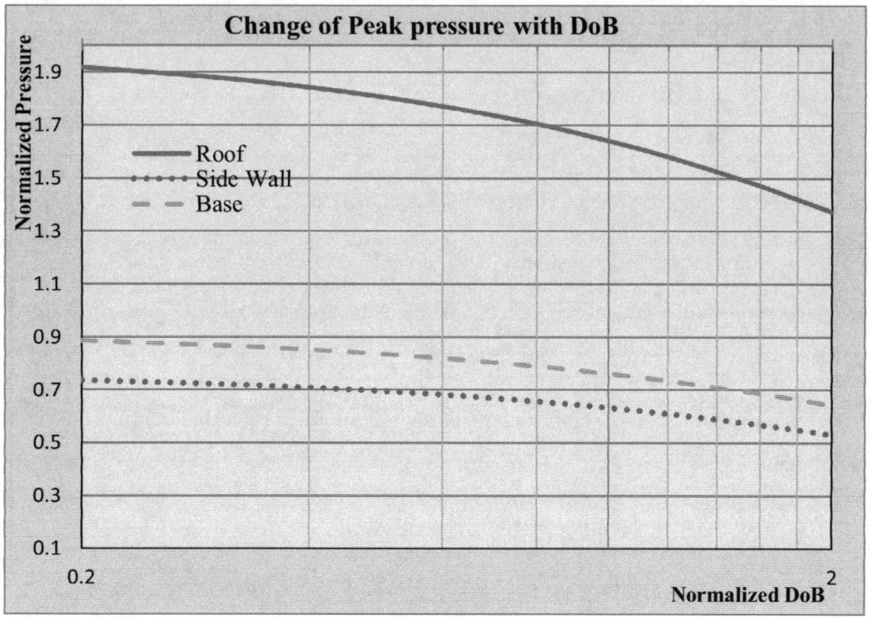

Fig (13) effect of depth of burial on peak loading on roof, side and base of buried structure

For Roof: $P_{peak}/P_{so}=0.0438*(DoB/L)^2 - 0.398*(DoB/L) + 2.0$ (9.a)

For Base: $P_{peak}/P_{so}=0.0188*(DoB/L)^2 - 0.176*(DoB/L) +0.919$ (9.b)

For Side: $P_{peak}/P_{so}=0.0162*(DoB/L)^2 - 0.15*(DoB/L) +0.766$ (9.c)

IOP Conf. Series: Materials Science and Engineering **974** (2020) 012007 doi:10.1088/1757-899X/974/1/012007

7. Discussion of Results

This paper presents a program called "**BLAST**" based on a collection of equations from literatures to predict the shock wave from explosion and to calculate the load affecting any structure or target above or under the ground located at a distance from the centre of explosion. The results obtained for different weapon yields at different distances were compared well to those obtained from "ConWep". The load on the structure is also compared to that obtained from "AUTODYN" software. It can be concluded that the results obtained from the program "BLAST" ***-which are easy to use and save effort and time-*** are compared well to those obtained from sophisticated software. A parametric study to investigate the effect of depth of burial of structure on the load affecting the different faces was carried out. The pressure affecting the roof decreased by (17%) from $1.8065P_{so}$ to $1.497P_{so}$ as the DoB increased (300%) from 0.5L to 1.5L, while for the base the pressure decreased by (16.5%) and for the side it decreased by (16.8%).

8. REFERENCES

1- Speicher S. J. and Brode H. L. "Air Blast Overpressure Analytic Expression for Burst Hight, Range, and Timeover an Ideal Surface", PSR Note 385, Pacific Sierra Research, Santa Monica, Calif., (1981).

2- Brode H. L. "Analysis Approximations to Dynamic Pressure and Impulse and Other Fits for Nuclear Blast". PSR Note 529, Pacific Sierra Research Corporation, Santa Monica, Calif. (1983)

3- ASCE-Manuals and Reports on Engineering Practice-No.42 "Design of Structures to Resist Nuclear Weapons Effects" (1985).

4- Farag H. M. "The Transient Analysis and Non Linear Behaviour of Reinforced Concrete Element", PhD Thesis Salford University (1995).

5- Schenker, A., Anteby, I., Gal, E., Kivity, Y., Eyal, N., Sadot, O., Michaeilis, R., Levintant, O., Ben-Dor, G., "Full scale field tests of concrete slabs subjected to blast loads". International Journal of Impact Engineering 35, 2008, pp. 184-198.

6- ConWep "Conventional Weapons Effects" U S Army Corps of Engineers.

7- TM5-855-1 "Fundamental of Protection Design for Conventional Weapons" U S Army Corps of Engineers.

8- Improvement performance of reinforced concrete structures under blast loads using composite materials, Ph D Thesis, by: Mohamed Mustafa Mahmoud, Defence Systems Studies Centre Military Technical College, Cairo, 2015.

9- Effect of air gap on the structural response of aluminium foam protected reinforced concrete panels, Capt. Eng. Mohamed M. Mahmoud, Assoc. Prof. Sameh Y. Mahfouz and Maj. Gen. (Ret.) Dr. Hassan M. Farag, *Proceedings of the 12th Conference* on Civil and Architecture Engineering *ICCAE*-12, 3-5 April, 2018.

10- TM-5-585-3, "Design Facilities to Resist Nuclear Weapon Effects". Technical Manual, U S Army Corps of Engineers.

MilitaryTechnicalCollege
Kobry El-Kobbah,
Cairo, Egypt

13th International Conference on Civil and Architecture Engineering ICCAE-13-2020

Simulation and Optimizationof aMetallic Ferry under MLC70Loading

Mohamed N. Lotfy[1], Yasser A. Khalifa[1], Abdelrahim k. Dessouki[2]and Elsayed Fathallah[1]

[1]Department of Civil Engineering, Military Technical College, Cairo, Egypt.
[2]Ain Shams University, Cairo, Egypt.

Abstract. Floating ferries are used for both civilian and military purposes. This study concerned with a ferry composed of sixteen connected floating pontoons. This ferry is simulated and optimized to carry Military Load Capacity MLC70 (Tank load).Consequently to the increasing demand of evolution and cost optimization, the design optimization is performed in this paper to obtain the optimum minimum weight which minimizesboth the cost and the buoyancy factor. The simulation of the ferry is performed using the finite element program ANSYSsoftware. Furthermore, different grades of the structural steel, hybrid materials (steel stiffeners covered with aluminium plates) and aluminium alloy are incorporated in this study. Thesimulation is verified with both practical and mathematical results. The performance of the ferry is investigated. In addition to the design parameters, constraints and objective functions are determined. The optimum weight of the ferry is obtained, followed by a reduction in the buoyancy factor; accordinglythe capacity of the ferry can be increased. Comparison between the behaviour of the different ferries using different materials is operated considering stresses, deformations and weight. Conclusions and recommendations are then stated.

Keywords: Floating; Pontoon;Optimization; Ferry; Steel; Aluminium; Modelling; ANSYS

1. Introduction

Transport is of vital importance in the flexibility of traffic and the affluence of the economy, delivering people, goods and vehicles from one place to another. The usage of floating bridges is essential when facing relatively deep water obstacles. Floating bridges usage can be dated back to 2000 BC.[1]. The floating bridges are used in peace and war times as they have a great importance in emergency situations. There are many types of floating bridges according to the conditions of the site to be constructed in and according to the types of obstacles to cross [2].The most common types of the floating bridges are the continuous pontoon floating bridge and the discrete pontoon floating bridge [3, 4]. There are two main features that judge the efficiency of the pontoon bridges; the safety and the speed of erection over the obstacles [5]. As a result of the great need to the usage of the floating bridges, many upgrades and develops are being applied to the floating bridges. Many researchers interested in studying floating bridges usinganalytical and numerical methods in a companion with software programs. At the beginning of the modelling of the floating pontoon bridges many simplifications are applied to be able to simulate the required models with the available software programs. First they considered the pontoon as a beam rested on an elastic foundations andsubjected to the applied loads [6].Shahrabi et al. [7]developed a method to analyze the motions of a floating pier which was represented as number of floating pontoons. These floating pontoons were simulated as rigid bodies and connected to each other's by rigid and flexible connectors. After that, the development of the finite element programming resulted in the complete simulation of the models. Khalifa et al.[8]studied the performance of the floating bridges under longitudinal and transverse eccentric static load using ANSYS. Fu and Cui [9], numerically studied the effect of the dynamic and the static loading on the hydro elastic responses of connected pontoons and introduced a numerical method for analysis of a ribbon floating bridge [10]. Sun et al. [11] studied the dynamic response of a floating bridge which was consisting of multi-modules. They determined the bending moment and the mooring force of the floating bridge. Zhang et al. [3] investigated the dynamic response of two analytical models for the two types of pontoon bridges subjected to moving loads for different depths of water. They concluded that the water depth has a little effect on the dynamic responses of floating bridges.Hirono et al. [12] divided the floating bridge to floating units and developed a measurement system to determine the vertical displacement of these floating units to determine the displacement of the floating bridge under the applied loads. The design of floating bridges to meet the requirements with an economic cost is a new challenging issue. The floating bridges design must consider several environmental loads such as waves, wind and current loads [13]. These loads in addition to the applied loads from the carried vehicles must be investigated and analyzed very carefully. Cheng et al. [13] studied the effect of the environmental loads. Halvor et al. [14] investigated the floating bridges under the same environmental loads and presented a numerical solution to get the required stresses, deformations and moments. Sha et al. [15] studied the dynamic response of the floating bridges under these environmental loads.

2. Optimization

Economic design of structures without exceeding the constraints is a challenge for the designers. Optimization is defined as doing something or design something as well as possible[16]. There are many techniques of optimization which is differing in applications, advantages and disadvantages. The development of optimization techniques and optimization softwareprograms are in a continuous increase. Genetic Algorithms are used in many optimization experiments with different techniques a long time ago. These genetic algorithms are used for the single objective optimization experiments at the first, and thenthey are expanded to deal with multi-objective optimization. Pareto genetic algorithm is a developed technique that can deal with the multi-objective optimization experiments depending on the weight factor of variables to get the optimum solution as well as possible [17]. Pareto optimization technique is used till our moment due to its advantage of showing the best desired solution with a great weight factor from many other solutions [18, 19]. The researchers and designers continued their development in such a challenging topic and applied many developed optimization techniques to many structures. In 1989 Torn et al [20] presented a guide book that explained the global optimization and introduced examples of many optimization problems and how to get the proper technique of the optimization and the solution of the problem. Carlos and Coello[21] made a comprehensive survey of evolutionary-based multi-objective optimization techniques. Koumousis andArsenis[22] performed a detailed optimization design of reinforced concrete members of a multi-story building. Hongand Adeli[23] studied the genetic algorithm optimization which was applied by binary representation of parameters. They used floating-point genetic algorithm in studying the cost optimization of composite floors to get rid of the disadvantages of the binary method. Philippe Rigoa and Claude Fleury[24] used the Stiffened panelssoftware as a development of a new methodology to determine the preliminary design of floating hydraulic structures and naval structures. They presented an algorithm for optimization based on convex linearization. Mehdi and Aidini[25] defined the optimum mooring systems. They used a genetic algorithm for the optimization of the design of

IOP Conf. Series: Materials Science and Engineering **974** (2020) 012008 doi:10.1088/1757-899X/974/1/012008

mooring platforms and presented new procedures to solve these floating structures in a quick way to get the best mooring system. Matthias et al [26] presented a new tool of optimization for mooring systems for floating structures foundations. They used MATLAB in their optimization experiments. Fathallah et al [27, 28] applied optimization for a composite submerged hull composed of composite layers with different layers orientations. Their objective function was to minimize the buoyancy factor by using ANSYS. Brittani R. Russell et al [29] developed the characteristics of the deployable floating causeways due to the increasing need to this type of bridges. They studied the reconceptualization and the optimization of floating causeways.
In this study, the steel ferry is simulated and optimized to achieve the minimum weight for the ferry using finite element modelling by ANSYS software. Additionally, different grades of the structural steel, hybrid materials (steel and aluminium) and aluminium alloy are incorporated in this study.

3. Model definition

3.1. Geometry

The studied floating ferry is analysed to carry the Military Load Capacity MLC-70 (Tank load). It is composed ofa sixteen floating pontoons connected together. Each pontoon is composed of upper deck, lower deck and round side sheets. These sheets are modelled as shell elements, which aresupported by internal stiffeners in vertical, longitudinal and transverse directionswith different cross sections. These stiffeners are designed as beam elements. The pontoon's dimensions and characteristicsareillustrated in **Table 1**. The pontoon is simulated in finite element program ANSYS shown in Figure 1

Table 1: The floating pontoon dimensionsand characteristics

Length	5.25m
Width	2.4 m
Depth	1.5 m
Weight	3.356 ton

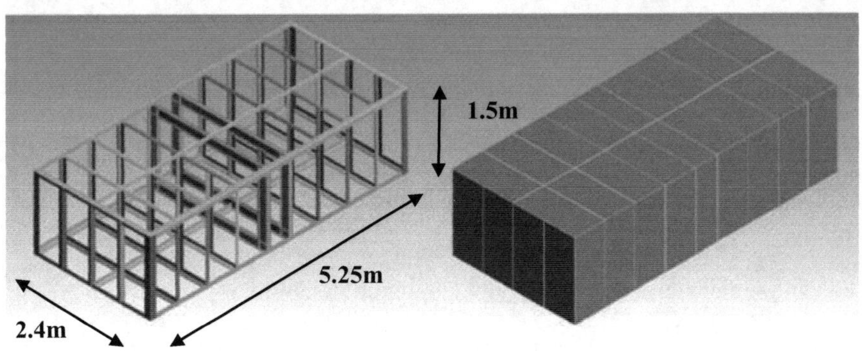

Figure 1: The Isometric view of the floating pontoon.

These floating pontoons are made of steel 37 with properties illustrated in **Table 2**[8, 30].

Table 2: The properties of structural steel

Density	7850 kg/m^3
Tensile yield strength	2.4×10^8 (Pa)

Compressive yield strength	2.4×10^8 (Pa)
Tensile ultimate strength	3.6×10^8 (Pa)

The ferry is composed of sixteen floating pontoon with a total length of (21.3 m), a total width of (9.9 m) and a depth of (1.5 m). These pontoons are arranged as follows in **Figure 2**; four pontoons in the transverse direction are separated by 10 cm and these pontoons are repeated another three times in the longitudinal direction.

Figure 2: The floating ferry

These pontoons are connected together by upper and lower connections. The connection system between pontoons is simulated to be 8 connections at the long directions divided on the upper and the lower levels and 6 connections in the short direction also, divided also on the upper and the lower levels. The characteristics of the ferry are as shown in **Table 3**.

Table 3: The dimension of the floating ferry

Length	21.3 m
Width	9.9 m
Depth	1.5 m
Weight	53698 Kg

3.2. Loads

The applied load is the load of the tank MLC70 and the characteristicsof the tank are as shown in**Table 4**.

Table 4: The floating ferry properties

Tank weight (W)	63.5 ton
Track length (L)	4.5 m
Track width (B)	0.5 m

The pressure from the tank tracks over the ferry (P)is calculated as:

$$P = \frac{W}{2 * L * B} \quad (1)$$

3.3. Boundary conditions

As shown in **Figure 3** the ferry is subjected to the tank load simulated as the two paths illustrated in red colour. The own weight of the ferry is taken into consideration. The elastic supports are assigned to the lower deck to represent the effect of the water, usingcalculated specific elastic stiffness (K).

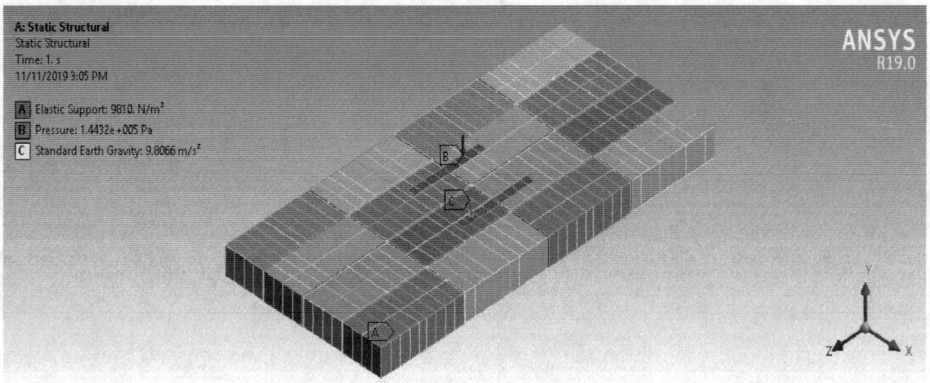

Figure 3: The boundary conditions

Following the principles of floating; the weight of the ferry is equal to the weight of the displaced volumeof water by the ferry[31]. The stiffness of the elastic supports differs from one case to another according to the density of the fluid, the stiffness of the submerged structure and its mass.

3.4. Elastic support stiffness calculation

In this model the lower deck of the ferry is supported with elastic supports with stiffness (K) so,

$$F_w = w_w \tag{2}$$

Where, (F_w)represents the weight of the ferry, and (w_w)represents the weight of the displaced water. The own weight of the steel ferry (F_w) is calculated using ANSYS software and(w_w) must be equal to that weight to keep the ferry floating. The weight of the displaced water is computed as follows:

$$w_w = F_b \times F_L \times \Delta \times \gamma_w \tag{3}$$

Where, (F_b) represents the breadth of the ferry without connections = 9.6 (m), (F_L)represents the length of the ferry without connections = 21 (m), (Δ)represents the draft of the ferry under its own weight, (γ_w) represents the density of water = 1000 (kg/m^3)
By substituting in equation (3), (Δ) is calculated. The stiffness of the elastic support (K) must make the deformation of the ferry under its own weight equal to (Δ). The stiffness is computed as follows:

$$w_w = k \times F_b \times F_L \times \Delta \tag{4}$$

By substituting in equations (3) and (4).

$$k = \gamma_w \tag{5}$$

4. Validation

The draft of the ferry from the simulation by ANSYS software is compared to the practical and mathematical results. The practical result is shown in **Figure 4**, showing the draft of the ferry under its own weight. The draft of the ferry under its own weight measured to be (26.8 cm) from field.

IOP Conf. Series: Materials Science and Engineering **974** (2020) 012008 doi:10.1088/1757-899X/974/1/012008

Figure 4: The practical value of the draft of the steel ferry

Using mathematical equations, the draft is determined as follows:

$$F_w = w_w \tag{6}$$

The own weight of the steel ferry equals (53698 Kg), so w_w must be equal to that weight to keep the ferry floating.

$$w_w = F_b \times F_L \times \Delta \times \gamma_w = F_w \tag{7}$$

$$\Delta = \frac{F_w}{F_b \times F_L \times \gamma_w} \tag{8}$$

By substituting in equation (8), (Δ) is calculated to be (26.59 cm).The numerical results from the simulation in ANSYS are based on the mathematical equations therefore; the results are the same as shown in **Figure 5** and matches with the practical results with a matching percentage of (99.2%).

Figure 5: The draft of the ferry under its own weight

5. Simulation of the ferry

5.1. Beam and shell elements simulation

The simulation of the ferry begins with determining the coordinates of every member. These coordinates are simulated as the key points in the graphical user interface of the software program. The stiffeners of the pontoon are simulated as beam elements. There are five different cross sections for the beam elements used in this model. On the other hand, the upper, lower and rounded side sheets are simulated as shell elements which are created from edges with the required thicknesses. After that, the pontoon is repeated by generatingtwo patterns; the first pattern repeats the finished pontoon in the X direction another three times to get the required four pontoons. The second pattern repeats these four pontoons in the Z direction another three times to get the required ferry which is composed of 16 floating pontoons as shown in**Figure 2**.

5.2. Elements contact

The elements are now generated, but still work individually, thereforethe elements must be connected together. To connect the elements there are a target and contact bodies. The shell elementsare considered the target and the beam elements (stiffeners) are considered the contact bodies.

5.3. Meshing

The smaller the element size is the more accurate the results are, but also it takes more time in solving the model. The smart meshing with different elements sizesis performed more than one time until the results have almost the same values. The meshing is refined at the critical position(tank tracks and the deck of the ferry around the tank tracks) to get more accurate results as shown in **Figure 6**.

Figure 6: The meshing of the ferry

5.4. Solution

The analysis is performed according to the previous simulation and the required results are determined. The total deformation and Von Misses stresses aretabulated in
Table *5*.

Table 5: The results of the ferry under MLC70

Total deformation	0.59407 m
Von Misses stresses	8.7215×10^7 Pa

Figure 7shows that the total deformationof the ferry under the static load of MLC70 is symmetric with a maximum value of (0.594 m).

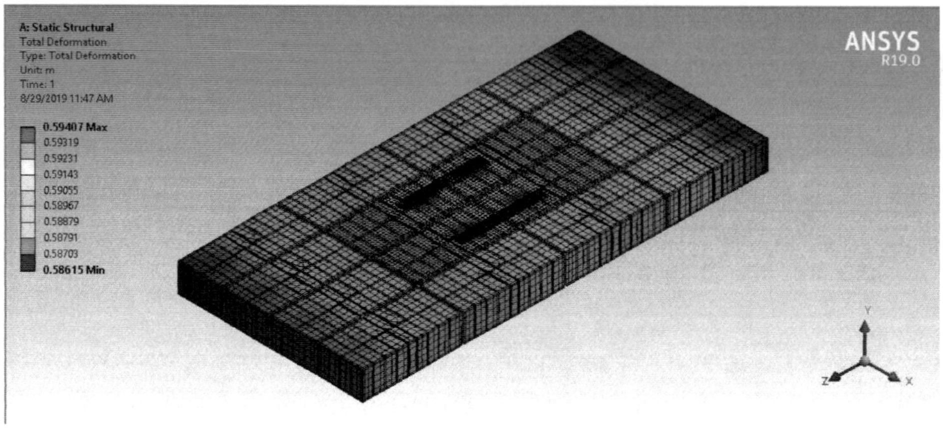

Figure 7: The total deformation of the ferry under MLC70

Figure 8 illustrates the Von Misses stresses. The ferry is affected by a symmetric manner of the stresses over the ferry under the load of the MLC70 with a maximum value at the tank tracks. The connections between the pontoons are affected by these applied loads, specially the transverse connections. The maximum value is (8.7215 $\times 10^7$ Pa).

Figure 8: The von-Misses stresses under MLC70

6. Optimization Process

There are many techniques of optimization and many are supported by ANSYS such as screening, multi-objective genetic algorithm,non-dominated sorting genetic algorithm (NSGA), non-linear programming by quadratic lagrangian, etc. According to the number and type of input parameters the used technique is determined automatically [32]. In our study the screening optimization is used to get the optimum weight of the ferry as well as possible as a rapid process. The Screening optimization method uses a simple approach based on sampling and sorting. This method supports multiple objectives and constraints as well as all types of input parameters. The screening method is the first process that is used in the new developed optimization strategy which is called the multiphase optimization strategy (MOST)[33, 34]. Usually, it is used for preliminary design, which may lead the designers to reach to the predicted results in a quick way instead of try and error methods. This flow chart illustrate the used technique to get the optimum design.

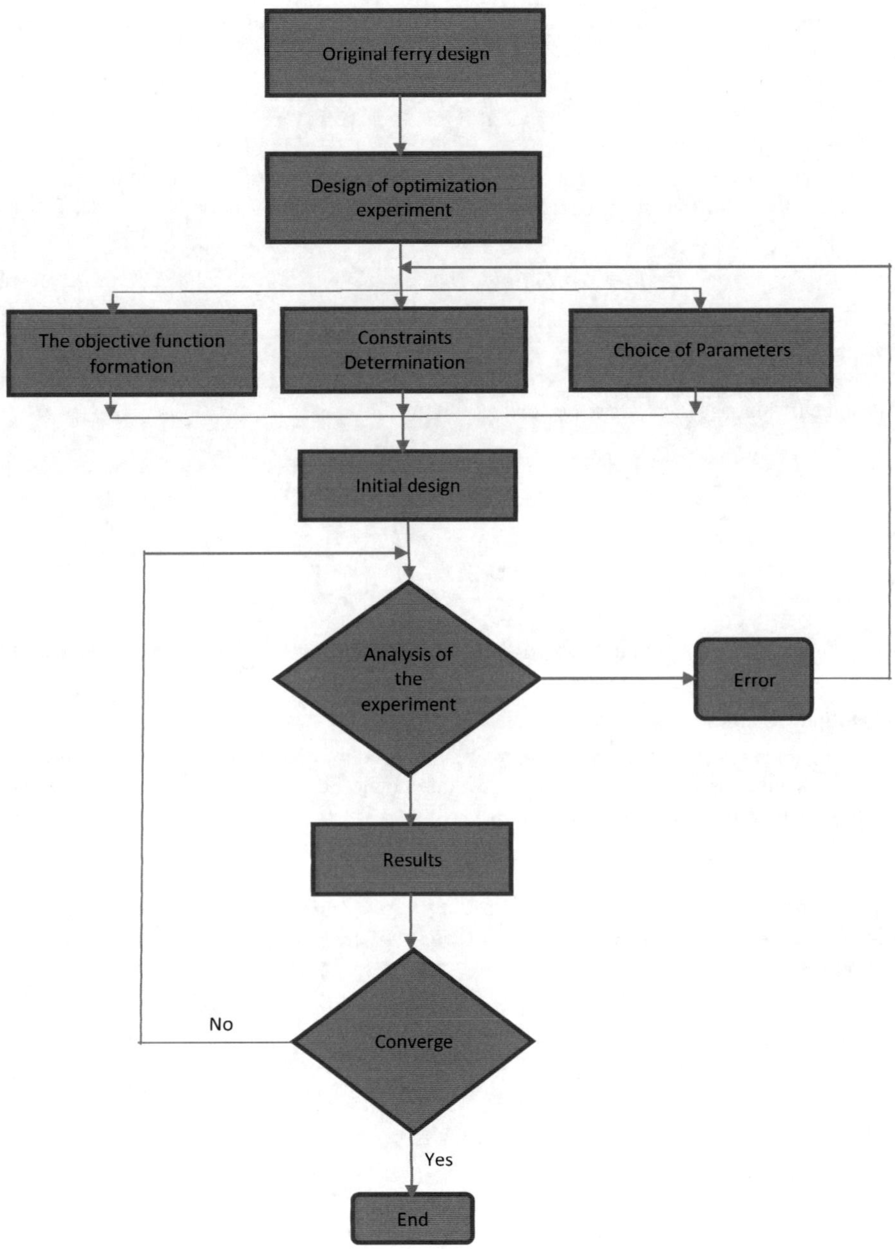

Figure 9: Optimization Flow Chart

6.1. Objective function

In order to apply the optimization process to the ferry, it is obligatory to determine in advance the objective function to be able to determine the parameters and the constraints. In this study, the optimization is performed at two stages; the cross sections are optimized first, then the shell thicknesses in the second stage. The objective function is to minimize the buoyancy factor (B F) that results in the reduction of the ferry weight. The following equation defines the buoyancy factor[27, 35-37].

$$The\ buoyancy\ factor\ (\text{B F}) = \frac{W_t}{V_t \times \gamma_w} \tag{9}$$

Where, (W_t)represents the total weight of the structure and(V_t)represents the total volume of the structure.Hence, the objective function is to minimize the buoyancy factor. Therefore, the objective function is defined as:

$$F_1(X): minimize\ B\ F = \frac{W_t}{V_t \times \gamma_w} \tag{10}$$

6.2. Constraints

The optimization problem is constrained under two constraints; the stress constraint (g_1) and the deformation constraint (g_2) which are defined as follows:

$$g_1 = \frac{f_{act}}{f_y} \leq 1 \tag{11}$$

$$g_2 = \frac{\Delta}{D_t} \leq 0.8 \tag{12}$$

Where, (f_{act}) represents the actual stress from the applied loads and (f_y) represents the strength of the materials which equals (2.4 t/cm^2) for steel 37.

6.3. Design variables

The design variables are the parameters, which are supposed to be optimized. These parameters are the cross section dimensions of the three angles sections (angle 80, angle 90 and angle 120) and two I-beams sections (I-beam 160 and I-beam 180).

$$W_i^L \leq W_i \leq W_i^u,\ i = 1,2,\ldots,6 \tag{13}$$

Where, (W_i) represents the cross sections of the angles (the 2 legs dimensions) for every angle (angle 80, angle 90 and angle 120) respectively, (W_i^L) represents the lower bound value for every parameter and(W_i^u) represents the upper bound value.

$$W_{fi}^L \leq W_{fi} \leq W_{fi}^u,\ i = 1,2,3,4 \tag{14}$$

Where, (W_{fi}) represents the flanges of the two I-beams (lower flanges, then upper flanges) respectively,(W_{fi}^L) represents the lower bound and (W_{fi}^u) represents the upper bound.

$$W_h^L \leq W_h \leq W_h^u,\ i = 1,2 \tag{15}$$

Where, (W_h) is the height of the two I-beams (I-beam 160 and I-beam 180) respectively, (W_h^L) represents the lower bound and(W_h^u) represents the upper bound.These parameters are optimized together with a thousand trial for each parameter to reach the optimum solution as well as possible. Every parameter resultsare determined through the optimization processand these resultsare chosen to be convergent to the objective function within the limits of the constraints. The results have a theoretical value that are not easy to be manufactured, but these values may be rounded to the values that can be manufactured.

IOP Conf. Series: Materials Science and Engineering **974** (2020) 012008 doi:10.1088/1757-899X/974/1/012008

7. Results

The results are generated from the chosen samples composing a thousand result of the optimization process as shown in **Figure 10**.

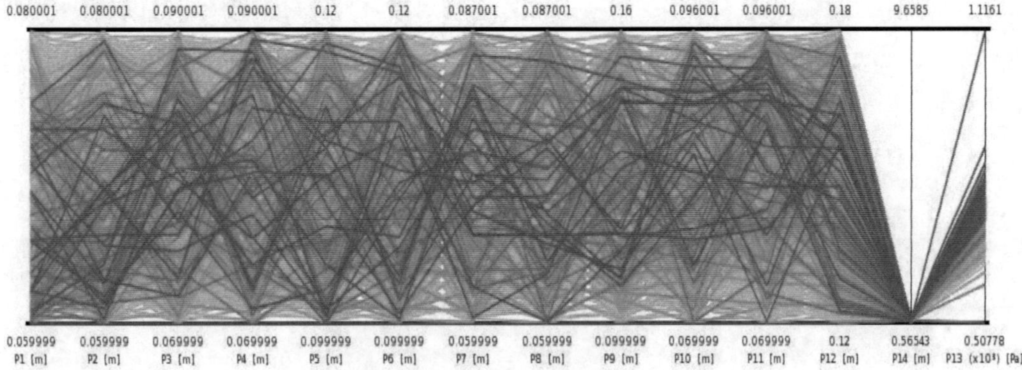

Figure 10: The whole optimization results

According to the design of the experiment these are the best three candidate results which are the most approached results to the optimum result. These candidate points achieve the objective function without exceeding the constraints limits. These results are tabulated in **Table 6**.

Table 6: The parameters values for the three candidate points

Parameter number	Candidate result 1	Candidate result 2	Candidate result 3
1	0.0657 (m)	0.06657 (m)	0.06905 (m)
2	0.0661 (m)	0.06146 (m)	0.06278 (m)
3	0.076 (m)	0.079 (m)	0.08526 (m)
4	0.08474 (m)	0.082554 (m)	0.078586 (m)
5	0.11375 (m)	0.11914 (m)	0.11197 (m)
6	0.11379 (m)	0.11756 (m)	0.1032 (m)
7	0.06814 (m)	0.682 (m)	0.08209 (m)
8	0.06825 (m)	0.6815 (m)	0.07674 (m)
9	0.11822 (m)	0.11864 (m)	0.14807 (m)
10	0.08698 (m)	0.7748 (m)	0.0879 (m)
11	0.07785 (m)	0.7842 (m)	0.08572 (m)
12	0.1627997 (m)	0.15549 (m)	0.15574 (m)

Figure 11 illustrates the candidate results with the constraints.

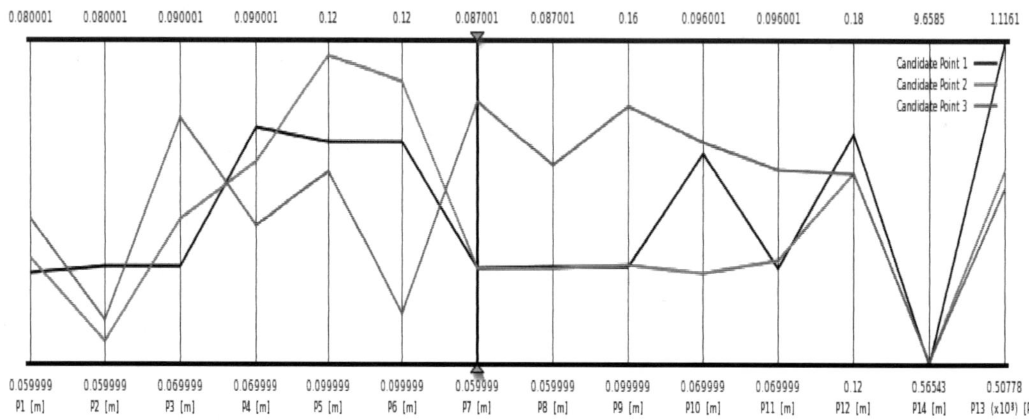

Figure 11: The candidate points

The lower horizontal axis represents the input parameters and the design variables and the upper horizontal axis represents the Von Misses stresses. The maximum value of Von Misses stresses is (1.1161×10^8Pa). The first candidate resultis used to apply the shell thickness optimization to reach the optimum solution. The upper, lower and side shells are used as design variables.

$$W_s^L \leq W_s \geq W_s^u,\ i = 1,2,....,6 \qquad (16)$$

Where, (W_s) represents the shell thickness for upper, lower and side shells respectively, (W_s^L) represents the lower bound and (W_s^u) represents the upper bound. The objective function is the same in stage 1. The values of the shell thicknesses before and after optimization are shown in **Table 7**.

Table 7: Shell parameters values before and after optimization

Parameter number	Original value	Optimized value
T1	0.006 (m)	0.006 (m)
T2	0.004 (m)	0.002 (m)
T3	0.004 (m)	0.002 (m)
T4	0.004 (m)	0.002 (m)
T5	0.004 (m)	0.002 (m)
T6	0.004 (m)	0.002 (m)

The thickness of the upper deck is not reduced as the stresses would increase. The lower and side shells are reduced. Figure illustrates the difference of weight before and after optimization.

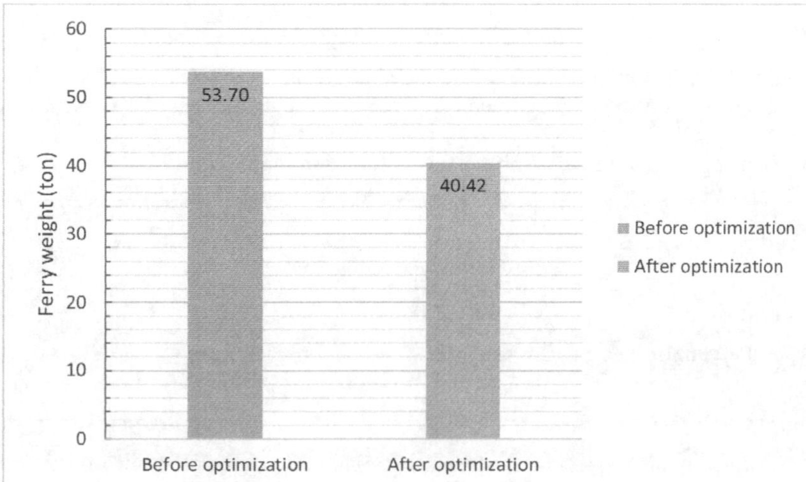

Figure 12: The comparison between weight before and after optimization

The optimization process of the ferry under the studied constraints and objective function reduces the weight to be optimum with a reduction of 24.7% from the original weight. The total deformation is also improved due to the reduction of weight. **Figure 13**shows the comparison of the total deformation before and after optimization.

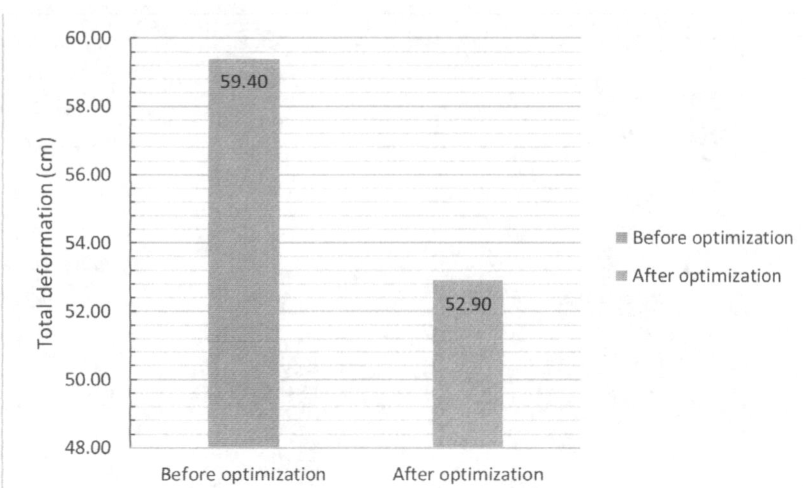

Figure 13: The comparison of total deformation

This chart illustrates that the buoyancy factor is minimized that results in the reduction of the ferry weight and the total deformation. It also illustrates that the effective constraint is the strength of the used material not the draft. If the material is changed with another one with a higher strength there will be better optimization results and more reduction in weight and cost of the ferry.

8. The redesign of the ferry

It is obvious that using material with more strength for the ferry results in more reduction in cross sections that result in further weight optimization. Different grades of the structural steel are used instead of steel 37 in addition to both the hybrid materials and the aluminium alloy (Al 6061) [38] also for the simulation of the ferry. The hybrid design is composed of the original steel stiffeners without optimization covered with aluminium sheets from all sides. This hybrid design reduces the weight in addition to the simplification of the manufacturing process. The properties of the used materials are shown in **Table 8**.

Table 8: The properties of the used materials

	Steel 44	Steel 52	Al 6061
Density	7850 kg/m^3	7850 kg/m^3	2770 kg/m^3
Tensile yield strength	2.8× 10^8 (Pa)	3.6×10^8 (Pa)	2.8×10^8 (Pa)
Compressive yield strength	2.8×10^8 (Pa)	3.6 ×10^8 (Pa)	2.8 ×10^8 (Pa)
Tensile ultimate strength	3.6 ×10^8 (Pa)	4.6 ×10^8 (Pa)	3.1 ×10^8 (Pa)

The material is changed in the model and the new constraints values are determined for every material. The new values of the constraints are the yield strength values for the used materials.
Figure 14shows the comparison between the ferry weight for the used materials after optimization and the original weight of the ferry.

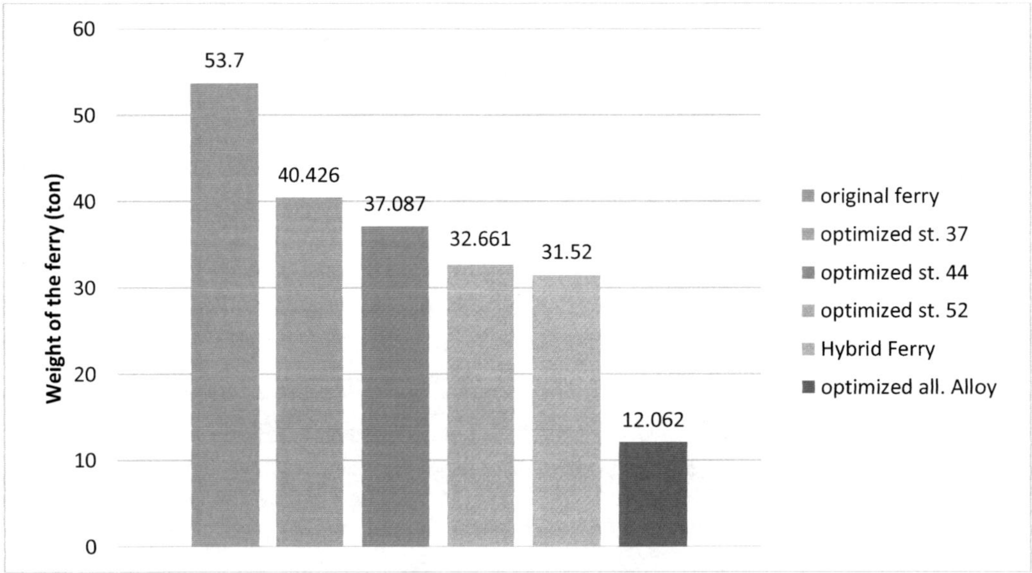

Figure 14: The Weight before and after optimization with different materials

For steel 37 the reduction in weight is about 24.7% as shown in**Figure 12**. For steel 44 the reduction of weight is about 30.9% of the original load, for steel 52 the reduction of weight is about 39.2% of the original load, for the hybrid design the reduction of weight is about 41.3% and for aluminium alloy the reduction of weight is about 77.5% of the original weight. This optimization values reduce the buoyancy factor and make it possible to increase the capacity of the ferry. This optimization values affect the total deformation as shown in **Figure 15**.

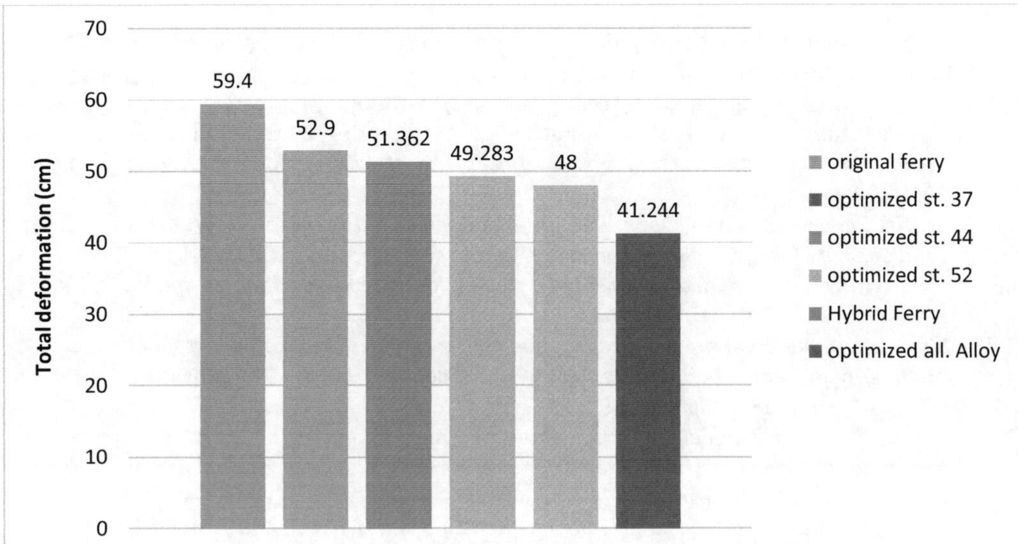

Figure 15: Total deformation before and after optimization with different materials

The deformation values are reduced as the total weight of the ferry is reduced.

9. Conclusion

In this study, the considered metallic ferry is simulated using ANSYS software. The ferry is loaded with a tank load (MLC-70). The model is verified with both practical and mathematical results. Optimization process is performed to obtain the minimum buoyancy factor and the optimum weight. Based on the aforementioned results, the buoyancy factor of the ferry is minimized and the ferry weight is reduced with a percentage of (24.7%) of the original weight for steel 37. Changing the grade of the structural steel to become steel 44 results in a reduction in weight of about (30.9%) of the original weight. The reduction in weight for steel 52 equals (39.2%) of the original weight. The hybrid design reduced the weight by about 41.3% from the original weight. Using aluminium alloy (Al 6061) instead of steel 37 results in a reduction of about (77.5%)of the original weight. Additionally, the total deformations are reduced for the whole optimized cases due to the reduction in weight. The strength of the used material has the effective constraint so, increasing the strength of the used material results in increasing the strength of the material and results in more reduction in weight.

References

1. Watanabe, E. and T. Utsunomiya, *Analysis and design of floating bridges.* Progress in Structural Engineering and Materials, 2003. **5**(3): p. 127-144.
2. Watanabe, E. and T. Utsunomiya, *Analysis and design of floating bridges.* Prog. Struct. Engng Mater, 2003: p. 127-144.
3. Zhang, J., et al., *Analytical Models of Floating Bridges Subjected by Moving Loads for Different Water Depths.* Journal of Hydrodynamics, Ser. B, 2008. **20**(5): p. 537-546.
4. Zhang, J., et al. *Simulation of a discrete pontoon floating bridge induced by moving loads.* in *Mechanics and Architectural Design: Proceedings of 2016 International Conference*. 2017. World Scientific.
5. Kim, T., et al., *A Study on Motion Control of Multiple Floating Units.*

6. Wu, J.-S. and P.-Y. Shih, *Moving-load-induced vibrations of a moored floating bridge.* Computers & structures, 1998. **66**(4): p. 435-461.
7. Shahrabi, M. and K. Bargi, *Numerical simulation of multi-body floating piers to investigate pontoon stability.* Frontiers of Structural and Civil Engineering, 2013. **7**(3): p. 325-331.
8. Khalifa, Y.A., *Study the structural system effect on the stability of floating metallic bridges.* 2008.
9. Erdal, O. and F.O. Sonmez, *Optimum Design of Composite Laminates for Maximum Buckling Load Capacity Using Simulated Annealing.* Composite Structures, 2005. **71**(1): p. 45-52.
10. Fu, S. and W. Cui, *Dynamic responses of a ribbon floating bridge under moving loads.* Marine Structures, 2012. **29**(1): p. 246-256.
11. Sun, J., et al., *An experimental investigation on the nonlinear hydroelastic response of a pontoon-type floating bridge under regular wave action.* SHIPS AND OFFSHORE STRUCTURES,, 2018. **13**: p. 233-243.
12. Hirono, Y., et al. *Positional displacement measurement of floating units based on aerial images for pontoon bridges*. in *International Conference on Advanced Engineering Theory and Applications*. 2016. Springer.
13. Cheng, Z., Z. Gao, and T. Moan, *Numerical Modeling and Dynamic Analysis of a Floating Bridge Subjected to Wind, Wave, and Current Loads.* Journal of Offshore Mechanics and Arctic Engineering, 2019. **141**(1): p. 011601.
14. Lie, H., et al. *Numerical modelling of floating and submerged bridges subjected to wave, current and wind*. in *ASME 2016 35th International Conference on Ocean, Offshore and Arctic Engineering*. 2016. American Society of Mechanical Engineers.
15. Sha, Y., et al., *Numerical investigations of the dynamic response of a floating bridge under environmental loadings.* Ships and Offshore Structures, 2018. **13**(sup1): p. 113-126.
16. Salem, A.I., *Weight and cost multi-objective optimization of hybrid composite sandwich structures.* 2016, University of Dayton.
17. rey Horn, J., N. Nafpliotis, and D.E. Goldberg. *A niched Pareto genetic algorithm for multiobjective optimization*. in *Proceedings of the first IEEE conference on evolutionary computation, IEEE world congress on computational intelligence*. 1994. Citeseer.
18. Zitzler, E. and L. Thiele, *An evolutionary algorithm for multiobjective optimization: The strength pareto approach.* TIK-report, 1998. **43**.
19. Corne, D.W., J.D. Knowles, and M.J. Oates. *The Pareto envelope-based selection algorithm for multiobjective optimization*. in *International conference on parallel problem solving from nature*. 2000. Springer.
20. Törn, A. and A. Žilinskas, *Global optimization*. Vol. 350. 1989: Springer.
21. Coello, C.A.C., *A comprehensive survey of evolutionary-based multiobjective optimization techniques.* Knowledge and Information systems, 1999. **1**(3): p. 269-308.
22. Koumousis, V.K. and S. Arsenis, *Genetic algorithms in optimal detailed design of reinforced concrete members.* Computer - Aided Civil and Infrastructure Engineering, 1998. **13**(1): p. 43-52.
23. KIM, H. and H. ADELI, *Discrete cost optimization of composite floors using a floating-point genetic algorithm.* Engineering Optimization, 2001. **33**(4): p. 485-501.
24. Rigo, P. and C. Fleury, *Scantling optimization based on convex linearizations and a dual approach—Part II.* Marine structures, 2001. **14**(6): p. 631-649.
25. Shafieefar, M. and A. Rezvani, *Mooring optimization of floating platforms using a genetic algorithm.* Ocean Engineering, 2007. **34**(10): p. 1413-1421.
26. Brommundt, M., et al., *Mooring system optimization for floating wind turbines using frequency domain analysis.* Energy Procedia, 2012. **24**: p. 289-296.

27. Fathallah, E., et al., *Design optimization of lay-up and composite material system to achieve minimum buoyancy factor for composite elliptical submersible pressure hull.* Composite Structures, 2015. **121**: p. 16-26.
28. Fathallah, E., et al., *Multi-Objective Optimization of Composite Elliptical Submersible Pressure Hull for Minimize the Buoyancy Factor and Maximize Buckling Load Capacity.* Applied Mechanics and Materials, 2014. **578**: p. 75-82.
29. Russell, B.R., et al., *Reconceptualization and optimization of a rapidly deployable floating causeway.* Journal of Bridge Engineering, 2013. **19**(4): p. 04013013.
30. *ANSYS, Inc. Products Release 19.0.* 2019.
31. King, R., *Principles of flotation*. 1982: South African Institute of Mining and Metallurgy Johannesburg.
32. *ANSYS, Inc. Release Notes*. 2018.
33. Collins, L.M., et al., *A strategy for optimizing and evaluating behavioral interventions.* Annals of Behavioral Medicine, 2005. **30**(1): p. 65-73.
34. Collins, L.M., S.A. Murphy, and V. Strecher, *The multiphase optimization strategy (MOST) and the sequential multiple assignment randomized trial (SMART): new methods for more potent eHealth interventions.* American journal of preventive medicine, 2007. **32**(5): p. S112-S118.
35. Elsayed, F., et al. *Optimal Design Analysis of Composite Submersible Pressure Hull*. in *Applied Mechanics and Materials*. 2014. Trans Tech Publ.
36. Helal, M., et al., *Numerical Analysis of Sandwich Composite Deep Submarine Pressure Hull Considering Failure Criteria.* Journal of Marine Science and Engineering, 2019. **7**(10): p. 377.
37. Fathallah, E., *Finite Element Modelling and Multi-Objective Optimization of Composite Submarine Pressure Hull Subjected to Hydrostatic Pressure.* Materials Science Forum, 2019. **953**: p. 53-58.
38. Hellier, A., P. Chaphalkar, and G. Prusty, *Fracture Toughness Measurement for Aluminium 6061-T6 using Notched Round Bars*. 2017.

IOP Conf. Series: Materials Science and Engineering **974** (2020) 012009 doi:10.1088/1757-899X/974/1/012009

Military Technical College

Kobry El-Kobbah,

Cairo, Egypt

13th International

Conference on Civil and

Architecture Engineering

ICCAE-13-2020

Multi-objective optimization of sandwich composite pressure hull for decreasing weight and drag force and increasing buckling load capacity

Mahmoud Helal[1,2] and Elsayed Fathallah[3,4*]

[1]Department of Mechanical Engineering, Faculty of Engineering, Taif University, Taif 21974, Saudi Arabia

[2]Production and Mechanical Design Dept., Faculty of Engineering, Mansoura University, Mansoura 35516, Egypt

[3]Department of Civil Engineering, Military Technical College, Cairo, Egypt

[4]Ships and Submarines Engineering Department, Military Technical College, Cairo, Egypt.

* saidhabib2000@mtc.edu.eg & saidhabib2000@hotmail.com

Abstract:

In underwater applications, space vehicles, and aircrafts, the weight becomes an important factor. Additionally, the design of composites structures greatly depends on the number of layers and the fiber orientation angle. Therefore, this work presents the optimization of sandwich composite pressure hull in order to minimize (weight and drag force) and maximize the buckling load capacity using ANSYS Parametric Design Language (APDL). Tsai-Wu and maximum stress failure criteria were incorporated for predicting the first-ply failure. The major and minor radius of the pressure hull, the ring and long beams dimensions, the fiber orientation angle and layer thickness are taken as design variables. The results illustrated that, core thickness (T_{core}) has a great effect to resist the shell buckling. While, has a little effect on both Tsai-Wu and maximum stress failure index.

Keywords: Multi-objective optimization; Buckling; drag force; pressure hull; Tsai-Wu.

IOP Conf. Series: Materials Science and Engineering **974** (2020) 012009 doi:10.1088/1757-899X/974/1/012009

1. Introduction

The pressure hulls are one of the most main structures of the submarine which provide high load capacity for electronic systems and buoyancy [1]. The optimization and buckling behavior have been attracted some recent attention [1-17]. Additionally, Zhang et al. [18], investigated the egg-shaped pressure hulls to improve the low buckling resistance, the difficult interior arrangement and the poor hydrodynamics of the spherical pressure hull. Vosoughi et al. [19], investigated the optimum stacking sequences of laminated composite plate to maximize the buckling load by achieving the optimum fibers orientations. Significant research work had been presented so far in this field. Among them, Mian et al. [20], presented the design optimization procedure for a composite pressure vessel, considering both maximum stress and Tsai-Wu failure criteria. Pan et al. [21], studied the optimization of composite cylinder due to hydrostatic pressure. The ply orientation and thickness are studied. Furthermore, Fathallah and Helal [22], optimized deep cross elliptical pressure hull taking into consideration both buckling and failure criteria. Imran et al. [23], optimized a composite pressure hull to minimize the buoyancy factor. Lund [24], Optimized a laminated composite structures using failure criteria. Moreover, Fathallah et al. [25], investigated the optimization of elliptical composite pressure hull for minimizing buoyancy factor. In this wok, a sandwich composite pressure hull is proposed. The view of the model is shown in **Fig. 1**. The multi-objective optimization methodology for the pressure hull is presented. The optimization is established for maximizing the buckling load capacity and minimizing the drag force and (weight/displacement) ratio, of the pressure hull.

IOP Conf. Series: Materials Science and Engineering **974** (2020) 012009 doi:10.1088/1757-899X/974/1/012009

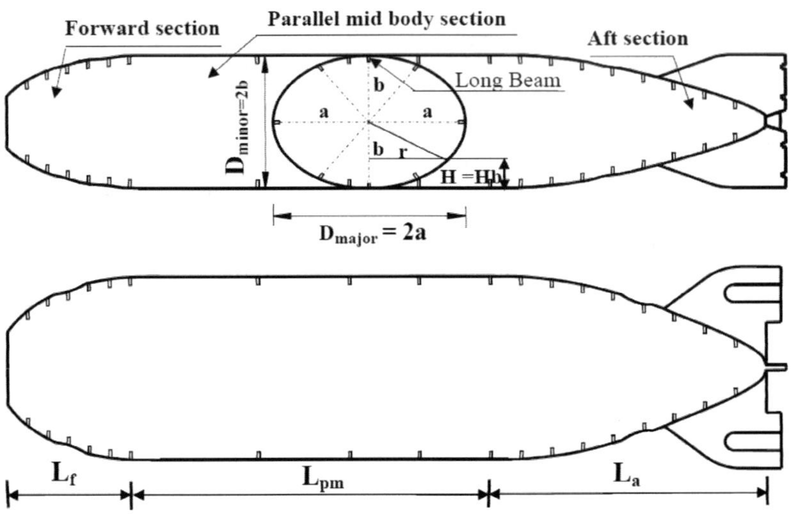

Fig. 1 Parameterization of the submersible pressure hulls geometry.

1.1 Buckling in multilayer-sandwich composite pressure hull

In this work, a scalar multiple λ of the design load, the so-called buckling load strength factor, is introduced to identify pre-buckling of the multilayer sandwich submersible pressure hulls and defined as [1]:

$$\lambda = \frac{P_S}{P_{act}} \quad (1)$$

Where: P_{act} is the actual load and P_s is the structural critical buckling strength. The buckling will occur when $\lambda < 1$ and (P_s) less than (P_{act}). The critical value of the pressure P_{cr} that will cause buckling is determined using the following equations [22, 26]:

$$P_{cr} = \left(\frac{R}{\left(n^2 + 0.5\left(\frac{m\pi R}{L} \right)^2 \right)} \right) \frac{\begin{bmatrix} C_{11} & C_{12} & C_{13} \\ C_{21} & C_{22} & C_{23} \\ C_{31} & C_{32} & C_{33} \end{bmatrix}}{\begin{bmatrix} C_{11} & C_{12} \\ C_{21} & C_{22} \end{bmatrix}} \quad (2)$$

Where: L, R, m and n are the length, the radius, the number of buckle half waves in the axial direction and the number of buckle waves in the circumferential direction, respectively. The coefficients C, A, B and D are defined as follows:

$$C_{11} = A_{11}\left(\frac{m\pi}{L}\right)^2 + A_{66}\left(\frac{n}{R}\right)^2, C_{22} = A_{22}\left(\frac{n}{R}\right)^2 + A_{66}\left(\frac{m\pi}{L}\right)^2$$

$$C_{33} = D_{11}\left(\frac{m\pi}{L}\right)^4 + (4D_{66} + 2D_{12})\left(\frac{m\pi}{L}\right)^2\left(\frac{n}{R}\right)^2 + D_{22}\left(\frac{n}{R}\right)^4$$

$$+\frac{A_{22}}{R^2} + 2\frac{B_{22}}{R^2}\left(\frac{n}{R}\right)^2 + 2\frac{B_{12}}{R^2}\left(\frac{m\pi}{L}\right)^2, C_{12} = C_{21} = \left(A_{12} + A_{66}\right)\left(\frac{m\pi}{L}\right)\left(\frac{n}{R}\right) \quad (3)$$

$$C_{23} = C_{32} = \left(B_{12} + 2B_{66}\right)\left(\frac{m\pi}{L}\right)^2\left(\frac{n}{R}\right) + \frac{A_{22}}{R}\left(\frac{n}{R}\right) + B_{22}\left(\frac{n}{R}\right)^3$$

$$C_{13} = C_{31} = \frac{A_{12}}{R}\left(\frac{m\pi}{L}\right) + B_{11}\left(\frac{m\pi}{L}\right)^3 + \left(B_{12} + 2B_{66}\right)\left(\frac{m\pi}{L}\right)\left(\frac{n}{R}\right)^2$$

Where: A_{ij}, B_{ij} and D_{ij} are the stiffness components including extensional stiffness, coupling stiffness and bending stiffness coefficient matrix, respectively. The elements of the stiffness matrices are:

$$A_{ij} = \sum_{k=1}^{n}(\bar{Q}_{ij})_k(Z_k - Z_{k-1}), B_{ij} = \frac{1}{2}\sum_{k=1}^{n}(\bar{Q}_{ij})_k(Z^2{}_k - Z^2{}_{k-1}), D_{ij} = \frac{1}{3}\sum_{k=1}^{n}(\bar{Q}_{ij})_k(Z^3{}_k - Z^3{}_{k-1}) \quad (4)$$

Where: n is the number of different plies in the stacking sequence and Z_k, Z_{k-1} are the upper and lower (Z) coordinate of the k_{th} ply layer as shown in Fig. 2. $\bar{Q}_{ij}$ are the elements of the transformed reduced stiffness matrix $[\bar{Q}]$ and are defined as in [27]: The optimization can be achieved to obtain the highest buckling strength factor. The number of plies can be varied as well as their orientation. Since the fiber orientation has no influence on the mass function.

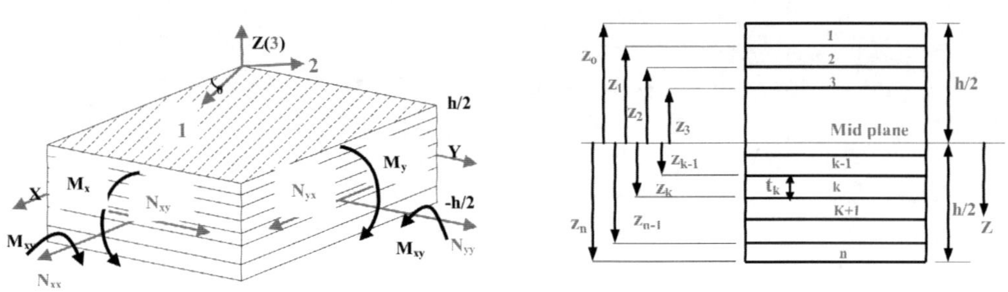

a) A composite laminate subjected to forces and moments. b) Enlarged view of laminate cross-section

Fig. 2 Laminated composite shell and coordinate locations of plies in a laminate.

The successful design requires an efficient and safe use of materials. Therefore, theories are needed to develop and compare the state of the stresses and the strains in the material [28]. In this study both Tsai-Wu and maximum stress failure criteria were incorporated for predicting the first-ply failure. and are defined as in [1]. The von Mises yielding criteria is employed here to assess the capability of the core materials to withstand the yielding failure.

1.2 Drag Estimation and power requirements

The proper hydro-dynamic design is necessary to achieve the effective performance of the submarine. An improper shape can cause excessive noise, drag and instability. Drag is the hydro-dynamic force exerted on the hull body in a direction opposite to its velocity. The propulsive power requirement is proportional to the drag and times the velocity divided by the propulsive efficiency. The drag is consist of two components, the form drag (function of the shape and frontal area) and friction drag (function of the speed and wetted surface area) [29]. For drag estimation, the following formula has been used [30]:

$$D = \frac{1}{2}\rho V^2 C_V S \tag{5}$$

Where: D denote the submarine drag in Newton (N), ρ denote the density of the fluid in kg/m^3, V denote the velocity in m/s, S denote the wetted surface area of the vehicle in m^2 and C_V denote the coefficient of viscous resistance for the smooth hull. There are three methods used to compute the coefficient of viscous resistance (C_V) namely; Virginia Tech, MIT and G&J method [31]. According to G&J method, the coefficient of the viscous resistance, C_V, can be calculated as [32]:

$$C_V = C_F\left[1 + 0.5\left(\frac{d}{L}\right) + 3\left(\frac{d}{L}\right)^3\right] \tag{6}$$

In MIT method, C_V defined as:

$$C_V = C_F\left[1 + 1.5\left(\frac{d}{L}\right)^{\frac{3}{2}} + 7\left(\frac{d}{L}\right)^3 + 0.002(C_P - 0.6)\right] \tag{7}$$

Where: d denote the maximum body diameter in meters, L denote the total length in meters, C_F denote the bare hull skin friction drag coefficient and defined as a function of Reynolds number (R_n) as [33]:

IOP Conf. Series: Materials Science and Engineering **974** (2020) 012009 doi:10.1088/1757-899X/974/1/012009

$$C_F = \frac{0.075}{(\log_{10} R_n - 2)^2} \tag{8}$$

Where: C_P denote the prismatic coefficient and can be defined as:

$$C_P = \frac{V_d}{\pi \left(\frac{d}{2}\right)^2 L} \tag{9}$$

Where: V_d denote the displacement volume of the submarine in m^3, R_n denote the Reynolds number and can be defined as:

$$R_n = \frac{\rho V L}{\mu} \tag{10}$$

Where: ρ is the density of the fluid in kg/m^3, V is the velocity in m/s, L is the overall length of the submarine pressure hull in meters and μ is the dynamic viscosity of the fluid in kg/ (m.s). In this study, (V_d, S) will be calculated using volume and surface measurement function in ANSYS.

2. Submarines pressure hulls materials and simulation

The model constructed from Boron/Epoxy B(4)/5505 composites with the lay-up composed of sixteen plies (n = 16) having equal thicknesses (t_1) and fiber orientation angle (α) and one core layer with thickness (T_{core}). Low density PVC foam is used for the core layer. The material properties and strength parameters of B(4)/5505 Boron/Epoxy and PVC foam are given in **Table 1** [25, 34, 35]. **Fig.3** illustrates the relationship between the collapse depth and buoyancy factor for a visual comparison of different structural materials. [36]. The sandwich structure is modeled using Shell 281 element. The Beam 189 element is used for ring and long beams [37, 38].

Table 1. Strengths of unidirectional composites and material properties of the sandwich components [34, 35, 39, 40].

Material	Material and strength properties
B(4)/5505 Boron/Epoxy	E_{11}= 204GPa, E_{22}= 18.5GPa, , E_{33}= 18.5GPa, G_{12}= 5.59GPa, G_{13}= 5.59GPa, υ_{12}=0.23, X_t=1260MPa, X_c=2500Mpa, Y_t=61MPa, Y_c=202MPa,S=67 MPa, ρ= 2000 kg/m^3
H200	E=250MPa, G=73MPa,υ=0.3, X_t=7.1MPa, X_c=5.4Mpa, Y_t=7.1MPa, Y_c=5.4MPa,S=3.5MPa, ρ= 200 kg/m^3

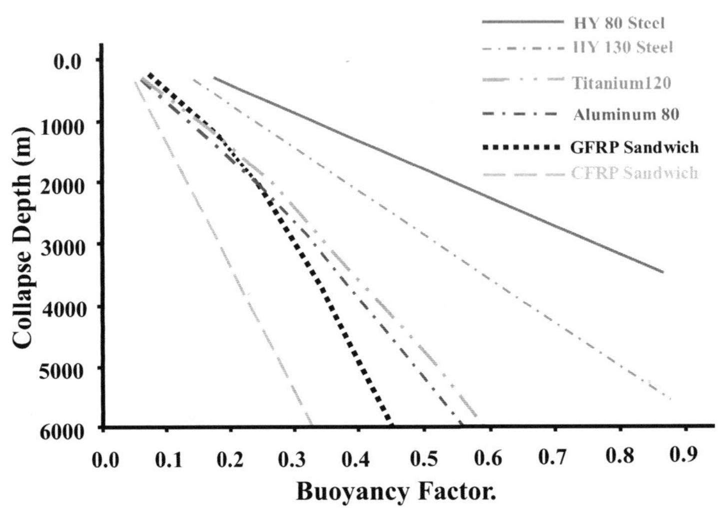

Fig.3 Weight to displacement ratio Vs Collapse depth for stiffened cylinders.

The boundary conditions applied here as in [1, 41]. The pressure hull is loaded by external pressure ($P = \rho gh$). Where ρ denotes the density of the sea water, g is the acceleration due to the gravity and *h,* represents the operating depth. Two failure criteria, Tsai-Wu and maximum stress are used to evaluate the failure of the pressure hull [25]. **Fig. 4** shows the final layer stacking for the global model.

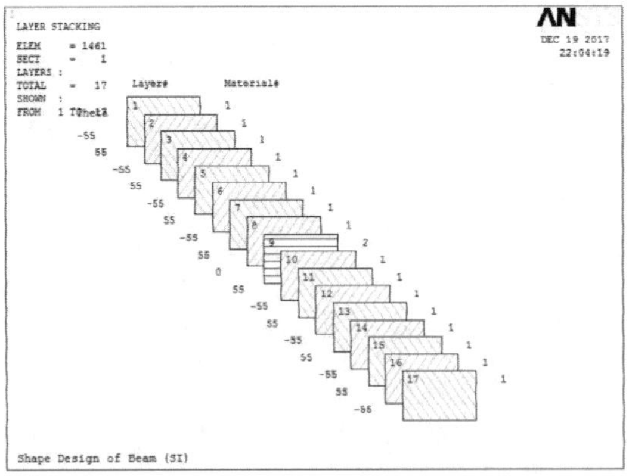

a) Layer stacking

Fig. 4 The composite pressure hull layer stacking.

3. Structural design optimization

Fig.5 shows the flow chart of the multi-objective optimization procedure. The optimization design is described as:

Objective function: F(X): Minimize (weight/displacement ratio (B.F) and drag force *(D_F)*) and maximize (buckling capacity (λ)).

Design constraints:

$$FS(i) \geq 1, \; i = 1,2,3,......n \tag{11}$$

Where: FS represents the factor of safety and must be greater than one to avert the first-ply failure of the angle-ply laminated for each ith layer and equals to the inverse of the failure index for both Tsai-Wu and maximum stress failure criteria and *(n)* is the number of layers.

$$\frac{\sigma}{\sigma_y} - 1 \leq 0 \tag{12}$$

Where: σ and σ_y are the actual stress and yielding strength in the core layer, respectively.

$$H \leq H_{\max} \tag{13}$$

Where: *H*, and H_{max} represent the operating depth and the maximum operating depth, respectively.

Design variables:

The composite pressure hull diameters are taken as:

$$D_i^L \leq D_i \leq D_i^U, i = \min, \max \tag{14}$$

Where: D_i, D_i^L and D_i^U are the ith submersible pressure hull diameters and its upper and lower limits, respectively.

IOP Conf. Series: Materials Science and Engineering **974** (2020) 012009 doi:10.1088/1757-899X/974/1/012009

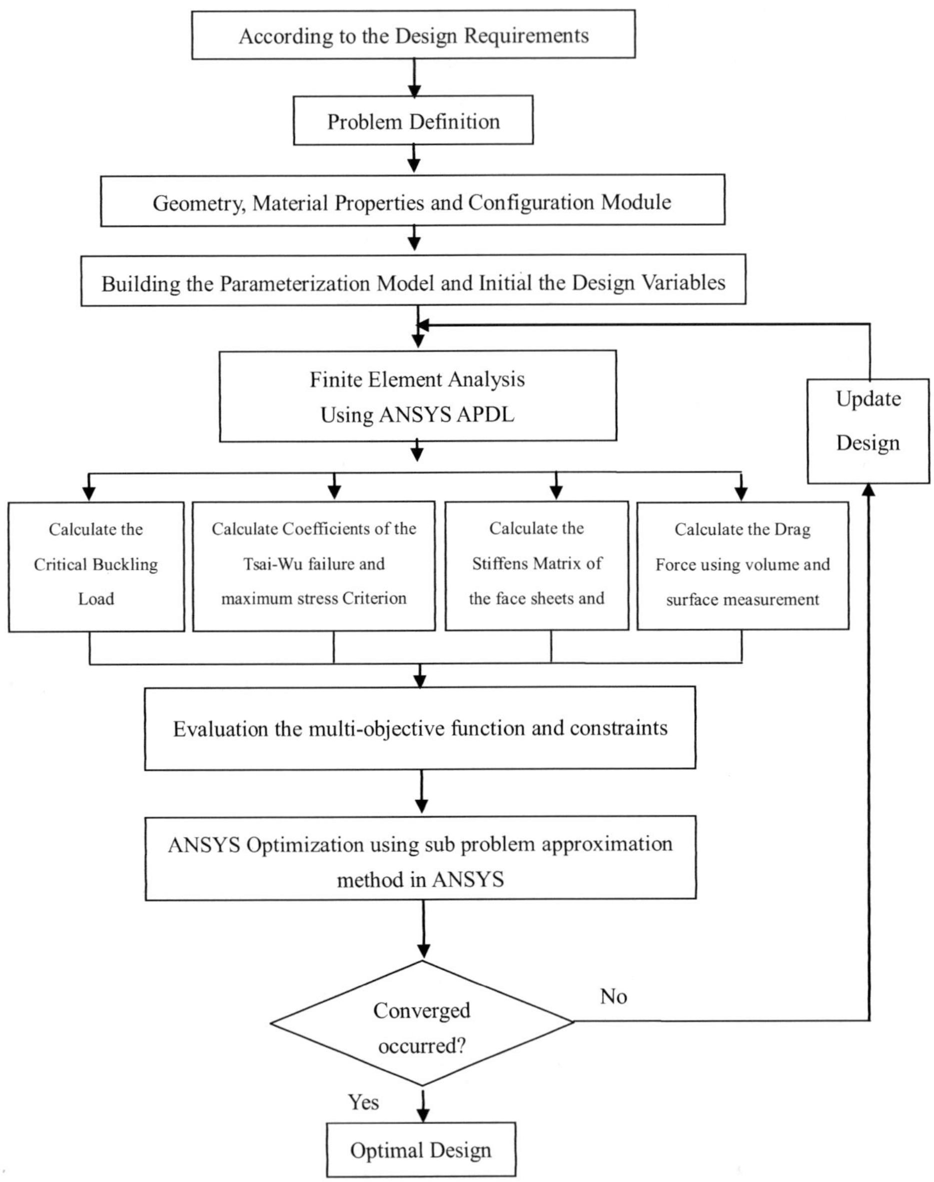

Fig.5 Flow chart layout of multi-objective optimization procedures

For core thickness:

$$T^L \le T_{core} \le T^U \tag{15}$$

Where: T_{core}, T^L and T^U represent the core thickness of pressure hull, and its lower and upper limits, respectively. For ply thicknesses:

$$t^L \le t_i \le t^U,\ i = 1,2,3,\ldots,n \quad (16)$$

Where: t_i, t^L, t^U represent the ith thickness for the lower and upper bounds of the individual ply thicknesses, respectively.

$$\alpha^L \le \alpha_i \le \alpha^U,\quad i = 1,2,3,\ldots,n \quad (17)$$

Where: α_i, α^L, α^U represent the ith orientation angle of each layer and their lower and upper limits, respectively.

4. Results and discussions

The results of Multi-objective optimization are summarized in **Table 1**. The drag force is equals to 2.19925×10^{12} N, buckling strength factor (λ) equals 118, at (*B.F*) equals 0.47.The maximum magnitude of deflection value (δ_{max}) equal to 2.18 mm. **Fig. 6** illustrates the displacement distribution in *X*, *Y* and *Z* directions for the model. The maximum displacement occurs at the centroids regions of the composite pressure hull and occurs in vertical direction with value equals to 2.14 mm. The optimum configuration of the sandwich composite pressure hull can overcome all structural failures until operating depth (*H*) equals to 255 m.

Table 2. Results of the optimal design of the sandwich composite deep pressure hull.

FS_{TWSR_1}	FS_{TWSR_2}	FS_{TWSR_3}	FS_{TWSR_4}	FS_{TWSR_5}	FS_{TWSR_6}	FS_{TWSR_7}	FS_{TWSR_8}
1.59	1.63	1.60	1.64	1.61	1.65	1.62	1.65
FS_{TWSR_10}	FS_{TWSR_11}	FS_{TWSR_12}	FS_{TWSR_13}	FS_{TWSR_14}	FS_{TWSR_15}	FS_{TWSR_16}	FS_{TWSR_17}
1.73	1.72	1.70	1.69	1.67	1.66	1.65	1.64
FS_{MAXF_1}	FS_{MAXF_2}	FS_{MAXF_3}	FS_{MAXF_4}	FS_{MAXF_5}	FS_{MAXF_6}	FS_{MAXF_7}	FS_{MAXF_8}
1.76	1.78	1.77	1.79	1.78	1.80	1.78	1.81
FS_{MAXF_10}	FS_{MAXF_11}	FS_{MAXF_12}	FS_{MAXF_13}	FS_{MAXF_14}	FS_{MAXF_15}	FS_{MAXF_16}	FS_{MAXF_17}
1.88	1.86	1.84	1.82	1.80	1.79	1.77	1.75
Maximum deflection (δ_{MAX})	LAY9_S_{XMAX}	LAY9_S_{XMIN}	LAY9_S_{YMAX}	LAY9_S_{YMIN}	D_{max}	D_{min}	t_1
0.00218 (m)	158772(Pa)	-28654 (Pa)	219556.6 (Pa)	-69869 (Pa)	2.0838(m)	1.8842(m)	1.59(mm)
Buckling strength factor (λ)	L_{pmb}(m)	T_{cor} (m)	α	Operating depth (*H(m)*)	LAY9_von Mises (Pa)	Drag Force (N)	B.F
118	2.80	0.091	55°	255.00	1636834.95	2.19925×10^{12}	0.47

IOP Conf. Series: Materials Science and Engineering **974** (2020) 012009 doi:10.1088/1757-899X/974/1/012009

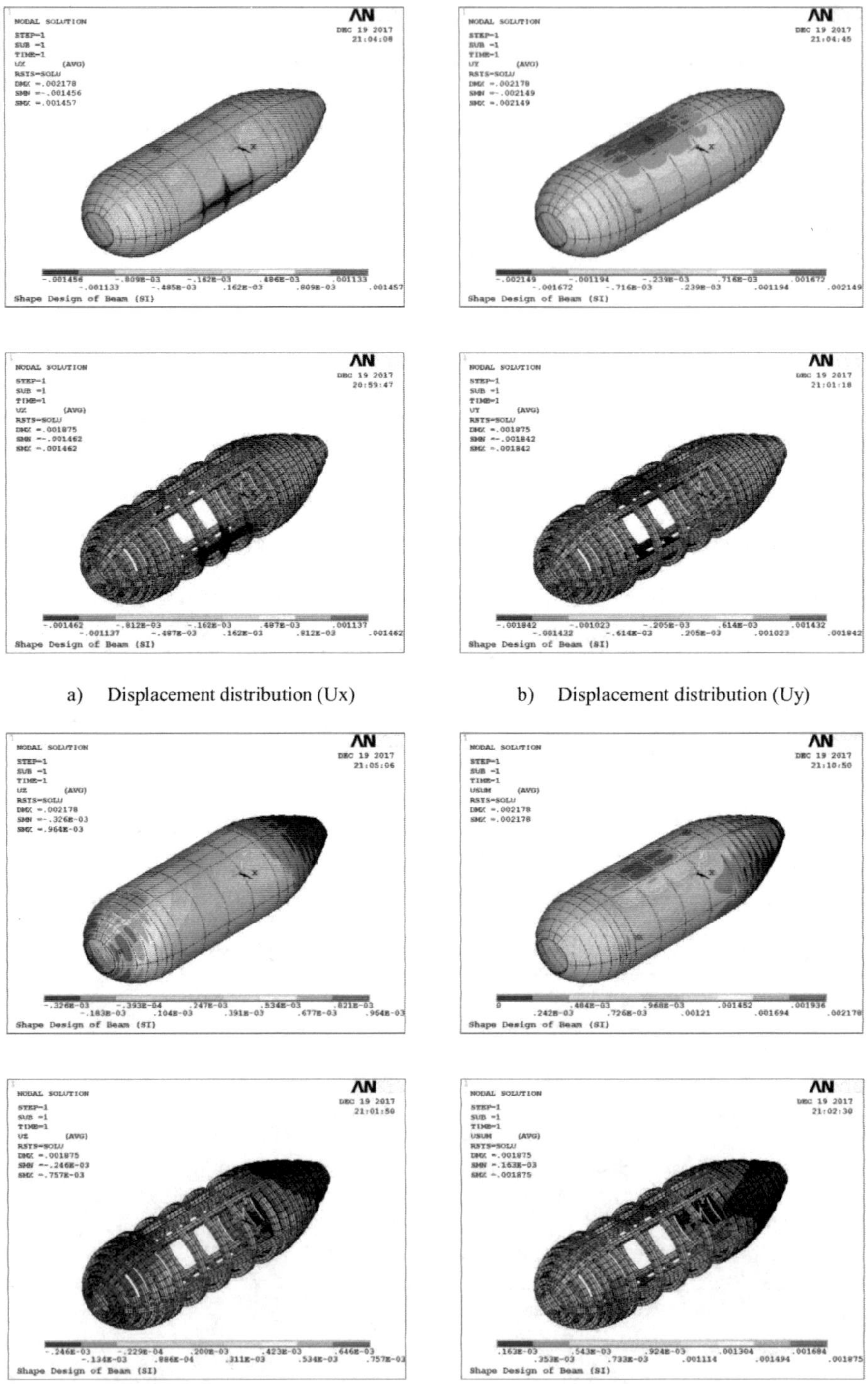

a) Displacement distribution (Ux) b) Displacement distribution (Uy)

c) Displacement distribution (Uz) d) Displacement distribution ($U_{magnitude}$)

Fig. 6 Displacement distribution (Ux,Uy, Uz and $U_{magnitude}$) in composite pressure hull.

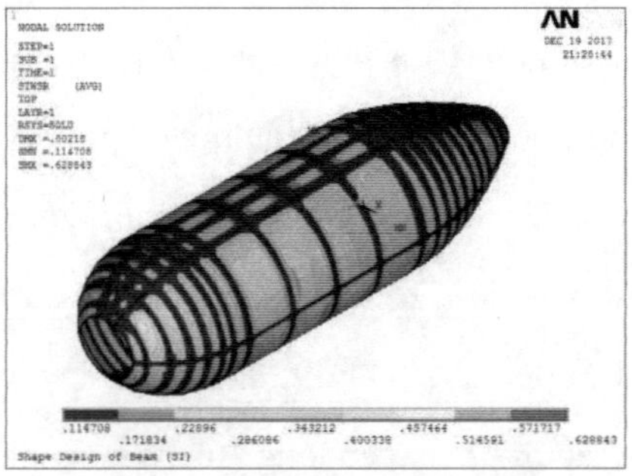

a) Tsai-Wu failure distribution at Ply-1

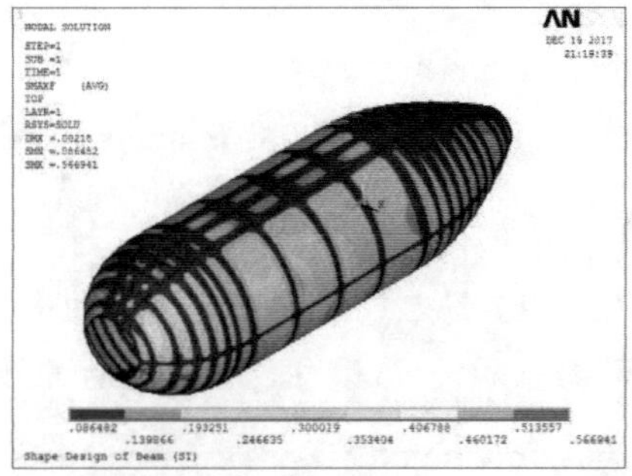

b) Maximum stress failure distribution at Ply-1

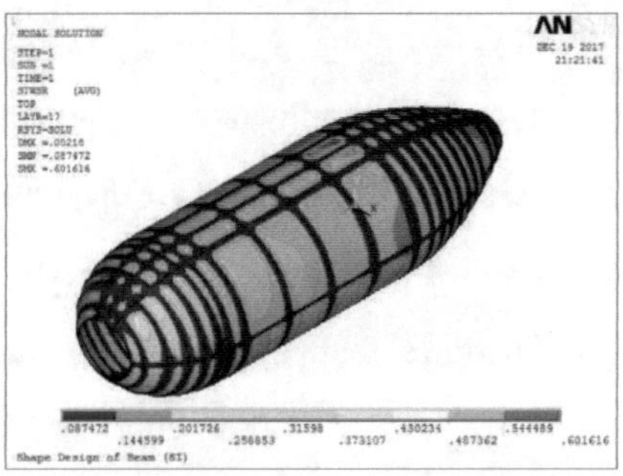

c) Tsai-Wu failure distribution at Ply-17

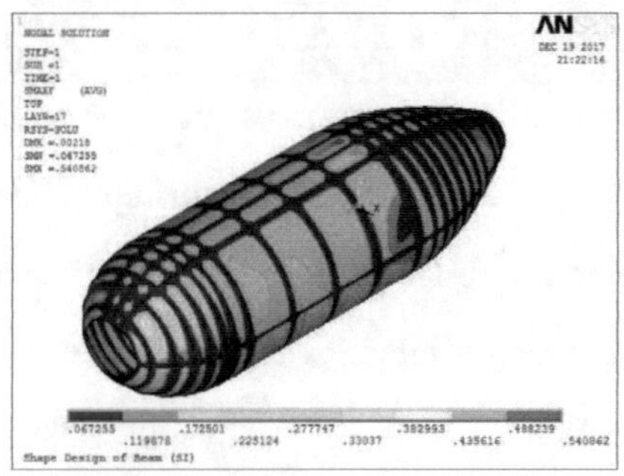

d) Maximum stress failure distribution at Ply-17

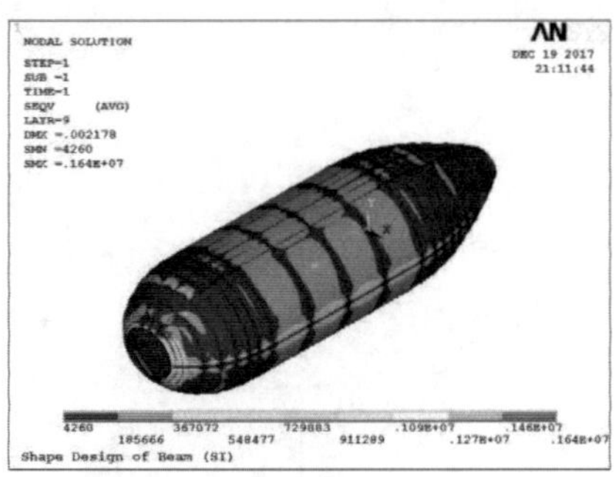

e) von Mises stress distribution at core layer (Ply-9)

Fig. 7 Tsai-Wu, maximum stress and von Mises distribution in composite pressure hull.

Also, **Table 2** illustrated that, material failure incorporating both Tsai-Wu and maximum stress failure criteria were considered together for predicting the first-ply failure. The minimum factor of safety (FS_{TWSR} and FS_{MAXF}) are occurring in the lower face at ply-1 due to maximum tensile stresses and equal 1.59 and 1.76, respectively. At upper face, the minimum values of F_{TWSR} and F_{MAXF} are occurring at ply-17 due to maximum compressive stresses with value equal to 1.64 and 1.75, respectively. The optimized orientation fiber α equals to (55°) with T_{core} of about 91 mm. **Fig.** 7 (a, b, c and d) shows the Tsai-Wu and maximum stress failure distribution at critical Plies (Ply-1 and pl-17). The maximum Tsai-Wu and maximum stress failure at ply-1 will be initiates first at the middle of the left and right sides of the composite pressure hull. For ply-17, the failure will be initiates first at the upper and lower regions of the composite pressure hull. **Fig.** 7 -e shows the von Mises stress distribution for core layer (Ply-9) and illustrated clearly that, the maximum von Mises stresses occurs at the regions between long and ring beams.

4.1 Effect of fiber orientation on design variables and design constraints

Fig. 8-(a and b) shows the effect of (α) on maximum deflection value ($\boldsymbol{\delta_{MAX}}$) and buckling strength factor (λ). It can be seen that, ($\boldsymbol{\delta_{MAX}}$) equals to 4.63 mm and occurs when α equals to 0°. As (α) increases the maximum deflection value decreases and reaches to the minimum value when α equals to 90° with value equals to 2.04 mm. **Fig.** 8-b shows the variations of buckling strength factor with fiber orientation α. It illustrates that, the minimum buckling strength factor value is equal to 16.5 and occurs when α equals to 0 °. After that, as (α) increases (λ) increases and reaches to maximum value when α equals to 60° with value equals to 157.2. With increasing α, (λ) decreases again and reaches to 45 at α equals to 90°.

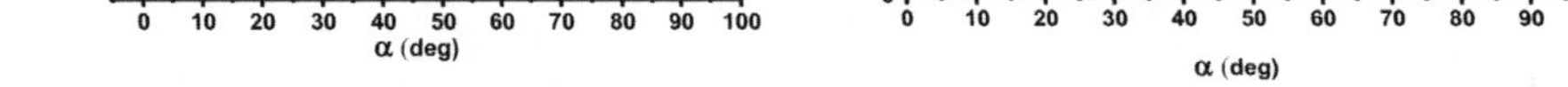

a) α on δ_{MAX} b) α on Buckling strength factor

Fig. 8 The effect of fiber orientation (α) on the buckling strength factor and maximum deflection value (δ_{MAX}).

IOP Conf. Series: Materials Science and Engineering **974** (2020) 012009 doi:10.1088/1757-899X/974/1/012009

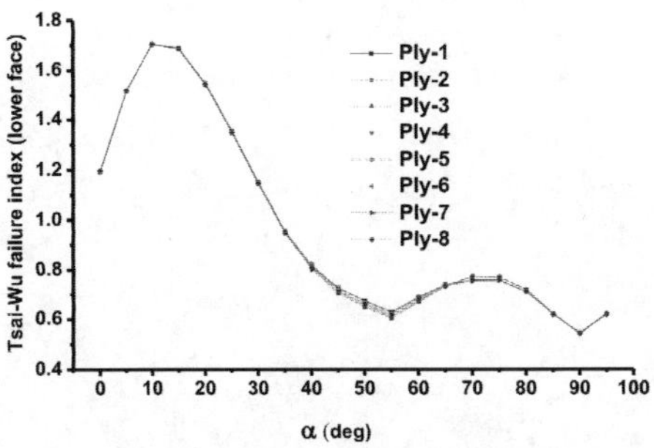

a) α on Tsai-Wu failure index (lower face)

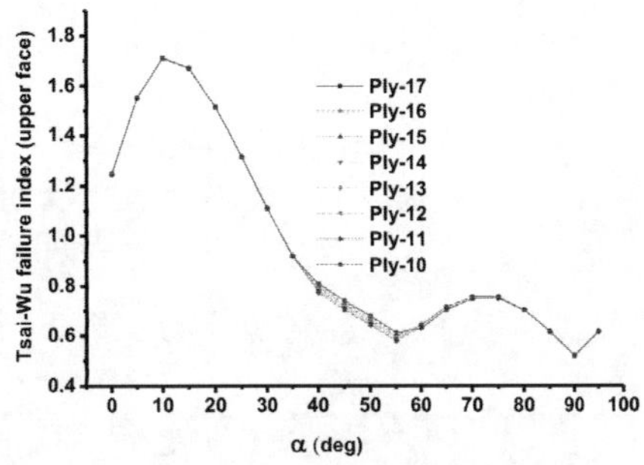

b) α on Tsai-Wu failure index (upper face)

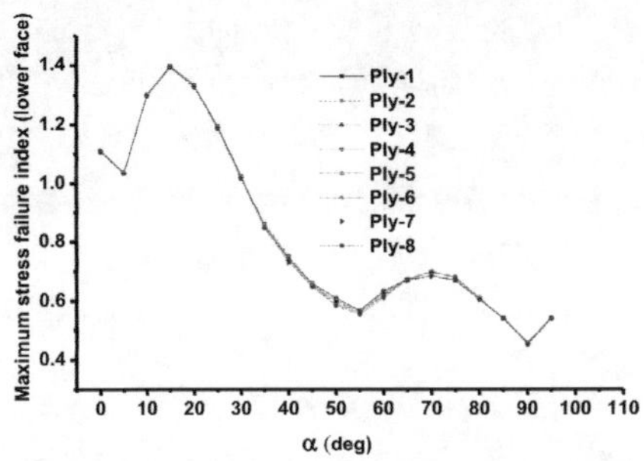

c) α on maximum stress index (lower face)

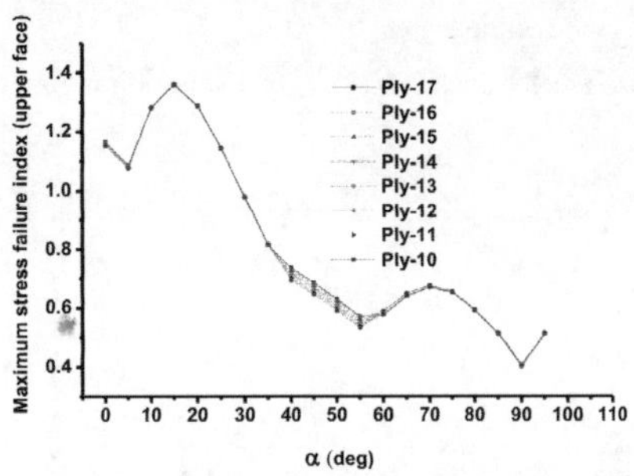

d) α on maximum stress index (upper face)

Fig. 9 The effect of fiber orientation (α) on the Tsai-Wu and maximum stress failure index.

Fig. 9 presents the effect of (α) on Tsai-Wu and maximum stress failure index for both upper and lower faces of the sandwich composite pressure hull. The maximum failure indices for Tsai-Wu and maximum stress failures occur when α equals to 10° at lower face with values equal to 1.7 and 1.29, respectively and the maximum occurs at (Ply-1) due to the high tensile stresses. With increasing α, the maximum failure indices for Tsai-Wu and maximum stress decrease and reach to the minimum values when α equals to 55°. After that, the maximum failure indices increase until α equals to 70° with values equal to 0.76 and 0.69 for Tsai-Wu and maximum stress failures, respectively. The maximum failure indices for the lower face decrease until α equals to 90° and reach to the minimum with values equal to 0.55 and 0.52 for Tsai-Wu and maximum stress failures, respectively. For the upper face, the minimum values of Tsai-Wu and maximum stress failure are equal to 0.52 and 0.5, respectively.

IOP Conf. Series: Materials Science and Engineering **974** (2020) 012009 doi:10.1088/1757-899X/974/1/012009

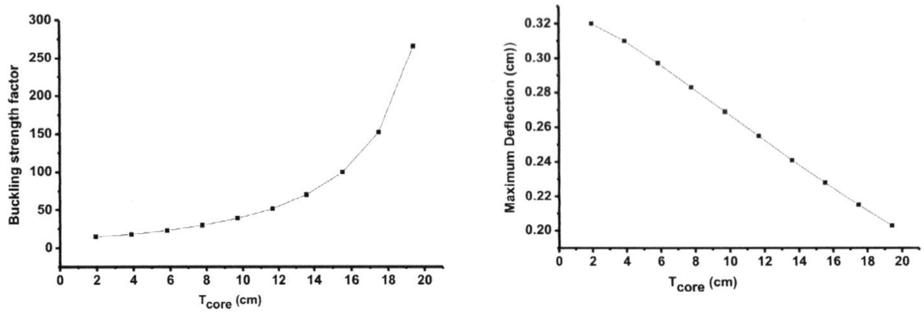

a) T_{core} on Buckling strength factor *b)* T_{core} on δ_{MAX}

Fig. 10 The effect of core thickness (T_{core}) on the buckling strength factor and maximum deflection value (δ_{MAX}).

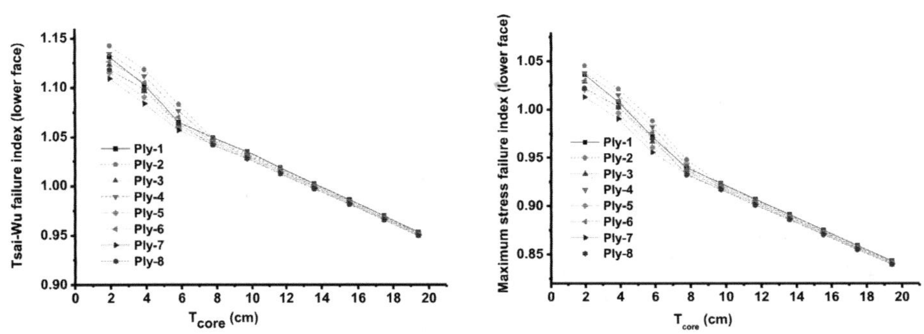

a) T_{core} on Tsai-Wu failure index (lower face) *b)* T_{core} on maximum stress failure index (lower face)

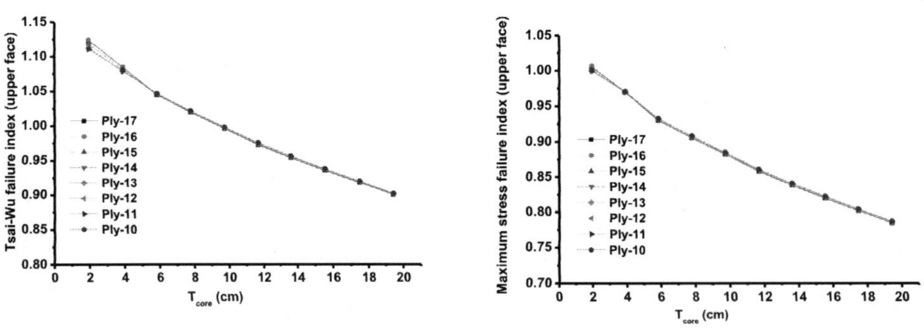

c) T_{core} on Tsai-Wu failure index (upper face) *d)* T_{core} on maximum stress failure index (upper face)

Fig. 11 The effect of core thickness (T_{core}) on Tsai-Wu and maximum stress failure index.

4.2 Effect of core thickness (T_{core}) on design variables and design constraints

Fig. 10 illustrates the effect of core thickness (T_{core}) on buckling strength factor and maximum deflection value (δ_{MAX}). The figure shows that, as T_{core} increases the buckling strength factor increases and δ_{MAX} decreases. The results emphasize that, T_{core} has great effect on buckling strength factor. Fig. 11 presents the effect of T_{core} on maximum Tsai-Wu and maximum stress failure indices. It reveals that, increasing T_{core}, decreasing the failure index and the core thickness has moderate effect on both Tsai-Wu and maximum stress failure index.

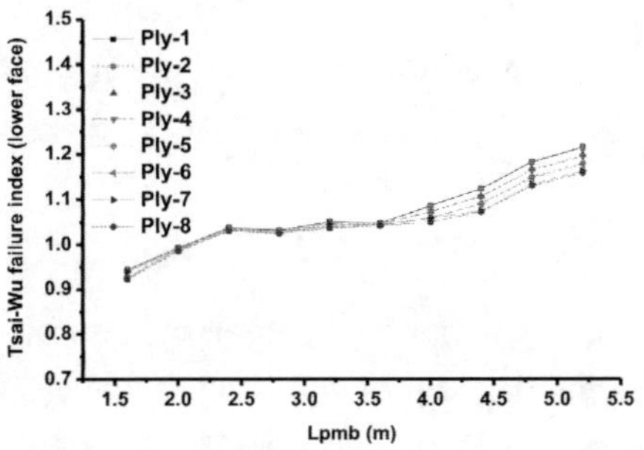

a) L_{pmb} on Tsai-Wu failure index ((lower face)

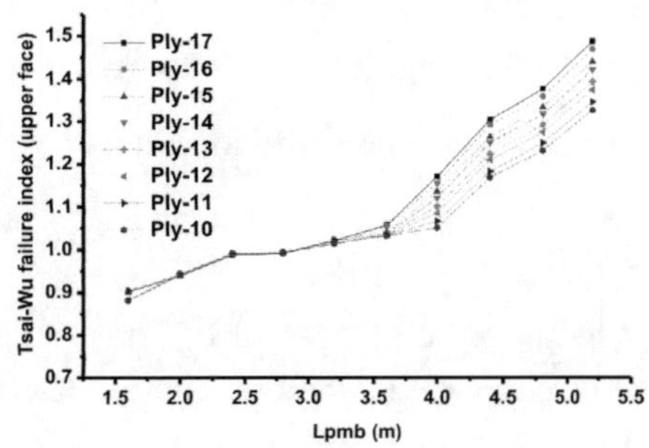

b) L_{pmb} on Tsai-Wu failure index (upper face)

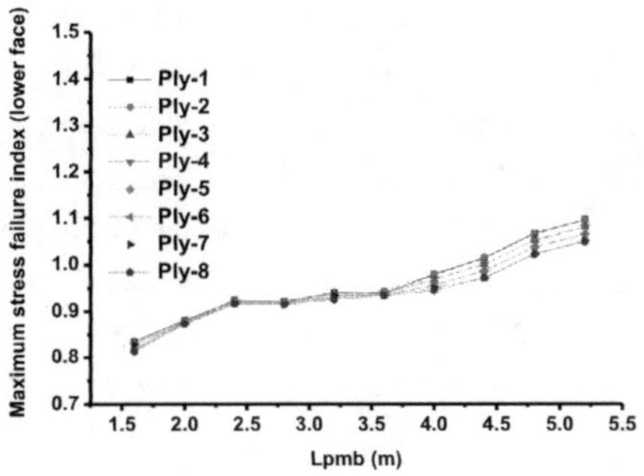

c) L_{pmb} on maximum stress failure index (lower face)

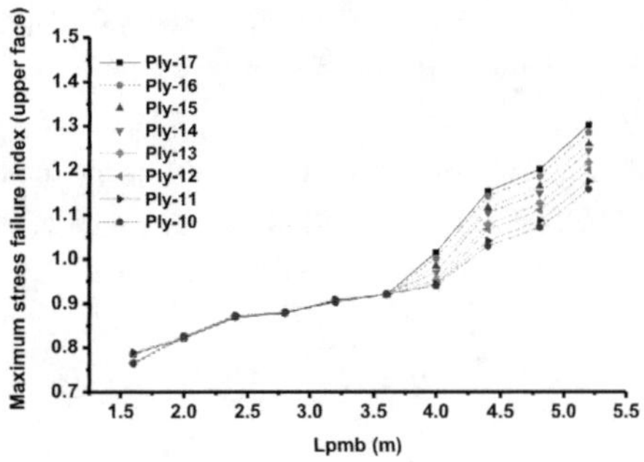

d) L_{pmb} on maximum stress failure index (upper face)

Fig. 12 Effect of (L_{pmb}) on Tsai-Wu and maximum stress failure index.

5. Conclusions

In this study the finite element simulation and optimization were presented for increasing the buckling load capacity, minimizing the drag force and the weight/displacement ratio using ANSYS (APDL). The following observations were made:

- The pattern of optimum points is presented which is used in fabricating and designing the composite pressure hull under hydrostatic pressures.
- At lower face sheet the failure will be initiate first at (ply-1), for upper face sheet the failure will be initiate at the upper most layer (ply-17).
- The core thickness is important to resist the shell buckling.

- The fiber orientation angle (α) has a great effect on buckling strength factor, maximum deflection value, Tsai-Wu and maximum stress failure index of the sandwich composite pressure hull.
- The results suggest that, in the design of composite pressure hull, both buckling and material failure should be considered.

References

[1] Helal M, Huang H, Wang D, Fathallah E. Numerical Analysis of Sandwich Composite Deep Submarine Pressure Hull Considering Failure Criteria. Journal of Marine Science and Engineering. 2019;7(10):377.

[2] Fathallah E, Qi H, Tong L, Helal M. Multi-Objective Optimization of Composite Elliptical Submersible Pressure Hull for Minimize the Buoyancy Factor and Maximize Buckling Load Capacity. Applied Mechanics and Materials. 2014;578:75-82.

[3] Zhang J, Zhang M, Tang W, Wang W, Wang M. Buckling of spherical shells subjected to external pressure: A comparison of experimental and theoretical data. Thin-Walled Structures. 2017;111(Supplement C):58-64.

[4] Zhang M, Tang W, Wang F, Zhang J, Cui W, Chen Y. Buckling of bi-segment spherical shells under hydrostatic external pressure. Thin-Walled Structures. 2017;120(Supplement C):1-8.

[5] Pranesh SB, Kumar D, Subramanian VA, Sathianarayanan D, Ramadass GA. Non-linear buckling analysis of imperfect thin spherical pressure hull for manned submersible. Journal of Ocean Engineering and Science. 2017.

[6] Zhang J, Zhang M, Cui W, Tang W, Wang F, Pan B. Elastic-plastic buckling of deep sea spherical pressure hulls. Marine Structures. 2018;57(Supplement C):38-51.

[7] Topal U, Uzman U. Multiobjective Optimization of Angle-ply Laminated Plates for Maximum Buckling Load. Finite Elementsin Analysis and Design. 2010;46(3):273-9.

[8] Błachut J, Smith P. Buckling of multi-segment underwater pressure hull. Ocean Engineering. 2008;35(2):247-60.

[9] Carvelli V, Panzeri N, Poggi C. Buckling strength of GFRP under-water vehicles. Composites Part B: Engineering. 2001;32(2):89-101.

[10] Błachut J. Buckling of composite domes with localised imperfections and subjected to external pressure. Composite Structures. 2016;153(Supplement C):746-54.

[11] Fathallah E, Qi H, Tong L, Helal M. Optimal Design Analysis of Composite Submersible Pressure Hull. Applied Mechanics and Materials. 2014;578:89-96.

[12] Fathallah E, Qi H, Tong L, Helal M. Design Optimization of Composite Elliptical Deep-Submersible Pressure Hull for Minimizing the Buoyancy Factor. Advances in Mechanical Engineering. 2014.

[13] Fathallah E. Finite Element Modelling and Multi-Objective Optimization of Composite Submarine Pressure Hull Subjected to Hydrostatic Pressure. Materials Science Forum. 2019;953: 53-8.

[14] Helal M, Fathallah E. Multi-objective optimization of an intersecting elliptical pressure hull as a means of buckling pressure maximizing and weight minimization. Materials Testing. 2019;61(12):1179-91.

[15] Helal M, Huang H, Fathallah E, Wang D, ElShafey MM, Ali MAEM. Numerical Analysis and Dynamic Response of Optimized Composite Cross Elliptical Pressure Hull Subject to Non-Contact Underwater Blast Loading. Applied Sciences. 2019;9(17):3489.

[16] Fathallah E, Helal M. Optimum Structural Design of Deep Submarine Pressure hull to achieve Minimum

Weight. The International Conference on Civil and Architecture Engineering. 2016;11(11PthP International Conference on Civil and Architecture Engineering):1-22.

[17] Alogla A, Helal M, Abdal Allah AA, Fathallah E. Design Optimization of a Flexible Hinge Compliant Micro icro-Gripper ripper Mechanism with Parallel Movement Arms Using Pseudo-Rigid -Body Model. International Conference on Aerospace Sciences and Aviation Technology. 2017;17(AEROSPACE SCIENCES & AVIATION TECHNOLOGY, ASAT - 17 – April 11 - 13, 2017):1-18.

[18] Zhang J, Wang M, Wang W, Tang W, Zhu Y. Investigation on egg-shaped pressure hulls. Marine Structures. 2017;52(Supplement C):50-66.

[19] Vosoughi AR, Darabi A, Dehghani Forkhorji H. Optimum stacking sequences of thick laminated composite plates for maximizing buckling load using FE-GAs-PSO. Composite Structures. 2017;159(Supplement C):361-7.

[20] Mian HH, Wang G, Dar UA, Zhang W. Optimization of Composite Material System and Lay-up to Achieve Minimum Weight Pressure Vessel. Applied Composite Materials. 2013;20:873–89.

[21] Pan G, Lu J, Shen K, Ke J. Optimization of Composite Cylindrical Shell Subjected to Hydrostatic Pressure. Springer International Publishing Switzerland 2015:81-90.

[22] Fathallah E, Helal M. Finite element modelling and multi-objective optimization of composite submarine pressure hull subjected to hydrostatic pressure. IOP Conference Series: Materials Science and Engineering. 2019;683:012072.

[23] Imran M, Shi D, Tong L, Waqas HM. Design optimization of composite submerged cylindrical pressure hull using genetic algorithm and finite element analysis. Ocean Engineering. 2019;190:106443.

[24] Lund E. Discrete Material and Thickness Optimization of laminated composite structures including failure criteria. Structural and Multidisciplinary Optimization. 2018;57(6):2357-75.

[25] Fathallah E, Qi H, Tong L, Helal M. Design optimization of lay-up and composite material system to achieve minimum buoyancy factor for composite elliptical submersible pressure hull. Composite Structures. 2015;121(Supplement C):16-26.

[26] Vinson JR, L R. The Behavior Of Structures Composed Of Composite Materials: Springer, Dordrecht; 2008.

[27] Kaw AK. Mechanics of Composite Materials. Second Edition ed: CRC Press Taylor & Francis Group; 2006.

[28] Soden PD, Hinton MJ, Kaddour AS. A comparison of the predictive capabilities of current failure theories for composite laminates. Composites Science and Technology. 1998;58(7):1225-54.

[29] Paster D, Co R, Portsmouth, RI. Importance of Hydrodynamic Considerations for Underwater Vehicle Design. In: Proceedings of MTS/IEEE Oceans. Washington, DC, USA, Conference, Conference 23-25 Sept. 1986. p. 1413-22.

[30] Alam K, Ray T, Anavatti SG. Design of a Toy Submarine Using Underwater Vehicle Design Optimization Framework. IEEE. 2011.

[31] M.A.Martz. Preliminary design of an autonomous underwater vehicle using a multiple-objective genetic optimizer [Master'sthesis]. Blacksburg,VA,United States: Virginia Polytechnic Institute and State University; 2008.

[32] Alam K, Ray T, Anavatti SG. Design and construction of an autonomous underwater vehicle. Neurocomputing. 2014;142(Supplement C):16-29.

[33] Alvarez A, Bertram V, Gualdesi L. Hull hydrodynamic optimization of autonomous underwater vehicles operating at snorkeling depth. Ocean Engineering. 2009;36(1):105-12.

[34] Zheng JY, Liu PF. Elasto-plastic Stress Analysis and Burst Strength Evaluation of Al-carbon Fiber/Epoxy Composite Cylindrical Laminates. Computational Materials Science. 2008;42(3):453-61.

[35] Panahi B, Ghavanloo E, Daneshmand F. Transient Response of a Submerged Cylindrical Foam Core Sandwich Panel Subjected to Shock Loading. Materials & Design. 2011;32(5):2611-20.

[36] Smith CS. Design of submersible pressure hulls in composite materials. Marine structures. 1991;4:141-82.

[37] Inc A. ANSYS Theory Reference Release 14.5. October ed2012.

[38] M. Erdogan GI. The Finite Element Method and Applications in Engineering Using ANSYS. New York: Springer; 2006.

[39] Liang CC, Chen HW, Jen CY. Optimum Design of Filament-wound Multilayer Sandwich Submersible Pressure Hulls. Ocean Engineering. 2003;30(15):1941-67.

[40] Ross CTF. A conceptual Design of an Underwater Vehicle. Ocean Engineering. 2006;33(16):2087-104.

[41] Ma S, Mahfuz H. Finite Element Simulation of Composite Ship Structures with Fluid Structure Interaction. Ocean Engineering. 2012;52:52-9.

MilitaryTechnicalCollege
Kobry El-Kobbah,
Cairo, Egypt

13th International Conference
on Civil and Architecture
Engineering
ICCAE-13-2020

Optimization of the Panels Used in Free-Form Buildings and Its Impact on Building Cost.

Sara N. Elkabany[1], Ahmed M. Elkordy[2], Hesham A. Sobh[3]

[1]Assistant lecturer at Architecture Department, Faculty of Engineering, Al-Azhar University, Cairo, Egypt.

[2] Professor & Head of Architecture Department, Faculty of Engineering, Al-Azhar University, Cairo, Egypt.

[3] Professor of Architecture & Vice Dean of Faculty of Engineering, Al-Azhar University, Cairo, Egypt.

Corresponding author: Sara.alkabany@azhar.edu.eg

The described study presents the context of the process of optimization curved panels and the direct impact on reducing high cost, especially in this type of free-form buildings. The main motivation for this study is the avoidance of the traditional style of design where it needs to produce expensive designs because free shapes require a huge amount of specialization, the use of digital technology can overcome the problem of high cost through the production of optimized designs in addition to the ease of implementation of these forms which can be complicated in some cases as the number of freeform buildings that use building information modeling technology has increased recently. The importance of research is shown in guiding designers to optimize curved panels used in free formation during the early design stage so that design solutions can be correctly identified. This paper presents the results of analyzing the types of curved panels by comparing the cost before and after the optimization process, using algorithms for the selected materials (FRP, steel, and aluminum) that aim to reduce the value of curve coordinates while preserving the curve from deformation. In order to further devote attention to methodological frameworks to optimize curved panels besides cost control. the primary goal of the research is innovation at the initial design stage by designers through the use of algorithm-based assistance programs to optimize curved shapes that can produce contemporary project optimization curved panels are less expensive while maintaining the desired curved shape.

Keywords: Free-Forms, Cost control, Optimization, Panel, Algorithms.

1. Introduction

In the manufacture and design of freeform shapes in architecture, one constantly faces design challenges and often high costs [1]. Composite materials also greatly influenced how many forms of engineering systems used in the automobile, medicinal, aerospace, construction, civil and mechanical industries have been developed and studied in recent decades [2]. Therefore, the optimization of the forms of this type of building has become a hot topic since its widespread use. Since then, a lot of curved shape optimization research has been undertaken that has revealed many new optimizations and their direct impact on reducing high cost [3-9]. In parallel, research is being done to build panels that are not only aluminum but it's from different materials such as the height of architectural construction with glass fiber reinforced plastic (GRP) was the era 1960-1972 [10], and lightweight and sustainable wood panels using laminated veneer lumber, where optimization plays a major role in the ability to quickly and accurately assemble curved shapes to maintain the shape of deformation with the help of numerically controlled machines for integrating joints into panels engineering [11]. As higher panel curvature leads to higher cost, the panel optimization method is required to modify curved panels with double curvature into planar panels or single curvature panels. The desired esthetic quality of free-forms and their surface smoothness shall be without damage. According to the breakdown of costs for recent free-form building projects worldwide, the costs for panel construction accounted for between 15 and 20 percent of the total project cost [12]. The study suggests a panel optimization process from different materials for free-form buildings by using a worldwide application that free-form designers can easily use. The study measures the construction cost changes during the panel optimization process by connecting the optimization results with panel construction cost estimates of the free-form building projects. A case analysis on the free-form building projects with different materials is performed to check the applicability of the research findings.
Throughout the analysis, free-form buildings Heydar Aliyev Cultural Center, Weatherhead School and DC New Logic III have cost assessed. Differently from the optimization levels, construction cost control is found for each selected case. In the case study, the most curved parts were chosen for the facades of the free-form projects, and as a result, the optimization was achieved without distorting the shape at about 20.2% of the budget by comparison with the initial cost before optimization process, by choosing the appropriate optimization point for each case while maintaining the continuity of the curved line.

2. Methodology

The five stages of the study were carried out as linear methods to optimizing free-form panels: (1) configuration of the algorithm for optimization analyzes, (2) Script creation which can parametrically monitor panel conversion information, (3) Analysis of continuity and curvature, (4) Optimizing and analyzing panel shape transformation, And (5) preliminary estimates of costs to minimize distortions and save production costs for every range of optimization. Further sophisticated applications can be created in potential studies depending on step-by-step processes see (Figure 1).

3. Material and Software Selection

Grasshopper has been chosen for use in the modeling process which is an algorithm graphical publisher that is closely incorporated with the Rhino 3D modeling instruments for parameter modeling tool for free-form structures., while Microsoft ExcelTM is converted by the Para Cloud modeler to a strong parameter modeler. Furthermore, the study assumed the panels are made of Fiberglass Reinforced Polyester (FRP), steel and aluminum as the most popular materials for curved facade panels.

IOP Conf. Series: Materials Science and Engineering **974** (2020) 012010 doi:10.1088/1757-899X/974/1/012010

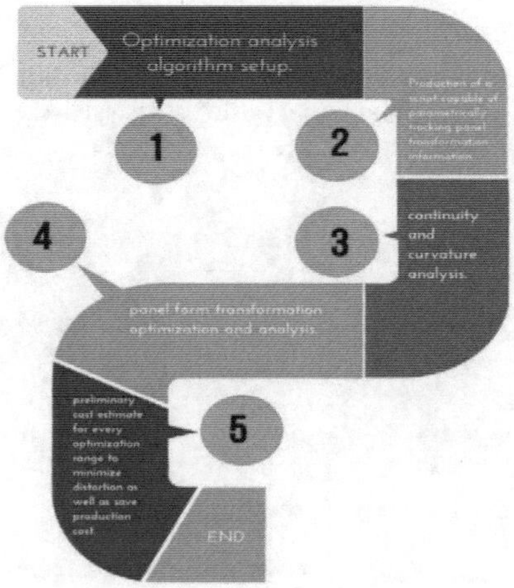

Figure 1: The Study Steps for Optimizing Free-Form Panels.

4. Case Projects

The three case projects this study referred to for inducing free-form surfaces. The second line of (Table 1) indicates the surfaces modeled for the optimization test. The surfaces intended to reflect the biggest curved portions of each case project the red circles in the case projects.
The model for each surface of free form is a 100 m2 area with an error margin of ± 5 m2 so that that the curvature range of each case project can well be reflected. The free-form surfaces were scheduled for a quadrangle that was commonly used in previous free-form building projects.

Table 1. Test Optimization Subjects of the Panel.

	Selected Projects	Materials	(The surface modeled for test)		Panel Type
Case 1	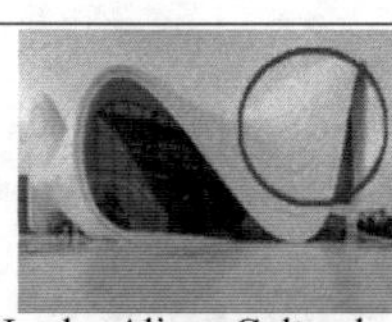 Heydar Aliyev Cultural Center (2012) [13]	Fibreglass Reinforced Polyester (FRP)	Area: 103.8 m2	Planar 5 Single Curvature 9 Double Curvature 86	
Case 2	 Weatherhead School (2002)[14]	steel	Area : 99.7 m2	Planar 3 Single Curvature 9 Double Curvature 88	
Case 3	 DC New Logic III (2019)[15]	Aluminum	Area : 104.4 m2	Planar0 Single Curvature 62 Double Curvature 38	

IOP Conf. Series: Materials Science and Engineering **974** (2020) 012010 doi:10.1088/1757-899X/974/1/012010

The three test surfaces were then tested with Grasshopper. The number of panels should be restricted to 100 in each situation. The highest curvature of aluminum, FRP and steel panels that can be produced using MPF processes is 30 cm. The panels whose curvature surpassed the value were therefore divided into smaller parts to half the size of the panel above see (Figure 2).

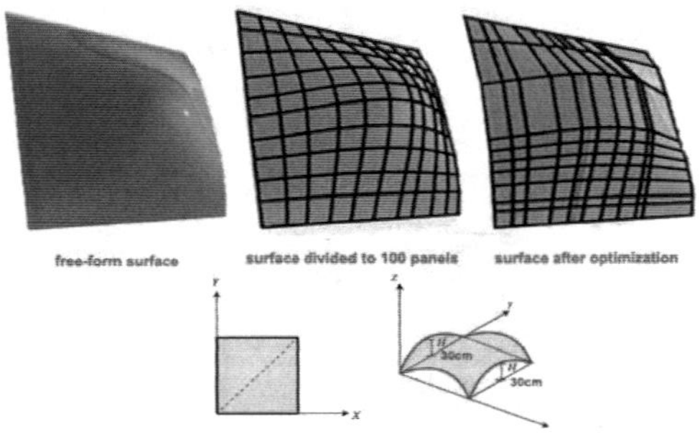

Figure 2: Divide Curved Panels.

The study created, as shown in (Figure 3), a panel optimization algorithm. Panel Arc height, from P1 to P2, was divided into 30 cm parts and curved to six different points was optimized.

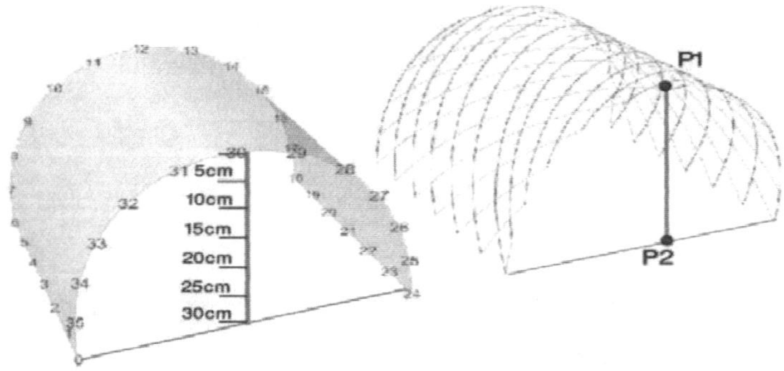

Figure 3: The Process Determines The Circular Parts Of Each Surface.

This study developed the continuous transformation process of the curved surfaces as shown in (Figure 4). The method is chosen as the circled parts of each surface that demonstrate the conversion of the surfaces from double curvature panels through the optimization method to single curvature and planar panels. The study also created scripts using Grasshopper's mark curve start & end feature to measure the quality level of keeping curved surfaces. Coordinate points are extracted, identified as optimization points, and used to maintain the original plan (coordinate points).

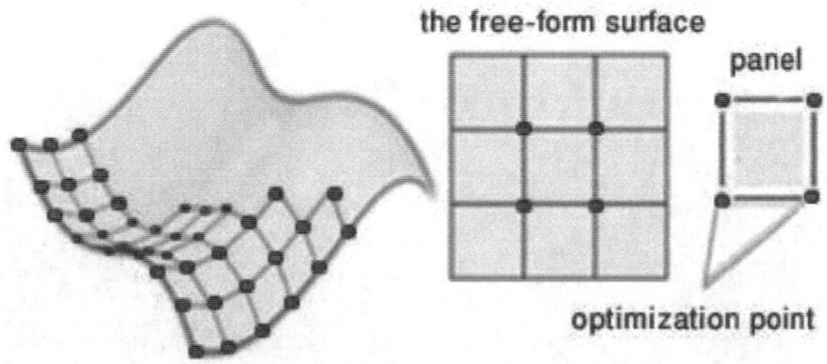

Figure 4: The Process Determines The Circular Parts Of Each Surface.

In each panel is four edges differ in height. The study created a Grasshopper Script, allowing automatic measurement in the optimization of the maximum location value and the height value of all four edges of each panel see (Figure 5).

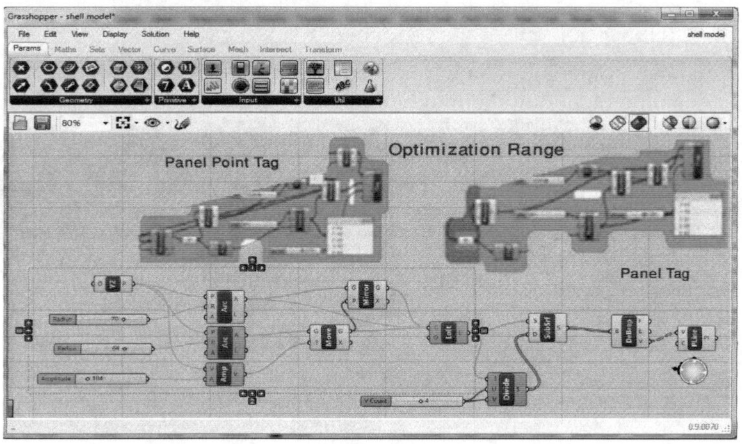

Figure 5: Panel Optimization Grasshopper Script.

5. Panel Optimization Analysis

Developable surfaces were commonly used as part of these methodologies owing to their geometric efficacy in distributing bent panels over planar substrates. A surface that can be developed is the material that can be applied without deformation to a plane (i.e. bending or compression). The use of developable surfaces is suitable for sheet metal panels which can be deformed to developable or almost developable shapes at reasonable expenses [16].

(Figure 6) is the graphic representation of data in (Table 2). It shows a decrease of double curvature panels at each optimization point of the cases and an increase in single curvature and planar panels. The issues with the optimization results are evaluated by test results of Curvature Analysis in (Table 3).

IOP Conf. Series: Materials Science and Engineering **974** (2020) 012010 doi:10.1088/1757-899X/974/1/012010

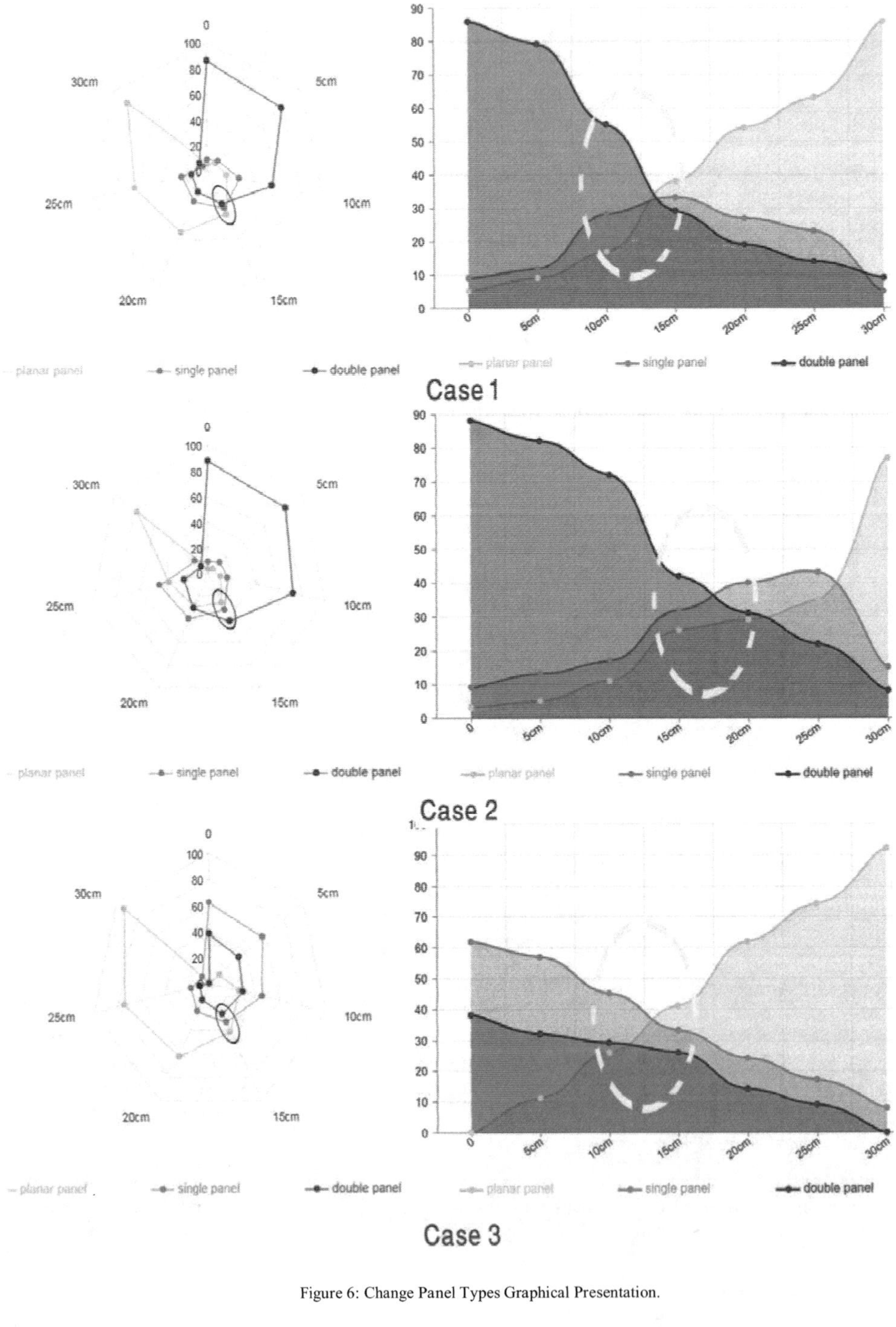

Figure 6: Change Panel Types Graphical Presentation.

IOP Conf. Series: Materials Science and Engineering **974** (2020) 012010 doi:10.1088/1757-899X/974/1/012010

Table 2. After Optimization Test, the variation of the panel types.

case (1)

1		0	5cm	10cm	15cm	20cm	25cm	30cm
2	planar panel	5	9 ▲	17 ▲	38 ▲	54 ▲	63 ▲	86 ▲
3	single panel	9	12 ▲	28 ▲	33 ▲	27 ▲	23 ▲	5 ▼
4	double panel	86	79 ▼	55 ▼	29 ▼	19 ▼	14 ▼	9 ▼

case (2)

1		0	5cm	10cm	15cm	20cm	25cm	30cm
2	planar panel	3	5 ▲	11 ▲	26 ▲	29 ▲	35 ▲	77 ▲
3	single panel	9	13 ▲	17 ▲	32 ▲	40 ▲	43 ▲	15 ▲
4	double panel	88	82 ▼	72 ▼	42 ▼	31 ▼	22 ▼	8 ▼

case (3)

1		0	5cm	10cm	15cm	20cm	25cm	30cm
2	planar panel	0	11 ▲	26 ▲	41 ▲	62 ▲	74 ▲	92 ▲
3	single panel	62	57 ▲	45 ▲	33 ▲	24 ▲	17 ▲	8 ▲
4	double panel	38	32 ▼	29 ▼	26 ▼	14 ▼	9 ▼	0 ▼

Table 3. Test results of Curvature Analysis.

ptimization point	0	5cm	10cm	15cm	20cm	25cm	30cm
Case 1	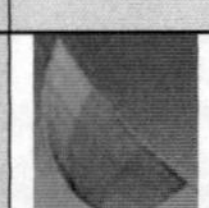	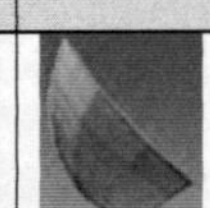	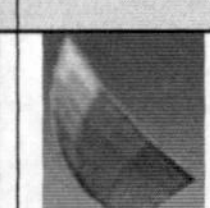	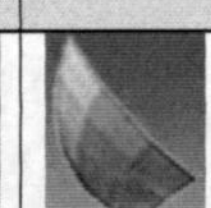	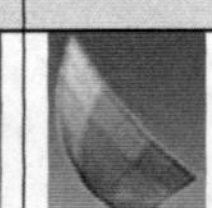	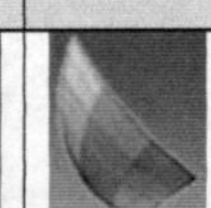	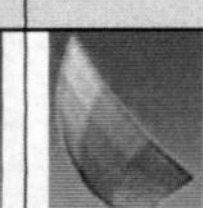
Case 2	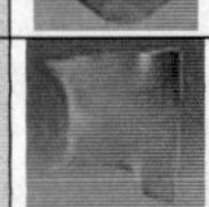		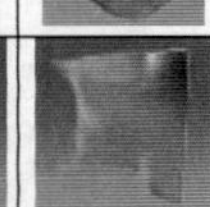	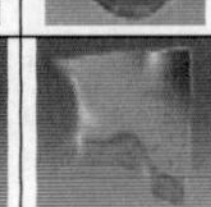		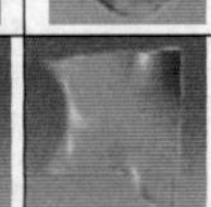	
Case 3							

The area change for each panel type's with the procedure of panel optimization at 6 optimization points and cost comparison of the manufacturing of the panel together. To calculate the cost rate in (Table 4) multiply each area of panel types by cost rate then the values are collected for each panel type used in each case study to obtain production cost rate. The same method was followed in the second and third cases by using different values of the Cost Rate between the kinds of the panel materials for each case as illustrated in (Table 5).

Table 4. Free-Form Facade Panels Costs Rate.

Division		Planar	Single Curvature	Double Curvature
Heydar Aliyev Cultural Center (2012) (Fiberglass Reinforced Polyester)	Contract Price	$ 201	$ 301	$ 482
	Cost Rate	**1**	**1.5**	**2.4**
Weatherhead School (2002) (Steel)	Contract Price	$ 232	$ 441	$ 626
	Cost Rate	**1**	**1.9**	**2.7**
DC New Logic III (2019) (Aluminum)	Contract Price	$ 178	$ 196	$ 392
	Cost Rate	**1**	**1.1**	**2.2**

Table 5. Each Optimization Point Cost Estimated Panel Types.

case (1)

Division	0	5cm	10cm	15cm	20cm	25cm	30cm
planar m²	2.26	4.55	7.68	17.17	41.61	52.09	85.15
single m²	13.84	18.94	39.34	55.53	41.52	35.84	7.68
double m²	87.70	80.5	56.08	29.57	19.37	14.27	9.17
total m²	103.8	103.5	103.1	102.8	102.5	102.2	102
production cost rate	**233.5**	**226.1**	**201.2**	**171.4**	**150.3**	**140.0**	**118.6**

case (2)

Division	0	5cm	10cm	15cm	20cm	25cm	30cm
planar m²	3.72	6.72	13.64	32.24	35.96	43.4	95.48
single m²	18.75	20.82	22.28	29.71	35.14	34.2	6.08
double m²	77.23	71.96	63.18	36.85	27.20	19.30	7
total m²	99.7	99.5	99.1	98.8	98.3	96.9	96.4
production cost rate	**247.8**	**201.0**	**226.5**	**188.1**	**176.1**	**160.4**	**125.9**

case (3)

Division	0	5cm	10cm	15cm	20cm	25cm	30cm
planar m²	0	10.65	29.29	41.89	63.31	74.85	102.3
single m²	71.8	66	52.11	38.21	27.79	19.63	0.14
double m²	32.6	27.45	22.30	23.30	12	7.72	0
total m²	104.4	104.1	103.7	103.4	103.1	102.2	102.5
production cost rate	**149.7**	**142.8**	**135**	**134.4**	**119.9**	**113.1**	**102.5**

6. Results

Changes in panel types were examined at each facade and the cost of production was assessed during the process of optimization of the panel with the three facades. Where we notice after the improvement process, the biggest decrease was in the first case, which is Fiberglass Reinforced Polyester (FRP) material; the rate fell by 26.5 percent from 233.5 to 171.4. And then the second case that used steel material, the rate fell by 23.9 percent from 247.3 to 188.1. The least is the third case, which used Aluminum material, it decreasing by 10.2 percent from 149.7 to 134.4, whereas the low-cost average decreased by 20.2 percent. A model designed with the utmost optimization against deformation was obtained see in (Figure 7).

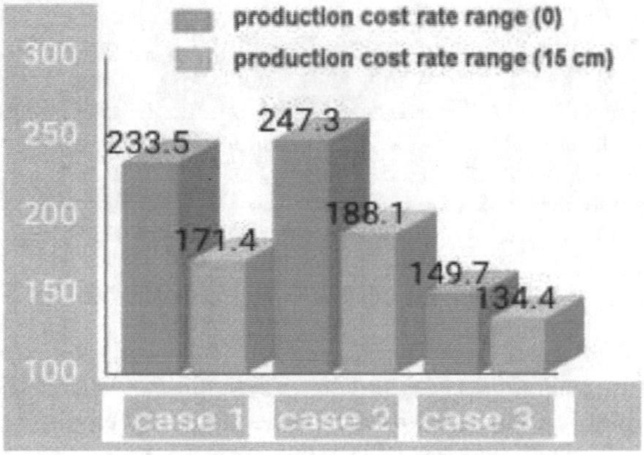

Figure 7: Low-Cost Projects after Process Optimization.

7. Conclusion

In conclusion, the design and manufacturing approach developed by algorithms in this work has become a guide for designers to focus on architecture and make design decisions. It changes the traditional design methodology in such a way that reduces the design complexity for the designer which is optimized by the algorithm resides in the potential of creating a wide range of solutions quickly and effortlessly, developing various approaches to design and gives more flexibility for the designers to focus on the more important issue in the context of architecture and the built environment design.

A structure of free-form building is a sloping, deformed building of design. The buildings outer facades include the panels with very complicated curvature. While the panels display impressive building forms in the urban environment, so the forming of the panels is extremely complex in reality. The higher the curvature of the panels, the higher the cost, For this, the designer must take into account optimization process by reducing the height of the curvature during the early design stages so as not to affect the interface functions without prejudice to the aesthetics of the curved shapes and preventing them from deformation. If this method can be related to the cost estimate of panel construction, cost variation can be observed by panel curvature change so that a feasible array of panel curvatures can be defined that fulfill both the construction cost limit and the free-form model purpose.

Throughout the analysis, free-form buildings Heydar Aliyev Cultural Center, Weatherhead School and DC New Logic III, Were cost assessed. Differently from the optimization levels, construction cost control is found for each selected case. In the case study, the most curved parts were chosen for the facades of the free-form projects, and as a result, the optimization was achieved without distorting the shape at about 20.2% of the budget by comparison with the initial cost before optimization process, by choosing the appropriate optimization point for each case while maintaining the continuity of the curved line.

8. References

1. Jan-Ruben F.(2012) Optimizing Digital Organic Freeform Modelling for Fabrication by Using Parameterization With Glass Fibre Reinforced Plastics (GRP): A Case Study.
2. Reddy, J. N. (2004). Mechanics of laminated composite plates and shells. CRC Press.
3. P. Hao, B. Wang and G. Li, (2012) Surrogate-based optimum design for sti®ened shells with adaptive sampling, AIAA J. 50(11), 2389–2407.
4. B. Wang, P. Hao, G. Li, K. Tian, K. F. Du, X. J. Wang, X. Zhang and X. H. Tang, (2014) Twostage size-layout optimization of axially compressed sti®ened panels, Struct. Multidiscip.Optim. 50(2) 313–327.
5. B. Wang, P. Hao, G. Li, X. J. Wang, X. H. Tang and Y. Luan, (2014) Generatrix shape optimization of sti®ened shells for low imperfection sensitivity, Sci. China. Technol. Sci.57(10),2012–2019.
6. Eigensatz, M, Deuss, M, Schiftner, A, Kilian, M, Mitra, NJ, Pottmann, H and Pauly, M 2010, (2010) 'Case Studies in Cost-optimized Paneling of Architectural Freeform Surfaces', Advances in Architectural Geometry: 49–72.
7. Kiesow, G and Friedrich, B (1987), Naturwerkstein in Der Denk¬malpflege, Ulm: Ebner Verlag GmbH & Co. KG.
8. Wallner, J, Schiftner, A, Kilian, M, Flöry, S, Höbinger, M, Deng, B, Huang, Q and Pottmann, H (2010), 'Tiling Freeform Shapes With Straight Panels: Algorithmic Methods', Ad¬vances in Architectural Geometry: 73–86.
9. Hyung Keun Park, Jong-Ho Ock, (2015), Developing the preliminary cost estimate for the free-form building facade in conjunction with the panel optimization process.
10. Voigt, P (2007), Die Pionierphase Des Bauens Mit Glasfaserverstärkten Kunststoffen (GFK) 1942 Bis 1980, Vol. Doctor Phil., Weimar: Bauhaus-Universität Weimar.
11. Robeller C (2015). Integral Mechanical Attachment for Timber Folded Plate Struc-tures. PhD thesis, EPFL ENAC, Lausanne.
12. Lee, K. and Kim, S. (2012). "Case study of mass customization of double-curved metal façade panels using a new hybrid sheet metal processing technique." Journal of Construction Engineering and Management, ASCE, Vol. 138, No. 11, pp. 1322-1330.
13. Heydar Aliyev Center / Zaha Hadid Architects, https://www.archdaily.com/448774/heydar-aliyev-center-zaha-hadid-architects, Accessed: May 2019.
14. Weatherhead School, https://www.azahner.com/works/weatherhead, Accessed: May 2019.
15. DC New Logic III, https://www.reynaers.com/en/get-inspired/aluminium-project-references/dc-new-logic-iii, Accessed: May 2019.
16. An, J. (2011). "A study on the lessons learned for improving efficiency of free-form facade construction." A Master of Science Thesis, Seoul National University of Science and Technology, Korea, pp. 22-35.

MilitaryTechnicalCollege
Kobry El-Kobbah,
Cairo, Egypt

13th International Conference on Civil and Architecture Engineering
ICCAE-13-2020

Optimization of Load-Bearing concrete Wall Using Genetic Algorithm To achieve Mechanically Integrated Behavior.

Sara N. Elkabany[1], Ahmed M. Elkordy[2], Hesham A. Sobh[3]

[1]Assistant lecturer at Architecture Department, Faculty of Engineering, Al-Azhar University, Cairo, Egypt.

[2] Professor & Head of Architecture Department, Faculty of Engineering, Al-Azhar University, Cairo, Egypt.

[3] Professor of Architecture & Vice Dean of Faculty of Engineering, Al-Azhar University, Cairo, Egypt.

Corresponding author: Sara.alkabany@azhar.edu.eg

This paper proposes mechanical and growth characteristics of nature that inspire shapes to be applied to load-bearing walls. The problem is that this type of buildings needs a high cost due to its design without using optimization processes that have a big role in reducing cost while maintaining the required aesthetic shape. In this study, this is performed by analyzing the case studies of O-14 Building, Victoria Gate and The Broad Museum where the building materials used differ from case to case. A model was created through the SolidThinking Inspire program using the morphogenesis tool. The model design is based on the algorithm for topology optimization and will be compared in terms of structural and material performance with the selected case studies. Optimization results are found to be effective and designed with maximum stiffness in order to prevent deformation which is clearly reflected in the cost estimate in the initial design stages.

Keywords: Genetic Algorithm, Topology Optimization, Load Bearing Walls, Generative Design, 3D printing.

1. Introduction

The use of structural optimization has rapidly increased over the last decade. The upstream phases of the design process account for 5 percent of the time involved in product development, but they represent 75 percent of the global development costs [1]. This is very important to integrate optimization into the early stages of a project. The use of genetic algorithms to optimize products has become necessary to test various materials and forms. Structural optimization design has become one of the most important ways of obtaining lightweight and high-performance structures with advances in computer science and technology. Structural optimization is generally divided into size, shape and topology optimization, depending on various design variables. Topology optimization is considered the most generic of these three methods of optimization because it can provide new and sometimes even unforeseen design ideas to engineering designers without requiring a pre-established design. Topology optimization, in general, utilizes optimization techniques to try to find out where to position content in the design realm .
Over the past four decades, topology optimization has achieved rapid development and has been successfully applied to structural design in many industrial sectors, including the automotive, aerospace and biomedical industries [2][3]. Many specific methods of topology optimization are proposed, including the density method [4][5], evolutionary approaches bi-directional evolutionary structural optimization (BESO) and Bi-directional Structural Optimization (BESO) and Structural Evolutionary Optimization (ESO) evolutions [6]. Of these methods, the density method, utilizing element-constant density to characterize the structural topology, is the most advanced technique due to its computational efficiency and stability.

2. Methodology

Model design based on the optimization algorithm for structural and physical performance, they were compared with the three case studies of the selected buildings O-14 Building, Victoria Gate and The Broad Museum, Where the results of the Optimization were found to be effective and designed with the utmost rigidity In order to prevent deformation Which resulted in reducing the cost in the initial design stages based on the characteristics of each material, which differ from one to other.
The model input parameters (design volume, materials, and loading conditions) interact as they affect each other. Integral systems are inseparable and are given a direct integration result. The aim of this study is possible to produce shapes with distributed loads directly related to the shape and materials used. The methodology followed in this work is illustrated on Fig. 1.

3. Material and Software Selection

Technological developments in parametric design make it possible to shape the large-scale production of curved forms. Through a generative sequence and relations between geometric objects, the parametric design approach presents logic into a geometric model. Applications available on the market include Grasshopper TM, Para Cloud TM, SOLIDTHIKING INSPIRETM, TOPOSTRUCTM and CATIA TM for parameter modeling tool for free-form structures.
As for the modeling instruments to implement parametric models, geometric factors need to be easily applied and easily reflected by designers throughout their modeling phase. In Table 1. the instruments were compared the tools with a variety of user comfort views. Every instrument is shown by levels in each category.
Reinforced concrete offers tensile strength such as steel plates, and is resistant to metal pressures such as concrete [7]. Where both substrates are composites and isotropically performed, the software for modeling load-bearing wall SolidThinking Inspire 2019 is the chosen software. The use of an instrument of morphology based on optimization of topology and algorithms which attempt to imitate the growth and weight of natural shapes. Morphogenesis produces effective lightweight structures that fulfill the weight, stiffness and strength efficiency requirements.

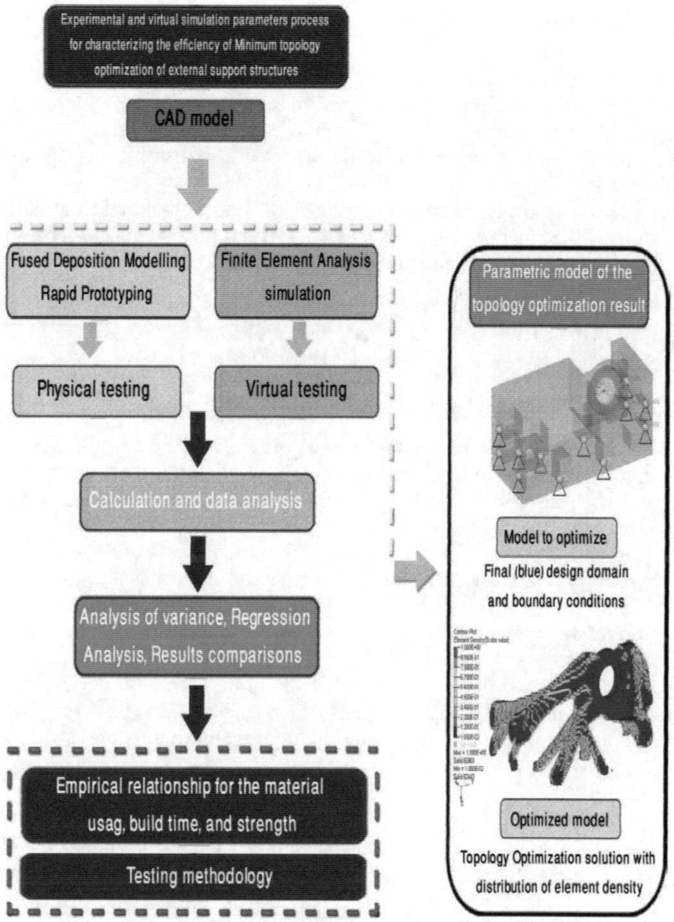

Figure 1: Load-Bearing wall Process Optimization.

Table 1. Free-Form Analysis Tools Comparison. [8]

Division	Usability	Learnability	Popularity	Compatibility	Comprehensive Evaluation
Grasshopper	medium	medium	medium	High	High
ParaCloud	medium	medium	Low	High	medium
SOLIDTHIKING INSPIRE	High	High	medium	medium	High
TOPOSTRUCT	medium	medium	medium	medium	Low
CATIA	High	medium	medium	Low	medium

4. Formulation of Problem

The optimization of form, structure and material in architectural models as distinct fields has resulted in inefficient use of structural materials in designs of architecture [9]. Improved material and structural performance is important in order to combine form, material and structure into one single feature-the load-bearing wall- and to solve the question of facade placement and stop utilizing substructures and structural obstacles within facades.
The goal of Application is applying mechanical growth and behavior in the early stages of architectural design to a bearing wall is to build its physical and structural efficiency. The strategy used to solve the problem is the chosen metrics method (known parameters problem - loads - support conditions - design size).

5. Applied Forces [10]

5.1. Loads of Gravity (Vertical Loads)

(Live Loads) are the loads that can alter in magnitude, including all products that are found in a construction during its lifetime (individuals, furnishings, safes, libraries, vehicles, machines, equipment or deposited equipment). Construction codes generally specify the magnitudes of building design live loads. Usually defined live loads for houses as evenly spread surface loads. (Dead loads) are steady magnitude loads and set locations that operate on the structure continuously. These stresses comprise the weight of the structural system itself and the rest of the structural system's materials and machinery.

5.2. Loads of Lateral (Horizontal Loads)

Lateral loads provide live loads that serve as the main component with horizontal force on the structure. A normal lateral load becomes a wind load against an exterior or storm. Most horizontal stresses differ in strength based on the place of the building, building equipment, height and structure of the building.

6. Selected Case Studies

The focus of this research is a rectangle 11 meters broad and 7 meters on the case study building's structure. A model size for the Maximum 3D printing machine is 14 x 22 cm, use a scale of (1:50). A model part of the same dimension is to be investigated from weight and resistance comparative directions compared with the following selected case studies structures:

- Victoria Gate in Leeds City Centre.
- O-14 Folded Exoskeleton in Dubai.
- The Broad Museum in Los Angeles.

7. Selected Case Studies Portion Weight Calculations

Table 2, Table 3, and Table 4. Show typical for O-14 and Urban Hive buildings, a 7 X 11 m of the external skin portion will be studied.[11] in (VICTORIA GATE) white Concrete density calculations is 3150 kg/m3, in (O-14 FOLDED) calculations Reinforced Concrete density is 2500 kg/m3, While in (Broad Museum) concrete of density 1800 kg/m3 [12] is used for the panels are made of fiberglass-reinforced concrete.

IOP Conf. Series: Materials Science and Engineering **974** (2020) 012011 doi:10.1088/1757-899X/974/1/012011

Table 2. Calculation of Victoria Gate Building in Leeds City Centre.

	VICTORIA GATE IN LEEDS CITY CENTRE	
SECTION OF THE STUDY	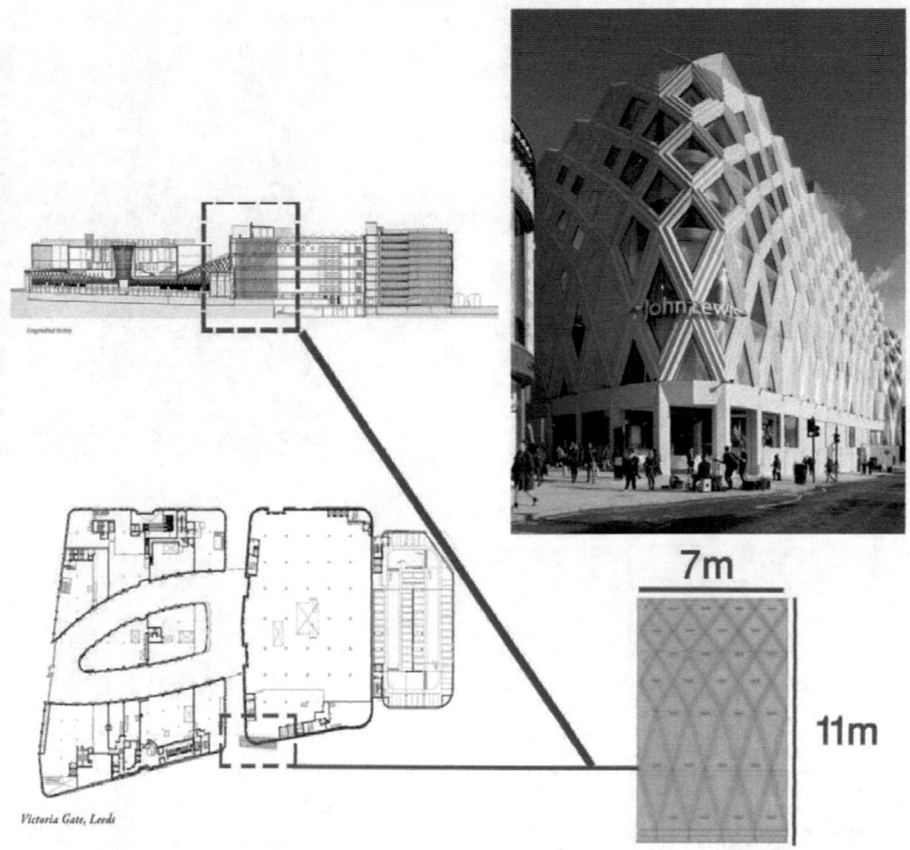	
CALCULATIONS	Openings percentage	34% of the facade
	Wall thickness	0.4m
	Total Concrete Area	7 X 11 = 77 m2
	Total Openings Area	26.18 m2
	Gross Concrete Mass	Density X Volume = 3150 X (7 X 11 X 0.4) = 97020 kg
	Gross Openings Mass	Density X Volume = 3150 X (26.18 X 0.4) = 34986 kg
	Net Concrete Mass	97020 – 34986 = 62034 kg = 62 ton

IOP Conf. Series: Materials Science and Engineering **974** (2020) 012011 doi:10.1088/1757-899X/974/1/012011

Table 3. Calculation of O-14 Folded Building in Dubai.

	O-14 FOLDED IN DUBAI	
SECTION OF THE STUDY	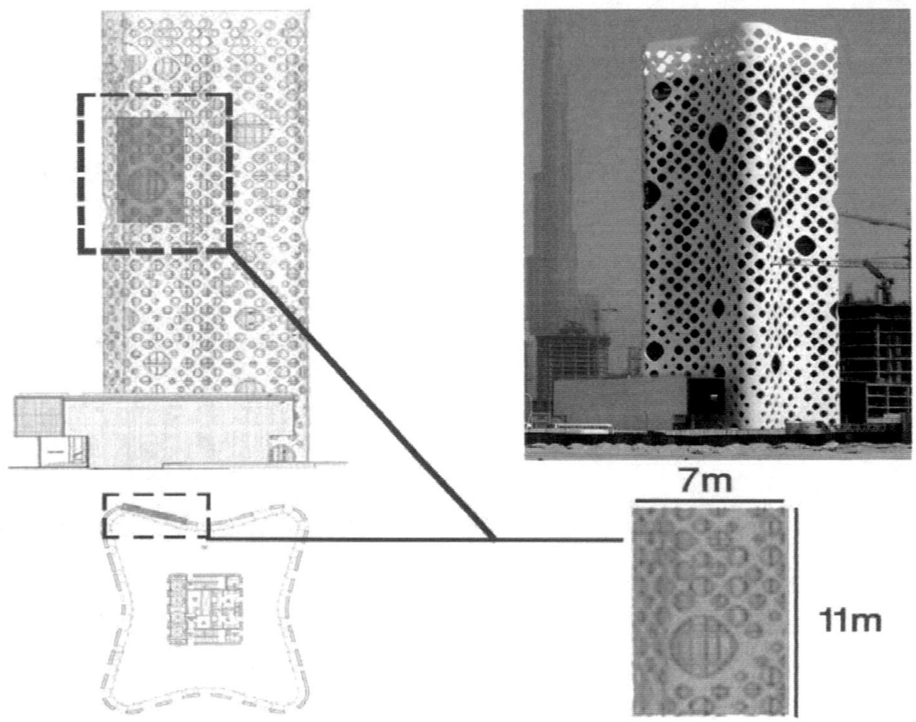	
CALCULATIONS	openings percentage	45% of the facade
	Wall thickness	0.6m
	Total Concrete Area	7 X 11 = 77 m2
	Total Openings Area	34.65 m2
	Gross Concrete Mass	Density X Volume = 2500 X (7 X 11 X 0.6) = 115500 kg
	Gross Openings Mass	Density X Volume = 2500 X (34.65 X 0.6) = 51975 kg
	Net Concrete Mass	115500 – 51975 = 63525 kg = 63.525 ton

Table 4. Calculation of the Broad Museum Building in Los Angeles.

	THE BROAD MUSEUM IN LOS ANGELES	
SECTION OF THE STUDY	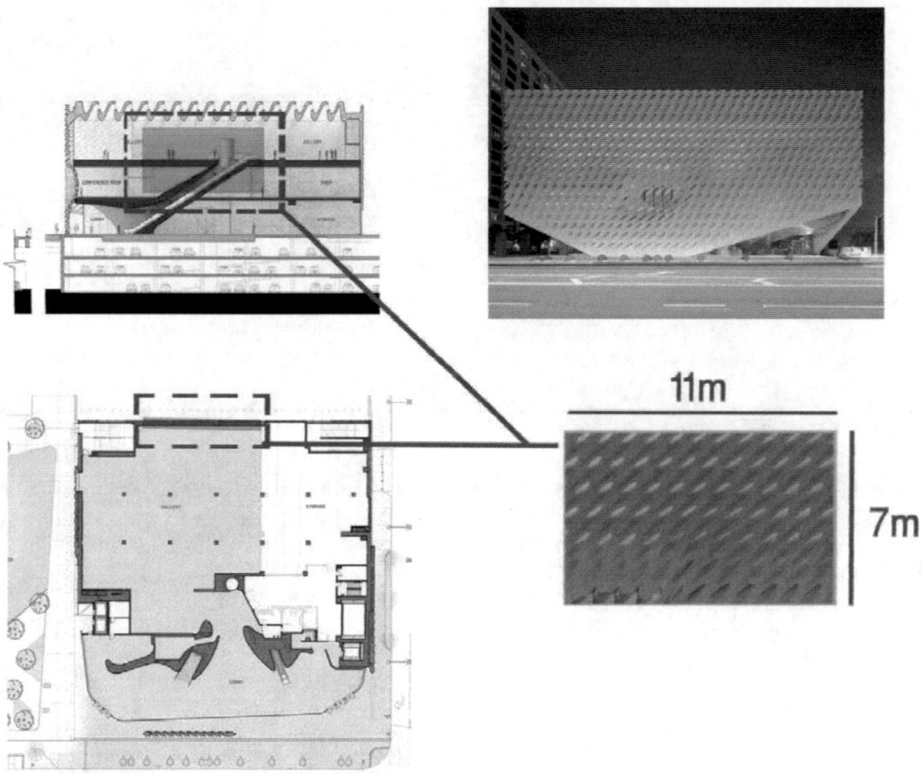	
CALCULATIONS	3 X 6 m panel weighs	6.8 ton
	panel Area	3X 6 = 18 m2
	Total Concrete Area	7 X 11 = 77 m2
	number of panels	77/ 18 = 4.2 panel
	Net Concrete Mass	4.2 X 6.8 = 28.56 ton

8. Process of Generating Model of Solidthinking Inspire

This program optimization is executed as indicated in Fig. 2. In order to generate an objective structure (maximum stiffness) with the following parameters:

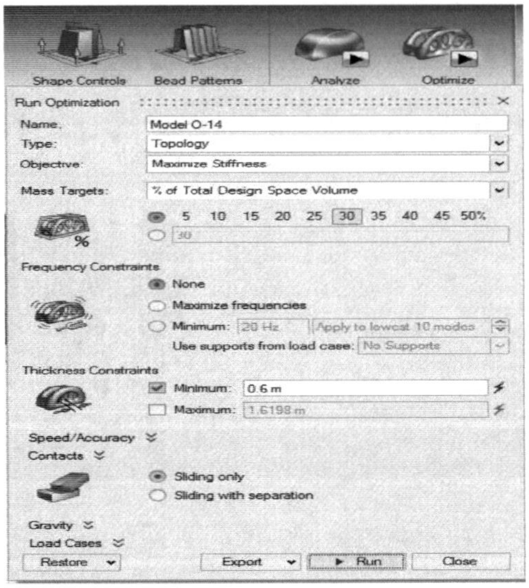

Figure 2: Program Run Optimization.

9.1. Target of Mass and Thickness Minimum

(30% of the total volume of design space) When mass objectives are used to define the amount of material to be kept, this target can either be defined as a percentage of the total design area volume or as the entire weight of the entire model. In a (Run Optimization) window, a preliminary minimum and (0.6 m) thickness may be regulated by specifying the minimum and/or maximum density.
The thickness of the design can be modified by the slider bar in the form explorer after optimization shown in Fig. 3, (Topology slider in the design space adding or extracting material from the design area in the form explorer) [13].

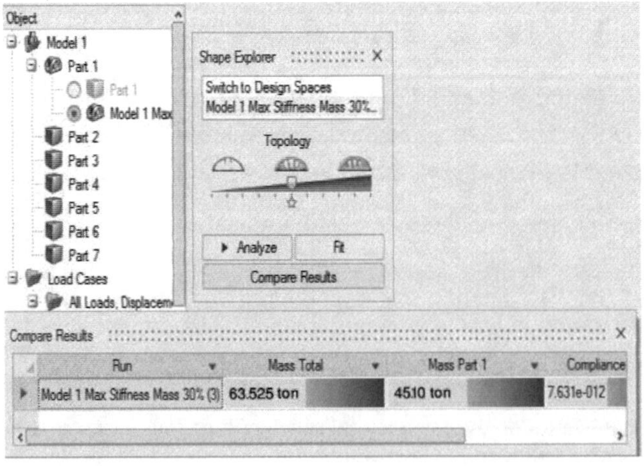

Figure 3: Shape Explorer and Compare Results.

IOP Conf. Series: Materials Science and Engineering **974** (2020) 012011 doi:10.1088/1757-899X/974/1/012011

9.2. Distribution of Forces and Supports

Structural skin forces and supports are shown in Fig. 4. Where the conditions of load are a lateral forces operating in the direction of the (+ X-axis) and the gravity loads operate in the position of the (–Y-axis). [14]

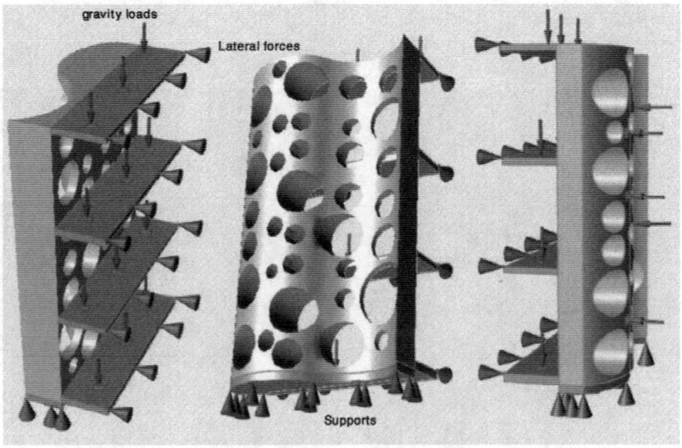

Figure 4: Distribution of Forces and Supports.

9.3. Selection of Materials

In the library of the program's materials, the characteristics of reinforced concrete, white concrete and fiberglass-reinforced concrete, as shown in Fig. 5. Where the input values of material properties are E: Young's Modulus of Elasticity (Young's modulus) or modulus of elasticity is a number which measures the resistance of an object or material to elastic deformity such as non-permanently when a force is applied to it Elastic [15], Poisson's Ratio When a material is compressed in one direction, It always tends to expand in two other directions parallel to the compression direction. This technique is called the Poisson effect. (nu) is a measure of this effect. The Poisson ratio is the expansion fraction separated by the compression fraction [16], in engineering and materials science (Density and Yield Stress) or yield point of a material is defined as the stress at which the material begins to deform plastically. When the tension added is withdrawn, the substance bends elastically before the yield point, and returns to its original shape. After reaching the yield stage, any proportion of the deformation would be persistent and non-reversible [17].

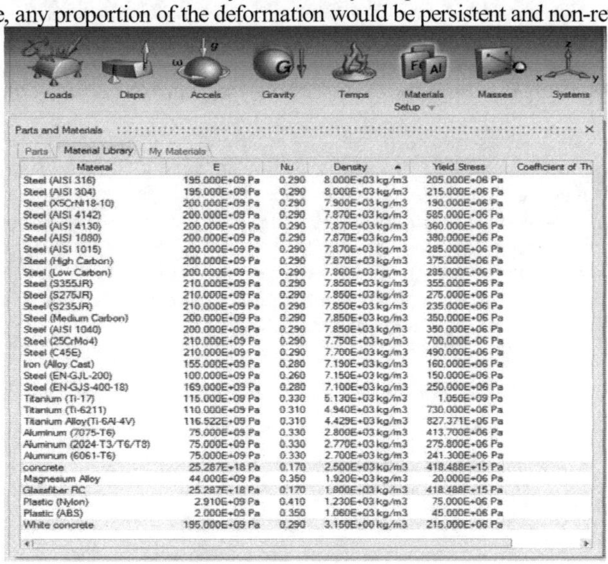

Material	E	Nu	Density	Yield Stress	Coefficient of Th
Steel (AISI 316)	195.000E+09 Pa	0.290	8.000E+03 kg/m3	205.000E+06 Pa	
Steel (AISI 304)	195.000E+09 Pa	0.290	8.000E+03 kg/m3	215.000E+06 Pa	
Steel (X5CrNi18-10)	200.000E+09 Pa	0.290	7.900E+03 kg/m3	190.000E+06 Pa	
Steel (AISI 4142)	200.000E+09 Pa	0.290	7.870E+03 kg/m3	585.000E+06 Pa	
Steel (AISI 4130)	200.000E+09 Pa	0.290	7.870E+03 kg/m3	360.000E+06 Pa	
Steel (AISI 1080)	200.000E+09 Pa	0.290	7.870E+03 kg/m3	380.000E+06 Pa	
Steel (AISI 1015)	200.000E+09 Pa	0.290	7.870E+03 kg/m3	285.000E+06 Pa	
Steel (High Carbon)	200.000E+09 Pa	0.290	7.870E+03 kg/m3	375.000E+06 Pa	
Steel (Low Carbon)	200.000E+09 Pa	0.290	7.860E+03 kg/m3	285.000E+06 Pa	
Steel (S355JR)	210.000E+09 Pa	0.290	7.850E+03 kg/m3	355.000E+06 Pa	
Steel (S275JR)	210.000E+09 Pa	0.290	7.850E+03 kg/m3	275.000E+06 Pa	
Steel (S235JR)	210.000E+09 Pa	0.290	7.850E+03 kg/m3	235.000E+06 Pa	
Steel (Medium Carbon)	200.000E+09 Pa	0.290	7.850E+03 kg/m3	350.000E+06 Pa	
Steel (AISI 1040)	200.000E+09 Pa	0.290	7.850E+03 kg/m3	350.000E+06 Pa	
Steel (25CrMo4)	210.000E+09 Pa	0.290	7.750E+03 kg/m3	700.000E+06 Pa	
Steel (C45E)	210.000E+09 Pa	0.290	7.700E+03 kg/m3	490.000E+06 Pa	
Iron (Alloy Cast)	155.000E+09 Pa	0.280	7.190E+03 kg/m3	160.000E+06 Pa	
Steel (EN-GJL-200)	100.000E+09 Pa	0.260	7.150E+03 kg/m3	150.000E+06 Pa	
Steel (EN-GJS-400-18)	169.000E+09 Pa	0.280	7.100E+03 kg/m3	250.000E+06 Pa	
Titanium (Ti-17)	115.000E+09 Pa	0.330	5.130E+03 kg/m3	1.050E+09 Pa	
Titanium (Ti-6211)	110.000E+09 Pa	0.310	4.940E+03 kg/m3	730.000E+06 Pa	
Titanium Alloy(Ti-6Al-4V)	116.522E+09 Pa	0.310	4.429E+03 kg/m3	827.371E+06 Pa	
Aluminum (7075-T6)	75.000E+09 Pa	0.330	2.800E+03 kg/m3	413.700E+06 Pa	
Aluminum (2024-T3/T6/T8)	75.000E+09 Pa	0.330	2.770E+03 kg/m3	275.800E+06 Pa	
Aluminum (6061-T6)	75.000E+09 Pa	0.330	2.700E+03 kg/m3	241.300E+06 Pa	
concrete	25.287E+18 Pa	0.170	2.500E+03 kg/m3	418.488E+15 Pa	
Magnesium Alloy	44.000E+09 Pa	0.350	1.920E+03 kg/m3	20.000E+06 Pa	
Glassfiber RC	25.287E+18 Pa	0.170	1.800E+03 kg/m3	418.488E+15 Pa	
Plastic (Nylon)	2.910E+09 Pa	0.410	1.230E+03 kg/m3	75.000E+06 Pa	
Plastic (ABS)	2.000E+09 Pa	0.350	1.060E+03 kg/m3	45.000E+06 Pa	
White concrete	195.000E+09 Pa	0.290	3.150E+00 kg/m3	215.000E+06 Pa	

Figure 5: The Library of Program's Materials.

9.4. Analysis of Compression and Tension

Fig. 6 shows the resulting forces of compression and tension. This enables the allocation of the material where needed by the material characteristics and the forces of acting (metal plates are tension, concrete is compression) at further phases of the design.

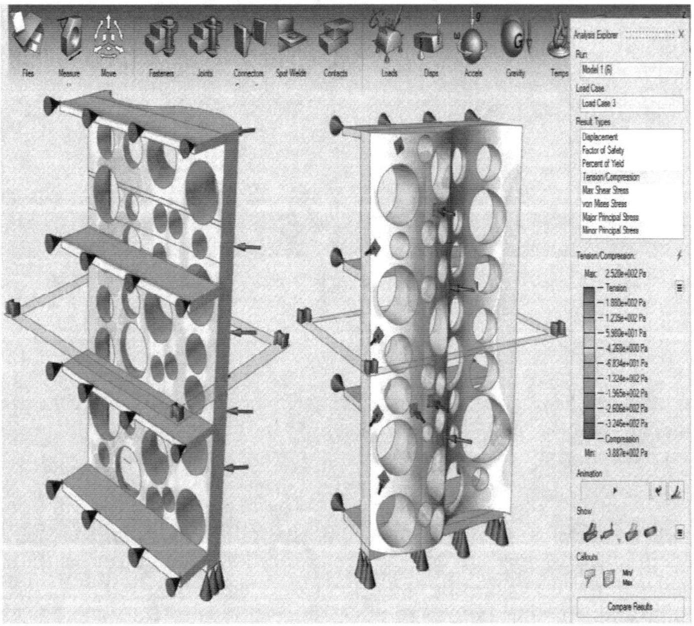

Figure 6: Analysis of Compression and Tension.

10. Results

The following Table 5. summarizes comparative outcomes from the material optimization and structural effectiveness factors of contrast between models generated from solidthinking inspire software and the case study buildings designs and That produces a 3D printed models Which produce forms with a lower density ratio for the materials used, Where we notice after the improvement process, the biggest decrease was in the second case, which is the traditional Reinforced Concrete material, and then comes the first case that used white concrete and the least of the third case, which used Fiberglass Reinforced Concrete material, which indicates that the use of improvement differs from one substance to another, where optimization is at the highest level concerning untreated traditional materials, which leads to a difference in the density of the materials and thus a decrease in cost.

Table 5. Comparison of material and structural efficiency with
The victoria gate, the o-14 building and the broad museum.

	Victoria Gate	O-14 Folded	The Broad Museum
Model to optimize	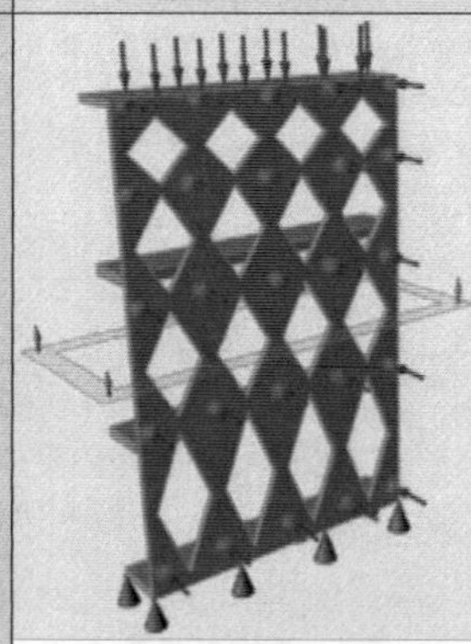	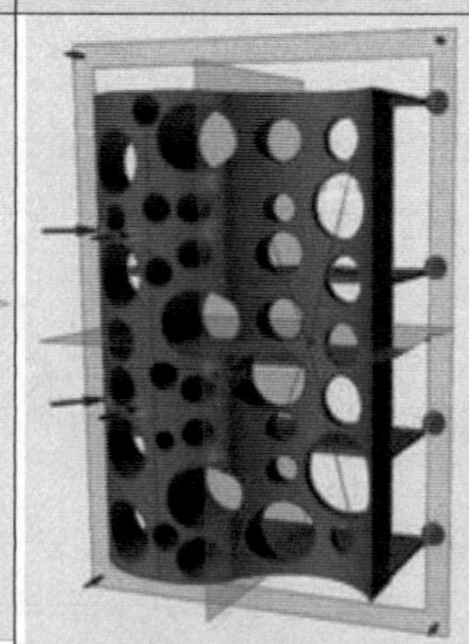	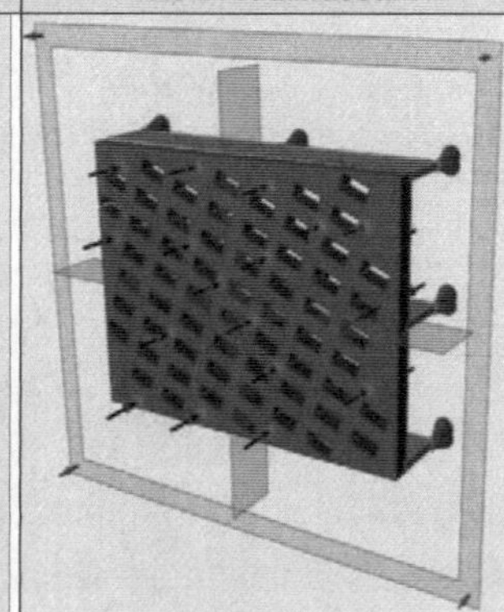
Optimized model by SolidThinking Inspire	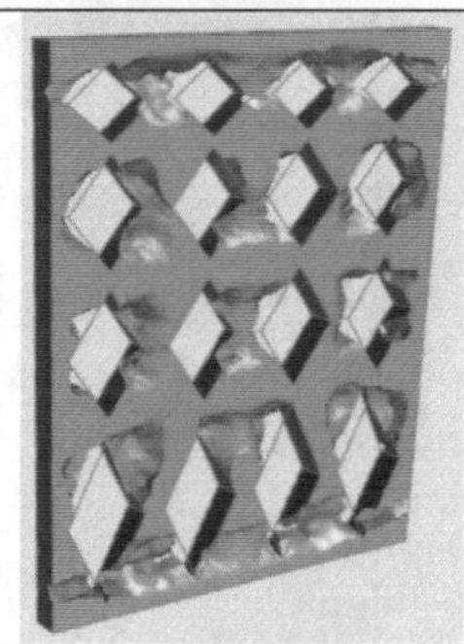	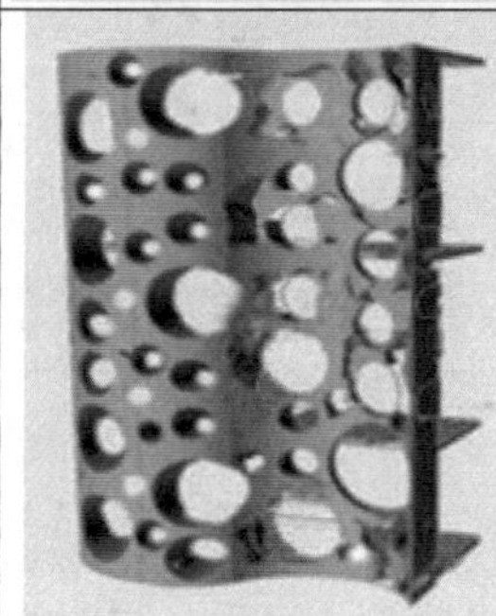	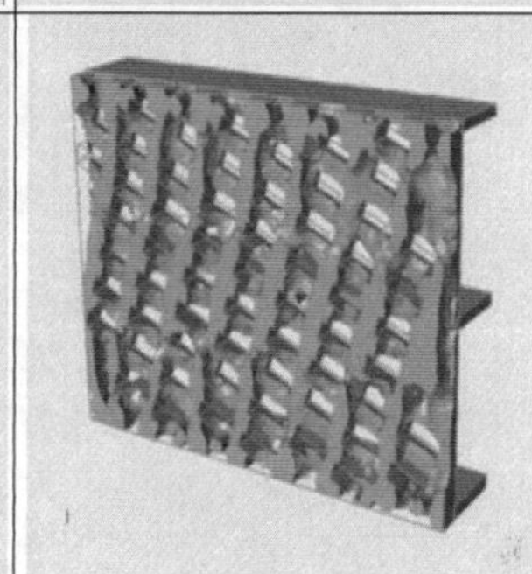
Structural System	Diagrid	Diagrid	Exoskeleton
Skin Material	White Concrete Panels (Density = 3150 kg/m^3)	Reinforced Concrete (Density = 2500 kg/m^3)	Precast Fiberglass Reinforced Concrete Panels (Density = 1800 kg/m^3)
Weight of (7m X 11m) Exterior Load-Bearing wall section before optimization	62 ton	63.525 ton	28.56 ton
Weight of (7m X 11m) Exterior Load-Bearing wall section after optimization	53.514 ton calculation by solidThinking Inspir	45.10 ton calculation by solidThinking Inspir	25.01 ton calculation by solidThinking Inspire
A 3D Printed model		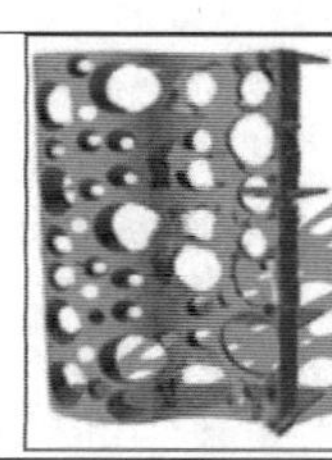	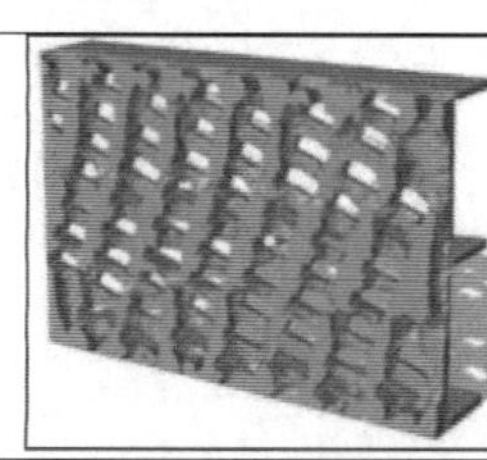
Average reduced percentage of material	The percentage of Weight reduce to **13.7%**	The percentage of Weight reduce to **29%**	The percentage of Weight reduced to **12.4%**

The improvement aimed to maximize hardness to obtain the lowest material density and thus the lowest cost while preserving the shape and preventing deformation. The optimization of the structural wall provides mechanical properties as designed with the parameter (maximum deviation hardness) where a structure weighs 18.36% less than similar cases before optimization, Where in the first case, the rate decreased from 62 ton to 53.514 ton by 13.7 percent, In the second case, the rate decreased by 29 percent from 63,525 ton to 45.10 ton and In the third case, it decreased by 12.4 percent from 28.56 ton to 25.01 ton As shown in Fig. 7.

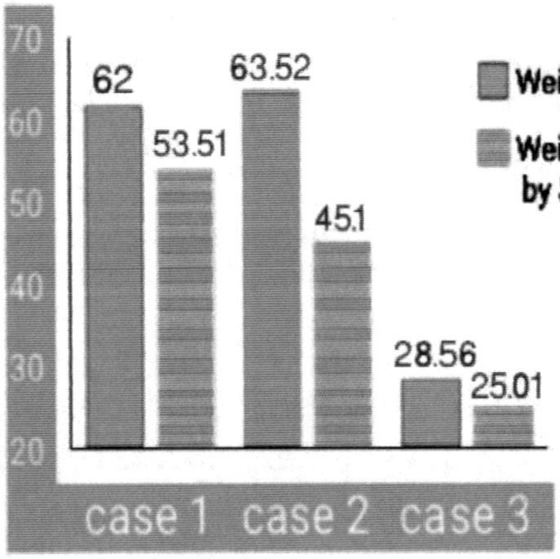

Figure 7: Reduce Weight of Load-Bearing Wall after Process Optimization.

The model input parameters (design volume, materials, and loading conditions) interact as they affect each other. Integral systems are inseparable and are given a direct integration result it is possible to produce shapes with distributed loads directly related to the shape and materials used which is the aim of this study.

4. Conclusion

The use of algorithms enables the creation of a wide range of solutions characterized by speed and effortlessly, in addition to allowing the development of different design methods.
Application of mechanically integrated behavior in the early stages of architectural design on the load-bearing wall, to build its material and structural efficiency, the strategy used to solve the problem is It is a method of integrating each of the known parameters such as load cases, supports, design volume and Materials:

- Structural wall skin is modeled and created using a topology optimization algorithm using software (solidthinking inspire).
- Model size constraints for the Maximum 3D printing machine are 14 x 22 cm, accordingly and to maintain a scale of (1:50); a 7 x 11 m portion of the structural skin was examined during the experimental study.
- The optimization on the structural wall skin of mechanical properties provides a structure which is 18.36 percent less in weight than the comparable cases.
- The use of algorithm programs to improve materials has a Clear on the relationship between form and functionality. As for the improvement of conventional reinforced concrete, it resulted in lower density, hardness and better surface quality than white concrete, followed by fiberglass-reinforced concrete. Conventional reinforced concrete can be used to build lightweight structures after improvement which is more structurally functional.

IOP Conf. Series: Materials Science and Engineering **974** (2020) 012011 doi:10.1088/1757-899X/974/1/012011

5. References

1. Sharpe JEE. Computer tools for integrated conceptual design. Des Studies1995;16:471–88.
2. Zhu J, Zhang W, Beckers P. Integrated layout design of multi-component system. Int J Numer Methods Eng 2009;78(6):631–51
3. Liu S, Hu R, Li Q, Zhou P, Dong Z, Kang R. Topology optimization-based lightweight primary mirror design of a large-aperture space telescope. Appl Opt 2014;53(35):8318–25.
4. Zhou M, Rozvany GIN. The COC algorithm, Part II: topological, geometrical and generalized shape optimization. Comput Methods Appl Mech Eng 1991; 89(1–3):309–36.
5. Bendsøe MP, Sigmund O. Material interpolation schemes in topology optimization. Arch Appl Mech 1999;69(9–10):635–54.
6. Xie YM, Steven GP. A simple evolutionary procedure for structural optimization. Comput Struct 1993;49(5):885–96.
7. Wilfredo Méndez , Structuring Biomimicry, Improving Building's Resiliency, http://ieet.org/index.php/IEET/print/6168 , Accessed: October 2019.
8. Semra Arslan sulcus, a biomimetic approach, doctor of philosophy in building science in architecture, middle east technical university, 2009.
9. Neri Oxman, Structuring Materiality Design Fabrication of Heterogeneous Materials, Architectural Design Journal AD, The New Structuralism Design, Engineering and Architectural Technologies, Wiley, https://onlinelibrary.wiley.com/doi/abs/10.1002/ad.1110, Accessed: June 2019.
10. Chapter 16, Structural Design, International Building Code, New Jersey Edition, 2006.
11. Thomas A. Holm and John P. Ries, Specified Density Concrete; A Transition for the Concrete Industry, Expanded Shale, Clay & Slate Institute (ESCSI) - United States.
12. Pshtiwan N. Shakor, S. S. Pimplikar, Glass Fiber Reinforced Concrete Use in Construction, International Journal of Technology And Engineering System (IJTES), Jan – March, Vol.2.No.2, 2011.
13. solidThinking Inspire 2014, Tutorial: Sketching and Optimizing a Bridge, http://www.solidthinking.com/tutorials/2014- Tutorials/mac/en/sketching_optimizing.htm , Accessed: May 2019.
14. Loads were calculated according to "The Egyptian Code for the Calculation of Loads and Forces in Structural and Masonry Works, 2008.
15. Elastic modulus, https://en.wikipedia.org/wiki/Elastic_modulus , Accessed: October 2019.
16. Poisson's ratio, https://en.wikipedia.org/wiki/Poisson's_ratio , Accessed: October 2019.
17. Yield (engineering), https://en.wikipedia.org/wiki/Yield_%28engineering%29 , Accessed: October 2019.

MilitaryTechnicalCollege
Kobry El-Kobbah,
Cairo, Egypt

13th International Conference on Civil and Architecture Engineering ICCAE-13-2020

Deflection of Thin-Walled Panels Loaded in Shear with Different Types of End Stiffeners

E A Mousa[1], M H El-Boghdadi[2], N M Yossef[2,3]

[1]PhD student, department of Structural Engineering, Faculty of Engineering, Tanta University, Tanta,
[2]Associate Professor, Dept. of Structural Engineering, Faculty of Engineering, Tanta Univ., Tanta,
[3] Corresponding author (Mobile: +201225151616; Fax: *+20403315860*; Email: *nashwa.yossef@gmail.com*, *nashwa_abdeltawab@f-eng.tanta.edu.eg*

Abstract, Numerical analysis for the load/deflection behavior of thin-web plate girders have been studied in this paper using a finite element program. An analytical approach presented in literature was discussed, and hence, it was used to calculate the load/deflection values in pre-bucking and post buckling stages. The validated finite element model is used to discuss the meaningful of the analytical approach. The effect of different parameters on load-deflection curves is proposed. Stiffener thickness (t_s), end post type (no end post (NEP), rigid end post (REP), and non-rigid end post (NREP)) and end distance (e) are the main parameters considered in this paper. The analytical approach estimates well the load/deflection behaviors of REP and NREP plate girders with ($a/h_w < 2.2$). On the other hand, the analytical approach can't estimate either for the shear strength or the deflection of NEP plate girders.

KEYWORDS
Plate girders; Shear behaviour; In-plan deflection; End stiffeners; Finite element model.

1. Introduction

Plate girder is a structural element which contains steel plates assembled together by welded or bolted to one another. Rolled sections account for structures with relatively low loads and limited spans. However, when it comes to relatively heavy loads with large spans, as bridge girders with small to medium spans, designed plate girders are among the most widely used sections [1, 2].

Optimizing the plate girder dimensions presents deep thin web, that thin web tends to buckle. Hence transverse intermediate stiffeners are used to restrain the web and increase the buckling strength. At shear force, the thin web panels buckle. Therefore the shear behavior of thin web panel is one of the important design aspects of plate girder [3, 4].

A thin web panel subjected to shear was studied by several researchers [3-14]. The shear resistance of transversely stiffened plate girders was affected mainly by three factors [15]; 1) the buckling shear resistance, 2) the post-buckling shear resistance, and 3) flanges contribution to the shear resistance.

The buckling shear resistance was studied and formulated by Porter et al. [16], hence Lee and Yoo [4] propose design equations for the determination of ultimate shear strength of the web panels. In the meanwhile, Ajam and Marsh express the elastic shear buckling strength based on the Tresca yield criterion [17]. For all studied models, the shear buckling coefficient is the key parameter for the shear strength. As a simplification, and as an accurate approximation, all panel edges are assumed to be simply supported, although Galambos has present the shear buckling coefficient for panel with fixed rotational restraints at the top and bottom [18].

Wilson in 1886 was one of the first to discuss the post-buckling behaviour [19]. In a major advance in 1950, Basler and Thurlimann [20] presented extensive studies on the post-buckling behaviour of plate girder web panels. They assumed that the transverse stiffeners doing as anchors and the yield was far from flanges. Then later Basler [21, 22] presented different approach on the post-buckling behaviour of thin web panel of a plate girder. In 1970s, Rockey and his co-workers predicted the ultimate strength of webs subjected to shear [16, 23].

As noted by many researchers, increasing flange dimensions develops larger shear strength, which means flanges contribute the shear strength by the frame action. However, Höglund [24] calculated the shear contribution from the frame action.

A numerical study of nonlinear large deflection behaviour of plate girder under shear was presented by Alinia et al. [7]. It was observed that plastic hinges induced by shear in the end panel only. These hinged are induced by the shear deformation near the supports. The researchers state that the end-posts are important elements in the behaviour and design of plate girder under shear.

A number of experimental and analytical studies have been carried out to establish a mechanical model to represent the out-of-plane and the in-plane deformation characteristics of web panel under shear [25, 26]. However the proposed model by Qian and Tan [25] does not take the effect of end-post into account, there is still a need for general analysis and design methodology to combine the effect of end-post on the out-of-plane and the in-plane deformation characteristics of web panels under shear.

Hence, this paper focuses on the shear behaviour of the thin web panel of plate girders. The principal aim is to improve the current presented model of the structural behaviour for the thin panel under shear formed by Qian and Tan [25]. The key parameters that affect this panel in the current investigation are the stiffeners thickness (t_s), the type of end-post, the web aspect ratio (a/h_w) and the end-post distance (e). The scope of this study can be summarized as:

1. Discuss the current literature model for calculating the in-plane deflection of web panels loaded in shear.
2. Propose and validate a detailed finite element 3D model used to address the behaviour of full-scale plate girders.
3. Investigate factors affecting the in-plane deflection, and compare between the finite element results and current literature model.

2. LITERATURE APPROACH FOR IN-PLANE DEFLECTION

Qian and Tan [25] presented a mechanical approach to calculate the in-plane deflection of web panel loaded in shear. Qian had used Cardiff theory to calculate the shear strength in his analytical approach. In Qian approach the overall deflection is divided into two stages, which are deflection of pre-buckling stage (δ_{cr}) and deflection of post-buckling stage (δ_b). Such that the total deflection for the panel loaded in shear (δ_{pb}) could be calculated as follow;

$$\delta_{pb} = \delta_{cr} + \delta_b \qquad 1$$

Thus, the pre-buckling deflection can be calculated as:

$$\delta_{cr} = 2\,\tau_{cr}\, a\,(1+v)/E \qquad 2$$

in which:
a is the panel length;
v is the Poisson's ratio;
E is the elastic modulus;
τ_{cr} is the shear buckling stress of the web panel $\left(= K\,\frac{\pi^2 E}{12\,(1-v^2)}\left(\frac{t_w}{h_w}\right)^2\right)$;
t_w and h_w are the web panel thickness and height respectively;

K is the shear buckling coefficient in case of web panel with transversal stiffeners only and equals to

$$\begin{cases} 5.34 + 4/(a/h_w)^2 & a/h_w \geq 1.0 \\ 4 + 5.34/(a/h_w)^2 & a/h_w < 1.0 \end{cases}$$

In the meanwhile, the post-buckling deflection was derived for the shear case as follows:

$$\delta_b = \frac{(\sigma_t^y)^2 c^2 h_w t_w}{E\,(V_{ult} c - 4\,M_{pf})} \qquad 3$$

where:

c is the distance assumed between the plastic hinge and the end bearing stiffener

$$\left(= a\left(0.25 + \frac{1.6\, b_f t_f^2 f_{yf}}{t_w h_w^2 f_{yw}}\right)\right)$$

M_{pf} is the plastic moment capacity of flanges $(= 0.25 f_{yf} A_f\, t_f)$

σ_t^y and V_{ult} are the tensile yield stress of web panel under shear and the post-buckling shear force component, respectively, which can be calculated as:

$$\sigma_t^y = -\frac{3}{2}\tau_{cr} \sin 2\theta + \sqrt{f_{yw}^2 + \tau_{cr}^2[(1.5 \sin 2\theta)^2 - 3]} \qquad 4$$

$$V_{ult} = \sigma_t^y h_w t_w \sin^2\theta \left(\cot\theta - \frac{a}{h_w}\right) + 4\,\sigma_t^y t_w \sin\theta \sqrt{\frac{M_{pf}}{\sigma_t^y t_w}} \qquad 5$$

Thus, θ is the inclination of the tensile field, the angle θ may be calculated by iteration to give the maximum value of V_{ult} or can be approximated as $\left(\theta = 2/3 \tan^{-1}(\frac{h_w}{a})\right)$ as given by Rockey et al. [24].

However, equation 5 is subject to the following control [27]:

$$\frac{M_{pf}}{{h_w}^2 t_w f_{yw}} < 0.125 \left(\frac{h_w}{a}\right)^2 \left\{-\frac{3}{2}\frac{\tau_{cr}}{f_{yw}} + \sqrt{1 - 0.25\left(\frac{\sqrt{3}\tau_{cr}}{f_{yw}}\right)^2}\right\} \qquad 6$$

If the left-hand side is greater than the right-hand side stated in equation 6, the value of V_{ult} is given by:

$$\frac{V_{ult}}{\tau_{cr} t_w h_w} = 4\sqrt{3}\left(\frac{h_w}{a}\right)\frac{M_{pf}}{{h_w}^2 t_w f_{yw}} + \left\{-\frac{3\sqrt{3}}{4}\frac{\tau_{cr}}{f_{yw}} + \sqrt{0.75 - \frac{3}{16}\left(\frac{\sqrt{3}\tau_{cr}}{f_{yw}}\right)^2}\right\} \qquad 7$$

3. FINITE ELEMENT MODEL

A detailed finite element (FE) 3D model used to address the behaviour of transversely stiffened, full-scale plate girders was developed and verified. Nonlinear behaviour of the thin web panel was presented. The verified FE model is used to promote the current literature equations for in-plane deflection and study additional factors affecting the behaviour of thin web panels under shear force.

3.1 Model description

Transversely stiffened plate girders were modelled using reduced-integrated shell element S4R. The element S4R, from non-linear finite element analysis of the ABAQUS platform element library [28], has four nodes each with six degrees of freedom. The full-scale 3D model includes both geometric and material nonlinearity. The mesh convergence study verified that the element with size 30 mm x 30 mm gives reasonable accuracy.

Girders with no end post (NEP), with rigid end post (REP), and with non-rigid end post (NREP) were included. Plated girder dimensions notations and various end supports were introduced, according to EN 1993-1-5 [29], as shown in figure 1.

IOP Conf. Series: Materials Science and Engineering **974** (2020) 012012 doi:10.1088/1757-899X/974/1/012012

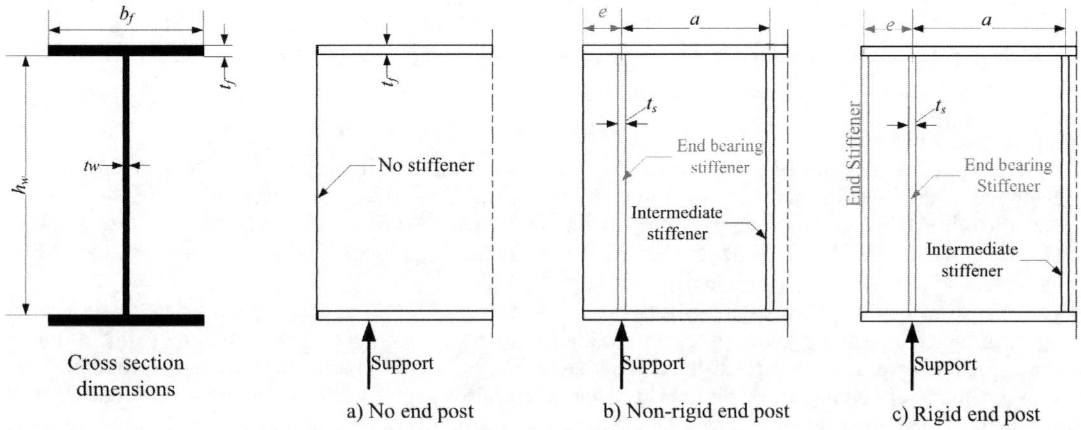

Figure 1. Cross section dimensions and types of end supports.

3.2 Boundary conditions and load procedures

The boundary conditions and load procedures follow what is indicated in [7], to assure quite small bending moment and constant shear force for the end panel. However, two-point loads were applied at the third points, and simply supported boundary conditions were applied to the end section. Both load and boundary conditions are presented in figure 2.

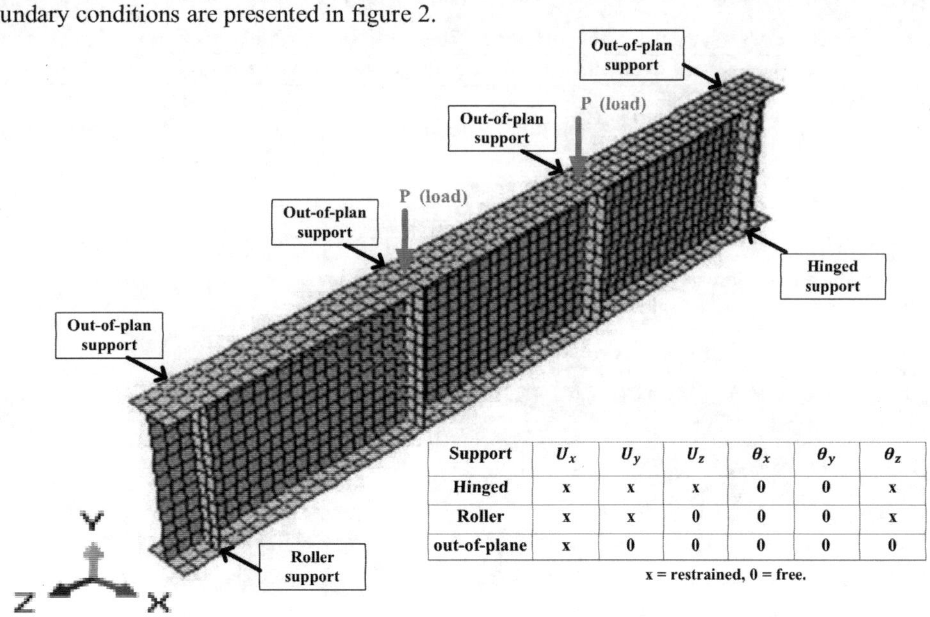

Support	U_x	U_y	U_z	θ_x	θ_y	θ_z
Hinged	x	x	x	0	0	x
Roller	x	x	0	0	0	x
out-of-plane	x	0	0	0	0	0

Figure 2. FE mesh, load and boundary conditions used.

3.3 Material properties

Bilinear elastic-plastic curve with linear strain hardening was used to simulate the behavior of the steel material for web, flanges and stiffeners. The slope of the linear elastic part of the curve, which mean

IOP Conf. Series: Materials Science and Engineering **974** (2020) 012012 doi:10.1088/1757-899X/974/1/012012

Young's modulus, was used as E = 200 GPa, and Poisson's ratio was taken v = 0.3. The slope of the hardening part was taken as 2 GPa. The used steel material is S355, which has ultimate stress 510 MPa and yield stress 355 MPa.

3.4 Model validation

To determine the accuracy of the FE model, the model was used to simulate the behaviour of specimens exhibited in the experimental tests of Zhu and Zhao [26]. As well as, the numerical model studied by Alinia et al. [7]. Table 1 presents the comparison between the available results in the literature and the FE model results. The mean value for the ratio of the available ultimate load in literature (P_u) to the FE ultimate load ($P_{u,FE}$) is 0.97 with standard deviation of 0.05.

The failure modes of plate girders in [26] were compared with the failure of FE model, as shown in figure 3, which indicates a good agreement. In the meanwhile, the failure modes stated by [7] for studied plate girders were also confirmed by the FE study, as shown in figure 4. Finally, the comparisons between load-deflection curves for the literature plate girders and the FE load deflection curves are presented. Where the FE model accurately predicts the load-deflection curves, as can be seen in figure 5.

Table 1. Comparison between available ultimate load in the literature ($P_{u,L}$) and the FE one ($P_{u,FE}$).

Plate girder	h_w [mm]	t_w [mm]	b_f [mm]	t_f [mm]	L [mm]	a [mm]	f_{yw} [MPa]	E [GPa]	$P_{u,L}$ [kN]	$P_{u,FE}$ [kN]	$P_{u,L}/P_{u,FE}$
Details of experimental tests of Zhu and Zhao [26]											
G1	600	8	180	16	2700	900	280	208	1200	1291	0.93
G2	600	8	180	16	3600	1200	280	208	985	1037	0.95
Details of the simulated plate girder studied by Alinia et al. [7]											
G(A)	1000	4	300	9	4000	1000	345	210	800	773	1.03
G(B)	1000	3.33	300	7.5	4000	1000	345	210	N/A	551	N/A
G(C)	1000	3.33	300	3.33	4000	1000	345	210	N/A	312	N/A
Mean*											**0.97**
Standard deviation											***0.053***

* The ultimate load $P_{u,L}$ for the plate girder G(B) and G(C) are unavailable, hence they were excluded from the mean and the standard deviation

a) b)

Figure 3. Lateral torsional buckling of tested specimen G2: a) Experimental results [26], b) F.E. results

IOP Conf. Series: Materials Science and Engineering **974** (2020) 012012 doi:10.1088/1757-899X/974/1/012012

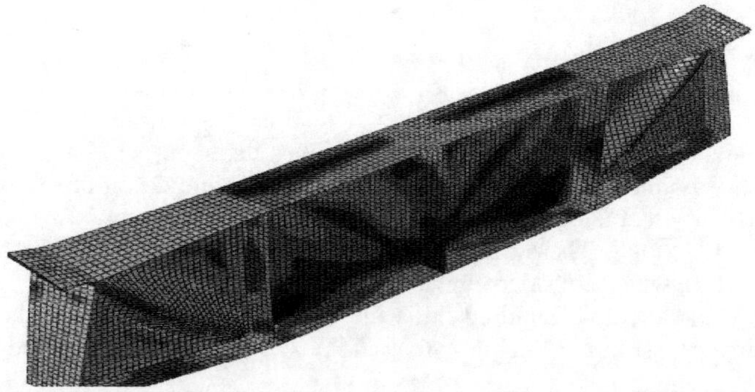

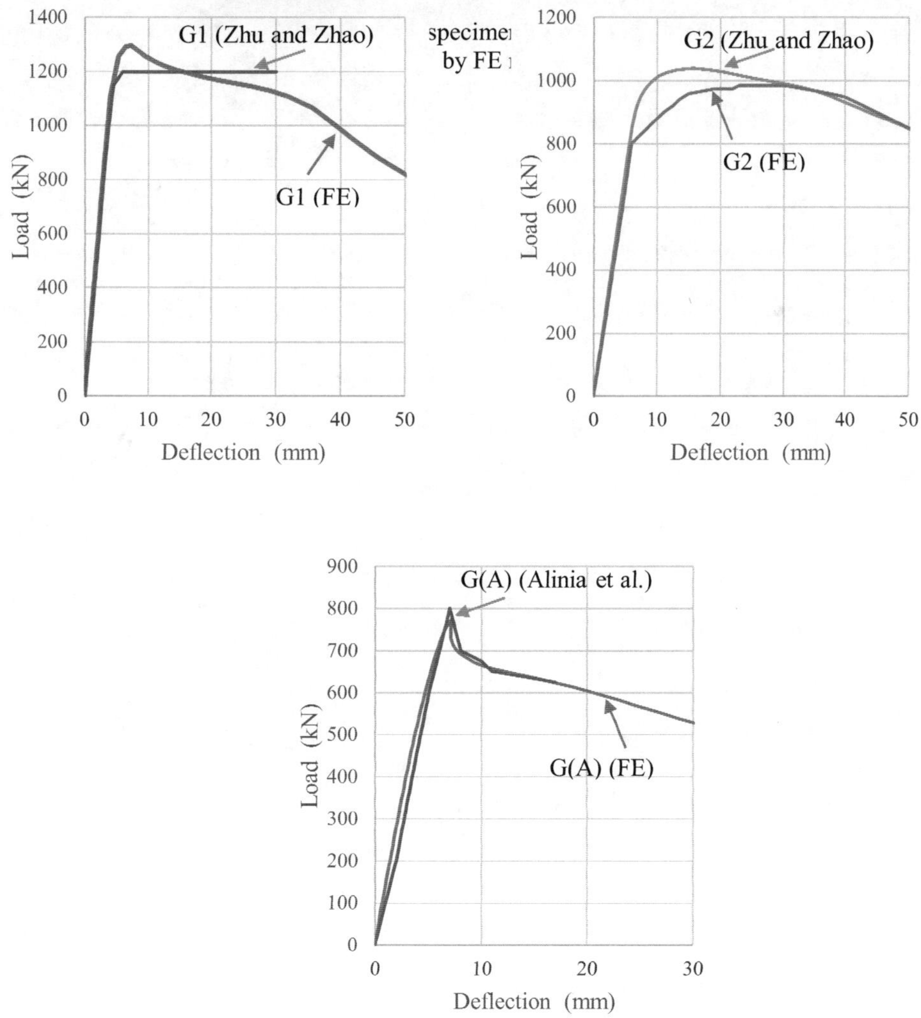

Figure 5. Comparison between literature girders and FE load-deflection curves.

4. NUMERICAL STUDY

IOP Conf. Series: Materials Science and Engineering **974** (2020) 012012 doi:10.1088/1757-899X/974/1/012012

4.1 Assumptions and parameters

In this section, the analytical and numerical load/deflection behaviour of thin-walled panel loaded in shear is investigated. Based on the main target of that section for studding the shear failure, the span length of the modelled plate girders was taken as 3 times the panel width ($3a$), so the bending effect on the studied panel was minimized. Regarding the geometrical limitation of thin-walled panel, h_w/t_w ratio was taken as 200 for all studied panels. Twenty (a/ h_w) ratios were investigated, such that (a/ h_w) ratio was varied from 3.5 to 0.8. The initial imperfections values (w_o) were introduced for the model with scale factors ($h_w/200$), which considers a greater value for the bow of the web.

The current study contains four hundred full scale thin-web plate girders, as shown in table 2, covering three main parameters; stiffener thickness (t_s), end-post type (REP, NREP and NEP) and end distance (e).

IOP Conf. Series: Materials Science and Engineering **974** (2020) 012012 doi:10.1088/1757-899X/974/1/012012

Table 2. Details of studied specimens.

Parameter	No. of specimens	h_w (mm)	a/ h_w	t_w (mm)	h_w/ t_w	span (L) (mm)	b_f (mm)	t_f (mm)	End stiffeners type	Initial imperfection w_o	t_s *(mm)*
Effect of stiffener thickness (t_s)	160	1000	3.5, 3.4, 3.3, 3.2, 3.1, 3.0, 2.8, 2.6, 2.4, 2.2, 2.0, 1.8 1.7, 1.6, 1.5, 1.4, 1.3,1.2, 1.0, 0.8	5	200	3a	300	25	Rigid end post (REP), Non-rigid end post (NREP)	$h_w/200$	5 10 15 20
Effect of end-post	60	1000	3.5, 3.4, 3.3, 3.2, 3.1, 3.0, 2.8, 2.6, 2.4, 2.2, 2.0, 1.8 1.7, 1.6, 1.5, 1.4, 1.3,1.2, 1.0, 0.8	5	200	3a	300	25	No-end post (NEP), Rigid end post (REP), Non-rigid end post (NREP)	$h_w/200$	10
Effect of end distance (e)	180	1000	3.5, 3.4, 3.3, 3.2, 3.1, 3.0, 2.8, 2.6, 2.4, 2.2, 2.0, 1.8 1.7, 1.6, 1.5, 1.4, 1.3,1.2, 1.0, 0.8	5	200	3a	300	25	e = 100 mm for NEP, REP, NREP e = 150 mm for NEP, REP, NREP e = 300 mm for NEP, REP, NREP	$h_w/200$	10

4.2 Results discussion

The specimens listed in table 2 were solved numerically using (ABAQUS), hence load-deflection curves for these specimens were determined numerically and analytically. Investigating the effect of different parameters on the finite element load-deflection curves, then conforming these results to the theoretical approach is the main goal of this study.

The results and the primary parameters are presented and discussed herein, where $P_{u,FE}$ is the ultimate load carrying capacity calculated numerically using the FE model, which is the maximum load stated, and was typically followed by a drop in the load value. $\delta_{Pb,FE}$ is the maximum FE vertical deflection under loading points for the post buckling stage, which is followed by failure stage (an increase in the deflection with nearly constant load).

In the meantime, the shear deflection $\delta_{pb,a}$ calculated by the analytical approach were expressed in equation 1, and the ultimate shear force $P_{u,a}$ is calculated such that;

$$P_{u,a} = \tau_{cr}(A_w) + V_{ult} \quad 8$$

4.3 Effect of stiffeners thickness

To investigate the effect of stiffeners thickness, 160 specimens were studied as listed in table 2. Specimens with REP and NREP were examined. The specimens had the same cross section dimensions, but different stiffeners thickness (5 mm, 10 mm, 15 mm, and 20 mm). Twenty ratios of (a/ h_w) were studied separately and combined, based on the aspect ratio of the web panels to evaluate the elastic shear buckling strength and post buckling strength, [4].

Table 3 & 4 present the comparison between the analytical results of the presented approach and the FE results for girders with stiffeners thickness 5 and 20, respectively. When the thickness of stiffeners were 5 mm, the mean values of $P_{u,a}/P_{u,FE}$ for REP and NREP are 0.96 and 1.00, with standard deviations of 0.17 and 0.19, respectively. Then the mean values $P_{u,a}/P_{u,FE}$ are 0.9 and 0.95 with standard deviations 0.11 and 0.15, for REP and NREP girders respectively, with stiffeners thickness 20 mm. Which indicates a good agreement with the analytical results.

For girders with thin stiffener (t_s = 5 mm), the mean values of $\delta_{pb,a}/\delta_{pb,FE}$ for REP and NREP are 0.80 and 0.82, with standard deviations of 0.11 and 0.12, respectively. Hence for girders with thick stiffeners (t_s = 20 mm), the mean values of $\delta_{pb,a}/\delta_{pb,FE}$ for REP and NREP are 0.82 and 0.87 respectively, with standard deviations of 0.10, for both. Deflection values indicate slight discrepancies between the FE and analytical approach. The authors believe that the presented approach underestimate the deflection values, since it only considers the shear deflection. But the FE model were subjected to shear and bending effects. Furthermore, results confirmed the small impact of the stiffener thickness on both shear strength and deflection of thin girders.

Figures 6, 7 and 8 present comparisons between the analytical and the FE load-deflection curves for different specimens, where it can be seen that the analytical approach accurately predicts the load-deflection curves for the studied specimens, especially for girders with panel aspect ratio ranged from a/h_w =1.7 to a/h_w = 2.2, hence the performance has some discrepancy for $a/h_w > 2.2$. That was probably as discussed, such that for $a/h_w > 2.2$, the effect of bending was significant as the panel and the girder length increased. Although these slight deviations between analytical predictions and numerical results, the given approach still exhibits acceptable accuracy compared with these FE especially for $a/hw < 2.2$. In the meantime, the stiffeners thickness has negligible effect on the load-deflection behavior.

Table 3. Analytical and FE loads/deflections of the studied specimens with ($t_s = 5\ mm$) values.

h_w. t_w (mm)	a/h_w	a mm	b_f x t_f (mm)	t_s mm	Analytical Eq. $P_{u,a}$ kN	Analytical Eq. $\delta_{pb,a}$ mm	FE (REP) $P_{u,FE}$ kN	FE (REP) $\delta_{Pb,FE}$ mm	FE (NREP) $P_{u,FE}$ kN	FE (NREP) $\delta_{Pb,FE}$ mm	Ratio (REP) $P_{u,a}/P_{u,FE}$	Ratio (REP) $\delta_{pb,a}/\delta_{Pb,FE}$	Ratio (NREP) $P_{u,a}/P_{u,FE}$	Ratio (NREP) $\delta_{pb,a}/\delta_{Pb,FE}$
1000 x 5	3.5	3500	300 x 25	5	701	19.14	954	23.20	939	23.00	0.73	0.83	0.75	0.83
	3.4	3400			714	18.12	950	21.41	843	21.16	0.75	0.85	0.85	0.86
	3.3	3300			727	17.12	952	19.84	942	19.74	0.76	0.86	0.77	0.87
	3.2	3200			742	16.15	961	24.28	946	23.90	0.77	0.67	0.78	0.68
	3.1	3100			757	15.21	969	22.40	951	22.26	0.78	0.68	0.80	0.68
	3.0	3000			773	14.30	976	20.64	952	18.36	0.79	0.69	0.81	0.78
	2.8	2800			808	12.57	987	15.37	962	13.04	0.82	0.82	0.84	0.96
	2.6	2600			848	10.95	994	12.53	973	12.62	0.85	0.87	0.87	0.87
	2.4	2400			894	9.46	1005	12.73	971	11.40	0.89	0.74	0.92	0.83
	2.2	2200			947	8.08	1005	10.91	989	10.52	0.94	0.74	0.96	0.77
	2.0	2000			1009	6.83	1027	8.29	992	10.01	0.98	0.82	1.02	0.68
	1.8	1800			1081	5.69	1083	6.73	1033	7.07	1.00	0.85	1.05	0.80
	1.7	1700			1123	5.16	1096	6.02	1032	5.71	1.02	0.86	1.09	0.90
	1.6	1600			1168	4.67	1095	5.07	1044	5.03	1.07	0.92	1.12	0.93
	1.5	1500			1217	4.20	1092	4.45	1063	4.88	1.11	0.94	1.14	0.86
	1.4	1400			1272	3.77	1135	3.98	1083	4.02	1.12	0.95	1.17	0.94
	1.3	1300			1332	3.36	1164	3.46	1106	3.50	1.14	0.97	1.20	0.96
	1.2	1200			1399	2.99	1193	4.42	1132	3.02	1.17	0.68	1.24	0.99
	1.0	1000			1556	2.33	1282	3.33	1203	3.36	1.21	0.70	1.29	0.69
	0.8	800			1769	1.82	1408	3.26	1314	3.53	1.26	0.56	1.35	0.52
							Mean				**0.96**	**0.80**	**1.00**	**0.82**
							Standard deviation				**0.17**	**0.11**	**0.19**	**0.12**

Table 4. Analytical and FE loads/deflections of the studied specimens with (t_s = *20 mm*) values.

h_w. t_w (mm)	a/h_w	a mm	b_f x t_f (mm)	t_s mm	Analytical Eq. $P_{u,a}$ kN	Analytical Eq. $\delta_{pb,a}$ mm	FE (REP) $P_{u,FE}$ kN	FE (REP) $\delta_{Pb,F}$ mm	FE (NREP) $P_{u,FE}$ kN	FE (NREP) $\delta_{Pb,FE}$ mm	Ratio (REP) $P_{u,a}/P_{u,FE}$	Ratio (REP) $\delta_{pb,a}/\delta_{Pb,FE}$	Ratio (NREP) $P_{u,a}/P_{u,FE}$	Ratio (NREP) $\delta_{pb,a}/\delta_{Pb,FE}$
1000 x 5	3.5	3500	300 x 25	20	701	19.14	946	23.00	946	22.86	0.74	0.83	0.74	0.84
	3.4	3400			714	18.12	945	21.22	951	21.05	0.76	0.85	0.75	0.86
	3.3	3300			727	17.12	966	19.66	947	19.66	0.75	0.87	0.77	0.87
	3.2	3200			742	16.15	970	24.20	958	18.19	0.76	0.67	0.77	0.89
	3.1	3100			757	15.21	978	22.36	962	22.42	0.77	0.68	0.79	0.68
	3.0	3000			773	14.30	983	20.46	966	20.46	0.79	0.70	0.80	0.70
	2.8	2800			808	12.57	994	15.10	975	12.95	0.81	0.83	0.83	0.97
	2.6	2600			848	10.95	1002	12.91	985	11.52	0.85	0.85	0.86	0.95
	2.4	2400			894	9.46	1016	11.70	995	12.64	0.88	0.81	0.90	0.75
	2.2	2200			947	8.08	1023	10.45	1018	10.25	0.93	0.77	0.93	0.79
	2.0	2000			1009	6.83	1054	8.26	1020	7.62	0.96	0.83	0.99	0.90
	1.8	1800			1081	5.69	1119	6.65	1074	6.73	0.97	0.86	1.01	0.85
	1.7	1700			1123	5.16	1115	5.66	1077	5.70	1.01	0.91	1.04	0.91
	1.6	1600			1168	4.67	1151	5.00	1097	5.02	1.01	0.93	1.06	0.93
	1.5	1500			1217	4.20	1200	4.45	1124	4.47	1.01	0.94	1.08	0.94
	1.4	1400			1272	3.77	1251	3.92	1153	3.92	1.02	0.96	1.10	0.96
	1.3	1300			1332	3.36	1311	3.40	1185	3.41	1.02	0.99	1.12	0.99
	1.2	1200			1399	2.99	1387	4.32	1229	2.94	1.01	0.69	1.14	1.02
	1.0	1000			1556	2.33	1577	3.12	1351	3.13	0.99	0.75	1.15	0.74
	0.8	800			1769	1.82	1833	2.81	1497	2.17	0.97	0.65	1.18	0.84
							Mean				**0.90**	**0.82**	**0.95**	**0.87**
							Standard deviation				**0.11**	**0.10**	**0.15**	**0.10**

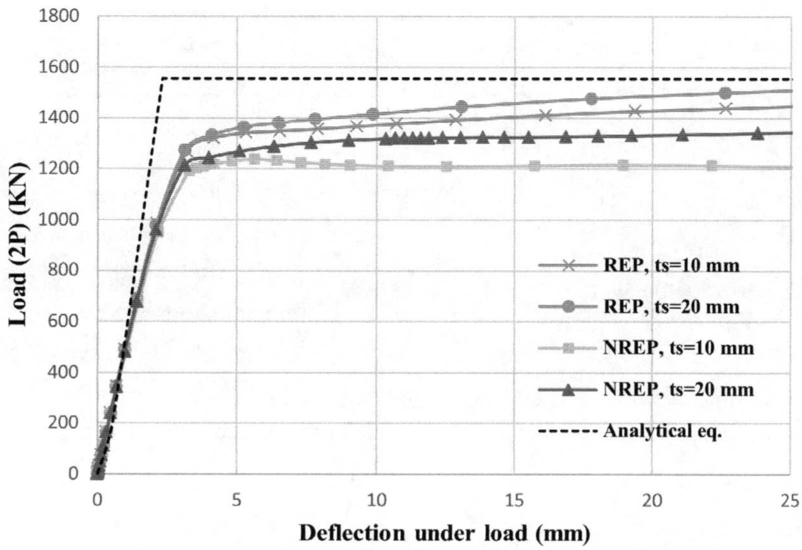

Figure 6. Comparison between analytical equation and FE load-deflection curves for REP and NREP plate girders with ($a/h_w = 1.0$).

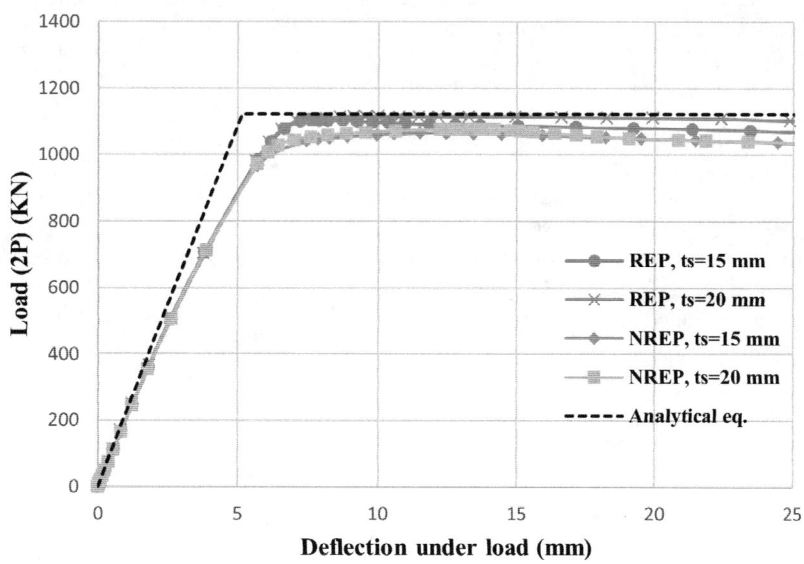

Figure 7. Comparison between analytical equation and FE load-deflection curves for REP and NREP plate girders with ($a/h_w = 1.7$).

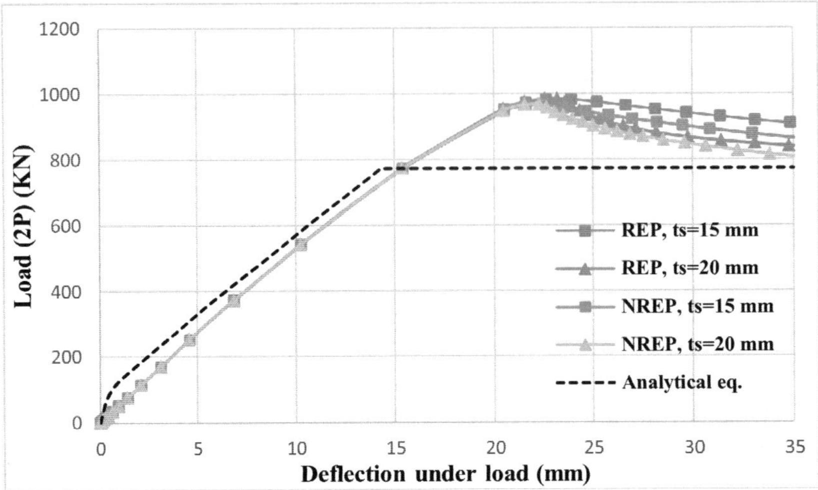

Figure 8. Comparison between analytical equation and FE load-deflection curves for REP and NREP plate girders with ($a/h_w = 3.0$).

4.4 Effect of end-post

Sixty plate girders having the same cross-sections and the same spans but with different end-post profiles were studied. Table 5 presents a comparison between analytical and FE load/deflection results for the studied plate girders. Since the analytical approach of Qian and Tan [25] neglected the effect of end-post, the analytical load/deflection results did not confirm any difference with different end-posts. Unlike the analytical results, the FE results highlight the effect of end-post rigidity on the load-deflection behaviour. The results show that providing a REP increases the ultimate shear strength ($P_{u,FE}$) of the girder, if compared with NREP. In the meantime, changing end-post from REP to NREP has insignificant effect on maximum vertical deflection ($\delta_{Pb,FE}$), in the most studied cases. It also shows that, ($P_{u,FE}$) of the same plate girder considerably decrease by providing NEP. However NPE profile has irregular effect on ($\delta_{Pb,FE}$).

The column charts presented in figure 9 shows the ratio between the analytical approach and finite element results for REP, NREP and NEP plate girders with a/h_w = 0.8, 1, 1.4, 1.7, 2, 2.2 and 3. The analytical approach estimates well the shear strength of REP and NREP plate girders, hence it can't estimate the shear strength or deflection for NEP plate girders.

In the meantime, the presented analytical approach gives underestimates deflection results for both REP and NREP plate girders.

4.5 Effect of end-distance (e)

The current sub-section studies the effect of end-distance (e) in the load-deflection curves, note that the cross-section dimensions of the studied plate girders remained constant. However, a/h_w ratio was varied from 0.8 to 3.5, as listed in tables 5, 6 and 7 for *(e)* equals 100, 200 and 300 mm, respectively.

Figure 10 presents the load-deflection curves of plate girders ($a/h_w = 2$) with different end-post profiles. Accordingly, the end distance (e) are 100 mm, 150 mm and 300 mm. the FE results show that increasing the end-distance (e) has no effect on the initial stiffness of the plate girders.

However, it slightly increases the ultimate shear strength of the girders. That was expected, since increasing the end-distance influence the rigidity of end-post and hence the position of plastic hinges on flanges at failure stage. Hence, that explains the insignificant change of the maximum deflection at the post buckling stage for the studied girder. In other word, the end-distance becomes effective only in the failure stage. Figure 10 also states the misestimating of the analytical approach for the NEP girders.

The same conclusions can be seen clearly form tables 5, 6 and 7, increasing the end-distance has insignificant effect on the deflection of the studied girders. Given that, for *(e)* equal 100, 150 and 300 mm, the mean values of $\delta_{pb,a}/\delta_{pb,FE}$ are 0.84, 0.85 and 0.85 respectively for REP girders. And equals 0.86, 0.87 and 0.86 for NREP girders.

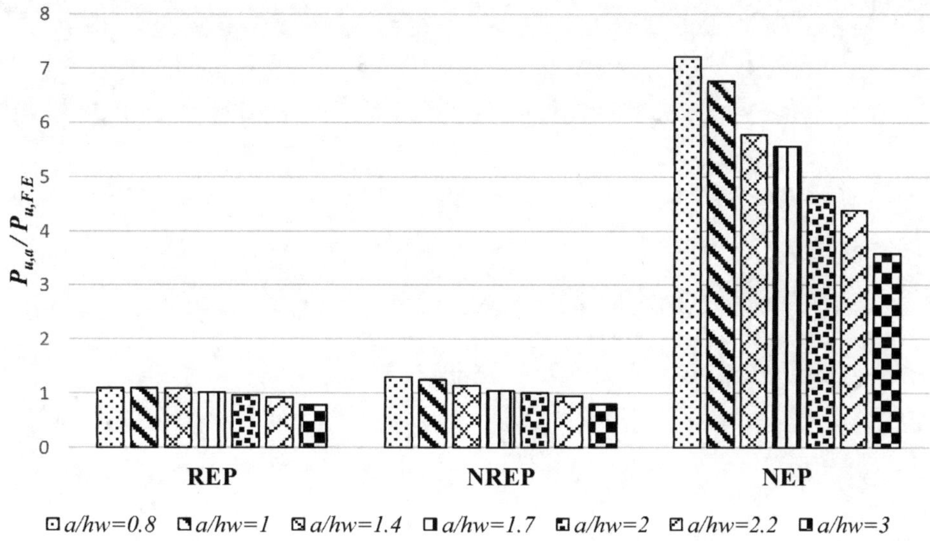

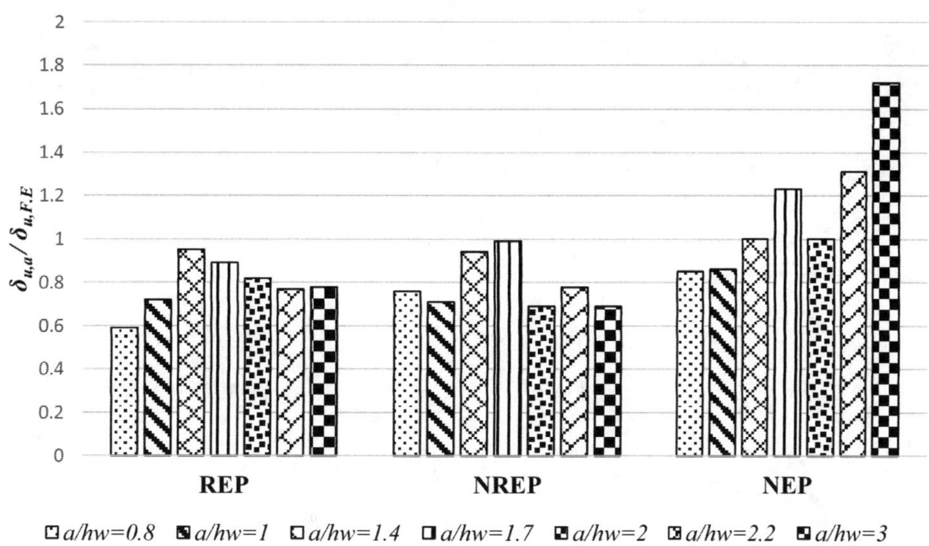

IOP Conf. Series: Materials Science and Engineering **974** (2020) 012012 doi:10.1088/1757-899X/974/1/012012

Figure 9. Comparison between the analytical and FE results (loads/deflections) with different end-post profiles.

IOP Conf. Series: Materials Science and Engineering **974** (2020) 012012 doi:10.1088/1757-899X/974/1/012012

Table 5. Comparison between analytical and FE loads/deflections results for specimens (ts = 10 mm, e =100 mm).

a/h_w	Cross section	Analytical Eq.		FE (REP)		FE (NREP)		FE(NEP)		Ratio (REP)		Ratio (NREP)		Ratio (NEP)	
		$P_{u,a}$ kN	$\delta_{pb,a}$ mm	$P_{u,FE}$ kN	$\delta_{pb,FE}$ mm	$P_{u,FE}$ kN	$\delta_{pb,FE}$ mm	$P_{u,FE}$ kN	$\delta_{pb,FE}$ mm	$P_{u,a}/P_{u,FE}$	$\delta_{pb,a}/\delta_{pb,FE}$	$P_{u,a}/P_{u,FE}$	$\delta_{pb,a}/\delta_{pb,FE}$	$P_{u,a}/P_{u,FE}$	$\delta_{pb,a}/\delta_{pb,FE}$
3.5	$h_w.t_w$ = 1000 x 5 mm, b_f x t_f = 300 x 25 mm	701	19.14	951	23.01	942	22.90	216	10.12	0.74	0.83	0.74	0.84	3.25	1.89
3.4		714	18.12	943	21.28	945	21.09	215	9.72	0.01	0.85	0.01	0.86	0.03	1.86
3.3		727	17.12	961	19.68	950	19.64	216	9.20	0.76	0.87	0.77	0.87	3.37	1.86
3.2		742	16.15	968	18.17	951	18.21	216	8.85	0.77	0.89	0.78	0.89	3.44	1.82
3.1		757	15.21	974	16.80	955	18.23	216	8.61	0.78	0.91	0.79	0.83	3.50	1.77
3.0		773	14.30	979	18.34	961	20.62	216	8.33	0.79	0.78	0.80	0.69	3.58	1.72
2.8		808	12.57	991	15.20	966	13.00	216	7.51	0.82	0.83	0.84	0.97	3.74	1.67
2.6		848	10.95	998	12.18	971	11.40	217	7.18	0.85	0.90	0.87	0.96	3.91	1.53
2.4		894	9.46	1004	12.64	980	12.80	217	6.67	0.89	0.75	0.91	0.74	4.12	1.42
2.2		947	8.08	1015	10.47	1001	10.36	216	6.17	0.93	0.77	0.95	0.78	4.38	1.31
2.0		1009	6.83	1037	8.28	1003	9.91	217	6.83	0.97	0.82	1.01	0.69	4.65	1.00
1.8		1081	5.69	1096	6.64	1046	7.00	216	5.13	0.99	0.86	1.03	0.81	5.00	1.11
1.7		1123	5.16	1102	5.81	1065	5.20	202	4.18	1.02	0.89	1.05	0.99	5.56	1.23
1.6		1168	4.67	1097	5.00	1064	5.03	217	5.31	1.06	0.93	1.10	0.93	5.38	0.88
1.5		1217	4.20	1130	4.47	1082	4.52	272	1.93	1.08	0.94	1.12	0.93	4.47	2.18
1.4		1272	3.77	1155	3.98	1103	3.99	220	3.77	1.10	0.95	1.15	0.94	5.78	1.00
1.3		1332	3.36	1207	3.46	1130	3.47	225	3.78	1.10	0.97	1.18	0.97	5.92	0.89
1.2		1399	2.99	1218	4.42	1158	3.00	220	1.66	1.15	0.68	1.21	1.00	6.36	1.80
1.0		1556	2.33	1407	3.22	1237	3.28	230	2.70	1.11	0.72	1.26	0.71	6.77	0.86
0.8		1769	1.82	1596	3.10	1353	2.41	245	2.15	1.11	0.59	1.31	0.76	7.22	0.85
						Mean				**0.94**	**0.84**	**0.98**	**0.86**	**4.69**	**1.43**

Table 6. Comparison between analytical and FE loads/deflections results for specimens (ts = 10 mm, e =150 mm).

a/ h_w	Cross section	.Analytical Eq		FE (REP)		FE (NREP)		FE(NEP)		Ratio (REP)		Ratio (NREP)		Ratio (NEP)	
		$P_{u,a}$ (kN)	$\delta_{pb,a}$ (mm)	$P_{u,FE}$ (kN)	$\delta_{pb,FE}$ (mm)	$P_{u,FE}$ (kN)	$\delta_{pb,FE}$ (mm)	$P_{u,FE}$ (kN)	$\delta_{pb,FE}$ (mm)	$P_{u,a}/P_{u,FE}$	$\delta_{pb,a}/\delta_{pb,FE}$	$P_{u,a}/P_{u,FE}$	$\delta_{pb,a}/\delta_{pb,FE}$	$P_{u,a}/P_{u,FE}$	$\delta_{pb,a}/\delta_{pb,FE}$
3.5	h_w. t_w = 1000 x 5 mm , b_f x t_f = 300 x 25 mm	701	19.14	952	23.06	952	23.11	234	10.40	0.74	0.83	0.74	0.83	3.00	1.84
3.4		714	18.12	944	21.29	947	21.28	234	9.94	0.76	0.85	0.75	0.85	3.05	1.82
3.3		727	17.12	967	19.68	956	19.59	234	9.30	0.75	0.87	0.76	0.87	3.11	1.84
3.2		742	16.15	972	18.18	960	18.17	234	8.91	0.76	0.89	0.77	0.89	3.17	1.81
3.1		757	15.21	977	16.78	965	18.17	234	8.84	0.77	0.91	0.78	0.84	3.24	1.72
3.0		773	14.30	983	20.64	967	18.37	234	8.48	0.79	0.69	0.80	0.78	3.30	1.69
2.8		808	12.57	994	15.21	978	12.86	232	6.96	0.81	0.83	0.83	0.98	3.48	1.81
2.6		848	10.95	1004	11.29	986	11.46	235	5.39	0.84	0.97	0.86	0.96	3.61	2.03
2.4		894	9.46	1030	12.66	998	12.72	234	6.57	0.87	0.75	0.90	0.74	3.82	1.44
2.2		947	8.08	1060	10.25	1023	10.31	235	6.34	0.89	0.79	0.93	0.78	4.03	1.27
2.0		1009	6.83	1049	8.29	1050	8.32	234	6.00	0.96	0.82	0.96	0.82	4.31	1.14
1.8		1081	5.69	1114	6.68	1079	6.73	233	5.66	0.97	0.85	1.00	0.85	4.64	1.01
1.7		1123	5.16	1114	5.84	1077	6.00	234	9.95	1.01	0.88	1.04	0.86	4.80	0.52
1.6		1168	4.67	1129	5.00	1066	4.04	234	10.40	1.03	0.93	1.10	1.16	4.99	0.45
1.5		1217	4.20	1157	4.48	1113	4.48	227	3.75	1.05	0.94	1.09	0.94	5.36	1.12
1.4		1272	3.77	1205	3.98	1138	4.00	230	3.78	1.06	0.95	1.12	0.94	5.53	1.00
1.3		1332	3.36	1259	3.47	1167	3.49	236	3.91	1.06	0.97	1.14	0.96	5.64	0.86
1.2		1399	2.99	1326	3.00	1202	3.02	243	3.84	1.06	1.00	1.16	0.99	5.76	0.78
1.0		1556	2.33	1498	3.29	1297	3.31	236	3.72	1.04	0.71	1.20	0.70	6.59	0.63
0.8		1769	1.82	1723	3.17	1443	2.43	251	2.28	1.03	0.57	1.23	0.75	7.05	0.80
Mean										**0.91**	**0.85**	**0.96**	**0.87**	**4.42**	**1.28**
Standard deviation										**0.15**	**0.10**	**0.18**	**0.10**	**1.24**	**0.43**

Standard deviation 0.13	0.11	0.17	0.11	1.24	0.51

Table 7. Comparison between analytical and FE loads/deflections results for specimens (ts = 10 mm, e =300 mm).

a/h_w	Cross section	.Analytical Eq		FE (REP)		FE (NREP)		FE(NEP)		Ratio (REP)		Ratio (NREP)		Ratio (NEP)	
		$P_{u,a}$ (kN)	$\delta_{pb,a}$ (mm)	$P_{u,FE}$ (kN)	$\delta_{pb,FE}$ (mm)	$P_{u,FE}$ (kN)	$\delta_{pb,FE}$ (mm)	$P_{u,FE}$ (kN)	$\delta_{pb,FE}$ (mm)	$P_{u,a}/P_{u,FE}$	$\delta_{pb,a}/\delta_{pb,FE}$	$P_{u,a}/P_{u,FE}$	$\delta_{pb,a}/\delta_{pb,FE}$	$P_{u,a}/P_{u,FE}$	$\delta_{pb,a}/\delta_{pb,FE}$
3.5	h_w. t_w = 1000 x 5 mm, b_f x t_f = 300 x 25 mm	701	19.14	955	23.09	953	23.09	275	9.21	0.73	0.83	0.74	0.83	2.55	2.08
3.4		714	18.12	946	21.30	947	21.29	276	8.84	0.75	0.85	0.75	0.85	2.59	2.05
3.3		727	17.12	970	19.68	965	19.66	276	8.10	0.75	0.87	0.75	0.87	2.63	2.11
3.2		742	16.15	975	18.18	972	18.17	275	7.69	0.76	0.89	0.76	0.89	2.70	2.10
3.1		757	15.21	982	16.78	978	18.16	275	8.40	0.77	0.91	0.77	0.84	2.75	1.81
3.0		773	14.30	987	20.06	981	20.65	276	6.75	0.78	0.71	0.79	0.69	2.80	2.12
2.8		808	12.57	998	19.43	992	12.86	276	5.86	0.81	0.65	0.81	0.98	2.93	2.15
2.6		848	10.95	1014	11.31	1001	11.37	275	5.17	0.84	0.97	0.85	0.96	3.08	2.12
2.4		894	9.46	1044	12.68	1020	12.69	276	4.62	0.86	0.75	0.88	0.75	3.24	2.05
2.2		947	8.08	1075	10.28	1046	10.29	276	4.04	0.88	0.79	0.91	0.79	3.43	2.00
2.0		1009	6.83	1062	8.30	1039	8.29	270	3.14	0.95	0.82	0.97	0.82	3.74	2.18
1.8		1081	5.69	1133	6.69	1105	6.70	275	3.35	0.95	0.85	0.98	0.85	3.93	1.70
1.7		1123	5.16	1153	5.85	1095	5.74	276	8.64	0.97	0.88	1.03	0.90	4.07	0.60
1.6		1168	4.67	1191	5.01	1110	5.00	275	9.21	0.98	0.93	1.05	0.93	4.25	0.51
1.5		1217	4.20	1240	4.50	1140	4.48	273	3.04	0.98	0.93	1.07	0.94	4.46	1.38
1.4		1272	3.77	1299	4.00	1169	4.00	277	2.72	0.98	0.94	1.09	0.94	4.59	1.39
1.3		1332	3.36	1366	3.50	1201	3.49	277	2.77	0.98	0.96	1.11	0.96	4.81	1.21
1.2		1399	2.99	1449	3.03	1237	3.03	278	2.68	0.97	0.99	1.13	0.99	5.03	1.12
1.0		1556	2.33	1647	3.36	1338	3.29	282	3.37	0.94	0.69	1.16	0.71	5.52	0.69
0.8		1769	1.82	1893	2.48	1477	2.44	294	2.47	0.93	0.73	1.20	0.75	6.02	0.74

IOP Conf. Series: Materials Science and Engineering **974** (2020) 012012 doi:10.1088/1757-899X/974/1/012012

Mean	**0.88**	**0.85**	**0.94**	**0.86**	**3.76**	**1.61**
Standard deviation	**0.10**	**0.10**	**0.16**	**0.09**	**1.06**	**0.60**

IOP Conf. Series: Materials Science and Engineering **974** (2020) 012012 doi:10.1088/1757-899X/974/1/012012

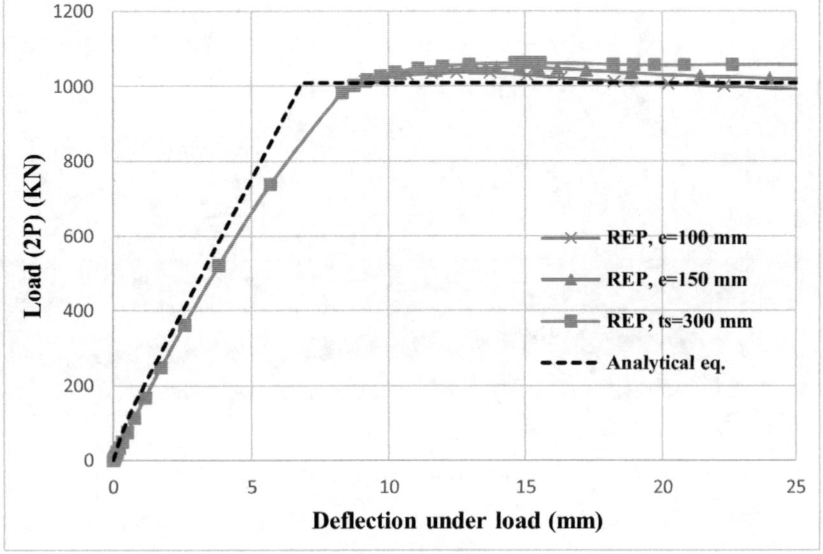

(a) REP

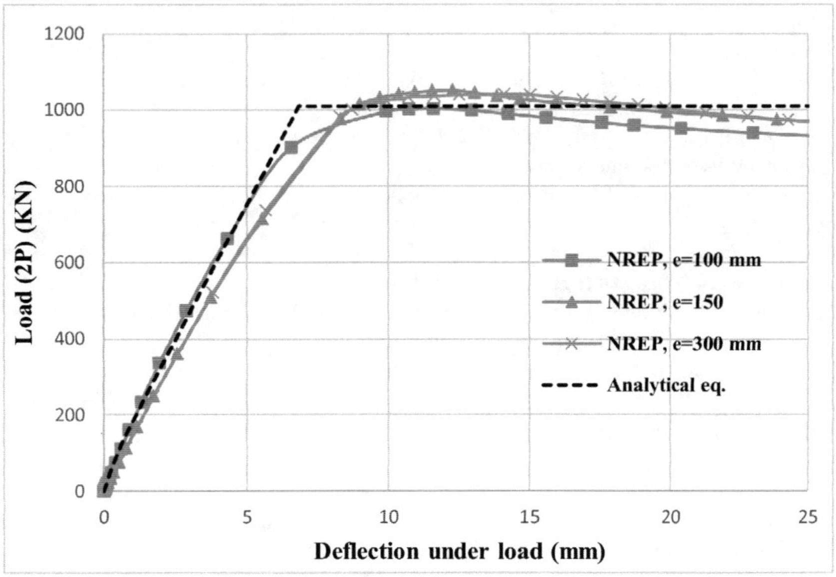

(b) NREP

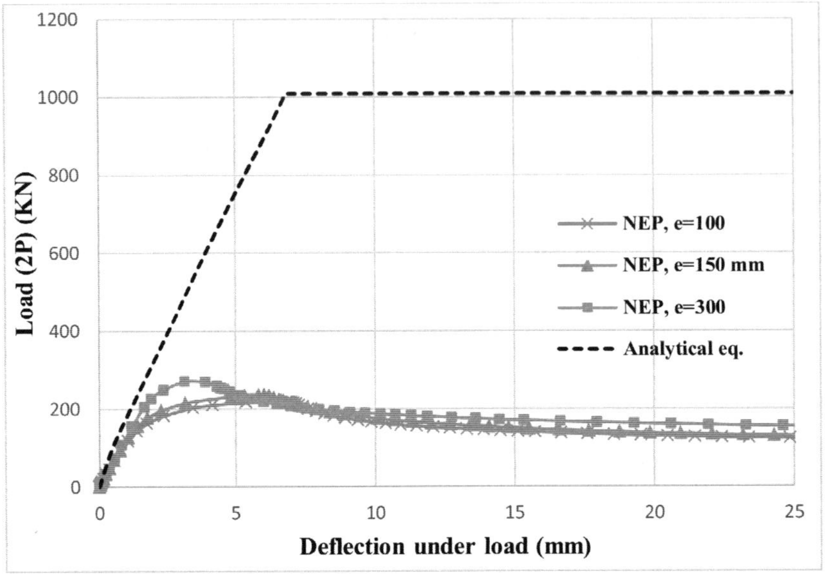

(a) NEP

Figure 10. Analytical and FE load-deflection curves for plate girder ($a/h_w = 2.0$) with different end-distances.

5. Conclusions

The load/deflection of thin-walled panel loaded in shear were studied using numerical and analytical methods. The Qian and Tan [25] mechanical approach, to calculate the in-plane deflection web panel loaded in shear, has been applied to calculate pre- and post-buckling deflection. And hence the non-linear large deflection FE analysis of the full-scale plate girders were studied. The comparison between the results of two methods concluded that:

- The analytical approach presented by Qian and Tan [25] can be used to simulate the load/deflection behaviour of thin-walled panel loaded in shear for $a/h_w < 2.2$, with REP and NREP. Such that, it estimates well the shear strength of REP plate girders and slightly overestimates the shear strength for NREP plate girders.
- The presented analytical approach can't be used for calculating the deflection of NEP plate girders. And more studied is needed to find an accurate approach for NEP panels.
- The stiffeners thicknesses have insignificant effect on the load-deflection behaviour.
- The type of end-post directly affects the ultimate shear strength(P_u). The results show that providing a REP increases the ultimate shear strength ($P_{u,FE}$) of the girder, if compared with NREP.
- Changing end-post from REP to NREP has insignificant effect on maximum vertical deflection ($\delta_{Pb,FE}$).
- The FE results show that increasing the end-distance (e) has no effect on the initial stiffness of the plate girders. However, it slightly increases the ultimate shear strength of the girders.

Conflict of interest

On behalf of all authors, the corresponding author states that there is no conflict of interest.

6. References

[1] M. Hirt, J.-P. Lebet, Steel Bridges: Conceptual and Structural Design of Steel and Steel-Concrete Composite Bridges, EPFL Press 2013.
[2] N.M. Yossef, S.Taher, Cost Optimization of Composite Floor Systems with Castellated Steel Beams, Practice Periodical on Structural Design and Construction 24(1) (2018).
[3] S.C. Lee, J.S. Davidson, C.H. Yoo, Shear buckling coefficients of plate girder web panels, Computers & Structures 59(5) (1995) 789-795.
[4] S.C. Lee, C.H. Yoo, Strength of Plate Girder Web Panels under Pure Shear, Journal of Structural Engineering 124(2) (1998).
[5] C.H. Yoo, S.C. Lee, Mechanics of Web Panel Postbuckling Behavior in Shear Journal of Structural Engineering 132(10) (2006).
[6] M.M. Alinia, H.R. Habashi, A. Khorram, Nonlinearity in the postbuckling behaviour of thin steel shear panels, Thin-Walled Struct. 47(4) (2009) 412-420.
[7] M.M. Alinia, M. Shakiba, H.R. Habashi, Shear failure characteristics of steel plate girders, Thin-Walled Struct. 47(12) (2009) 1498-1506.
[8] J. Loughlan, N. Hussain, The in-plane shear failure of transversely stiffened thin plates, Thin-Walled Struct. 81 (2014) 225-235.
[9] G. Quan, S.-S. Huang, I. Burgess, An analytical approach to modelling shear panels in steel beams at elevated temperatures, Eng. Struct. 85(15) (2015) 73-82.
[10] Y.B. Kwon, S.W. Ryu, The shear strength of end web panels of plate girders with tension field action, Thin-Walled Struct. 98(January) (2016) 578-591.
[11] Y. Xiao, X.Y. Xue, F. FeiSun, G.Q. Li, Postbuckling shear capacity of high-strength steel plate girders, Journal of Constructional steel research 150(November) (2018) 475-490.
[12] A. Reis, N. Lopes, P.V. Real, Ultimate shear strength of steel plate girders at normal and fire conditions, Thin-Walled Struct. 137(April) (2019) 318-330.
[13] N.M. Yossef, J. Fogarty, S. El-Tawil, Residual Compressive Strength of Flange-Damaged Steel Columns, Journal of Structural Engineering (2015).
[14] O.F. Zaher, N.M. Yossef, M.H. El-Boghdadi, M.A. Dabaon, Structural behaviour of arched steel beams with cellular openings, Journal of Constructional steel research 148 (2018) 756-767.
[15] D.W. White, M.G. Barker, Shear Resistance of Transversely Stiffened Steel I-Girders, Journal of Structural Engineering 134 (2008) 1425-1436.
[16] D.M. Porter, K.C. Rockey, H.R. Evans, The collapse behaviour of plate girders loaded in shear, The Structural Engineer 53(8) (1975) 313 - 325.
[17] W. Ajam, C. Marsh, Simple model for shear capacity of webs, Journal of Structural Engineering 117(2) (1991) 410-422.
[18] T.V. Galambos, Guide to stability design criteria for metal structures, 5th ed., McGraw-Hill, New York., 1998.
[19] J. Wilson, On specifications for strength of iron bridges, Trans. Am. Soc. Civ. Eng. 15(1) (1886) 489-490.
[20] K. Basler, B. Thurlimann, Plate Girder Research, AISC Nat. Engrg. Conf., American Institute of Steel Construction, New York, N.Y., 1959.
[21] K. Basler, Strength of plate girders under combined bending and shear., Journal of the Structural Division, ASCE 87(2) (1961) 97-181.
[22] K. Basler, Strength of plate girders in shear, Transaction of ASCE, American Society of Civil Engineers., New York, 1961.
[23] K.C. Rocky, M. Skaloud, The ultimate load behaviour of plate girders loaded in shear, The Structural Engineer 50(1) (1972) 29 - 47.
[24] T. Höglund, Strength of steel and aluminum plate girders shear buckling and overall buckling of plane and trapezoidal webs comparisons with test, Dept. of Structural Engineering, Royal Institute of Technology, Stockholm, Sweden., 1995.
[25] Z.-H. Qian, K.-H. Tan, Deflection behaviour of plate girders loaded in shear at elevated temperatures, Journal of Constructional steel research 65 (2009) 991 - 1000.
[26] Y. Zhu, J. Zhao, Experimental and numerical study on member classification of I-section plate girders, Journal of Constructional steel research 137(October) (2017) 192-200.

[27] K.C. Rockey, E. H.R., D.M. Porter, A design method for predicting the collapse behavior of plate girders Proceedings of the Institution of Civil Engineers 65(1) (1978) 85-112.
[28] ABAQUS Standard User's Manual, The Abaqus Software is a product of Dassault Systèmes Simulia Corp, DassaultSystèmes, Providence, RI, USA, 2008.
[29] CEN, EN 1993-1-5, Eurocode 3. Design of steel structures - Part 1-5, Plated structural elements., 2007.

Nomenclature

σ_t^y	tensile yield stress of web panel under shear;
a	panel length;
a/h_w	web panel aspect ratio;
A_w	area of the web;
b_f	width of the flange;
c	distance assumed between the plastic hinge and the end stiffener;
E	elastic modulus;
f_{yf}	yield stress of the flange plate;
f_{yw}	yield stress of the web plate;
h_w	height of the web;
h_w/t_w	web slenderness ratio;
K	shear buckling coefficient;
L	span of the plate girder;
M_{pf}	plastic moment capacity of flanges$(= 0.25 f_{yf} A_f\, t_f\,))$;
$P_{u,a}$	ultimate shear force
t_f	thickness of the flange;
t_s	thickness of the stiffeners;
t_w	thickness of the web;
V_{ult}	post-buckling shear force component;
w_o	initial imperfection
δ_b	deflection of web panel from pre-buckling to the post buckling stage;
δ_{cr}	deflection of web panel at the pre-buckling stage;
δ_{pb}	total deflection of web at the post buckling stage for a panel loaded in shear;
v	Poisson's ratio;
τ_{cr}	shear buckling stress of the web panel
θ	inclination of the tensile field;

Page | 1Military Technical College
Kobry El-Kobbah,
Cairo, Egypt

13th International Conference on Civil and Architecture Engineering
ICCAE-13-2020

Elements of Safety Management System in the Construction Industry and Measuring Safety Performance – A Brief

Mostafa Elsebaei[1]*, Omar Elnawawy[2], Ayman Othman [3], Mohamed Badawy[4]

[1] Teaching Assistant at The British University in Egypt, Post Graduate Student at Ain Shams University; Mostafa.adel@bue.edu.eg
[2] Professor, Structural Engineering Department, Ain Shams University; nawawyomar@hotmail.com
[3] Professor, Head of Architecture Department, The British University in Egypt; Ayman.othman@bue.edu.eg
[4] Ph.D., Structural Engineering Department, Ain; Mohamed.badawy@eng.asu.edu.eg

* Correspondence: Mostafa.adel@bue.edu.eg

Abstract. Construction industry is considered to be one of the most hazardous industries in the world. The reason could be attributed to its hazardous nature as it is an accident-prone industry. Thus, a need for better understanding of safety management system is essential for improving safety performance in this sector. This paper discusses briefly the elements of safety management by presenting different systems (such as Oregon OSHA Occupational Health and Safety Administration, and OTAR Overseas Territories Aviation Circle) and elaborating their elements. It also discusses two types of measuring safety performance the first is the lagging indicators and the second is the leading indicator. In addition, a field study was conducted to explore contractors' perception on safety management. A questionnaire was distributed to construction firms. 200 responses were collected and analyzed. All of the results showed positive answers which indicate that safety in performance in Egypt is slightly above average as all means were close to average.

1. Introduction

Construction industry is considered one of the most dangerous industries in the world. By looking at accidents records, it turns out that the construction industry scores the highest among other industries. For example, it was reported that 9209 accidents occurred in 2012 in the construction industry by the Turkey Statistical Institute, and 568 of them were permanent disability, and 256 resulted in fatality[1]. Moreover, a recent study in Nigeria revealed that the incident rate (IR) in the construction industry in Nigeria is 2 (which means that for every 100 workers, 2 of them encounter an accident)[2]. The reason could be attributed to its unique characteristics which reflect how hazardous it is. The characteristic of the construction industry has been analysed in different studies[3,4]. Briefly, construction industry creates one-of-a-kind product which creates a difficult in safety fixtures as the have to be replaced continuously, and being a temporary multi organization industry makes it difficult to communicate between different parties.Thus, it is important to implement a safety management system in this industry to improve its safety indicators.

There is no specific definition that can best describe safety management system as this term may have different interpretation among different countries and even among different organizations. Table (1) shows different definitions of safety management system by different organizations and studies. Another reason for having different definitions could be that most studies done on safety management, in the past as noted, have been done in the field of psychology and sociology, as noted in the previous section about human behavior[5].

Table 1: Definitions of safety management systems by different organizations.

Organization or Study	Definition
SMIC (Safety Management International Collaboration Group) [5,6]	"A safety management system is a series of defined, organization-wide processes that provide for effective risk-based decision-making related to your daily business."
ILO (International Labour Organization) [7,8]	"A set of interrelated or interacting elements to establish OSH policy and objectives, and to achieve those objectives."
ICAO (International Civil Aviation Organization) [9,10]	"A systematic approach to managing safety, including the necessary organizational structures, accountabilities, policies, and procedures."

Most of the studies have showed over the years that safety management systems have a positive relationship with safety performance.Bottani et al. [11] have investigated the influence of safety management system between adopting and non-adopting companies. The results showed companies adopting safety management systems exhibit higher performance in training employees and assessing risks. The benefits of safety management system according to [5] are briefly:

- Reducing the number of accidents and minimize the risk accidents in the workplace by controlling the workplace hazards.
- Improving the employee morale and enhancing their productivity by minimizing production interruptions.
- Reducing the cost of employees' absence and the cost of their insurance as well.
- Reducing the cost of legal litigation in court, and reducing investigation time for accident.

2. Elements of Safety Management System

Elements of safety management systems varies from one organization to another [12]. Some organizations have drawn the elements of safety management by making a comparison between high and low accident rate in companies. Whereas, others have drawn the elements of safety management based on actual case studies of highly safe and reliable organizations with good safety performance [5]. The following section will cover different safety management systems according to different organizations in different countries from various studies. In each safety management system, elements proposed will be discussed and elaborated.

A study aimed to detect successful elements of safety management by comparing high and low accident firms revealed five major successful elements of safety management [13]:

- The first and the most important element is the strong top management commitment towards safety. This commitment can appear in various ways. For instance, it was found that in low-accident companies top management was highly involved in safety activities on a regular basis, whereas, this involvement was on the other hand absent in high-accident companies. It has been reported that top management take safety issues seriously in meetings in all low-accident companies.
- The second element of safety management according to [13] is safety training of workers. Safety training in low-accident companies has been addressed as an essential part of new workers training, as well as, as a following up for old personnel.

- The third element is establishing well communication link between management and workers. This may include periodic safety inspection by professionals.
- The forth safety element is good housekeeping which may include high usage of the latest safety devises.
- The fifth element is maintaining low turnover rates. It is found that low turnover rates reinforce work relations and enhance the personal development practices. Finally, the last element of safety management is incentives. This could be implemented in a variety of ways such as praising individuals through recognitions for safety performance.

A more developed safety management system was developed and created by Oregon OSHA (Occupational Health and Safety Administration) consists of seven elements of safety management [14]:

- The first element is management commitment to protecting employees. In fact, this element is the most important element in any safety management system as it determines how much the top management are committed on the safety of their workers.
- The second element of Oregon OSHA is accountability. One of the most important methods which can improve the accountability is stating in the employees' job description their safety responsibility.
- The third element is employee involvement in safety such as: allowing employees to participate and be part of the committee, and posting safety policies and guidelines in workplace where all of them can see, and promoting recognitions awards for best safety awards. It was found that safety incentives and rewards are one of the most effective techniques for improving safety performance [15].
- The fourth element is hazard identification and control. There are a variety of ways to identify hazard (such as maintaining a periodic inspection, providing an efficacious reporting system for the employees). As whole, well-developed hazards identification techniques can be classified into two categories: First is the Reactive Approach which is identifying hazards that may lead to an accident event before it occurs. Second is the Proactive Approach which is identifying the hazards before an accident occur depending on historic data and previous experiences [16,17].
- The fifth element of Oregon OSHA is accident analysis. Accident analysis is very important to prevent future accidents and improve the safety performance.
- The sixth element is educating and training. It is very important to train the employees about the risks and the hazards and teach them how to protect themselves.
- The last element is evaluating and reviewing the safety program [14].

Another organization the Overseas Territories Aviation Circle OTAR has developed another safety management system consisting of six elements[18]:

- The first element is objective which includes the organization's mission and vision towards safety. It acts as a motivation for the whole organization.
- The second element is defining roles and responsibilities, which is similar to accountability (from the previous system).
- The third element is hazard identification which is similar to the previous system.
- The forth element is risk assessment.
- The fifth element is monitoring and evaluation such as safety audits and reviews.
- The last one is safety documentation like safety manuals, and accident records.

3. Safety Performance Measurements

Measuring safety performance allows the organizations to take important decisions and appropriate actions towards their adopted safety management system. Measurements are very important as it determines the effectiveness of the safety management system on the overall safety performance, which can be either accident prevention strategies and/or safety practices and activities [19]. However, is safety performance measurable? For example, the standard unit for measure lengths is "meter" and a measure tape can be used for that. The standard unit for measuring mass is "kilogram" and a balance can be used for that. So, how to measure the safety performance?

In fact, there are two types of measuring safety performance actually based completely on two different concepts. The first is "lagging indicators", and the second is "leading indicators".

Lagging indicator is safety performance measurement based on failures in the past. So, it is a retroactive measurement which only record incidents in the past [19]. Lagging indicator measure the incident after it occurs, that is why it was described as measuring the absence of safety rather than the presence of safety [5,19]. According to [20] the lagging indicators are widely use as easy to collect and understood, comparable with each other, and useful in the identification of a trend.)

The most common lagging indicators were discussed briefly by [5,19]. Table (2) shows the most common lagging indicator used for measuring safety performance and Table (3) shows a brief description for each of them. Most of them are lagging indicators developed by OSHA organization which has a standard. It bases its calculation on 200,000 labour hours in a year, and this should be equivalent to 100 employees working 40 hours per week, and 50 weeks per year.

Table 2: The most common lagging indicators.

Lagging Indicator Name	Equation
OSHA Recordable Incident Rate (TRIR)	$TRIR = \frac{Total\ recorded\ incidents\ a\ year * 200{,}000}{Total\ labor\ hours\ worked\ a\ year}$
OSHA Lost Time Case (LTC)	$LTC = \frac{No. of\ lost\ time\ cases * 200{,}000}{Total\ labor\ hours\ worked\ a\ year}$
OSHA Lost Work Day Rate (LWD)	$LTC = \frac{No. of\ lost\ days\ a\ year * 200{,}000}{Total\ labor\ hours\ worked\ a\ year}$
OSHA Days Away, Restricted or Job Transfer (DART)	$DART = \frac{No. of\ DART\ incidnts\ a\ year * 200{,}000}{Total\ labor\ hours\ worked\ a\ year}$
Incident Rate (IR)	$IR = \frac{Total\ Incidents * 100}{Total\ workers}$
Severity Rate (SR)	$SR = \frac{Total\ lost\ days}{Total\ incidents}$
Risk Rate (RR)	$RR = \frac{Total\ lost\ days * 100}{Total\ workers}$

Table 3: Description of the lagging indicators.

Lagging Indicator Name	Description
OSHA Recordable Incident Rate (TRIR)	The number of OSHA recordable incidents per year for every 100 workers working the standard OSHA working hours a year. For example, if the TRIR for an organization is 8, this means that for every 100 workers working (the standard OSHA working hours) in that organization, 8 of them had recordable accident/injury in a complete year.
OSHA Lost Time Case (LTC)	The number of incidents in a year which results in sick leaves for every 100 workers working the standard OSHA working hours. For example, if the LTC is 8, this means that 8 workers for every 100 workers (working the standard OSHA hours) were unable to go to work for some time because of an incident.
OSHA Lost Work Day Rate (LWD)	The number of lost work days in a year for every 100 workers (working the standard OSHA working hours). For example, if LWD is 8, this means that 8 complete days were lost in a year for every 100 workers (working the standard OSHA hours).
OSHA Days Away,	It is the number of incidents in a year for every 100 workers which

Restricted or Job Transfer (DART)	results in workers being restricted,
Incident Rate (IR)	Total number of incidents per 100 workers (not necessarily working the standard OSHA hours) in a certain time (not necessarily a year).
Severity Rate (SR)	The total number of lost days for every incident within a period of time.
Risk Rate (RR)	The total number of lost days for every 100 workers (not necessarily working the standard OSHA hours) in a certain time (not necessarily a year).

Although lagging indicators are widely used in almost all organization in all sectors, yet they have many problems and defects. The first problem is luck. Some workers might have worker for a long period of time without encountering incidents. That doesn't mean that they were safety, they might have been exposed to extreme hazards but without and accident. Such traditional measurements will not detect this kind of events. The second problem is susceptibility to manipulations and lack of precision. Many organizations tend to not reveal information concerning injuries occurred in their workplace. Moreover, some attempt to play with injury records and description to reduce the compensation costs. The third problem is as stated before that these measurements are used to project the absence of safety not the presence of safety as they measure past incident which have already occurred. So, they cannot be used efficiently to determine hazards or eliminate them, or even to improve the safety [21].

There is a more improved type of safety performance measurement than lagging indicators, which is leading indicators. Leading indicator are safety measurement which provide a future forecast of the safety performance based on the activities and practices implemented not incidents. So, it is proactive measure to what might happen in the future [22]. Leading indicators were developed from the root causes of accidents, that is why they are very practical in improving safety performance.

There are many examples of leading indicators. In fact, each industry has its own leading indicators developed as mentioned before from the nature of the accidents in that industry. All safety practices and activities are considered as leading indicators. For instance, all workers in an organization should have an OSHA safety certificate. Another example is providing a certain number of training hours in a certain period of time [5]. Table 4 shows different examples of leading indicators.

Table 4: Examples of different leading indicators

Leading Indicator	Description
Worker Observation Process [23]	• Common techniques used to evaluate ongoing tasks in construction. • Unsafe conditions and acts that contribute to injury, property damage, or equipment failure can be identified, recorded and used to monitor and predict safety performance.
Near Miss Reporting [23]	• Defined as an incident where no property damage and no personal injury was sustained, but where, given a slight shift in time or position, damage and injury easily could have occurred. • Near misses are measurements of processes, activities and conditions that assess safety performance and can predict future results. • Near miss reporting is used as a safety management tool in many other industries within the U.S. private sector.
Project Management Team Safety Process Involvement[23]	• Demonstration of leadership and commitment via active management walking around. • Senior management and supervisors are encouraged to participate in site safety walks. • Management plays a key role in promoting a positive safety culture. • Allocating resources, time, and inspections.

Job Site Audits [24]	• Systematic measurement and evaluation of the way in which an organization manages its health and safety program against aseries of specific and attainable standards. • Conducted to identify problem areas including unsafe conditions and unsafe behaviors. • The results can predict trends to show that safety is improving or that jobsite safety is decreasing.
Stop Work Authority [23]	• Workers are expected to stop any work they consider to be unsafe until they feel it is safe to proceed. • Stop work authority is to be clearly communicated to workers in initial orientation and at regular intervals throughout each project.
Housekeeping Program [23]	• Helps achieve a further reduction in the occurrence of job site accidents. • The level of housekeeping at a given site is an indicator of safety at that site.
Safety Orientation and Training [24]	• Helps workers become aware of project hazards. • The nature of the orientation will help to determine the probable success of delivering a safe project. • The orientation training should be provided to all individuals who will be working on site, including the field employees, subcontractors' employees, and all salaried personnel on site.

4. Methods.

4.1. Data Collection.

Data collection in this research was done in two stages. First stage is gathering relevant data through literature review. The second stage was a field study which was done in the form of a questionnaire.

The questionnaire aimed to investigate contractors' perception of their evaluation of their companies in terms of safety performance. It consists of nine Likert-scaled questions about evaluating safety management adopted from [24]. It consists of nine questions and is expected to be answered in 2:30 minutes. The questionnaire consisted of two sections (general information and evaluation of companies). All questions were designed on 4-point closed-end Likert-type scale measuring the level of agreement of each variable, except for an open-end question. The reason why even number of (4-point) alternatives was used instead of the most common 5 alternatives explained by [25] is: first to avoid selecting the middle alternative if the respondent felt bored, second some respondents tend to choose the middle alternative despite of having a solid answer. This would be because some respondents would not exert the cognitive effort to form an opinion about the question and this is called "satisficing". The only disadvantage for this type of scale is that those respondents, who actually do not have any idea about the question or the topic, will be forced to select one of the two-directional answer categories [26]. Because this questionnaire was distributed to experts, this problem did not exist.

4.2. Techniques of analysis

All the quantitative questions in this questionnaire were analysed using two methods. The first method is done by measuring the central tendency. Central tendency presents a certain value for a variable or probability distribution. Its most used measures are the arithmetic mean, median and mode. The second method used in this questionnaire is dispersion. Dispersion specifies the nature of data whether it is homogeneous or heterogeneous. It also assesses how much differences and variations are presented in the data and are apart from the presented value in the sample. The dispersion measures, which were used, are standard deviation and variance [27]. Excel was used to get both central tendency (mean, median and mode) and dispersion (standard deviation and variance) and relative importance index RII.

4.3. Sampling and questionnaire distribution

The statistics of contractors for complete building works were obtained from the Egyptian Federation for Construction & Buildings Contractors (EFCBC). Table 5 shows the number of contractors (for complete buildings works) in Egypt and some of its governates. It shows all the seven grades of contractors as well. According to [28] the large firms are the first 3 grades, the medium firms are the fourth and fifth grades, and the sixth and seventh grades are the small firms. This study targeted the large and medium firms as the large firms must be registered in the International Project Management Association IPMA. The medium firms were also selected as one of the terms required for their registration that they should have an experience for at least 2 years, which is suitable for this study to cover a wide spectrum of opinions.

Table 5: Number of contractors for complete buildings works in different governates in Egypt

Class	1st	2nd	3rd	4th	5th	6th	7th	**Total**
Egypt	138	144	182	453	653	565	4884	**7019**
Cairo	79	57	62	154	161	100	787	**1400**
Giza	23	21	20	44	68	66	389	**631**
Alexandria	6	6	10	24	32	35	296	**409**
Sohag	4	4	3	11	22	20	368	**432**
Minya	2	1	3	13	15	17	178	**229**
Dumyat	2	3	0	4	15	6	120	**150**
Asyut	2	4	7	10	16	19	175	**233**
Fayoum	1	5	8	21	22	31	172	**260**
Suez	0	2	5	13	27	11	119	**177**
Port Said	3	4	6	10	16	13	44	**96**
Others	16	37	58	149	259	247	2236	**3002**

Source: [29]

The total number of the large and medium categories was 1570 which is the population size. The sample size was calculated using the simplified formula by Yamane [30]:

$$n = \frac{N}{1 + N(e^2)}$$

Where:
n is sample size
N is the population size
e is the level of precision (5% in this case)

$$n = \frac{1570}{1 + 1570(0.05^2)} \approx 319$$

This takes into consideration that the confidence = 95%... If the confidence is a value different from 95%, this equation becomes invalid.

Another equation could be chosen for that instead of the simplified equation which is the most common fundamental equation for large population (which should be adjusted if the population is small) by Cochran [30]:

$$n_o = \frac{z^2pq}{e^2}$$

Where:
no is sample size for large population
p is an estimated proportion of an attribute that is present in the population
e is the level of precision (5% in this case)
q is equal to 1-p
z is a point on the abscissa of the standard normal curve that specifies the confidence level

It is recommended to take the product p*q (which is called the variance) as the maximum value. Therefore, p = 0.5 and q = 0.5. For a confidence percentage of 95% z value should be 1.96

$$n_o = \frac{z^2 pq}{e^2} = \frac{1.96^2 \times 0.5 \times 0.5}{0.05^2} = 384.16$$

If no (sample size) is greater than 10% of the population, then the population is not considered a large one. Therefore, no should be subjected to adjustments for true sample value. The true sample value can be calculated using the following equation:

$$n_R = \frac{n_0}{1 + \frac{(n_0 - 1)}{N}} = \frac{384.16}{1 + \frac{384.16-1}{1570}} \approx 309$$

309 companies were contacted. The questionnaire was distributed in large construction sites in Cairo, as well as, it was sent on Linkedin to reach firms in Cairo and other governates which are inaccessible. 200 of the 309 respondents, answered the questionnaire.

5. Results and Discussion

Table 6: Results of the questionnaire

Questions	Mean	Median	Mode	SD	Variance
Q1: When a risk is detected, management in my company (the company you work in) ignores it without an action	1.995	2	1	0.97	0.95
Q2: Management in my company ensures that safety problems discovered are corrected immediately	2.99	3	3	0.93	0.86
Q3: Management in my company places safety before production	2.64	3	3	0.99	0.98
Q4: Management in my company ensures that everyone receives the necessary information on safety	2.875	3	4	1.01	1.02
Q5: Workers have confidence in the management's ability to deal with safety in my company	2.735	3	3	0.94	0.89
Q6: Management in my company encourages employees to participate in decisions which affect their safety	2.72	3	3	0.93	0.86
Q7: Management in my company never considers employees' suggestions regarding safety	2.245	2	2	0.96	0.92
Q8: Fear of negative consequences from management discourages employees from reporting near-miss accidents	2.275	2	2	0.95	0.90
Q9: Management always blames employees for accidents	2.45	2	2	0.98	0.95

It is noticed from Table 6 and the radar diagram that more than half of results showed positive responses (level of agreement). All the nine questions showed positive responses, however almost close to the average. The Mean showed that respondents pointed out that management in their

IOP Conf. Series: Materials Science and Engineering **974** (2020) 012013 doi:10.1088/1757-899X/974/1/012013

companies did not ignore risks without an action, ensures that safety problems discovered are corrected immediately, places safety before production, ensures that everyone receives the necessary information on safety, encourages employees to participate in decisions which affect their safety, and considers employees' suggestions regarding safety. In addition, they believe that workers have confidence in management, and they do not fear to report near-miss accidents. They also believe that their management don't blame workers for accidents.

In the professional opinion of the author, that these results may not be as they seem, and there may be problems related to management whom respondent did not point out. Respondent might have been afraid to answer this section as it directly asks them to assess the company, they work in. So, results were good. The author believes that results are over-assessed because some of the respondents refused to answer this questionnaire.

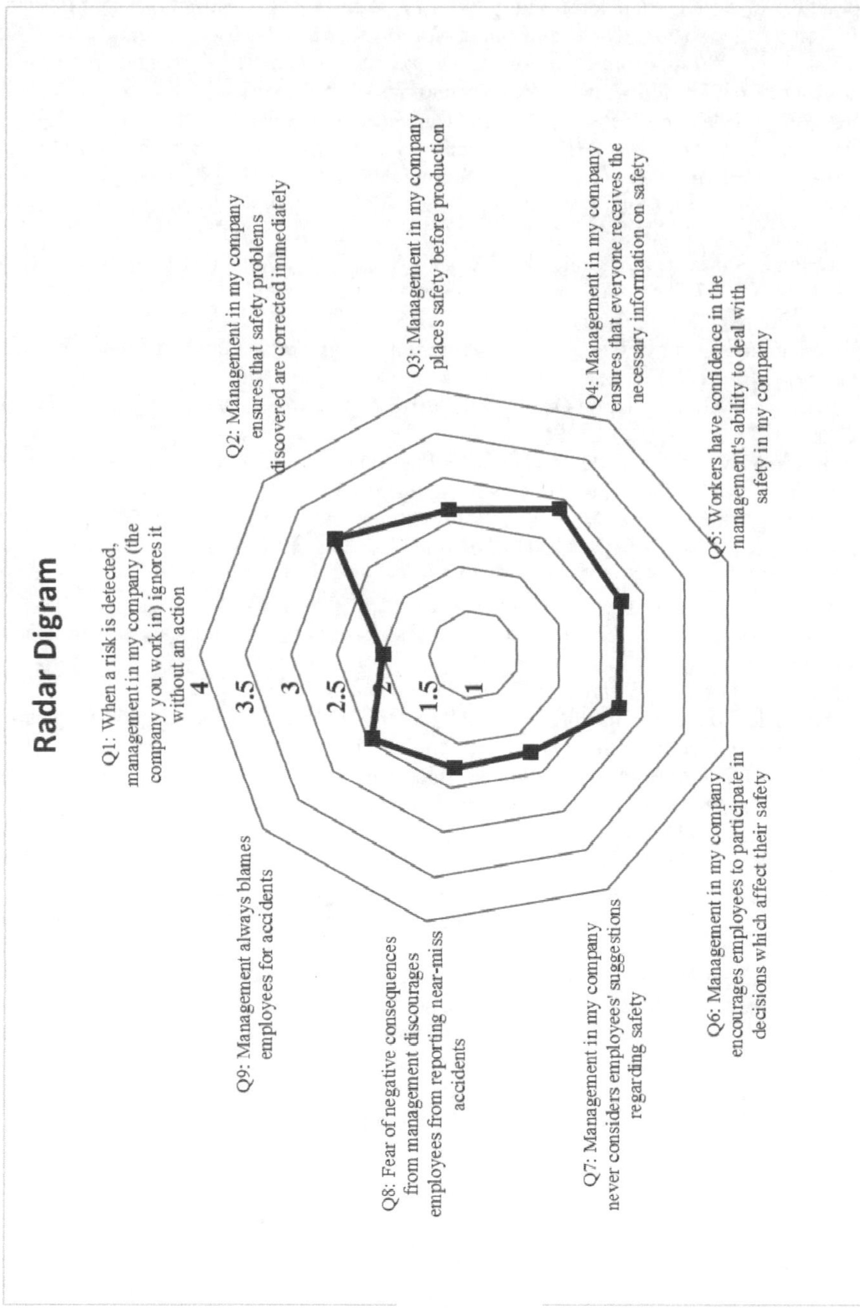

6. Conclusion

The continuous increase in the accident rates in the construction industry makes it a necessity to start implementing safety management systems. Thus, this paper has presented a brief review in some of the most important safety management systems such as: Oregon OSHA, and OTAR in a better way of improving the safety performance. In addition, this paper has elaborated tools used for measuring safety performance for more accurate performance assessment.

The paper has also explored contractors' perception on their safety management. Most of the respondents showed positive answers to the questions.

References

[1] Yilmaz F 2014 Analysis of Occupational Accidents in Construction Sector in Turkey *J. Multidiscip. Eng. Sci. Technol.* **1** 421–8

[2] Agbede J O, Manu P, Agbede O A and Mahamadu A-M 2016 Health and safety management practices in the Nigerian construction industry: A survey of construction firms in South Western Nigeria *Proc. CIB World Build. Congr. 2016* **2** 293–304

[3] Koskela L 2000 *An exploration towards a production theory and its application to construction* (Helsinki University of Technology)

[4] Koskela L 1992 *Application of the new production philosophy to construction* vol 72

[5] Jazayeri E and Dadi G B 2017 Construction Safety Management Systems and Methods of Safety Performance Measurement : A Review *J. Saf. Eng.* **6** 15–28

[6] Safety Management International Collaboration Group (SM ICG) 2010 10 THINGS YOU SHOULD KNOW ABOUT SAFETY MANAGEMENT SYSTEMS (SMS)

[7] Robson L S, Clarke J A, Cullen K, Bielecky A, Severin C, Bigelow P L, Irvin E, Culyer A and Mahood Q 2007 The effectiveness of occupational health and safety management system interventions : A systematic review *Saf. Sci.* **45** 329–53

[8] International Labor Office 2001 *Guidelines on occupational safety and health management systems* (GENEVA)

[9] International Civil Aviation Organisation 2013 *Safety Management Manual (SMM)* (Quebec, Canada)

[10] Batuwangala E, Silva J and Wild G 2018 The Regulatory Framework for Safety Management Systems in Airworthiness Organisations *Aerospace* **5**

[11] Bottani E, Monica L and Vignali G 2009 Safety management systems: Performance differences between adopters and non-adopters *Saf. Sci.* **47** 155–62

[12] Ismail Z, Doostdar S and Harun Z 2012 Factors influencing the implementation of a safety management system for construction sites *Saf. Sci.* **50** 418–23

[13] Zohar D 1980 Safety Climate in Industrial Organizations : Theoretical and Applied Implications *J. Appl. Psychol.* **65** 96–102

[14] Oregon OSHA 2002 *Safety and Health Management – the Basics*

[15] Alarcón L F, Acuña D, Diethelm S and Pellicer E 2016 Strategies for improving safety performance in construction firms. *Accid. Anal. Prev.* **94** 107–18

[16] Willquist P and Marianne T 2003 Identifying and analysing hazards in manufacturing industry — a review of selected methods and development of a framework for method applicability *Int. J. Ind. Ergon.* **32** 165–80

[17] Khanzode V V, Maiti J and Ray P . 2012 Occupational injury and accident research : A comprehensive review *Saf. Sci.* **50** 1355–67

[18] Suan A 2017 A Mini Review on Efficacy of Safety Management Systems in Construction *Int. J. Eng. Sci. Comput.* **7** 14997–5001

[19] Antillón E I 2010 *A research synthesis on the interface between lean construction and safety management* (University of Colorado at Boulder)

[20] Arezes P M and Sérgio Miguel A 2003 The role of safety culture in safety performance measurement *Meas. Bus. Excell.* **7** 20–8

[21] Blair E H and Spurlock B S 2007 Leading measures for improving safety performance *ASSE Professional Development Conference* (Orlando, Florida: American Society of Safety

Engineers)

[22] Alexander D C 2016 *Using Precursor Analysis To Predict and Prevent Fatal* (University of Colorado Boulder)

[23] Hallowell M R, Hinze J W, Baud K C and Wehle A 2013 Proactive construction safety control: Measuring, monitoring, and responding to safety leading indicators *J. Constr. Eng. Manag.* **139** 1–8

[24] Ramirez L S M 2014 *Safety climate, safety hazards and organizational practices in the construction industry in Colombia (Unpublished doctoral dissertation)* (UNIVERSITY OF MASSACHUSETTS LOWELL)

[25] Sturgis P, Roberts C and Smith P 2014 Middle Alternatives Revisited: How the neither/nor Response Acts as a Way of Saying "I Don't Know"? *Sociol. Methods Res.* **43** 15–38

[26] DeMars C E and Erwin T D 2005 Neutral or Unsure: Is there a Difference? *Poster presented at the annual meeting of the American Psychological Association* (Washington, DC) pp 1–12

[27] Deshpande S, Gogtay N J and Thatte U M 2016 Measures of central tendency and dispersion *J. Assoc. Physicians India* **64** 64–6

[28] El Ehwany N 2009 The Construction and Related Engineering Services in Egypt: Challenges and Policies *Egypt. Cent. Econ. Stud.* **146**

[29] Egyptian Federation for Construction & Buildings Contractors 2019 (EFCBC)

[30] Israel G D 1992 Determining Sample Size *Univ. Florida Coop. Ext. Serv. Inst. Food Agric. Sci. EDIS, Florida*

**Military Technical College
Kobry El-Kobbah,
Cairo, Egypt**

**13th International Conference
on Civil and Architecture
Engineering
ICCAE-13-2020**

Cairo Airport Between Origination And Future Vision

Mostafa Mahmoud 1, Alaa El Din Abdel Rahman 2 and Salah ZakiSaeid 3

[1]PhD researcher, Department of Architecture, Al-Azhar University.

[2]lecture , Department of Architecture, Al-Azhar University.

[3] Professor, Department of Urban Planning, Al-Azhar University.

Corresponding author:**mostafa2ms@yahoo.com**

ABSTRACT
Airports are the latest means of transportation that reflect the progress of human civilization and reflect the distinctive identity of each country. The passenger building is the main component of the airport, which in turn contains many dynamic and constantly evolving facilities depending on the increase in passenger traffic and the continued entry of new aircraft and modern operating and service systems that are constantly heading to try to facilitate and accelerate the travel procedures, as the development of airports in recent years is a remarkable development that promotes the level of booms as a result of increased air traffic in the world. The airports have become one of the most important international projects.
In this research, Cairo Airport has been chosen because it is considered the largest airport used in Egypt, and the interface that arrives from the outside world to Cairo. It accommodates all foreign air movements for all foreign airlines operating in Egypt. In this research, a detailed explanation about Cairo Airport will be provided by studying its components from the ground side, passenger buildings and the air side, and studying the area of Cairo Airport and how to benefit from unexploited places at the airport by studying the elements of the airport city and how to benefit from the application of the components of the airport city at Cairo International Airport.

Key Words: Airport - Airport City –landside –Terminal Building - Airside.

Introduction: The airport with its various elements is considered a city within the city, and this city with its components, the way of work inside it, the distribution system and the relationship of these elements to each other, which aims first and foremost to achieve the requirements of travelers and arrival. The consequences of the movement of these departures and arrivals to and from the state on which the airport is located are precautions for security and securing the airport's entrances and exits. In this research, Cairo Airport is studied in terms of location, access roads to the airport, a study of the ground side and passenger buildings, methods of distributing passenger buildings and assessing their locations in relation to each other and their relationship to the air side with the study of the preparation of the airstrips and their relationship to each other, as well as the study of vacant places and how to benefit from them by studying the components of the airport city and adding its elements in these spaces and making use of the entire surface of the airport land.

1-1 Research Hypothesis: If the efficiency of Cairo airport performance increases through the development of the airport and the exploitation of vacant flats and its transformation into an airport city, that development affects the airport's income and consequently the Egyptian economy is affected by that growth.

1-2 Research Structure:

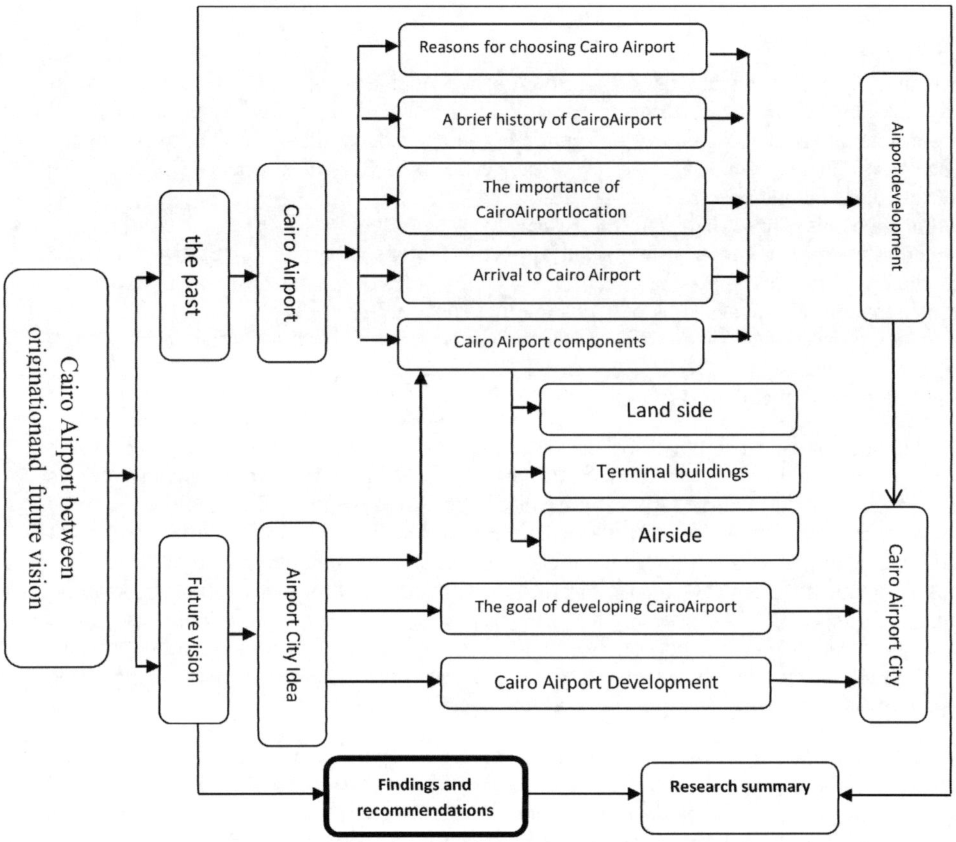

Figure (!) Shows the research structure

1-3 Airport Definition:

A specified area on the surface of the earth or water that includes any buildings, installations or equipment that are wholly or partly designated for the arrival, departure and movement of aircraft on the surface[1].

1-4 Determining The Problem:

The research problem lies in developing Cairo Airport and studying the factors that have a significant impact on prosperity and economic progress, and the growth of the airport with the aim of achieving the desired benefits from them through the concept of linking the airport with the surrounding environment and how to take advantage of the whole airport space in the development work, which results in large numbers of users (passengers & non passengers), which makes designers keep this in mind during the planning process in order to improve, develop and convert the airport into a small city to attract investors.

1-5 Research Objectives:

1- Introducing Cairo Airport.

.[1]Annex 14 to the Convention on International Civil Aviation –A erodromes- Volume I- pp 1-2

IOP Conf. Series: Materials Science and Engineering **974** (2020) 012014 doi:10.1088/1757-899X/974/1/012014

2- How to take advantage of the entire surface of the airport land and benefit from the airport's location in the economic aspects through studying the development of the airport to reach the airport city of Cairo, which needs to add some services to passengers and non-passengers such as (entertainment - logistical - administrative - commercial - educational - residential - Healthy - etc.) services.

2 Cairo Airport:

Cairo International Airport is an international airport away from downtown Cairo, the capital of Egypt, about 22 kilometers in the northeastern direction. The land area of the airport is about 40 million square meters and is the air gateway to Egypt and to the continent of Africa. The airport is the second largest airport in the continent in terms of congestion and passenger density. In 2017, the airport served about 31 million passengers.

2-1 Reasons for choosing Cairo Airport in the research:

• Cairo International Airport is the most important airport in the Arab Republic of Egypt and one of the most important airports in Africa.
• Identifying the strengths and weaknesses of Cairo International Airport.
• The ability to define the strategy followed in the Cairo Airport growth process.

2-2 Historical overview of the airports(2):

The airport's construction dates back to 1942, when the US Air Force, in cooperation with the British Army, built a military airport 5 kilometers in the north of Almaza Airport to serve the coalition forces participating in World War II, and the airport was called "Pine Field Airport" in relation to the name of the American pilot soldier "John Payne", who was the first American pilot killed in the battles of World War II, and the airport was very large if compared to the standards that prevailed in the airports at that time as it included two runways for the aircraft, an air traffic control tower, four hangars, and many buildings . On April 22, 1945, the Egyptian Civil Aviation Authority was established after it was a small department in the Ministry of War, and after the end of World War II, the airport administration and all Egyptian airports that were under British administration, aviation facilities and their responsibilities moved to the Egyptian side on December 15, 1946.

The new administration began to equip an international civil airport to accommodate the largest number of traffic, so the travel and arrival halls were expanded to accommodate the incoming and outgoing passenger traffic to the Egyptian lands, while Almaza Airport was designated for domestic flights. In 1946 the name of the airport was changed from Pine Field Airport to Farouk I Airport, and the number of passengers traveling through the airport in that year reached nearly 200 thousand passengers, and after the revolution of July the name of the airport was changed from "Farouk First Airport" to "Cairo Air Port". In 1955, some studies were conducted to build a new terminal building instead of the old one in order to keep pace with the increasing travel movement, and the location of the new building was chosen between the two main runways. Construction began in 1957, and the building was officially opened on March 18, 1963, and the capacity of Cairo Airport at that time reached 5 million passengers annually.

The increase in travel traffic through Cairo Airport and its arrival to the maximum absorptive capacity led to thinking about the expansion of the airport and the creation of new halls in building (1) and the establishment of a new terminal building, between 1977 and 1979 two travel & arrival halls NO.2were established in the main terminal building, and a new third runway for aircraft was established, in 1980 the terminal hall NO. 3 was established , and in 1986 the Terminal (2), located in the southern direction of Building (1), was opened, and the capacity of the new building reached about 3.5 million passengers annually. This was followed by the expansion and development of the passenger lounges, which increased the airport's capacity to 11 million passengers annually. In 2005 work began on the construction of Terminal (3) near Terminal (2) in order to reach the airport to its maximum operational capacity- the capacity of the new building is 11 million passengers annually.

(2) Cairo Air Port History, the official website of Cairo Air Port

IOP Conf. Series: Materials Science and Engineering **974** (2020) 012014 doi:10.1088/1757-899X/974/1/012014

The building was officially opened on December 18, 2008and it was actually put into operation on April 27, 2009bringing the capacity of Cairo International Airport in 2010 to 23 million passengers annually.

Due to the revival of the travel and arrival movement and its severe intensity in some times of the year excluding others such as the Hajj and Umrah season, the seasonal trips building was opened in 2011 to be allocated to serve the guests of Rahman in travel and access to EgyptAir flights flying to and from Jeddah& Madinah to alleviate the pressure on the travel hall No. 1 from which Egypt's Hajj and Umrah flights were taking off, along with other companies, and the passenger terminal yardsNo.1were also developed.

In 2012, the multi-storey garage was opened with a capacity of 3700 cars. The automated train project has also been opened to connect all terminals to the airport in record time and facilitate the travel and transit process. A firefighting station has also been established.

In 2013, the "Meridian Cairo Airport" hotel was opened with a capacity of 350 rooms. A new air cargo village was also developed, which provides technical facilities to support the growth in freight traffic through the airport.

The year 2014 came to witness the opening of the new cargo village, which is located on an area of 17 thousand meters, which represents a boom for the export and import movement in Egypt. The DHL building at Cairo Airport was opened on an area of 10,000 square meters. The center is the main hub for distribution in Africa and the Middle East, where the center works on Promoting sorting operations and re-exporting of incoming shipments from Europe to the countries of the region. The new center includes the latest international technologies that ensure receiving and delivering shipments of the highest quality and speed.

In addition, Terminal 2 was opened after its development and the increased capacity to 7.5 million passengers annually, raising the total capacity of Cairo International Airport to more than 30 million passengers annually.

2-3 The importance of Cairo Airport location[3] :

Cairo International Airport has a great importance, especially after Egypt Air Company joined the Star Alliance and has transformed Cairo Airport into a hub connecting Africa, the Middle East and Europe together in order to collect transit passengers and take them to the world's airports.

In view of the situation of Cairo Airport with the importance of its location, we find that it does not have a significant impact on air traffic in Africa, the Middle East and Europe due to several reasons including (mismanagement, planning, lack of customs exemptions on products and high ticket prices, especially in Egypt Air Company and the lack of improvement services that qualify the airport to increase the reception of transit passengers, which results in fewer passengers at Cairo Airport).

2-4 Arrival at the airport:

The airport can be reached through three roads: the first road is Salah Salem Street which passes in the middle of the Heliopolis area, the second

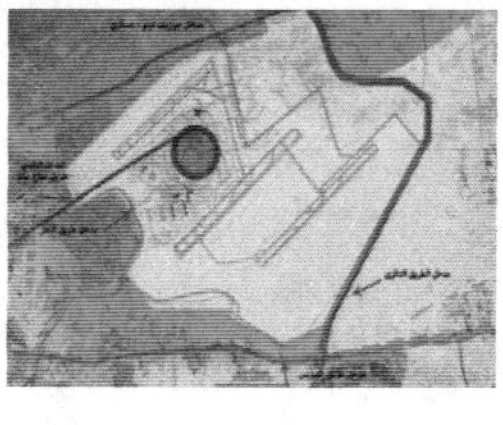

[33] Eng. Ahmed Nabil - Design Foundations of Terminal Buildings- Master Thesis - Al-Azhar University - 2014 AD p.
Figure No. (2) shows the way to get to Cairo Airport 216.

IOP Conf. Series: Materials Science and Engineering **974** (2020) 012014 doi:10.1088/1757-899X/974/1/012014

road is Al-Nasr Highway, and the third road is the road that runs between the Cairo-Suez road and the western ring road and has two entrances, there is another entrance on the one hand of Joseph Tito Street (New Nozha), but it is a special entrance to Sharq base (Military Airport), and the following figure shows the entrances of Cairo Airport, The airport will be connected to the neighborhoods of central Cairo and Giza Governorate through the third line of the Cairo subway. The new third metro line will carry passengers traveling and arriving through the airport to the neighborhoods of the capital, then an extension to Dokki and Giza.

By studying the roads leading to Cairo International Airport, we find the main roads of the airport work as car roads, and this is not sufficient for entry and exit of the airport; therefore it is recommended to increase railways (train - metro) linking the airport and the city center so that the load on the highways is reduced and the benefit becomes greater for travelers.

2-5 Cairo Airport components:

2-5-1 Landside[4]

: It means all uses surrounding the airport connected to and from the city, private parking lots, bus and taxi stops, and other services related to the part surrounding the airport affecting or affected by it; in general it is the area of the movement of cars and buses carrying passengers, their luggage and depositors, concerning that the path of the tracks must be appropriately expanded for the sake of the movement outside the airports.

When we talk about the landside, it is necessary to define what the landside is and what its limits are. The landside is the external area in front of the terminal building starting from the terminal building pavement, services, garages, and so on. Therefore, Cairo International Airport has landsides according to the number of terminal buildings inside it, which are the landside of the terminal building (1, 2, 3, 4 and the seasonal hall) ;and we will talk in detail about each part separately.

After completing the study of the landside of Cairo International Airport, there is some difficulty in using the parking lots in front of the three buildings of Cairo International Airport and in a very bad condition, but the worst thing is the case of Building «2» because it doesn`t accommodate the number of cars and it doesn`t fit with the intensity of the cars movement, as well as after adding the multiple storey garage next to the terminal building (2), but the distance is far, especially for baggage holders. It was preferable to make multi-storey car parks under the garage currently standing next to the terminal building (3) to achieve proximity to the terminal building and increase the area of parking lots whether in the horizontal direction or the vertical one; and the following figure shows the car parking lots in Terminal 2 and 3.

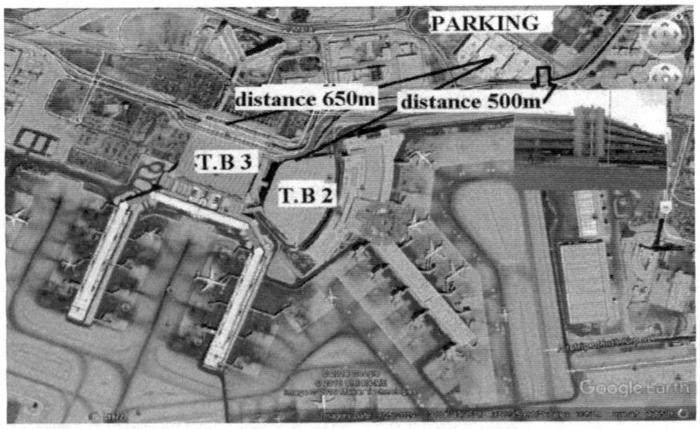

Figure(3) shows car parking lots in Terminal 2&3

2-5-2The Terminal Building[5]:

The terminal building area is the most important one in the airport, as it contains the terminal building with all its planning and designing problems, its conflicting requirements, and its multiple services, etc. Consequently, this area of the airport is the scope of work for architects and

-topic134[4]http://gacaksa.forumarabia.com/t

[5]Airport Passenger Terminal , Planning and Design, ACRP,REPORT, Volume 1: Guidebook, p.25

IOP Conf. Series: Materials Science and Engineering **974** (2020) 012014 doi:10.1088/1757-899X/974/1/012014

engineers for(acoustics, lighting, ventilation, air-conditioning, different treatments, and other various disciplines).

2-5-2-1 The importance of the terminal building (1):

A- The terminal building is the contact point between the airside and the landside. Thus it contains gates and loading bridges connected to the plane.

B- The terminal building is the one that achieves the concept, processes and procedures for separating travelers and arrivals.

C- The terminal building is the one that receives the travelers` relatives, and provides them with all means and services.

D- The terminal is the airport building through which travel and immigration processes are controlled.

E- As a result of the modern services provided by the terminal building including restaurants, hotels, business centers, etc., the importance of these buildings increases even for non-travelers themselves who may use these services.

2-5-2-2 Terminal buildings at Cairo International Airport: Cairo International Airport consists of more than one terminal building as follows: terminals No. 1, 2, 3, 4 for VIPs (very important persons) and the seasonal terminal for Hajj and Umrah trips. Each building will be explained in detail with its components and services from its inception until the present time.

2-5-2-3 Terminal Building No. (1) :

The construction history of the building dates back to 1963, which is the oldest terminal in the airport, and Egyptian citizens call it the "old airport", which includes three halls as follows: the internal hall No. (1), the international travel hall No. (2), the international arrival hall No. (3) and the hall No. (4) that was dedicated to Egypt Air Company for its flights to the cities of Hurghada and Sharm El-Sheikh, in addition to a hall for air services for catering aircraft. The landside of Terminal No. (1) contains an open car garage with a capacity of more than 5,000 cars and connects to Terminal 2 and 3 through a train with a top bridge and access to it directly from Al-Oruba - Salah Salem Road, and next to the building there is a large commercial market called "Air Mall". The following figure shows Terminal 1 at Cairo Airport.

Figure No. (4) shows terminal building No. (1) at Cairobefore development

Figure No. (5) shows terminal building No. (1) at Cairo Airport after development

IOP Conf. Series: Materials Science and Engineering **974** (2020) 012014 doi:10.1088/1757-899X/974/1/012014

2-5-2-3 Terminal Building No. (2)[6]:

Terminal (2) was opened in 1986, Egyptian citizens call it the "new airport", and it is primarily dedicated to European, Gulf and Far East airlines such as Korean Air, Italian Airlines and Saudi Arabian Airlines. The building contains only 7 bridges to move passengers to the planes, and when more than four trips are met at one time, the terminal becomes in a marked congestion; that`s why the Cairo Air Port Company announced a project to develop the terminal in July 2008. The terminal development project aims to increase the building area and modify an area Loading the aircraft, which are bridges, to double it to 14 instead of only 7 in order to receive the largest number of passengers during peak times[7]. The officials expected that the building capacity will increase from 3.5 million passengers annually to 7 million passengers, so that the airport capacity after completing that development will reach 23 million passengers annually, as well as that terminal will be linked to terminal No. (3) by a bridge in order to facilitate the movement of transportation between the two terminals. The terminal development process has already started in February 2010, and it has been declared that the building was closed for a period of three years. During that period, all terminal (2) airlines have been operated in terminal (1).

On September 28, 2016, the Ministry of Civil Aviation celebrated the first actual operation of the terminal after its development and its increased capacity to 7.5 million passengers annually, which raised the total capacity of Cairo International Airport to more than 30 million passengers annually. The terminal contained 28 loading bridges, and the number of passport counters reached 78 counters. It also contained 11 hotel rooms inside the customs department highly equipped and used by transit passengers without the need to obtain entry visas to Egypt. the following figure shows terminal 2 before and after the development.

Figure No. (6) shows terminal building No. (2) at Cairobefore development

Figure No. (7) shows terminal building No. (2) at Cairo Airport after development

The two previous figures show the impact of the terminal No(3) construction at Cairo Airport upon terminal No. (2), whereas the terminal building hall has been modified and a new wing for the old terminal has been built to increase the building capacity and the numbers of travelers; so that it has used the parking area located near the new terminal, which represents a remedial adjustment for the concept of the airport growth. That remedial amendment is an additional economic burden, resulting from the failure to study the concept of the terminal capacity and the way in which these buildings grow in an integrated manner with the previous and future stages of growth, and through this it becomes clear to us that the parameters of the remedial adjustment are unclear and not planned. Consequently, the parameters for future growth are unobvious

1Eng. Ahmed Ibrahim - Safety Standards in Terminal Buildings - Master Thesis - Ain Shams University - 2013, p. 257.

2 Dr.Nafisa Mahmoud Al-Nasharti Transfer Airports - A case study of Cairo International Airport and the problem of transferred passengers - 2006 - p. 168

IOP Conf. Series: Materials Science and Engineering **974** (2020) 012014 doi:10.1088/1757-899X/974/1/012014

2 - 5 - 2 - 4 - Terminal Building No. (3)

Building No. (3) is the latest building to be constructed at the Airport, which is allocated primarily to Egypt Air flights and Star Alliance's Members Airlines. The idea of constructing the third terminal building started in the early nineties of the last century after the growing need to construct a building to accommodate the expected increase in the number of passengers during the next twenty years. However, the idea did not come into effect, especially since whenever one of the studies concludes the importance of the building to accommodate the increasing movement during the coming years, some incidents that affect the volume of traffic in the region, such as the second Gulf War and the events of September eleventh, after which the rates of passenger flights decreased. However, the establishment of the Ministry of Civil Aviation in March 2002, and the establishment of the Egyptian Holding Company for Airports and Air Navigation, and the Egyptian Airports Company in addition to converting the Cairo Air Port Authority into a commercial company created the motivation to realize that idea. Work began on the building in 2005, with an area of 190,000 m².

The building consists of a basement, three main floors, and two mezzanine floors, with a maximum height of 122.6 meters above sea level, as follows: Basement which includes the building's services such as workshops, maintenance, warehouses, and planes supplies, etc.; the first floor serves international arrival flights; the second floor serves international departure flights; the third floor serves domestic arrivals and departure flights; as for the mezzanine floor, it includes administration offices, Airlines Offices and Air Traffic Services, in addition to parking lots, and 63 airplanes parking areas. The building can accommodate 15 large model aircraft at one time, and it is equipped with 23 loading bridges for landing and boarding of passengers from and to airplanes serving thus 23 parking lots adjacent to the building. As for the 40 remote parking lots, passengers will be transported thereto by bus. The Departure and Arrival terminals are also designed to accommodate all aircraft models through 15 waiting halls, passenger security devices, in addition to follow-up and security systems that are consistent with the highest standards of the global code. The building includes five commercial areas and restaurants with an area of 3300 square meters, in addition to the equipped waiting areas and VIP halls. The Landside of Terminal Building No. 3 contains a parking area that serves the building to accommodate 2360 private cars and 200 buses. There is also the Meridian Hotel, which is connected directly with Terminal 3 by a bridge

Figure No. (8) shows linking of terminal building No. (2) to terminal building No. (3)

2 - 5 - 2 - 5 - Services Inside the Terminal Building

The research sheds light on the concept of the terminal building as a means of generating income for the airport, as well as a building for public and social facilities, since the relative balance between the two functions is what directly affects the personality of the terminal. In airports designed to generate

IOP Conf. Series: Materials Science and Engineering **974** (2020) 012014 doi:10.1088/1757-899X/974/1/012014

the maximum possible income, passengers are directed to banks, shops, and customs exemption areas (free zones) while aviation offices take a secondary role. On the other hand, in other airports such as John Kennedy Airport in New York, which is primarily concerned with public and social service, it contains organized spaces and uncrowded lobbies that give priority to passengers and their needs for comfort and relief from the hassle of traveling. One of the significant trends in the design of passenger terminals is the exploitation of the time factor in activities that entertain passengers once they arrive at the airport until the moment the plane takes off.

Considering the Cairo International Airport and the services provided therein, we did not find any kind of integration between the design and planning of spaces in the terminal buildings, so that the passengers are intentionally directed to the commercial areas and are encouraged to wait and spend some time between the check period and leaving the terminal building (waiting hall). Some airports also provide some services, as "USA today" newspaper listed a number of the strangest activities provided by airports for passengers to ease waiting times.

For example, "Munich" Airport in Germany includes a covered "Surf & Style" unit that provides passengers with the opportunity to surf, as the unit produces a set of the highest artificial waves in the world. The following figure shows surfing at Munich Airport.

Figure No. (9) shows linking of terminal building No. (2) to terminal building No. (3)

In Japan, Narita Airport in Tokyo offers oxygen sessions, during which the passenger inhales the oxygen with different scents and flavors on demand, such as cinnamon, menthol and other smells that help relax and calm the nerves. The session lasts ten minutes and costs 7 dollars.

Figure (10) shows the oxygen sessions at Tokyo Airport

In the United States of America, "Pittsburgh" Airport offers a unique service that allows passengers to hunt wild deer in the early morning. The airport raises deer on a plot of land estimated at 906 acres, and issues 60 hunting licenses.

Figure (11) Forest Airport "Pittsburgh"

Those wishing to get a license submit applications at the airport's website that draws lots to choose winners. Applicants are not allowed to attend the lottery, but winners are notified via email. Hunting tools are limited to arrows because the airport law does not allow carrying or using firearms.

IOP Conf. Series: Materials Science and Engineering **974** (2020) 012014 doi:10.1088/1757-899X/974/1/012014

Also, Istanbul Airport offers massage service, food, beverages and various games in addition to cinematic shows for free at the frequent passenger suite.

2 - 5 - 2 - 6 Shopping Services inside the Terminal:

In some airports, such as Paris airports, passengers who reside in a country that is not a member in the European Union are entitled for leisure purpose to the privilege of deduction of value-added tax on purchases bought from France. **In Cairo Airport, however, the prices are increased for the passenger with the addition of value-added tax and thus the prices are exaggerated significantly resulting in limiting shopping process except for necessity, not to mention that there are no major suites inside the waiting hall for shopping or entertainment to amuse passengers as mentioned in the previous examples.**

2 - 5 - 2 - 7 How to Connect Terminals Together:[8]

Terminal 2 is near Terminal 3 and they connect through a hanging aisle. The two suites of Terminal 3 in which the travel or departure gates are located extend near the aircraft aisles, and they are also located near the aircraft parking spaces. Terminal 2, Terminal 3 and Terminal 1 are connected via the Automated People Mover- (APM)

Figure No. (12) shows connection of terminal No. (1) with terminal Nos. 3&2

Figure No. (13) shows connection of terminal No. (2) with terminal No. (3)

By studying the terminal buildings at Cairo International Airport, we notice that terminal building 2 and 3 are close and are connected through a covered corridor. It would have been better to keep more distance between terminal building no. 3 and terminal building no. 2 to avoid congestion, especially in the parking lots outside the building. If we take a look at the airport as a whole and review the location of the terminal building No. 3 at the left side of C23 / C05 and L23 / R05 passageways, while it would have been better for terminal 3 building to be located between the C23 / C05 Airfield and L23 / R05 Airfield as in the following figure in order for the C23 / C05 Airfield not to be interrupted. In which case

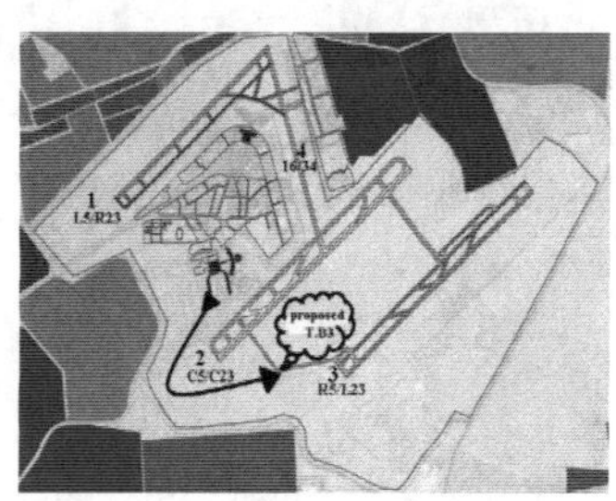

Figure No. (14) shows the ideal location for the terminal 3

[8] Eng. Ahmed Nabil – (الأسس التصميمية لمباني الركاب) Foundations of Designing Passenger Terminals - Master Thesis - Al-Azhar University - 2014, p. 235.

the Cairo airport buildings distribution is as follows Terminal 1 and 4 for R23 / L05 Airfield, terminal 2 for C23 / C05 Airfield, and terminal 3, and the seasonal hall for L23 / R05 Airfield.

2 - 5 - 3 - The Air Side:[9]

The air side is the area extending from the exit phase of the terminal building to the boarding of the plane and is divided into two sections: The Airfield, referring to the movement of the plane on the ground (waiting yards, parking, airstrips, aisles, maintenance, supplies, control tower and hangars); and the Airspace that refers to the plane and its field of movement in the air during taking off and landing. The Air side of Cairo Airport includes four runways for takeoff and landing of aircraft, the first runway is R23 / L05 with length of 3300 meters and width of 60 meters, located in the north area of the airport; while the second runway C23 / C05, is 3500 meters long and 60 meters wide as well, located in the south area of the airport; the third runway L23 / R05 has a length of 4000 meters and a width of 60 meters, and this runway is able to accommodate giant Airbus A382 aircraft, and is located in the south area of the airport next to runway 2; and the fourth runway (16/34) is currently suspended due to some contraindications. The following figure shows the runways at Cairo International Airport.

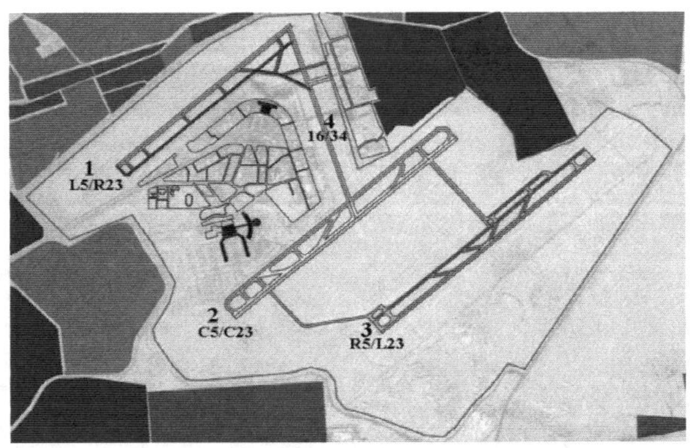

Figure No. (15) shows the number of runways at Cairo International Airport

Having studied the Air Side of Cairo Airport, it is clear to us that there are four runways, as previously mentioned, one of them is closed due to some barriers, and the reason for this is the absence of applying the legislation for airports by allowing the residents near the airport to increase the height of buildings, especially the new Nozha area and Al Salam area, which led to closing the runway and reliance on the three other runways 5/23. Terminal 3 is located to the left of runways C23 / C05 and R23 / L05 and it would have been better to increase the length of runway C23 / C05 to 4 km instead of constructing a new runway, especially that in case of using runway R23 / L05 runway C23 / C05 shall be interrupted which comes in violation of runway specifications according to the requirements of the International Civil Aviation Organization.

3 - Cairo Airport Development to be an Airport City

When aiming at developing Cairo Airport to be an Airport City, the following should be studied:

A - The vacant spaces in Cairo Airport and the extent of the possibility of expanding the airport buildings.

B - Studying the current components, mentioned above, i.e. the Landside, Terminals, and the Air Side.

C - Studying the components of the Airport City and how to develop the airport through these components.

---

[9] Dr. Nafisa Mahmoud Al-Nasharti - (مطارات التحويل - دراسة حالة مطار القاهرة الدولي ومشكلة الركاب المحولين) Transfer Airports - A case study of Cairo International Airport and the problem of transferred passengers - 2006 - p. 19.

IOP Conf. Series: Materials Science and Engineering **974** (2020) 012014 doi:10.1088/1757-899X/974/1/012014

A detailed explanation will be provided regarding Cairo International Airport, a study of the buildings at the airport, and identification of places that can be expanded. The following figure indicates the use of land at Cairo International Airport, including the basic buildings, which are terminal buildings 1, 2 and 3, in front of which the Air Side appears, which consists of aircraft parking lots and then airstrips to arrive at the runways as the cargo village appears in the area between Terminals 1 and 2. There is also a hotel area near Terminal 1, such as Airport Movenpick Hotel and the Meridian Hotel connected to Terminals 2 and 3. There is also a manufacturing area used as aircraft maintenance workshops in the eastern area. All these elements are shown in the following figure with keys to read the details

Figure No. (16) shows land use at Cairo International Airport

3 - 1 Aims of Developing Cairo International Airport

- Taking full advantage of the land areas (plots) at the airport
- Enhancement of the aesthetic appearance of the State.
- Revitalization of tourism.
- Increasing foreign investment in our country.
- Reducing unemployment and creating job opportunities for youth
- Expanding resource development as a result of job opportunities created by the project.
- Airports around the world directly support the economic movement by providing access to global markets and supporting trade. The global air traffic increases significantly, which directly and greatly affects the economy[10]

2 - 3 Cairo Airport City:

The airport city may be defined as "the final form or configuration that gives a picture of the entire airport area", that is, what relates to aviation affairs such as the land runways and airspace over this area, as well as what is not related to aviation affairs meaning (what is related to the airport buildings itself, along with its

Figure No. (17) shows the components of the airport city

[10] Dr. Sahl Abdullah Saad EddinWaheeb - The possibility of benefiting from the concepts of sustainability in the future of airport design in the Kingdom of Saudi Arabia - Umm Al-Qura University, Makkah Al-Mukarramah, Saudi Arabia - 2015 - pp. 12 and 13.

facilities, services, roads, parking lots, and the rest of the region within the area or cordon of the airport itself). Future airports, which some countries have begun to work on their designs and study their application, depend on the so-called small cities, which are an independent and self-sufficient entity that includes all the services that include shopping centers, retail, entertainment, gardens, and restaurants, and making it available to passengers, and that is what we refer to by Airport City. The city also includes different models of housing, schools, universities, hospitals [11]and business administrative and residential areas, conference centers, exhibitions and logistics areas. Figure no. 13 on your right illustrates the airport area and the surrounding services that helps to upgrade the airport into an airport city.

Having concidered the use of lands in the airport area, we find some areas where there are no buildings or facilities and therefore these areas must be defined and its area to be measured and a study shall be made concerning the possibility of adding activities to the airport to help the growth and improvement of airport revenues, how to integrate these elements with the current airport components, and how to link these components with each other. These components are a set of services presented at the airport for passengers and non-passengers, including (logistical - commercial - administrative - medical - entertainment - ... etc.) services.

After mentioning the elements of the airport city, the vacant areas in Cairo International Airport were identified and divided into 4 main areas, and proposals for each area were made separately regarding how to benefit therefrom through the activity that is appropriate to the site. The following figure indicates areas that can be used at Cairo International Airport without obstacles.

Figure No. (18) shows the vacant areas in Cairo International Airport

Having studied the unused areas at Cairo International Airport, a proposal is put in place to use each region in a way that suits it and that accomplishes the primary goal of developing the airport, which is to bring in revenues to improve the economy. These regions were identified, and they are represented in five main regions. We start with the first area, which is the airport entrance area from Al-Oruba Road. This region is one of the best areas in terms of location, it is at the front of the airport, and thus it is considered the interface of Egypt. Thus, it is possible to take advantage of the site by using this area as administrative, commercial, medical and educational complexes with exhibition and conference centers. The second area is the area adjacent to the New Nozha neighborhood on the side of Joseph Teto Street. And that region is preferable to establish a commercial residential area to be used to attract tourists and provide commercial malls near it. The third region is the area next to Wadi Degla Club, which is the area between the Suez Road and the Autostrad Road. It is better to use it as a sports recreational area. The fourth zone is the area above Runway 3, which is the area between the Ring Road and the Suez Road. This area from the research point of view should be made a cargo village, especially

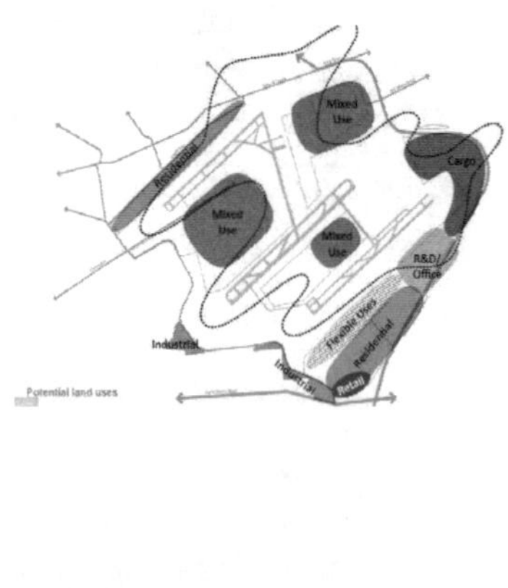

[11] Global Airport Cities-- London –2009- p no 39 - John D. Kasarda

IOP Conf. Series: Materials Science and Engineering **974** (2020) 012014 doi:10.1088/1757-899X/974/1/012014

because of its connection to the ring road, where a new entrance is created from the ring road to the cargo village directly. This village has been linked to a logistical area next to it in the area between the Ring Road and the Suez Road. The following figure shows a sketch of the areas proposed for use at Cairo International Airport.

Figure No. (19) shows a proposed sketch of land use at Cairo International Airport[12]

Conclusion:

This study, entitled (Cairo Airports Between Creation and Future Vision), dealt with a simplified explanation of the concept of airports, with a historical overview of the establishment of Cairo International Airport and its components, starting from the Landside, through the Terminal buildings, a stage study reaching the Air side. The study also examined the design criteria for the components of the airport, and the concept that development at the airport creates a new term for us, which is the city of the airport, which consists of several complex components, through which, through its components, attracts investments and competes with transcontinental services. And, as many economists say, without strong regional airports, a comprehensive civilization renaissance will not be achieved in countries. In order to achieve effective planning, there must be a framework for the development of areas near Cairo Airport based on the (economic) dimensions. Cairo airport should be linked with the surrounding areas

and take advantage of its distinctive geographical location and link it with new cities such as New Cairo, the city of transit, the city of sunrise and the administrative capital, and that by linking it to the logistical areas near Ain Sukhna, with the aim of achieving comprehensive economic development through developing the airport internally, with the elements that supports the economy, and externally with other countries.

Findings:

A - Cairo Airport is considered Egypt's international interface, and therefore management must be improved to achieve the required good reputation.

B - The entire area of the airport land should be used in order to achieve comprehensive development in the area around the airport and create new job opportunities.

C - The airport has a civilizational, economic, commercial and environmental impact through its location in the middle of the world.

D - In the event that Cairo Airport is converted into an Airport city, the Cairo Airport City after development shall be considered an independent and self-sufficient entity that includes all commercial, administrative, residential, recreational, logistical services ... etc. Thus, a noticeable change will occur in the airport's revenues, whether from aviation or non-aviation business.

E - There is randomness in choosing the locations of the terminal buildings and their connection to each other, which causes problems for visitors in moving from one terminal to another.

F- Lack of sufficient number of car parks, which creates a crisis in the official seasons.

[12] Cairo Aerotropolis | Interim ReporTMasterplanRepoRT November 2012 |2 May 2013pp 167

IOP Conf. Series: Materials Science and Engineering **974** (2020) 012014 doi:10.1088/1757-899X/974/1/012014

G - A strategic plan for the growth of the airport and the terminal buildings has not been developed yet. Development must be carried out according to a previously studied and determined plan, with the aim of avoiding the process of developing the passenger buildings in a preventive manner, as happened in the plan to develop terminal 2 at Cairo Airport after the construction of the terminal building 3.

Recommendations

A - A future plan for the development of Cairo International Airport should be established until the completion of all phases of development, explaining the use of the airport's lands and traffic paths for visitors, and it shall be implemented within specific time plans and schedules without prejudice to the general location of the airport.

B - Overcoming bureaucratic obstacles in developing the airport.

C - Taking into consideration the principles of airport planning and its relationship with cities when selecting new airport sites.

D - Taking into account the principles of the environmentally compatible architectural design for airport buildings.

E - Taking into consideration that the size and quality of services and facilities provided by the airport (commercial, tourism, businessmen services, etc.) ... are commensurate with the quality of passengers through the building, and are closely related to the general activity of the state and its strategic location.

F - Using the help of international offices with expertise in airport science to study the modernization and development of the airport.

Sources and References:

Arabic References

1 - Muhammad Majed Khulusi - Airport Design - Dar Qabes for printing, publishing and distribution - 1998.

2 - Dr. Nafisa Mahmoud Al-Nasharti Transfer Airports - A case study of Cairo International Airport and the problem of transferred passengers - 2006.

3 - Eng. Ahmed Nabil - Design Foundations of Terminal Buildings - Master Thesis - Al-Azhar University - 2014 CE.

4 - Eng. Ahmed Ibrahim - Safety Standards in Passenger Buildings - Master Thesis - Ain Shams University - 2013

5 - Dr. Sahl Abdullah Saad EddinWaheeb - the possibility of benefiting from the concepts of sustainability in the future of airport design in the Kingdom of Saudi Arabia - Umm Al-Qura University, Makkah Al-Mukarramah, Saudi Arabia - 2015

Foreign References

1 - Annex 14 to the Convention on International Civil Aviation - Aerodromes- Volume I -

IOP Conf. Series: Materials Science and Engineering **974** (2020) 012014 doi:10.1088/1757-899X/974/1/012014

2 - www.wikipedia.com

3 - John D. Kasarda - Global Airport Cities-- London –2009- p no. 39 -

4 - Cairo Aerotropolis | Interim Report MasterplanRepoRT November 2012 |2 May 2013

5 - John D. Kasarda- The Evolution of Airport Cities and the Aerotropolis - London –2008.

6 - John D. Kasarda - Airport Innovation _New Planning Models- London –2007.

7 - John D. Kasarda - AEROTROPOLIS-AIRPORTS AS THE NEW CITY CENTER - London –2007.

8 - The Emergence of the Airport City in the United States - California - 2013.

IOP Conf. Series: Materials Science and Engineering **974** (2020) 012015 doi:10.1088/1757-899X/974/1/012015

Military Technical College
Kobry El-Kobbah,
Cairo, Egypt

13th International Conference on Civil and Architecture Engineering ICCAE-13-2020

The Reflective Impact of Different Teaching Strategies in Teaching Architecture and Interior Design Practical Courses

Marwa Abdelalim[1], Noura Aji[2], Sherif A. Mahmoud[3], Mohammad Fahmy[4]

[1] Assistant Professor, Architecture Engineering Department, Faculty of Engineering, Prince Sultan University, Riyadh, KSA.
[2]Lecturer, Prince Sultan University, Interior Design Department, Riyadh, KSA
[3]Assistant Professor, Architecture Engineering Department, Military Technical College, Cairo, Egypt.
[4]Associate Professor, Architecture Engineering Department, Military Technical College, Cairo, Egypt.

Corresponding email address: mabdelalim@psu.edu.sa

Abstract. Achieving learning outcomes in a creative manner, actually considers a vision for all instructors whether theoretical courses or practical courses, in a reflective way where such reflection interfere to boost students' self-confidence while enhancing their knowledge. The research investigates some effective teaching strategies that attempt to bridge the gap between theories and real application trying to close the loop of getting both interior/architectural design projects, and construction technology integrated. The research adopts an ontological assumption of applying questionnaire and surveys to recognize this silent war, is it in the students' mind as just fear of MAKE NO MISTAKE or fear from the future career. According to Tony Robbins. He said if you want a better result then ask yourself a better question. The main research problem is thinking about the rational beyond the reflection of effective teaching strategies as well as solving the problems that fear junior architects and interior designers, to know exactly where is the battle field? A whole mechanism of some teaching and learning strategies as well as direct assessment tools, in accordance with well-defined rubrics were applied.The researcher presents a considerable hypothesis that assures that believing in the architecture and interior design students' abilities and the instructor's variable teaching strategies can bring out examples of best practices in teaching philosophies.

Keywords: Flipped classroom strategy, One to one basis, circle talk, assessment tools

IOP Conf. Series: Materials Science and Engineering **974** (2020) 012015 doi:10.1088/1757-899X/974/1/012015

1. Introduction

The UK Professional Standards Framework supports the initial and continuing professional development of staff engaged in teaching and supporting learning, which can be tracked in a sequential activity accordingly mapped to the core knowledge and its values.[Fig.1]Starting from this understanding, the research emphasizes on the design and planning of the learning activities in practical courses' labs, which will further be discussed.

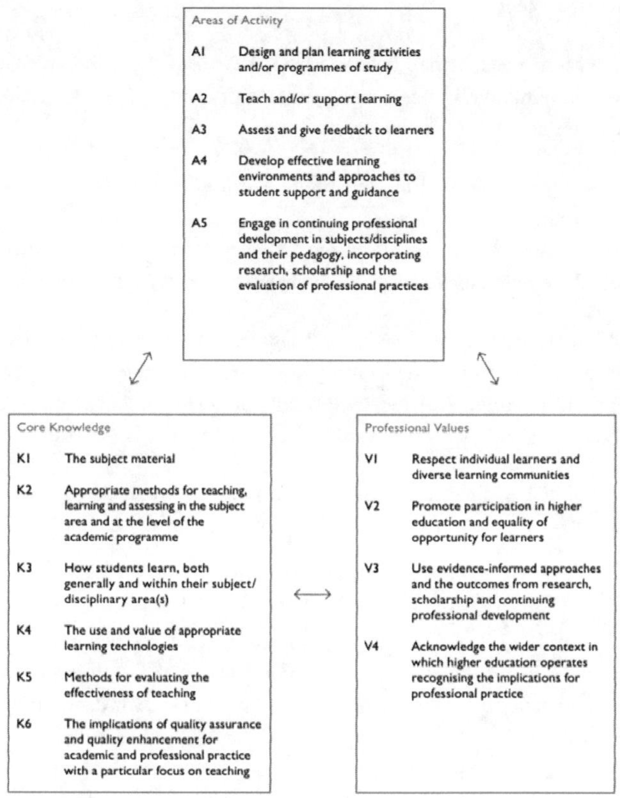

Figure 1: UK professional Standards Framework

The main problems identified revolved around how students encounter difficulties when trying to link a theoretical detail learnt in class with real life projects. Students tend to have a pre-set fear of incorporating an idea they are imagining in their projects because of their worry of not being able to actually execute it in the right way. Therefore, the main challenging task can be viewed as how to grab the students' attention when delivering the theoretical part of a studio course while making sure the students' confidence in their ideas remains stable.The matter which puts some obstacles in the way of the whole learning and teaching cycle, raising the question of WHY junior architect or interior design students feel they are part of a long lasting *silent war*; in particular, students feel they are part of a jury throughout the entire time leading to involuntary defensive character.

The researchers planned to apply several teaching strategies, such as flipped classroom where the lecture is devoted to discussions and activities, in order to test the students' skills in applying knowledge and interacting with one another throughout hands-on activities. In studio sessions, the researchers tried to address different students' learning styles and needs, adopting one-to-one basis to accommodate individual differences.

In regards to students' sharing ideas, it has been noticed that once they are given opportunities to participate and share their thoughts with their peers, good results can be achieved; circle talk teaching strategy has therefore proven itself as a successful one.

An additional strategy employed, revolved around the self-critique concept. This idea improves the students' visualization of their own weaknesses, which stimulates one's self-enhancement. As a strategy, self-critique can be allocated as one of the most successful ones as its results derive from the students' themselves; defensive behavior towards the instructor is therefore evitable in this scenario.

Moreover, examining field trip experiences was a very interesting and effective application placing it at the top of the list. During a field trip, students are given important moments in learning in a shared social experience that provides them with the opportunity to meet and discover new things in an authentic setting.

Surveys and charts were conducted to find out how close or far the relationship between the applied strategies is, allowing measurable outcomes to show the success or failure of the applied strategies.

2. Methodology

The research adopts both quantitative methods as surveys and qualitative methods as interviews with both students and instructors. The feedback of all stakeholders was gathered to come up with a SWOT analysis that highlights some alternatives that have been set as solutions for developing more successful learning activities.

Learning and teaching strategies will be triangulated to their direct effects on the students, which will lead to suggestions for improvement resulting in the best practices.

3. Applications of the different teaching strategies on students

An effective teaching is not solely linked to the course content, it is rather measured by the process that an instructor follows to deliver that content. With the availability of such a big variety of teaching strategies, every instructor has the freedom to pick the ones proven to be the most successful: online searching, book related research, in-class activities. The key to be able to decide on the usefulness of a teaching strategy resides in understanding the background of the students along with the environment they are coming from and their learning goals. Students have different abilities when it comes to demonstrating their knowledge, which opens the door to the urge of using various learning styles to accommodate all the students' needs and equalize their opportunities to learn. Nevertheless, no matter the differences among the students, praise will always result in positive outcomes.

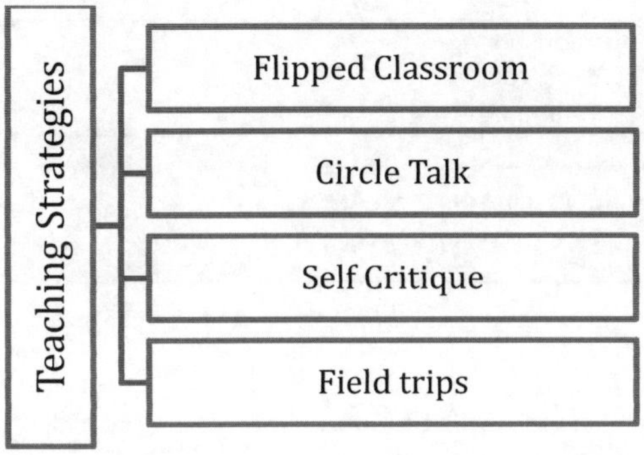

Figure 2: Different applied Teaching Strategies

3.1 Strategy: Flipped classroom

As a first teaching strategy, students are given the opportunity to share their own understanding and knowledge within the classroom. Students have the freedom to gather information from whatever sources they have in relation to a certain topic; after that they would meet with their instructor and peers and discuss what they have found. This method allows to broaden students' knowledge along with receiving deeper insights from the instructor and enhancing their problem solving skills. Thus, the classical roles of students and teachers, as well as the chronology of a course or lecture are 'flipped'. To develop our flipped-classroom's structure, we followed the suggestions of Prober and Khan (2013) which helps create an innovative design of students' learning environment,[Fig.3] along with an evaluation of the students' experience[1]

To help ensure student preparation for class, students were expected to prepare drafts for their chosen architectural and interior details to be applied in their projects, the challenge was they were suppose to search for specific details that they inspired from other implemented famous projects, seeing how this can be fit with their projects. Their efforts were randomly collected and graded. The students got benefits from their instructors' feedback,[Fig.4] which was positively influencing helping in understanding and connecting corresponding courses together and in the same time help saving time for a good results.

IOP Conf. Series: Materials Science and Engineering **974** (2020) 012015 doi:10.1088/1757-899X/974/1/012015

Flipped classroom strategy[1] does in fact allow better use of class time, increases the one-on-one time between instructors and students at the same time as it gives the opportunity for active learning techniques.

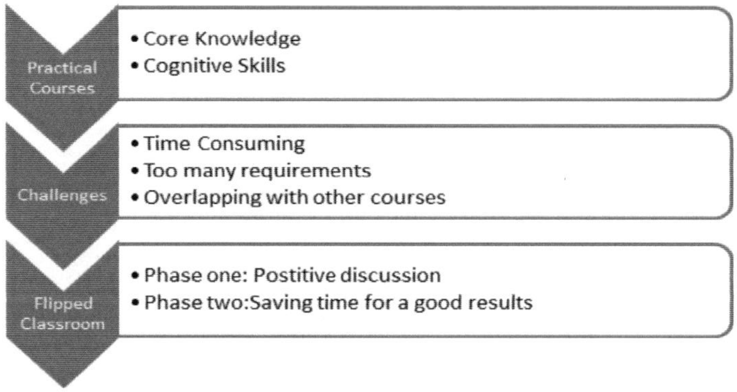

Figure 3: Flipped classroom's Structure, Source: the researchers

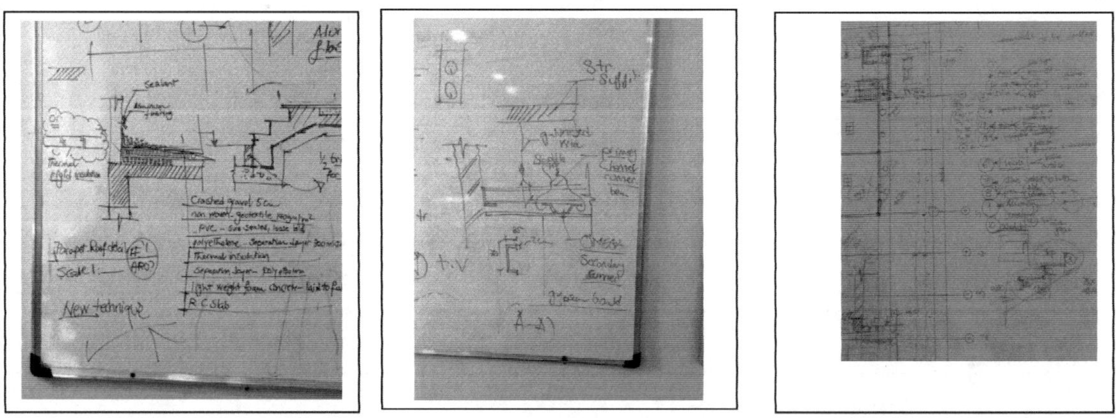

Figure 4: Sequence of given feedback to students in accordance with the approached flipped class, Source: researchers

[1]Flipped classroom frameworks improve efficacy in undergraduate practical courses – a quasi-randomized pilot study in otorhinolaryngology, Tobias Dombrowski, Christian Wrobel, Stefan Dazert&Stefan Volkenstein

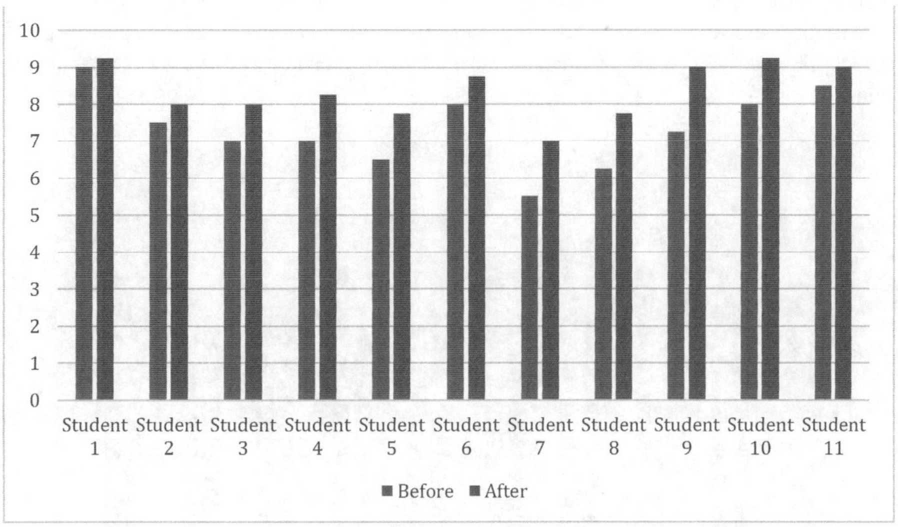

Figure 5: Showing the comparison of grades before implementing the flipped classroom strategy versus after implementing it
Source: Researcher

Analysis:
By looking at the above graph, the average of grades before implementing the strategy was 7.3/10, whereas after implementing the flipped classroom strategy it became 8.4/10. These results show an incremental rate of 15%. As a result, these numbers clearly show that students have evolved since the flipped classroom strategy was implemented, which proves its successfulness.
Since the researches' introduction of the flipped classroom model, the role has changed; most of the time was spent in class walking around helping the students who struggle the most. This may be the single most important reason students thrive for the flipped model.

3.2 Strategy: circle talk[2]

Logically speaking, there is always a certain gain when sharing ideas among peers, or when asking for more than an opinion in regards to a certain idea. Throughout experiencing this second teaching strategy, all of the students along with their instructor were gathered to discuss all the projects together. The main challenge of this strategy was to make sure every student gave their opinion objectively and freely. What was noticed during this experience, is that students shared their opinions wisely while trying to help their classmates come up with better outcomes. The atmosphere residing was not like a typical classroom anymore, it became much more enjoyable for the students as they forgot about grades and cared only about discussing each other's work as if they were doing so outside of class[Fig.5].

---

2 Brame, C., (2013). Flipping the classroom. Vanderbilt University Center for Teaching. Retrieved [todaysdate] from http://cft.vanderbilt.edu/guides-sub-pages/flipping-the-classroom/.

IOP Conf. Series: Materials Science and Engineering **974** (2020) 012015 doi:10.1088/1757-899X/974/1/012015

In fact, students always find it easier to handle feedback given from their peers rather than from their instructor; therefore allowing students to critique each other's work and give out positive and/or negative feedback helps to promote their critical thinking, opening the door for continuous development.

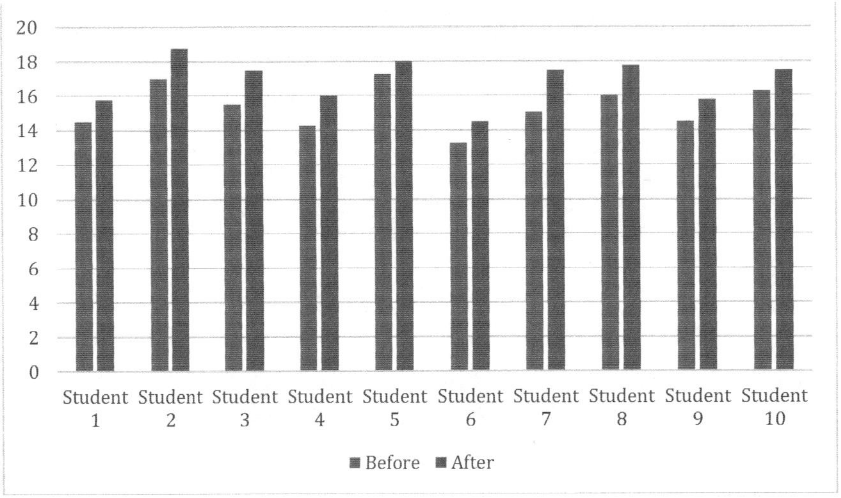

Figure 6: Comparison of grades before implementing the circle talk strategy versus grades after implementing it
Source: researchers

Analysis:
By looking at the above graph, the average of grades before implementing the strategy was 15.25/20, whereas after implementing the circle talk strategy it became 16.9/20. These results show an incremental rate of 10.8%. As a result, these numbers clearly show that students performed better after having implemented circle talks in the classroom.
Discussing in circle talks can be used for critiques, problem solving, and/or decision making. The basic purpose of a talking circle is to create a safe, non-judgmental place where each participant has the opportunity to contribute to the discussion of difficult and/or important issues.

3.3 Strategy: Self-critique

*"Involving students in this live marking activity engages both them and the tutors in further understanding the criteria..."*Dr. Darrall Thompson, UTS.

In order to increase the students' self-esteem, letting them criticize their own work is a thoughtful strategy. Self-critique does in fact help every student reflect on their own work, which might very often end up with the students correcting their own work for themselves. In fact this teaching strategy gives the students the opportunity to dig deeper into their own thoughts and their outcomes to understand why they have done so and so and maybe reflect on a better way to having solved a certain issue.

IOP Conf. Series: Materials Science and Engineering **974** (2020) 012015 doi:10.1088/1757-899X/974/1/012015

It has been noticed that self-critique is an overall very beneficial strategy not only for the students, but for us instructors as well. Taking time to think about a certain course syllabus, or about a certain grading criteria or even about projects' requirements, allows instructors to thoroughly organize their work for better learning/teaching outcomes' achievements.

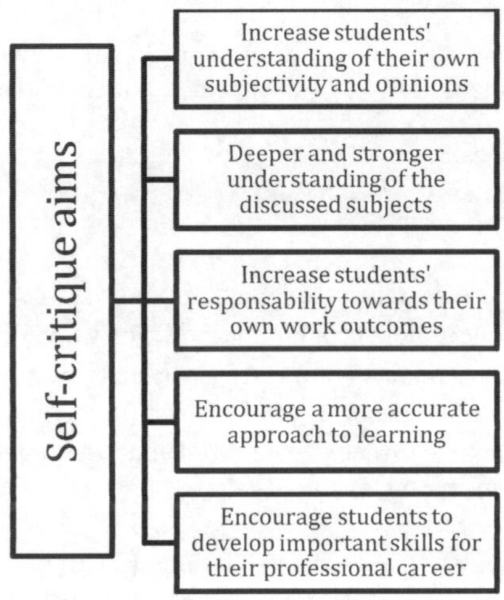

Figure 7:Self critiques objectives

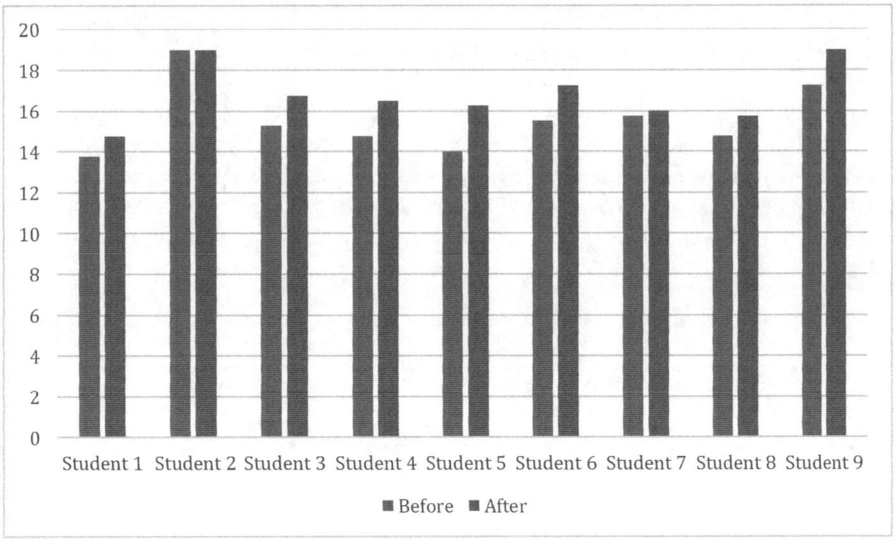

Figure 8: Showing the comparison of grades before implementing the self-critique strategy versus after implementing it
Source: researches

IOP Conf. Series: Materials Science and Engineering **974** (2020) 012015 doi:10.1088/1757-899X/974/1/012015

Analysis:
By looking at the above graph, the average of grades before implementing the strategy was 15.5/20, whereas after implementing the circle talk strategy it became 17/20. These results show an incremental rate of 9.7%. As a result, these numbers clearly show that students performed better after having implemented circle talks in the classroom.

In fact, students felt that sharing ideas out loud in a professional setting, helped them perform better during their projects as they had more ideas to implement, and more feedback to achieve better outcomes.

3.4 Strategy: Field Trip

'LEARNING FROM FIELD TRIPS' researchers have had varying degrees of success in measuring cognitive learning resulting from a school field trip, but the evidence generally suggests that such trips can have a positive impact on learning of facts and concepts (Anderson, 1999; Anderson & Lucas, 1997; Bamberger & Tal, 2006; Beiers&McRobbie, 1992; Feher& Rice, 1985; Flexer&Borun, 1984; Gottfried, 1980; Knapp, 1996; Mallon & Bruce, 1982; Miglietta, Belmonte, &Boero, 2008; Orion &Hofstein, 1994; Stronck, 1983; Tuckey, 1992a).

In regards to this other teaching strategy which involves off-campus activities, students were taken to a specific location related to their major and were asked to provide a report. The report requested had to include 20 negative and/or positive points of the visited site, which students had to analyze and prove with images taken from the site. An interesting fact during the visits was the amount of questions asked by the students to the personnel in charge, in order to understand every structural detail which lead to a specific outcome in the architecture/interior design of the site.

Once submitted, the reports were assessed by the instructors, and very positive feedback was remarkable from the students. The purpose of these trips was to assess their results in the projects given afterwards and compare them to the ones previously assessed. The astonishing outcomes were noticed right away in the projects given, where students showed great link to real life sites. Students acknowledged great development in their way of problem solving, and demonstrated outstanding improvement in achieving the intended learning outcomes of the new projects.

IOP Conf. Series: Materials Science and Engineering **974** (2020) 012015 doi:10.1088/1757-899X/974/1/012015

Figure 9: Real photos from a field trip to ALWAH ALKHALEG TECHNOBOND CLADDING, PSU

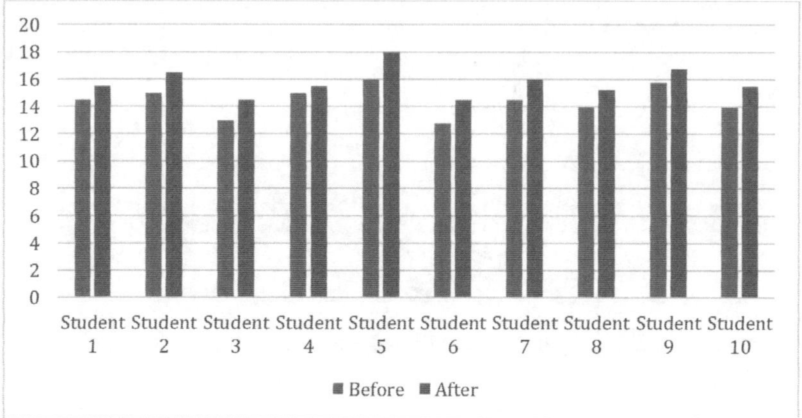

Figure 10: Showing the comparison of grades before implementing the field trip strategy versus after implementing it. Source: researchers

Analysis:

By looking at the above graph, the average of grades before implementing the strategy was 14.45/20, whereas after implementing a field trip it became 15.8/20. These results show an incremental rate of 9.34%. As a result, these numbers clearly show that students performed better after having a field trip added to their class schedule. Students have recognized the positive aspects of a field trip as they better imagined a detail in their project or an interior element. Moreover by pointing out the negative points of their visited site, students made sure not to include such undesirable detail in their projects.

4. Surveys/Interviews

The research debates the satisfaction levels of students towards which teaching strategy they feel more comfortable to be applied through quantitative methodology where a survey consisted of 5 multiple-choice questions and 10 questions that the respondents had to answer with a 5-point scale.

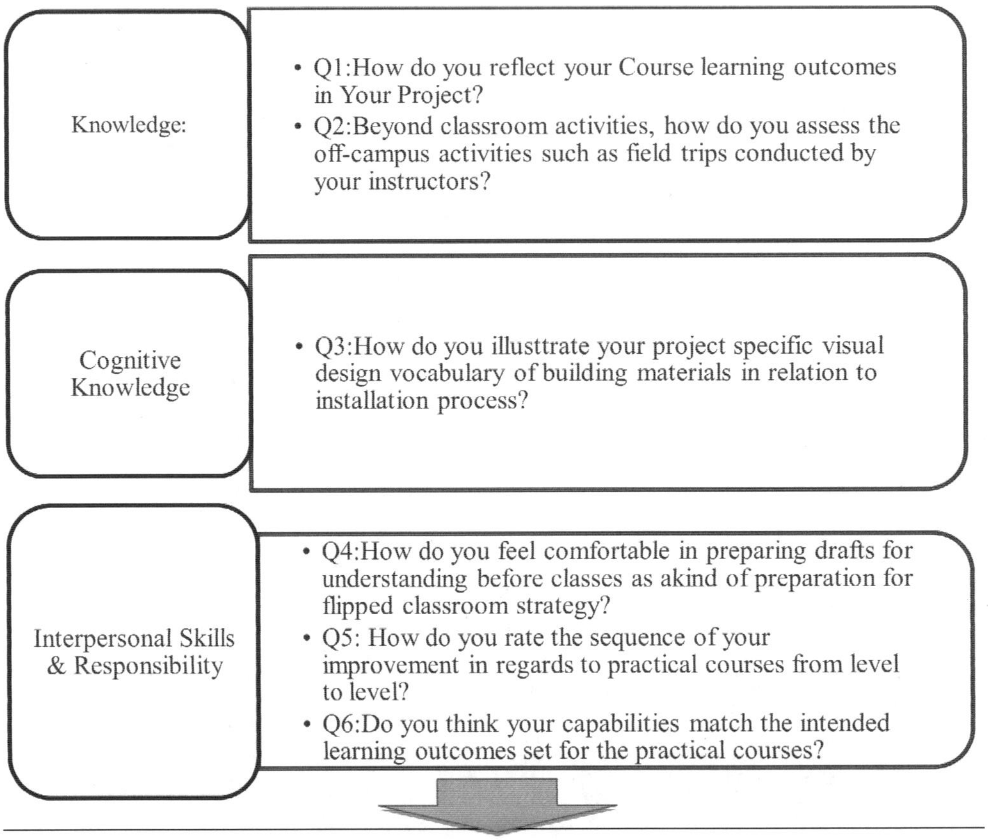

Q7: Which teaching strategy motivates you more: Flipped class strategy, Circle Talk Strategy, Self critique strategy or Field Trip Strategy?

Figure 11: Scheme of Questions adopted in the questionnaire in accordance with NCAAA main Domains

5. Findings

Students were able to observe, understand and recognize how building materials can be assembled on site in a comprehensive way. The matter, which was very useful and had been reflected in their

IOP Conf. Series: Materials Science and Engineering **974** (2020) 012015 doi:10.1088/1757-899X/974/1/012015

drawing sheets; they got the researchers' feedback in accordance with their submissions, which gave them more opportunities to modify their work, trying to get their projects more understandable

The research will present some of the questions answered by students, showing the analysis for each. Which helps understanding and stating the improvement that have been recorded by the researches, assuring that the research hypothesis whereas, believing in the architecture and interior design students' abilities and the instructor's variable teaching strategies can bring out examples of best practices in teaching philosophies.

Q1: Beyond classroom, how do you assess the off-campus activities such as field trips conducted by your instructors?

For question(1) about 80% of surveyed students found that off-campus activities were very effective as it helped them better imagine their projects' outcomes in real-life and be able to assess whether a specific idea can actually be achieved in a successful way or not.

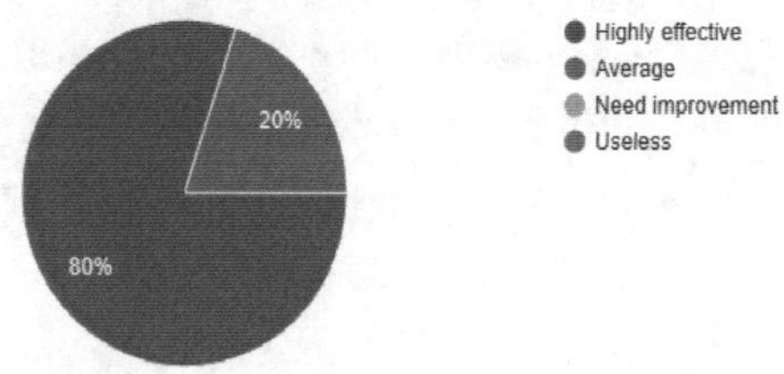

Q2:Do you think your capabilities match the intended learning outcomes set for the practical courses

For question(2) about 80% of surveyed students found that their capabilities match the intended learning outcomes, which means that the practical courses' level of difficulty is acceptable.

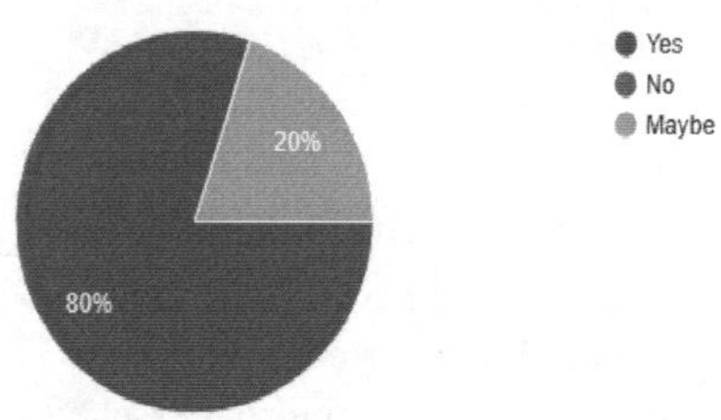

Q3: How do you rate the sequence of your improvement in regards to practical courses from level to level?

IOP Conf. Series: Materials Science and Engineering **974** (2020) 012015 doi:10.1088/1757-899X/974/1/012015

For question (3), 60% of surveyed students rated their sequence of improvement in regards to practical courses from level to level as very good.
This shows that practical courses evolve fairly in regards to the students' capabilities and performances.

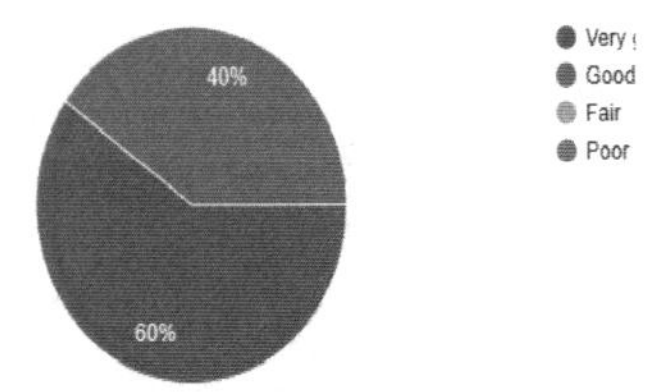

Q4:How do you feel comfortable in preparing drafts for understanding before classes as a kind of preparation for flipped classroom strategy?

For question (4), majority of the surveyed students rated their feeling comfortable in preparing drafts for understanding before classes as akind of preparation for flipped classroom strategy.
This shows that their own perception towards this startegy is positively recorededties and performances.

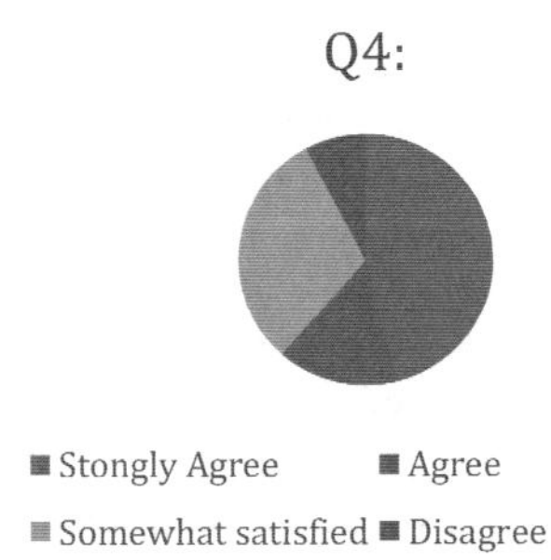

Q5:How do you rate the sequence of your improvement in regards to practical courses from level to level?

For question (5), majority of the surveyed students ratedsequence of your improvement in regards to practical courses from level to level in an ascending progress.
This shows that their own performance and acquisitiona has been improved positively because of the applied strategy.

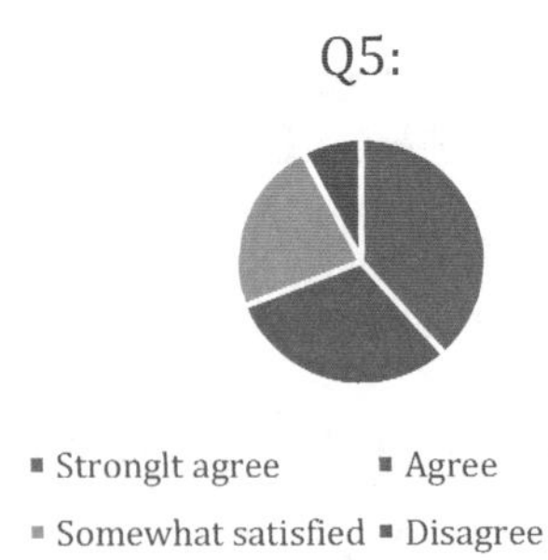

References

[1] Flipped classroom frameworks improve efficacy in undergraduate practical courses – a quasi-randomized pilot study in otorhinolaryngology, Tobias Dombrowski, Christian Wrobel, Stefan Dazert&Stefan Volkenstein

[2] Brame, C., (2013). Flipping the classroom. Vanderbilt University Center for Teaching. Retrieved [todaysdate] from http://cft.vanderbilt.edu/guides-sub-pages/flipping-the-classroom/.

[3] Blended Learning: A Flipped Classroom Experiment by William R. Slomanson Journal of Legal Education Vol. 64, No. 1 (August 2014), pp. 93-102

[4] Pranis, K. (2005). The little book of circle processes: A new/old approach to peacemaking. Intercourse, PA: Good Books.

[5] Mills, H., & Jennings, L. (2011). Talking about talk: Reclaiming the value and power of literature circles. The Reading Teacher, 64(8), 590-598. doi: 10.1598/RT.64.8.4

[6] The Importance of Student Self-Assessment- http://academ.com.au/importance-student-self-assessment/

[7] Panadero E., Jonsson A., Strijbos JW. (2016) Scaffolding Self-Regulated Learning Through Self-Assessment and Peer Assessment: Guidelines for Classroom Implementation. In: Laveault D., Allal L. (eds) Assessment for Learning: Meeting the Challenge of Implementation. The Enabling Power of Assessment, vol 4. Springer, Cham

[8] Panadero, E., Brown, G. T. L., & Strijbos, J. W. (2016). The future of student self-assessment: A review of known unknowns and potential directions. Educational Psychology Review. doi: 10.1007/s10648-015-9350-2

[9] A Short Review of School Field Trips: Key Findings from the Past and Implications for the Future Jennifer DeWitt a & Martin Storksdieck b a People Science & Policy Ltd, London, UK b Institute for Learning Innovation, Edgewater, Maryland, USA Version of record first published: 16 Oct 2008

[10] Dewey, J. (1938), Experience and education, NY. Macmillan

[11] Gillett, J. (2011). The use of experiential education and field trips for learning. Journal of Educational Multimedia and Hypermedia, 20(2), 173-177.

[12] Krepel, W. J., & DuVall, C. R. (1981). Field trips: A guide for planning and conducting educational experiences. Washington, D. C.: National Education Association

Military Technical College
Kobry El-Kobbah,
Cairo, Egypt

13th International conference on Civil and Architecture Engineering ICCAE-13-2020

Study the Social Needs for Proper Housing

Maged Moneer Gad[1], Ahmed Galal El Din Aly[2]

[1] Assistant Professor, Architecture Engineering Department, Obour high Institute for Engineering and Technology, Cairo, Egypt.
[2] Assistant Professor, Civil Engineering Department, Modern University for Technology and Informations, Cairo, Egypt.

Corresponding Author: gadmaged@yahoo.de

Abstract:
The housing industry is an essential part of the urban development that requires full cooperation of the country's different institutes including public and private sectors' institutions, individuals, and groups. This is to achieve the sustainable development in all different aspects; economic, social, and urban developments. As the basic citizen needs of "everyone's has the right to have a suitable home" made it essential requirement to offers many housing solutions to fulfill different people needs.

The human concepts changed through the years and evolved the housing patterns, methods, forms and uses. As the human needs, concepts and cultures changed - from the need of only a shelter above the head - to protect from weather, into home that provides not only protection but also physiological and psychological comfort and recreational facilities to practice various activities. The industrial revolution, and the development of mechanization with the technology development, in addition to economic and social changes, leading to lack of areas, which helped the change of housing features and its meanings. As it used to be consisted of two or three floors to hold one family only, it transformed into the existing towers and muti-story buildings to conclude large numbers of residents.

The research relied mainly on surveys to measure the population's preference for housing. He also used field visits (for construction companies that play the role of owner and sometimes contractor and development) and documentary analysis to study the nature of the residential product mainly as well as to measure the level of preference for types and requirements of housing, as well as the use of interviews and panel discussions. The criteria, recommendations and proposals contained in this research represent a picture of the needs of a sample of the community, but the results of this research cannot be indicated merely or considered decisive, but must be analyzed to suit each case, and use it in the best possible way to contribute to the design decision.

The study reached several results; the most important is to provide privacy and tranquility, due to the low population density with the provision of services, based on the following:
1- The preference for a residential building without other services was the largest.

2- The residential buildings' height should not exceed 5 floors.
3- The residential building should not contain more than 10 apartment units.
4- There should be some special services for residential buildings, as a suitable number of parking spaces, as well as the elevator in the buildings and a terrace.

Key words: residents, residential needs, populations, multistory buildings, residential buildings, office buildings, commercial buildings.

Introduction:

The architectural design reflects many aspects; the human being wishes with regard to the inhabitant of the place, human crafts-man-ship, the natural material that exists. The architectural design is the result of the combining professional, ideological, cultural, social, and behavior aspects that affect human behavior. Architectural design is art of self-influence. The surrounding environment affects the human being. The architect is influential as well as it is a direct reflection of human behavior. If we look to old patterns and architectural styles we find out that they are the ones that contributed to making society monolithic, it participate and active member of society and participated architecture in the formation of the personality of the individual. (Wdh, 2005)

The responsibility for architectural planning lies on the project's owner, the revising authorities, the designer, and the contractor in solidarity.
Residential communities are not just buildings and condominiums, but are a social connection and a common interest for all residents, through which their social and humanitarian needs meet. (Kamas, 2006)

In the 20th century, enormous changes established in various science and human arts, the most important of which is architecture that had rapid high rates changes. The evaluating of post-usage performance of buildings is one of those developed branches that contribute to a better understanding of the needs of users and the advancement of architecture both functionally and aesthetically. The concept of evaluation still overlaps with many other concepts used in architecture and the most important of these concepts is architectural criticism. However, the prevailing concept of the evaluation process means judging the validity and efficiency of the design product to meet the general requirements of the beneficiary in accordance with the available resources and these judgments are often made after the establishment of specific and clear criteria whether these standards are technical related to the profession itself or functional.

The city must meet the current and future needs of work and people requirements in general and housing in particular as housing is very important. (Moulin, 1995)

Putting forward the idea of a city for all through the "right to the city", the right to the city manifested by living there, and considering the city is the right of all. This process carried out through meeting the civil forces with the state, where the state controls the laws of buildings and their regulation, to solve the problem of the applications submitted and this through the reconciliation of the various housing offers submitted by those involved in construction with the demands and needs of the community. Decent housing is a fundamental right for all. (H. Lefebvre, 1974)

The study (Shaukat, 2014) analyzed housing policy through government projects and plans and compared it with social and economic studies and human rights standards to find the pros and cons of this policy, and ended with some recommendations for a more comprehensive and efficient policy in providing adequate housing. The designs have varied in terms of providing housing for a family such as the Two House Projects or a joint housing project for a group of families. The study discussed the modular presentation method as well as the long-term rental method, which subsequently became ownership. The study took more account of the number of units for the population. However, it is not clear how well the quality of the housing provided by the densities and their relationships with the requirements or wishes of the inhabitant.

The study (Faraj, 2013) addressed the role of services in urban development, as it is one of the components of the success of urban development and a key factor in the population attraction and represents one of the three main pillars of development (housing, employment opportunities and services). The topic addressed in terms of distribution methods and the social and economic impacts of the service sector, as services continue to suffer from incomprehensible disparities caused by the levels of success, and this disparity affects the satisfaction and interaction of the population with these services.

The research (Ben Adawa, 2016) discusses the housing needs in Algeria at densely populated cities. Based on the increased need, demand for housing, in the absence and scarcity of land and the lack of possibility of horizontal expansion in cities with high population density. The option of vertical constructions in the field of housing programs, with the goal of establishing as many units as possible on the lowest possible area, was the best solution for such cities. While for us in Greater Cairo the possibility of study, While, of course, places with high population density excluded, while the legal and cultural planning nature of the area allows for the vertical development of the property as well as the multiplicity of uses of the same property.
Some major companies in Egypt are studying the citizen's need for housing types by studying buyer preference for units for population projects, but the study was limited to analyzing customer data, which refused to disclose the results in general. However, as reported by the analysis of the data, the population's preference for higher units starting from the second floor, the preference for places that do not allow a large number of units in the same building.

Research Problem:

Different types of residential buildings, including those that do not meet the needs of the residents due to the lack of the designer's clear view regarding the resident's design requirements.

Research Aim:

1. Identify the preferred residential pattern that meets the actual needs of the occupants, in terms of the usage and population density according to the criteria of a questionnaire.
2. Reaching appropriate planning for the comfortable citizen's housing needs.

Methodology:

The study followed the descriptive analytical approach. In order to achieve the theoretical framework research formation, the importance of studying the population and their needs presented by using data collection, organizing and analysis of data research plan.

The study also followed the study and analyze approaches through surveying work, then evaluates and analyzes data to reach results to support design decision-maker.
Different regions and neighborhoods selected to represent different cultural, social and financial levels segments of Egyptian society, .in order to gain knowledge about the community special design requirements needed to help the architect to consider those aspects when designing residential buildings to conform the society culture and needs and to help making the best decision when designing residential buildings. The greater Cairo region is the geographical scope of the study.

The study focused on the importance of studying the needs of population so that the designer engineer would consider it when designing housing patterns that are compatible with the culture of society and its requirements.
The study followed the theoretical part in the theoretical description in describing the components of housing and determining the basic needs for housing, and the practical part through evaluating a questionnaire form and analyzing it and reaching the results and recommendations.
Hence, it was necessary to know some details about the special design requirements that society needs, thus allowing to form an idea about these needs, which must be reflected in the future to form a clear vision for the designer.

A questionnaire was prepared and the various social, cultural, and income levels and segments of society were taken into consideration and the appropriate age groups to collect information about the suitable housing for the population and thus to devise their requirements for a comfortable housing that provides the necessary needs for the population.

Several different areas and different regions were chosen in the Greater Cairo region, those regions included; Heliopolis, Hadaeek El Zatoon, Nasr City, Ain Shams, Ghamra, The First Settlement in New Cairo.
The nature of the region differs from one region to another. As does the societal culture of the residential community for those areas, as well as the level of living that reflects the changing economic level. The shape and composition of the housing of main entrance and three floors in New Cairo to buildings with higher density in Heliopolis and Nasr City to a higher density In Hadaeek El Zatoon and Ghamra area.
The questionnaire designed to conclude:

- Type of housing in terms of preference for private ownership property or government agency ownership.
- The percentages of individuals who prefer private housing or housing units
- Type of building in terms of use
- The preferred number of units in the building.
- The number of rooms required in the housing unit
- Preferred building height
- Favorite services in the housing unit
- The floors that the resident prefers to live in
- How close commercial services to residential units
- Preference for housing in Greater Cairo or in new cities
- Prefer housing in the suburb or down town
- The housing unit is on a main or secondary street
- Prefer housing over a specific street's width

Results:
The questionnaire distributed to 500 people, but only 420 responses obtained (84%). The survey lists valid for analysis and coding were 386 questionnaires (91.90%). The collected primary data analyzed.
From the questionnaire, conclusions reached on the general form of the type of comfortable housing for the population in terms of:

1. Ownership preference:
Table 1 and figure 1 illustrate that the type of housing in terms of preference for private ownership or government agent housing ownership, the result is as follows:
Most people prefer living in private sector owned buildings (80%), as the rest (20%) people prefer government agent ownership.

IOP Conf. Series: Materials Science and Engineering **974** (2020) 012016 doi:10.1088/1757-899X/974/1/012016

Figure (1) Ownership preference

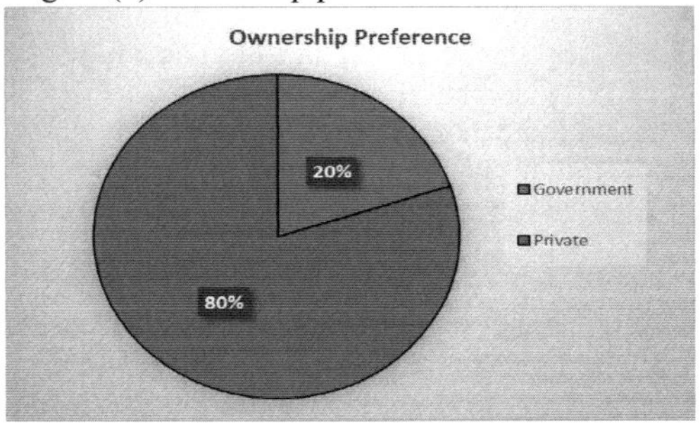

Table (1) Ownership preference

Type of Building	No.
Government	78
Private	308

2. Private Residential:

Table (2) and figure (2) illustrate that The preference of people resident ownership type results are as follows: 69% of the people prefer Private residential buildings, while 38% prefer resident apartment units, the rest (13%) gives irrelevant concern.

Figure (2) Private Residential

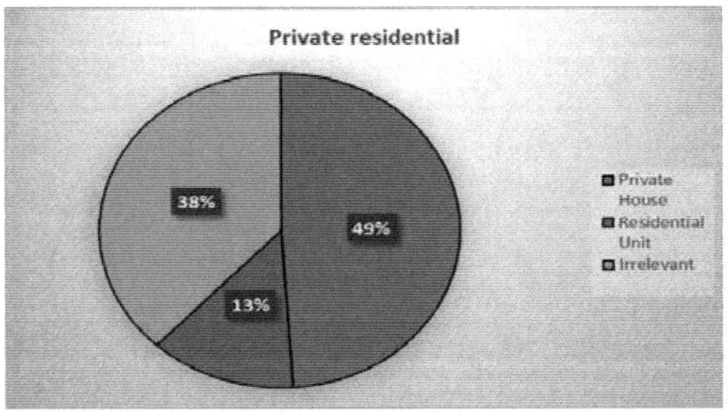

Table (2) Private Residential

Type of Building	No.
Private Resident	187
Residential Unit	148
Irrelevant	51

3. Type of building in terms of use:

Table (3) and figure (3) illustrate that the type of building usage gives the following:

53% of the people prefer to have a dedicated residential building, then 15% would tolerate the building to be residential commercial, 13% of the people agreed to have a residential administrative building, 9% like to have residential commercial administrative building, and 10% acts with irrelevant reaction.

Figure (3) Type of building in terms of use

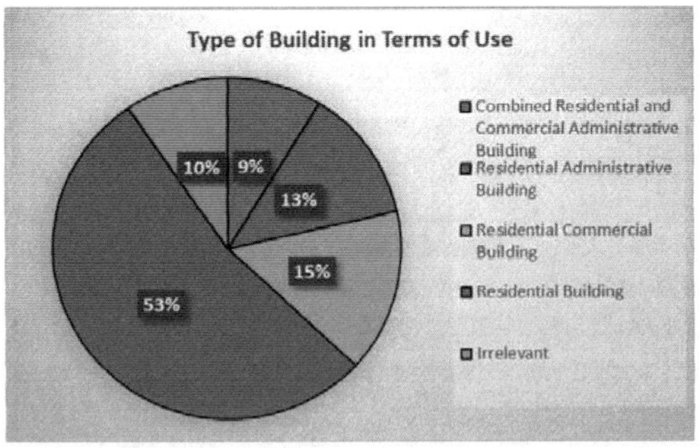

Table (3) Type of building in terms of use

Type of Building	No.
Combined Residential and Commercial Administrative Building	34
Residential Administrative Building	49
Residential Commercial Building	60
Residential Building	206
Irrelevant	37

4. Preferred number of units in a multi-unit building:

Table (4) and figure (4) illustrate that the Preferred residential buildings is that contain 10 units at most as it came with 75%, while residential buildings that contain 20 units occupy a maximum of 16%, and residential buildings that contain more than 20 units alone have a ratio 9%.

Figure (4) Preferred number of units in a multi-unit building

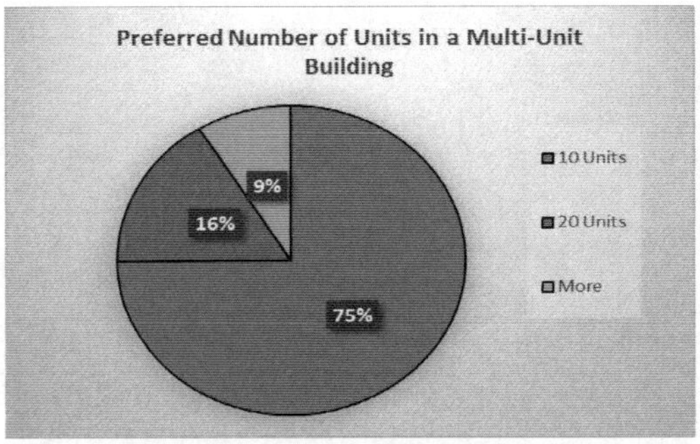

Table (4) Preferred number of units in a multi-unit building

Preferred Number of Units in a Multi-Unit Building	No
10 Units	289
20 Units	63
More	34

5. Preferred Number of Rooms of The Residential Units:

Table (5) and figure (5) illustrate that most of the people wants to have 3 rooms apartments (54%), 31% wants to have an apartment with more than 3 rooms, people wants 2 rooms comes with 14%, and at last comes people who wants only one room with 1%. This result is due to the culture of the Egyptian people that doesn't want to move to another area when financially capable.

Figure (5) Preferred Number of Rooms of the Residential Units

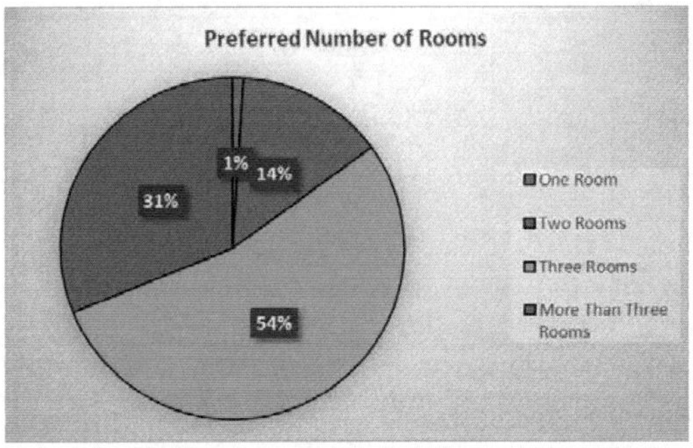

Table (5) Preferred Number of Rooms of the Residential Units

Preferred Number of Rooms	No.
One Room	4
Two Rooms	54
Three Rooms	208
More Than Three Rooms	120

6. Preferred Building Height:

Table 6 and figure 6 illustrate that Most of the people prefer 5 story buildings (65%), 25% accept up to 10 story buildings, while 10% accept more than 10 story buildings.

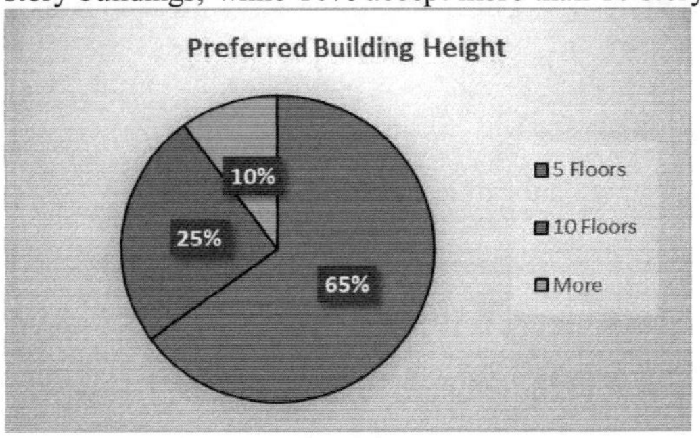

Figure (6) Preferred Building Height

Table (6) Preferred Building Height

Preferred Building Height	No.
5 Floors	251
10 Floors	96
More	39

7. Preferred Services Availability in the Housing Unit:

Table 7 and figure 7 illustrate that Most of the people (73%) prefer to have all types of facilities (balconies, elevators, and cars parking plots), 11% are settled with elevators only, as for the cars parking plots only and balcony only requirements are 8% each.

Figure (7) Preferred Services Availability in the Housing Unit

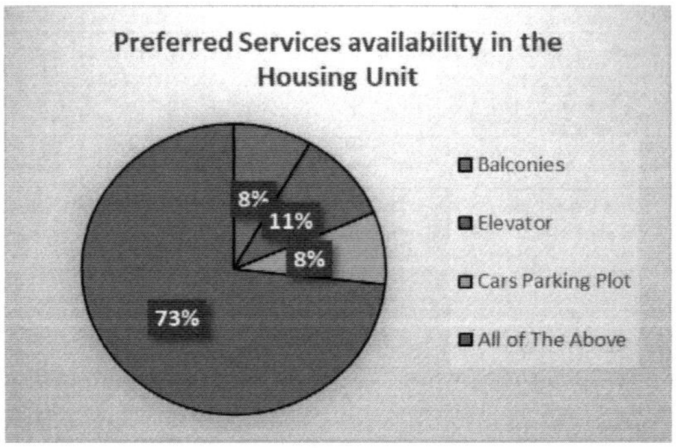

Table (7) Preferred Services Availability in the Housing Unit

Preferred Services availability in the Housing Unit	No.
Balconies	32
Elevator	40
Cars Parking Plot	31
All of The Above	283

8. Preferred Inhabitant Floor:

Table 8 and figure 8 illustrate that 23% of peoples prefer to live on the 3rd floor, 22% prefer to live on the 4th floor, 18% prefer to love on the 5th floor, 15% prefer to live on the 2nd floor, the last floor comes next with 10%, and at last the ground or first floor comes with 4%.

Figure (8) Preferred Inhabitant Floor

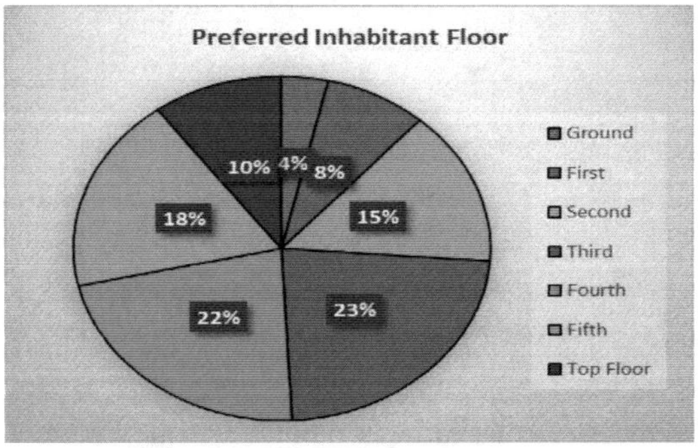

Table (8) Preferred Inhabitant Floor

Preferred Inhabitant Floor	No.
Ground	14
First	31
Second	56
Third	89
Fourth	86
Fifth	71
Top Floor	39

9. Preferred Inhabitant Close to or Far from Commercial Centers:

Table 9 and figure 9 illustrate that 77% of the sampled people prefer to live close to the commercial centers, as 13% prefer to be far from commercial centers, the rest are not interested or it doesn't matter to them.

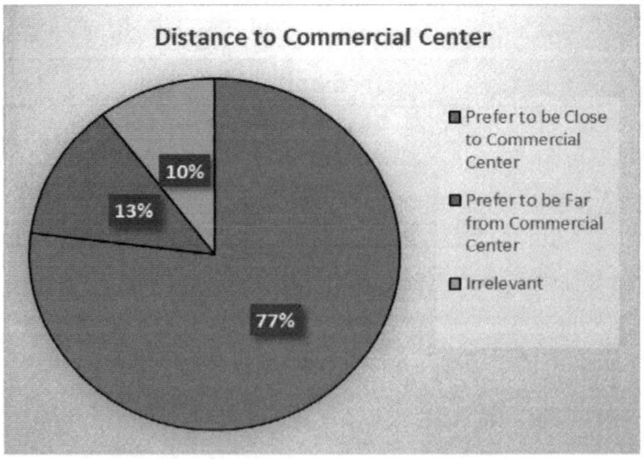

Figure (9) Preferred Inhabitant Close to or Far from Commercial Centers

Table (9) Preferred Inhabitant Close to or Far from Commercial Centers

Distance to Commercial Center	No.

IOP Conf. Series: Materials Science and Engineering **974** (2020) 012016 doi:10.1088/1757-899X/974/1/012016

Prefer to be Close to Commercial Center	297
Prefer to be Far from Commercial Center	49
Irrelevant	40

10 Preferring to Live in the Greater Cairo Region or in the New Cities Region:

Table 10 and figure 10 illustrate that 58% of the people prefer to live in new cities, as 42% prefer to live in Cairo region.

Figure (10) Preferring to Live in the Greater Cairo Region or in the New Cities Region

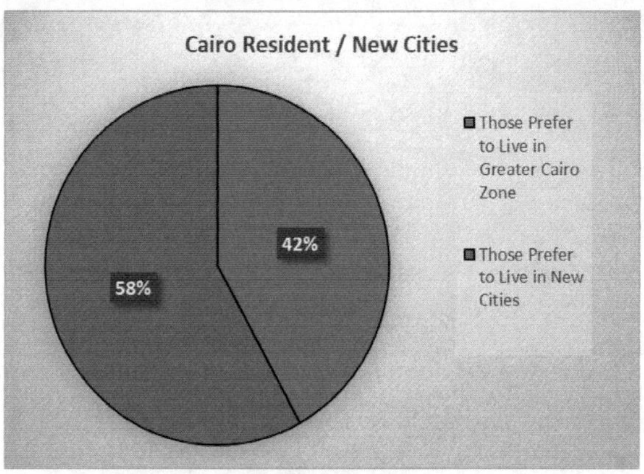

Table (10) Preferring to Live in the Greater Cairo Region or in the New Cities Region

to live in Greater Cairo or Prefer New Cities	No.
Those Prefer to Live in Greater Cairo Zone	162
Those Prefer to Live in New Cities	224

11. Preference of Living Downtown or City Suburb:

Table 11 and figure 11 illustrate that 46% of the people prefer to live in the suburb, as 36% prefer to live downtown.

Figure (11) Preference of Living Downtown or City Suburb

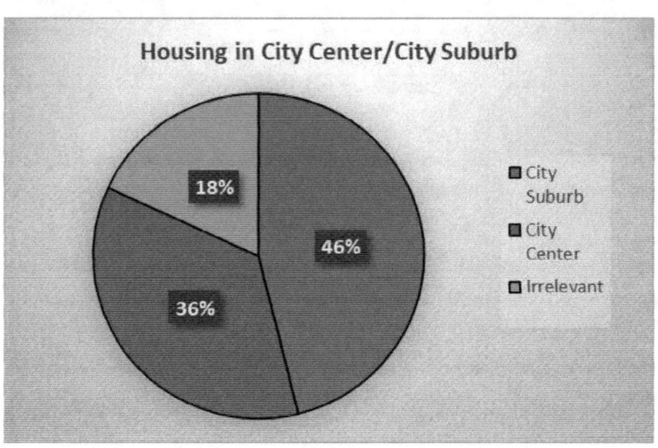

Table (11) Preference of Living Downtown or City Suburb

Housing in Down Town/City Suburb	No.
City Suburb	178
Down Town	138
Irrelevant	70

12 Preference to Overlook On Main Or Secondary Street:

As illustrated in table 12 and figure 12, 40% of the people prefer to live on main streets, 37% prefer to live on Secondary Street and the rest has irrelevant response.

Figure (12) Preference to Overlook On Main Or Secondary Street

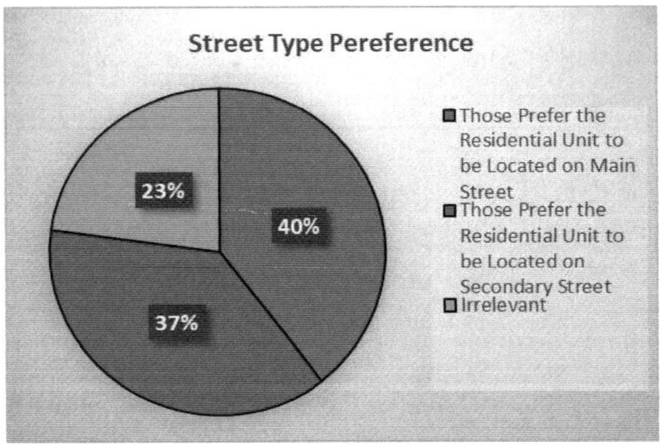

Table (12) Preference to Overlook On Main Or Secondary Street

Street Type Preference	No.
Those Prefer the Residential Unit to be Located on Main Street	153
Prefer the Residential Unit Those to be Located on Secondary Street	145
Irrelevant	88

13 Preference to Live on Special Width Street:

Table 13 and figure 13 illustrate the analysis result of the Egyptian citizen requirements for the housing of the wide streets, starting from 15 m street width. The major demands are for 20 m street width with 31% preference, 19% prefer to live on 15m street width, and the rest of the people sample percentage are distributed over the rest of streets with widths of 6m and 8m.

Figure (13) Preference to Live on Special Width Street

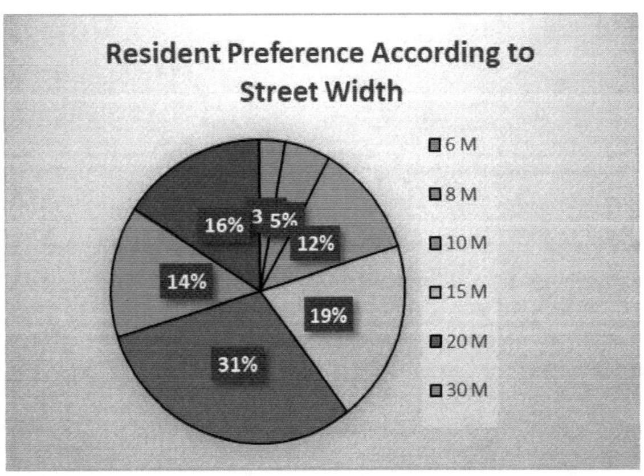

Table (13) Preference to Live on Special Width Street

Resident Preference According to Street Width	No.
6 M	11
8 M	20
10 M	47
15 M	75
20 M	118
30 M	54
Irrelevant	61

Conclusion:

The results of the questionnaire show that the inhabitant prefers low occupants density in the building; these shown as to prefer the short height buildings as well as few units' occupancy.

It also shows the preference of the largest percentage for buildings with residential units only which shows the preference for privacy.

It also shows that there is a large number of segments tends for the quietness, whether it occurs at the city suburbs or at the new urban communities providing services.

The following is a detailed result explanation for each questionnaire criteria analysis collected from different inhabitants' categories.

1. Most of the selected survey sample prefers private housing instead of the government housing, as the ratios of people who prefer to live in private housing 80% against 20% of the sample prefer governmental residential buildings. Those who prefer the government-owned housing related this as to guarantee that the buildings are safe as they have opinion that some private sector buildings are bad constructed.
2. People who prefer private residence are 49%, while those who prefer residential buildings units are 38% and the rest are not concerns. The reason is to have privacy in their house property.
3. As shown in table 3, 53% of the chosen sample prefer residential only building with no other activities in it. While 37% of the sample prefer to have a multi usage building. The reason of the choice of one usage is to have safety and quietness.
4. It is therefore clear that 75% of the residential buildings are to be with 10 residential units, as for the 20 residential units buildings at most comes with ratio 16%, and the residential buildings that consists of more than 20 units is 9%. The reason for that is to have more privacy and to make it easier to communicate to their neighbors.
5. The requirement of the Egyptian society questionnaire for three-room housing was 54% of the Egyptian housing demands. This result is (what the researcher believes) based on the culture of the Egyptian citizen more than needs cause. The Egyptian citizen culture is searching for housing not only to cover the current need but to be a place fulfill his future needs to not to be in-need to move to another larger place when he is financially capable. While the ratio of the need for more than three rooms is 31%, and then the demand for two rooms comes to be 14% and the demand for only one room is very small 1%.
6. The people prefer buildings up to five floors, as the preference for buildings up to five floors is 65%, while the preference for buildings up to ten floors is 25%, and more than ten floors is 10%.
7. Most of the demands are for all the facilities are required such as balcony, elevator, and parking plots, as the results are 73% required all the previously mentioned facilities, while elevators only are 11%, parking plots only and balcony only are with the same results with 8% each. This illustrates the importance of the availability of services such as elevators and parking plots.
8. The Egyptian people mostly prefer to have apartments on the third floor level with 23% result, then comes the fifth floor level with 18%, then the second floor level with 15%, then the top floor of the building with 10%, then the first floor level with 8%, and at last the ground floor level with result of 4%.
9. The Egyptian people prefer to live close by the commercial center with 77% result, as for those who prefer to live far away from commercial centers are 13%, the rest of the sample gives irrelevant preference.
10. People who prefer to live in new cities are 58%, while people who prefer to live in greater Cairo zone are 42%. As in new cities the population density is less, the buildings' spacing gives more privacy.
11. The percentage of the people who prefer to live in the suburb are 46% where more quietness is provided, while the percentage of the people who prefer to live in the downtown are 36%, the rest gives irrelevant opinion.
12. People who prefer their apartment to locate on Main Street are 40%, while 37% prefer to have their apartments locate on secondary streets, the rest gave irrelevant preference.
13. People who prefer to live on wide street with more than 15 m width (20 m and more) are 31%, then who prefer the 15 m width comes with 19%, then the rest of the result are distributed between 6m and 8m width streets.

Recommendations:

IOP Conf. Series: Materials Science and Engineering **974** (2020) 012016 doi:10.1088/1757-899X/974/1/012016

1. The government must provide the requirements and the needs that provide the psychological and physiological comfort for the people with no contradicting the planning of the cities. The questionnaire shows the need to provide services close to the new residential communities, which encourages individuals to get out of crowded places.
2. The civil society and the concerned private sectors that contribute to the construction industry should fulfill the community's needs for residential buildings, whose architectural design matches the requirements of the number of rooms and services.
3. Universities and educational institutes should document, confirm and publish the community's actual needs for housing requirements, such as interior design and surrounding housing services.

References:

1. Ahmed Mahmoud Abdel Hamid Faraj, "Impact of Urban Services Planning and Management on The Prosperity of New Residential Communities", Cairo University 2013.
2. Ben Adawa, Suad, Manasr, Emad, "Residential Vertical Construction and Its Role in Maintaining and Managing The Property", Um al-Bawaqi University, 2016.
3. Craig W. Kelsey,Thesis , "Smart Growth Planning Principles", California State University, USA , 2006.
4. Dalila Zarqa, Housing and Housing Policies between Speech and Reality, Ph.D. thesis dissertation for Science in Sociology, University of Oran- Algeria, 2016.
5. David Walters," Designing Community", London, Elsevier Ltd., 2007.
6. El Ghoneim, Osman, "Planning Standards Philosophy, Types, Methodology of Preparation and Applications in the Field of Urban Planning", Amman, Safaa Publishing and Distribution House, 2011.
7. Ginearalta & Caighdeain, Chapter 15, "Development Standard-General Site",Dublin , 2005.
8. Hani Hashem Wdh, "An Analytical Study of Architectural Facades", Tishrin University Journal of Scientific Studies and Research - Volume (27) Issue (2), 2005.
9. Iraqi Republic, Ministry of Architecture and Population, General Population Authority, Division of Studies, Urban Housing Standards, 2010.
10. Larbi Messaoud .S, L'art de L'espace: du moi-peau à la maison-peau. Problème de congruence entre rêve et réalité, Mémoire de Magister en Architecture, Université d'Oran Usto, 2014.
11. Messahel. A, Les Mécanismes de productions foncières et immobilières en Algérie, Thèse de doctorat es sciences en Urbanisme, Université d'Oran , Usto, 2010.
12. Mohammed Abbas Mustafa, Ph.D. thesis dissertation, "Natural, Social and Economic Factors affecting Housing Planning in The Arab Republic of Egypt", Al-Azhar University, Engineering College, Planning Department, 1972.
13. Moughtin, "Urban Design, Green Dimensions", London , 2005.
14. Qamas, "Urban Residential Complexes in Constantine, Reality and Planning Requirements", University of the Mantori Brothers, Constantine 2006.
15. Raed Mohammed Saleh Yousef, "Design Standards of housing for Low-Income peoples", Master of Urban and Regional Planning thesis dissertation, Graduate School of Al-Najah National University, Nablus, Palestine, 2002.
16. Said A, Le logement social urbain à Oran, nouvelle politique de l'habitat et retombées locales, Mémoire de Magister en géographie, Université d'Oran, 2002.
17. Sally, L," Front to Back – A Design Agenda for Urban housing", London: Architectural Press, 2005.
18. Salma Ahmed, "The problem of the theory of the contemporary Arab city, the residential neighborhood between the two and the expansion", Al-Azhar 9th International Engineering

Conference, Urban Planning Department, Faculty of Engineering, Al-Azhar University, Cairo, 2007.

19. Trache S. M, Mobilités résidentielles et périurbanisation dans l'agglomération oranaise, Thèse de doctorat d'Etat en géographie, Université d'Oran, 2010.
20. Yahya Shaukat, "Egypt's Housing Policy", Economic and Social Justice Unit, First Edition/December 2014.

**MilitaryTechnicalCollege
Kobry El-Kobbah,
Cairo, Egypt**

13th International Conference on Civil and Architecture Engineering ICCAE-13-2020

The history of monasteries in Egypt as self-sustained settlements

G Romel, L Sherif, and S Ashour

Department of Architecture, Faculty of Engineering, Arab Academy for Science, Technology & Maritime Transport, Cairo, Egypt

gina_romel@aast.edu

Lobna.sherif@aast.edu

shaimaa.ashour@aast.edu

Abstract. Monasticism initiated in Egypt and has spread worldwide. It introduced a new self-sustained architectural innovation called “monasteries”. Monasteries can be considered as homogeneous self-sustained settlements through history. This paper aims to document the origins of monasticism, leading to the development of self-sustained monasteries: conceptually, economically, and architecturally. It relies on observations, maps, and historical references. The main challenges in collecting data for this paper were the lack of published references about origins of monasticism. Most of the resources are rare documents stored in the Coptic Clerical and Theological College of Cairo, and libraries of monasteries. Thus, the paper analysis present monasteries as self-sufficient, self-sustained settlements that survived through history since the fourth century to the current state through: (zoning, architectural elements, building materials, and buildingtechniques. The architectural analysis shows that the architectural solutions of monasteriesevolved in respect to the geographical location, and context. Monasteries architecture relies on trials and errors, until it reached an applicable architectural model. Keywords: History of Monasteries, Monasteries architecture, Monasteries zoning, self-sustained settlements.

1. Introduction

“Coptic monasticism started as the embodiment of the ascetic philosophical intellectual theory of celibacy and devotion in the wilderness” [1]. In other words; monasticism relies on loneliness and optional poverty. This term was first introduced in the second half of the third century as Eremiticálmode (solitude) of life, then evolved to Cenobitic(communal) life [2]. Monk is a Greek word derived from “Monos”, which means solitary or loneliness. On the other hand, in Arabic, it is called “Râheb”, which means fear of God [3]. This paper will document monasteries built by the main two monastic systems through history; founded by Father Anthony (semi-Cenobitic) and Father Pachomius (total-Cenobitic). Both models have architectural existing references, that led to the establishment of these self-sustained settlements. The chronological evolution of monasteries will be covered conceptually, architecturally, and sustainably, influenced by evolution and traces of monasticism since the fourth century. This study will cover two monasteries dating to the fourth century (Saint Simon, and Abu Makâr Monasteries), reflecting how these systems of Cenobitic life affect architecture.

This paper will introduce monasticism and its main pillars to explain architectural development of monasteries. The analysis includes historical architectural elements, and clarifies the two monastic systems of the fourth century. Furthermore, the paper will explain the main architectural elements in monasteries design in the current state, in respect to zoning. The methodology of this paper lies in understanding Coptic monasticism idea and translating the historical text into diagrams. This paper will be constructed vertically, starting with the origins of Coptic monasticism, the main two monastic models (Eremitic and Cenobitic), then the geographical location of these models, and their differences especially in the cell design, using saint Simon and Abu Makar Monasteries as an example of each monastic system.

2. The origins of Coptic monasticism in Egypt

The basis of monasticism was the life of Jesus; taking Him as the role model. Also, following many figures like Paul the Tarsus, who was a great inspiration for monasticism [1]. In contrast to the claims that monasticism was inspired by different cultures; like Indians, ancient Egyptians, Jewish, and Greeks, all monastic rituals were extracted from the Bible [1, 4]. Coptic monasticism stands on three pillars which are solitary life in the desert, optional poverty, and obedience [1]. "They tried to reach a state of a pure heart. Through ascetic control over their passions, Monks were meant to reach a purity of soul, in which state they could gain so-called spiritual knowledge" [5].

Egypt's rich history and special geographical location helped in initiating Monasteries. Egypt's desert had a great impact on religious life since the ancient Egyptians. In the hard life of the desert, they found a spiritual connection to their gods. Furthermore, it was documented in "Oxyrhynchus papyri" the third century, that was discovered in the late nineteenth century, that religion has dominated the culture in almost every aspect of their daily life, and in their building of tombs and temples [6]. Many historians claimed that after the end of the era of martyrdom for Christians, by the Romans at the beginning of the fourth century, living in the desert with all its suffering, a kind of martyrdom without bloodshed [7].

2.1. Monastic models

There were two main attempts to Coptic monasticism, named after each Father established it, and reflect the idea of monasticism.

The first attempt for monasticism was established in the age of the emperor Antoninus Pius by a wealthy man called Frontonius in the mid of the second century (138-161 A.D). Frontonius stayed with seventy men in Nitra desert in Egypt, although this attempt failed to complete due to the lack of supplies[8] .Coptic monasticism could be divided into two main models, Eremiticál mode of life, known as the solitary hermit life, and Cenobitic mode of life, known as the communal life [1,3].

Saint Paul of Thebes, the first hermit, started the Eremiticál mode of life in the middle of the third century. He settled nearby a well and a palm tree in a cave, in the Egyptian desert near the red sea. He settled there when he was fifteen years old, and it was said that he spent almost ninety years in the desert without seeing any human, until he was found by Father Anthony and passed away around (ca. 340/341) [9]. Eremiticál mode was known as hermit model, in which each hermit stayed in total loneliness, in a cave beside a source of water. This life was very hard to maintain due to lack of food, and wild animals attacks; therefore, Cenobitic mode appeared to endure the hard life of solitary hermits [10].

There were two parallel movements in the fourth century towards Cenobitic mode. The first movement was semi-Cenobitic by Father Anthony who formed small hermitage communities in Egypt by the red sea (ca. 313). Having some rules, the monks split the desert into small monasteries, known as "Manshoubia". Each monastery is ruled by the eldest monk "Sheikh", known for his wisdom. They built in the middle of each Manshoubia a church, in which they gather for worship and prayer on Sundays. After Saint Anthony passed away, his students built the Monastery of Saint Anthony by the red sea [1].

The second community bloomed on the Nile valley in Thebaid (ca. 320) by Father Pachomius (290/346 AD), inspired by the imperial army [11]. Father Pachomius led to a major movement in communal monasticism in Egypt by constructing the first monastic enclosure on the east bank of the Nile river substituting the solitary hermit life by a total-Cenobitic life, through a daily based program for praying and working. Father Pachomius had a different perspective for monasticism, allowing monks to share each and every aspect of life. Every three monks shared the same room, they were divided according to their specialization into groups (workers, farmers, chefs...etc.). All monks in this new enclosure pray together in one church, eat in a refectory, and no one was allowed for any solitary activity, nor break any rules. This was the first uniform communal regime known worldwide [12].

After Father Pachomius passed away, his ideological successor was Father Shenoute (348/466 AD). Father Shenoute found that the two Cenobitic systems should be merged into one model, as both do not contradict but helps to develop each other. He believed that Pachomius rules for the closed Monastery helped the self-sustained community to survive with all its architectural elements, but every monk shall have his own solitary private life as Father Anthony believed. Thus, in Father Shenoute Monastery, every monk has his own room, where he could have his private loneliness life in prayer. While, monks share work and special prayers every day as a regime. Also, some elder monks were allowed for a total solitary life in the desert, but in certain known locations [1].

Hence, Monasteries great evolution as an economical self-sustained community bloomed in the twentieth century by Pope Shenoute III (1923/2012 AD), the 117th Pope of Alexandria and patriarchal evangelism of Saint Mark. Pope Shenoute III believed that education is the keynote of any community to evolve. That's why he had added new regulations for any man to be a monk. Importantly, anyone who is interested to enter the Monastery must finish his/her bachelor's degree. This ensures a new well-educated generation, which will help the development of Monasteries in all aspects. Nowadays, this new generation of (engineers, doctors, scholars...etc.) helps Monasteries to be listed as a successful self-sustained settlement. Today Monasteries became productive settlements having poultry farms, livestock farms, apiaries, fish farms, citrus farms, and automobile repair workshops using the latest technologies, that not only serve the Monastery, but also enhance the economic aspect by selling their goods outside. Coptic monasticism timeline was analyzed through historical references, to understand the chronological order and the origins of monasticism idea, that was reflected architecturally into Monasteries seen today (Figure 1).

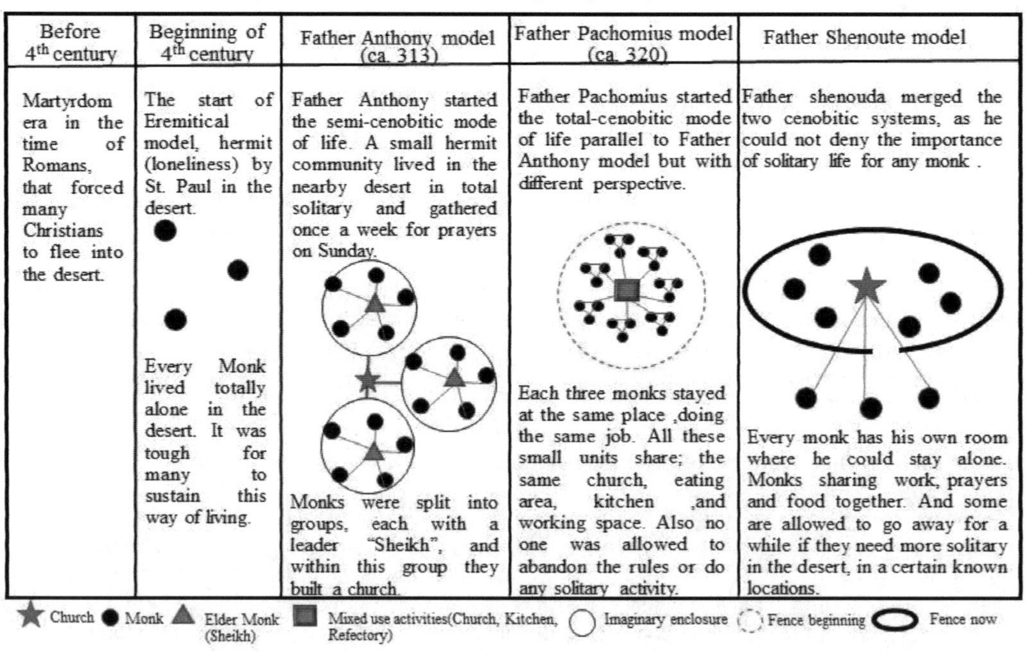

Fig. 1:Timeline analysis of Coptic monasticism.

2.2. Geographical location of Cenobitic modes

Through analyzing historical references, there was a chronological development for Monasteries to reach the self-sustained community of today. The north semi-Cenobitic mode of life established by Father Anthony, consisting of the Red sea (ca.318), Nitria (ca. 330), Scetis (ca. 330), and Kellia (ca.

338). The south total-Cenobitic mode of life, started by Father Pachomius, was around the Nile valley in Thebaid (ca. 320) all the way down to the south (Figure 2) [11].

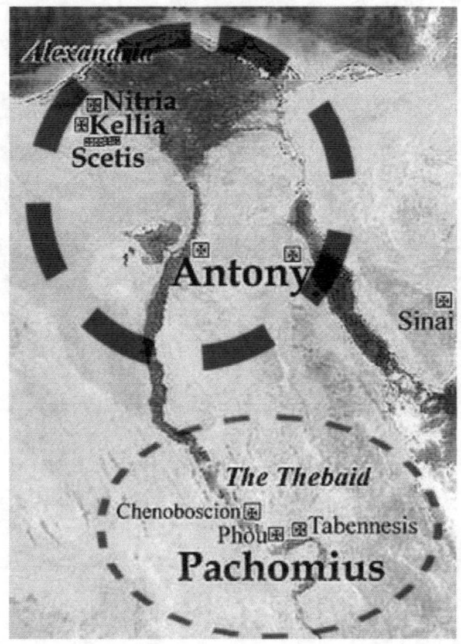

Fig. 2: Location of Anthony and Pachomius monasteries [13].

There were no architectural remains of Nitria, as the historical site was destroyed by irrigation water that covered the whole site [1]. The first architectural vestiges for Anthony monasteries were found in Kellia, which was made for three, then ten monks[14] .

There are no traces of architectural records for Father Pachomius monasteries. Monks destroyed the old buildings to replace them with new buildings, to accommodate their need for life, and to house the increasing number of monks in Pachomius monasteries.

3. Differences between Father Anthony, and Father Pachomius monasteries

Based on historical references, Coptic monasteries could be distinguished according to architectural relationships. Understanding the architectural differences between the two models of Cenobitic life will facilitate recognizing whether the monastery belongs to Anthony or Pachomius from the architectural plan. There are two Cenobitic modes of life, the first was semi-Cenobitic by Father Anthony while the second was total-Cenobitic by Father Pachomius.

3.1. Semi-Cenobitic mode of life by Father Anthony

The first architectural evidence of Anthony monasteries appeared in Kellia excavations that divided Father Anthony monastery into periods that reflect the architecture of each era. In the fourth century, this hermit's community lived in separate built cells, besides a water source, in a solitary life. Every cell had two spaces, one for prayer and sleeping, while the other was a reception for visitors. A kitchen was added to this cell and a staircase leading to the second floor, where toilets were added. This community built a church and a refectory in its center, to share Sundays prayers (Figure 3).

In the fifth century, the Keep (fortress) was added to this community due to multiple Bedouins and Barbarian attacks that cause the death of many hermits. The Keep was known for its large walls, with only one door on a high level for entrance using a removable staircase, and a reachable water source from the inside (Figure 4).

IOP Conf. Series: Materials Science and Engineering **974** (2020) 012017 doi:10.1088/1757-899X/974/1/012017

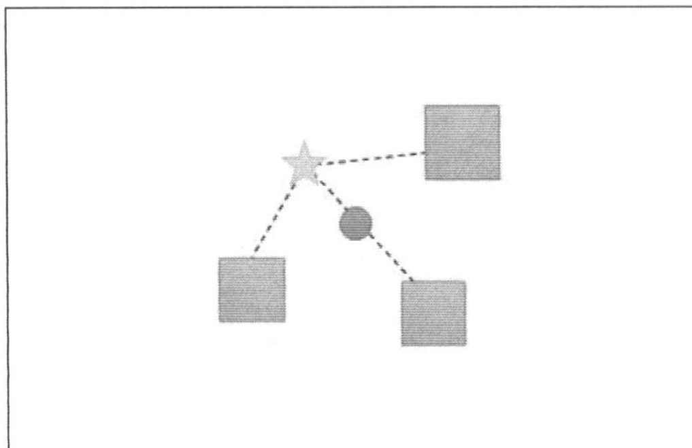

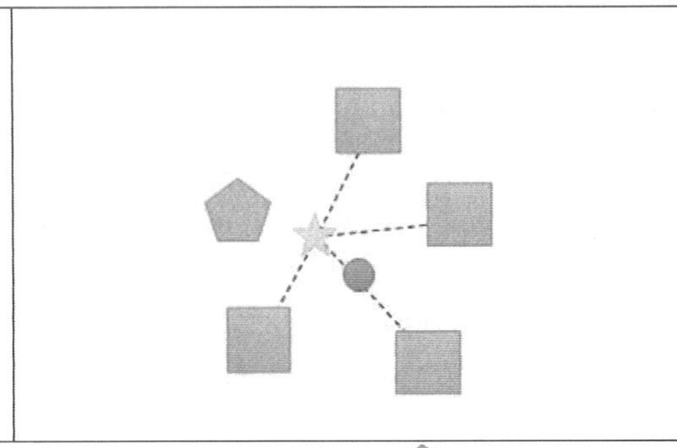

cell Water Source Church, Refectory, Kitchen Keep

Fig. 3. The fourth century diagram shows Father Anthony's monastery components.

Fig. 4. The fifth century diagram shows Father Anthony's monastery components.

Since the middle of the fifth century, until the beginning of the sixth century the primary monastery known as "*Manshoubia*" appeared with all its elements that fit from three monks until it reached ten monks. This new monastery was fenced and contained multiple separate cells, *Keep*, water source, church, refectory, and a big court for vegetation (Figure 5).

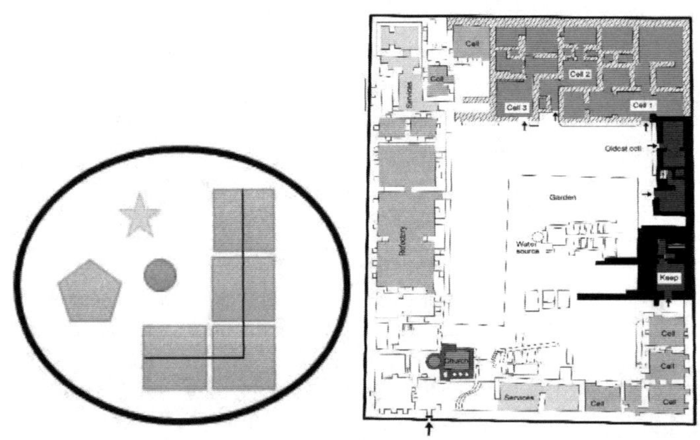

cell Water Source Church, Refectory, Kitchen Keep Fence

Fig. 5. The right figure shows the Manshoubia prototype that fits a maximum of ten monks in the fifth century, Kellia[1], while the left figure is an analytical diagram of the Manshoubia contents.

The monastery of Abu Makâr

This monastery is located in "Scetis" Wadi El-Natrûn, south of Alexandria. The monastery of Abu Makâr is one of the oldest surviving monasteries following the semi-Cenobitic model of Father Anthony. This Monastery was built in the fifth century after the third barbarian attack 444 A.D which caused the death of many monks [15]. This was the first Monastery that has been visited by people after the Pope of Alexandria moved to live there 451 A.D. It has evidence of the old and the modern monasteries.monasteries of Wadi El-Natrûn were built as an example of a small fortress, its nucleus is the church; furthermore, fences function were changed nowadays from defensive to the private idea [16]

The plan drawn by Evelyn-White for the Abu Makâr monastery in 1933 A.D (Figure 6) shows that the cells were designed on multiple longitudinal axes. Every cell is totally separated from the next cell, to respect the solitary life of each monk. The cells overlook an intermediate court, vegetation with a well in its center, the kitchen and the refectory were connected. There are three churches built in different eras in the monastery. The Keep was part of the walls of the monastery, with huge walls approximately 3.5 meters of depth. The height of the monastery's walls was 14 meters, while the Keep was 16 meters. Also, there was a mill, and stable used for the daily work of monks. Monks came together on Sundays to share prayers, and eat together.

Each architectural element in Father Anthony's monastery is independent. It is shown that cells are away from the Keep, services (kitchen, refectory, and guesthouse) and the church. This type encouraged the solitary life of monks. Workshops didn't appear as an important architectural element in the plan, although this could not deny the importance of work for any monk in Anthony's monastery. They worked as farmers to plant their food, and manufacturing baskets out of palm.

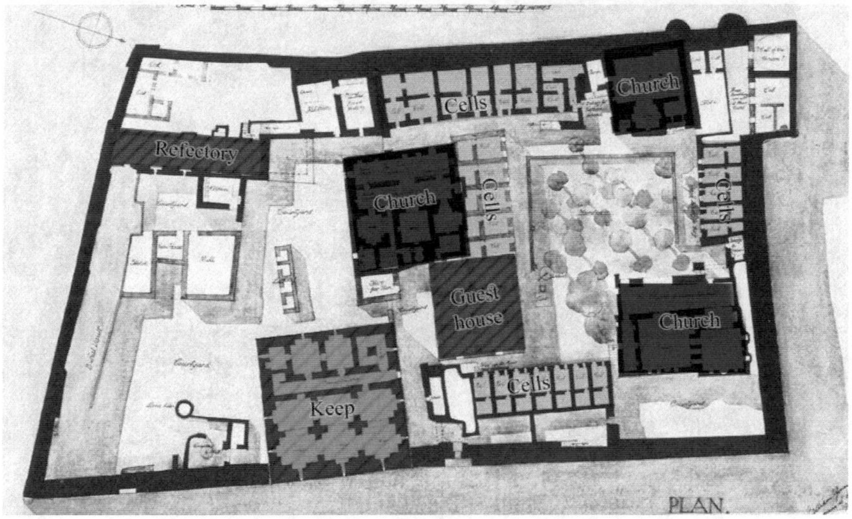

Fig. 6. Abu Makâr old monastery in Scetis since 1933 [15].

3.2. Total- Cenobitic mode of life by Father Pachomius

Father Pachomius built his monastery in the fourth century. His monastery was fenced and divided into two main parts: living area which was the Keep that included (cells, refectory, and kitchen), and workshops. In all Pachomius monasteries, workshops occupied a large area of the plan. As Father Pachomius believed that work is important for any monk. Monks were divided according to their work into groups, they shared daily activities, and they strictly followed the monastery rules.

Saint Simon monastery in Aswan

This monastery is one of the rare Pachomius monasteries that survived without changes in architectural form or function, as it was abandoned by monks for a long period. This monastery was located on the Nile river bank. Due to the special topography of the land, the Monastery was levelled on two terraces. The upper terrace is accessed from the land side, while the lower terrace is accessed from the Nile riverside. There are three components in the plan; Fig. 7a the keep, the church, and the workshops. This emphasizes that the bigger part of the plan was the workshops. The upper terrace of the plan; Fig. 7b is divided into two parts. The first part is the keep: where the living area, kitchen, and the refectory are located. While the second part has workshops. The church is on the lower terrace. The cells were divided according to monks' work in a linear alignment; they shared everyday life. The monastery organization reflects total-Cenobitic life. work was a basic life regime [12].

Fig. 7. "A" Saint Simon monastery master plan, Aswan [17], analyzed by Author." B" Saint Simon monastery upper terrace, Aswan [18], analyzed by Author.

By understanding the two main concepts of Cenobitic life it is obvious that Pachomius and Anthony's monasteries share the same architectural elements, but with different plan organization. The main elements of monasteries are: the water source which was important for the survival of any community, the cells, where monks live, and had their spiritual solitary life, the keep (fortress), where they shelter of attacks, the fence appeared in the fifth century after the third barbarian attack [3], the church was the base of this Cenobitic life, where monks had their communal weekly prayers, the refectory and the kitchen were of great importance to this Cenobitic life, where they shared food and prayers after the holy mass.

Work has been important to the monastic system, it helped occupy the empty time of monks, and provided the monks with economic sustainability, and probability to improve their community. The cell design is one of the most important differences that could differentiate Father Anthony and Father Pachomius monasteries.

Shenoute the Great (347- 465 A.D), Father Pachomius successor, was the abbot of the White monastery in Egypt. Shenoute found that the two Cenobitic systems should be merged together in one model, as both models do not contradict but help develop and sustain each other. He believed that Pachomius rules for the total-Cenobitic monastery helped this community to survive sustainably through rules and economics. Furthermore, every monk shall have his own solitary private life as part of his daily regime, to ensure the sustainability of the essence of Coptic monasticism, as Father Anthony believed. Thus, in Father Shenoute monastery, every monk lived his private life in prayer, while sharing other prayers and work with the whole group [19]. Thus, the second part will concern studying the living quarters' design in a monastery, according to Anthony and Pachomius monastic systems.

4. Living quarters

Pachomius monastery cells were arrayed linearly. Monks share every aspect of life from food, work, and prayer. The refectory and the kitchen were connected to the end of the group of cells; to emphasize the total Cenobitic life; Fig. 8a. While Anthony monastery cell design was based on the solitary life of each monk; their daily needs were all in the same cell (living area, kitchen, and toilet on the second floor). The monastery complex had one refectory and kitchen connected to the church,

IOP Conf. Series: Materials Science and Engineering **974** (2020) 012017 doi:10.1088/1757-899X/974/1/012017

used once weekly after the holy mass; Fig. 8b, [1]. The new Shenoute (347- 465 A.D) cells in monasteries respect the solitary life of each monk. Each cell contains all its functions which allow the monk to live privately in the monastery; Fig. 8c.

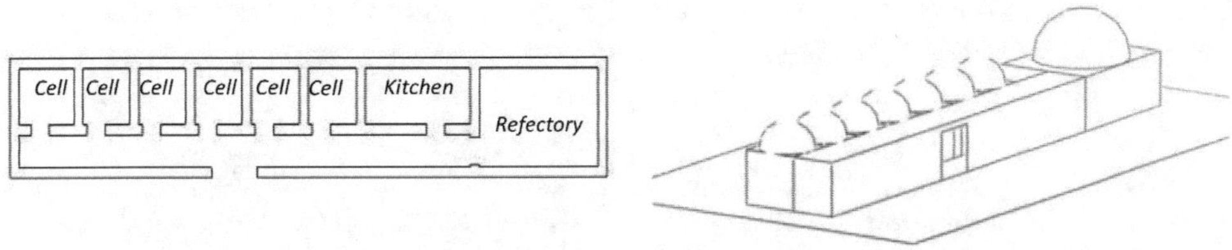

A

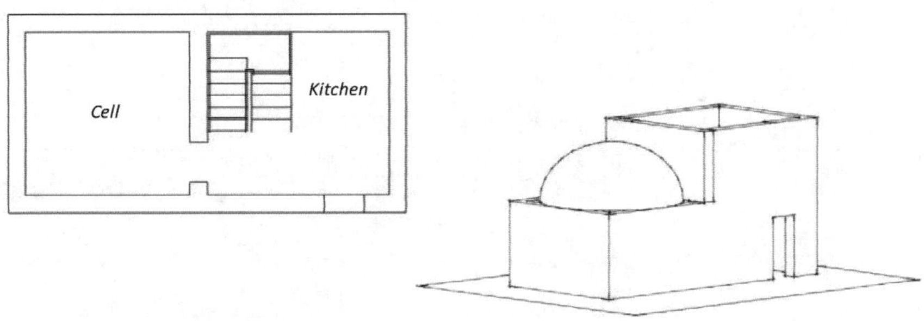

B

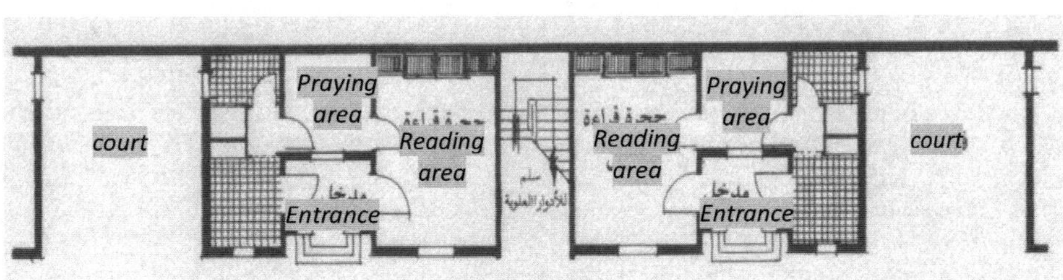

Fig. 8. Different cells design. "A" Father Pachomius total Cenobitic cell design (plan, and perspective). "B" Father Anthony prototype solitary cell in Kellia (plan, and perspective). "C" Abu Makâr cell plan [1], analyzed by Author .

5. **Abu Makar monastery as an example of Self-sustained settlements**

Sustainability is the ability to survive over a period of time, causing no damage to the environment [20].

Abu Makâr monastery is one of the oldest monasteries that survived sustainably since the fourth century. This monastery was built on the semi-Cenobitic mode of life by Father Macaruis. Abu Makâr was the first monastery visited by people after the patriarchal evangelism of Saint Mark moved to it. This monastery is one of the leading examples of self-sustained settlements through time. Analyzing Abu Makâr monastery since the fourth century till nowadays, sustainability was found in all aspects

IOP Conf. Series: Materials Science and Engineering **974** (2020) 012017 doi:10.1088/1757-899X/974/1/012017

of the monastery. Abu Makâr monastery could be considered as a self-sustained community in its architectural elements, building materials, zoning, architectural design, and economics.

Comparing the twentieth century 1933 plan referenced by Evelyn-White (Figure 6) and the renovation new plan of the monastery (Figure 9), the main architectural elements remain the same, but new elements were added to sustain this new life. All elements respect the solitary life of a monk. While workshops became of great importance.

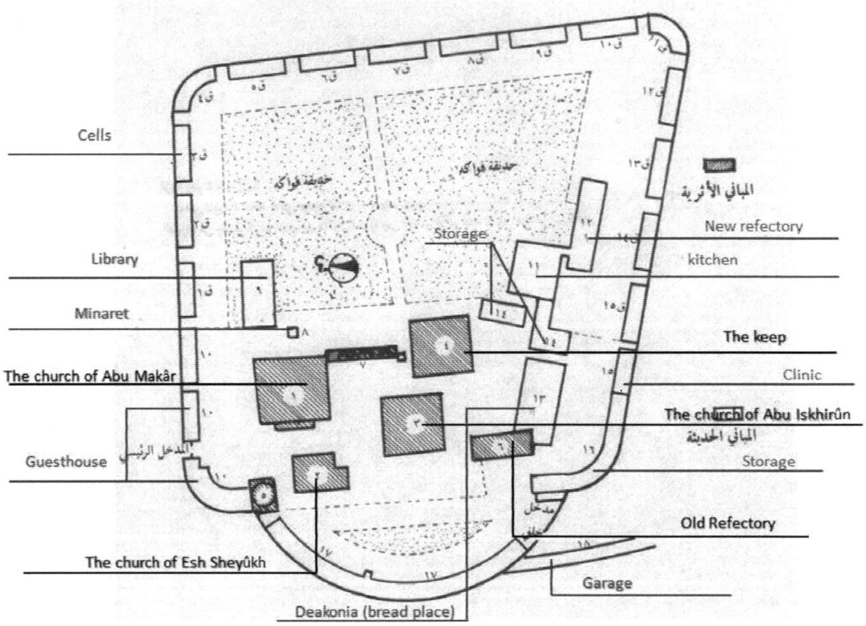

Fig. 9: Abu Makâr new plan analysis [1], analyzed by Author.

Abu Makâr monastery used vernacular materials like limestone. Bricks were made from Tafla mixed with sand and sometimes with sand and lime. Their sizes were (40x20x7 cm) or (42x42x10 cm), and the mortar was clay taken from some lower layers of the ground. The floor of the cell was a deck of gravel, gypsum, and ochre. Monks used insulation materials from nature to create cool and comfortable interiors [1]; Fig. 10a.

Abu Makâr was the first visited monastery by people, zoning was of a great importance for these monks to sustain their solitary spiritual life. The new layout of Abu Makâr monastery emphasizes the separation of public areas of visitors from private areas of monks, to maintain their solitary life. Due to climatic changes, the levelling of the monastery has changed. The new monastery level was higher than the old by six meters. Also, Traces of the old fence act as the entrance to the old monastery, which serves as a natural separator; Fig. 10b. The new design of the monastery was made respecting to the primary designs, and to sustain climatic changes. The fence was built on the trace of the old monastery. Cells were built to reinforce the fence and prevent its destruction. Every group of cells was made of three floors facing the fence, which act as pillars supporting it [1].

IOP Conf. Series: Materials Science and Engineering **974** (2020) 012017 doi:10.1088/1757-899X/974/1/012017

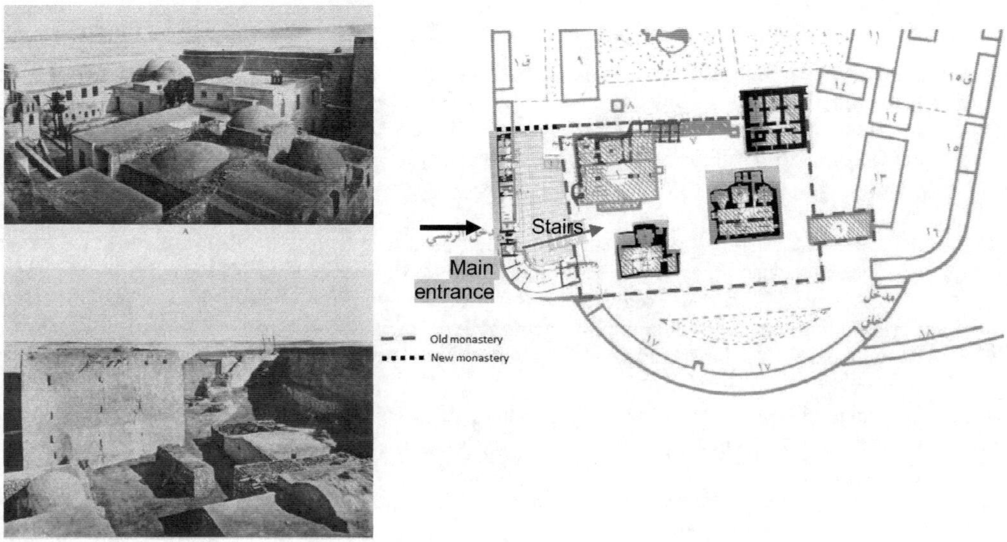

A ***B***

Fig. 10. Abu Makâr monastery. "A" interior photo[15]. "B" New Zoning monastery, analyzed by Author.

Conclusion

This paper presents monasteries as self-sufficient, and self-sustained settlements survived through history. The main architectural elements to form a monastery since the fourth century are: church, refectory, kitchen, water source, cells, Keep and fence. The fence main function changed from a protective to a space defining element that ensures privacy.

The architecture of father Anthony monastery provides privacy for all the elements in the main plan, where the cells were far away from the church and the kitchen. This monastery design ensures the solitary life of each and every monk.

On the contrary, the architecture of father Pachomius monastery enhances the community collaboration, where monks share every aspect of life, therefor all the architectural elements aligned together.

The main difference between father Anthony and father Pachomius monasteries is the cell design. The cell design in Anthony's monastery stands alone with all its functions in it: reception, sleeping area, kitchen and toilet; where a monk could stay in privacy for a long period of time. The cell design of Pachomius monastery was on a longitudinal axis where all cells are aligned sharing same kitchen, toilets, sleeping and prayer areas.

Monasteries of both models used vernacular materials from the site and applicable architectural techniques. These monasteries evolved through history with respect to the identity and authenticity of the place. Furthermore, the topography of the land acts like natural barriers or defender in the monastery design.

Nowadays, the main architectural elements of monasteries did not change but new elements were added to accommodate the new aspects of life as clinics, guests house and workshops, in respect to the solitary life of each monk. Zoning was applied on old and new monasteries in their architectural design where visitors may enter any monastery without accessing private areas of monks.

References

[1] Mattá, Coptic Monasticism in Saint Macarius era (al-Rahbanah al-Qibṭīyah fī ʻaṣr al-Qiddīs Anbā Maqqār), 2nd ed. Barrīyat Shīhīt [Egypt]: Saint Macarius Monastery, 1984.
[2] J. Cassian, C. Luibhéid, and E. Pichery, Conferences. New York: Paulist Press, 1985.
[3] P. Georgy, “Encyclopedia of Pop Georgy-Coptic Monasticism,” Encyclopedia of Pop Georgy, vol. 3. Library of Pop Georgy- Monastery of Father Ruwais in Abbasiyah, Egypt, p. 192, 2003.
[4] I. G. Smith, Christian monasticism: from the fourth to the ninth centuries of the Christian era (classic ... reprint). Place of publication not identified: FORGOTTEN Books, 2016.
[5] D. Kasprzak, “Monastic Exegesis and the Biblical Typology of Monasticism in the Patristic Period,” The Journal of Systematic Theology, vol. 2, pp. 85–103, 2014.
[6] R. A. Coles, N. Gonis, and B. P. Grenfell, The Oxyrhynchus papyri. 45, London: Egypt Exploration Soc., 1977.
[7] E. Budge, The paradise of the holy fathers II, vol. 2, 2 vols. Putty, N.S.W.: St Shenoute Coptic Orthodox Monastery, 2008.
[8] P. Bedjan and C. Detienne, Acts of martyrs and saints: acta martyrum et sanctorum. Piscataway: Gorgias Press, 2008.
[9] L. Regnault, The day-to-day life of the Desert Fathers in fourth-century Egypt, 1st English ed. Petersham, Mass: St. Bede’s Publications, 1999.
[10] E. Budge, The paradise of the holy fathers, vol. 1, 2 vols. Putty, N.S.W.: St. Shenoute Coptic Orthodox Monastery, 2008.
[11] B. Ward, The sayings of the Desert Fathers: the alphabetical collection. London: A.R. Mowbray, 1984.
[12] A. Veilleux, Pachomius, and A. de Vogüé, Eds., The life of Saint Pachomius and his disciples, 2. Aufl. Kalamazoo, Mich: Cistercian Publ, 1985.
[13] Saint Andrew’s Abbey, “Monks and nuns of the Egyptian desert,” NITRIA, KELLIA, and SCETIS, 1990. [Online]. Available: http://ldysinger.stjohnsem.edu/@texts/0400_apophth/02_nit-kel-scet.htm. [Accessed: 07-Oct-2019].
[14] N. H. Henein, M. Wuttmann, and P. Ballet, Kellia. 2: l’ermitage copte QR 195. Le Caire: Institut français d’archéologie orientale, 2000.
[15] H. G. Evelyn-White, The Monasteries of the Wadi ’n Natrûn: Part III, The architecture and archaeology, vol. 3. The Metropolitan Museum of Art, 1933.
[16] W. Youssef, “Evolution of the design of the Coptic churches in Egypt”, Unpublished master's thesis, Ain Shams University, Egypt, 1974.
[17] G. Gabra, S. Shaker, and F. M. El Muharraq, Monastery of Anba Hatre (St. Simeon). Aswan, Egypt: The St. Mark Foundation, 2011.
[18] G. Gabra and T. Vivian, Coptic Monasteries: Egypt’s monastic art and architecture. The American University in Cairo, 2002.
[19] Besa and D. N. Bell, The life of Shenoute. Kalamazoo, Mich: Cistercian Publications, 1983.
[20] Cambridge University Press, Ed., Cambridge academic content dictionary, 1st ed. New York: Cambridge University Press, 2009.

Military Technical College
Kobry El-Kobbah,
Cairo, Egypt

13th International Conference on Civil and Architecture Engineering
ICCAE-13-2020

Flexible Design: An Innovative Approach for Achieving Sustainability in Primary Public Schools in Egypt

Yara A Fahmy [1] And Ayman A E Othman [2]

[1]Senior Student, Architecture Engineering Department, Faculty of Engineering, The British University in Egypt, Cairo-Suez Desert Road, El-Shorouk, Cairo, Egypt. E-mail[1]: Yara141478@bue.edu.eg

[2]Professor of construction and project management, head of architectural engineering department, faculty of engineering, The British University in Egypt, Cairo-Suez Desert Road, El-Shorouk, Cairo, Egypt. E-mail[2]: ayman.othman@bue.edu.eg

Abstract: Public schools in Egypt are considered a valuable asset that should be preserved and sustained, as they play a paramount role for the development of a sustainable community. Whereas, currently, the schools in Egypt are not sustainable from social, economic and environmental perspective as they are generally heavy, fixed and normally irreversible once construction has been completed. This is due to the traditional construction method used in the construction of schools which consumes time and generates a vast amount of waste. In addition, the changing demands of the occupants may confront the need for future expansion or complete changeover. With the increase in labour costs, lack of skilled labours for different trades, demand for faster construction, increased considerations for safety and environment and above all meeting these demands at an economic and reasonable price requires us to change the approach used in construction. Hence, these faults could be sufficed with the introduction of green prefabrication methods as a flexible design approach for achieving sustainability for primary public schools in Egypt. Therefore, this paper aims to formulate a matrix that correlates the flexible design criteria with suistainability in order to achieve suistainability in Primary Public schools in Egypt.A research methodology consists of literature review to investigate the topics of primary public schools, flexible design, Green prefabrication and sustainability.

Keywords: Green prefabrication, Sustainability; Primary Public Schools; Egypt; Flexible Design; Matrix

1. Introduction

Public schools are considered one of the pillars of development in Egypt. Not only does a school provide students with foundational skills but also creates a sustainable community. Current status of schools in Egypt is not sustainable in terms of poor maintenance, increased costs and overpopulation. Subsequently, this negatively impacts the performance of schools, students and the educational process. Although there are around 6,000 public schools in Egypt, there is still a demand to build more schools in order to provide students with better education [1]. Consequently, Egypt was ranked 133rd out of 137th countries for the quality of the education in primary schools according to the World

Economic Forum Global Competitiveness Index [1]. Substantially, the traditional construction method of schools consumes time and produces a vast amount of waste [2]. Hence, flexible design is a proposed approach to achieve sustainability of primary public schools through green prefabrication. Prefabrication is defined as the offsite manufacturing of components or section of a building. These sections are then transported or delivered to the site for complete assembly [3]. However, there are some limitations for prefabrication .For instance, as the sections or the components are delivered from the manufacturing location to the onsite location, there is a restriction on the maximum tolerable proportions and sizes of these sections that rely on the ergonomics of the highway through which the components are transported. Consequently, a prefabricated structure cannot be constructed in one piece, and accordingly, they are fragmented into smaller sections or components that could be managed allows the execution of a building or a structure with higher quality and at a reasonable cost [3]. According to the Education Funding Agency (EFA), a school design is suggested to be green prefabricated as it is more environmentally friendly when compared with the traditional building method. Not only does the prefabrication reduces the maintenance costs but also reduces the construction time.

2. Literature Review

2.1 Public Schools in Egypt

The quality of education is an essential tool for achieving a sustainable environment [4]. Education for sustainable development (ESD) endorses the development of several aspects including knowledge, skills understanding, actions and values required to ensure environmental protection (environmental sustainability), to promote social equity (social sustainability), and encourage economic sustainability. The development of the concept of ESD was established from environmental education, which pursued to develop the knowledge, skills and values in people to maintain their environment. The goal of ESD is to enable people to carryout decisions and actions to enhance the quality of life. In addition, ESD also aims to assimilate the values of sustainable development into all levels of learning [4].

2.2 The General Authority for Educational Buildings (GAEB)

The general authority for educational buildings is a public building that is present in each governorate in Egypt. The authority is responsible for the planning of the educational buildings, developing the standards and specifications of the design of public schools and determining the building regulations and rules in the light of a scientific plan and taking into account the objectives of the development plan and the general policy of the state. The GAEB has typical designed prototypes of schools. These designed prototypes can be either a single linear form through which the floor encompasses four classrooms or an L-shape or U-Shape building with eight classrooms per floor [5].

2.3 Traditional Construction

2.3.1 Definition of traditional construction

The term traditional means conventional or customary. Forbes added that cast-in-situ is a commonly used term which is conventional method through which concrete is poured into a formwork on the site. This conventional method encompasses the use of common or traditional materials such as brick or concrete in the construction of buildings [6].

2.3.2 Characteristics of traditional construction

The traditional design and construction method are considered a waste generating process which uses materials that produce inordinate and excessive waste during the construction of a building [7]. In this context, bricks produce a vast amount of waste attributable to the broken units resulting from transportation [8]. Also, the wooden strips that are used in the traditional formwork to fit the required dimensions for the cast-in-situ could be ruined and eventually will not adapt to other uses. All these aspects can contribute to waste production which is detrimental to the environment. Unlike prefabrication, the traditional construction methods produce an inflexible building (Mahayuddin, 2016). A flexible building must be adaptable. However, the traditional construction method hinders the flexibility of building [6]. This means that the building is not adaptable with the user's changing requirements. Hence, future extensions of spaces in a building cannot be applied.

2.3.3 Traditional vs Prefabrication

Although both prefabrication and traditional construction commence the same way-planning, designing, approvals, site preparation and development, there are various differences between the two approaches in terms of quality, cost, required workers and machinery, construction time, environmental considerations, safety and flexibility [9]:

- Unlike traditional construction, quality in prefabrication construction can be controlled and maintained as the prefabricated elements or components are manufactured by using laser measured machines with precise code tolerances
- Although the maintenance cost of a traditional structure is less when compared to the prefabricated structure, the overall maintenance and lifecycle costs of the structure are decreased over the long term.
- The traditional construction involves massive number of workers using simple machinery. However, prefabrication requires less labor with heavy cranes and machinery for handling large prefabricated elements.
- The process of prefabrication is faster than the traditional construction as prefabricated units are produces in mass modules through which elements or components are easily assembled on the site. Apart from this, the prefabricated elements are delivered in few weeks after the drawings are verified as illustrated in figure 1.
- Prefabrication allows for recycling and reusing of materials, thus generating less waste. On the other hand, a traditionally constructed building produces waste as the remains of the building components are demolished and therefore the materials cannot be reused.
- In the traditional construction, workers may be exposed to severe weather conditions which may cause serious accidents and hinder the construction schedule. Conversely, in prefabrication constriction, an indoor controlled environment for workers is maintained. This indoor environment ensures that workers are manufacturing in a safe environment with reduced risk of accidents.
- Once traditional construction has been completed, a heavy, fixed and irreversible building is constructed Akin to this, flexibility is suppressed in traditional construction. However, a prefabricated building offers flexibility in two main ways: Firstly, the parts or components could be disassembled and reused in another location. Secondly, the building can adapt with the users' needs, thus allowing for future extension to occur in the building.

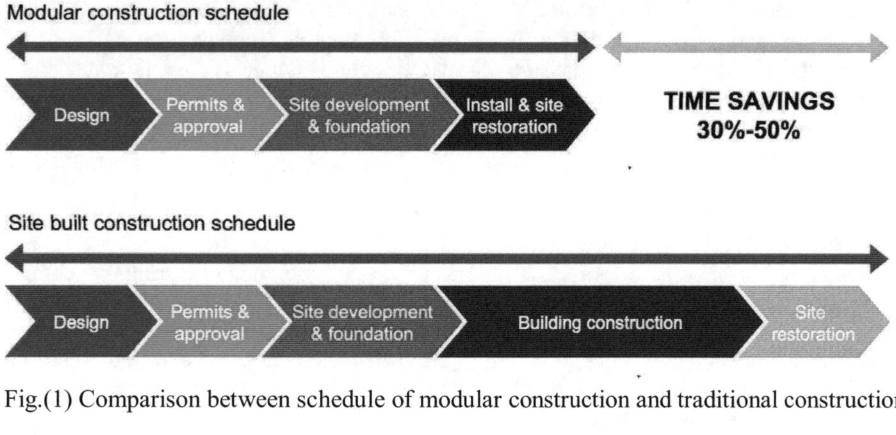

Fig.(1) Comparison between schedule of modular construction and traditional construction

IOP Conf. Series: Materials Science and Engineering **974** (2020) 012018 doi:10.1088/1757-899X/974/1/012018

2.4 Flexible Design

2.4.1 Definition of flexible design

The concept of flexible design was defined by researchers in different ways [10]. Flexibility was defined as the ability of buildings' features to adapt to the needs of the users. On the other hand, a flexible building must be adaptable which means the ability to change in order to meet the user's changing requirements as well as the physical environment [6]. These changes mean that the building is designed in a way to allow for modifications of a certain part without affecting other parts of the building. A balance between adaptability and durability while designing the building creates a flexible design [10].

2.4.2 Design for flexibility

There are three changes that can happen through the lifetime of the building. These changes or transformations could occur in the structure, the space and elements and material transformation [11]:

- Structural transformations occur when the capacity of the building changes. This means that the building components are reused, replaced or recycled.
- Spatial transformations occur due to changes in the function or the use of the building, thus requiring disassembly and adaptability in order to preserve the integrity of spaces.
- Spatial as well as elemental and material transformation occur due to changes in flow within the building.

2.4.3 Design criteria for increasing building flexibility

There are two main indicators that should be considered in the initial phase of the building design in order to reach an optimal flexible design: design for durability and design for adaptability [6].

2.4.3.1 Design for durability

Durability is the ability of a building to fulfill its functions over a period of time [11]. Also, a building element will tolerate changes if they have a long life time. This ability entails a durable structure in order to allow for changes. A durable design increases the productivity of resources as the materials could be reused and optimized [11]. In addition, quality and lifetime of building components are important determinants for building durability [11].

2.4.3.2 Design for adaptability

Designing for adaptability means the ability of the building to adapt or change while fulfilling various functions during its lifetime [13]. Using modular systems and prefabricated components could simply increase a building's flexibility. In addition, modular systems or the use of prefabricated elements simplify the process of disassembly in order to change the function of a space. This paradigm reduces the construction time and energy consumption which produces an adaptable and sustainable building [13].

2.5 Prefabrication

2.5.1 Definition of prefabrication

Prefabrication or precast construction refers to the offsite manufacturing of components or segments of a structure. These segments are then transported or delivered to the site for complete assembly [9]. The components or the segments of the prefabricated structure can be fabricated of different materials such as steel, concrete, reinforced concrete, wood and Aluminum composite. Prefabrication can have several terms: off-site prefabrication (OSP), off-site manufacture (OSM) and as a modern method of construction (MMC)[9].

2.6 Types of Prefabrication

IOP Conf. Series: Materials Science and Engineering **974** (2020) 012018 doi:10.1088/1757-899X/974/1/012018

There are three main systems that are used in the prefabricated structural systems [9]: Modular prefabrication, panelized prefabrication and hybrid prefabrication as illustrated in figure 2. All the types are divided according to the used materials, methods and structural configuration.

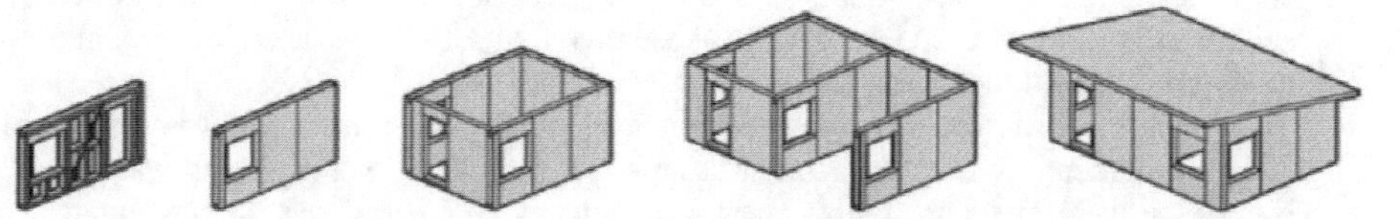

Fig. (2) Prefabrication systems (Dunlop, 2010)

Hence, Prefabricated systems are divided in three different types

- Modular Prefabrication (volumetric module)
- Panelized Prefabrication (component or panel)
- Hybrid Prefabrication (volumetric panel)

2.6.1 Modular Prefabrication

Modular buildings are volumetric 3d elements or components that are constructed entirely in a building facility and then transported to the final location for assembly [14]. These components are comprised of walls, ceiling and floor panels that are assembled to form the 3d volumetric module. The volumetric modules could be used for any building that with repetitive units such as hotels, hospitals, educational facilities and many other building uses.

2.6.2 Panelized Prefabrication

The panelized prefabricated system as a 2D-panel or components through which components are produced in masses and then delivered to the site for the three- dimensional assembly. These components or panels can be made of several materials including mainly concrete panels, timber panels or recycled timber[14].Furthermore, the panelized prefabricated system can be either an open-panel system in which panels are not fully insulated in the site or a closed-panel in which panels are insulated and completely manufactured in the factory.

2.6.3 Hybrid Prefabrication

The Hybrid prefabrication system is an integration of both the modular prefabrication system and the panelized prefabrication system. This means that some of the building components could be 2D panels or 3D volumetric modular construction [14].

2.7 Advantage of prefabrication

There are numerous advantages of Prefabrication [15]:

- Prefabrication reduces the construction time as the prefabricated units are delivered in few weeks after the approval of drawing.
- Prefabrication allows flexibility in the construction industry such that the prefabricated components can be easily dismantled and relocated to different sites. In addition, prefabrication units can blend in with any building type as the units can be used in different spaces. Moreover, prefabricated units offer future expansions in the building.
- The assembly process of the sections or components of prefabricated units are produced in a factory-controlled environment which creates a safer working

environment. This means that any complications associated with moisture and environmental hazard is reduced. In addition, an indoor construction environment ensures that workers are less likely to be exposed to accidents or other liabilities.

- Not only does a controlled environment provides safety for workers but also produces units with consistent quality. Specific standards are followed to have assemblies constructed to a uniform quality. Moreover, precise machines are used to ensure adherence to building codes as well as quality control checks are made to monitor quality throughout the whole process.
- The technologies associated with prefabrication produces a building that is sustainable and environmentally friendly. Kumar asserts that the overall water saving was reduced by 70%, therefore enhancing energy efficiency by 20% per square meter of construction.
- Although the initial costs of capital and transportation costs from factory to the specified site may be high, the maintenance and life cycle costs of the building are reduced over the long term.
- Prefabrication produces a building that is flexible, safe, and environmentally friendly with a consistent quality, reasonable cost and reduced construction time.

2.7.1 Prefabrication offers flexibility

Prefabrication offers flexibility in terms of portability or movability, reprocessing or reusing components and expansions [6]. Portability or movability means that the prefabricated components could be easily disassembled and relocated to different sites. In addition, the elements also could be reused or repurposed for other functions. Moreover, a modular design allows for future expansions in the building [11].

2.8 Green prefabrication

2.8.1 Green prefabrication definition

The green prefabricated building as one which consumes less water, generates a reduced amount of waste, conserves or protects natural resources and enhances energy efficiency of a building [16]. Not only does a green prefabricated building provide a healthier environment for the occupants or the users of the building but also helps in using processes that are environmentally friendly. These processes aid the design, construction, operation and renovation of the building's life cycle by utilizing resources efficiently. Hence, green prefabrication expands the concerns of environmental, social and economic sustainability, thus creating a sustainable or "high performance building" [16].

2.8.2 Characteristics of green prefabrication

Green prefabrication incorporates a smart design that is energy efficient by using eco materials that are environmentally friendly while conserving water. High performance building focuses on sustainability [16]. Green prefabricated buildings consume less energy when compared to conventional or traditional buildings. This means that the amount of water, energy and raw materials used in the green prefabricated building during construction and operation supports the creation of a high- performance building. In addition, less emissions are generated due to the efficient use of resources. All these features create a better and healthy environment for users, reduced the cost of design and emphasizes the recycling and reusing of materials [11].

2.9 Sustainability

2.9.1Definition of Sustainability

Sustainability is defined as the efficient distribution of resources intra-generationally with the process of socio-economic activities within the borders or limitations of a finite ecosystem [17].On the other hand, sustainability is perceived as a dynamic process through which there is an interaction between the carrying capacity and the population of its environment such that the population expresses its full potential without creating or producing irretrievable impacts on the carrying capacity of the environment[18]. From this standpoint, sustainability emphasizes human actions or activities and their capabilities to meet the expectations, needs or desires of human in order to satisfy their needs and wants [18]. These needs and wants are sufficed without exhausting or diminishing the prolific resources at their disposal. This, consequently, incites thoughts on the manner through which people must lead their economic and social lives on the existing resources for human development.

2.9.2 Sustainability and sustainable development

The main concept of sustainability is to focus on the conditions of the environment in order to attain a designed product or output with maximum internal qualities of the environment such that it can reduce the undesirable features or aspects of these constructions [19]. Consequently, buildings must respond to the environment initially from the design stage and settling when they are to lessen or decrease confronting with nature. Sustainability was depicted as a three-legged stool through which the environment, the economy and the society are depicted as legs. Hence, the fundamental concept of the stool is that if any leg is missing or shorter (less important), the stool will be unbalanced. However, if these legs have the same dimension, this means that each pillar has an equal weight. Consequently, there would be a balanced stool that will support sustainable development [19].

2.9.3 The Three Pillars of Sustainability

There are three main attributes of sustainable development which are environmental protection, economic expansion and social integrity. Based on this, talyor,2016 stated that the idea of sustainable development rests, primarily, on three key pillars. These key pillars are "economic sustainability", "social sustainability" and "environmental sustainability" [20].

2.9.4 Environmental Sustainability

The first pillar of sustainability by defining the concept of environmental sustainability such that it supports human life while remaining productive and resilient [8]. In addition, environmental sustainability recounts to the carrying capacity and ecosystem integrity of natural environment. This requires that the natural capital should be used sustainably as a source of economic input and as a disposal of waste [9]. As the earth systems have limitations or boundaries within which equilibrium is sustained, the process of regenerating and harvesting should be balanced. On the other hand, Kumar stated that the waste should be disposed in a way that the environment can maintain.

2.9.5 Social Sustainability

The second pillar is social sustainability which represents any aspect that is related to human rights. In this context, social sustainability is not only about ensuring that human needs are met but also aiming at providing conditions for humans to have the capacity to realize their needs. This means that anything that impedes or hinders this capacity is considered a barrier [19]. However, if human needs are met, a socially sustainable environment will exist.

2.9.6 Economic Sustainability

The third pillar of sustainability is economic sustainability which requires the use of resources efficiently [21]. The economic sustainability is used to outline strategies that promote the consumption of socio-economic resources to their maximum benefit. In addition, there should be an efficient allocation of resources as well as an equitable distribution in order for the economy to operate in s sustainable manner [21]. However, during the design and construction phase, for example, aspects such as building materials should be considered in terms of their cost effectiveness. This means that using recycled materials during the design phase and construction phase can help lower the costs, leading to a long-term cost efficiency.

2.9.7 Relationship between Environmental, Social and Economic Sustainability

In order to achieve sustainability, there must be a balance of economic, social and environmental factors in equal harmony [22]. As shown in figure (3) there are three main aspects of sustainability which include environmental, social and economic. The interaction between these aspects created three new aspects, specifically: Social- Environmental, Environmental-Economic and Economic-Social, which explains that each individual has a right or share of natural resources of the environment at both national and international levels. This ensures that resources are not consumed by a portion of the society leaving the of the society with needs that cannot be met by remaining resources [22].

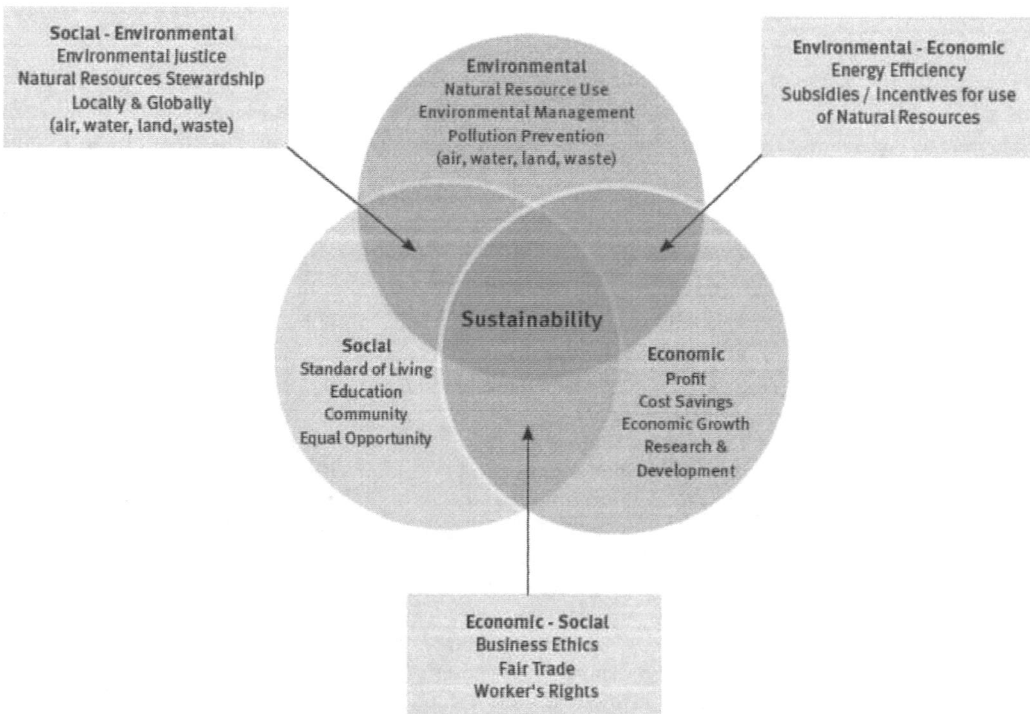

Fig. (3) Aspects of Sustainability

2.9.8 Benefits of Sustainability

The application of sustainability in construction is reflected in every aspect of the construction process. It includes, for instance, sustainable design, sustainable planning, sustainable architecture, sustainable landscaping and sustainable construction. By applying the concepts of sustainability in the construction industry, a number of benefits will be achieved. Environmentally, sustainability helps reducing the use of non-renewable resources, minimizing environmental risk and uncertainty, reducing waste and pollution, and increasing the reuse and recycling of building materials [22]. Socially, sustainability focuses on identifying stakeholders' requirements and ensuring that the developed project fulfils their needs and meets their expectations. Moreover, it encourages the involvement and getting feedback from all parties affected by the built environment, ensures health and safety requirements are achieved, considers people with special needs and provides support and adding value to communities and the supply chain. Economically, sustainability supports growth in the construction industry through increasing gross domestic product (GDP), providing more job opportunities, raising the client's profit and investment return [22].

3. Conclusion

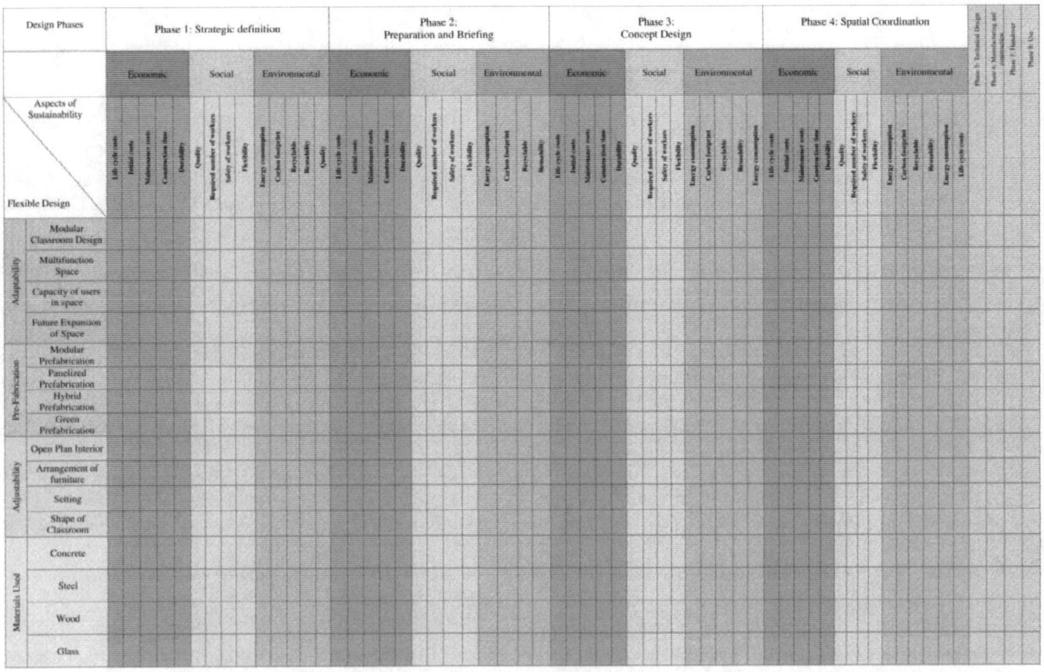

Fig. (4) Matrix showing correlation between Flexible design criteria and sustainability

Primary Public schools have a vital role towards developing a sustainable community. Despite its important role, the inflexibility of the traditional methods used in construction create social, economic and environmental challenges that makes the school unsustainable. Hence, flexible design is a proposed approach to achieve sustainability in Primary Public schools. The paper's methodology consists of qualitative data gathered from several reliable and valid sources in order to formulate a comprehensive literature review that would help to investigate the role of flexible design for achieving sustainability in primary public schools in Egypt through Green Prefabrication. Finally, the output is formulated in the form of a matrix which shows the correlation between flexible design criteria and the three pillars of sustainability as illustrated in figure (4).

Recommendation

Implementing green prefabrication in Primary Public schools in Egypt entails the creation of flexibility which includes both adaptability and durability in order to suffice the future needs of the users. Both green prefabrication and flexibility are essential tools that, if applied in schools, will help in shaping an environmental-friendly and sustainable community.

References

[1] Patil, J. (2019). Prefabrication Technology -A Promising Alternative in Construction Industry. *International Journal of Science and Research*, 220 - 224.

[2] Blanc, D. L. (2015). Towards integration at last? The sustainable development goals as a network of targets. *Department of Economic & Social Affairs*, 141-160.

[3] Manley, k. (2015, July 9). Retrieved from https://theconversation.com/prefab-revolution-factory-houses-are-the-secret-to-green-building-42622

[4] UNESCO. (2002). *Education for Sustainability*. Paris.

[5] Sewilam, H. (2014). Introducing education for sustainable development into Egyptian schools. *Environment Development and Sustainability*, 1-18.

[6] Crowther, P. (2000). *DEVELOPING GUIDELINES FOR DESIGNING FOR DECONSTRUCTION* . Watford, England: Deconstruction- Closing the loop, BRE Proceedings.

[7] Drolet, J. (2015). *International Encyclopedia of the Social & Behavioral Sciences (Second Edition).* Edmonton: Science Direct.

[8] Morelli, J. (2011). Environmental Sustainability: A Definition for Environmental Professionals. *Journal of Environmental Sustainability, 1*(1), 1-10.

[9] Keshavarzi, A. (2018). *Prefabricated and traditional construction methods.* Larestan: Islamic Azad University of Larestan.

[10] Zuo, J. (2014). Green building research–current status and future agenda: A review. *Renewable and Sustainable Energy Reviews*, 271–281.

[11] Durmisevic, E. (2008). *Construction materials stewardship, the status quo in selected countries.* Rotterdam: Centre for building performance research, CIB.

[12] Sassi, P. (2008). Defining closed-loop material cycle construction. *Building Research and Information*, 509-519 .

[13] Mahayuddin, S. A. (2016). Assessment of building typology and construction method of traditional longhouse. *Procedia Engineering* , 1015 – 1023 .

[14] Macozoma, D. S. (2001). *Building Deconstruction.* South Africa: CIB Publication.

[15] Kumar, N. (2018). Environmental Sustainability—necessary for survival. *Environmental Sustainability, 1*(1), 1-2.

[16] Wilson, J. (2015, January). Retrieved from https://www.buildinggreen.com/feature/potential-prefab-how-modular-construction-can-be-green

[17] Mensah, J. (2019). Sustainable development: Meaning, history, principles, pillars, and implications for human action: Literature review. *Cogent Social Sciences*, 21.

[18] Malagnino, A. (2017). From Traditional Construction Industry Process Management to Building Lifecycle Management. *Product Lifecycle Management and the Industry of the Future* , 335-344.

[19] Vallance, S. (2011). What is social sustainability? A clarification of concepts. *Geoforum*, 342-348.

[20] Santra, S. (2012). The social pillar of sustainable development: a literature review and framework for policy analysis. *Sustainability: Science, Practice, & Policy*, 15-29.

[21] Thomas, C. F. (2015). *Naturalizing Sustainability Discourse: Paradigm, Practices and Pedagogy of Thoreau, Leopold, Carson and Wilson.* Arizona: Arizona State University.

[22] Wanamaker, C. (2018, April 24). The Environmental, Economic, and Social Components of Sustainability. *Components of Sustainability*, 25-28. Retrieved from Soapboxie.

IOP Conf. Series: Materials Science and Engineering **974** (2020) 012019 doi:10.1088/1757-899X/974/1/012019

**MilitaryTechnicalCollege
Kobry El-Kobbah,
Cairo,Egypt**

**13th International Conference
on Civil and Architecture
Engineering
ICCAE-13-2020**

Paths Towards Energy Efficient Buildings Using Nano Architecture

Nader Ghariband Basant Ahmed[1, 2,*]

[1] Head of Architectural Engineering and Environmental Design Department, Engineering & Technology College, Arab Academy of Science, Technology and Maritime Transport, Port Said 42511, Egypt.

[2]Architectural Engineering and Environmental Design Department, Engineering & Technology College, Arab Academy of Science, Technology and Maritime Transport, Alexandria 21913 P.O. Box 1029, Egypt.

*Arch_Nader@gmail.com

*Basantsaleh8@gmail.com

Abstract. Nanotechnology has been in a massive use in the last two decades in many fields specially the architecture and construction field. Applying nano technology in architecture is taking place and has become the biggest challenge in our time. The challenge of nano architecture is to improve the material properties, adding new properties or production new materials which used in architecture design of the building or in construction phase. The study aims to reduce the energy usage of the building by using nano architecture and to achieve one of the environmental sustainability factors.The research adopts the descriptive theoretical approach of nano technology and nano material in field of architecture and its impact on the energy use of the building. This followed by the second analytical part which examines the nano materials and nano devices through examples of global sustainable projects and nano architecture applications which leads us to a huge difference in the architecture design and reduces the total energy usage of the building. In conclusion, the research shows the effect of using nano material and devices on the building energy use which can help to save the environment by reducing the energy consumption.

Keywords: Nano architecture – Nano material – energy efficient building – environmental sustainability

IOP Conf. Series: Materials Science and Engineering **974** (2020) 012019 doi:10.1088/1757-899X/974/1/012019

1. Introduction

Nanotechnology is one of the new emerging technologies ofthe contemporary time, as a result of the international focus onnano sciences. Such technology has made it possible tomanipulate the matter on an atomic basis; this is expected totransform and revolutionizes the way of live. The nano world isa convergence of a real mix of scientific and technologicaldomains which once were separate.The Applications of nanotechnology include almost all aspects of our life, in medicine, industry, communications,transportation, and building façades. The application ofnanotechnology in architecture is wide and varies from theearly stages of sketching up to the final touches of finishing,especially in materials selection.

Nano architecture is the integration of nanotechnology in architecture, by using nano-products, nano-materials, nanotelecommunication, or even nano-shapes. In introducing nanotechnology to architecture, one should examine the benefits it can bring, such as additional functionality, value added, and market demand with regard to product development. Good design is based on demand, and the contributions to the evolution of both nanomaterial and the resulting nano-product in the long term.

1.1Problem Definition

The energy consumption is one of the largest sectors seen negatively on the environment. so consideringbuilding efficiency and recycling to preserve the environment one of the most important principles of sustainable architecture, and the methods usedSo far, there's a great need for a Re-studyProperties of materials based on modern technologies like nanotechnology to identify new possibilities of these materials reduce the negative returns are in construction and sustainability.

1.2 Aim and Objectives

This study aims to try using nano technology to achieve the principles of sustainability and reducing energy consumption of buildings by:

1. Monitoring and documenting the role of nano technology in achieving the principles of building sustainability.
2. Examine the application of nanotechnology in architecture to minimize the energy used in any building.
3. Clarification of nanotechnology techniques for achieving an appropriate living environment.

1.3 Nano Technology and Nano Sclae

Nanotechnology is the manipulation of matter at the scale of individual atoms and molecules. It includes processes for making materials, systems and structures. Some types of nano-materials exist in nature, but nanotechnology, the deliberate engineering of nanostructures and materials, largely began in the 1980s with the invention of new, more powerful types of microscopes. “Nanotechnology is an enabling technology that allows us to develop materials with improved or totally new properties. **(Figure 1) Development of nano products**

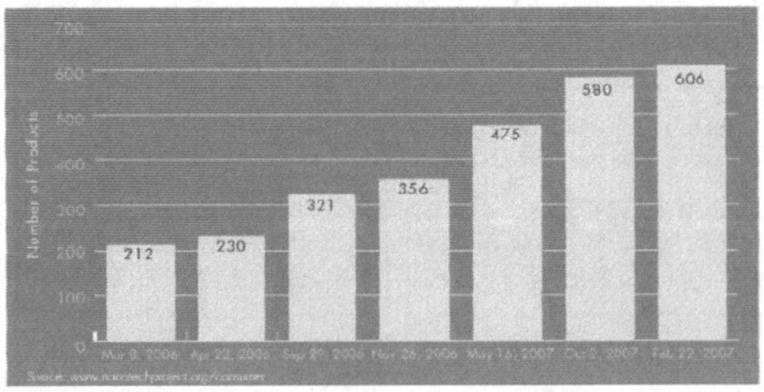

(Figure 1) development of total products listed for nanotechnology Source (1)

Nanoscale science, engineering, and technology are fields of research in which scientists and engineers manipulate matter at the atomic and molecular level in order to obtain materials and systems with significantly improved properties. Nanomaterials are usually defined as materials that have at least one dimension smaller than 100 nanometers. A nanometer is approximately 1/80,000th the width of a human hair or 1/7,000th the size of a single red blood cell. Materials at the nanoscale often exhibit physical, chemical and biological properties that are very different from those of their normal-sized counterparts.

2. Nano Technology and Architecture

The applications of nano technology depends on the potential inherent within the materials that have been examined to produce different materials with new properties and capabilities, in addition to the new technology of devices that have been rapidly produced that have a marked difference with many changes in the way of thinking in architecture and to lead us to a new architecture forms which helps us improve our indoor and outdoor environment.[1]**Figure 2. Applications of Nano Technology in Architecture**

2.1 Nano Material in Architecture

Nano technology enables us to work within the small particles of the material to improve its properties to give more different potentials which can make the architect face more challenges and have a lot of solutions for the design and construction problems.[2] Nano material can be classified as follows:

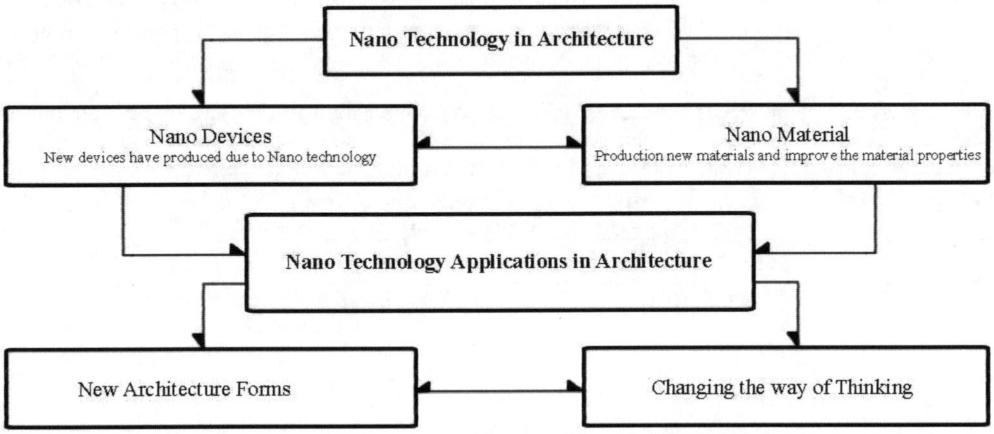

Figure 2.Applications of Nano Technology in Architecture (Source (1) editdede by the researcher)

- **Structure Material**

Nano technology in construction field can develop and improve the main structure material like concrete, steel by producing a new structure materials like Nano Carboon tubes.Structure material can do a lot like: Instead of cement in concrete mixture, Adding 3% increases pressure strength and bending from concrete mortar, Reduces CO2 emissions. **Table [1]**

- **Non – Structure Material**

Nano technology can help the non-structure material used in any building to be more effective by increasing the material's properties efficiency like Glass, wood and dry walls. They have a lot of benefits by adding SOI2 to the glass became a Fireproof glass. **Table[2]**

- **Insulation Material**

The progress has been made by nano technology in the insulation materials have made a huge difference in the lightness, firmness and flexibility of this materials like: Lumira, solar absorbing windows and thin-film insulation used in the indoor materials.There is a lot of technologies approved that insulation material have a very good effect on energy building consuming it helps transmission of light and CO2 emissions, reducing heat gain and noise. **Table[3]**

- **Nano Coating Material**

Coating takes a big place in nano technology field of research to achieve a huge possibility in the material properties to be more effective coating materials like: self-cleaning, easy to clean and anti-finger print coating.Coating materials and its benefits like:[1]Self-Cleaning (Photo Catalytic) & Easy To Clean it is Non-sticking impurities on paint, easy to clean and reducing maintenance costs, expelling oil and water coatings (ETC) and there is a lot of other coatings like Anti-Finger Print, Anti -Graffiti, Anti - Scratching, Anti Reflection and Bacteria they have a lot of benefits like:

Anti-finger print on the outside coating, easy to clean by water, graffiti and paint resistant adhesion, adding (TIO2) it makes coating more anti bacteria and germs, improve light permeability to glass by 98% and plastic by 99%, improve lighting efficiency by 15% and reduce the reflected lights from 1% to 8%. Ultra Violet Protection have another different benefit to the indoor environment by adding (TIO2) (ZNO) (CEO) to the coating it helps to absorb the harmful rays without blocking good light.

2.2 Nano Devices

The revolution of nano technology in the improvement of nano devices and its applications in lighting, air and water filtration and in energy storage has become more effective to our environment.**Table[4]**It had a lot of development across the twenty past years because the new technologies and it begins to be used a lot by all the means in any field of work and study.

This tables shows the new materials and technologies started from beginning till now. they have a new technology in nano technology and new devices for energy in lighting in heating using them to consume the energy used in the public buildings or in the small residential apartments. Starting with table one till table four.

Table[1]. Structure Materials(Source By Researcher)

Nano Materials	Nano Applications	Descriptions
Concrete	Nano Silica (NS)	Instead of cement in concrete mixture, Adding 3% increases pressure strength and bending from concrete mortar, Reduces CO2 emissions.
	Nano Titanium Dioxide (TIO2)	Adding from 5% to 10% from (TIO2) increases concrete strength by 26% to 35%, self cleaning concrete.
Steel	Nano Fillers	Adding nano fillers to concrete mixture increase bending strength from 15% to 20%
Carbon Tubes	Nano Carbon Tubes	Stronger than steel more than 100 times and more lighter, flame resistance by adding nano tubes to the glass, frost resistance increased from 150 to 400 degrees Fahrenheit 46%

Table[2]. Non – Structure Materials (Source By Researcher)

Glass	Nano Titanium Dioxide (TIO2)	Blocks ultra violet rays, expel water and anti reflection by adding zinc oxide ZNO + TIO2 and self cleaning glass plus controls the air pollution.
	Silicon Dioxide (SIO2)	By adding SOI2 to the glass became a Fireproof glass
Dry walls	Dehydrated Calcium Sulfate	Light interior walls and ceilings, water and mildew resistant, less energy consumption in construction
Wood	Nano Wood	Water and oil expelling and easy to clean, wood doesn't change its color and protect it from temperature changes

Table[3]. Insulation Materials (Source By Researcher)

Dry Silica	Lumira (Aerogel)	Helps transmission of light, less consumption of energy and CO2 emissions, reducing heat gain and noise
Thin – Film Insulation	Thin – Film Insulation Sheets	Low room temperature from 2-3C, reducing electricity cost for air conditioning, blocking sunlight if needed
Solar Absorbing Windows	Solar Absorbing Nano Material	Heat gain coefficient is 0.55, visible light permeability up to 70% and less outside noises

Table[4]. Nano Devices (Source By Researcher)

Lighting	Nano LED	More energy efficient and smaller than normal lamps up to 93%, economically good for the long term use , long life more than 50.000 hours, have more speed of light
	Organic Light Emitting Diodes (OLED) – Quantum Dot Lighting (QLED)	(OLED) Reducing energy efficient usage from 2 to 10 Volt and (QLED) is more energy efficient than the OLED from 30% to 40%, reducing energy

Filtration	Indoor Air and Water Filtration	Eliminate pollutants such as viruses and bacteria without releasing oxidation in the air, detoxification of the groundwater by adding nano particles, elimination of industrial pesticides in water.

3. Environmental Sustainability

As a general term refers to the capacity to endure. Sustainable development is the development thatmeets the present needs without compromising the ability of future generations to meet their own needs [5].Currently it is considered as an infinite process through the ecosystem, a dynamic Evolutionary way towards theimprovement of management of human and natural resources. In the construction field, sustainability is a matter ofgrave importance. In architecture sustainability describes environmentally conscious design techniques, minimizingnegative environmental impact and enhancing efficiency in the use of materials, energy and space.

The principles for sustainable environmental design include:

- Low-impact materials (non-toxic, sustainably produced or recycled)
- Energy efficiency (less energy in manufacturing and production)
- Quality, durability,
- Reusability and Recyclability [6]

3.1 Building sustainability with Nanotechnology

Nanotechnology is an enabling technology that is opening a new world of materials functionalities, and performances. Therefore, it is opening also new possibilities in construction sustainability. It could lead to a better use of natural resources, obtaining a specific characteristic or property with minor material use. It can (also) help to solve some problems related to energy issues in building (consumption and generation), or improve internal environmental conditions to mention only a few matters. [7]

3.2 Energy Efficient Buildings with Nano Architecture Application

The use of nanotechnology in architecture is strongly linked to sustainability, to improve energy efficiency and reducegreenhouse gases. Architects are to find innovative solutionsfor climate changes, by combining ambitious architecture withenergy efficiency. The use of materials and surfaceproperties that have now become possible throughnanotechnology offer architecture, interior architecture and related disciplines a mean of achieving greater energy efficiency and sustainable architectural design through innovations.

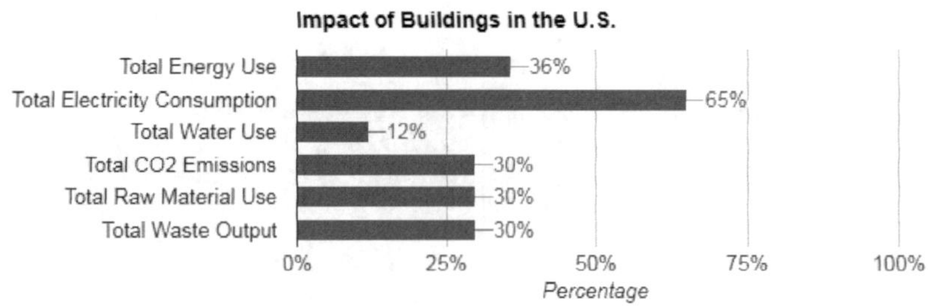

Source: EPA, USGBC

IOP Conf. Series: Materials Science and Engineering **974** (2020) 012019 doi:10.1088/1757-899X/974/1/012019

3.3 Efficient Use of Energy

Another key aspect of sustainability is the efficient use of energy. The most advanced nanotechnology projects related to energy are: storage, conversion, manufacturing improvements by reducing materials and process rates, energy saving by better thermal insulation for example, and enhanced renewable energy sources like solar cells. Efficient use of energy in building can be achieved through the following aspects:

3.3.1 Reduction of Energy Consumption

- **Insulation:**

In the EU, over 40% of total energy produced is consumed by buildings. Insulation is an obvious solution to reduce some of this energy use; however, limited space for installation is a major problem for building renovation. As a possible remedy, work by Aspen Aerogels has produced an ultra-thin wall insulation which uses a nano porous aerogel structure which is hydrophobic and repels water so it is mould free. Another intriguing application of aerogels is silica-based products for transparent insulation, which leads to the possibility of super-insulating windows. [4]**Figure.3 OLED Structure and Nanogel Aerogel System**

- **Lighting:**

Nanotechnological approaches like light-emitting diodes (LEDs) or quantum caged atoms (QCAs) could lead to a strong reduction of energy consumption for illumination. Reducing energy use of lighting can be the most essential field of research can be made nowadays. [15]

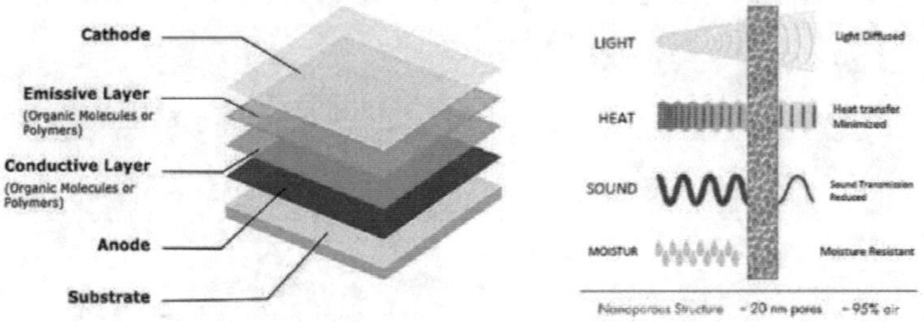

Figure.3 OLED Structure and Nanogel Aerogel System

3.3.2 Close Definition of OLED Lighting Structure Material:

An OLED (organic light-emitting diode) is a light-emitting diode (LED) in which the emissive electroluminescent layer is a film of organic compound that emits light in response to an electric current. OLEDs are used to create digital displays in devices such as television screens, computer monitors, portable systems such as mobile phones, handheld game consoles and PDAs. A major area of research is the development of white OLED devices for use in solid-state lighting applications.

An OLED is a solid-state device consisting of a thin, carbon-based semiconductor layer that emits light when electricity is applied by adjacent electrodes. In order for light to escape from the device, at least one of the electrodes must be transparent. The intensity of the light emitted is controlled by the amount of electric current applied by the electrodes, and the light's color is determined by the type of emissive material used.

4. Analytical Nano Technology Studies

There is an opportunity for OLED Nano lighting to have a significant impact in short term. A combination of Recent

IOP Conf. Series: Materials Science and Engineering **974** (2020) 012019 doi:10.1088/1757-899X/974/1/012019

LED and OLED lighting will enable the greatest energy and cost savings. What is required is collaboration. OLED panel performance has been demonstrated, but the high cost of panels is slowing the market development of OLED lighting. By thinking creatively and collaborating together, the industry can come up with lower-cost solutions. Dealing with nano technology and all of its new terms can make the future more advanced and glorious. By mixing insulation and lighting.

4.1 Example 1.

Sports hall, Carquefou, ZAC du Souchais, France	
Architecture Agency	MA, Murail Architectures, Nantes,France
Client	City of Carquefou
Manufacture	Multi-wall panels with nano Aerogel filling
Opened	2006
Area	3,360 m² with Surface l,450 m²
CO2 Emissions	Greenhouse gases reduction by energy saving and natural lighting

Figure 4 Sports hall, Carquefou, ZAC du Souchais, France (internal and external view)

4.1.1 Description:

All elevations of this sports complex have been clad withaerogel-filledmulti-wall polycarbonate panels. With thisconstruction the architects voluntarily comply with theguidelines set down by the French "green" environmentalinitiative Haute QualityEnvironmental (HQE).

Additionalsolar protection is unnecessary, allowing a clean and unifiedappearance uninterrupted by brise-soleils or louvers. Naturaldaylight provides an even and glare-free illumination of theindoor space and additional indoor lighting is not necessaryduring the day. There are no cast shadows that could bedistracting for certain sports. The thermal insulating effect ofthe aerogel panels also reduces the heat demand: a 25 mm thickpanel has a U-value of 0.89 W/ m² K and is available in 1.05 m
wide panels of up to 6 m in length.

4.2.2 Close Definition of Aerogel Insulation Material:

Aerogel currently holds the record as the lightest known solid material and was developed back in 1931. It is relatively banal: it is simply ultra-light aerated foam that consists almost 100% of air. The remaining foam material is a glass-like material, and silica. The nanodimension is of vital importance for the pore interstices of the foam: the air molecules trapped within the minute nanopores –each with a mean size of just 20nm – are unable to move, lending the aerogel its excellent thermal insulation properties. **The Nanogel makes up a translucent panel, which achieves**

IOP Conf. Series: Materials Science and Engineering **974** (2020) 012019 doi:10.1088/1757-899X/974/1/012019

a remarkable level of energy saving while providing indoor spaces with natural light.

4.2 Example 2.

LightHouse Tower Project	
Architecture Agency	Mikou Design studio
Client	Brazilian city of Rio de Janeiro
Manufacture	NanoLED
Opened	2011
Type	paths and public venues
CO2 Emissions	Illuminated with bright light with little emission heat -save energy

4.2.1Description: Infusing new life to conventional lighthouses, installed to mark dangerous coastlines, hazardous shoals and reefs in and around the sea, Mikou Design Studio has planned a tower to build in the Brazilian city of Rio de Janeiro. Entitled the “Lighthouse Tower,” the mammoth structure is rooted on the island of Cotunduba and makes an arched gateway to the capital city. Accessed through a large gate from the sea, the modern lighthouse provides enough space for a number of observation points, an auditorium, skywalk, bungee jump platform and climbing tower, together with a gyro drop, cafeteria, souvenir store, urban balconies and multi-usage space.

Figure 5 multi-usage space in lighthouse tower

Illuminated with bright (possibly Nano LED) lights, the tower does not only look good at night but also provides a mesmerizing view of the “samba” city. Nano LED Makes the building More energy efficient and it's smaller than normal lamps up to 93%, economically good for thelong-termuse, long life more than 50.000 hours, have more speed of light.

4.2.2 Examples Conclusion

Nanomaterialsand nanotechnologies are expected to contribute to greatly improve different environments.

- Nanotechnology give great facilitates to better use of natural resources.
- Nanotechnology not only resolves many problems associated with various construction materials, but it also reduces the impact of the construction industry on the environment and improves sustainability.
- Nanotechnology offers low cost solutions for demanding problems such as energy consumption in buildings.

IOP Conf. Series: Materials Science and Engineering **974** (2020) 012019 doi:10.1088/1757-899X/974/1/012019

5. Conclusion and Recommendation

Potentiality - benefits and risks

A central aim of nanotechnology is to consistently use the minimum amount of raw material and energy: from an ecological point of view "nano" is a winning factor. From the point of view of the client or the user, the most realistic and sensible application of nanotechnology focuses on aspects of functionalityand sustainability. The use of nanotechnologies in the design and construction disciplines usually involves the optimization of existing products or common materials. Nanotechnology brings us a step closer towards customized materials with specific individual properties and represents a shift away from the catalogue of standard materials. Similarly to Le Corbusier, who once quoted that "a house is a machine to live in"; we are able now to describe ahouse as a living machine or even a living organism. There are nevertheless two important challenges for future research: achieve to combine properties in real life and manage to produce large scale structures within viable cost limits.

The importance of taking into consideration what Nanotechnology has to offer concerning energy consumption in buildings and its sustainability.

- To Keep with and share experiences with all nations that have started to interact with these technologies in different fields by establishing nanotechnology research centers.

- Nanotechnologies may be directly used for new forms of architectural art, and architectural urban design, but it will be an effective tool to give many functional alternatives in design process.
- The Nanotechnology has possibilities to help many developing countries solve problems. It also present an opportunities to bridge the gap between the east and west.

- Egypt has a very big market for Nano products in order to help solve many of our urban architectural problems.

References

Books:

[1] NanoWerk. (2014). Nanotechnology Introduction. Retrieved from NanoWerk: http://www.nanowerk.com/nanotechnology/introduction
[2] An Introduction to Nanoscience and Nanotechnology by Alain Nouailhat(2008) ISTELtd.
[3] UNEP (2010) Assessing the Environmental Impacts of Consumption and Production: Priority Products and Materials
[4] Johansen, J. (2002). NanoArchitecture: A New Species of Architecture. New York: Princeton Architectural Press.(2002)

IOP Conf. Series: Materials Science and Engineering **974** (2020) 012019 doi:10.1088/1757-899X/974/1/012019

[5] World Commission on Environment and Development. (1987). Our Common Future. Oxford: Oxford University Press.
[6] Anastas, P., & Zimmerman, J. (2003, March 1). Design through the 12 Principles of Green Engineering. Environmental Science &
Technology, 94-101A.
[7] The Biomimicry Institute. (2014). What is Biomimicry? Retrieved from Biomimicry Institute: http://biomimicry.org/what-is-biomimicry/

Reports:

[8] F. A. Hemeida, "GREEN NANOARCHITECTURE" (M. Thesis), University of Alexandria, Arch. Dpt., 2010
[9]Nanomaterialhttp://www.gitam.edu/eresource/nano/nanotechnology/nano_materials.htm
[10] Nanomaterials for Today (Narita-International-Airport-Corporation) http://es.cyclopaedia.net/wiki/Narita-International-Airport-Corporation

Websites:

http://www.pvupscale.org/IMG/pdf/Schlierberg
http://www.nano.gov/search?keys=facts&searchButton.x=11&searchButton.y

Military Technical College Kobry El-Kobbah, Cairo, Egypt

13th International Conference on Civil and Architecture Engineering ICCAE-13-2020

PV Grid-Connected Systems: Performance and Architectural Integration Aspects

Hala Alaa[1], Nader Gharib[2]

[1]Architectural Engineering and Environmental Design Department, Engineering & Technology College, Arab Academy of Science, Technology and Maritime Transport, Alexandria, Egypt.

[2]Head ofArchitectural Engineering and Environmental Design Department, Engineering & Technology College, Arab Academy of Science, Technology and Maritime Transport, Port Said, Egypt.

Corresponding author: eng.hala.alaa1@gmail.com

Considering the high global rates of carbon dioxide emissions, there is a great concern about renewable energy especially solar energy in buildings. Despite its growth, there has been limited acceptance of architects and stakeholders concerning the use of solar energy technologies in buildings, and subsequently, insufficient architectural integration. Here, the research aims to develop better understanding of architects for the design stages, considerations, and possibilities regarding solar PV grid-connected systems in buildings using inductive methodology, thus analyzing the technology specifications and good integration practices. The research determines the factors that influence the early design and conception phase of the building. It develops a framework for architects concerning the technical and physical aspects affecting the design process, through building needs, cell types, system performance, site conditions, space and environmental requirements, and BIPV integration possibilities. In conclusion, this study will help in the implementation of PV grid-connected powered buildings in new projects by addressing previous design challenges and providing data for architects and researchers on the system design and performance.

Keywords:solar energy, photovoltaics,grid-connected system,architectural integration,BIPV.

1. Introduction

Considering that the major component of greenhouse gases (GHGs) is carbon dioxide, there is a global concern about enhancing renewable energy especially solar energy to reduce building's carbon footprint. Solar energy total installed capacity at the end of 2015 was 227 GW with more than 50 GW installed just in 2015. Despite this growth only about 1.3% of the world's electricity is produced by PV [1].
At regional scale, efforts are made to develop solar energy solutions or to bridge the gap of limited acceptance from architects through sharing knowledge between architects and specialists of the photovoltaic sector. Surveys reveal that one of the main barriers to the spread of solar power systems is the high cost, perceived by architects and contractors, who are considered the main stakeholders. On the other hand, the drastic cost breakdown of solar energy over recent years has decreased system prices. Thus, it is essential to build trust of architects, investors, and financial stakeholders in solar power. The language how the technology components are presented physically and virtually is one of the main keys to this acceptance.
The study aims to develop better understanding of architects for designing solar PV grid-connected systems in architectural integrated buildings, that act as micro energy hubs consuming and generating energy with a reduced carbon footprint. Therefore, the objectives of this paper include;

- To explore the solar technologies and needs.
- To understand types of photovoltaics (PV) cells.
- To study the factors affecting PV energy yield, space requirements, and on-site conditions.
- To study architectural integration solutions of BIPV systems.
- To conclude a framework for architects regarding the factors affecting the design process.

The methodology of this research implies an inductive approach, through exploring the various design solutions and investigating the factors affecting it, focusing on the technical and physical aspects.

2. Photovoltaics Technology

Today, solar technologies have fulfilled various needs in buildings, ranging from passive technologies and active technologies (Figure 1.). In the case of active systems, solar energy can be transformed into heat using solar thermal collectors, or into electricity using photovoltaic (PV) modules [2]. Solar thermal collectors can provide domestic hot water (DHW), space heating, and recently transform solar heat into cold to fulfil building cooling needs; while PV modules can provide building cooling, light, and electricity needs [3].
Solar thermal collectors include two systemsdepending on the medium used for energy transformation; air collector systems, and hydraulic collector systems. Hydraulic collectors are more cost effective, and require more sophisticated technology. Furthermore, hydraulic collectors are divided into three types; glazed flat plate collectors, unglazed flat plate collectors, and evacuated tubes collectors [4]. However, the research only covers the photovoltaics technology in further detail.

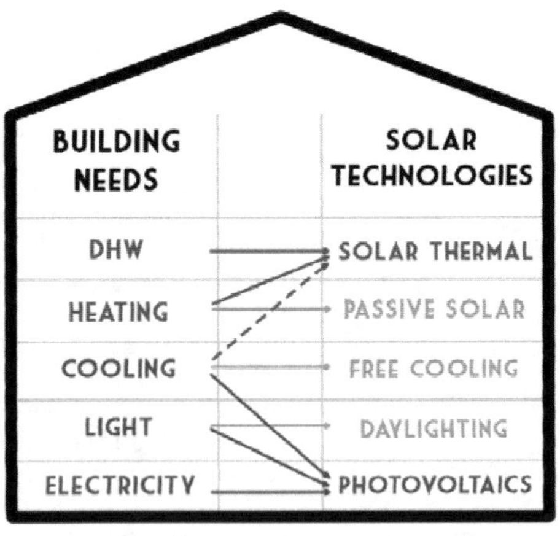

Figure 1: Different solar technologies covering different building energy needs [3]

2.1. PV cell types

A PV system is formed by a number of cells combining to make a module, and modules combining to make an array. Cells can differ according to their manufacturing process. Basically, they can be divided into three main categories; crystalline silicon cells, thin film cells (Table 1), and third generation cells [3]. The third category is still emerging with less sufficient data and results, while the first two categories are the main established technologies that the research would focus on.

Table 1. PV cells categories[2, 3].

Category	Crystalline silicon solar cells		Thin film cells
Types	Monocrystalline silicon	Polycrystalline silicon	• Amorphous or micromorphous silicon • Gallium- Arsenide (GaAs) • Cadmium-Telluride (CdTe) • Copper-Indium-Diselenide (CIS)
Photo Example			
Description	They are cut into thin wafers from a singular continuous crystal grown for this purpose, and has uniform colour, ranging from blue to black. The cells may be fully round or they may be trimmed into other shapes.	They are melted and poured into a mold. As the material cools, it crystallizes in an imperfect manner, forming random crystal boundaries. This forms a square block that can be cut into square wafers.	They have a low-cost potential because its manufacture requires only a small amount of material and less energy. They are available in range of colors.
Average production in European weather	900-1000 kWh per each kW installed	750-850 kWh per each kW installed	600-800 kWh per each kW installed

2.2. PV systems

Photovoltaic systems can be either stand-alone or grid-connected systems. Stand-alone/off-grid systems are more suitable for isolated or desert settlements without access to the government electricity grid [5]. They supply energy independently and thus, energy must be balances and energy surplus must be stored in batteries. On the other hand, grid-connected systems can be a very efficient solution that avoids costly batteries and reduce high losses, therefore, it is the simplest and cheapest solar power setup available [6].

In this system, electricity is generated using PV modules as direct current (DC), and as the electricity supply in the grid is alternating current, an inverter is used to convert DC to AC (Figure 2.). Whenever the electricity generation from PV modules does not cover the building consumption, the shortage is compensated from the public grid. However, if generation exceeds consumption, the excess energy goes to the grid [3],[6]. The study will focus on grid-connected systems performance and applications.

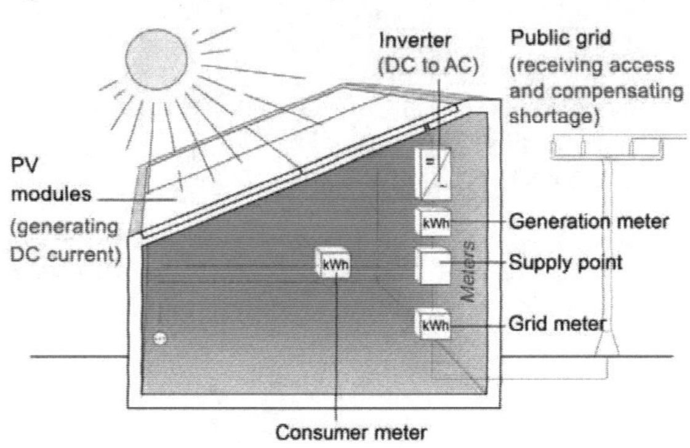

Figure 2: Scheme of a common grid-connected PV installation. [2] edited by researcher.

IOP Conf. Series: Materials Science and Engineering **974** (2020) 012020 doi:10.1088/1757-899X/974/1/012020

3. PV Performance

For designing a solar system, it is crucial to study its production and the energy required to fulfil the building needs, as well as the factors affecting the production and causing efficiency losses. The study of the technical side on the early design phase leads the architect to design the optimum conditions for the system to deploy the most out of the it.

3.1. Energy yield

The production of a PV system can be roughly estimated through a simple calculation in predesign stage, as followed [3];

- FinalYield = *G* [kWh/ m2]·*OrientationFactor*[%]·*Area*[m2]·*eff* [%]·*PR*[%]

***G*:** Global irradiation is the amount of solar energy on the horizontal surface. Irradiation maps can be found online, and they calculate the irradiation for specific locations as an annual long-term average (Figure 3.).
***OrientationFactor*:** The orientation factor is the effective radiation reaching the PV modules and is determined by the module's orientation.
***Area, eff*:** Cell efficiency depends on the technology of the cell types, and it is specified in the PV module datasheet provided by manufacturer. By multiplication of it with module area, the nominal power of the system is obtained.
***PR*:** The performance ratio of the systems depends on the ratio of losses due to other factors like type of installation, poor ventilation… etc.

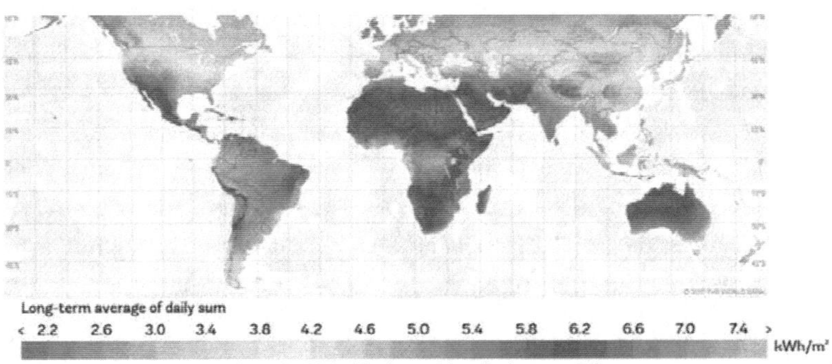

Figure 3: Global horizontal irradiation map [7]

3.2. Space requirements

The space required (Sr) to install 1 kWp PV system can be obtained with a simple calculation depending on cell efficiency, as follows [3];

- *Sr* [m2] =1/*eff*

IOP Conf. Series: Materials Science and Engineering **974** (2020) 012020 doi:10.1088/1757-899X/974/1/012020

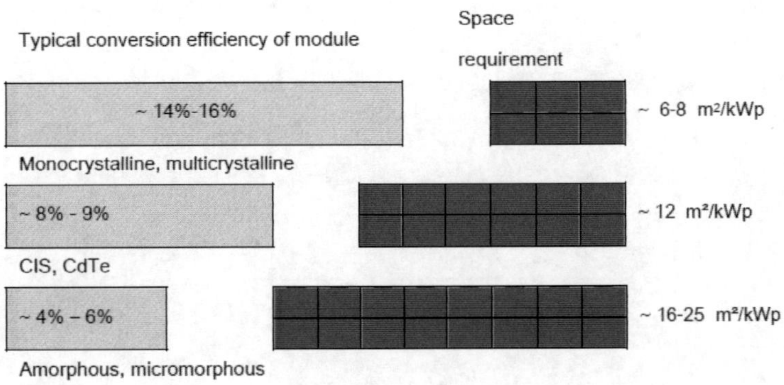

Figure 4: How cell efficiency affects the modules space required [3]

3.3. On-site considerations

3.3.1. Orientation and tilt. Solar power gain is affected by on the module orientation and tilt for each site. The best orientation (azimuth angle) is south facing at the northern hemisphere, and north facing at the southern hemisphere. While, the tilt (elevation angle) is influenced by the latitude, as it is found that the best module tilt is facing the equator with a tilt equal to the latitude angle [1]. Studies has showed that a system installed at the vertical façade gained an annual solar energy with 30% losses compared to a system installed at the altitude angle. While a system installed at horizontal angle losses were as high as 10%. [8]

3.3.2. Temperature dependency. Losses in the cell performance highly depend on the module temperature, especially mono- and multi-crystalline cells. Depending on the cell technology (Table 2.), each have a different temperature coefficient (%/°C); which is defined as *"the dependence of the electrical characteristics of PV on the operating temperature"*[1]. It indicates the percentage of efficiency loss for every degree Celsius temperature above STC conditions -Standard Test Conditions; a radiation of 1 kW/m2, a cell temperature of 25°C, and 1.5 air mass- (Figure 5.), and can be found at the module datasheet as well.

Table 2. Temperature coefficients for different photovoltaic technologies.[1]

Technology	Temperature coefficient (%/°C)
Mono-c-Si	-0.40
Multi-c-Si	-0.45
a-Si	-0.20
a-Si/μc Si	-0.26
CIGS	-0.36
CdTe	-0.25

IOP Conf. Series: Materials Science and Engineering **974** (2020) 012020 doi:10.1088/1757-899X/974/1/012020

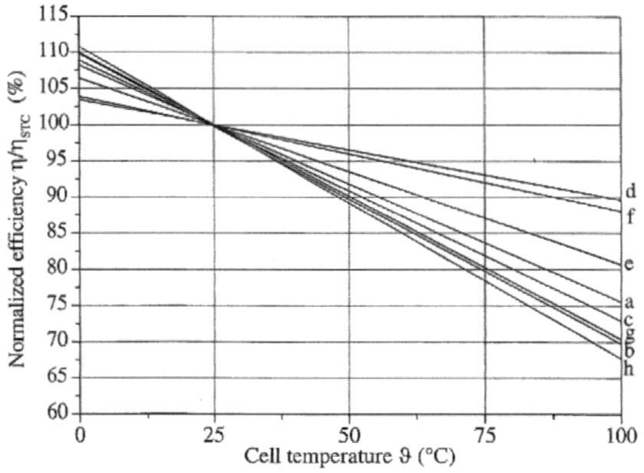

Figure 5: A representation of the influence of cell temperature on the normalized efficiency of the modules for different cell technologies. (a) high efficiency crystalline silicon; (b) monocrystalline silicon; (c) polycrystalline Silicon; (d) amorphous silicon; (e) "micromorph" tandem; (f) CdTe; (g)-(h) CIGS. [3]

3.3.3. Self-shading. Shading greatly affects solar PV production. Shading the bottom part of the module causes losing the entire module production as they are interconnected ribbons (Figure 6.- Left), while side shades only causes losing those particular cells production (Figure 6.- Right). [9]Strategies to minimize self-shading require careful design and placing of the modules (Figure 7.). The calculation of row inter-spacing can be simplified as follows;

- d = (h ÷ tan a)

d: minimum distance between rows
h: height differential between the top of the panel and the ground
a: solar altitude angle

Figure 6: Interconnect cell ribbons, (Left: side shading, Right: bottom shading). [10]

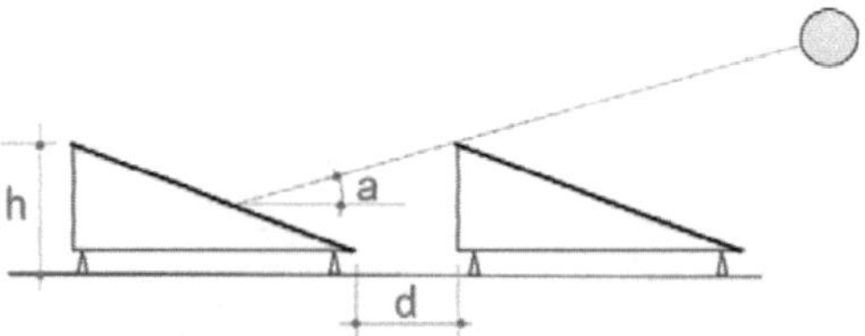

Figure 7: Design of modules inter-spacing. [9] edited by researcher

The solar altitude angle can be calculated using online software can be used, such as NOAA Solar Calculator (https://www.esrl.noaa.gov/gmd/grad/solcalc/) [10], using the following steps;

- Firstly, the location is determined by online map, or by entering latitude and longitude values,
- Also, the time zone, specific date and time are determined (Figure 8.),

- Eventually the software calculates the altitude angle for the time and area.

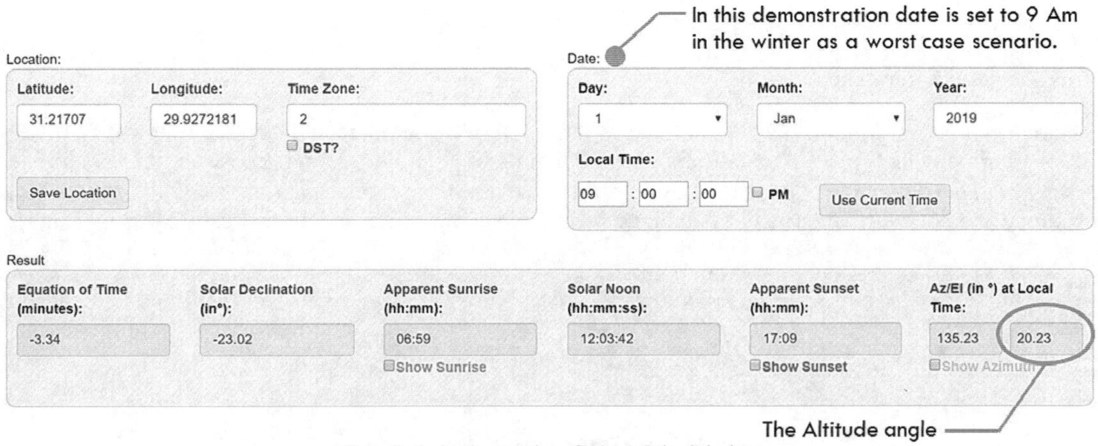

Figure 8: An implementation of NOAA Solar Calculator
Source: Researcher, 2019

3.3.4. Soiling and system losses. Soiling indicates dust accumulation above the surface of PV modules (Figure 9.). It is a factor of losses of the system performance depending on the dust type, time period since previous rainfall, and the cleansing schedule. To overcome soiling issues, studies are exploring solutions such as developing soiling sensors, cleaning strategies, and anti-soiling coatings.
A study that has taken place in Egypt compared three PV systems production; a clean PV module, a module that was exposed to dust for a year, and a module that was exposed to dust as well but was cleaned every two months. Results showed that the one-year soiled module and the two-months soiled module produced 35% and 25% less energy, respectively, when compared to the clean module. [1]

Figure 9: Interconnect cell ribbons, (Left: side shading, Right: bottom shading). [1]

Other system losses can be caused by inverters, wires, glass reflection, snow, deviation from STC, …etc. (Table 3.). All these parameters have an influence on the final Performance Ratio (PR) of the system. [3]

Table 3. Common losses percentage [3]

Common losses factors	Mean losses
Glass reflection losses	2 – 4%
Deviation from STC	2 – 4%
Temperature effect	3 – 6%
Snow. Dust/soiling	1 – 2%
Shadows	≥ 0%
Tolerance and mismatching	2%

Inverter losses	≥ 5%
Cables and line losses	1 – 2%

4. Architectural integration

The architectural integration quality is defined as the result of a controlled and coherent integration of the solar system simultaneously from all architectural points of view, functional, constructive, and formal (aesthetic) [3]. This implies that the PV system that would replace a building element shall carry out its functional and constructive quality, while being be in good composition and aesthetically pleasing (Figure 10.).

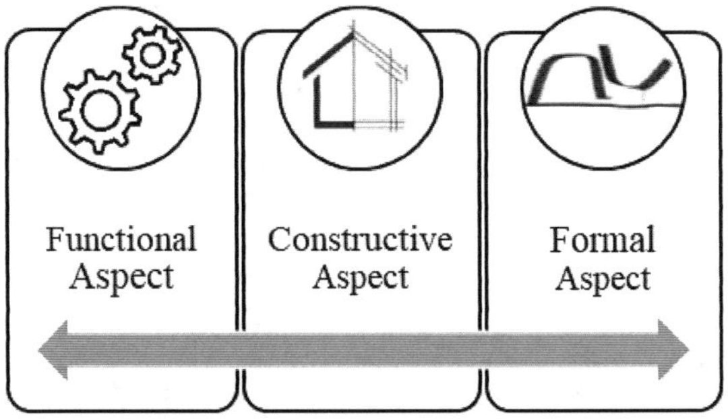

Figure 10. Architectural integration aspects.
Source: Researcher, 2019

4.1. Integration terminology

- Building added photovoltaics (BAPV); In this type the modules are only added on top of the building envelope without replacing any building elements. Therefore, there are no savings of materials nor cost, but furthermore, additional mounting systems are used. [2]
- Building integrated photovoltaics (BIPV); Photovoltaics modules are integrated within the building envelope to be used as an architectural element and for energy generation. This could result in savings when the cost of the PV elements is below the traditional elements. [2]

There are various advantages of integrating photovoltaics into buildings [11];

- No need for additional land, as PV is installed on a part of the building, which is optimum for dense cities.
- The cost of PV elements is offset against the cost of the replaced building elements.
- Avoiding losses during distribution as the power is generated on-site.
- Architectural elegant, which helps increasing the market acceptance (Figure 11.).

IOP Conf. Series: Materials Science and Engineering **974** (2020) 012020 doi:10.1088/1757-899X/974/1/012020

Figure 11: Inclined PV integrated glazed façade of Vocational College in Tirol, Austria [12]

4.2. BIPV solutions

In case of BIPV, the relationship between the system and the building envelope can be categorized into three main categories (Table 4.); roof, façade, and external elements solutions. Roofs usually are the most efficient solution to achieve the optimum angle. They can be tilted roofs, flat roofs, or skylights. Façade solutions can be in the form of cladding or glazing PV panel. Furthermore, external elements can take the form of shading canopies and louvers, or the form of railing and parapet systems.

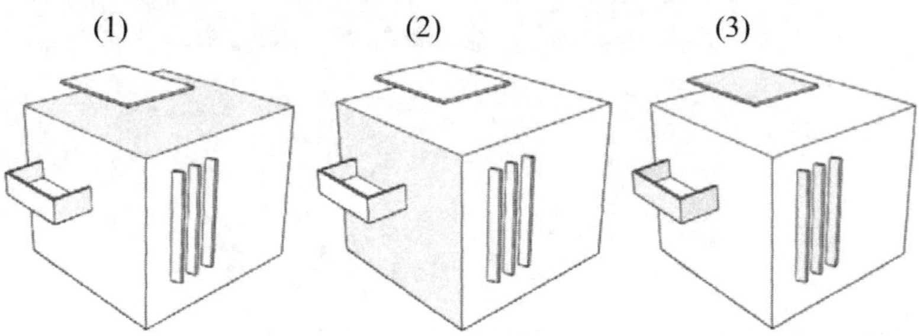

Figure 12: PV integration within the building envelope. (1) Roof solutions. (2) Façade solutions. (3) External devices solutions. Source: Researcher, 2019

Table 4. Solutions of PV integration within the building envelope[2, 3]

Category	Solution	Description

(1) Roof	Flat roof modules	Special racks might be used to create tilt angles. Crystalline modules may also be used on plastic substrates and with adhesive backing. Another trend is using thin film laminates above waterproof membranes.  Crystalline modules for flat roof. Left: standard module on rack mounting system. Right: special rack system for thin film cells on stainless steel substrate.  Integration of PV laminates on plastic substrates in flat roof. Left: mono-crystalline modules. Right: thin film modules.
	Tilted roof tiles	There have been many products developed for roof tiles, shingles, and slates; using both crystalline and thin film cell technologies. They can match traditional products with great aesthetics.  Crystalline tilted roof systems  Thin film tilted roof systems
	Skylights	In case of skylights, PV modules could be the only element covering the roof. Mostly, semi-translucent crystalline modules are custom made so the architect must ask for a simulation or a test about the performance. While for thin films, translucent modules are standard with detailed datasheets available.  Semi-translucent skylights. Left: crystalline modules. Right: thin film modules.
(2) Facade	Cladding	PV modules can be used as cladding installed on a load-bearing wall. The modules are usually back ventilated to avoid efficiency loss due to temperature. Both crystalline and thin films can be used, with various fastening systems. 

		Façade cladding solutions
	Glazing	Translucent PV modules could be integrated into curtain wall systems; post-beam structure, structural glazing, and spider system. It also can be used for the outer layer of double façades. PV glazing solutions can greatly affect indoor climate, and can transfer heat into the space. Therefore, products with good thermal insulation have been developed with low-E value, which could be used as double or triple glazing elements.  Façade glazing solutions. Left: Translucent thin film module curtain wall. Middle: Frameless modules with spider glazing system. Right: Double glazing module with thermal insulation.
(3) External devices	Shading devices	PV modules can be used as sun shades in a form of canopies, louvers, or movable shutters. They can be semi-translucent crystalline or thin film modules. Also, opaque modules are used, which require that the upper part is made without cells to avoid shading.  Shading devices solutions. Left: sliding shutters. Right: shading louvers.
	Railings	PV can be integrated in railing and parapet systems as well. There are mostly semi-translucent modules that are made with security glass (SSG).  Railing and parapet solutions

4.3. Physical characteristics of PV cells

For the design and implementation of PV technology in buildings, it is important to take the aesthetic value into consideration. This requires from the architect to harmonise the physical and visual characteristics of the PV modules with the architectural language in the building [13]. Various products have been developed for the sake of integration within the buildings with many physical properties that provide variety in solutions. It is essential to explore these properties which include; colour, texture/pattern, translucency/reflection, and flexibility.

Table 5. Physical properties in BIPV and their characteristics

Property	**Characteristics**

Colour	There are many colour variations for PV modules. Polycrystalline cells are usually black or dark blue, but other colour variations can be produced by changing the thickness of the anti-reflective coating [13]. Darker cell colours maximize sunlight absorption and are the highest in efficiency, so the change in colours must compromise the efficiency [14]. Thin film cells provide various ranges of colours as well. Amorphous silicon cells or CIS are usually available in black, grey, or dark brown, while CdTe cells have greenish shades [15].  Coloured crystalline silicon cells manufactured by Solartec[13]
Texture and pattern	The cells texture varies depending on the technology they are made of. Monocrystalline cells have a solid finish and a homogenous texture, while polycrystalline cells have a marble-like texture and multifaceted reflections [18]. Variations of patterns of polycrystalline cells can be produced by the grey metallic grid which collects the produced energy [13]. Amorphous silicon cells mostly have a homogenous texture with fine lines, but it is also possible to produce semi-transparent cells with a variety of patterns by laser cutting (point, stroke, stripes). 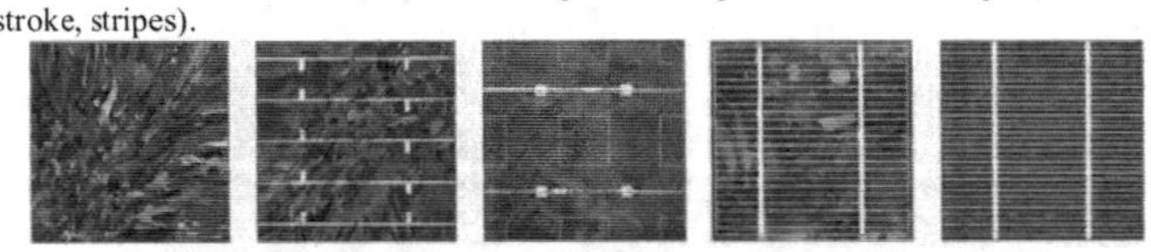 Variations of the pattern of metallic grid on polycrystalline cells [3] 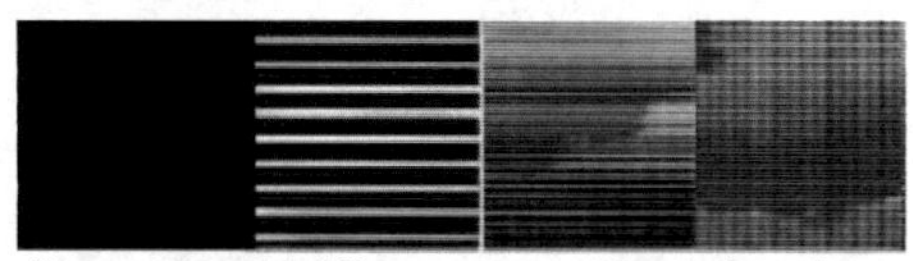 Thin film modules with different pattern and level of translucency [3]
Translucency and reflection	Both crystalline silicon cells and thin film cells can be produced as translucent or semi-translucent cells [16]. There is a variety of products in the market with different transparency rates such as 10%, 20% and 30%. These products maximize sunlight indoor spaces and avoid infrared radiation [6]. The module cover is commonly made of glass, which has a reflective surface causing annoying glare. Some products overcome this by sandblasting the glass cover creating a matt finish [15].  Semi-transparent c-Si cells with 22% transparency [17]
Flexibility	Thin film technologies provide flexible solar cells which creates PV foil as thin and lightweight building element for architectural integration [18]. The flexible CIGS and CdTe cells can be laminated to the building elements as the flat roof membranes or tiles without adding an extra weight which highly reduces the cost and extend the opportunities for new PV concepts and applications [19].  Flexible thin film solar cells integrated in a curved rooftop [20]

5. Conclusion and recommendations

The early study of the design process and possibilities of the photovoltaics system has a huge effecton avoiding future challenges and achieving a cost-effective system. The research investigated the technical and physical aspects of the design phase that would be affected by the architectural decisions. The paper concluded four design determents that shall be considered by the architect during the design of the solar system which are; cell type (technical), the position of the modules in the building, the integration solution, and cell type from a physical/ visual aspect. Figure 13. illustrates a framework for the design determents of PV grid-connected solar systems.

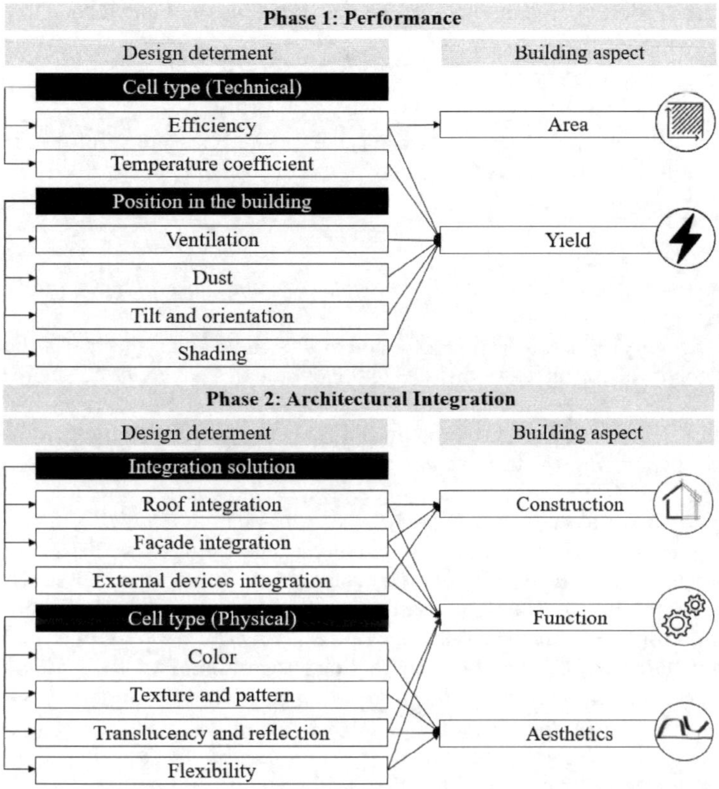

Figure 13: A framework for the design determents of PV grid-connected solar systems
Source: Researcher, 2019

Phase 1 concludes the design determents by the technical characteristics of the solar system. The first determent is the technical cell type, where the architect choose the optimum efficiency and temperature coefficient that affects the facing building aspects. The second determent is the position of the solar system in the building, where the architect studies ventilation, dust, tilt, orientation, and shading results.The building aspects that are affected in phase 1 are the area of the building space needed to place the system, and the final yield produced by the system to fulfil the building needs.

On the other side, phase 2 concludes the design determents by architectural integration of the solar system. The first determent is the integration solution, where the architect determines the relationship between the system and the building envelope; roof solutions, façade solutions, or external devices solutions. The second determent is the cell type in terms of physical and visual characteristics; where the architect chooses the cell that suits the desired design in terms of colour, texture and pattern, translucency and reflectivity, and flexibility. The building aspects that are affected in phase 2 are construction, function, and aesthetics.

The research concludes some recommendations for architects, as follows:

- Starting to direct awareness towards PV grid-connected solar systems for proving to be the best cost-effective and less complicated system.
- Making preliminary estimated building electrical need analysis using simplified equations or software, and estimating the required area in early design phase.
- Studying the precise location of the PV system and analysing various alternatives to check on ventilation, dust, solar angle, and shading of the panels.
- Reviewing the products in the market with various efficiencies to make a cost-effective analysis.

IOP Conf. Series: Materials Science and Engineering **974** (2020) 012020 doi:10.1088/1757-899X/974/1/012020

- Reviewing the products in the market that are developed for roof and façade integration, as well as external devices products.
- Experimenting with different design solutions for asuitable architectural quality in terms of colour, texture and pattern, translucency and reflectivity, and flexibility.
- Co-operating with manufacturers regarding the desired architectural needs and visions for developing more state-of-art products.

6.References

1. Kalogirou, S. *McEvoy's handbook of photovoltaics: fundamentals and applications*; Academic Press: **2017**.
2. Basnet, A. Architectural integration of photovoltaic and solar thermal collector systems into buildings. Faculty of Architecture and Design, Norwegian University of Science and Technology, **2012**.
3. Munari Probst, M.C.; Roecker, C.; Frontini, F.; Scognamiglio, A.; Farkas, K.; Maturi, L.; Zanetti, I. *Solar Energy Systems in Architecture-integration criteria and guidelines*; Solar Heating and Cooling Programme: International Energy Agency, **2013**.
4. Munari Probst, M.C.; Roecker, C. *Designing solar thermal systems for architectural integration*; Solar Heating and Cooling Programme: International Energy Agency, **2013**.
5. Dabaieh, M.; Johansson, E. Building Performance and Post Occupancy Evaluation for an off-grid low carbon and solar PV plus-energy powered building. A case from the Western Desert in Egypt. *Journal of Building Engineering* **2018**, *18*, 418-428.
6. Achenza, M.; Desogus, G. Guidelines on building integration of photovoltaic in the Mediterranean area. *ENPI CBC Mediterranean Sea Basin Programme* **2013**.
7. Global solar atlas. *World Bank Group [US]* **2018**, *9*.
8. Eiffert, P.; Kiss, G.J. *Building-integrated photovoltaic designs for commercial and institutional structures: a sourcebook for architects*; DIANE Publishing: **2000**.
9. U.S. Department of Energy. Architectural Integration of Solar PV into Building Design.
10. NOAA Solar Calculator. Availabe online: https://www.esrl.noaa.gov/gmd/grad/solcalc/(accessed on 30/12/2019).
11. Reijenga, T.H.; Kaan, H.F. PV in Architecture. *Handbook of Photovoltaic Science and Engineering, second ed., John Wiley & Sons, Chichester, UK* **2011**, 1043-1077.
12. Solarfassade. Imst vocational school. Availabe online: http://www.solarfassade.info/en/project_examples/at/imst_vocational_school.php(accessed on 21/11/2019).
13. i Florensa, R.S.; Cueva, R.L. Architectural Integration of Solar Cells. In *Practical Handbook of Photovoltaics*, Elsevier: **2012**; pp. 917-941.
14. Solarfassade. Format and colour. Availabe online: http://www.solarfassade.info/en/fundamentals/components/format_colors.php(accessed on 16/1/2020).
15. Hermannsdörfer, I.; Rüb, C. *Solar design*; **2005**.
16. Jelle, B.P.; Breivik, C. State-of-the-art building integrated photovoltaics. *Energy Procedia* **2012**, *20*, 68-77.
17. Heinstein, P.; Ballif, C.; Perret-Aebi, L.-E. Building Integrated Photovoltaics (BIPV): Review, Potentials, Barriers and Myths. *Green* **2013**, *3*, doi:10.1515/green-2013-0020.
18. Pagliaro, M.; Ciriminna, R.; Palmisano, G. Flexible Solar Cells Vol. *ChemSusChem* **2008**, *1*, 880-891, doi:10.1002/cssc.200800127.
19. Buecheler, S.; Chirila, A.; Perrenoud, J.; Kranz, L.; Gretener, C.; Blösch, P.; Pianezzi, F.; Seyrling, S.; Tiwari, A. Flexible and lightweight solar modules for new concepts in building integrated photovoltaics. *Proceedings of the CISBAT, Lausanne, Switzerland* **2011**, 14-16.
20. Sinapis, K.; van den Donker, M. *SEAC BIPV Report 2013: state of the art in building integrated photovoltaics*; **2013**.

MilitaryTechnicalCollege
Kobry El-Kobbah,
Cairo, Egypt

13th International Conference
on Civil and Architecture
Engineering
ICCAE-13-2020

Comparison of egyptian innovation in affordable housing with global models.

Hebatalrahman [1]

[1]Dr.eng. Consultant in material Science, Egypt.

Corresponding author: hebatalrahman11@yahoo.com & hebatalrahman11@gmail.com

In these work Survey about different types of affordable housing was dond with the advantages and disadvantages of each type. The combtability of each type with sustainable in construction according to national and international standards was considered. The risk assessment related to this type of construction was evaluated. The Egyptian innovation in the affordable housing which is registered as patent in the egyptian patent office is studied. The new building has one floor with spherical external shape for energy conservation , it is consists of multilayers movable walls moving on the streams. The metalic structure is installed in the soil surface by fixation elements, it is equipped with sensors which is controlled automatically to detect climate changes. the movement of successive layers can be controlled manual or automatic according to climate changes, The movement of insulated layers is done by sliding on dedicated channels, It is equipped with upper and lower ducts located on several axes positioned at the top and bottom of the ball so that the upper and lower positions are located on one axis. The unit has conical top contain the layers folded to the top and prevent the accumulation of dust and water. The layers can be stored inside the conical top. The Egyptian patent model is compared with the common global models of affordable housing to explain its advantages and disadvantages

Keywords: Affordable Housing, Risk, Sustainable Building, Innovation

IOP Conf. Series: Materials Science and Engineering **974** (2020) 012021 doi:10.1088/1757-899X/974/1/012021

1. Introduction

Affordable housing is a set of building models which is considered as type of low-income housing primarily by providing residential products that are within their reach and difficult living conditions. The models of affordable housing are often linked to rapid and constructive housing in the face of disasters.

Recent studies reveal that green buildings with good overall environmental quality can reduce the rate of respiratory disease and enhance worker performance,(1). The green buildings measures include building shape and orientation, passive solar design, and the use of natural lighting(2),(3). **The conservation of enery is main topic ingreen buildings design and codes. Affordable housing** is defined as housing which is deemed affordable to those with a median household income or below[1] as rated by the national government or a local government by a recognized housing affordability index. In other words housing for poor people specially incase of emergency. The hard and critical question in the construction field is how to achieve the balance between affordable housing and green buildings. Figure (1) shows the main elements in green building and affordable housing.

Figure (1) The main elements in green building and affordable housing

Technical problems with affordable housing constructions

1.Construction time is relatively long and not appropriate for emergency situations and disaster situations
2.The insulation layers in the buildings are fixed and do not suit climate change and temperature difference between different seasons of the year
3. Energy conservation increase in buildings have sharp edges. Figure (2) shows the rate of energy conservation according to the building shape.
4.affordable housing with conventional techniques is relatively expensive and take long time so green buildings in the future technique for thermal comfort in low cost energy-efficient buildings.

IOP Conf. Series: Materials Science and Engineering **974** (2020) 012021 doi:10.1088/1757-899X/974/1/012021

5.All Green building codes and techniques which are environmentally compatible are very complicated and expensive.

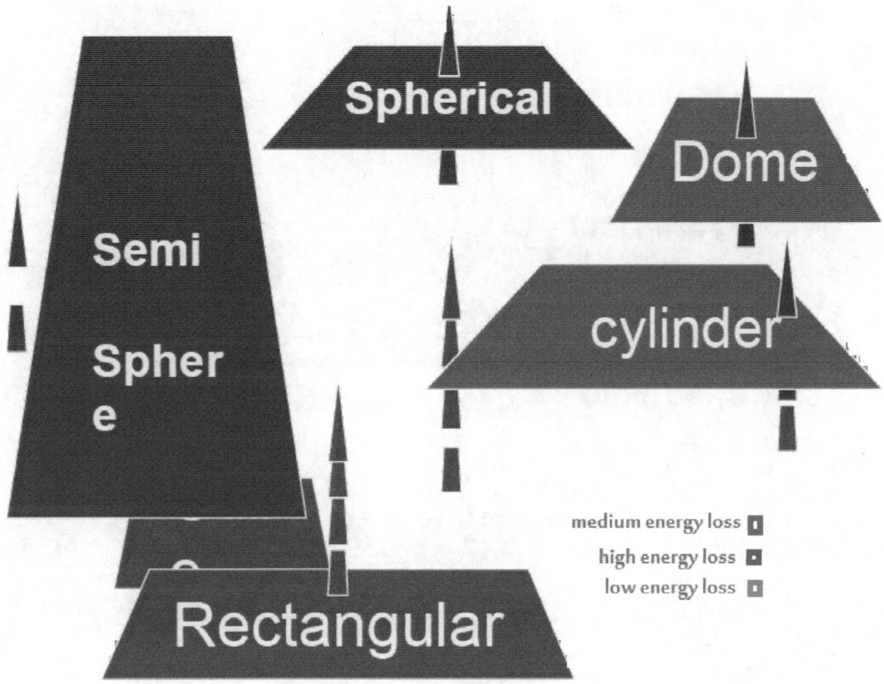

Figure (2) the rate of energy conservation according in different building shapes.

6.The conventional building consists of only one layer and fixed walls that lack the ability to cope with climate changes

7.Prefabricated buildings are suitable only for very limited types of soil

8.Connections of electricity, water, sanitation and communications are part of the building and are adjacent to the walls, which causes the building to be rapidly damaged and difficult to maintain

9.Insulation layers , doors, windows , coatings and protections are connected to the main building body.

2. Methodology

2.1 Egyptian innovation technique in affordable housing

The Egyptian green buildin for affordable and low income housing is patent registered and granted from the Egyptian patent office. patent number is 27118/2015. The invention is a design for building element in the form of incomplete sphere, The design save energy. It consists of multiple layers on moving streams equipped with sensors, The movement of the working layers were controlled automatically or manually according to climate changes in the surrounding atmosphere. The top and bottom of the layers are installed on several axes centered together. The movement of insulated layers are done by sliding on the certain channels. Figure (3) The section elevation for main elements in the building.

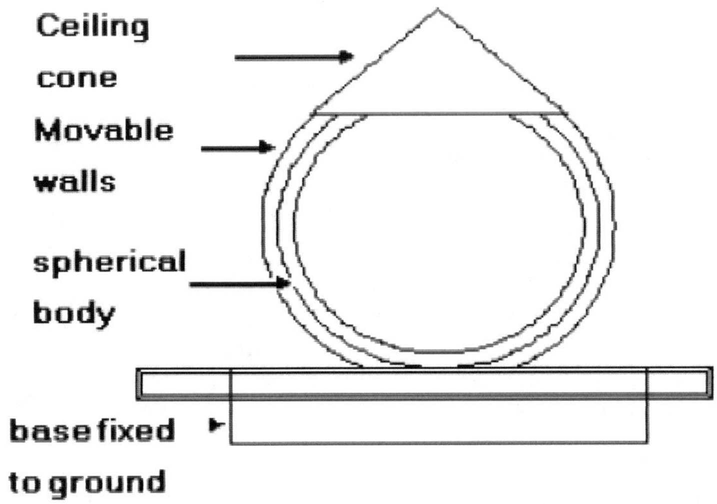

Figure (3) The main elements in the building

2.2 Green building design

In this work green building is designed in spherical shape with moving walls. figure (4) and Figure (5) show the main construction of the new green building . Figure (4) shows the exterior shape of the building, while Figure (5) shows a building block showing the general structure which are a general shape of a semi-spherical structure

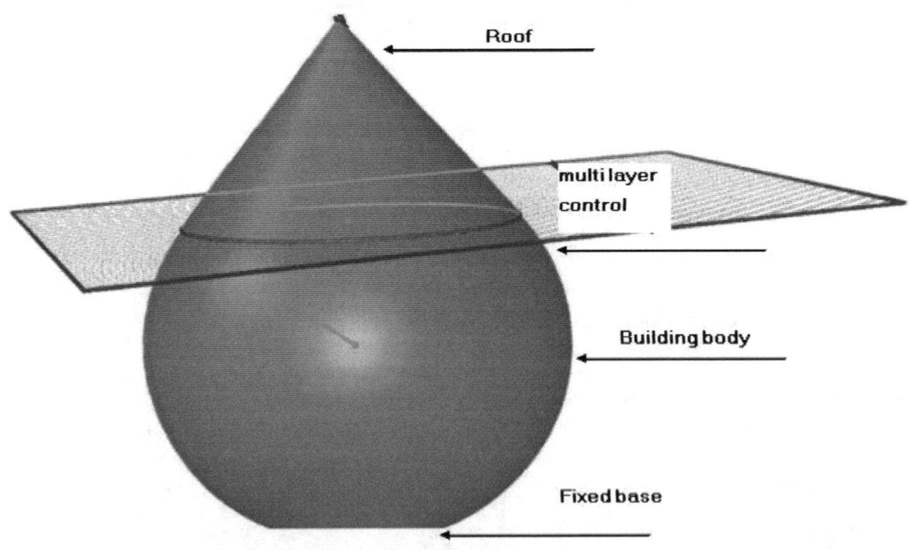

Figure (4) 3D view of the building

IOP Conf. Series: Materials Science and Engineering **974** (2020) 012021 doi:10.1088/1757-899X/974/1/012021

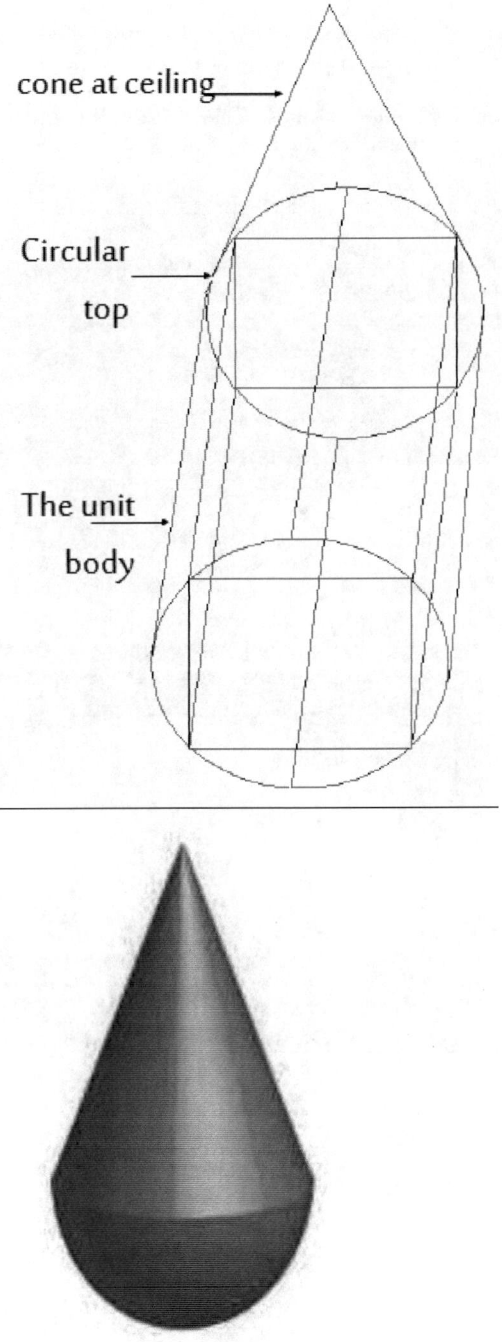

Figure (5) The main construction of the new green building

IOP Conf. Series: Materials Science and Engineering **974** (2020) 012021 doi:10.1088/1757-899X/974/1/012021

2.3 Building shape

The shape is in the form of a great ball, the ball is not completely circular. The circular shape is the best engineering shape can preserves energy and prevents leakage, it is also contains the lowest heat transfer rates. The building shape is free from any right angles,it is achieve ideal distribution of energy inside the building. Figure (6) shows the energy distribution inside the structure .

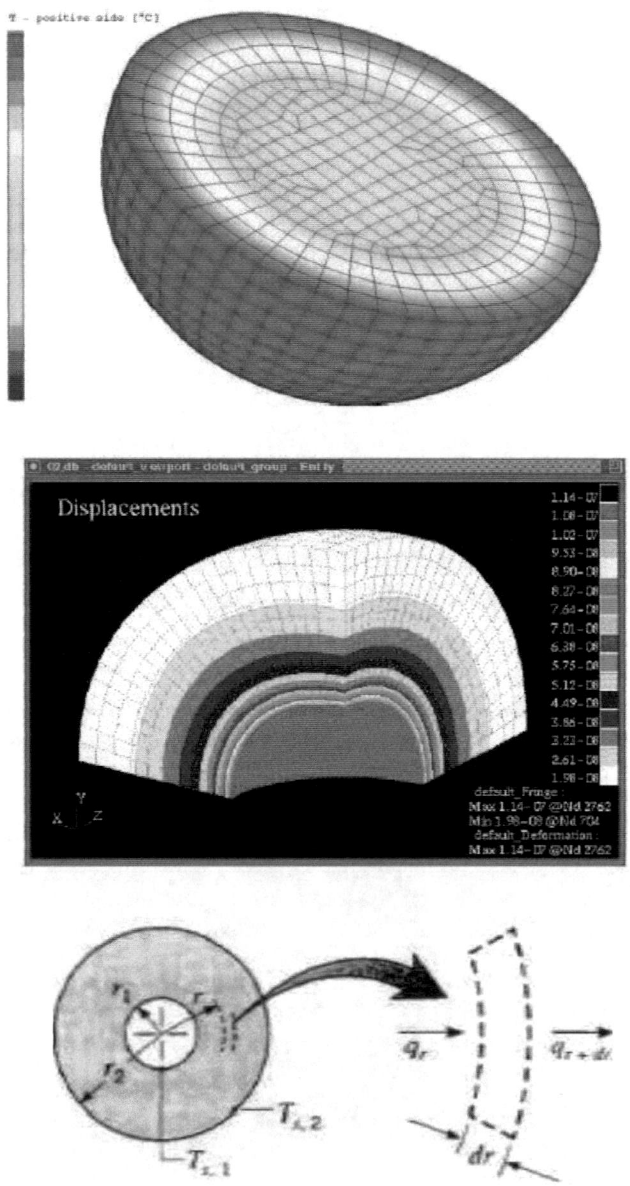

Figure (6) The energy distribution inside the structure

The new building consists of the following main parts:-

1. The conical top
2.The upper cavity
3.The lower cavity
4.The transparent layer
5. Insulating layers
6. Install the unit
7. The upper base
8. Eight fixing bars
9. A layer of heat transfer insulation
10. A layer to achieve acoustic insulation
11. A layer of reinforced plastic with additives to achieve light weight and integration of building elements.
12. An additional layer when isolating lead or radiation for special applications
13. Side places where the layers are combined
14. Bottom base
15. Laying down the layers while folding
16. The window that penetrates the insulation layers

Conical top to prevent the accumulation of dust and water on the roof surface, The upper cavity is regarded as a suspended ceiling for lighting, air conditioning, and other facilities. the upper cavity is used to store the layers of the walls for long periods by pushing the layer (body) up [26]. The upper cavity follows the upper base (7), which contains interlocking circular rings with variable diameters, where the diagonal spacing depends on the required distance between the layers of the movable and folding walls. Each ring fixed to the upper base corresponds to a ring at the bottom base (14)

The lower cavity is considered as raised floor containing cables, water pipes and electricity connections. The upper base is connected to the roof of the building adjacent to the base of the conical top. Fixation of the unit in the soil is done by means of metal columns and anchors, The eight fixation coloums extend from the conical top to under ground to fix the building in the subsurface of the soil to achieve stabilization of the building.

The Egyptian patent in affordable housing consists of Spherical structure with conical tops and movable multi-layered walls. The outer layer is transparent layer to allow penetration of sunlight. The Insulating Layers is multi layes all of them are movable and separated from each others. Figure (7) shows the plain of the different wall layera and the windows. One of the building layers can be used and the other layer is folded through side places . In the side places the layers (13) are collected and fixed from the top and from the bottom in the bottom base (14), cross-straps are installed by valves to control the fastening of the layers during the folding (15) and their lengths can be controlled. the size of the folded part and its height can be also adjusted.

IOP Conf. Series: Materials Science and Engineering **974** (2020) 012021 doi:10.1088/1757-899X/974/1/012021

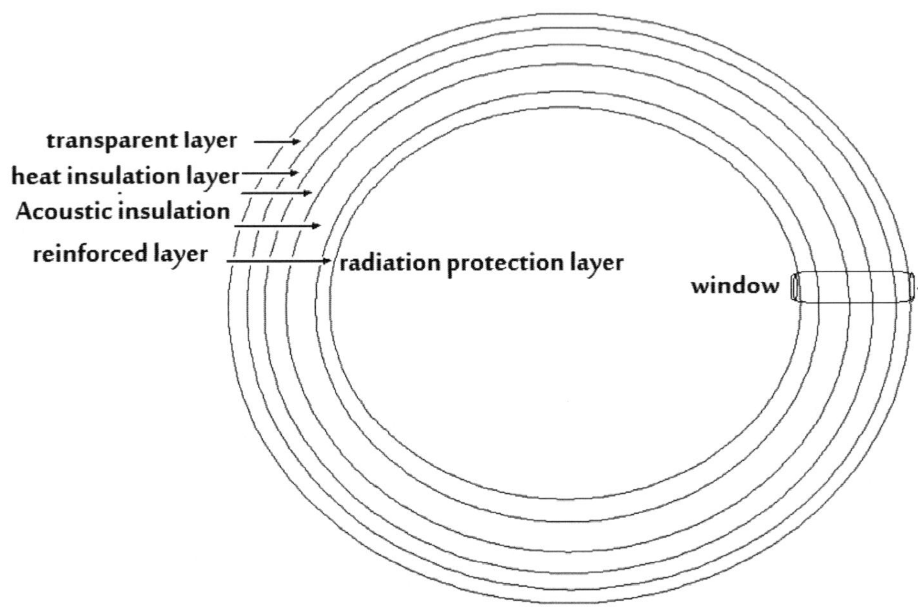

Figure (7) The different wall layers

Building walls are consists of several layers completely separated from each other, it is serves as completely separate building units, the arrangement of the layers from the outside to the inside is as follows:

A. layer of heat transfer insulation
B. layer to achieve acoustic insulation of the sintered felt on both sides
C. layer of protection from reinforced plastic with additives to achieve light weight and integration of building elements.
D. protection layer against radiation and Bullets for special applications
E. The windows are penetrates the insulation layers
F. The bottom base is the interior floor of the building

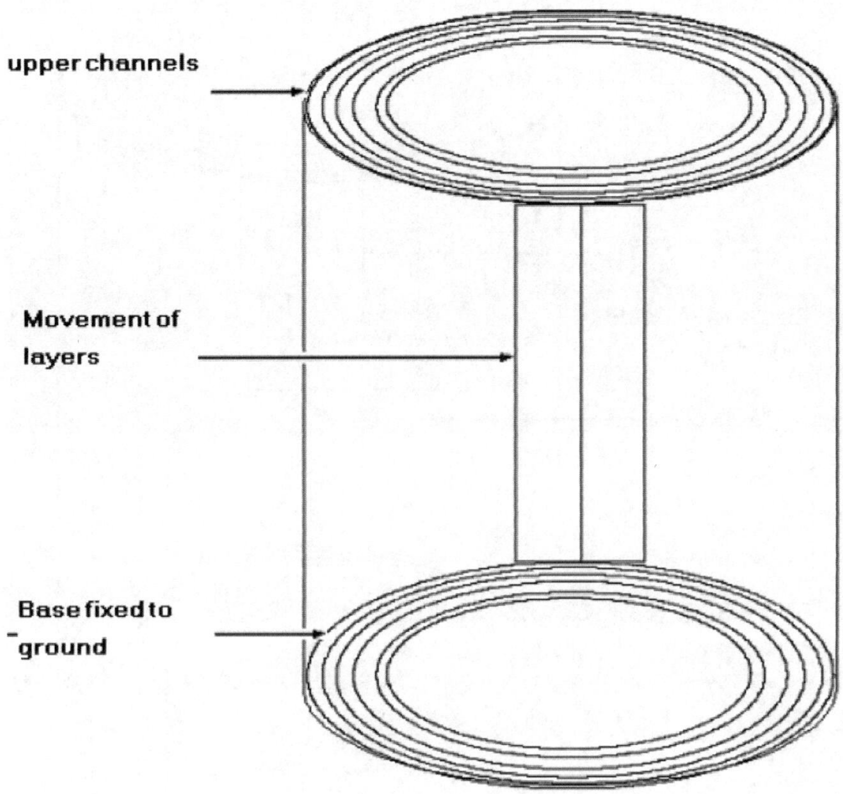

Figure (8) The movement mechanism of the different layers

Side places where the layers are combined, allowing the unused layers to fold, Elements for fastening elements while folding. Figure (8) shows the movement mechanism of the different layers relative to each others, while figure (9) shows the folding of layers and fixation techniques. The insulating layers (5) are attached to the fixed rings, each layer is installed to the upper and lower rings to fix the ducts moving of the circular layers, it is maintain their ability to slip and control the folded or visible amount of circular layer. The new design in spherical shape have the best energy distribution to achieve thermal comfort. The unit also contains an anchor base (6) topped by a lower cavity (3) and adjacent to the surface of the soil, the lower base (14) is considered as raised floors for construction, it is allow all ducts and cables such as electricity connection computer cables,web wires water and sewage pipes to connected from the base of the building to outside sources vice versa

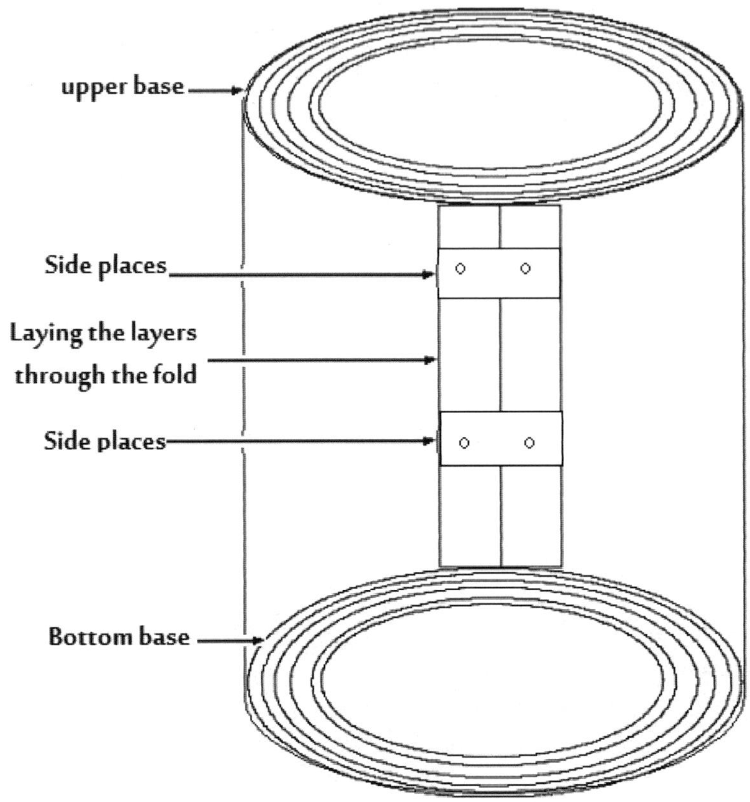

Figure (9) The folding of layers and fixation techniques.

the body of the internal building is almost completely isolated from the external climate. The movement of the successive layers is done manually or automatically with mechanical system connected to electrical circuit with remote control 26.] [

The building width can be controlled by increasing or reducing layer size and movement ,it is done by controlling the separation between the successive and second layers by folding some of the layers or storing them in the upper cavity (2), each of the previous layers forms a unit in itself closed from both sides with a wooden or metal casing to assemble the layers when folded and not needed.

2.4 Building installation method

The building can be moved from place to place by loosening the fixing screws represented by the eight (8) fixing columns and reinstalling them, Figure (6) showing the mounting columns and conical tops.
The installation columns are distributed regularly over the area of the building, three converging and concentrated columns are located below the conical top, while the other columns are distributed regularly on the outer circumference and along the body of the building, it is in the distances between the insulating layers (4), the columns are fixed centrally between the upper base (7) And the bottom base (14) by fastening nuts.

The distance between the two bases is about (2/5)the total length of the fastening columns, which constitute the exploited height of the building , the columns are extend up to penetrate the conical top (1) with a length of about (1/5) the total length, it is penetrates the bottom base (14) and the socket Bottom (3) so that it is about (2/5) the total length of the stabilization columns below the soil surface.

2.5 The internal division of the building

It is done through metal cutters that are fixed in the upper base (7) and the lower base (14) and the metal cutters are provided with doors to facilitate access between different regions inside the building , it is also control the rate of their division.

2.6 External entry door

The door is in the form of a metal plate followed by a panel of tent fabrics, it is fixed and wrapped on a cylinder, the cylinder is fixed on its sides in the upper base , the door is opened and closed by rotating the cylinder and wrapping the fabric on it , the bottom end of the door contains removable fastening rings that attach easily to the bottom base, when it removed the door move automatically up.

2.7 Windows

The window penetrates the insulation layers, it is a cylindrical pipe that can be opened and closed from both sides (16) The pipe can be pulled inside to return the insulation layers ti its original place. Each temporary opening can be covered with a layer of the same type of insulation present in the layer.

3. Results

3.1 Distinctive elements of the design

The design is distinguished by the following topics:-

1. The positions of the slots are variable due to the relative movements of the successive layers,(movement up and side movements).
2. Ease of disassembling and installation is done any number of times.
3. The quality and kind of the soil does not affect the movement process it is affected only by the depth of the plug and the mechanics of fixation
4. The elements of protection against thermal and acoustic insulation vary according to the movement of the sun during the day or during different periods of the year.
5. The semi spherical design is ensuring energy conservation by folding, changing, or storing of different the layers.
6. The joints of electricity , water, drainers and internet cables are located in special layers away from the body of the building .
7. The joints are connected to the ceiling and the floor, that is increases its life span and facilitates the dismantling, installation, and maintenance.
8. The Egyptian model is considered as the cheapest green building because It does not need any cement, sand or grit for building operations.
9. The Egyptian model does not need any advanced techniques to implement it.
10. The building consists of several moving layers each layer suitable for specific climatic or environmental conditions.
11. The building is equipped with sensors to detect external and internal light and heat change to choose the appropriate layer.
12. The process of changing the insulating layers can be controlled manually or automatically.
13. The building is equipped with layers of protection against risks such as electrical hazards and lightning strikes
14. Multipurpose design, easy to disassemble, install and move around.

3.2 Applications of the Egyptian model in affordable housing:-

1. Rapid construction in case of relief, crisis and disaster.
2. Industrial and domestic uses such as setting up industrial parks.
3. Construction for energy efficient purposes.
4. The egyptian model is multipurpose design which is easy to disassemble and install

IOP Conf. Series: Materials Science and Engineering **974** (2020) 012021 doi:10.1088/1757-899X/974/1/012021

5. The egyptian model is ease to move around from place to another places.
6. The egyptian model is ease and speed of construction and
7. The egyptian model has the ability to add any number of additional units because the units are easy to dismount, install and pre-equip
8. The egyptian model is provide protection against external hazards and enviromental factors. Figure (10) shows the prototype of the Egyptian patent and its main parts.

Figure (10) The prototype of Egyptian patent

4. Conclusion & recommendation

❖ In view of the importance of acquiring suitable housing for all segments of society, especially groups that are limited income or non-intervention or that are subjected to forced displacement,

❖ it is necessary to develop structural and construction systems that meet the needs of Arab societies for relief construction in disaster situations.

❖ Affordable housing design must be developed to resist the expected climate changes that It will lead to many environmental and natural disasters.

❖ **The Egyptian model that has been patented was invented and is characterized by the following- :**

1. The Egyptian model is a one-story building that has a metal structure and is easy to dismantle and install and can be moved from one place to another so it is a unique model for rapid construction work
2. The Egyptian model achieves cost savings in terms of construction and operation with the possibility of reuse and ease of storage
3. Achieving sustainability in the Egyptian model by linking housing design and design elements that lead to sustainability
4. The multiplicity of walls in the Egyptian model, and its ease of folding or expansion according to the external climate are the main factor achieve thermal comfort

5. The Egyptian model provide the energy depleted in its operation, which would achieve its sustainability.
6. The internal divisions of the Egyptian model are flexible and adapt to the needs of the user
7. The method is simple, inexpensive and environmentally compatible
8. The Egyptian model is in the form of a sphere in order to reduce the excess in internal energy loss.
9. The Egyptian model is asily change buildinge function without any cost or demolish by re-arranging layers.

Recommendations

Making industrial scale prototype for Egyptian patent number **2016-(27668)** in order to measure the following parameters:-

- o Thermal comfort during day and night hours
- o Thermal and acoustic insulation rate
- o Different parameters change during different periods of the year
- o Calculation of economical feasibility study
- o Study the compatibilty of the new egyptian patent with the enviroment in the arab region
- o Development of an Arabic code for unconventional building systems that support affordable housing
- o Mention the required developments to make the egyptian patent suitable for different regions with different climate

5. References

1. Hebatalrahman (Method for creating a spherical building unit with sliding walls) patent nmber (27118), Egyptian patent office , 2015.
2. UN-HABITAT, "A Practical Guide for Conducting: Housing Profiles," United Nations Human Settlements Programme, Kenya, 2010.
3. UN-HABITAT , "Affordable land and housing in Africa," United Nations Human Settlements Programme, Kenya, 2011.
4. Center for Affordable Housing Finance in Africa, "Housing Finance in Africa,," The Centre for Affordable Housing Finance in Africa (CAHF) , South Africa2016.
5. D. G. F. Hassan, "Review and Re-evaluation for successful housing policy in Egypt". [Performance]. Faculty Of Engineering - Ain Shams University, 2010.
6. The Center for Economic and Social Rights (CESR), "THE RIGHT TO HOUSING, URBAN GOVERNANCE AND ORGANIZATION," The Center for Economic and Social Rights (CESR), Egypt, 2014.
7. World Bank, "INCLUSIVE HOUSING FINANCE PROGRAM," World Bank, Egypt, 2015.
8. Centre for Affordable Housing Finance in Africa (CAHF), "HOUSING FINANCE IN AFRICA," Centre for Affordable Housing Finance in Africa (CAHF), South Africa, 2012.
9. Richard Hares , Malak Wahba, "The Urban Geography of low-income housing:Cairo (1947-96) Exemplifies a Model," International Journal of urban and Regional Research, 2002, vol. 26, no. 1, pp. 58-79.
10. University Press of America, "A Possible Way Out: Formalizing Housing Informality in Egyptian Cities," United states of America, USA, 2004.
11. S. E-Attia, "Rent control Delimma come back in Egypt governance,," Egypt Economic Research Forum , Egypt, 2012.
12. K. S. A. A. M. AFIFY, "Integrated strategies for improvement of housing conditions for poor communities in inner cities". [Performance]. Faculty of Engineering El Mataria Helwan University, 2006.
13. World Bank, "ANALYSIS OF HOUSING SUPPLY MECHANISMS," Wold Bank, Egypt, 2007.
14. World Bank, "memorandum and recommendation of the president of the international development associate to the excestive directors," Wold Bank, Egypt, 1991.
15. Naglaa Fahmy, Khaled Wahba, "The Utilization of System Dynamics in Concluding Policies for Greater Cairo Sustainable Development Egypt " CD Conference, Egypt, 2013.

IOP Conf. Series: Materials Science and Engineering **974** (2020) 012021 doi:10.1088/1757-899X/974/1/012021

16. Ministry of Housing, Utilities and Urban Communities, "Shelter programs and City Development Strategies in Egypt," Ministry of Housing, Utilities and Urban Communities, Egypt, 2001.
17. United Kingdom University Of Cambridge , "Housing Generated By (Re) generation," United Kingdom University Of Cambridge , United Kingdom, 2004.
18. Tornado Center for planning and community studies, "State versus civil society: An approach to affordable housing in Egypt: Policy and Practice," Tornado Center for planning and community studies, Egypt, 2004.
19. Dr. Edgar Gael, Dr. Nisreen El-Lahham, Dr. Waleed Hussein, Sara El-Khishin, Samoa Soliman , "Sustainable Cities in Egypt Learning from Experience: Potentials and Preconditions for New Cities in Desert Areas," The Egyptian Cabinet Information and Decision Support Center, Egypt, 2009.
20. World Bank , "A framework for housing policy reform in urban areas in Egypt," World Bank , Egypt,2008.
21. The Egyptian Initiative for personal Rights, "Mubarak"s Promise Social justice and the National Housing Program; affordable homes or political gain?," The Egyptian Initiative for personal Rights, Egypt,2014.
22. A. M. Shalaby, Implications from Recent Experience of an Incremental Housing Project in Egypt. [Performance], Faculty of Engineering |Cairo university, 2014.
23. The Affordability of the Provided Housing Units in Egypt; Case Study of the National Housing Program (NHP), The 1st International Conference:Towards A Better Quality of Life,Technische Universität Berlin Campus El Gouna, Egypt, 24 - 26 Novemeber 2017.
24. A framework for using urban indicators in planning and managing human settlements, General Organization for Physical Planning, Ahmed O. El-Kholei, DPSIR Matrix, 2013.
25. Community Indicators, American planning Association, Rhonda Phillips, report no. 517,2003.
26. Developing indicators "Lessons Learned from Central America", Lisa Segnestam, world bank, 2000.
27. Indicator for Urban and Regional Planning "the interplay for policy and methods",Cecilia Wong, Routledge,2006.
28. Urban Indicators for Managing Cities: Cities Data Book, 2001.
29. United Nations Development Program, Human Development Report, 2013.
30. United Nations Development Program, Human Development Report, 2014.
31. Republic of the Arab Republic of Egypt's experience in establishing urban observatories, Report of the Ministry of Housing, Utilities and Urban Development, Urban Planning Authority, 2013.
32. Strategic Growth Council, Affordable Housing and Sustainable Communities Program: Program Guidelines,2015, (pp. 1–52).
33. Law No. (119) for the year 2008 - The Unified Egyptian Cooperative Building Law, 2008.

MilitaryTechnicalCollege
Kobry El-Kobbah,
Cairo, Egypt

13th International Conference on Civil and Architecture Engineering ICCAE-13-2020

Design out waste as an approach for achieving sustainability in Egyptian housing projects

Laila A.Elsawaf[1] And Ayman A. E.Othman[2]

[1]Senior Student, Architectural Engineering Department, Faculty of Engineering, The British University in Egypt, Cairo-Suez Desert Road, El-Shorouk, Cairo, Egypt. E-mail[1]: Laila141068@bue.edu.eg

[2]Professor ofConstruction and Project Management,
Head of Architectural Engineering Department,
Faculty of Engineering, The British University in Egypt, Cairo-Suez Desert Road, El-Shorouk, Cairo, Egypt. E-mail[2]:ayman.othman@bue.edu.eg

Abstract:Housing projects have been increasing rapidly during the previous years to fulfil community needs and achieve the national and international sustainable development objectives. Due to their nature, housing projects generate a significant amount of waste that accounts for approximately 40% of the solid waste that the construction industry generates annually. Many of these wastes are a result of inappropriate decisions adopted during the design process such as using non-sustainable materials and inefficient use of natural recourses. Meanwhile, research has revealed that achieving sustainability within housing projects is a crucial challenge, yet an essential one. This called to think differently and adopting creative strategies for reducing construction waste during the design process. Therefore, this research aims to develop a matrix that correlates the aspects of sustainability with design out waste strategy as an approach within RIBA design phases for achieving sustainability in Egyptian public housing projects. A research methodologyconsists of literature review and case study was designed to achieve a working matrix by accomplishing three objectives. Firstly, literature review was used to investigate the topics of sustainability, design out waste strategy and housing projects. Secondly, a case study collected by literature presentedand analysedto identify the design out waste strategy for reducing waste as an approach to achieve sustainability. Finally, developing a matrix that correlates sustainability and design out waste strategy within RIBA design phases.

Keywords: Waste Reduction, Design out waste strategy, Sustainability, housing projects, Waste, RIBA design phases.

1. Introduction

Waste is one of the most devastating global dilemmas generated from the construction industry, particularly from housing projects.smith, [1] stated that the U.S. alone generates roughly 251 million tons of solid waste annually, of which 40% is produced from construction projects alone. According to the World Bank[2] only 4% of the waste generated from low-income countries is recycled and one third of the high-income countries. Due to the negative impact of waste economically, socially and environmentally, the construction industry aims to reduce this percentage down to 30% approximately by 2020[3]. Whilst, it may seem to be a low percentage, yet it is not assured to be done. The majority of waste and unusable materials as solid waste are generated from the construction industry. Which leads to the question of how we can minimize waste generated from construction projects? A plausible solution would be considering the utilization of the design out waste strategy within architectural design firms to reduce the waste from the early stages of the design of housing projects as an approach to achieve sustainability. Waste in the construction industry usually results from inefficient or limited guidelines that aid with resources[4]. Particularly, that generated from housing projects, which mainly begins to appear from the earliest stage of the design process[5]. Whereas, the decisions taken during the design phase have proven to decrease the overall waste generated from the housing projects[6]. Waste management is one of the main aspects that controls and regulates the avoidance of waste in housing projects. Nevertheless, architects responsible for the design of housing projects should take into consideration the types of wastes that are primarily a result of decision made during the design process. However, these wastes should be decreased by considering the multiple design out waste techniques for further sustainable development. Therefore, the issue is based on exploring the impacts of waste on construction and contributing causes which will help the practitioners to formulate ways of avoiding or reducing waste. It also highlights the sustainable approach in design out waste management as a strategy. Therefore, design out waste strategies would minimize the causes and effects of waste on environment, economics, and social impact. According to Wrap [7], design out waste is a principle of building using resources in an efficient manner. Usually, it is first introduced within the design stage, which in turn helps with reducing the amount of resources that are consumed throughout the project. On the other hand, it could also be considered a strategy that increases the overall quality of the project.However, with focusing on the resources not only human resources but energy consumption, water, health, education and environment as a whole leads to Egypt 2030 aspiration for developing sustainability. As well as, to keep the country participators in developments challenges facing it keeps initiative plans to determine these areas.

2. Methodology

The methodology adopted throughout the research for the purpose of attaining its aim and objectives. A research methodology consists of literature review and case study was designed to achieve a working matrix by accomplishing the objectives. However, the matrix developed will be in form of a checklist for easy use as it's the first of its kind as a model collaborating design out waste strategy , RIBA design phases and sustainability.

Aim

- The research aims to develop a matrix that correlates the aspects of sustainability with design out waste strategy as an approach within RIBA design phases for achieving sustainability in Egyptian public housing projects.

Objectives

To achieve the mentioned aim, a number of objectives have to be accomplished.

1. Investigating the waste in public housing projects in Egypt, design out waste strategy and sustainability listed in literature review.

2. Acase study collected by literature presented and analysed to identify the design out waste strategy for reducing waste as an approach to achieve sustainability.
3. Developing design matrix to assist ADFs with achieving sustainability through implementing the design out waste strategy within collaboration to RIBA design phases.

2.2 Research Methodology

The research methodology is based on data collection and the action demanded to achieve the proposed aim to develop a design matrix for achieving sustainability within Egyptian public housing projects through the implementation of design out waste strategy as an approach.

- **Data collection and gathered information:**

  Literature review which is based on the literature review that investigate the waste in public housing projects in Egypt, DOW strategy and sustainability.

  Case studyis researched to achieve objective two that uses design out waste strategy to reduce project waste within the construction projects and achieve sustainability.

- **Required Course of Action:**

  Based on the results of the previous objectives analysed developing a design matrix to achieve objective three to develop the proposed matrix and research conclusions for achieving sustainability in housing projects and meet with Egypt 2030 vision.

3. Literature Review

2.1 Waste in housing project in Egypt

The construction industry specially housing projects is one of the biggest industries worldwide. It is one of the essential sector for the development in countries worldwide. Moreover, the housing projects plays a significant role in increasing Gross Domestic Product (GDP), economic growth and urbanization development improvement of countries. In Egypt, the construction industry employs about 11% of the total population which contributes towards increasing[8] GDP by 5% through 2009/2010-2013/2014. However, the construction industry through housing projects provides a contribution toward meeting society needs and requirements. Furthermore, the housing project is a complex business in the construction industry. It is a risky business that is affected by some internal and external factors which impacts its performance.Housing projects are subject to a number of problems including low productivity, time and cost overrun, poor safety, lack of quality and lack of working conditions. Durgekar[9]stated that Egypt has increased its solid waste in housing projects by 36% since 2000. Globally, producing construction materials are using about 40% of the global economy. The 60% is about accountable construction sector of natural materials resources from nature, 23% of air pollution and 50% of water pollution, 50% of water and energy consumption and 50% of total generated waste which is a huge percentage. However, the problem has been released throughthe lack of direct guideline and effective legislations. According to Durgekar,[9] housing projects in Egypt waste has been generated among each structure participant working separately, that also lacks the vision of planning and collaboration in each construction phase. Furthermore, the limited reliable services providing cost-efficiently and funds provided by municipal authorities.As well as there is a lack in public awareness and citizen behavior. Those unconsidered lacks generates the waste in housing projects. A result of the dismiss performance of waste in Egypt during the last decades, the pollution and its impact has been highly increased [9,19]. The fishbone diagram shows different types of construction waste through the construction project, see figure (1). Development of construction wastes can becaused by various factors. It is very important to identify andunderstand those causes for controlling waste generation atsource. However, this diagram below explains the causes and effects

IOP Conf. Series: Materials Science and Engineering **974** (2020) 012022 doi:10.1088/1757-899X/974/1/012022

ofconstruction waste. Identification ofthe different factors can help in increasing awareness ofconstruction practicing to control construction waste.

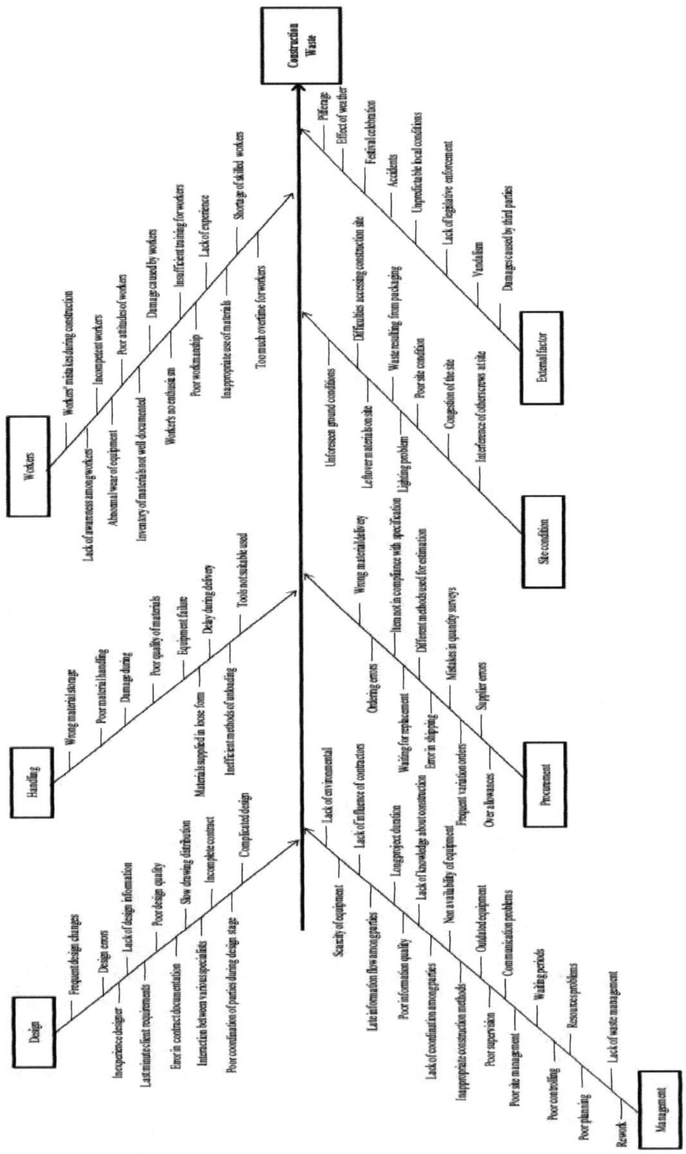

Figure (1) construction waste fishbone [9].

2.1.1 Types of waste

Construction waste is one of the most recycled materials and recovered waste according to its diversity in types and quality[10]. Construction waste generated from housing projects includes a wide variety of materials. Some of those materials may be recycled considered as hazards materials while other materials could not be recycled that considered asnon-hazardous in nature. For managing the construction waste effectively, should classify the type of waste to work with it as a recycled material or to used it carefully without generating any waste, see table(1). As for example, concrete separated means the separation of concrete from steel or other materials in the components [10].

IOP Conf. Series: Materials Science and Engineering **974** (2020) 012022 doi:10.1088/1757-899X/974/1/012022

Table (1) types of wasted materials [10] (developed by the author)

Materials	Hazards (could be recycled)	Non- hazardous (couldn't be recycled)
category 1	Concrete separated	Concrete
	Ceramic separated	Ceramic
	Bricks separated	Bricks
	Tiles separated	Tiles
Category 2	Wood	Treated wood
	glass	Gypsum
	plastic	paint
	Paint cans	
Category 3	Fibre glass	insulation
Category 4	Steel	Metallic cables containing oil
	Copper	

Construction waste also is divided into two groups including physical and non-physical waste as shown in figure (2). The physical waste represents the materials loss, while the non-physical waste represents waste in time and cost.

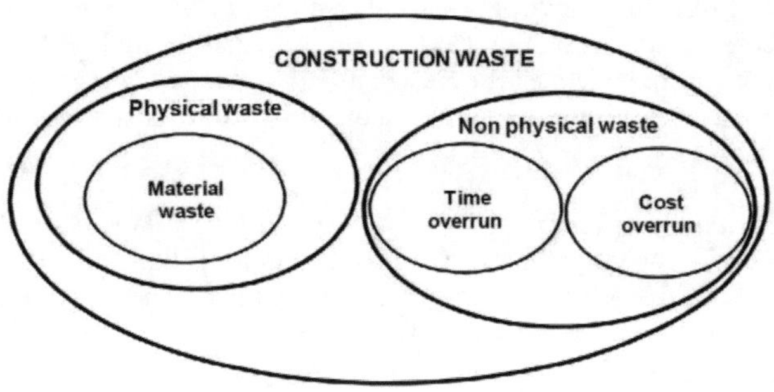

Figure (2) the waste occurs in the form of material, time and cost losses [9].

2.1.2. Waste Different impact of the housing project

2.1.2.1. social impact of the housing project
Dispute economic and environmental impacts, social impact is the least in the "triple bottom line" the principles of sustainability developing. However, it's an impact of housing construction projects in Egypt that refers to certain social aspects to human populations of housing projects as the change of the way people work, relate to one another live, to meet people needs and overall organize how to cope as members of the society. In construction housing projects the development may lead to positive or negative social impact and maybe both, resettlement of nearby resident provides a negative impact on society. Furthermore, land acquisition, disposals and resources depletion effects on a negative way to the surrounding settlements and residents. Social impacts of the construction waste under certain conditions could lead to social risks and evolves a conflicts if different stallholders didn't dealt with the waste problem carefully and properly. While society couldn't cope with the current situation for long period of the housing project lifecycle.

2.1.2.2. Environmental impact of the housing project
The successful construction projects should recognize that not only cost is the effective method but also environmental methods should be respected. The Environment is affected by the unethical activities that cause problems generated from the housing waste.The waste contributes environmentally disposal to pollution globally within the wrong deposit activity of physical wastes. However, the waste impact on the public generally and to the whole environment around the world.

Construction waste has an effect that requires an immediate attention to protect the environment globally. According to Nagapan[11]Studies shows that the adverse wasted materials of housing projects has significant highly impact on the environment.Compromising the Egyptian natural resources improper disposal of housing projects waste like waterways that contaminate the water supplies[9]. This contributions and current lack causing serious environmental impact. Various countries are blown down by neglecting the concept of waste impact on environment and its very important to understand the current situation to discover the solution.

2.1.2.3. Economic impact of the housing project

Construction housing projects waste increases the project cost above the budget of the project[11]. However, the cost overrun by different aspects as the transportation, the surplus materials and the repairing damages. The materials wasted in housing projects generated from the construction activities in different identifications as concrete leftovers, steel scrap and demolished debris [11].

2.2. Achieving Sustainability In Housing project

2.2.1. Definition of sustainability

Sustainability means the capacity to sustain, the current need and future generations for well and survival being depending on natural environment, either directly or indirectly way [12]. Furthermore, sustainability aims to consent humans within nature in present and future by creating and maintaining the environment, social and economic aspects. Sustainability is becoming a great variety term that can empiric to almost every aspect of life on earth, ambulatory from local to globally scale and over diversity of time period[13], see figure (3).

2.2.2 Aspects and dimensions of sustainability

Environmental, social and economic those are the main three pillars of sustainability. However, combining between those three pillars generate three new pillars, namely: Social Environmental, Environmental-Economic and Economic-Social[14],see figure 3.

2.2.2.1 environmental aspect

The environmental pillar of sustainability concern to use the natural resources in an efficient way within reducing the waste, effluent, pollution and emissions generation to the environment. Moreover, it aims to protect human health from the negative impact reduced, while exterminate the toxic substances and encouraging the use of renewable raw materials[13].

2.2.2.2 social aspect

The social pillar of sustainability is a fair and achievable society fairness and justice when conveying its resources within itself. According to BenzuJK,[13]the society would not demanding in the rights of each and every individual based on sequences of their age, sex, religious, ethnicity and social background. The rights to have sustainable is living in a good quality life, right to housing, right to social security, freedom of speech and right to facilitate property ownswith sustainable aspect according to[13,15].

2.2.2.3 economical aspect

An economical pillar means the self-sustaining society which it using the available resources provided in efficient way according to its individuals with its needs without help from the neighbouring countries. according to Othman et al.,[13]to develop an economically sustainable society, increasing productivity, adding value and minimizing waste is by creating new market share, reducing cost, energy through improving efferently and using raw material.

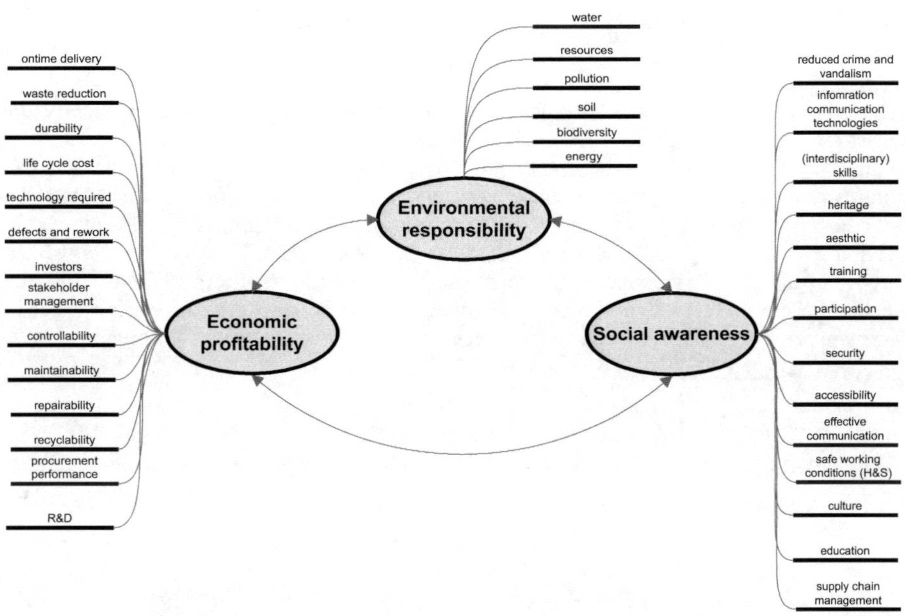

Figure(3) Qualitative data - the construction industry sustainability tri-polar[13].

2.2.3 Dimensions of Sustainability

Sustainability is a long period structure strategy for environment , economic and social system, which aims to minimize the burden on natural resources. However, balancing these three pillar is the aim for developing sustainability for long term. Otherwise, neglecting one of the three pillar will impact potentially danger on developing the whole process.

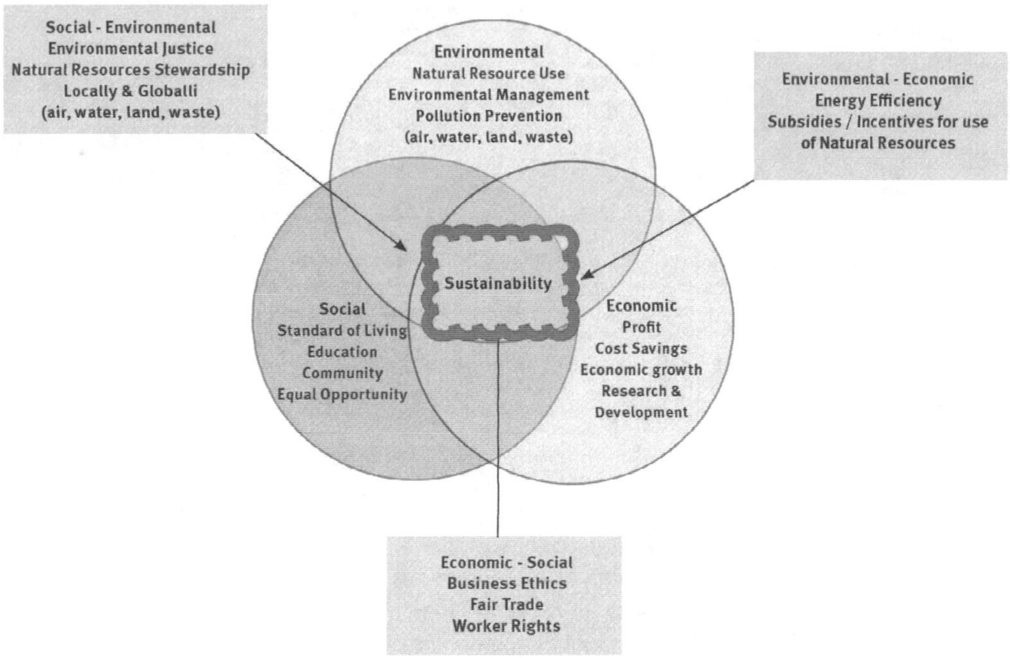

Figure (4) sustainability dimensions [15].

IOP Conf. Series: Materials Science and Engineering **974** (2020) 012022 doi:10.1088/1757-899X/974/1/012022

4. Design Out Waste Strategy for achieving sustainability in Egypt

3.1 Definition of design out waste

Design out waste (DOW) defined as building principles within using resources efficiently during design process of construction projects[16]. The concept of this strategy to design the use of each material with no value intrinsic, or inherent value discarding. Design out waste strategy can be integrated into a process matrix to evaluate the waste generated and identify how to be reduced within a buildability ,cost less, high quality and less time. It aims to plan the minimization of resources within use of available resources and materials efficiently as possible as can. Design out waste circular economy depending on pollution and optimizing usage of natural resources. The UK waste and resources action program (WRAP) designed this strategy to reduce waste during design process has provide five main principles[16], as follows the processes production serve several principles:

- Design for Waste Efficient Procurement
- Design for Materials Optimization
- Design for Off-Site Construction
- Design for Reuse and Recycling
- Design for Deconstruction

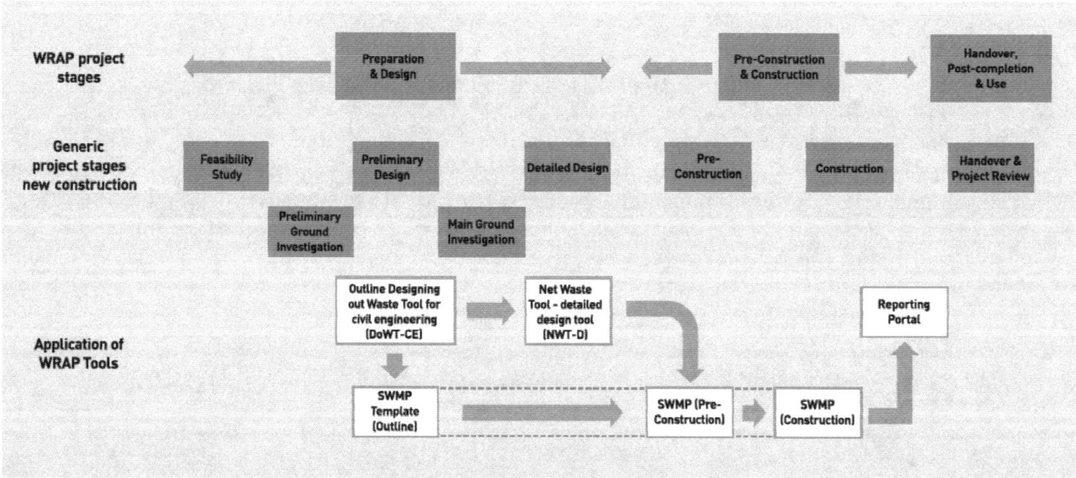

Figure (5) WRAP project stages [16].

3.2 Principles Of Design Out Waste

3.2.1. Design for reuse and recovery [16].

- This principle focuses on recycling used material (ex. reclaims products or crushed demolition materials) throughout project whole life cycle.
- Reuse demolition materials and building components.
- Reuse existing site buildings.

3.2.2. Design for offsite construction[16].

- Through early phase in construction project contractors should specify a concept of using prefabrication to improve the supply chain.
- Use modular volumetric construction.
- Use of-site prefabrication of structure elements.
- Use of-site prefabricated pods, bathrooms.

- Use of-site prefabricated and pre-cut building elements.

3.2.3. Design for material optimization[16].

- Focusing on the design process with an approach towards "good practice" that allows designers to consider materials usage to be more efficient, which will further reduce the overall waste in construction, by minimizing excavation and material.
- Facilitate the building layout, form and elements.
- Design within standard room sizes, ceiling high stand material sizes.
- Reduce number of material used using local material standard.
- Specify recycled content.
- Consider the services, maintenance and replacement required of each component for the long run life time.

3.2.4. Design for waste efficient procurement[16].

- Allows the designer to be involved within both the design and construction process, which aids with the considerations that dictate workflows.
- Use documents for setting the waste performance requirements.
- Supply chain collaboration.
- Identify the source responsible of recycled content and materials.
- Reduce the required packaging in using materials procurement.
- Prepare plans within contractors waste minimization and site waste management plan (SWMP).
- Consider logistics material and in time delivery.

3.2.5. Design for deconstruction and flexibility[16].

- The concept of deconstruction and flexibility is principle that on climate adaptation throughout the whole life cycle of the building, which is linked to the reuse and recovery material principle.
- Use precast steel frame.
- To simplify the future reuse of bricks use lime mortar.
- To change of use enabling method use flexible construction.
- Facilitate the design life time potentials once the project is done.

3.3. Benefits and Methods of Design Out Waste Strategy

Benefits of the design out waste strategy in projects, that it can be applied in any constructionphase throughout the project, while its effective action done mostly during the structure of the project as approached by[16]. Previous research found that poor design planning followed with failure, according to WRAP, [16]33%of all on-site waste is due to a failure to implement waste reduce on measured ways during the early design stages.However, to minimize waste through the initial design planning phase are the best opportunities using (DOW) in it.

3.3.2 Role Of Design Out Waste Towards Achieve Sustainability

If the ideas and principles of designing out waste strategy are to be successfully applied to any project, The effective impact is achieving sustainability across the project is absolutely vital. Every team member or stakeholder involved in the design, construction and ongoing management of the project should result a building to be engaged a with design out waste as early as possible[7]. This will ensure that the importance, benefits and implications of designing out waste are fully effective to achieve sustainable housing project. While, technical team members such as contractors, construction engineers and facilities management teams can all bring extremely valuable insight on the practical implications of the delivery of sustainability[17]. All team members and stakeholders should be encouraged to apply the strategy for achieving sustainability. It is also important to engage as early as

IOP Conf. Series: Materials Science and Engineering **974** (2020) 012022 doi:10.1088/1757-899X/974/1/012022

possible with the relevant planning, building regulation and other statutory compliance authorities. Communication with these bodies should include discussion of the planned designing out waste strategy and measures of sustainability meet together. This means they can be addressed and resolved before causing disruptions of work environment to reach the effective implementation of sustainability during housing projects plan.

3.4. Sustainable development in Egypt 2030

Egypt 2030 vision is known with sustainable development strategy, set a goal to be initiative with creating productive, modern, striving, happy society in addition to that is to be economically, environmentally and politically stable[18]. Maximizing the domestic resources is a need for developing the energy sector and the construction industry to accomplish an effective accommodation. Additionally, urban development is another Egyptian 2030 vision issue that admired of multiplying the current urban spaces within relocating both projects and citizens for introducing more effective use of numerous lands and resources. The key for developing urban spaces include creating developed 7.5 million housing units within finding kind of a solution for slums in Egypt. As the housing project is the main source of comfort and save accommodations[18]. This leads to the next level of initiative future plan, which is the consequently quality of life overall improvement. The construction industry combine both economic and environment for developing a sustainability approach, while balancing between both to of them adhesive the economical level in advance way that do not effect on the environment in negative way[18]. To be par with the environmental standard internationally, Egypt should follow this strategy to be on track, as well as establishing consumptions practice, sustainable approach and significant waste management methods. However, design out waste is a successful and efficient method for delivering sustainability of construction projects in Egypt, as it followed by the three pillar of sustainability environmental, economic and social to meet with Egypt 2030 futuristic vision.

4. Case-Study: Model D house

4.1 Background Information

Figure (6)Model D house [16].

- Location : Pitmanchie near Insch, Aberdeen
- Source : (zero waste Scotland, 2011)
- Architects : GokayDeveci ARB RIBA FRIAS
- Project Type : Residential
- Date Completed : October 2011

- Gross Internal Floor Area : 155 sqm

Case study objective: This case study uses Model D house located at Pitmanchie near Insch, Aberdeen as an example to investigate the compatibility of new uses design out waste strategy. The case study also takes into consideration several indicators such as architectural integrity, building form, building function and sustainability (considered as sustainable building values listed previously in the literature review) . This case study aims to identify qualitative issues and to understand the new features to provide a guidelines for using it during the investigation of waste that generated from housing projects and reduce it, based on the analysis.

Introduction: The Model D house was designed for Zero Waste Scotland at 2011 is an innovative project designed and managed as part of resources efficient for Scotland by GokayDeveci ARB RIBA FRIAS[16]. The house was opened in October 2011 is now a new feature and visitor attraction at it's a new iconic in the city. It shows how several of the ideas of designing out waste can be successfully incorporated in a building to have high potentials to be cost-effective

4.2 Project Brief

The house is located near Insch in Aberdeenshire, and is designed to determine the different alternatives of unaffordable products that offered by high developers. The design aims to offer a new approach of design that consist of large scale and related to the surrounded context rather than kit-built homes that are expensive.

Figure (7) model D house [16].

4.3 Case Study Target

Additional to that, the clients were focusing on to develop an alternative design to be providing an affordable, high quality within low-energy dwelling that adopts home-grown timber. The design main target is to stimulate a transformation in our old version of current thinking about architectural design and house design[16]. Here in, it focuses on the ecological awareness increasing and the concentrating on environmental and friendly designs and the use of local materials.

IOP Conf. Series: Materials Science and Engineering **974** (2020) 012022 doi:10.1088/1757-899X/974/1/012022

4.4 Case Study Process

4.4.1 Orientation

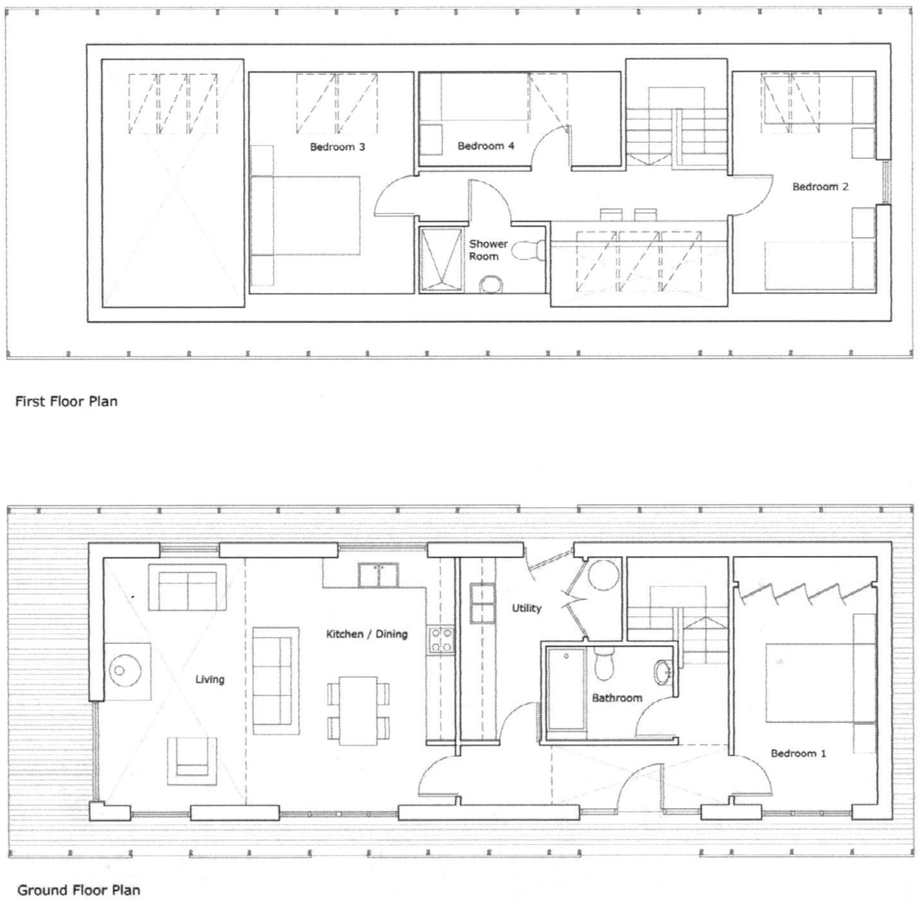

Figure (8) model D house plans [16].

The orientation of the building was to develop a flexible approach for the living space that proposed in summer. As shown in figure (8) the design of the house is designed within a very simple design within rectangular long plan, building including an open living, dining room, kitchen space and one double bedroom on the ground floor. There is a separated utility room that used as a recycling storage room and also a bathroom on the ground floor plan. On the first floor plan there is a two double bedrooms, an additional shower room is in the attic story and an extra open gallery space[16]. The bedroom that located over the living space (bedroom 3) can be used as an open studio or office space only by removing the alternatively wall. There is a main circulation space within the living space both designed double height space to give the sense of lighting and space in the interior[16]. The design incorporate a terrace in the house that offers a buffer space, protected within timber screen. Taking advantages of the solar heat gain by using large south facing floor-ceiling openings, while it also maximize the views.

IOP Conf. Series: Materials Science and Engineering **974** (2020) 012022 doi:10.1088/1757-899X/974/1/012022

Figure (9) model D house screens [16].

The large widows openings of the building is in the south façade that takes the desired solar gain and maximize the views. Durability and homogeneous appearance is provided by the meticulously detailed external rain screen which provides shading and privacy where required.

4.4.2 Materiality

Due to the smaller parts of timber panels sections of the wood used in the building is manufactured on site, as well the superstructure of Scottish timber was delivered on site in open panels[16]. The use of standardized timber kit construction was adopted to provide the required U-values and air tightness. The wall construction became two leaf's of 95x45mm C16 timber frame panels, with a layer of OSB sheathing between to accommodate the air-tight membrane, Similar to the roof arrangement [16]. As shown the materials and setting of the dwelling through the process that followed by the design out waste strategy application.

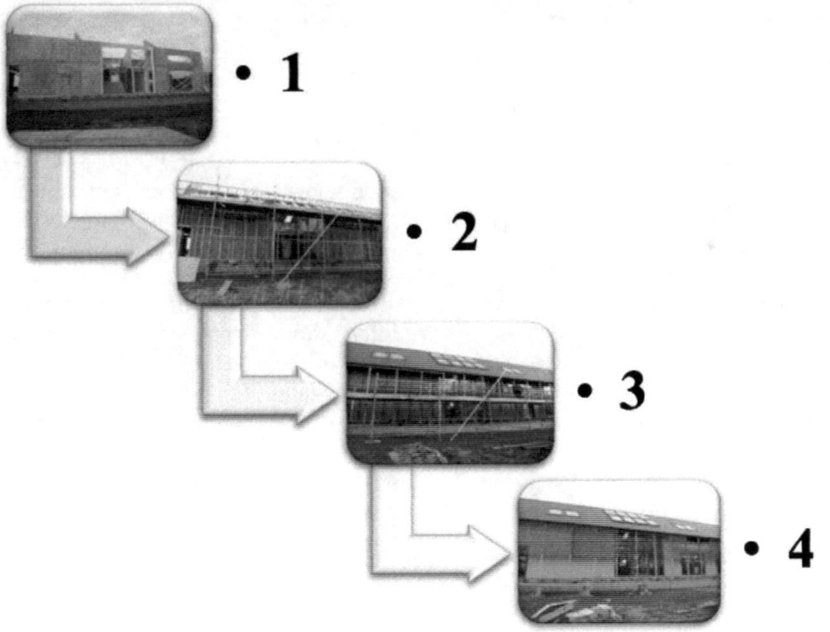

Figure (10)model D house building process[16].

4.4.3 Energy Efficiency

The dwelling was designed to meet the German passive house standard, where it strict that the houses should build with high thermal performance to minimize heating demand. The U-Values achieved throughout the House include:
Walls 0.15 W/m2/degC
Air tightness 0.6 m3/(h.m2) @ 50Pa

4.5Response achievement to sustainability

According to the stages of design out waste highlights the key points in that process where clear and unambiguous communications between the architect/design team, the client and other project team members are crucial. It provides a guide maps that would be recommended measures, interventions andconsiderations of designing out waste to the Working stages, see Figure (11).

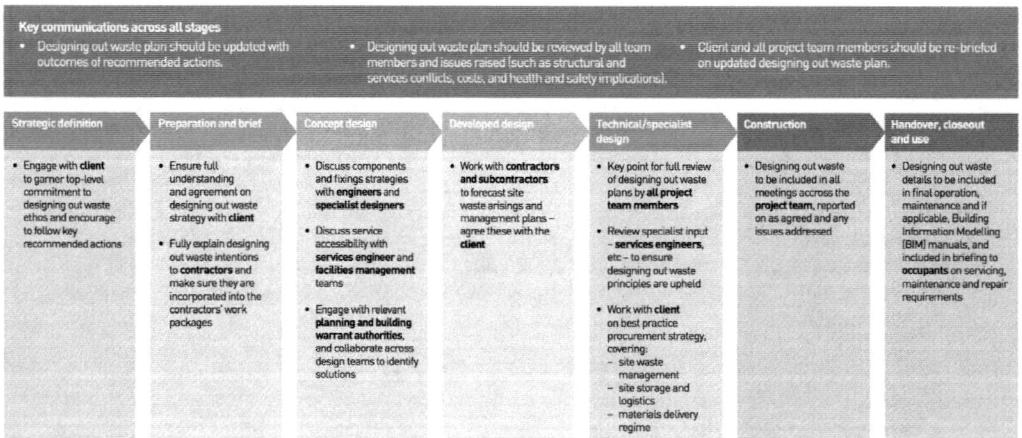

Figure (11) key communication points [17].

4.6Challenges with the implementation

Establishing targets and reporting structures needs to be supported and complemented bysuitable additions or alterations to standard procurement templates in respect of the clientorganization's engagement with others involved participant in the project. In large organizations, this canbe complicated by the fact that different departments or individuals are responsible for differentcontractual arrangements, so a concerted effort to integrate the subject is required.In construction projects, it can be useful to ensure the information is contained within thedocuments issued and that a separate briefing exercise, perhaps as part of the tenderingprocess, is undertaken to highlight and explain the aspirations of the client. Taking this approach gives a legal standing to the targets and will create a binding keyperformance indicator that gives the client recourse about overall the process to be aware of the entire success.

5. Discussion

A brief overview of the paper was introduced in the first part explaining research background stating the problem tackled and why it is important to solve and addressing its causes as well. The second part presented a thorough overview of the research pillars, concluding with the criteria will be illustrated in form of matrix. Developing a matrix that corelates sustainability and design out waste strategy to be applied on housing projects in Egypt in order to achieve the 2030 Egypt sustainability vision. However, the impact of current state working technique gives an impact on the environmental, social and economic aspects. Furthermore, analysing this impacts give opportunity to choose the will established strategy for applying it. While, Design out waste can achieve sustainability in public housing projects in Egypt. As mentioned before housing projects applied using the DOW strategy has

IOP Conf. Series: Materials Science and Engineering **974** (2020) 012022 doi:10.1088/1757-899X/974/1/012022

achieved critical point that meets sustainability approach. The matrix on table (2) shows the criteria that will be used to analyse buildings according to it as a checklist. However, The matrix developed within the correlation of RIBA design, design out waste strategy and sustainability. This matrix will also help in extracting some guidelines and recommendations for future studies to follow those successful steps for a beneficial outcome.

Table (2) DOW and sustainability matrix (developed by the author)

DOW strategy			Sustainability: Environmental							Economical							Social					
Dow principles	RIBA Design phases	Dow process	Recyclable	Reusability	Carbon footprint	Energy consumption	Water consumption	Life cycle	Pollution	Initial costs	Maintenance costs	Defects and rework	Recycle cost	controllability	Repairability	Procurement performance	Required number of workers	Safety of workers	Training of workers	User requirements	Supply chain management	User changes
Design for Materials Optimization	Strategic Definition	local material standard																				
		designer use of efficient materials																				
	Preparation And Brief	Specify recycled content																				
		Maintenance consideration																				
Design for Reuse and Recycling	Concept Design	Reuse existed materials building facilities																				
		Reuse site demolition																				
		Use precast steel frame																				
	Developed Design	recycling used material																				
		Preparation of recycling content																				
Design for Off-Site Construction	Technical Design	Off-site prefabrication elements																				
		Off-site prefabrication components																				
		Off-site prefabrication structure																				
Design for Waste Efficient Procurement		logistics material																				
		time delivery																				
		designer involvement																				
	Construction	supply chain collaboration																				
		Reduce the required materials																				
		SWMP																				
Design for Deconstruction and flexibility	Close out	Flexibility																				
		lime mortar																				
		Facilitate building																				

IOP Conf. Series: Materials Science and Engineering **974** (2020) 012022 doi:10.1088/1757-899X/974/1/012022

Design out waste strategy matrix application

Design out waste strategy matrix explanation of each principle as follows:

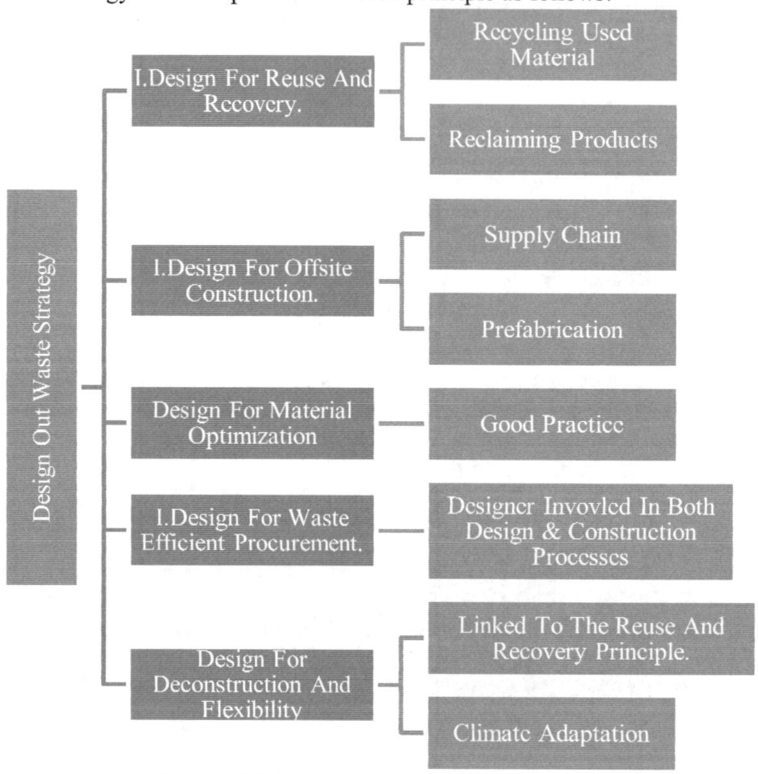

Figure (12) design out waste matrix application

Design For Reuse And Recovery.

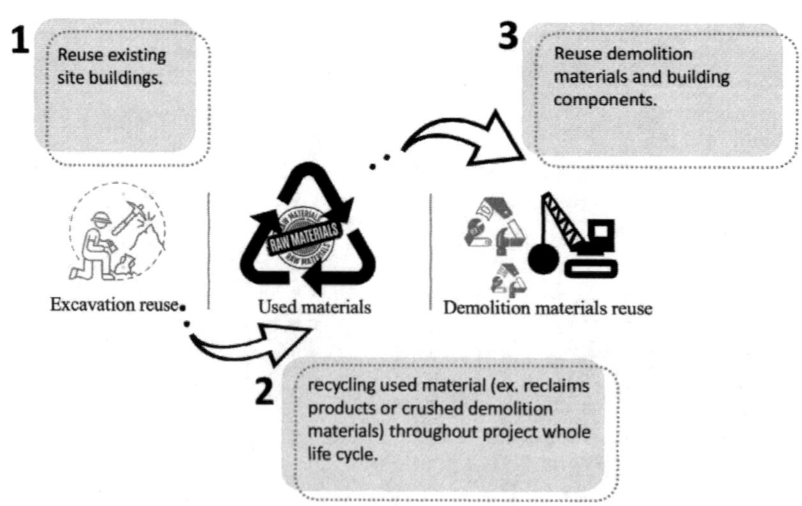

Figure (13) design out waste Design For Reuse And Recovery matrix explanation

Design for Deconstruction and Flexibility

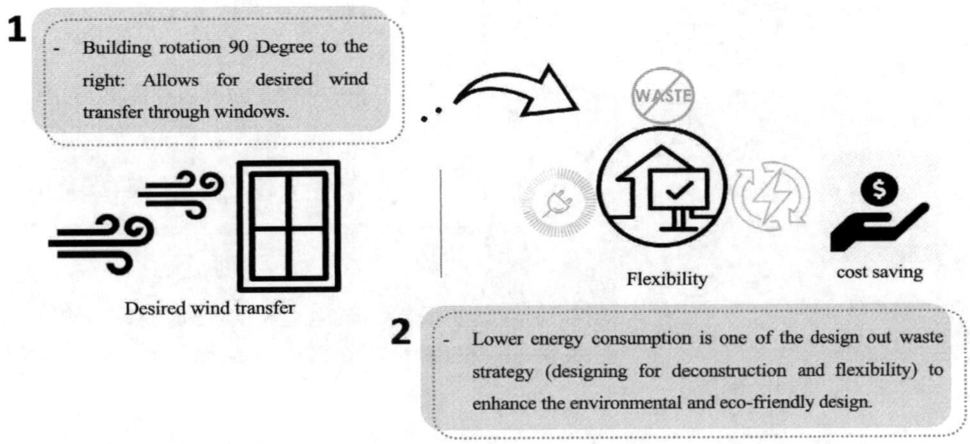

Figure (14) design out waste Design For deconstruction and flexibility

Design for Waste Efficient Procurement

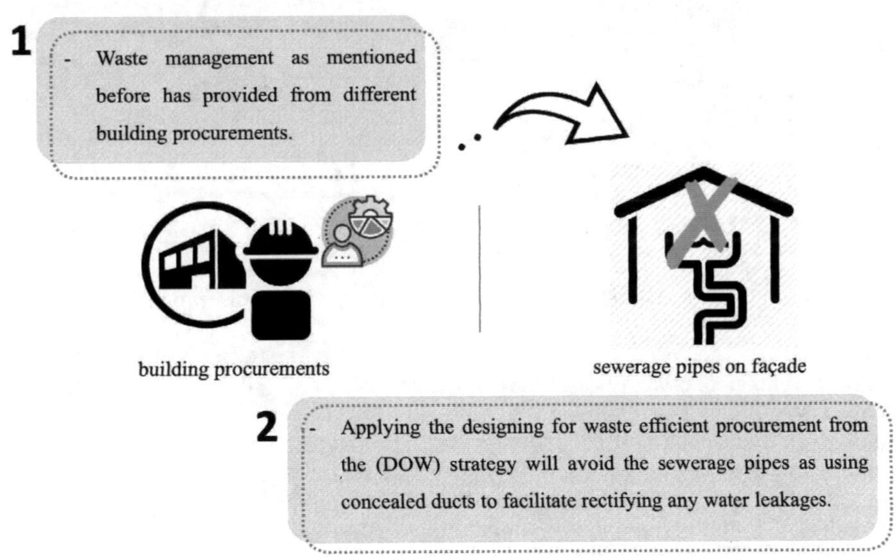

Figure (15) design out waste Design For waste efficient procurement

Design for Material Optimization

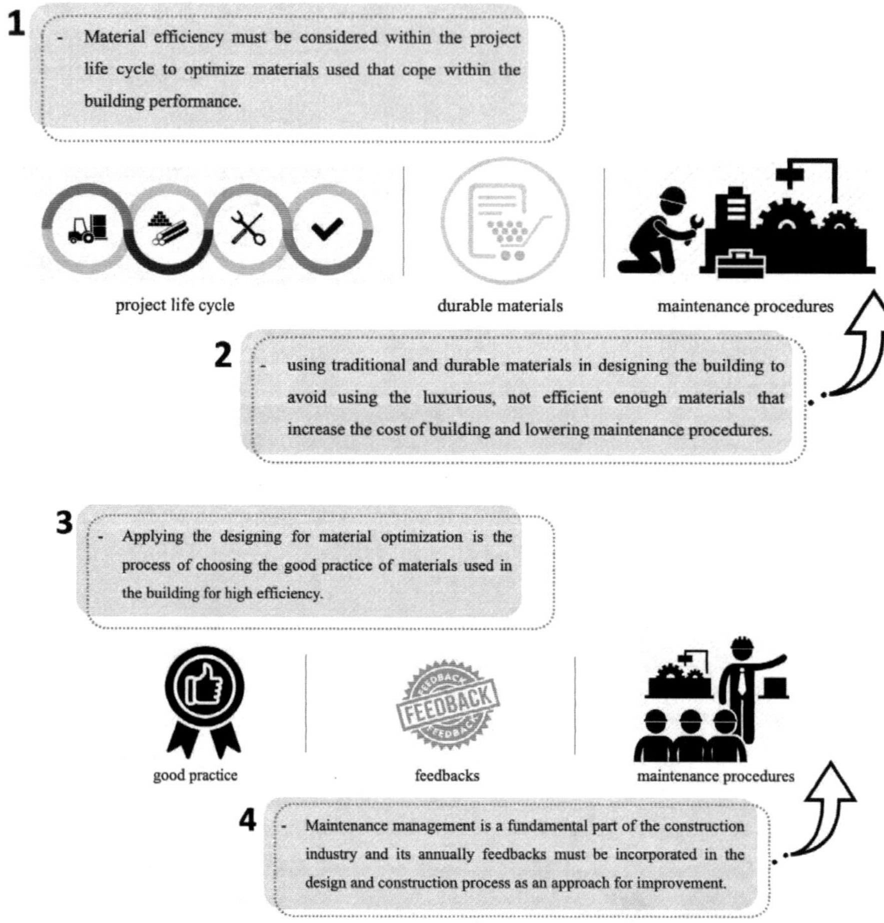

Figure (16) design out waste Design For material optimization

6. Conclusion

In conclusion, the literature review along within the case-study proposed have highlighted the need of a collaborative integration of management within design a housing project for avoiding waste through construction project. It emphasized the prominence of waste integrationin construction site through design process changes resilient design standards to promote sustainable development. Design out waste was introduced as a strategy where the matrix would properly position the role of architecture throughout the cycle towards waste prevention, while bridging the gap between the rest of the disciplines involved. The case study was predominately related to the waste reduction through implementing DOW strategy, which included a multitude of essential attributes to be present for each state, the sound practices prove that waste reduction could be achieved at any stage if the capability of compliance is available, as some practices have shown prominent success at the early stages. Finally, managing the project phases is to eliminate the consequences of the waste production, giving flash to the idea of design out waste strategy. While, existence of relative applicable resilient design standards could lead to a more waste reduction and achieving sustainability to meet with Egypt sustainable vision 2030. The collection of model D house is based off real-life case study which further support the value of the research and its aim to develop a design matrix for housing projects as an approach for achieving sustainability through implementing design out waste to meet with the Egypt sustainable development strategy vision of 2030.

IOP Conf. Series: Materials Science and Engineering **974** (2020) 012022 doi:10.1088/1757-899X/974/1/012022

7. Recommendation

The need to Adopt a DOW strategy in housing projects to enhance the architect's ability to develop more resilient solutions by implementing resilient design standards within everyday practice, in addition to integrating DOW strategies within their firms with resiliency as a key indicator for sustainable development. Moreover, raising awareness and communication through education and introducing the ideologies within architectural education and lastly, linking the resources with other disciplines.

References

[1] Smith, a, Brown, K., Ogilvie, S., Rushton, K., & Bates, J. (2001). Waste management options and climate change: Final report to the European Commission, DG Environment. In ... *At Www. Envirohelp. Co. Uk/France/ ....* https://doi.org/10.1016/S1352-2310(01)00532-5

[2] Mohan, G., Sinha, U. K., & Lal, M. (2016). Managing of Solid Waste through Public Private Partnership Model. *Procedia Environmental Sciences*, *35*, 158–168. https://doi.org/10.1016/j.proenv.2016.07.066

[3] Review, G., & Management, S. W. (n.d.). *68135-REVISED-What-a-Waste-2012-Final-updated.*

[4]Shooshtarian, S., Maqsood, T., Wong, P., Khalfan, M., & Yang, R. (2019). *Construction Waste Management in Natural Disasters in Australia*. 4–5. https://doi.org/10.33552/GJES.2019.03.000573

[5]Akash, Bath Kumar, M., & Rana, P. (2018). International Journal of Advance Engineering and Research. *International Journal of Advance Engineering and Research Development*, *5*(3), 434–439. https://doi.org/2348-4470

[6]Poon, A., &Jaillon. (2003). *Table 2.1-Waste generation in Hong Kong.* (2003).

[7] WRAP. (2007). Designing out Waste: a design team guide for buildings Halving Waste to Landfill Less Waste, sharper Design. Retrieved from https://www.modular.org/marketing/documents/DesigningoutWaste.pdf

[8]Ahmed, A., & Othman, E. (2019). *Lean Construction for Reducing Construction Waste in the Lean Construction for Reducing Construction Waste.* (February).

[9]Durgekar, V. (2016). Towards Sustainable Waste Management Through Technological Innovations, Effective Policy, Supply Chain Integration & Participation. *Procedia Environmental Sciences*, *35*, 140–149. https://doi.org/10.1016/j.proenv.2016.07.061

[10]What are the different types of Construction Waste? - GreenSutra™. (2017). Retrieved November 9, 2019, from team greensutra website: https://www.greensutra.in/question/what-are-the-different-types-of-construction-waste/

[11]Nagapan, S., Rahman, I. A., Asmi, A., Memon, A. H., & Latif, I. (2012). Issues on construction waste: The need for sustainable waste management. *CHUSER 2012 - 2012 IEEE Colloquium on Humanities, Science and Engineering Research*, (May 2014), 325–330. https://doi.org/10.1109/CHUSER.2012.6504333

[12] Emas, R. (2015). *The Concept of Sustainable Development: Definition and Defining Principles, Florida International University*. 1–3. Retrieved from https://s3.amazonaws.com/academia.edu.documents/43652555/5839GSDR_2015_SD_concept_d

efiniton_rev.pdf?response-content-disposition=inline%3B filename%3DThe_Concept_of_Sustainable_Development_D.pdf&X-Amz-Algorithm=AWS4-HMAC-SHA256&X-Amz-Credential=AKIAIWOW

[13] Othman, A. A. E., Ghaly, M. A., &ZainulAbidin, N. (2014). Lean Principles: An Innovative Approach for Achieving Sustainability in the Egyptian Construction Industry. *Organization, Technology and Management in Construction: An International Journal*, *6*(1), 917–932. https://doi.org/10.5592/otmcj.2014.1.2

[14] Kolekar, K. A., Hazra, T., & Chakrabarty, S. N. (2016). A Review on Prediction of Municipal Solid Waste Generation Models. *Procedia Environmental Sciences*, *35*, 238–244. https://doi.org/10.1016/j.proenv.2016.07.087

[15] Abidin, N. Z., Yusof, N., & Othman, A. A. E. (2013). Enablers and challenges of a sustainable housing industry in Malaysia. *Construction Innovation*, *13*(1), 10–25. https://doi.org/10.1108/14714171311296039

[16] WRAP. (2010). Designing out Waste Tool for Buildings. *A Full User Guide to WRAP's Designing out Waste Tool for Buildings*, (February), 1–42. Retrieved from http://dowtb.wrap.org.uk/

[17] Zero Waste Scotland. (2015). Designing Out Construction Waste. *Scotland*, 1–46. Retrieved from https://www.resourceefficientscotland.com/sites/default/files/Designing Out Construction Waste Guide.pdf

[18] El-Megharbel, N. (2015). *Sustainable Development Strategy: Egypt's vision 2030 And Planning Reform Deputy to the Minister of Planning, Monitoring and Administrative Reform Integrated Approaches to Sustainable Development Planning and Implementation New York*. Retrieved from https://sustainabledevelopment.un.org/content/documents/15262El-Megharbell, Egypt NSDS 20150527.pdf

[19] Ibrahim, M. I. M., & Mohamed, N. A. E. M. (2016). Towards Sustainable Management of Solid Waste in Egypt. *Procedia Environmental Sciences*, *34*, 336–347. https://doi.org/10.1016/j.proenv.2016.04.030

Military Technical College
Kobry El-Kobbah,
Cairo, Egypt

13th International Conference on Civil and Architecture Engineering ICCAE-13-2020

Simulation Study of Trombe Wall for Passive Solar Technique of buildings in Egypt

Ayah Mohamed Ramadan

Faculty of Engineering, Modern Academy for Engineering and Technology, Cairo, Egypt.

Abstract. The studies aimed to evaluate a Trombe wall, a solar wall, which is a critical component of passive solar strategies in energy usage. It is a very compact wall and can be south-facing or north-facing, which is painted black and made of exclusive material that absorbs a lot of heat. The main advantage of the Trombe wall is that they are often easily built from locally obtainable materials, very dependable, and having less repair, maintenance, and operation costs. In this study, a BASECASE model were simulate with "Design Builder" software based on "Energy Plus" with the passive solar technique of Trombe wall in Egypt of a model of residential building with two façade alternative base case. The research has shown that Trombe wall has the possibility of providing acceptable internal thermal comfort through "Mixed-Mode" ventilation strategy in hot climate, this is important in determining the possibility of Trombe wall in incorporating "Mixed-Mode" ventilation reducing energy usage in cooling demands. The conclusion of this research that Trombe walls can reduce the cooling load by 75% compared to the base case with annual energy savings.

Keywords: Trombe wall; Passive solar cooling; Energy efficiency; Thermal comfort.

1. Introduction
In Egypt, approximately 90% of the electricity produced depends on fossil fuel sources [1]. Similar to the elevated responsiveness of climate change in Egypt, the energy demand in buildings has also developed an increasing alarm in the challenge of the new power failure [2]. In 2015, residential electricity consumes the largest amount of the nationalized consumption of energy, nearly 52% (see Fig. 1) compared with 44.2% in 2014 [3].

2. Problem definition
Operation hours of air conditioners is rapidly increasing inside the residential homes sector through summer season, this is consequently responsible for a higher demand for heating and cooling, which already bills for 50% of power intake in Egypt [4]. The main problem in the design process is the lack of understanding of the passive solar techniques inside the space in the initial design stages.

3. Research aim
The aim of the research to achieve the standard Comfort Conditions under operation condition and decrease energy consumption as long as possible by adapting passive solar technique.

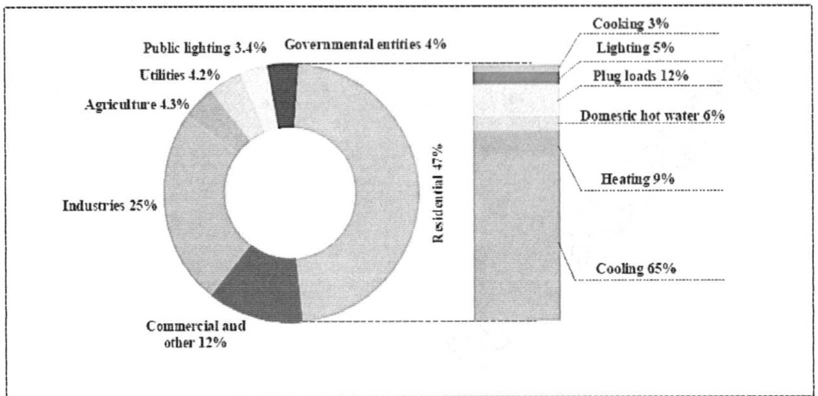

Fig. (1) Energy consumption sectors in Egypt (Ministry of Electricity and Energy) [5]

4. Passive Solar Techniques

4.1 Passive Solar Design Strategies Package

Passive Solar Architecture is a way of designing buildings that takes advantage of the benefits of the resident environment (for example sunlight) while reducing the unfavourable effects of the climate (for instance cold night-time temperatures) on the comfort level of the building.

The idea of passive solar is straightforward but applying it efficiently does need data and consideration to the details of design and construction. passive solar techniques are simple and low-budget and need only little modifications in a builder's standard preparation. A passive solar system can exactly remove a house's need [6].

4.2 Advantages of Passive Solar technique

The Advantages of Passive Solar technique can summarize in:

- **Energy performance:** Minimize energy bills all year-round.
- **Smart living environment:** huge windows and visions. Sunny interiors, open floor plans
- **Comfort:** noiseless, hard structure, cooler in summer, warmer in winter.
- **Value:** extraordinary owner approval, as customers buy more digital works, resale rights will become a bigger issue
- **Low Maintenance:** strong, reduced controlling and maintenances
- Investment: independence from future increases in fuel costs, will remain to keep money long after any initial costs have been improved
- **Environmental Concerns:** fresh, renewable energy to prevent growing alarms over global warming, acid rain and ozone reduction [7].

Passive solar technique provides a method of decreasing the amount of energy required to cool buildings to a suitable comfort level, by changing some of the heat resulting from biomass or fossil fuels with heat resulting from sunlight. Sunlight is permitted and has none of the negative economic, environmental or health properties of biomass and fossil fuel use [8].

4.3 Passive Solar Building Configurations

Passive Solar Design Strategies can be classified in five basic parts as shown in Fig 2. [9].

4.3.1 Passive Solar Concepts
There are four interconnected elements in passive solar buildings, which operate at the same time to build the structures well-organized of energy efficiency:
1. Assembly and absorption of the maximum amount of solar radiation during the day
2. Packing of the heat collected from the sun's radiation during the day
3. Emission of this heat into the internal spaces of the building during the night [10].
4. Insulation of the full construction to keep as much of the heat as imaginable in the building [11].

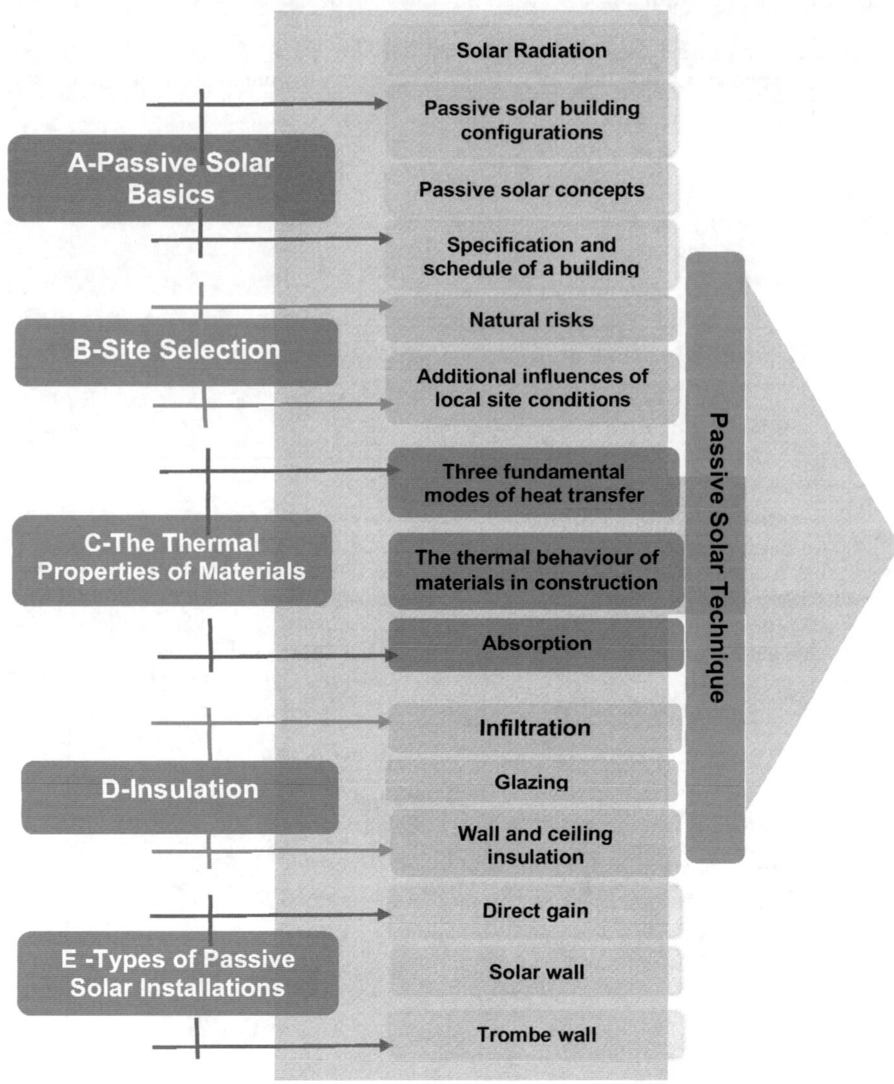

Fig. (2) Classification of Passive Solar Technique.

4.4 Wall Insulation

There are several different types of walls:

- **Partition Walls:** The closest area is a buffer zone where the temperature level is much greater than the outside temperature level. Minimal insulation is essential.
- **The wall is buried partially underground:** The ground is usually at a higher temperate than the outside air, so less insulation is required
- **The wall is external and east, west or south-oriented:** In this case, the wall must be carefully insulated [12].
- **Wall Design:** An insulated wall is consisting of three layers:
  1-An external wall to keep insulation from animals and rain
  2-Insulation layers
  3-An internal wall for thermal storage

This type of wall is identified as a cavity wall, and it is occupied with insulating materials. The external wall is the supporting wall and protects the insulation layer from moisture and animals [13].

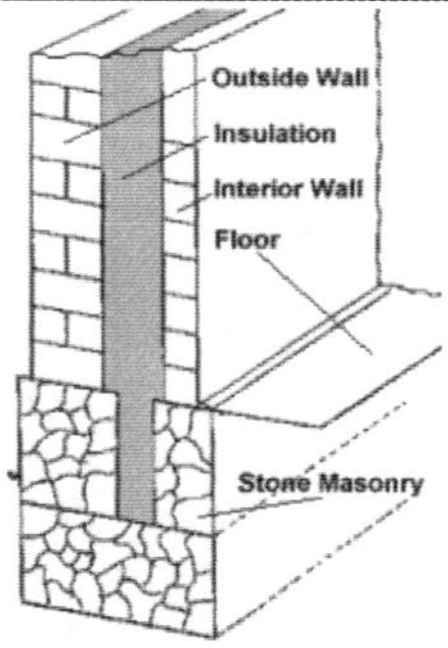

Fig. (3) Cavity Wall Cross Section

4.5 Solar Wall

Solar Wall: A sun wall is south oriented black painted a glazed wall. The black painted wall catches the solar radiation and by using the glass covering, the wall stays insulated from the weather outdoor so the heat is saved and migrates slowly to the inside [14]. A conceptual example of a few basic solar wall configurations is shown in Fig. 5.

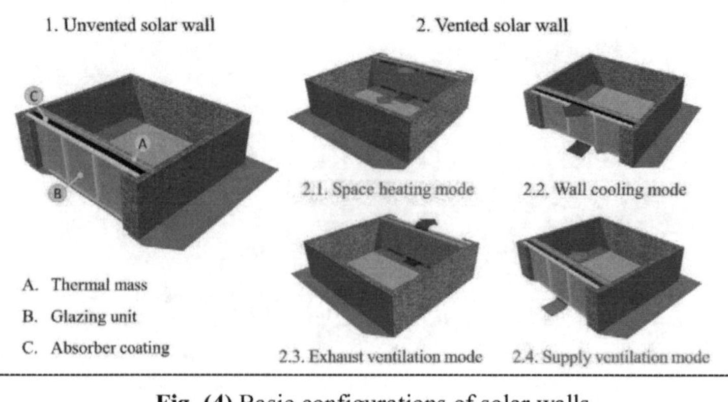

Fig. (4) Basic configurations of solar walls

A solar wall includes four major elements: (1) thermal storage mass, i.e. wall; (2) absorber coating on the exterior face of the wall; (3) glazing element installed in the external face of the wall; and (4) air gap between the internal wall and the external layer of the wall.

4.6 Trombe Wall

Trombe Wall: A Trombe wall is a ventilated solar wall. There is a vent in the upper and lowest of the wall. During the day, the airflow trapped between the external layer and the internal of the wall heats up and rises (warm air tends to go up) [15].

Practical Features of Trombe Wall: A familiar Trombe wall contains the following features/technical aspects:

- 10 – 41 cm thick masonry wall is painted with a dark, heat-absorbing color that is typically black in color.
- Covered with insulated glass (double glazing) or low emissivity glass (Low e-glass).
- The glass is laid at 2 – 5cm distant from the masonry wall that allows you to construct small air space.
- It must be South facing.
- For residential buildings in the southern facade, Trombe walls should be installed in the south- certain orientation for highest efficiency and extreme usefulness, as that is the side that will be exposed to the sunlight rays.
- Extraordinary transmission glass increases solar gains to the masonry wall.
- The glass layer avoids the leakage of radiant heat from the warm facade of the storage wall.
- The air gap prevents the heat radiated by the wall, controlling to additional heating of the wall surface [16].

5. Methodology

In this study, a BASE CASE model was simulated with Environmental simulation software (Design-Builder for thermal and energy) [17]. with the application of Trombe wall Techniques as a part of achieving sustainable buildings and increasing demands for energy in buildings. The results proved that the application of Trombe wall in Egypt had the main impact on enhancing performance and as a result, a reduction in energy usage. Moreover, a comparison between the base case and the final improved proposals in the two levels of technology will show the helpfulness and implantation of each case study.

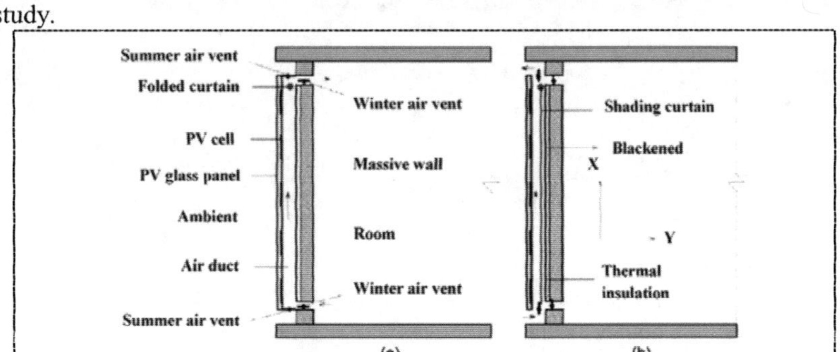

Fig. (5) Cross-section of a Trombe wall

5.1 Study Models design & Office Building model description

General villa building design data (see Table 1):

Table 1. Residential Building Architectural Data

Building type	Residential Buildings.
Building floors numbers	Ground floor + first floors 4.0 m (slab to slab), 3.70 m ~ 3.0 m (clear height)
Typical office floor	(4.0 m * 2 floors) + 2.0 m parapet of façade= 10.0m
Ventilation system	For "Base-Case" building: Artificial Mechanical ventilation (Full HVAC) system. For a Trombe wall case study: 'Mixed-Mode' (Hybrid) system all of the year.

A methodology with the flow chart short the investigation is proven in Fig.1

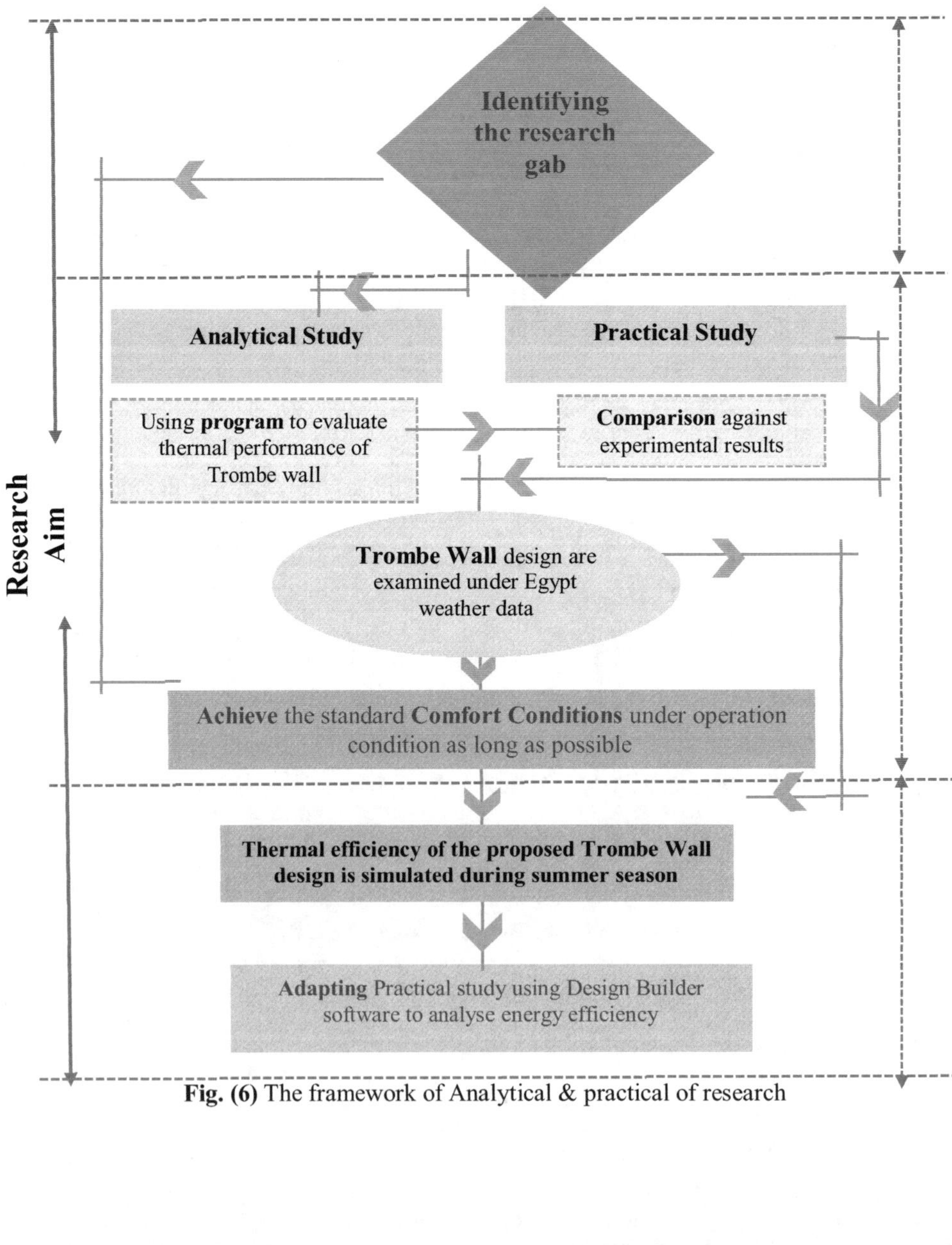

Fig. (6) The framework of Analytical & practical of research

IOP Conf. Series: Materials Science and Engineering **974** (2020) 012023 doi:10.1088/1757-899X/974/1/012023

5.2 Weather data file

The Egyptian Typical Meteorological Year (ETMY), the International Weather for Energy Calculations (IWEC) and Köppen categorized Cairo climate as a warm semi-arid climate with a hot to extremely warm dry summer and moderate to warm moist winter [18].

5.3 Standard Model Definition

The "standard model" was defined considering a one-story isolated room, with regular geometry 27 x 7 m (94.5 m²), a ceiling height of 3.5 m, and a total dimension of 15 x 3.5 m (52.5 m²) & 12 x 3.5 m (42 m²) for the façade glazing building block composition (Figure 8,9&10). For the Trombe wall module, the use of double glazing with a high shading coefficient was measured. The double glazing had two panes built of 6mm and 13mm Argon air space as shown in Fig 11. [19].

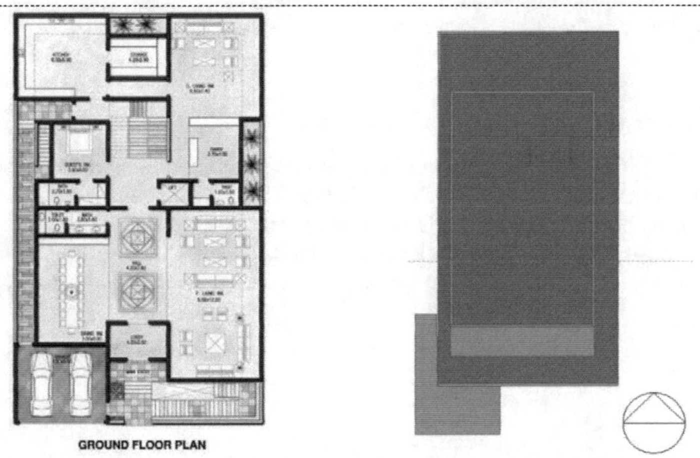

Fig. (7) A) Architecture plans of case study, b) view of base case Models, Design Builder Screen shoot.

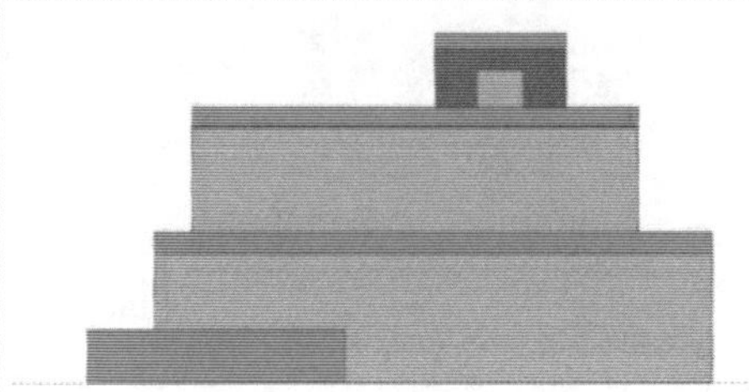

Fig.(8) Energy efficient Trombe walls, Design Builder Screen shoot.

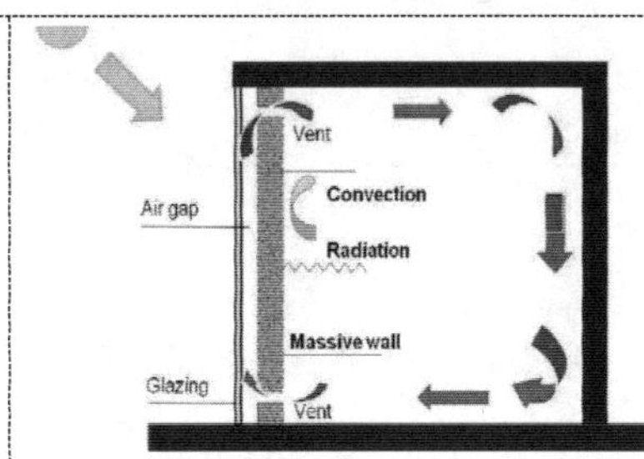

Fig. (9) Cross Section of Trombe walls

IOP Conf. Series: Materials Science and Engineering **974** (2020) 012023 doi:10.1088/1757-899X/974/1/012023

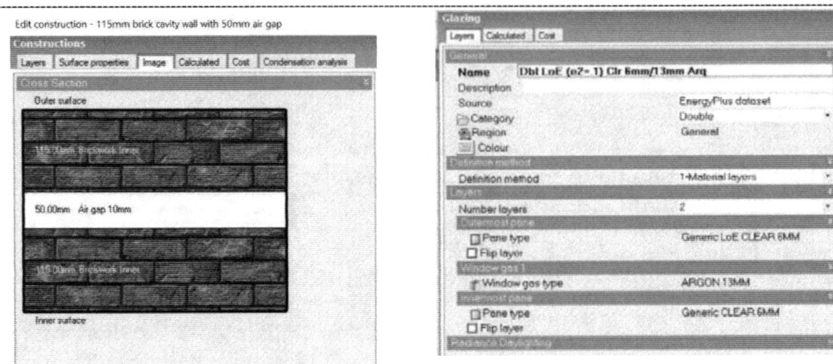

Fig. (10) Construction Details of a) Trombe wall, Screen shoot

6. Results and discussion

Figure 12 &13 shows the Zone sensible cooling for Hybrid Ventilation modes with the base case (without Trombe Wall) with Full (HVAC) Mechanical Ventilation (Fully Air-conditioned). This clearly shows that the best ventilation mode was the hybrid case (with Trombe Wall) which made a reduction in zone sensible cooling more than 50%. Figure. 12 &13 Shows the Effect of applying the Trombe Wall Strategy on annual sensible cooling with 247565 kWh compared to the base case 325947 kWh.

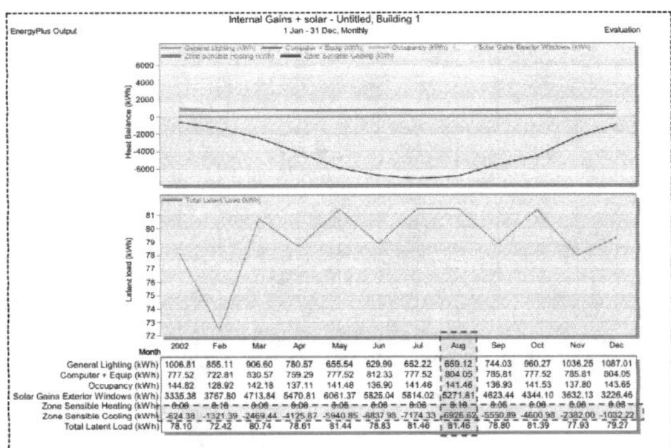

Fig. (11) Monthly results of Zone Sensible Cooling (kWh) for Trombe Wall case studies.

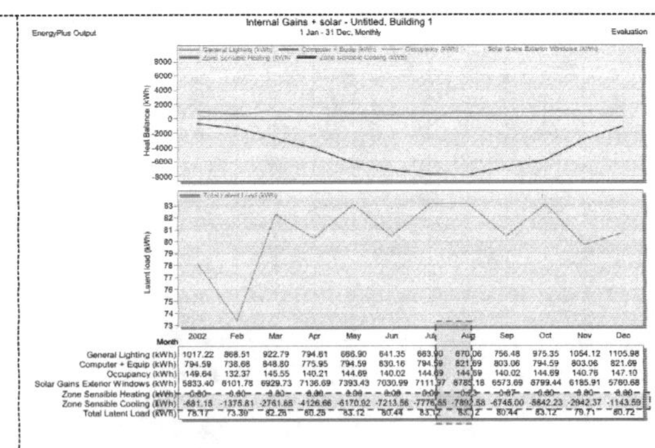

Fig. (12) Monthly results of Zone Sensible Cooling (kWh) for base case (without Trombe Wall).

Figures 14&15, In applying the Fanger PMV on the Trombe wall case study, it is clear that all lay between arranging of (-1 to 1), whereas the base case reached above 4.but still the most effective was the base case without Trombe Wall.

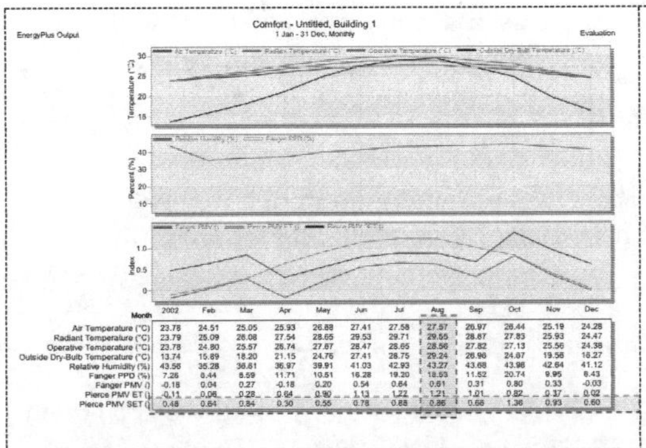

Fig. (13) Monthly results of PMV for Trombe Wall case studies.

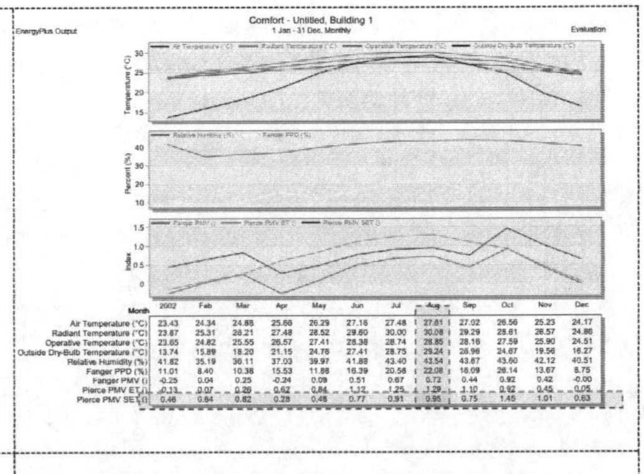

Fig. (14) Monthly results of PMV for base case (without Trombe Wall).

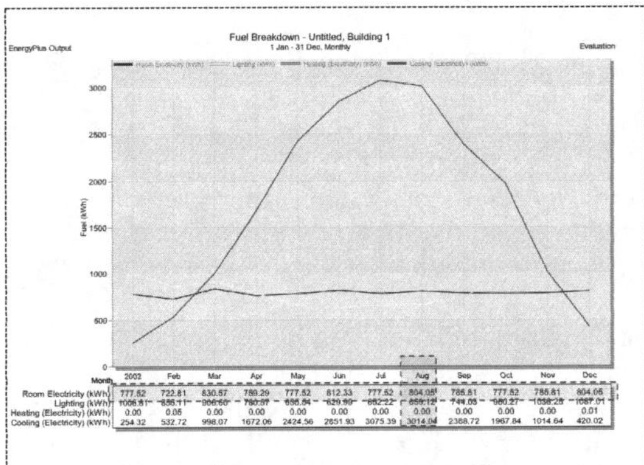

Fig. (15) Monthly results of Building Electricity (kWh) for Trombe Wall case studies.

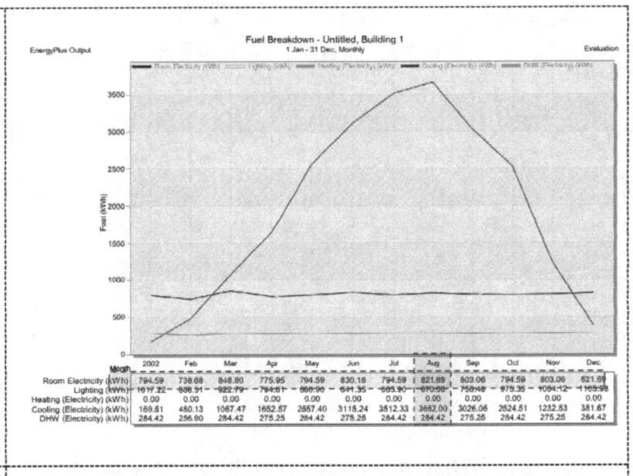

Fig. (16) Monthly results of Building Electricity (kWh) for Trombe Wall case studies.

Figure. 16 Shows the Effect of applying the Trombe Wall Strategy on room electricity with 284 kWh compared to the base case 3014 kWh.

7. Summary and conclusion

Obviously, the planned Trombe wall was presented to passively supply sensible comfort conditions in Egyptian residential buildings. The Trombe wall shows higher saving in total energy consumption compared with a conventional Trombe wall. The results, according to simulations, showed that the as-built base-case consumes 2730 kWh, which is greater with 75% than the modified-case that adopts passive design concepts. The Trombe wall can be observed as high energy conservation, improve thermal comfort, reduction in CO_2 emissions and producing ventilation in subfloor spaces. A Trombe wall can reduce more than 75% of zone sensible cooling. The economics of Trombe wall is dependent on many variable factors such as climatic condition, cost of materials, cost of power, cost of energy and type of construction.

11. References

IOP Conf. Series: Materials Science and Engineering **974** (2020) 012023 doi:10.1088/1757-899X/974/1/012023

[1] Ahmed M A (2012) Natural Ventilation Techniques as a Base for Environmental Passive Architecture With Special Reference to Residential Buildings in Greater Cairo. Cairo: Ain Shams University

[2] Egyptian Electricity Holding Company-Annual Report 2015/2016, Ministry of Electricity & Renewable Energy, Egypt, n.d

[3] Certain elements of the text have been extracted and modified from the CD-Rom Solar Bioclimatic Architecture, Lior Environmental, Charlierlaan 78, B1560 Hoeliaart, Belgium.

[4] NREA (New & Renewable Energy Authority), (2013). Ministry of Electricity & Renewable Energy Annual Report 2012/2013, Cairo.

[5] Ministry of Electricity and Energy , [online]. Available from: http://www.moee.gov.eg/english/e-frmain.htm

[6] Jaber, S., Ajib, S., 2011. Optimum design of Trombe wall system in mediterranean region. Sol. Energy 85, 1891–1898. http://dx.doi.org/ 10.1016/j.solener.2011.04.025.

[7] Abdallah, A., Yoshino, H., Goto, T., Enteria, N., Radwan, M., Eid, M., 2013. Integration of evaporative cooling technique with a solar chimney to improve indoor thermal environment in the New Assiut City, Egypt. Int. J. Energy Environ. Eng. 4, 1–15. http://dx.doi.org/10.1186/2251-6832-4-45

[8] A.A. Saleem, A.K. Abel-Rahman, A.H.H. Ali, Experimental Study on Thermal Comfort Conditions in Existing Public Primary Schools Buildings in Upper Egypt, in Sustain. Energy Build. Res. Adv., Future Technology Press, Cardiff, Wales, UK, 2014: p. 6. http://nimbusvault.net/publications/koala/sebr

[9] W. Weiss Editor “Solar Heating Systems for Houses - A design handbook for solar combi system” -James and James 2003.

[10] Norbert Lechner - Heating, Cooling, Lighting: Sustainable Design Methods for Architects, Wiley Publishers, 2015.

[11] MC Munari Probst, C Roecker, Architectural integration and design of solar thermal systems, PPUR -Routledge, Taylor&Francis, 2011

[11] B. Kundakci Koyunbaba, Z. Yilmaz, The comparison of Trombe wall systems with single glass, double glass and PV panels, Renew. Energy. 45 (2012) 111–118. doi:10.1016/j.renene.2012.02.026

[12] Z. Yilmaz, A. Basak Kundakci, An approach for energy-conscious renovation of residential buildings in Istanbul by Trombe wall system, Build. Environ. 43 (2008) 508– 517. doi:10.1016/j.buildenv.2006.11.033

[13] Marwa Dabaieha and Ahmed Elbablyb - Ventilated Trombe wall as a passive solar heating and cooling retrofitting approach; a low-tech design for off-grid settlements in semi-arid climates, Solar Energy, Volume 122, December 2015, Pages 820–833.

[14] Kreider, J.F., Kreith, F., 1982. Solar Heating and Cooling: Active and Passive Design. Hemisphere Pub. Corp., Washington.

[15] Chen, B., Chen, X., Ding, Y.H., Jia, X., 2006. Shading effects on the winter thermal performance of the Trombe wall air gap: an experimental study in Dalian. Renew. Energy 31, 1961–1971. http://dx.doi. org/10.1016/j.renene.2005.07.014

[16] Dowson, M., Harrison, D. & Dehouche, Z., Trombe walls with nanoporous aerogel insulation applied to UK housing refurbishments. International Journal of Smart and Nano Materials, 5(4), pp. 283–303, 2014.

[17] Design Builder 2018. Design Builder [Online], Available: http://www.designbuilder.co.uk/. Accessed 15/2/2020

[18] EMA (Egyptian Meteorological Authority), 2014. Meteorological Annual Report 2014. Egyptian Meteorological Authority, Cairo.

[19] Design Builder 2019. Design Builder Software [Online] (Gas glow Scotland: Design Builder Software Ltd).

Military Technical College
Kobry El-Kobbah,
Cairo, Egypt

13th International Conference on Civil and Architecture Engineering ICCAE-13-2020

The philosophy of visual perception of the external space Using the concept of biophilic design

Enjy Ibrahim[1], Khaled M. Khorshed[2], Sherif Sabri[3]

[1] PhD researcher, Department of Architecture, Al-Aznar University.

[2] Professor, Department of Architecture, Al-Azhar University.

[3] Professor, Department of Urban Planning, Al-Azhar University.

Corresponding author: enjyyusuf123@gmail.com

The research deals with the study of the integrated relationship between the elements of visual perceptions of the void and its interaction with the vital elements of it through the study and address the concept and philosophy of visual perceptions of the void and its development through the ages and the effect of these visual perceptions on the perception of life in the vacuum at its levels the three vitals It is the first level: belief in the existence of a life of visual formations and the sense of awareness of users in a dynamic interaction, and comes the second level: with regard to the mental and mental path of the user aware of these visual perceptions as physical elements with vital visions to convey the perception to the existence of real life and not Imagine it, level 3: The presence of the user in the scope of visual perceptions interactively with the characteristics of the formations that he realized between his mental perceptions and his inner feelings and his impression of the place and his human interaction towards it in the conscious and unconscious mind of his vital qualities surrounding the pulsating quality Life has vivid images that interact with the pulse of natural ingredients in the void The research has found that the interaction between human and visual perceptions in the void has exceeded the limits of time and space and has become more oriented towards biophilic design that enhances communication between nature and human, the architect has an important role in applying the concept of biophilic and expressing it to design elements space.

Key words: Quality of Life - biophilic Design - External space - Visual Perception

The problem of the study: The research problem lies in the loss of the population's dealing with the external space as a result of not feeling its belonging to it and benefiting from it and meeting their needs and feeling comfortable and gaining vital life energies from the design of this external space, which led to its neglect and neglect to its disappearance in some places and turn it into Bad or negative image. With human development, the environment around us is mostly made up of natural environments, but after the industrial revolution we find ourselves living and spending time in built environments, biophilic design translates the understanding of human convergence with natural systems and the so-called love of life into the design of the built environment. (According to the EPA, 90 of our days are spent at home EPA, 2004, rev2009), But as the world's population increases and urbanization expands as time is spent not in their homes in designed environments, research shows that attachment to nature is not just something that makes us feel comfortable, but vital to our health and well-being, understanding the love of life and how it teaches us how to design our environments more The importance of. (W, Browning, Ryan and Clancy 2014).

The hypothesis of the study: Through the research problem of loss of communication between the user and visual perceptions in space and even the extreme neglect and loss of these perceptions, the research assumes that there is a relationship between real life and imaginary perceptions in the external space and it relates to the human senses and his vision of those perceptions until they reach the level of love, hate, habituation and ten easily or simple through which the existence of a relationship dialogue of the conscious mind and the formation of the aware in a certain and imaginary reality exists and does not exist with the aim of moving from the abstract visual perception to the perception The bio-visual is most connected, co-existing and spacious with the user achieving the ultimate goal of quality of life.

The objective of the study: 1 - try to understand the role that the biophilic design of the external space can play in enhancing communication between nature and users, which is to develop an understanding of the subject and how it is applied - intentionally or not - such as (external spaces in some hospitals which are nature-oriented spaces aimed at Providing recreational, therapeutic or rehabilitation possibilities (theoretical study).
2- How can this understanding reach new biophilic design methods or ways of thinking that can be general and specific to the project in accordance with the quality of life requirements of the external space (analytical study)?

The methodology of the study: - first Theoretical approach: By addressing the previous studies that are related to the case of studying the biological design of external spaces and what are the elements of bio-visual perception in the space and related to the needs of the user of the vacuum environment with the recognition of the extent to use them to achieve quality of life For users.

Second Analytical approach: Through the analysis of new biophilic design methods and the development of visions to redesign the space vision of biophilic design and biophilia to achieve the method of quality of life of the user, and use some international experiences to reach the results of guidance for the designer in the future.

Introduction: Man by nature tends to nature with its scenic views that give serenity and comfort to the human soul and also positively affect his health and for this research touched on the biophilic design of the space that constitutes the place of nature in human society, where there are mutual relations, respect and enrichment relationships at all levels should appear as a rule and not an exception. The research dealt with the concept of quality of life biophilic to achieve happiness and belonging, and emphasizing the need for life outside the building to be more compatible and integrated with the human being, and to achieve the approach of biophilia. The human attachment to the natural environment of the space around him also makes him affect it positively by taking care of it and maintaining it and his sense of belonging to it and therefore the importance of the external environment came with the concept of biophilic design for the needs of users.

1-The concept and philosophy of visual perceptions of space and its evolution throughout the ages: The external space of the building block played a prominent role in human thought, as it was found in most human civilizations throughout history, taking into account the difference between the thought, the level of handling of each civilization, the extent of its development, the purpose and the role it played in achieving entertainment. **Tom Turner has known the space** as the planned place connected to the outside and organized the perceptions, the cultivation and enjoyment of plants and other natural formations and the combination of environmental natural and man-made materials, and its function is limited to aesthetics only. {1}
From the previous definition we have been looking at other definitions through successive civilizations so far, returning to the human cultural heritage we note that the external space is rooted (to the beginnings of the process of agriculture that paved the way for humans to begin to stabilize in the period of the Nedithic Neothetic revolution). Revolution that accompanied the end of the Stone Age more than ten thousand years ago {2}. This vision of the outer space is evolving through various original civilizations, from ancient Egyptian civilization and Mesopotamian, Greek, Persian, Roman, Chinese and Japanese Far Eastern civilization to Islamic civilization and its acquisition of deeper dimensions than symbolism. Social and architectural different from its predecessors.

2- The effect of visual perceptions on the perception of life in space:
Through the philosophical follow-up of some influential thinkers in human and social development with regard to human life, moving towards the depths of human and urban interaction and addressing the conscious mind aware of the orientation towards dynamic architectural and urban horizons, Until we reach a more spacious horizon for the unconscious interaction between the user and around him to increase his sense and sense of humor towards visual perceptions within the external space.
- Intellectuals and philosophers on the biophilic space: Intellectuals and philosophers in general have a variety of opinions about the perception of life in the space through the ages, both in the East and the West, and the reciprocal relationship between matter and meaning as a special degree of subjects of perception, Some architects have been concerned with the handling of life forms in external spaces and the diversity in feeling them, and among them;-

In another opinion of Christopher Alexander{3}: For his collection of works (The nature of order, the phenomenon of life), in his dealings about the feelings that many have about life in the perception of the cosmological system in general and in urbanization in particular, Through his studies over twenty years of thought and writing, the product of these books was that 'the sense of life in things is not fully compatible with the clear and well-known methods of scientific research to prove them despite our sense of existence', and therefore the thinker turned to rooting the concept of life in nature. By taking examples to convey to the reader what he feels to inspire his heart and his sense of purpose "By learning from the composition of life in nature and the overlapping elements of life that he referred to in the centers to integrate with each other in the form of the harmony of the composition of the urban ", he reached his vision that architecture has lost elements of life which led to the dynamics of composition and superficial formation and that it followed contemporary thought and its mechanisms At the end of his studies, he listed fifteen properties for the coherence of composition by its formative standards, one of which was simplicity and inner calm as a single moral element. {3}
Through these views about bio visual awareness in the external space from a natural perspective characterized by the belief that there is environmental and architectural compatibility and that urbanization is integral in this compatibility, and with some ideas of architects of the creation of innovation and plastic creativity and its association with the thought of realizing life in urbanization between reality and imagination, and by studying our vision of the physical architectural components in the space must acquire the vitality between reality and imagination, existence and absence, and these are realized In three levels{4}:

1- The first level: the user feels that there is life in the perceived elements.
2- Level two: Visualize things in the imaginary and mental dimension of vitally inspired elements.
3- Level three: Expressive projection between the properties of these elements and the characteristics of the perceived user, to fall on some perceptions that stimulate the perceived feelings of the user.

2.1. The method of bio visual perceptions in space {4}
2.1.1. Method 1: Inference of the existence of life: It can be inferred to prove the existence of life in the elements of visual perceptions in the biological environment by analyzing the sensory inputs of these perceptions in the mind of the viewer be a doctrine of the occurrence of the thing in the mind, and prove the existence of life must differentiate between the degree of life represented by awareness and will and the degree of expression of life represented by the manifestations of life and the ability to change things, While every sane organism has the will and purpose, its expressions vary in its gradual strength from being human to other living beings, Accordingly, the elements of visual perceptions in a vacuum are characterized by life according to the noble verses "The seven heavens and the earth and whatever is in them exalt Him. And there is not a thing except that it exalts [Allah] by His praise, but you do not understand their [way of] exalting. Indeed, He is ever forbearing and forgiving." (44) From Surat Al-Isra.
2.1.2. Method 2: Imaginary and mental level: The human mind, aware of the elements of emptiness in its imaginary and mental logic, realizes life in the perceptions around it to establish it in its conscious mind through feelings and sensations. But there remains a hypothesis related to the dependencies of these perceptions of the

IOP Conf. Series: Materials Science and Engineering **974** (2020) 012024 doi:10.1088/1757-899X/974/1/012024

fundamentals of life, like all creatures, and physiognomy is the link between the characteristics of life in its stereoscopic composition with features, that relationship is best known to Arabs for their ingenuity in imagination and perception about their analogy acquired the quality of life. The Arabs were like bending the branch of roses when wilted due to the emotional impact of the idea of weakness and bowing before death, To adapt and agree with it and come to mind an endless analogous picture that shares its formal vocabulary in nature with the emotional expression of the viewer like the palm of the palm, and the full moon's link in its illumination with the enlightening face, figure (1) shows a park bench in the shape of a human being with a dog, a children's play area with a human-shaped toy whose reputation is close to the ground to unleash children's imaginations, and the tree trunk on the ground formed to bear shape all these perceptions to unleash the user's imaginary and mental level.

Figure (1) - imaginary and mental level {10}

A case in point - th i.com)

One of the most dazzling scientific places in Dubai Green Planet The Green Planet is a miniature tropical jungle inside a origami style glass building with a biological dome and this vital dome provides a complete tropical system that includes more than 3000 species of plants and animals and The building is composed of four floors, air-conditioned and at a moderate temperature, to adapt to the diverse animals and birds in it. You will definitely notice a 25-foot tree, the largest man-made tree in the world, and you will see trees and three artificial waterfalls to give an air of coolness to a few animals and birds to provide them with a suitable natural environment, figure (2).

Figure 2 - Green Planet Dubai tropical forest

2.1.3. Third Level - Expressive Projection The process of simulation between a person and the surrounding visual perceptions is a product that generates spontaneous thoughts and mental perceptions in which a person shares with other beings the characteristics of life such as motherhood, affection, dependency, and containment as figure (3) Games for children in the form of bird's nest, in addition to projecting life to skill vocabulary in its rigid forms to give ambiguity to an understanding of a kind of life in the architectural composition and show the imaginative ability of the viewer to suggest expressing meanings and vital gains as the following image Containing the building for the external space, and these expressions include the curves that can be made in the perceived elements of the space in the horizontal and vertical dimensions and the third dimension (temporal), These expressions are the product of nature's

IOP Conf. Series: Materials Science and Engineering **974** (2020) 012024 doi:10.1088/1757-899X/974/1/012024

simulation and its characteristics and transfer of designer experiences to the perceived mind seen through the kinematic behavior of these elements.

Figure (3) - the level of expressive projection {10}

3-Vi ... **it** ... **phy:** {5}
With ... m ... all parts of the world, espe ... al ... what is used of materials in th ... en ... or the design of the outer spac... with nature that was lost in modern societies as a result of urbanization Which prompted some philosophers, including architects, to integrate the biophilic philosophy of space with design, which is known as (biophilic design) In this research, we will address evidence that confirms that there is a strong relationship between visual perceptions in space and that they retain vital qualities, The trend towards lively design aims to interact between interior and exterior design in addition to the free interaction within the outer space and its connection to natural elements, Philosophical architects have dealt with dynamic design patterns from many different perspectives. In one of them, Terrapin Bright Green 2014 book published the fourteen models of dynamic design by classifying them into three types as follows:
1. Nature in space. 2- Natural Analogues 3- The nature of space.

3.1: - Nature in space: {5}
3.1.1. Visual connection with nature (visual perception): Interactions through this pattern are made by visual contact between the viewer and the external nature, to bring about complete feeling, visual attention, motivation, and the acquisition of perceptions and other vital elements, This can be seen through the view between the interior space and the outer space through the windows, as figure (4), and watching the natural elements from trees, water and sky In other words, realize the plant elements, flowers and greenery by watching natural murals outside.

Figure 4 - Visual contact with nature with a view between the inner and outer spaces {5}

3-1-2 –Non-visual contact with nature (other senses): The interaction is through this pattern by hearing, such as the musical seat in figure (5), smelling, tasting and sensory feeling when touching and feeling things and other senses, To create a positive interaction with nature and its living systems through a sense of familiarity and comfort towards the perceived elements, their durability and balance, in addition to what it speaks of bird sounds and the scents of flowers and our feeling of being in a vital nature.

Figure 5: - Non-visual Communication with Nature {10}

3.1.3. Non-Rhythmic Sensory Stimuli: Interaction through this pattern is done by automatic, non-sensory communication, Through the rhythms nature occurs that can be perceived away from the senses, One of its characteristics is that it is suddenly aware and exciting and stimulates attention in a short time and always welcome, such as these interactions: - Kinetic rhythms that can be seen from different angles create an interactive design between the built mass and nature in space, and this is by the visual perceptions, Through other senses such as the rhythms of water fall or the buzzing of insects and the tweeting of birds as figure (6).

Figure 6: Non-Rhythmic Sensory Stimuli {5}

3.1.4. Thermal / Airflow Variability: This pattern is interacted by changing air currents loaded with temperatures and humidity between foreplay Collisions with the human body and surface temperatures that simulate the natural environment can be understood through a sense of freshness, activity, survival and freedom of movement Examples of this pattern include: air is interspersed with leaves and the rhythmic sounds that are visually recognizable as figure (7) The currents of the air approached the branches of the trees from each other form a bright heart and shaded on the rest of the corridor, and heard and felt through the rest of the senses.

Figure 7: Thermal / Airflow Variability {5}

3.1.5. Presence of Water: Interaction through this pattern is done by interacting with the 3D water visual perceptions and feeling of vision, hearing or touch The attraction, sound rhythm, stimulation, calmness, or both, through kinetic signal, such as: Dancing fountains or waterfalls, tranquil water pools and water murals as figure (8) are all reflected in the sky.

Figure 8: Presence of Water {5}

3.1.6. Dynamic and Diffused Light: Interaction through this pattern by providing variability diffuse light and the inter-distance of the observed elements in the biophilic space it is a movement and time, causing dramatic emotions, such as the spread of light and the creation of shadows on elements of visual perceptions outside the space Viewing them from several different angles creates visual cognitive multiplicity through the movement of the diffuse light that is reactive during the day as figure (9).

Figure 9: Dynamic and Diffused Light {5}

3.1.7. Connection to Natural Systems: Interaction through this pattern is done through visual and sensory communication with the conscious and unconscious natural environment system By summoning the relationship between the viewer to the visual perceptions of emptiness and elements of the natural living void, the life cycle of different seasons and its nostalgic and enlightening results, and the creation of depth and relaxation, Examples include: seating areas and celestial spaces added to buildings and their relationship to the biological environment, and elements of seasonal visual perceptions and their outcomes from the fall of leaves and flowers and their interaction with space and the creation of new visual perceptions as figure (10).

Figure 10: Seasonal interaction and visual trend {10}

2.3. Natural Anal of materials, colors and other patterns found in nature they appear in the form of perceptions of fixed and animated artwork and decorative elements of space in the form of the following patterns:

3.2.1 Biomorphic Forms and Patterns: - Interaction is achieved through this pattern by deriving symbolism from lines, textures, pluralism, autism and other forms embedded in the biological environment which makes us feel comfortable, twins, attractiveness and integration with visual perceptions of this space Examples include - devising the organic forms of the elements of visual perceptions in a vacuum, such as figure (11), the organic form of wood and lighting, and its harmonious, harmonious, and diverse plural colors in terms of shape and individuality of curves, refractions, and other geometric shapes.

3.2.2. Material Connection to Nature: The interaction is through this pattern by physical contact with the elements of visual perceptions in the space that are reflected through the environmental treatment of these elements To create a distinct and dynamic sense that increases human interaction with the vacuum environment that creates a sense of either warmth or originality and stimulates and inspires the feelings and feelings buried and unconscious in the viewer, It affects the psychological and physical state of the user, and this pattern is reflected in the visual perceptions through the use of wood, clay, leather, stones, wools, natural fabrics and other elements of the bio-environment as figure (12).

Figure 12: Material Connection to Nature {10}

3.2.3. Complexity and Order: Interactions through this pattern are made by means of the visual information and spatial observations that are caused by the dynamic space like hierarchical, rural, serial, intertwined, embracing, simplified, and other forms of spontaneous nature, to create a feeling in the viewer with participation and balance between systemic an etrical, overlapping, and embracing visual perc on, as figure (13).

Figure 13: Complexity and Order {5}

3.3. Nature of the space:

3.3.1. Prospect: The interaction is through this pattern by watching through depth, design, planning, opening and closing as figure (14) to gain the viewer a sense of security, control, familiarity and containment. Examples include visual perceptions on the floor of space and their orientation to depth, slopes, gradients, transparent elements, and scenes within sight and endless visual perceptions.

IOP Conf. Series: Materials Science and Engineering **974** (2020) 012024 doi:10.1088/1757-899X/974/1/012024

3.3.2. Refuge: Interaction through this pattern is carried out by escaping to the natural external environment and a place of resort to intimate compatibility with the elements of visual perceptions in space. To feel the twins and heal and gain positive energy and create separation and uniqueness with the ocean, for example: figure (15) of visual formations in the seating areas container such as pergolas and tree incubators, covered corridors and cultivated external terraces.

Figure 15: Refuge {5}

3.3.3. Mystery: The interaction is through this pattern by searching for more interactions with the vital external reality and devising concrete and intangible sensory depths to speak to the viewer, the visual perceived elements surprise, a kind of acquisition of ideas and elements of reward and exploration and penetration into the unknown Examples of these perceptions figure (16) of winding paths and labyrinths, research behind unknown audio and hidden scenes in other dimensions.

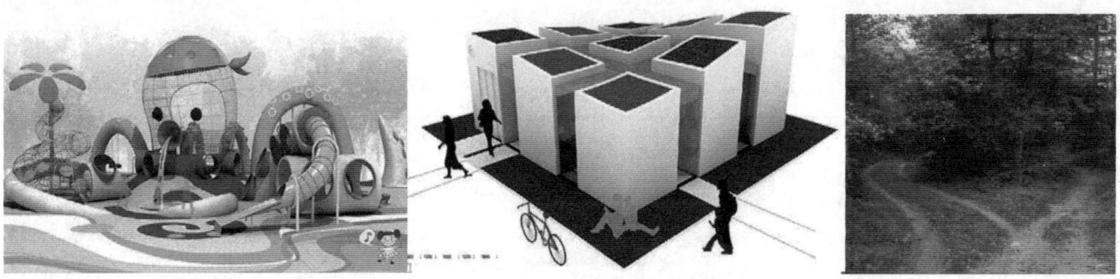

Figure 16: Mystery {10}

3.3.4. Risk/ Peril: - Interactions through this pattern are made by the viewer feeling that the perceived elements have a reliable risk of sudden ups or downs, Suspended, oblique, and tight forms, to make the viewer feel the effects of excitement and potential threat rhythm, but positively, which helps to explore and search for safety and resistance Examples of these elements - elevated formations, hanging passages, transparent barriers on heights as figure (17) and walking over water through transparent floors or the use of organic elements of their natural size posing a threat like spiders and snakes.

Figure17 Risk/ Peril {5}

IOP Conf. Series: Materials Science and Engineering **974** (2020) 012024 doi:10.1088/1757-899X/974/1/012024

4. Human interaction towards visual perceptions in the subconscious mind of the unconscious: {4} Figure (18) Life is the organizer of the elements of the vocabulary of the universe of humans, animals and inanimate objects, and that there is a relationship between these elements and life in terms of our awareness of consciousness and the unconscious. We notice that inanimate objects exist in life, but it does not realize awareness and the unconscious as for the animal, it realizes life with awareness only, so that man comes in a higher rank to realize the full aspects of life in the awareness of consciousness and the unconscious. This is what philosophers have dealt with about the concept of existence and life, and among them:

- **Reno Decartes (1650-1596)** where he discussed the dialectic correlation of reason and consciousness as a bilateral relationship on the basis that our awareness of things through the conscious mind and our own experiences what we realize with our conscious mind is nothing but the product of the contents of our minds and our cognitive experiences in the unconscious.

- **In another vision of Decartes,** he mentioned that every mental activity is by definition conscious activity and for the transition to a mindless subconscious that takes place through a self-contradiction in consciousness and subconscious.

- **In another opinion of the psychologist Freud**, the users realize things with the credibility of their existence and they realize other things with an unconscious conclusion of the true qualities of the things perceived in the unconscious, which confirms that the acquisition of life and the human qualities of visual perceptions take place in the Lao mind.

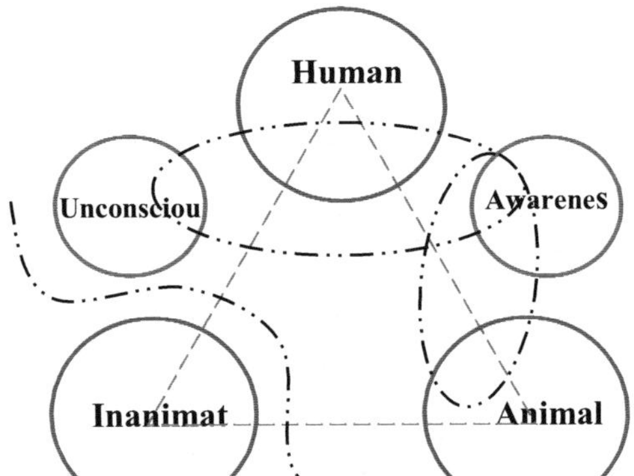

Figure (18) - The participation of humans, animals, and inanimate objects in the attribute of life with their difference in perception - Researcher Source

-From our faith background to perceive life in things, we derive from the Noble Qur'an. The Qur'an recognizes feelings for everything such as the weeping of heaven and earth {the heavens and the earth did not cry over them, and they were not seen.} (29) Surah Al-Dukhan, and the compassion of the earth, the sky, and the mountains, carry the faithfulness {Indeed, we offered the Trust to the heavens and the earth and the mountains, and they declined to bear it and feared it; but man [undertook to] bear it. Indeed, he was unjust and ignorant} (72) Surat al-Ahzab The reverence and rifting of the mountains of God fearing { If this Qur'an were revealed to us on a mountain, you would see him as submissive, rifting from the fear of God and those proverbs we give to people so that they may think} (21) Surat Al-Hashr, { Have you not seen that God worship him in the heavens and the earth, the sun, moon, stars, mountains, trees and animals and many of the people and many of the right to torment him and Lehne God He has no honor but God does what He wills} (Hajj 18) {You have never seen God be allowed to be in heaven, to be exposed, to the light and to fly. All his prays have been heard and allowed, and God is a day when they do} El-Nour 41{And from his signs that you see the earth as humbled, so when we descend upon the water on her, she is shaken, and she is raised, that the one who revives it to me} Phosellat 39, These are some Qur'an verses that give life everything, and this confirms the research hypothesis that it senses life in the rigid visual perceptions to create this consciousness with certainty of jealousy Feeling this effect and interacting with it on the psychological level despite its relative and stable variation in the subconscious mind affecting the visual

IOP Conf. Series: Materials Science and Engineering **974** (2020) 012024 doi:10.1088/1757-899X/974/1/012024

perception, and despite the difference between our awareness in the subconscious and the level of its expression in visual perceptions. Every living being has expressive intentions for its life and its administrative ability to act, graded in terms of strength and weakness between humans, animals, and inanimate objects. In light of this, the question arises whether each of the perceived elements is separate in its vital itself or is it in its comprehensive perspective within the space.

5- The vital characteristics of visual preceptors {6}

Visual perceptions in a vacuum have many characteristics that follow the mind map of the recipient, and through them we perceive these elements and explain them through human interaction towards these perceptions in the conscious mind of the unconscious and from these characteristics: The recipient's sense of its vitality due to his knowledge, observations, and experiences, and his ability to visualize, analyze, and produce, so that these characteristics range from being real to imaginary, from being to non-being, **Rapoport defined** properties in a vacuum with diversity and assimilation, according to specific laws , In Lironzaft's opinion, they are internal representations of the properties and relative locations of people and things in space , At a time when the viewer can see a house on the top of the mountain, at the edge of the river, or in the middle of the forest, he can transfer this image from the real reality to another imaginary reality and thus he has moved from being to non-existence and these characteristics have vital dimensions of it :

A - Bio-stability: where the mental image of visual perceptions interacts with any accident change due to the multiplicity of factors affecting this change and how to restore the true image with its vital characteristics.

B - Inaccuracy in symmetry: the visual perceptions are not the same in accuracy, but retain the subject figure (19).

Figure 19: Inaccuracy in symmetry in natural trees, unlike artificial trees {11}

C-Generaliza ... entioned differences and differences, which are sometimes clear and apparent, such as the composition of benches from tree wood and designing a general seat for a group and containing the trees as picture (19).

Figure (20) - Circular adjective forms {10}

D- Biasing: Adopting the idea of designing visual perceptions in their vital form, because their owners are biased and fanatical of that quality to preserve their nature and communicate with the vital environment.

E- Skip time and space: This is done through visual perception of these elements in the spatial space in a time between the past and the future, and vice versa, the perception of time in a realistic and imaginary place.

W-Tradition and simulation: the visual elements in space are perceived by a direct and indirect method, as directness can imitate and perceive a vital element, and the indirect is the inspiration for this living formation in the form of figure (21, 22).

IOP Conf. Series: Materials Science and Engineering **974** (2020) 012024 doi:10.1088/1757-899X/974/1/012024

Marwa Magdi, speaking about her artistic work, said: In my style of sculpture known as biomorphic, the artist extends his imagination to the fullest extent to find a relationship with the forms of the plant and human features which are images or forms, and refers to living forms such as plant, animal or human body or conjure up and see in the work of the sculptor As a mysterious cavity and organic lines inspired by nature is an irregular lyrical form of a living being but not specific or distinctive that the recipient can see according to his earlier fantasies, experiences and feelings of his own . {7}

Figure (22) - Simulation of a vital component and indirect awareness {7}

6- Due considerations in achieving the visual image and visual perceptions in architectural design

6-1- Achieving vibrant quality:

The visual perceptions of a vibrant space in order to achieve its goal must be a dominant force directed towards achieving the ability to live in a vibrant space, The vibrant elements are a spectrum of perceptions characterized by vitality, change and strengthening of the societal personality in the light of human interaction with them and achieving the concept of quality of life And that is by not directing the person to a feeling of happiness, love of life, and acceptance of the self-merging between personality, place and time in a harmonious and harmonious image that achieves the best ability to deal efficiently with the most severe conditions of life surrounding a person.

Figure 25: Accepting the self that is integrated between personality, time and space

6-2- Achieving vibrant quality visual perceptions :{ 8}

1- Integration, compatibility and overlap between functional elements in place and visual perceptions in a vibrant and dynamic approach.
2- Pictures of mixing the different activities of the population that gather throughout the day in an interactive environment, and the activities and visual perceptions in the void in a vibrant harmony.
3- Societal stability, despite its diversity and cultural and intellectual diversity, as a result of mixing and vital integration in the place and not feeling the difference and providing opportunities for interaction and awareness of the vibrant elements of emptiness.
4- A vibrant space has the potential and the ideals to exist despite differing self-interests and societal beliefs to achieve the dimensions of security, trust and belonging.
5- Space with vibrant perceived elements has the characteristic of sustainability as it can withstand the natural influences of sun, wind, rain and energy.

Results:-

1- The interaction between man and visual perceptions in a vacuum has transcended the boundaries of time and space, and these perceptions have become more oriented towards a dynamic design that strengthens the connection between nature and man.
2- The vitality and perceptions of the void must be preserved, the ability to attract attention, coexistence, and integration with the void and not be ignored.
3- The manufacture of biophilic space is a sequential process that starts from the use of space and then planning and then the stage of dynamic design and finally visualization of visual perceptions in the third dimension of space and its integration with the biological nature.
4- The human mind perceives life in visual perceptions with multiple images to give hidden emotion to it.
5- The unconscious mind, with its cultural and human background, and those who have an ideological background believe in life inanimate objects.
6- The study invites us to formulate the void with its visual perceptions on the basis of the interaction between the two parties, one of whom is alive and the human being and the other is the void with its vital components.
7- The architect has an important role in applying the concept of biophilic and expressing it to design elements of emptiness.
8- The process of coordinating visual perceptions can play a vital role in achieving environmental and physical balance.
9- Applying biophilic concept in the design of vacuum elements meets one of the requirements of quality of life.

Recommendations:-

1- The research recommends achieving a relationship between real and imaginary life in visual perceptions of the outer space.
2- The researcher recommends the transition from abstract visual perception to vital visual perception most relevant to the user.
3- The ultimate goal in voids must be achieved by achieving vital quality of life.
4- The necessity of fabricating innovation and visual creativity to achieve visual awareness that develops the imaginary and mental dimension of the viewer.
5- Visual effects and perceived elements must be created within the void to realize the concept of a natural life in the environment.
6- Ensuring the appearances of life and the ability to change in the perceptive elements of the viewer to form his belief that something happened in the mind.
7- The concept of biophilic is expected to transform as a developmental idea for the environment as a whole and the development of its various elements.
8- The architectural and urban designer must take advantage of the elements of the natural environment in its vital concept in the visual perceptions of space.

Recommendations for a future vision

1- The necessity of directing research on the concepts of residential quality of life and the biophilic design of spaces.
2- It is necessary to note the necessity of having a research center specialized in the field of biophilic-design in accordance with the foundations and standards of quality of life that contributes to housing decisions and serves as a house of expertise and has the authority to issue decisions according to new standards.

IOP Conf. Series: Materials Science and Engineering **974** (2020) 012024 doi:10.1088/1757-899X/974/1/012024

3- The necessity of communication between research centers and the residential community in order to develop renewed and continuous standards for the needs of society.

References: -

1- Turner, Tom," Garden history: philosophy and design, 2000 BC-2000 AD", New York, Spon Press, 2005.p5-23.

2- Al-Sabahi,Aref The effect of perception, imagination, and mental images on architectural design, College of Engineering and Architecture, Ibb University, Pen Magazine Issue 5 - (January/June 2016) p. 202.

3- Alexander, Christopher (2004), "The Nature of Order, The phenomenon of Life" (2nd book:" The process of creating life") (3rd Book:" A vision of a living world") (4th book: "The luminous ground"), an essay on the Art of Building and the Nature of the Universe", the center of Environmental Structure, Berkeley, California.p.25

4- Al-Hussein ,M 'Understanding life in the components of urbanization between truth and metaphor is an introduction to generating the influence ranges around it' - September 2012.p24

5- William Browning, Catherine Ryan & Joseph Clancy (2014) "14 Patterns of Biophilic Design", Terrapin Bright Green, New York. Available from: http://www.terrapinbrightgreen.com/report/14-patterns/.p5-12

6- Aref Abdullah Al-Sabahi "The Impact of Perception, Imagination, and Mental Image on Architectural Design". College of Engineering and Architecture - Ibb University - Al-Qalam Magazine (Scientific - Refereed), Fifth Issue - (January / June 2016).

7- https://www.youm7.com/story/2019/3/7.

8- Clarity and Visual Ratios of the Vibrant Place Making-Urban Planning Iraq Postmodern Urban and Regional Planning in Iraq Volume 2018.

9- Dickau, Ruth," Farming older than thought", University of Calgary Web site, 19/2/2007, http://www.ucalgary.ca/news/feb2007/early-farming.p1.

10- https://www.mazra3a.net/vb/showthread.php?t=19806 7-6-2019

Military Technical College Kobry El-Kobbah, Cairo, Egypt

13th International Conference on Civil and Architecture Engineering ICCAE-13-2020

Environmental Assessment of Coastal Resorts: Recreational Spaces as a Case Study

Mohamed Sobhi Hassan[1] , Wael Ahmed Sheta[2] , Ahmed Mohamed El Kordy[3]
[1] Lecturer Assistant of Architecture
Department of Architecture, Faculty of Engineering, Al-Azhar University- Cairo.
sobhieg@hotmail.com +201001007403
[2] Assistant Professor of Architecture,
Department of Architecture, Faculty of Engineering, Al-Azhar University- Cairo
wael.sheta@azhar.edu.eg
[3] Professor of Architecture
Department of Architecture, Faculty of Engineering, Al-Azhar University- Cairo
ahmed.elkordy@ymail.com

Abstract. Tourism is one of the most important resources of national income in Egypt. Tourism as a vital role in bringing economic benefits to local communities, helping with conservation efforts or placing value on aspects of cultural heritage. Egypt is one of the world's most attractive countries in terms of its natural and historical characteristics, which makes it eligible to play an important role in the international tourism market. The vision of Egypt 2030 urges all sectors of the state to adhere to the axes of sustainability. Thus, the tourism sector is studying the current situation in the hotel industry and exploring the possibility of applying sustainability to hotels and resorts. This paper studies some of the sustainable practices in outdoor recreational spaces (especially in landscape) affects positively on conservation and the reduction of operation cost. The research method applies the seven principals of Xeriscape, which is considered as one of the sustainable benchmarks to evaluate landscape practices. The research aiming at assessing the coastal resorts in the Red Sea, Egypt from an environmental standpoint, analysing the operation practices in the recreational landscape which affects positively resources, conservation and the reduction of operation cost.

Keywords: Environmental Assessment, Recreational Spaces, Sustainability, Tourism Development.

1-Introduction

According to definition, Recreational areas would include land that is designed, constructed, designated, or used for recreational activities. Examples are national, state, county, or city parks, other outdoor recreational areas such as golf courses or swimming pools and bodies of waters (coastal areas) when used by the public for swimming, diving or boating. Public and private areas that are predictably used for hunting, fishing, bird watching, bike riding, hiking, or camping or other recreational use also would be considered recreational areas. Most of the tourist activities in Egypt, particularly in Red Sea clustering in coastal areas. The coastal zone has gained a solid reputation as a place vacated for recreational activities and this is generally related to the presence of the sea. A resort is a full-service lodging facility that provides access to or offers a range of amenities and recreation facilities to emphasize a leisure experience. Resorts based on two sectors. The first is income-generating sector such as hotel rooms, conferences halls, outdoor restaurants and retail shops. The second is income-expensing such as swimming pools, gardens, sports yards and kids playing area. Resorts work as a primary provider of the guests' experience, often provide services for business or meetings, and are characteristically located in vacation-oriented settings. According to Brey [1] the minimum qualifications of a resort, related to recreation amenities, should provide one signature amenity or anchor attribute, provide five secondary recreation/leisure/entertainment experiences and emphasize a leisure or retreat-environment experience. According to Egyptian Tourism Development Authority (ETDA) resort ca be defined as an integrated tourist development project with an approximately total plot area of 500,000 sqm or more, that includes many hotels and tourist facilities, secured and surrounded by fences and managed by investor (developer or development company) who provides all project infrastructure and facilities of water and energy. Resort recreation offerings have grown over the years. Early resorts could focus on one amenity and be prosperous. Ski resorts didn't seem to need more than an alpine ski area and beach resorts could rely on the beach alone for visitors. But as the resort industry grew, multiple amenity offerings became the norm and consumers became conditioned to expect certain amenities as part of the resort experience [2]. Recreation participation changes, vacationer demand, and industry influence have swayed resort development over the years, so much so, that for each decade, a new seemingly "essential" recreation amenity appeared to emerge. A swimming pool was one of the earliest expected resort facilities, followed by tennis in the 70s, fitness in the 80s, golf in the 90s, and spa in the 2000s. Amenity facilities and related services can make or break the success of a recreation amenity department and is a way for resorts to stay competitive in the market [3]. According to recent studies, 40% of hospitality managers indicated they were ready to make 'recreation area' structural improvements, with twenty percent (20%) also specifying spa improvements [4]. Because today's travelers expect much from their recreation dollar and are offered more ways to spend that dollar, resorts must anticipate changing consumer preferences and cater to those needs [5]. During economic crises, vacationers are more discriminating when choosing a resort destination. Amenities often define a resort [6]. When making a vacation decision, tourism experts will cite resort offerings and amenities as one of the key deciding factors; that are 'any extra product or service found in the hotel [7]. This paper studies one of the new tourist developments in Red Sea region, Egypt. The purpose is to provide a comprehensive sight on the sustainable practices in coastal resorts from an environmental standpoint, particularly on resort recreational facilities and service offerings. The paper aiming at providing an environmental assessment to these facilities to establish a benchmark to enable appropriate solutions to be chosen at the early stages of a design to achieve more sustainable developments in tourism sector.

2- Literature review

A considerable amount of literature has been published on tourism sustainable development. In recent years, increasing attention has been paid to research regarding the evaluation of open spaces -including green areas- and their components as well as their effect on the environment. In Red Sea governorate, no such analytical research or evaluation relating to the effects of sustainability on open spaces and operating cost values has been made so far. This lack has been to the detriment of the development in all resorts in red sea governorate. However, the open spaces have increasingly greater roles in the sustainable city development processes because of the expansion of the tourist projects and also of the growing of the tourists' number. Globally, a large and growing body of literature has investigated environment friendly practices and cost reduction from many different standpoints. The United States Environmental Protection Agency (USEPA) provided the apply water-efficient landscaping principles using Xeriscaping method in landscape [8]. So State enacts rules put for landscapers including Group

IOP Conf. Series: Materials Science and Engineering **974** (2020) 012025 doi:10.1088/1757-899X/974/1/012025

plants according to their water needs, use native and low-water-use plants, limit turf areas to those needed for practical uses, use efficient irrigation systems, schedule irrigation wisely, make sure soil is healthy, remember to mulch and provide regular maintenance. Robert F Baldwin and others [9] studied practices in research (The Future of Landscape Conservation) and provide a wide vision for conservation in landscape and put points for Success of the landscape conservation cooperatives. A recent research provided a framework to include cultural resources in spatial conservation planning that acknowledges the different scales of importance and management of sites [10]. Eric W. Sanderson and others [11] outlined a conceptual methodology for landscape conservation being tested by the Wildlife Conservation Society at three sites in Latin America and Africa. Based on the biological requirements of an ecologically functioning population of a landscape species, the "biological" landscape is defined. This landscape is compared to the landscape of human activities through the use of Geographic Information Systems (GIS). Focal landscapes sufficient to meet species requirements are defined and threats from human activity evaluated with respect to biological requirements. A suite of landscape species may be selected depending on resources, leading to multiple, often overlapping, focal landscapes. A hypothetical example is presented [12]. Several different methods are used in the industry to classify resorts. One common technique is to identify the resort's signature amenity [13]. According to Brey [14] resort types should be including many elements and facilities such as beach or lakeside hotel, golf course, tennis playground, fishing and hunting facilities, marina, theme park, spa and health facilities, ski and mountain hiking, waterpark and aquatic resort and casino.

3- Case study

3.1 Soma Bay Resort

Soma Bay resort was selected as the location for the case study for several reasons. Rea Sea is considered as one of the world's most attractive destination for tourism, the resort located about 42 km from Hurghada International Airport. It spreads over an area of about 10 square kilometers as shown in Figure 1. The resort consists of two major zones; the first zone comprises 324 units owned for landlord and dedicated for investment. These units are distributed in clusters all over five main projects in the area, ranging from villas, chalets and housing units, with a maximum height of three stories. The second zone comprises five (5 stars) hotels, including some recreational outdoor spaces, central business, entertainment district and marina center.

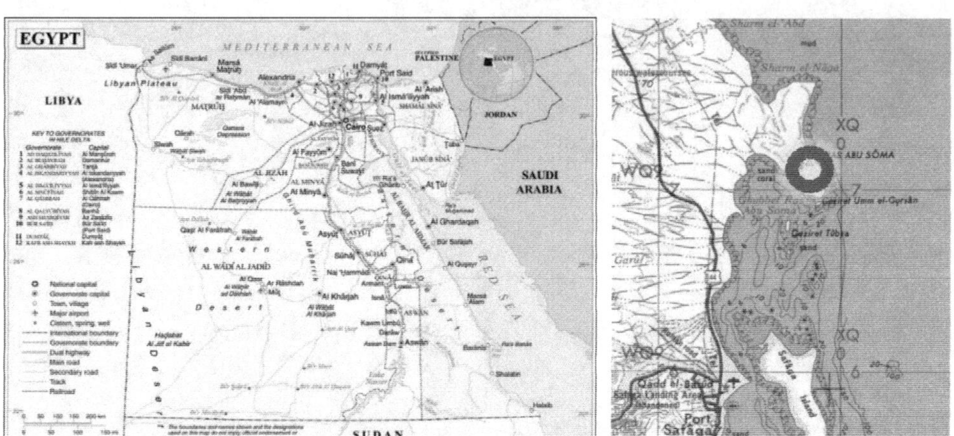

Figure 1: Rea Sea region location (in yellow) relative to Egypt and the location of selected case study (in red).

IOP Conf. Series: Materials Science and Engineering **974** (2020) 012025 doi:10.1088/1757-899X/974/1/012025

As shown in Figure2 and 3, all hotels guests have the ability to join sky-surfing, diving courses, marina activities, golfers globally and are home to a world-class spa and facility. Hotels are surrounded by charming coastal landscapes, entertainment and recreational zones offer luxurious gardens with a view of either the sea, pool, golf course, or a panoramic view of all. The resort serves an array of unique recreational activities creations to please every palate.

Figure 2: The master plan of Soma Bay resort located in Red Sea, Egypt.

Figure 3: An aerial view Soma Bay resort. (as planed)

IOP Conf. Series: Materials Science and Engineering **974** (2020) 012025 doi:10.1088/1757-899X/974/1/012025

3.2 Recreational Open Spaces

Recreation open spaces are the heart of resort attractions. The recreation and entertainment can be considered as the main aims of visiting the resort. Beach, gardens, golf, swimming pools, lakes, water-park facilities, outdoor activity center, kids' area and sports courts are dominant recreation open spaces in coastal resorts. They are huge areas and consume a lot of money during construction phase that is covered by the investor. The presence of recreational spaces (specially golf course and gardens) at coastal resorts is frequent and costly. For example, in some cases the amount of freshwater required to maintain them is very high. A study of several tourist facilities' water use in Zanzibar [15] showed that hotels with extensive gardens require an average of up to 50% of the total freshwater consumption. Different landscape design strategies aim to reduce the need for garden irrigation, which can greatly contribute to reducing the whole water demand of resorts. Therefore, one of the main aims of this study is to find sustainable solutions to reduce the cost of operation and overcome the side effect of the presence of recreational area.

3.3 Sustainable practices

Many practices can be implemented to improve recreational spaces sustainability and reduce the cost of operation process on different scales and capacities. These practices can be outlined in the following scales.

3.3.1 On Beach

- Conservation of marine life, stop over-fishing, protect the rare organisms and protect coral reefs.
- Improve waste separation through the use of color-coded bins for different kinds of waste: recycling, organic waste, composting
- No plastic waste.
- Fit water taps with flow controllers such as aerators, restrictors or percussion. Fit showers with flow restrictors or special low-flow shower heads.
- Put in place a towel and bed linen re-usage program with an incentive for guests
- Reduce waste by limiting disposable items such as small packages for food such as jam, butter where bulk dispensers could be used instead.
- Use refillable bottles and containers for items such as shampoo and soap.
- Place glass reusable water bottles in guest bedrooms rather than plastic ones.
- Offer a Sustainable Tourism Kit during the booking process filled with products such as a bamboo toothbrush, an ocean safe sunscreen, a water bottle or a reusable cotton swab.
- Offer a carbon footprint offset option like the one used by KLM Royal Dutch Airlines.
- Provide staff training on best environmental practices to reduce waste and create awareness around protecting local biodiversity
- Showcase recommendations for environment friendly behavior in orientation booklets available in each cabin, as well as through the company TV channel in cabins.
- Apply for a free sustainability certification by the Oceanic Global.

3.3.2 In Gardens, Sports Courts and Golf

- Develop clear understanding of species, habitats, hydrology and geomorphology.
- Ensure golf course and landscape design typology is clearly compatible with site and surroundings.
- Protect the most valuable landscape and ecological features
- Use native or locally important species in landscaping designs, and drought-tolerant varieties if applicable.

IOP Conf. Series: Materials Science and Engineering **974** (2020) 012025 doi:10.1088/1757-899X/974/1/012025

- Minimize pollution from maintenance facility.
- Plan an ecologically-rich6 landscape.
- Naturalize any water features as far as possible.
- Minimize the extent of intensively managed grass areas.
- Consider the visual impact of buildings.
- Maximize irrigation water efficiency.
- Use local materials.
- Select best-adapted turf species and cultivars for the local environmental conditions and to minimize resource requirements.
- Reduce fuel use in maintenance.
- Maximize the energy efficiency.
- Use recycled materials and materials with recycled content.
- Apply conservation methods in landscape (like xeriscape).

3.3.3 Swimming Pools, Lakes and Waterpark Facilities

- Harvesting rain water is as easy as putting a rain barrel under downspouts and using the water for interactive water play or site irrigation.
- Grey water harvested from sinks, showers and swimming pools can be recycled or used for site irrigation.
- Solar driven water pumps, water fountains and water features, like water curtains, can offer new dimensions in eco-friendly play.
- Alternative water playground designs allow children to pump water from collected rainwater and self-direct play opportunities in a low cost, environmentally friendly way.

4- Research Methodology

The context of this study concerns recreational spaces of tourist projects. The concept of sustainable touristic buildings includes the process of landscaping and gardening that reduces or eliminates the need for supplemental water from irrigation [16]. In some areas, terms as water-conserving landscapes, drought-tolerant landscaping, and smart landscaping are used instead. Plants whose natural requirements are appropriate to the local climate are emphasized and care is taken to avoid losing water to evaporation and run-off. The specific plants used in xeriscaping depend upon the climate. Xeriscaping approach is different from natural landscaping, because the emphasis in xeriscaping is on selection of plants for water conservation, not necessarily selecting native plants. Public perception of xeriscaping has frequently been negative as many assume that these types of landscapes are ugly expanses of cactus and gravel. However, studies have shown that education in water conservation practices in the landscape can greatly improve the public's perception of Xeriscaping [17].

In order to analyze and evaluate the impact of landscaping and gardening upon recreational spaces of tourist projects in Rea Sea region, the Xeriscaping approach was selected as a method in this paper. This approach was originally conceived by Denver Water, the seven design principles of xeriscaping have since expanded into simple and applicable concepts to creating landscapes that use less water. The principles are appropriate for multiple regions and can serve as a guide to creating a water conserving landscape that is regionally appropriate and since they were conceived for homeowners they are easy to implement [18]. Through the field visit and the meeting of officials, all operation details were recognized. The measurement ruler (Seven Principles of Xeriscape) was applied and the availability of this procedure was measured at the site. Every element will take a percentage score. Finally, research will provide the result to get accurate evaluation for existing situation in conservation field to be able to make a correct future plan to optimum environment-friendly and lowest operation cost.

4.1 Seven Principles of Xeriscape

4.1.1 Plan and Design

Create a diagram, drawn to scale, that shows the major elements of the landscape, including house, driveway, sidewalk, deck or patio, existing trees and other elements. Once a base plan of an existing site has been determined, the creation of a conceptual plan (bubble diagram) that shows the areas for turf, perennial beds, views, screens, slopes, etc. is undertaken. Once finished, the development of a planting plan that reinforces the areas in the appropriate scale is done.

4.1.2 Soil Amendment

Most plants will benefit from the use of compost, which will help the soil retain water. Some desert plants prefer gravel soils instead of well-amended soils. Plants should either fit the soil or soil should be amended to fit the plants.

4.1.3 Efficient Irrigation

Xeriscape can be irrigated efficiently by hand or with an automatic sprinkler system. Zone turf areas separately from other plants and use the irrigation method that waters the plants in each area most efficiently. For grass, use gear-driven rotors or rotary spray nozzles that have larger droplets and low angles to avoid wind drift. Spray, drip line or bubbler emitters are most efficient for watering trees, shrubs, flowers and groundcovers. If watering by hand, avoid oscillating sprinklers and other sprinklers that throw water high in the air or release a fine mist. The most efficient sprinklers release big drops close to the ground. Water deeply and infrequently to develop deep roots. To reduce water lost to evaporation, never water during the day. With the use of automatic sprinkling systems, adjust the controller monthly to accommodate weather conditions. Also, install a rain sensor to shut off the device when it rains.

4.1.4 Appropriate plant and zone selection

Different areas in a yard receive different amounts of light, wind and moisture. To minimize water waste, group together plants with similar light and water requirements and place them in an area that matches these requirements. Put moderate-water-use plants in low-lying drainage areas, near downspouts, or in the shade of other plants. Turf typically requires the most water and shrub/perennial beds will require approximately half the amount of water. Dry, sunny areas support low-water-use plants that grow well in the specific climate. Planting a variety of plants with different heights, color and textures creates interest and beauty.

4.1.5 Mulch

Mulch keeps plant roots cool, prevents soil from crusting, minimizes evaporation and reduces weed growth. Organic mulches, such as bark chips, pole peelings or wood grindings, should be applied 2 to 4 inches deep. Fiber mulches create a web that is more resistant to wind and rain washout. Inorganic mulches, such as rocks and gravel, should be applied 2 to 3 inches deep. Surrounding plants with rock makes the area hotter; limit this practice.

4.1.6 Limited turf areas

Native grasses (warm-season) that have been cultivated for turf lawns, such as buffalo grass and blue grama, can survive with a quarter of the water that bluegrass varieties need. Warm-season grasses are greenest in June through September and may go dormant during colder months. Native grasses (cool season) such as bluegrass and tall fescue, are greenest in the spring and fall and go dormant in the high heat of the summer. New cultivars of bluegrass,

IOP Conf. Series: Materials Science and Engineering **974** (2020) 012025 doi:10.1088/1757-899X/974/1/012025

IOP Conf. Series: Materials Science and Engineering **974** (2020) 012025 doi:10.1088/1757-899X/974/1/012025

Numbers Before and after applying Xeriscape to insure landscape sustainability.

5. Results

The land use of hotels was analysed to provide an accurate vision of hospitality capacity and open spaces. As shown in Table 2, Four hotels are sea front and one hotel is golf hotel. Findings show that according to the rooms (keys) capacity, the biggest hotel is Robinson (348 keys – 24% Resort keys). According to the hotel`s plot area, the largest hotel`s land is La Residence Des Cascades–golf (2020367 m2). And the largest area of used open space is Sheraton (190181.5 m2).

Table 1: Types of hotels and total plot area of Soma Bay resort.

Hotel Name	Hotel Type	keys	Plot Area (m^2)	Building Area (m^2)		Open Spaces (m^2)	
KEMPINSKI HOTEL	Beach	325	156153	20927.6	13%	135225.4	86.6%
ROBINSON CLUB	Beach	348	190807	25166.1	13%	165640.9	86.8%
SHERATON SOMA (+Island)	Beach	326	209499	19317.5	9%	190181.5	90.8%
THE BREAKERS	Beach (Diving)	173	82681	6811.4	8%	75869.6	91.8%
LA RESIDENCE DES CASCADES	Golf	249	2020366.6	19020	1%	2001347	99.1%

Table 2: Recreational spaces of all analyzed hotels (5 Hotels).

Hotel name	Kempinski hotel	Robinson club	Sheraton soma	The breakers	La residence des cascades
Beach	19843	39564	27893	3805	0
	14.7%	23.9%	14.7%	5.0%	0.0%
Gardens, Golf	4674	19950	8481	456	1368708
	3.5%	12.0%	4.5%	0.6%	68.4%
Pedestrians and Streets	1796	3032	3966	1837	9534
	1.3%	1.8%	2.1%	2.4%	0.5%
S. Pools, Lakes and Waterpark Facilities	13946	4153	3829	287	32088
	10.3%	2.5%	2.0%	0.4%	1.6%
Outdoor Children's Activity Centre	243	811	211	0	185
	0.18%	0.49%	0.11%	0.0%	0.01%
Sports Courts	2408	5432	2661	0	3641
	1.8%	3.3%	1.4%	0.0%	0.2%
Others (Un-used Areas)	92315.4	92698.9	143140.5	69484.6	587190.6
	68.3%	56.0%	75.3%	91.6%	29.3%
Total Open Spaces (m^2)	135225.4	165640.9	190181.5	75869.6	2001347

In this study recreation space have been analyzed, the percentage of its area were compared with other recreation spaces of all hotels. provide this analysis with the seven kinds of recreation open spaces and the total area of open spaces in each hotel. Research achieves that Robinson has the largest Beach area, Garden also, Kempinski has the largest Swimming Pools but Cascades has the largest Lakes and Golf areas. Resort has a lot of recreation kinds (See Table 3). Recreation open spaces shown in Figure 5 contain: Beach, Gardens, Golf Yards, Swimming Pools, Lakes, Waterpark facilities, Sports Courts and Outdoor Activities Center (Adults – kids areas). Every hotel has a special character in activities and recreation open spaces.

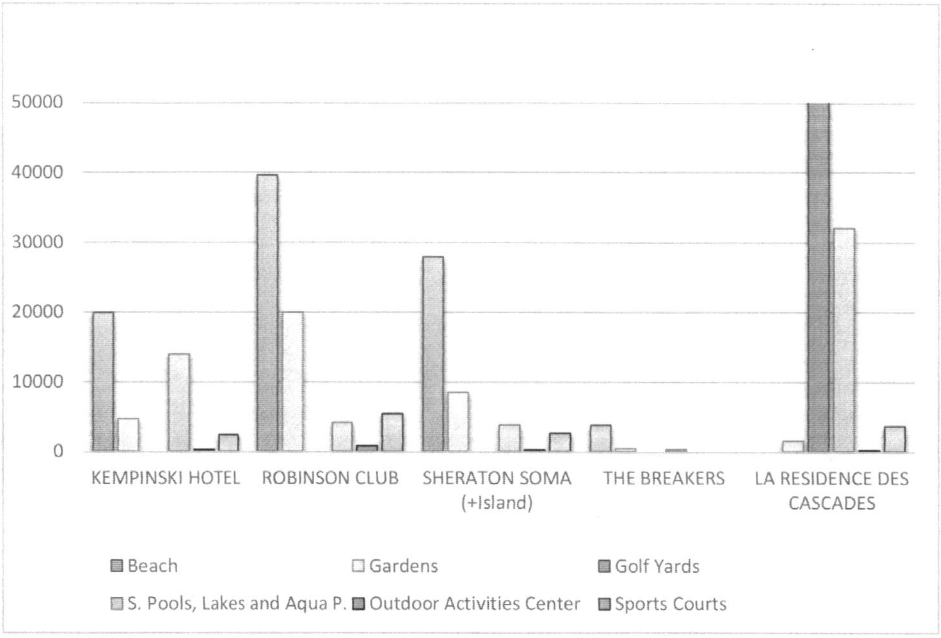

Figure 5: Comparison of the area of recreational spaces of 5 hotels in Soma Bay

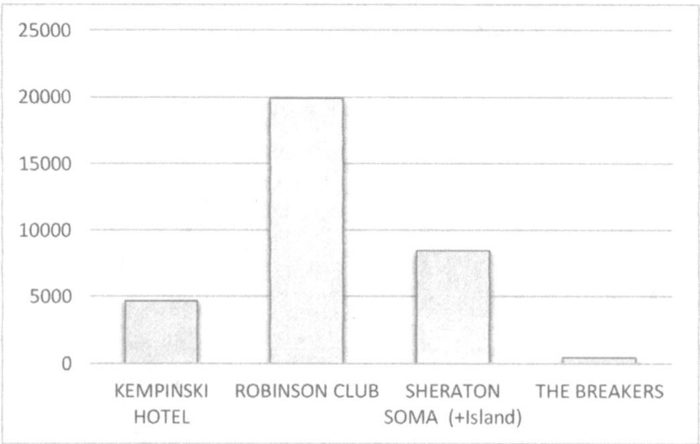

IOP Conf. Series: Materials Science and Engineering **974** (2020) 012025 doi:10.1088/1757-899X/974/1/012025

Figure 6: Comparison of the garden area (greenery places)

As shown in Figure 6, the papers also highlighted the Gardens areas in each hotel to be able to understand the operating cost of landscape (in following study). this paper made an assess for Soma Bay landscape existing situation (38561sqm). And measured the xeriscape principles as criteria shown in Figure 7. Some of landscape operating practices ware matching with the Xeriscape Principles:

- landscape was designed in the project design phase.it considers the climate, views, functions and the watering needs. But recently there are many modifications done for view reasons without any planning.
- Lawn is Limited with functions except rarely spaces.
- Some plants are native but most of plants are not water efficient.
- Irrigation systems are efficient at all resort gardens except stuff residence gardens. And most of gardens irrigated with Grey water and sea-water desalination.
- Soil of gardens (up to 75%) was prepared before cultivation.
- Mulches are not used at all.
- landscape maintenance is taken care regularly according to timelines and careful follow-up.

Findings show that landscape is Xeriscape up to 62%. that is meaning that Landscape is already conserving water and materials (and saving 37.2% of traditional gardens cost).

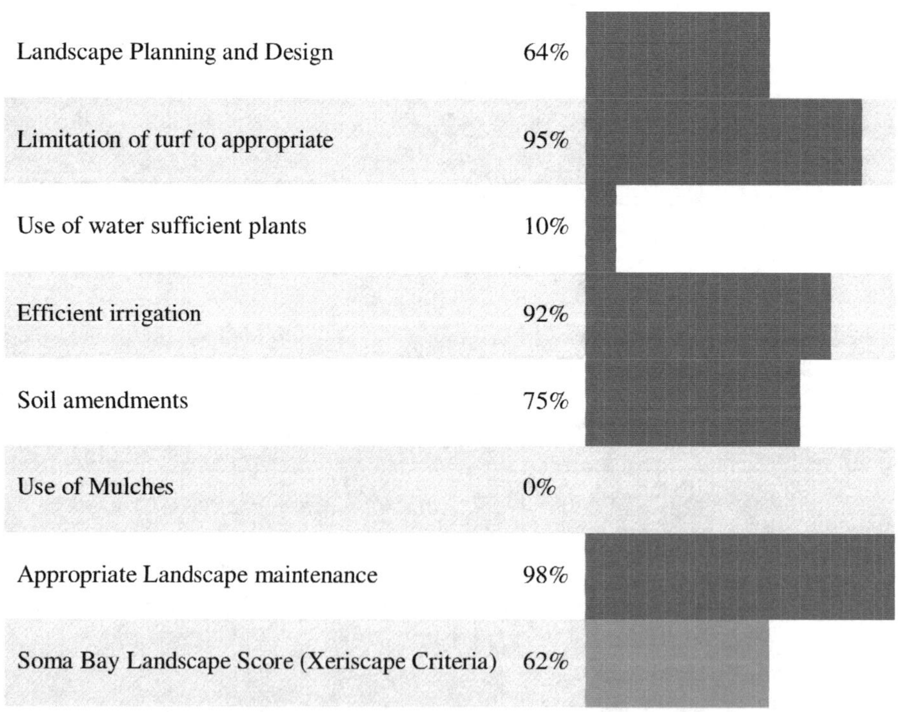

Figure 7: Results of applying Xeriscape evaluation approach.

Findings also revealed the amount of irrigation water used in resort gardens (grey and treatment water) as shown in Table 3 and the amount of irrigation water reduced by applying Xeriscaping illustrated in Figure 8. Sea-water desalination and gray water treatment stations provide 1000 m3/day of water covers all hotels, gardens and golf yard's needs. By using Xeriscape as a method makes a huge reduction in water consumption in gardens. So by extension cost, waste and manpower reduced.

IOP Conf. Series: Materials Science and Engineering **974** (2020) 012025 doi:10.1088/1757-899X/974/1/012025

Table 3: Daly Water Consumption in Garden (liters) per meter² (average).

Hotel Name	Gardens Area (m²)	Existing Irrigation Water(L)	Xeriscaping Reduction rate	Irrigation Water after using xeriscape (L)
KEMPINSKI HOTEL	4674	16359	37%	10306.17
ROBINSON CLUB	19950	69825	37%	43989.75
SHERATON SOMA (+Island)	8481	29683.5	37%	18700.61
THE BREAKERS	456	1596	37%	1005.48
Other Spaces (streets and Parking areas)	4956	17346	37%	10927.98
Total	38517	134810	37%	84929.99

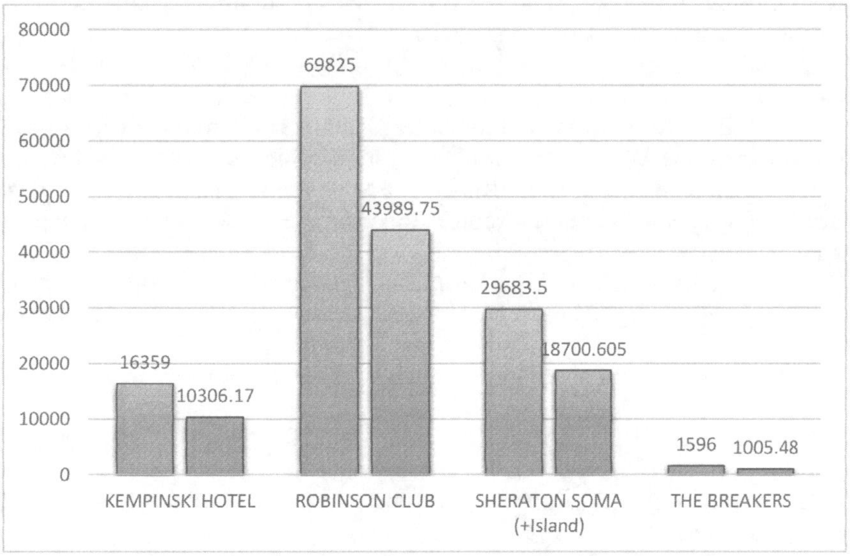

Figure 8: Xeriscaping reduces the amount of irrigation water

6. Conclusion

Some coastal resorts in Egypt are already applying some sustainable practices to reduce the running cost of landscape (20%-35%). But they need to apply more practices to achieve the optimum situation (Xeriscape = 60% saving). By using the aforementioned proposal system generated by the aim of the research, there are many Sustainable practices could reduce the cost of recreation open spaces. as well as serve the principles of sustainable development and contribute to the creation of more pleasant settlements. The proposal system connected to property and recreation open spaces developments could consist of the following three sections: A) Local Government Module.
B). Developer/Investor Module.
C). Public Space Developmental Contribution Module.

IOP Conf. Series: Materials Science and Engineering **974** (2020) 012025 doi:10.1088/1757-899X/974/1/012025

References

[1] Brey, E. T. & Choi, G.H. (2010), “Standard Resort Hospitality Elements: A Performance and Impact Analysis,” Journal of Tourism Insights, Vol. 1, No. 1, pp. 21-35.
[2] Stringam, B.B. (2008), “A Comparison of Vacation Owner Amenities with Hotel and Resort Hotel Amenities,” Journal of Retail and Leisure Property, vol. 7, pp. 186-203.
[3] Kandampully, J., Mok, D. and Sparks, B. (Eds.) (2001), Service Quality Management in Hospitality, Tourism and Leisure, Haworth Hospitality Press, New York, NY.
[4] Bohrod, G. (2010), “Resort Management Future Plan Initiatives Survey by Resort Trades,” Resort Trades, May 2010, p. 22.
[5] McElyea, J.R. and Cory, G.L. (1998), “Resort Investment and Development; An Overview of an Evolving Market,”.
[6] Mill, R.C. (2001), Resorts: Management and Operation, John Wiley and Sons, Inc., New York, NY.
[7] Vallen, G. K., & Vallen, J. J. (2005), Check-in, check-out: Managing Hotel
[8] Operations. Prentice Hall, Upper Saddle River, N.J.
[9] USEPA (United States Environmental Protection Agency).(2002), “Water-Efficient Landscaping” .
[10] Robert F Baldwin, Stephen C Trombulak, Paul B. Leonard and Reed Noss (2018), “The Future of Landscape Conservation” an article in Bioscience · January 2018.
[11] Scott Ogletree, Robert Powell, Robert F Baldwin, Paul B. Leonard (2019), “A framework for mapping cultural resources in landscape conservation planning“ an article. May 2019
[12] Eric W. Sanderson, Kent H Redford, Amy Vedder , Peter B. Coppolillo and Sarah E. Ward (2001), “A conceptual model for conservation planning based on landscape species requirements”.
[13] American Society of Landscape Architects " The Case for Sustainable Landscapes " (2014).
[14] Chon and singh, (1995), “Marketing resorts to 2000: review of trends in the USA”.
[15] "Xeriscape Plans." Xeriscape Plans. Denver Water. <https://www.denverwater.org/residential/rebates-and-conservation-tips/remodel-your-yard/xeriscape-plans>.
[16] S Gössling, (2001), “The consequences of tourism for sustainable water use on a tropical island: Zanzibar, Tanzania” an article. March 2001.
[17] "Xeriscape | Define Xeriscape at Dictionary.com". Dictionary.reference.com. Retrieved 2018-05-31.
[18] American Hotel and Lodging Association/Smith Travel Research (2010), Lodging Survey 2010: Lodging Services, Facilities and Trends. American Hotel and Lodging Educational Association, Washington D.C.
[19] Environmental Fact Sheet: Recycling Grass Clippings" (PDF). Environmental Protection Agency. United States. Retrieved 23 May 2017.

IOP Conf. Series: Materials Science and Engineering **974** (2020) 012026 doi:10.1088/1757-899X/974/1/012026

Military Technical College
Kobry El-Kobbah,
Cairo, Egypt

***13th* International Conference on Civil and Architecture Engineering ICCAE-13-2020**

Performative Design Optimization for Shading Screens

Mohga Ali Abdelslam Youssef
Assistant lecture - Department of Architectural Engineering– Al-Azhar University

Ahmed Mohamed ELkordy, Hesham Ahmed Sobh
Professor - Department of Architectural Engineering– Al-Azhar University

Corresponding author: mohga.ali@azhar.edu.eg

Abstract. Building façade plays a significant role in architecture; it is not only a mean to express the design concepts but it is the main moderator between exterior environment and internal spaces. Ecological facades are defined as the ability to response and adapt to the changes of the environmental conditions. The geometry of the proposed screen is not a focal point of this paper, **as the aim is to find a method for designing and evaluating non-conventional solar screens that improve indoor daylight quality**, while the specified shading screen is installed in front of a transparent full floor to ceiling glass facade. It is assumed that by using the shading screen, the required lighting levels are being maintained, while the heat gain is being reduced. In addition, the designed patterns play a role as an architectural feature of the building. Ultimately, **the presented shading screens are based on geometric ornamental patterns and arranged according to the lighting requirements.** Results are indicating a multi-disciplinary approach in the design of the shading screens, which can be employed in creating similar prototypes with different climatic and loading requirements.

1. Introduction

The design process is a set of actions that are taken in a hierarchal order aiming to reach an intended final output based on specific design requirements and objectives [1]. This set of actions is constantly changing through time trying to adapt with uprising design needs and respond to the demands of new design approaches. The design process can be described as an iterative process where a number of dependent or interdependent design tasks are performed in sequence till reaching a desired goal [2]. The effectiveness of this process can be revealed through this loop of action and assessment (activity and reflection) that is highly reliable on the design media. Design tools and techniques used influence the problem representation and how is perceived. Thus, they affect the designers conceptual thinking to develop ideas and reflect upon the results [3].

2. Generative systems in architectural design

IOP Conf. Series: Materials Science and Engineering **974** (2020) 012026 doi:10.1088/1757-899X/974/1/012026

In usual traditional design, the role of the designer is to explore a solution space. The key relationship between designer and product is a direct one 'Figure 1' (even if mediated via a third-party or medium). There is a direct relationship between the designer's intentions and that of the designed product [4].

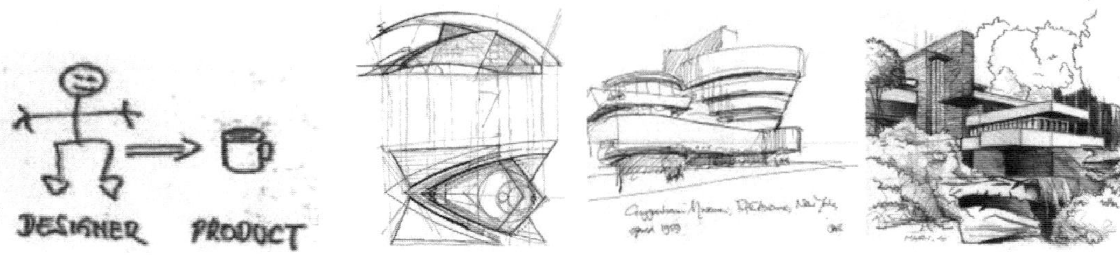

Figure 1. Traditional design approach [4].

In contrast, design using generative methods involves the creation and modification of rules or systems that interact to generate the finished design autonomously 'Figure 2' [5].

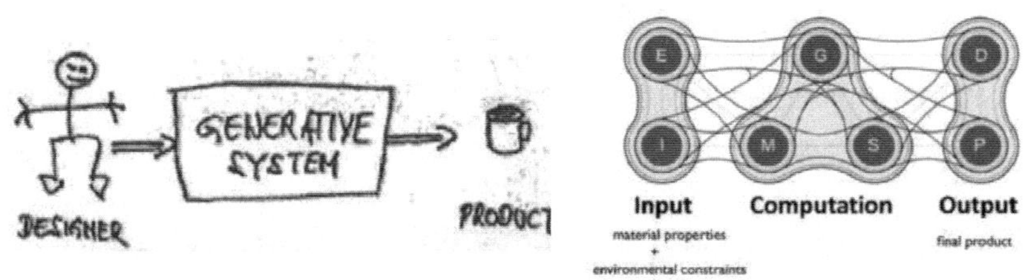

Figure 2. Form generation operation [4] [6].

"Generative design is not about designing a building, it's about designing the system that designs a building." Thus, Generative systems made a major shift in the concept of design modeling from the modelling of a pre-designed "object" to the modelling of the design "logic" [7].

Generally, the generative systems' process is based on four main components: the input parameters, the controlled rule, the output generations and the selection of the optimal solution as shown in 'Figure3'. The design artifact could only be obtained by the fourth phase.

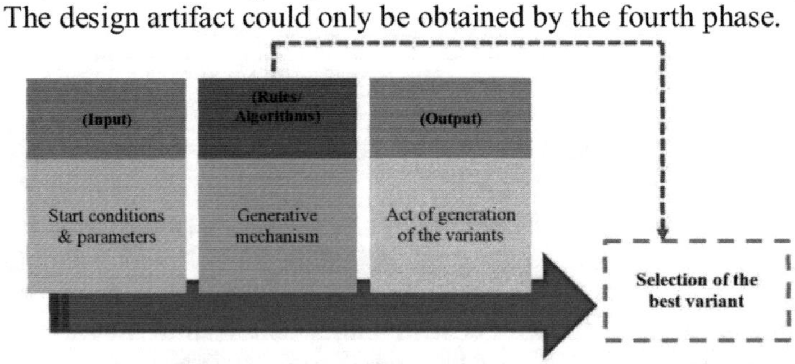

Figure 3. Generative system process
Source; by the researcher

2.1. Generative Performative Design

Performative design is an approach derived from performance models. It amalgamates form generation and performance, considering both through optimization algorithm and simulation technique as shown in 'Figure 4'. No longer are simulation tools utilized for analysis only, but they are used for both performing analysis and synthesis simultaneously; form is driven by generative processes guided by analytical simulation techniques that automatically modify the model [6].

IOP Conf. Series: Materials Science and Engineering **974** (2020) 012026 doi:10.1088/1757-899X/974/1/012026

Figure 4. performative design
Source; by the researcher

The concept of form making shifted to be form finding. Geometric models are needed to be formulated in a way that reacting to the stimulus of the evaluation process and complying with the modifications of generative process can be in a consistent manner. Hence, parametric modeling is essential to support the generative process informed by the performance evaluation. It needs three consecutive processes 'Figure 5':

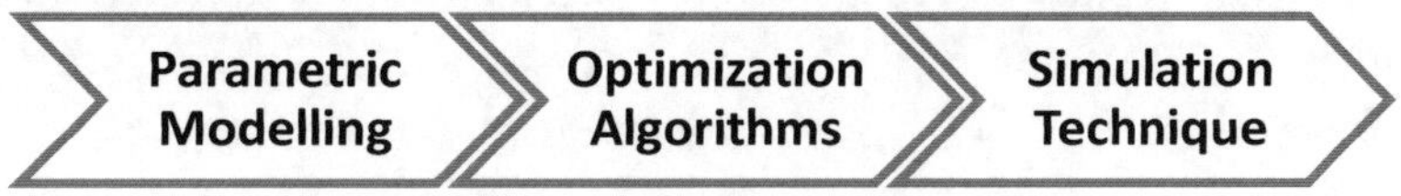

Figure 5. performative design process
Source; by the researcher

On the other hand, generative design captures the aesthetics qualities of the design through a rule-based process that inform the generation of large range of solutions. Combining both models in one approach gives the advantage of reaching the aesthetic quality aspired while respecting the performance criteria needed. This would be through 'Figure 6':

Figur e 6. Gener

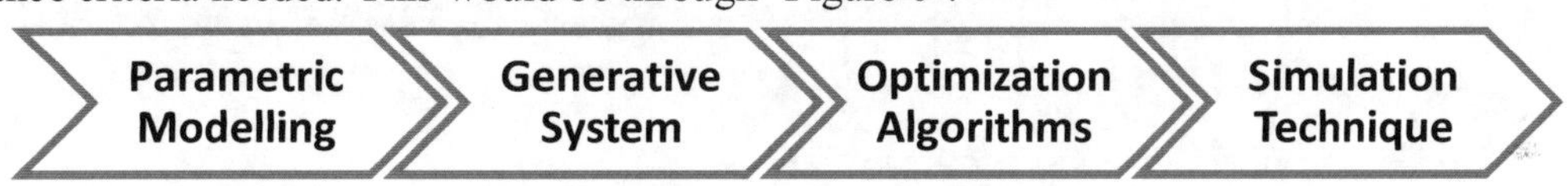

ative Performative Design process
Source; by the researcher

2.2. *Performative design exploration based on parametric aspects*

Current CAD and simulation tools only used as a representational tools for the final design case. Conversely, parametric modelling approach allows the modelling of unlimited set of design variations depending on the parametric definition which in turn opens a wide range of possibilities to explore, evaluate and even regenerate the design model to comply with targeted aim 'Figure 7'.

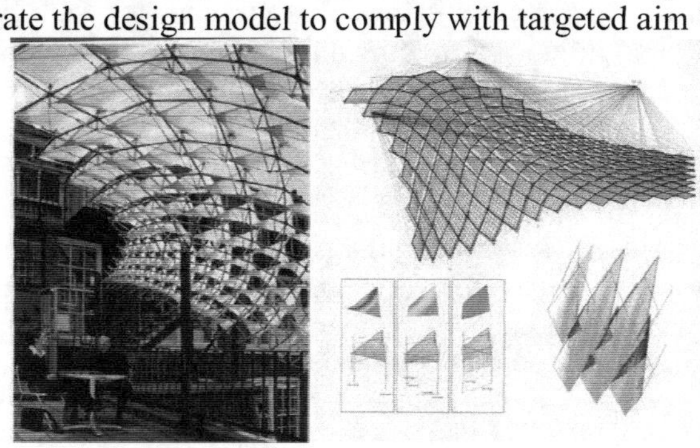

Figure 7. For the "AA Component Membrane", a terrace canopy designed at the Architectural Association, the form was simultaneously optimized for sun, wind, drainage and views using a parametric model combined with various computational analysis techniques including fluid dynamic wind flow analysis, precipitation analysis, stress analysis, and solar analysis [8].

Considering architectural field, performance simulation tools have been widely used to establish early design processes, which are effective as alternative solutions to isolate design problems. However, current simulation tools failed to address novel shading systems and façade designs due to its limitation in modeling and exploring wide range of solutions. Thus, integrating new digital generative systems such as parametric design systems as well as Genetic Algorithms with these simulation tools contributes in raising building performance through offering unlimited alternatives in specific ways that ensuring the simulation process is evaluating the resulted in a comprehensive cycle till reaching the optimal solutions. This paper will focus on the latter case.

3. Movement as performance

It is often the movement of people around and through a building that gives architecture its performative capacity. It is the experience of the building's materiality and spatial presence. In some recent projects, performativity is in the kinetic effects of the architecture. Rather than the subject that moves, the object itself creates an architecture of spectacle and architecture of performance [9] as shown in 'Figure 8'.

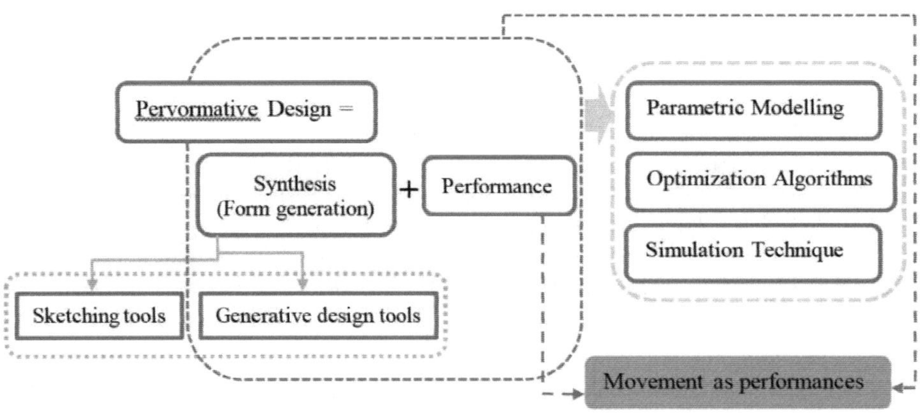

Figure 8. Schematic diagram suggested for approaching the performative design
Source; by the researcher

4. The Concept of Time in Design

Many scientists, also the architects addressed the concept of time as a term that has an impact on all aspects of life. Science until this moment still reveals to us that the time has great importance. This was at the level of physiological influence on human or on his environmental and urban surroundings or on his tools and uses. However, the virtual worlds through the means of communication can jump on the time, crossed the borders and exceeded the distances, and this forces us to understand the impact of time in the world in architecture now [10].

4.1. The importance of the role of Eye in the Design

The eye distinguishes between the types of movement, either dynamic or static in the design [11]. The architectural projects that give the sense of movement, give the design the pleasure and excitement. The movement represents the continuous change in the position of the image, as long as it does not interfere with the function required by the power of formation. The good movement design requires hundreds of the ways to read it, and this factor has a big role in distinguishing between the fertile form and the simple form.

It is the continuous change in a location of the body in relation to a location of another object are assumed it is constant. The movement in architecture 'Figure 9' depends on two main factors: [12]

The first factor: Benefiting from the human perception of dynamism in giving the spirit of movement in the still image, where the movement is only a developing concept to the conscious dynamism. The second factor: It relates to the same form where the dynamism does not use the usual coordinates of design (X, Y, Z), but depends on temporal time coordinates similar to the coordinates of energy [13].

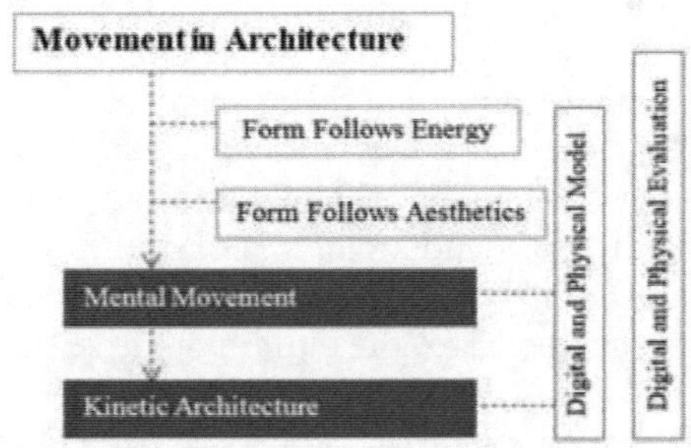

Figure 9. Diagram of action research and investigation
Source; by the researcher

4.2. Kinetic architecture

Kinetic architecture is a type of architecture that includes the concept "movement as performance", while approaching the concept of Adaptive architecture [14]. However, not all Kinetic architecture can be included in Adaptive architecture. Kinetic architecture allows parts of the building's structure to move, without decreasing the overall structure integrity. This skill for motion can be used, not only to improve the building's aesthetic qualities, respond to environmental conditions, but also to add functions that would not be possible in a static structure [15]. This type of architecture existed already in the past, but since the end of the 20th century its presence has increased due to the evolution in the fields of mechanics, robotics and electronics, which, consequentially, promoted more possibilities for the practical implementations of this architecture. Within Kinetic architecture, only the buildings that use movement to adapt to external conditions in order to optimize the energetic costs, can also be classified as Adaptive architecture.

Table 1. Typological Classification of Kinetic Architecture.

Kinetic	□Totally -Dynamic		▪ Partially –Dynamic		
	□ Dynamic the new dimension Architecture tower 	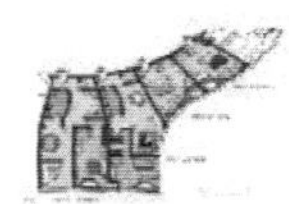	▪ Partially – Components The Suite Vollard 	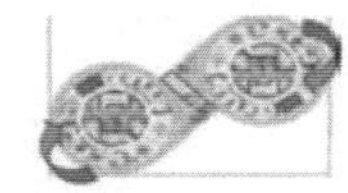	▪ Kinetic Wind- driven Façade Brisbane airport 
Partially - Elements	▪ Expanding Research Laboratory 	▪ Retracting Al Bahr towers 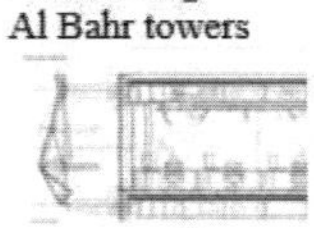	▪ Folding Kiefer Showroom 	▪ Sliding Fosun Foundation 	▪ Transforming One Ocean building 
Basic Motion Mechanisms	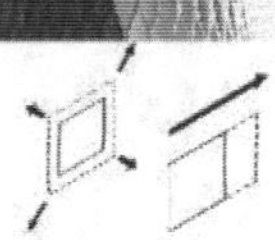			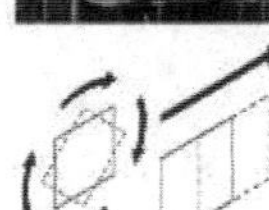	

4.3. The mental movement -Dynamic configuration

It exists in all aspects of the perception, represents the idea of movement in a simple and formal manner and what including constant formative relationships resulting from the simulation of movement and time. There are many potentials and relationships, such as the formation, juxtaposition, overlay, friction, exaggeration, absence as a few of the techniques and potentials used by the architect to express the apparent movement without the existence of time [16]. The main idea of the static movement does not depend on the direct physical movement, but is indirectly represented by the expressive form of the movement that depends on the nature of form, where each shape has values different in its mental movement or in parts of it in the same field [17]. The movement through the stability in design is the dynamic of architectural form of the building means the dazzling with the movement, which simulates the dynamic movement, but at rates lower than the changes resulting from the dynamic movement [18]. Therefore, the line and curved surface of the highest variables that give a continuous change in the building mass and architectural space that includes several components such as the position, direction and size [19]. It is the movement by inspiring through the arrangements of the elements, lines, basic forms and levels of the building to give the sense with the movement.

Table 2. The mental movement in the process of perception "The latent dynamism"

	The outer lines of the building	Architectural form including shape and mass	shaping the building	The Composition of the Building Surface	Foldable surfaces
	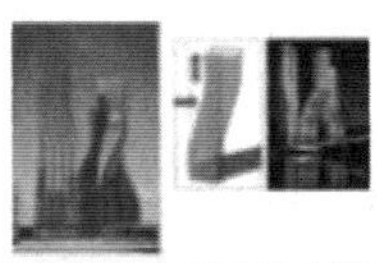	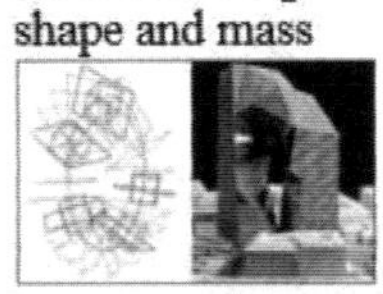		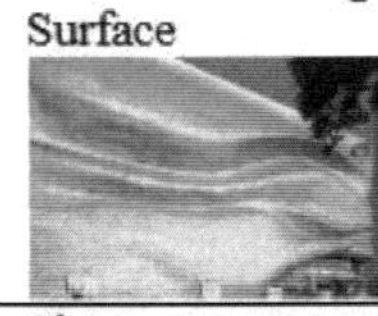	
façade configuration	Elements' Distribution	Elements' Size	Elements' Deformation	Shape Grammar	Shape Grammar Structure
					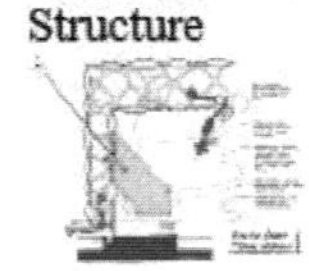

Generally, ecology is one branch of environmental science which is more concerned about the available natural resources scattered throughout an environment. Ecological impacts together with economic and social impacts of buildings on their surroundings are the main three principles of sustainable building design. Thus, ecological façade could be defined as the building facade that relates to climate conditions and surrounding environment resources while responding to conflicting needs for heating, cooling, daylight and ventilation.

5. Shading screens

Shading screens have traditionally evolved to control solar penetration in the buildings of the middle-east such as the Mashrabeya 'Figure 10'. Generally, shading screen systems can be considered as an essential integral part of ecological facade systems. They can be used to improve the indoor daylight quality [20].

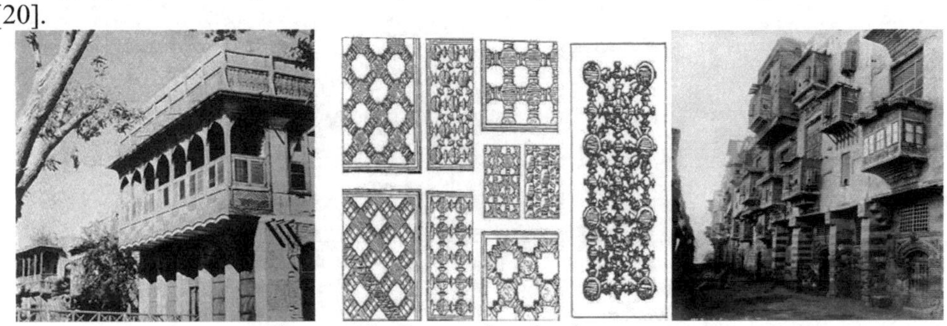

Figure 10. Conventional solar screens

IOP Conf. Series: Materials Science and Engineering **974** (2020) 012026 doi:10.1088/1757-899X/974/1/012026

Natural daylighting is desired in building envelopes, however thermal discomfort and the incidence of glare is unwanted [21]. One type of shading system that is used to permit daylight while controlling solar penetration is "shading screens". Shading screens are defined external perforated panels that are fixed in front of windows.

An additional advantage of these screens lies in their provision of privacy, which is a social-cultural need in its original region. During the past decade the traditional patterns of shading screens has been transformed to new patterns, by varying some of the basic mathematical rules like symmetry of the patterns or equal dimension of a replicated geometry. Furthermore, some active solar screens with such patterns has been developed. There is a close connection between the geometric patterns and the design of shading screens which play a role as daylight control systems [22].

Screens became a common and rich architectural device that can separate spaces, while maintaining a certain visual. In contrast to glass, screens have a strong presence and offer the possibility to vary their materials, color, texture, etc. They can assume other functions such as passive shading on facades 'Figure 11'.

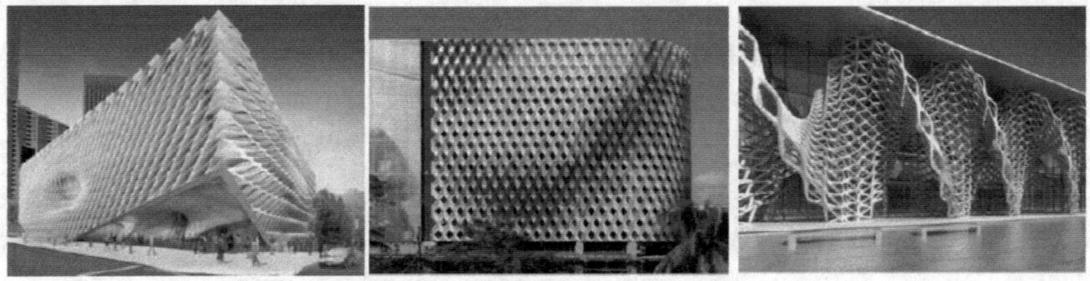

Figure 11. Generative Design Systems for Patterns Formation

6. Façade geometry

Pattern was designed and then replicated in the plane. A hexagonal grid 'Figure 12'was the underlying grid of the pattern. After the basic diamond pattern was designed, the diamonds were offset towards inside in order to create apertures. The offset value is dependent on the distance between hexagon's centers and a defined attraction point. This pattern provides the opportunity that the size and ratio of the apertures change, which leads to a change in the perforation ratio of the screen. This ultimately affects the daylight illumination.

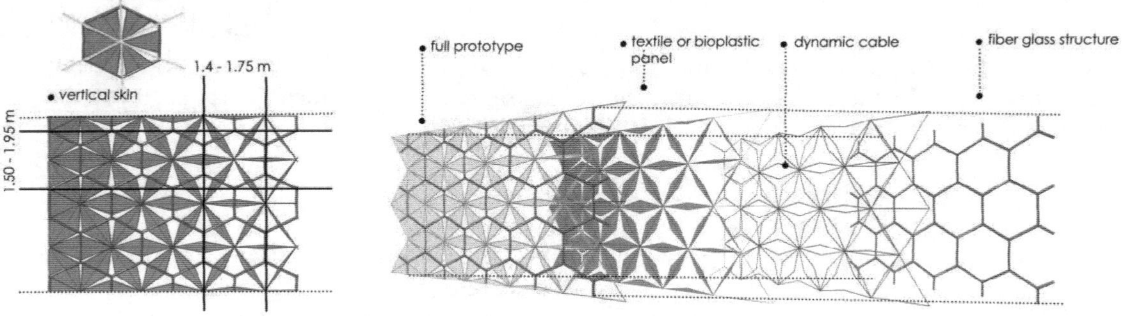

Figure 12. Skin design: a possible development of the skin system for the environmental rehabilitation of buildings.

All configurations have a dynamic surface on the front side of the wall. This provides the opportunity to change the perforation ratio of each module depending on the specific program of the space and the required daylighting levels of that space. On top of this, the screen walls can change their role from shading screen to the main building facade, by placing the glass panes in the apertures instead of adding a layer to the glass envelope.

7. Pilot study

The aim of this study is to offer a simple method that could be embedded within the typical design process of building façade, in order to evaluate and optimize façade daylighting performance during the design process. The proposed method could be applied to different oriented facade to offer a guide line for façade design.

8. Methodology:

To design the geometric patterns, Rhino has been opted as the modelling platform and Grasshopper, which is a parametric modelling plugin for Rhino, has been used. Grasshopper provides the opportunity to parametrically design and then alter the variables, in order to create various iterations of the design downstream as show in 'Figure 13'. By using Rhinoceros and Grasshopper, a model was created for a hypothetical indoor office space of **60** m2 area (6 m (w) x 1**0** m (l) x 4 m (h)). The space is a sidelit space facing the south direction at the ground floor, located in Cairo, Egypt. The designed shading screen is applied to a base case open-plan office space with minimum 500 lux illumination daylighting requirement [23].

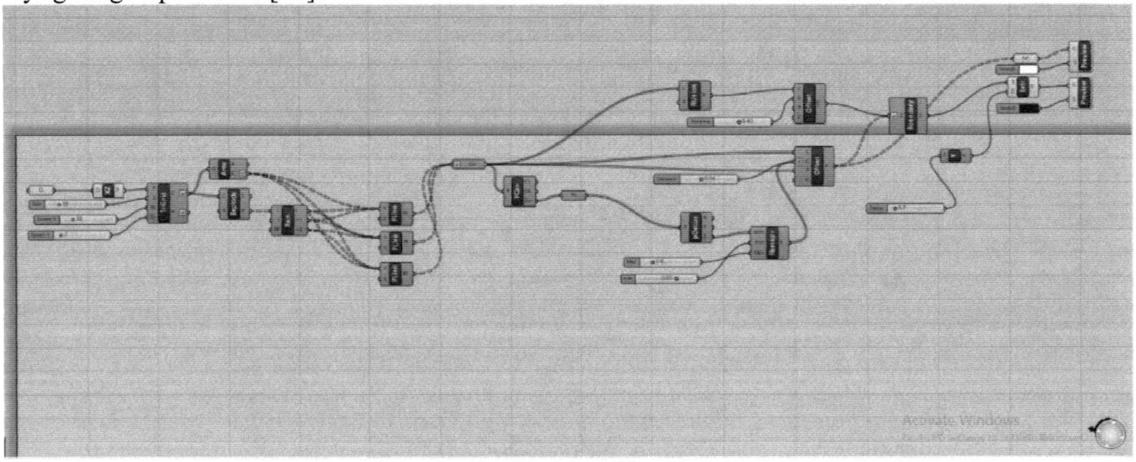

Figure 13. The Modeling Process
Source; by the researcher

Since Rhinoceros has been chosen as the modeling platform, DIVA-for-Rhino, which is a highly optimized daylighting and energy modeling plug-in for Rhino, has been chosen for daylighting simulations. The default DIVA materials have been assigned to the geometry. Year round performance is addressed by using a Dynamic Daylight Performance Metrics (DDPMs). Climate- The "Daylight Autonomy" index (DA) is defined as the percentage of the occupied hours of the year when a minimum illuminance threshold is met by daylight alone.

The process will follow two successive stages:

Stage one: this stage will evaluate daylighting performance for the hypothetical space with a mere glass façade.

Stage two: this stage will use optimization process to guide façade design and define the perforation of shading screen that is needed to optimize daylighting uniformity in the space. The aim of this stage is to evaluate the effect of changing the Shading Screen perforation percentage on annual performance, in order to conclude the best perforation percentage.

9. Results

Next table illustrates the four different stages and its results, all three configurations have been simulated and the results have been compared with a base case model with no shading screen as show in (Table 3).

IOP Conf. Series: Materials Science and Engineering **974** (2020) 012026 doi:10.1088/1757-899X/974/1/012026

Table 3. Visual daylighting simulation results.

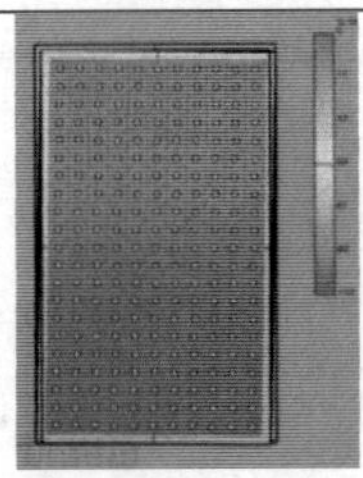	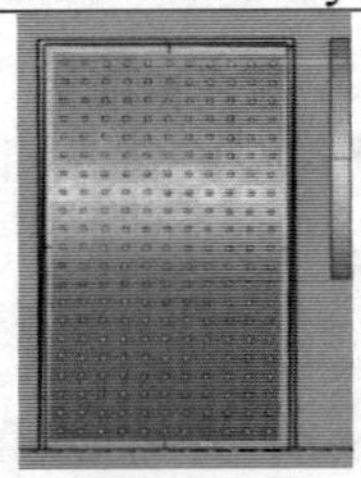	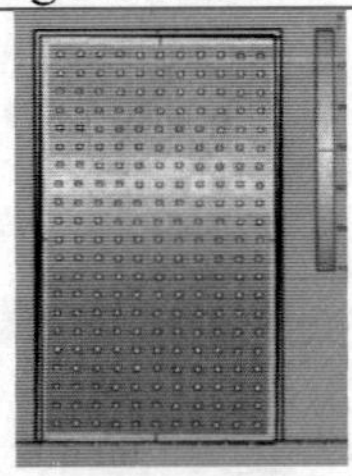	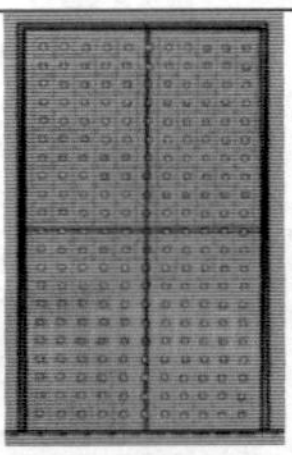
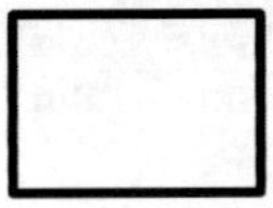	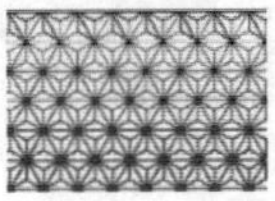		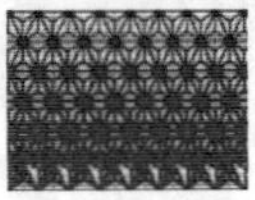
Case 1 Space with a mere glass façade- no screen.	Case 2 70% perforation	Case 3 50% perforation	Case 4 30% perforation

Point in time visualizations are included in the table for September 21st 2019 at noon, to better understand the shading effect of each configuration. Case 1 which is the window with no shading screen is included in order to have a base case for comparison. It was expected that as the perforation ratio gets lower, the daylighting performance becomes lower too. Therefore, the decreasing trend in mean daylight autonomy from case 2 (70% perforation) and case 3 (50% perforation) compared to case 4 (30% perforation), can be clearly explained. The optimum solution is using shading screen (70% perforation) that offer 69% daylit.

10. Conclusion

Perfromative design as an important design paradigm in architecture intended mainly to meet building performance requirements. This research looks at a multi-objective performance evaluation of a screen wall, inspired by geometric patterns. The study contributes to the design of screen walls, in order to control daylighting and to provide shading for the buildings. The perforation ratio of each module can differ to accommodate the unique daylighting needs of a specific space.

11. References

[1] Broadbent G and Ward A, 1969. *Design Methods in Architecture*: Lund Humphries London.
[2] Kalay Y E, (1999). "Performance-Based Design." Automation in construction 8, no. 4: 395-409.
[3] Anderson J, 2010. Basics Architecture 03: Architectural Design. Vol. 3: AVA Publishing.
[4] El Iraqi, A. (2008). FORMGENERATION IN ARCHITECTURE. Cairo: Ain Shams University .
[5] McCormack, J. & A. Dorin 2001, 'Art, Emergence and the Computational Sublime' in Dorin, A. (ed), Second Iteration: a conference on generative systems in the electronic arts, CEMA, Melbourne, Australia, pp. 67-81.
[6] Oxman R, (2009). "Performative Design: A Performance-Based Model of Digital Architectural Design." Environment and planning. B, Planning & design 36, no. 6: 1026.
[7] Leach N, (2009). "Digital morphogenesis." Architectural Design 79(1): 32-37
[8] Graham P C, "The Parametric Façade Optimization in Architecture through a Synthesis of Design, Analysis and Fabrication by," 2012.
[9] Albayrak C, "Performative Architecture as a Guideline for Transformation of the Defence Line of Amsterdam", Master of science, Middle East Technical University Delft University of Technology, 2011

[10] Abdelhamid A, "Tools of Activating the Time in the Design Process - a Study in the Paths of Movement", M. Sc., Faculty of Engineering, Cairo University, Egypt, pp.7,2010.
[11] Klaus R S, "The Dynamic Architecture of Emotion: Evidence for the Component Process Model, Cognition and Emotion", 23:7, pp. 1307-1351, DOI: 10.1080/02699930902928969, 2009.
[12] Saber A M, "The Static Dynamic in Architecture between the Concept and the Practice", 12th Al-Azhar International Eng. Conference, Faculty of Engineering, Al-Azhar University, Egypt, 2012.
[13] Karvigh S A, Ghahramani A., Gerber B.B., and Soibelman L., " One Size does not Fit all: Understanding User Preferences for Building Automation Systems", Elsevier, Energy and Buildings, Vol. 145, pp. 163-173, 2017.
[14] Ibrahim, Islam Ghonimi. (2015). The Role of Nature Form Versus Life Principles In Achieving Sustainability Of Bio-Mimic Architecture Measuring The Gap Of Contemporary Egyptian Practice Of Bio-Mimic Architecture. Journal of Engineering Sciences, Assiut University, Faculty of Engineering, Vol. 43, No. 6, November 2015, p.40. Retrieved from
[15] Caetano I, Leitão A. DrAFT: An algorithmic framework for facade design, generative design applications. Vol. 1. In: 34th CAADe; 2016; Oulu. pp. 465-474
[16] Hashim B H, and Al-Tai, M.M., "The Movement Architecture - The Impact of Objective Movement of Architecture on the Recipient", Iraqi Journal of Architecture, Vol. 28, No. 1-2, College of Engineering, University of Technology, Baghdad, Iraq, 2014.
[17] Antonio C, "Dynamic Reconfigurable Architectures and Transparent Optimization Techniques", Springer Science, London, 2010.
[18] Sulaiman A, "Evolution of the Concept of Movement and its reflection on architectural output", 12th Al-Azhar International Eng. Conference, Faculty of Eng., Al-Azhar University, Egypt, 2012.
[19] Baghaei D A, and Safari, H., "Factors Affecting Creativity in the Architectural Education Process based on Computer-Aided Design", Elsevier, Frontiers of Architectural Research, DOI.org/10.1016/j.foar.2017.09.001, 2018.
[20] Bellia L, Marino C, Minichiello F. and Pedace A (2014). "An Overview on Solar Shading Systems for Buildings." Energy Procedia **62**: 309-317.
[21] Freewan A, (2014). "Impact of external shading devices on thermal and daylighting performance of offices in hot climate regions." Solar Energy 102: 14-30.
[22] Sherif A, Faggal A and. Arafa R (2010). "External perforated solar screens for thermal control in desert environments: the effect of perforation percentage on energy loads." Proceedings of Renewable Energy.
[23] Illuminating Engineering Society of North America, The IESNA lighting handbook, 9th ed. NewYork, 2000.

Military Technical College
Kobry El-Kobbah,
Cairo, Egypt

13th International Conference on Civil and Architecture Engineering ICCAE-13-2020

Comparison of Occupant Thermal Comfort with and without Passive Design for a Naturally Ventilated Educational Building: a case study in Cairo, Egypt

Mohammad Fahmy[1], Sherif A. Mahmoud[2], Ibrahim M. Olwy[3], Marwa Abdelalim[4]

[1]Associate Professor, Architecture Engineering Department, Military Technical College, Cairo, Egypt.
[2]Senior Lecturer, Architecture Engineering Department, Military Technical College, Cairo, Egypt.
[3]Assistant Lecturer, Architecture Engineering Department, Military Technical College, Cairo, Egypt.
[4] Assistant Professor, Architecture Engineering Department, Faculty of Engineering, Prince Sultan University, Riyadh, KSA.

Corresponding author: md.fahmy@mtc.edu.eg

Occupant thermal comfort in educational buildings is important for students as it support the educational process itself. The passive design of naturally ventilated buildings plays a crucial role in achieving occupant thermal comfort but the passive design itself might be of higher cost to convince the owner even with proofs from the pre and while design simulation outputs. In this paper, the as built of a 1000 students' educational building is simulated using Design Builder and compared with the same thermal zones of the as built but with passive cooling applications. Two wind catchers, shading devices, low e-glass windows, double skin façade, and double roof were applied as amelioration strategies. The status of student comfort was improved in terms of both Predicted Mean Vote and air temperature, compared with the base case of the building. It is concluded that shading the roofs and southern facade of the building envelope were the most efficient scenarios for the passively modified version of the building.

Keywords: passive cooling, educational buildings, occupant thermal comfort, design builder, shading, PMV.

1. Introduction

Human thermal comfort is an important sustainability, wellbeing and productivity measure for buildings indoor spaces, [1]. Speaking about thermal comfort as a sustainability measure is crucially correlated to the energy consumption of mechanically cooled building. It is used as an indicator for the HVAC systems set points so that the cooling loads can be estimated efficiently, and the whole system can be designed [2]. On the other hand, indoor thermal comfort represents how occupants feel indoors which mean that their productivity can be estimated and maximized for better outcomes of the employer. Therefore, the assessment of thermal comfort status became a hot topic since the Fanger Predicted Mean Vote (PMV) have been adopted [3]. Since then, a lot of thermal comfort research has been undertaken which revealed many new and improvements for existing indices such as the Standard Equivalent Temperature (SET) and others [4-8]. In this concern, natural ventilation in buildings as a passive design strategy is characterized with wind flow stimulators like wind catchers or courts or so, in addition to applying shading as main heat gain prevention strategy, [9]. Moreover, the shade generation by fences, green roofs [10], double roofs and double skin facades are getting much interest in the recent years attributed to its efficiency in minimizing the heat gain and improving indoor thermal comfort in the excessive heat conditions of naturally ventilated buildings but subject to the optimization between the need to solar prevention and the need to daylighting even in the outdoor spaces, [11]. Further, improving and/or providing acceptable occupant thermal comfort levels is not only subject to the building envelope passive applications but extends to the outdoor environment that has to be coupled with indoor one not only for the sake of energy efficiency as discussed by Fahmy et al., [12,13], but also for the sake of occupant thermal comfort. However, assessment of post occupancy or existing buildings in terms of sustainability measures is gaining more platform [14-16]. From this standing point, the post occupancy of a naturally ventilated but not comfortable educational building in Cairo, Egypt using building performance simulation will be assessed in comparison to different scenarios of passive treatments in order to improve indoor thermal comfort.

2. Methodology

The methodology is divided into two main subsections: 1) case study and passive improvement strategies; and 2) design builder calibration and simulation. The first subsection describes the case study and the passive improvement strategies that were implemented in the simulations, whereas the second one describes the calibration and the setting of the design builder tool through the various simulations. The dynamic simulation tool DesignBuilder V4.2, [17], was used to estimate the occupant thermal comfort PMV levels as it incorporates an easy and simple CAD interface for the EnergyPlus core calculations. Site data is based on WMO meteorological measurements of station no.623660 at Cairo international airport, which is classified by Energy Plus conversion tool [8] statistically as mixed dry/semiarid whereas the extreme hot week typically lies between June 26th and July 2nd with a maximum dry bulb temperate of 44.0 °C. the weather data file that has been used is modified one by METEONORM [18] to adjust the exact data for the site which is far of about 5km from the above mentioned WMO station.

2.1. Base case and passive improvement strategies for thermal comfort

The case study building is a higher education building in Cairo, Egypt, which comprises of five above grade floors, ground floor and four typical floors, with a foot print area of 1950 m^2 as shown in Figure 1. The building comprises lecture halls, classrooms, and computer labs, as well as, W.C.s in each floor. The building is oriented 30 degrees in direction of the north east. Ten simulations were conducted; one simulation for the base case and other nine simulations for modified scenarios of the same building in order to improve passively the thermal comfort inside the building and decrease the air temperature in the summer months. There are seven correlated strategies are modified in each simulation to achieve the maximum achievable passive thermal improvement, which are: 1) walls; 2) openings; 3) window airflow control; 4) louvers; 5)double roofing; 6) wind catcher; and double skin façade.

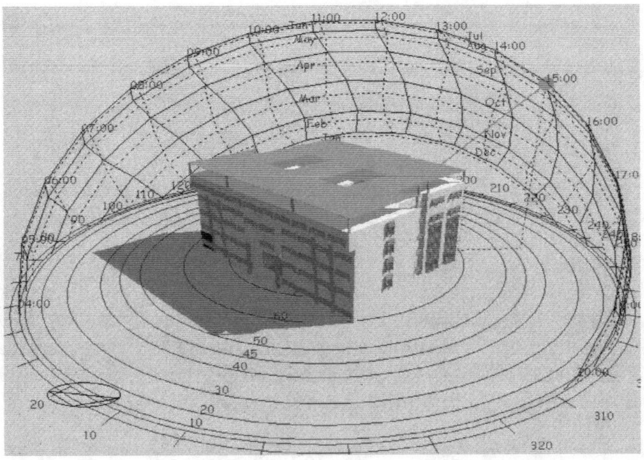

Figure 1: Design builder case study model

Table 1. Passive improvement domains for simulation scenarios

Simulation Scenarios	Walls	openings	Window air flow control	Louvers	Double ceil	Wind catcher	Double skin Façade
Base case	Single wall 250 mm brick	Single clear glass	No window air flow control	No louvers	No double roofing	No wind catchers	No double skin facade
Scenario (S1)	Double layer wall 250 mm brick-foam insulation-125mm brick	Double glazing Low-E 6mm glass- argon filled gap	No window air flow control	0.5 overhang louvers	Double roofing with holes over inner courts	No wind catchers	No double skin façade
Scenario(S2)	Double layer wall 250 mm brick-air gap-125mm brick	Double glazing Low-E 6mm glass- argon filled gap	No window air flow control	0.5 overhang louvers	Double roofing with holes over inner courts	No wind catchers	No double skin façade
Scenario(S3)	Single wall 250 mm brick	Double glazing Low-E 6mm glass- argon filled gap	No window air flow control	0.5 overhang louvers	Double roofing with holes over inner courts	No wind catchers	No double skin façade
Scenario(S4)	Single wall 250 mm brick	Double glazing Low-E 6mm glass- argon filled gap	No window air flow control	0.5 overhang louvers	Double roofing with holes over inner courts	No wind catchers	No double skin façade
Scenario(S5)	Single wall 250 mm brick	Single glazing	No window air flow control	0.5 overhang louvers	Double roofing with holes over inner courts	No wind catchers	No double skin façade
Scenario(S6)	Single wall 250 mm brick	Single glazing	No window air flow control	0.5 overhang louvers	Double roofing with holes over inner courts	No wind catchers	double skin Inclined fins at south facade
Scenario(S7)	Single wall 250 mm brick	Double glazing Reflective 6mm glass- air filled gap	No window air flow control	0.5 overhang louvers	Double roofing with holes over inner courts	No wind catchers	double skin Inclined fins at south facade
Scenario(S8)	Single wall 250 mm brick	Double glazing Reflective 6mm glass- air filled gap	Existence of air flow control	0.5 overhang louvers	Double roofing with holes over inner courts	No wind catchers	double skin Inclined fins at south facade

Simulation Scenarios	Walls	openings	Window air flow control	Louvers	Double ceil	Wind catcher	Double skin Façade
Scenario(S9)	Double layer wall 250 mm brick-foam insulation-125mm brick	Double glazing Low-E 6mm glass- argon filled gap	No window air flow control	0.5 overhang louvers	Double roofing (inclined panels)	8 wind catchers	double skin Inclined fins at south facade

2.2. Design builder calibration and simulation

Design Builder software was used for simulations of the base case and the nine scenarios. There are two main tabs in the software interface have the same setting through all the simulations: activity tab and HVAC tab. In the activity tab the schedule of the occupancy was chosen to be educational classrooms with time schedule from 8 am: 2 pm and from 5 pm: 10 pm. In the environmental control, the cooling set back temperature is set to be 20 degree Celsius. For the indoor minimum temperature control in the ventilation set point temperatures, the minimum temperature definition was set to be by schedule and the minimum temperature schedule was set to be 20 degree Celsius through six seven days of week except Friday is of for all the twelve months. Moreover, the office equipment and computers are checked on. In the HVAC tab, the template was selected to be natural ventilation with no heating nor cooling. The mechanical ventilation is checked on. Besides, the natural ventilation was activated with the same operation schedule as the occupancy in the activity tab. Further, in the seventh and the eighth simulation, the air flow controlled was activated to force keeping some selected windows opened and the source of air was outdoor, and the destination was the indoor.

3. Results

Figures 2 and 3, and tables 2 and 3 show the graphical and numerical representation of the resulted simulation outputs for both averaged PMV and air temperature (T_a) for the whole thermal zones of the examined building.

As illustrated in figure 2 and 3, the monthly distribution of PMV for all scenarios (column 2-9) compared to the base case results of column 1 show that generaly summer aleviation was limited which can be owed to the original compact design of the building. In addition, despite some scenarios have made a slight alleviaion in summer, all scenarios increased the PMV values in winter. However, only the wind flow stimulation through the long corridor side windows was effective whencompaered to using of wind catchers in S9 that hasn't been effective at all as shown in both tables 2 and 3. In summer months of June and July, the most improvement for PMV happened S7 and S8 (double roofing and double skin at southern façade with reflective glass) that reached a reduction of 0.1 and 0.15 respectively and corresponding to about 0.8°C reduction of T_a for both June and July months.
All building envelope design scenarios that has either double wall with cavity or double wall with insulation didn't reveal any improvments to the indoor conditions which can be attributed to same thermal effect of the insulation that it makes. It insulates and lags the conductive heat transfer from outside to inside and the same applies from inside to outside which means capturing the heat inside building and prevention of heat release.
In winter, the most effective scenarios were S5 and S6 which provided a slight heating with an increase of about 0.6 PMV in both January and February values corresponding to about 1.6°C and 1.9°C.

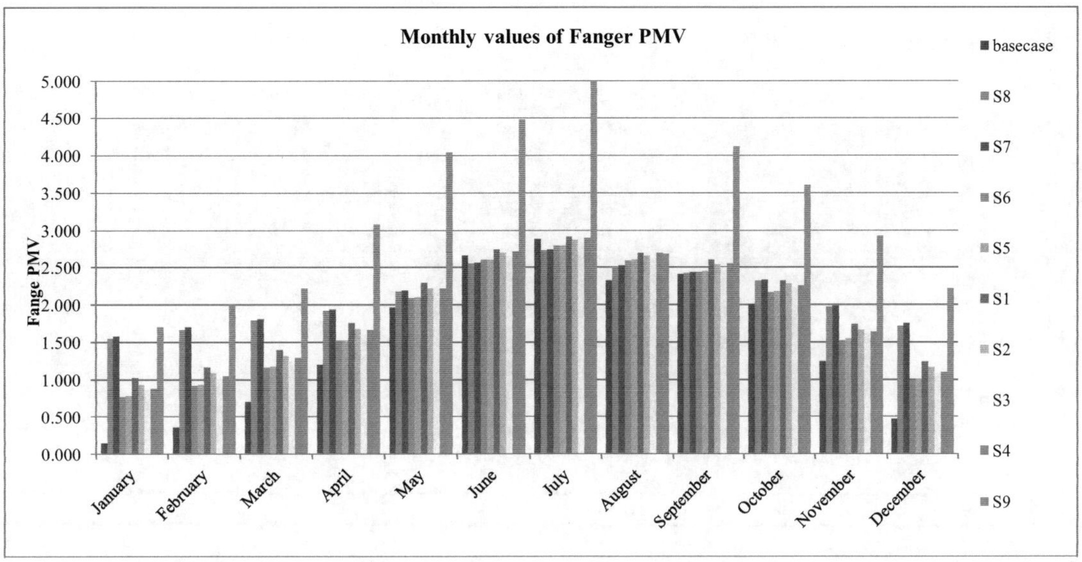

Figure 2: Fanger PMV monthly simulation results distribution for the base case and nine scenarios

Table 2: Fanger PMV monthly simulation numerical results for the base case and nine scenarios

Date/Scenarios	Basecase	S1	S2	S3	S4	S5	S6	S7	S8	S9
January	0.144	1.024	0.937	0.862	0.885	0.780	0.760	1.579	1.548	1.696
February	0.353	1.167	1.083	1.011	1.037	0.930	0.916	1.696	1.667	2.011
March	0.695	1.397	1.320	1.254	1.288	1.180	1.169	1.810	1.789	2.219
April	1.200	1.762	1.676	1.604	1.658	1.525	1.518	1.932	1.914	3.081
May	1.963	2.285	2.214	2.158	2.216	2.098	2.094	2.193	2.180	4.038
June	2.655	2.742	2.689	2.647	2.713	2.607	2.604	2.563	2.551	4.482
July	2.882	2.915	2.869	2.832	2.897	2.797	2.794	2.742	2.731	5.019
August	2.324	2.689	2.650	2.620	2.692	2.595	2.587	2.521	2.510	2.683
September	2.406	2.595	2.539	2.493	2.550	2.443	2.431	2.436	2.423	4.113
October	2.005	2.326	2.273	2.228	2.262	2.179	2.165	2.332	2.321	3.598
November	1.245	1.735	1.669	1.612	1.636	1.548	1.530	1.985	1.968	2.920
December	0.463	1.239	1.157	1.084	1.105	1.005	0.987	1.749	1.720	2.210

IOP Conf. Series: Materials Science and Engineering **974** (2020) 012027 doi:10.1088/1757-899X/974/1/012027

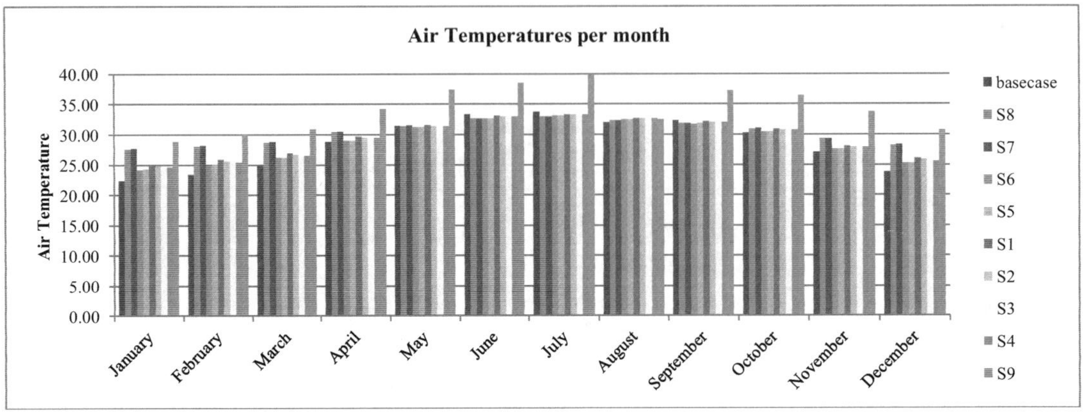

Figure 3: Air temperature monthly simulation results distribution for the base case and nine scenarios

Table 3: Air temperature monthly simulation numerical results for the base case and nine scenarios

Date/Scenarios	Basecase	S1	S2	S3	S4	S5	S6	S7	S8	S9
January	22.37	25.04	24.75	24.50	24.57	24.21	24.15	27.66	27.54	28.79
February	23.39	25.80	25.52	25.28	25.37	25.01	24.96	28.26	28.14	30.09
March	24.88	26.93	26.67	26.46	26.57	26.20	26.17	28.82	28.73	30.88
April	28.81	29.67	29.48	29.31	29.43	29.11	29.10	30.44	30.39	34.33
May	31.39	31.60	31.44	31.32	31.44	31.17	31.16	31.50	31.47	37.32
June	33.38	33.04	32.92	32.83	32.97	32.73	32.72	32.67	32.65	38.56
July	33.77	33.35	33.25	33.17	33.31	33.08	33.07	32.97	32.95	39.81
August	31.92	32.70	32.61	32.54	32.71	32.48	32.46	32.32	32.30	32.51
September	32.29	32.18	32.05	31.96	32.08	31.83	31.81	31.88	31.85	37.21
October	30.33	30.91	30.74	30.60	30.71	30.43	30.39	31.07	31.03	36.36
November	27.12	28.16	27.95	27.76	27.84	27.55	27.49	29.38	29.32	33.74
December	23.77	26.04	25.77	25.53	25.60	25.25	25.20	28.42	28.30	30.80

4. Conclusion

This research aimed to make a spot light on evaluating the application of different passive scenarios for an educational building in Cairo, Egypt after it has been already constructed to improve the post occupancy thermal comfort. The different scenarios showed in table 1 applied all available interventions of thermal insulation, ventilation and shading such as wind catchers, louvers, double skin facades, double roofing, double glass windows, etc. There was evident results for the effectiveness of the shading strategy (with reflective glass) in S7 and S8 compared to other strategies especially in the hot summer months of June and July. Thermal insulation using either double wall and cavity or double and filling insulating material didn't conclude any improvement but in fact made the results worse for the examined building as it depends on natural ventilation and needed to release the heat gained during day which hasn't been available because of the insulation. It was clear that the already existence of such a compact educational building and trying to modify its as built is much harder than constructing a pre and while designed passive cooling building which might has necessitated a different educational building zoning form to allow wind flow for a maximized ventilative cooling in addition to the shading.

5. References

1. Humphreys, M.A.; Nicol, J.F.; Raja, I.A.; . Field Studies of Indoor Thermal Comfort and the Progress of the Adaptive Approach. *Journal of Advances in Building Energy Research* **2007**, *1*, 55-88.
2. Mahmoud, S.; Fahmy, M.; Mahdy, M.; Elwy, I.; Abdelalim, M. Comparative energy performance simulation for passive and conventional design: A case study in Cairo, Egypt. In Press. https://doi.org/10.1016/j.egyr.2019.09.052. *Energy Reports* **2020**.
3. Fanger, P.O. *Thermal comfort; Analysis and applications in environmental engineering*; McGraw-Hill, first published in 1970, Danish Technical Press, Copenhagen: NewYork, 1972.
4. Hensen, J.L.M. Literature review on thermal comfort in transient conditions. *Building and Environment* **1990**, *25*, 309-316.
5. de-Dear, R.J.; Brager, G.S. Developing an Adaptive Model of Thermal Comfort and Preference. *ASHRAE Transactions* **1998**, *104* 145-167.
6. Nicol, J.F.; Humphreys, M.A. Adaptive thermal comfort and sustainable thermal standards for buildings. *Energy and Buildings* **2002**, *34*, 563-572.
7. Olesen, B.W.; Parsons, K.C. Introduction to thermal comfort standards and to the proposed new version of EN ISO 7730. *Energy and Buildings* **2002**, *34*, 537-548.
8. de Dear, R.J.; Brager, G.S. Thermal comfort in naturally ventilated buildings: revisions to ASHRAE Standard 55. *Energy and Buildings* **2002**, *34*, 549-561.
9. Al-Hafith, O.; B K, S.; Bradbury, S.; de Wilde, P. The Impact of Courtyard parameters on its shading level An experimental study in Baghdad, Iraq. *Energy Procedia* **2017**, *134*, 99-109, doi:https://doi.org/10.1016/j.egypro.2017.09.539.
10. Kumar, R.; Kaushik, S.C. Performance evaluation of green roof and shading for thermal protection of buildings. *Building and Environment* **2005**, *40*, 1505-1511.
11. Elwy, I., Yasser, I., Fahmy, M., Mahdy, M. (2018). Outdoor microclimatic validation for hybrid simulation workflow in hot arid climates against ENVI-met and field measurements. *Energy Procedia* **2018**, *153*, 29-34, doi:https://doi.org/10.1016/j.egypro.2018.10.009.
12. Fahmy, M.; Mahdy, M.; Mahmoud, S.; Abdelalim, M.; Ezzeldin, S.; Attia, S. Influence of urban canopy green coverage and future climate change scenarios on energy consumption of new sub-urban residential developments using coupled simulation techniques: A case study in Alexandria, Egypt. *Energy Reports* **2020**, https://doi.org/10.1016/j.egyr.2019.09.042, doi:https://doi.org/10.1016/j.egyr.2019.09.042.
13. Fahmy, M.; El-Hady, H.; Mahdy, M.; Abdelalim, M.F. On the green adaptation of urban developments in Egypt; predicting community future energy efficiency using coupled outdoor-indoor simulations. *Energy and Buildings* **2017**, *153*, 241-261, doi:https://doi.org/10.1016/j.enbuild.2017.08.008.
14. Sherif Mahmoud; Tarek Zayed; Fahmy, M. Development of sustainability assessment tool for existing buildings. *Sustainable Cities and Society* **2019**, *44*, 99-119.
15. Hwang, R.-L.; Lin, T.-P.; Kuo, N.-J. Field experiments on thermal comfort in campus classrooms in Taiwan. *Energy and Buildings* **2006**, *38*, 53-62.
16. Feriadi, H.; Wong, N.H. Thermal comfort for naturally ventilated houses in Indonesia. *Energy and Buildings* **2004**, *36*, 614-626.
17. DesignBuilder. *DesignBuilder [Online], Available: http://www.designbuilder.co.uk. Accessed 21/4/2018.*; 2018.
18. METEONORM. Global Meteorological Database for Engineers, Planners and Education, [Online]. Availabe online: www.meteonorm.com/pages/en/meteonorm.php. (accessed on 22/4/2009.).

**Military Technical College
Kobry El-Kobbah,
Cairo, Egypt**

13th International Conference on Civil and Architecture Engineering ICCAE-13-2020

Life cycle assessment of a residential building in Egypt: A case study

Dalia M.A. Morsi [1], Walaa S.E. Ismaeel [2] and A. E. ABD EL-HAMED [3]

[1] Teaching Assistant, Higher Institution of Engineering, Sherouk Academy, Egypt.
[2] Assistant Professor, Architecture Engineering Department, The British University, El-Sherouk City, Egypt.
[3] Assistant Professor, Civil Engineering Department, Higher Institution of Engineering, Sherouk Academy, Egypt.

Corresponding author: PG.dalia91630009@bue.edu.eg

The present study discussed the significant environmental impacts of a residential building located in New Cairo, Egypt. This covered all life cycle steps from cradle to cradle with a projected 60-year life span: (i) an inventory of all the construction materials were analysed, covering the building structure as well as the energy consumption; (ii) three types of functional units were defined; (iii) the two top building materials were examined, and a sensitivity analysis was conducted to investigate the impact associated with the choice of building materials. The result shows that two life cycle phases concerning, manufacturing and operation, were more significant in all of the impact categories. It also shows that building structure and flooring result in most of the environmental loads. In terms of the sensitivity analysis, it was found the structural concrete had the largest impact, dominating all selected impact categories except ODP. Finally, limitations and challenges are discussed to explore better design decisions for selecting buildings' structural systems in future studies.

Keywords: Environmental impact assessment, Life cycle analysis, Residential building, Tally plugin.

IOP Conf. Series: Materials Science and Engineering **974** (2020) 012028 doi:10.1088/1757-899X/974/1/012028

1. Introduction

In 2018, United Nations Environment Programme reported that the buildings worldwide are responsible for 40% of the world's total energy use, 25% of global water, 40% of global resources and they emit about one-third of greenhouse gases emissions, thus witnessed a prioritization to reduce the environmental impact on the built environment [1]. Saving energy in buildings has various benefits including reducing costs, increasing the security of supply and saving the environment [2]. Buildings account for 19 % of energy-related CO_2 emissions, 32 % of global energy use and 57 % of world electricity consumption, according to the report of the Intergovernmental Panel on Climate Change. Thus, buildings are one of the most significant infrastructures in modern society in terms of their energy consumption. The residential and commercial buildings contributed towards 41.1% to total primary energy consumption, because of their operation to primary energy use, 74% to electricity consumption, and 40% to CO_2 emissions [3]. Also, it is found that for the greenhouse emissions related to buildings, 40–95% of these emissions are caused by the operational energy use, with the remainder being caused by the construction and demolition processes. Yet, buildings also offer the greatest potential in developing countries to achieve significant reduction of emissions of GHGs. In addition, the consumption of energy in buildings can be reduced by proven and commercially available technologies about30 to 80 percent [1].

Any change in the structural system plays a role in the environmental impact assessment. Hence, this study aims at studying the environmental impact of concrete structural systems in residential buildings. Accordingly, the method adopted is; 1) compiling an inventory for the building's structural system and its energy consumption, 2) estimating environmental loads in several impact categories through life cycle stages, construction elements and conducting a sensitivity analysis of the two main construction materials within the building's life cycle.

2. Reviewing previous literature

It presents a recent review of the literature on Science Direct database covering the period 2000-2020. This literature review provides an overview of the Life Cycle Analysis and its application in the building industry.

2.1 Life cycle assessment framework

Life cycle assessment (LCA) is a technique for evaluating the environmental aspects and potential impacts associated with a product's life cycle from acquisition raw material to final disposal [4]. It yields reliable assessment results which practitioners may reply on for the decision-making process on both the building and urban levels [5]–[7]. The Life Cycle Assessment includes four phases: the goal and scope definition, life cycle inventory (LCI), impact assessment, and interpretation of results.

The goal and scope definition represents the functional unit, system boundaries, and the quality criteria for the inventory data. The life cycle inventory analysis (LCIA) is analyzing the inputs and outputs for a product are compiled and quantified for its life cycle. Finally, in respect of the defined objective and scope, the life cycle interpretation phase contains results of either the inventory analysis or the impact assessment (or both). Previous studies have followed the environmental impact categories characterization scheme reported by the 'Tool for the Reduction and Assessment of Chemical and Other Environmental Impacts' (TRACI) 2.1, developed by the United States Environmental Protection Agency (US EPA) to quantify environmental impact risk associated with emissions [8]–[10].

IOP Conf. Series: Materials Science and Engineering **974** (2020) 012028 doi:10.1088/1757-899X/974/1/012028

There have been various studies on complete LCAs as Ramesh, Prakash, and Shukla (2010) reviewed the life cycle energy analyses of 73 cases across 13 countries. It was found that the life cycle energy demand of office buildings varies from 45 GJ to 99 GJ per gross floor area over 50 years with 10-20% for embodied energy and 80-90% for operational energy. Also, Itard (2009) presented the results of an energy flow investigation, based on LCA considerations, on the relative values of embodied and operational energy of dwellings and dwelling renovation. The result after comparing the primary operational energy use and the new-built variants with the primary energy embodied in the variant itself showed that the embodied energy (EE) in renovation variants is very low, opposite to the EE in "low energy" new-built that can amount up to 60 years of operational energy use. Furthermore, the cumulative energy demand (CED) and the global warming potential (GWP) have examined by Bawden and Williams (2015), for a portfolio of 10 low, mid and high-rise multi family residences. The CED is increasing from 30, 34, to 39 GJ/m^2 for low, mid and high-rise multi-family residences, respectively and GWP is found to increase from 1.8, 2.2, to 2.5 CO_2eq/m^2, respectively. As well as another study by Gong et al. (2012) who compared the total life cycle energy and greenhouse gases (GHG) emissions of three multi-family residences of similar size but varying in commonly used framing materials (wood, steel and concrete). The ratios of energy consumption of concrete framework construction (CFC), light gauge steel framework construction (SFC) and wood framework construction (WFC) to that of the life cycle in the embodied materials phase are 13%, 23%, and 27%, respectively, and the ratios in the operation phase are 87%, 76%, and 71%, respectively.
The study also showed through the life cycle, the energy consumption of CFC is almost the same as that of SFC, and each of them is about 30% higher than that of WFC. Moreover, Lu et al. (2017) compared quantitative measurement for the environmental impact of using (engineered wood, concrete and steel) along 1m section of a structural beam in a continuous beam system with 6 m spans. The GWP of concrete had the greatest environmental burden (17.43 kg CO_2-eq), mainly due to significant transportation, fossil fuel consumption as well as the use of energy-intensive materials during the manufacturing process. The steel has reached (12.83kg CO_2-eq), so it has better environmental performance due to its lighter weight, which led to a reduction in the energy and material consumption, as well as transportation while the engineered wood beam had much better performance on GWP of (5.22kg CO_2-eq). Furthermore, Xia, Ding, and Xiao (2020) reviewed the reuse and recycling strategies for the design for deconstruction (DfD) and recycled aggregate concrete (RAC). The results were when applying only one strategy; the maximal impact indicator reductions induced by DfD are 1.8–2.8 times compared to that by RAC. When applying two strategies simultaneously, the benefit of either DfD or RAC shall decline compared to that of the corresponding strategy applied alone, but the overall benefits shall increase. The study also showed that using fully reusable structure with 100% (RCA) replacement of the conventional design will achieve the maximal environmental benefits, with percentage, 15.0% and 13.3% reductions for GWP and ADP, respectively

It is noted that the previously referenced studies describe various environmental loads and energy consumption for residential and commercial buildings worldwide, nevertheless, there is a lack of studies in Egypt. Hence, this study applies LCA for a residential building in New Cairo, Egypt.

3. Methodology

The research method follows three consecutive stages; 1) designing a solid slab structural systems and compiling an inventory of materials for an LCA, 2) developing 3D modelling for the structural building design and 3) integrating Revit 3D modelling and Tally environmental impact Plug-in, to define materials in the Tally plug-in for measuring the Life Cycle Categories according to TRACI 2.1 characterization scheme in the construction stage.

3.1 *Developing a Revit Structural 3D-Model*

This 3D model allows for the visualization and coordination of building components into an early design process that will dictate changes and modifications to the actual process of construction. This aims at obtaining an accurate estimate of material quantities for all structural elements in the building to be able to use the environmental impact assessment plug-in. The concrete solid slab 3D structural system model applied, and its plan view shown in figures (1&2).

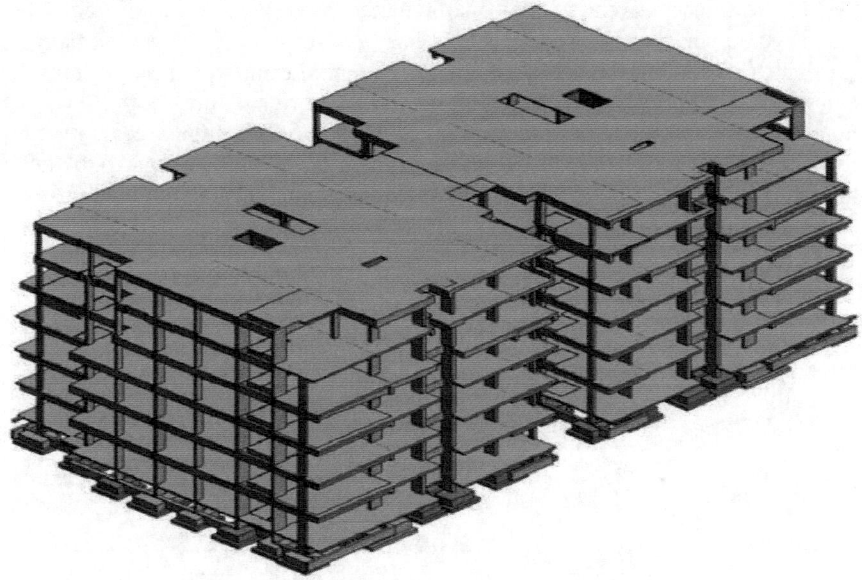

Figure 1. 3D view of the Concrete Structural Building

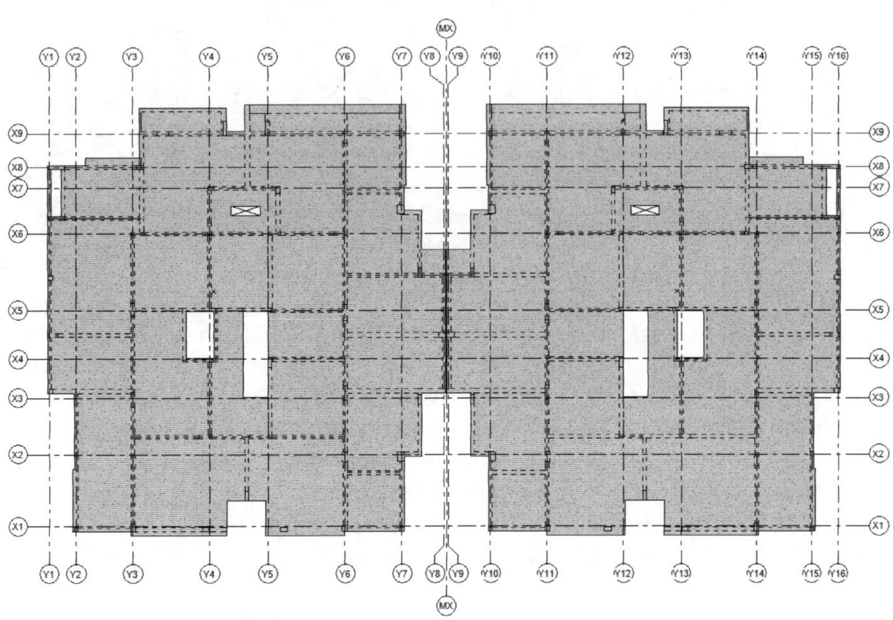

Figure 2. Plan view for the Concrete Structural Building

3.2 Tally Environmental impact assessment

The Tally plugin is integrated with Revit 3D structure model to allow the user to assign materials from its environmental product database to the BIM software; so, the user can view and follow the embodied environmental impact results based on the material selection decision-making process.

In this study, the system boundary in Tally plugin was defined as the process beginning from resource extraction and manufacturing of building products, to site preparation and the construction process, then moving to the operating energy and maintenance phase, the building demolition process and ending with recycling or reusing materials. Table (1) shows the system boundary defined for this study. To evaluate the goal and scope of LCA for residential buildings, it is essential to determine the functional unit. In building research there are several functional units for LCA. In this study, the functional equivalent is introduced as the square meter size of residential areas with a 60-year life span according to the Athena Sustainable Materials Institute (2016). In regard to the purpose of the building and the functions that it serves; the functional unit can be broken down into:

- Total impacts through building life cycle
- Total impacts per construction elements
- Total impacts per construction materials

Table 1. Life-Cycle Stages as defined by EN 15978, source [16]

Building life cycle																Supplementary information (Module)
Product / Manufacture Stage [A1-A3]			Construction Process Stage [A4-A5]		Use stage [B1-B7]							End-of-life Stage [C1-C4]				Benefits and load beyond the system boundary
					Building Fabric					Operation of the Building						
A1	A2	A3	A4	A5	B1	B2	B3	B4	B5	B6	B7	C1	C2	C3	C4	D
Raw materials supply	Transport	Manufacturing	Transport	Construction	Use	Maintenance	Repair	Replacement	Refurbishment	Operational energy use	Operational water use	Demolition	Transport (to disposal)	Waste processing	Disposal	Reuse Recycling Energy recovery
Cradle-to-Gate			Gate-to-Grave													
Cradle-to-Grave																
Cradle-to-Cradle																

In the life cycle impact assessment stage, seven impact categories have been selected: Acidification Potential (AP) (kg SO_2eq), Eutrophication Potential (EP) (kg Neq), Global Warming Potential (GWP) (kg CO_2eq), Ozone Depletion Potential (ODP) (kg CFC-11eq), Smog Formation Potential (SFP) (kg O_3eq), Primary Energy Demand (PED) (MJ), Non-Renewable Energy Demand (NRED) (MJ) and Renewable Energy Demand (RED) (MJ).

It is noted that the following life cycle processes are excluded from this study:

- The transportation to factory and transportation to disposal.
- Personnel- related activities (accessories, furniture, office supplies).
- Energy and water use related to company management and sales activities, which may be located either within the factory site or at another location.

IOP Conf. Series: Materials Science and Engineering **974** (2020) 012028 doi:10.1088/1757-899X/974/1/012028

3.3 *Description of the case study building*

The case study is a residential building in New Cairo. It consists of a ground and 5 floors with clear height for each floor equals 2.70 m. The building is under construction, the statically system was solid slab concrete structural system with various thickness (12-25 cm), the beams and columns were with several cross sections. The isolated and combined footing were linked by grade beams. the external view of the building shown in Figure (3). The case-study building has a total gross area 8676.25 m^2. Each floor level has 8 apartments with 3 bedrooms per apartment, two living-room areas, four bathrooms, a kitchen and a balcony (terrace) across the private garden of the building. In addition, the information on residential energy consumption was provided by Electrical Egyptian code, including the consumption of electricity 10 Kilo-volt ampere KVA for luxurious residential. The total consumption of electricity approximate was 281,880 annual kilowatt-hour kWh. On-site construction using 2 generators with capacity 100 (KVA) for each, so the total electricity uses 1,800 (KWh). These two data were taken into Tally plugin as factors of energy usage.

Figure 3. Front view of the case study.

3.3.1 Description of the construction materials

3.3.1.1 Reinforced Concrete

The concrete structural system was designed based on The Egyptian Code for Design and Executing Reinforced Concrete Structures. A bill of Materials report was produced by the Revit structure quantities takeoff sheet. The largest amounts of materials used in concrete solid slab structural system were concrete and steel reinforcement rod in floor elements 1,456.22 m^3 and 179,719.9 kg, respectively. Further information regarding the quantities of each material is shown in Table (2).

Table 2.Material quantities for each element in the Solid slab concrete structural system model.

The reference unit	Concrete solid slab structural system Elements		Quantitative (m^3) or (kg)
Materials required	Structural concrete, 3001-4000 psi, 0-19% fly ash and/or slag	Rectangular-Footing + Combined footing	770.98 m^3
		Combined footing Plain concrete	75.34 m^3
		Rectangular-Column	305.53 m^3
		Rectangular Beam & smells	541.43 m^3
		Floors	1456.22 m^3
		Core	73.98 m^3
		Shear walls	49.44 m^3
	Steel, reinforcing rod	Rectangular-Footing + Combined footing	22448.1 kg
		Rectangular-Column	76,991.70 kg
		Rectangular Beam	136,435.3kg
		Floors	179719.9 kg
		Core	5442.4 kg
		Shear walls	3637.6 kg

The LCI data on concrete is sourced in accordance with GaBi databases and modelling principles. Reinforced concrete grade in the project is with cubic characteristics strength of 25 N/mm^2. Ordinary Portland cement (Type I) has been used in the project. Cement is designed in accordance with the Egyptian Standard Specifications (ESS) 373, ESS. 541. Reinforcing bars satisfy the E.S.S 262-1974 for grades 24/35 & 36/52. Also, Table (3) provides a list of the building characteristics and the scope boundaries for each material in different stages of the LCA for solid slab concrete structural systems.

Table 3. The data describing scope boundaries for Concrete structural system

Material Name	Cast-in-place concrete, structural concrete, 3001-4000 psi	Steel, reinforcing rod
Description	Structural concrete, 3001-4000 psi, 0-19% fly ash and/or slag. Mix design matches the National Ready-Mix Concrete Association (NRMCA) Industry-wide Environmental Product Declaration (EPD).	Common unfinished tempered steel rod suitable for structural reinforcement (rebar)
Used in	All structural elements as (Foundation, Floors, Columns…etc.)	All structural elements as (Foundation, Floors, Columns…etc.)
Life Cycle Inventory	16% Cement, 7% Batch water 44% Coarse aggregate, 33% Fine aggregate	100% Steel rebar
Product Scope	Cradle to gate, excludes mortar Anchors, ties, and metal accessories outside of scope (<1% mass)	Cradle to gate
Transportation Distance	By truck: 24 km	By truck: 24 km
End-of-Life Scope	55% Recycled into coarse aggregate 45% Landfilled (inert material)	70% Recovered 30% Landfilled (inert material)
Module D Scope	Avoided burden credit for coarse aggregate, includes grinding energy	Product has a 16.4% scrap input while the remainder is processed and credited as avoided burden.
Service life	60 Years	60 Years

4. Results

A comprehensive LCA was carried out for a residential building in New Cairo, Egypt. The study included the entire life cycle of the building, including the manufacturing, construction, operating energy, maintenance, end-of-life and module D phases.

4.1 The results per building life cycle

The results show that the highest environmental impacts during the building's life cycle occurred during the operating energy phase, was approximately 62%–98% (except for the ODP), while product stage was approximately 16%-34% (except for the ODP was 70%), end of life was approximately 1%-4%. The construction phase and the last life cycle phase, module D, clearly had a lesser impact, less than 0.3% to -2%, respectively in all of the studied categories. This ratio corresponds with the study carried by Zhang et al. (2014) which computed the environmental impact of a residential building located in Vancouver, Canada. The study indicated that two life cycle phases had the greatest impact, these were; manufacturing and operation, with approximately 7%-51% and 30%-90%, respectively, while the end of life stage caused less than 1% impact in all of the studied categories.
The GWP in operation stage for this case study, was 7.05 E106 kgCO_2 eq, so the GWP was 1.19 per square meter. These results are consistent with the results of Bawden and Williams (2015) who examined the total life cycle energy, and GWP for a portfolio of 10 difference family residences. The GWP is found to increase from 1.8, 2.2, to 2.5 CO_2eq/m^2 for low, mid-rise and high-rise buildings, respectively. The proportion of the embodied energy and operation energy in the case study was 10-17% and 80-90%, respectively. This is similar to Ramesh, Prakash, and Shukla (2010) and (Wang, Yu, and Pan (2018) who analyzed the life cycle energy with 50 years and found that it was 10-20% for embodied energy and 80-90% for operational energy. The result of the environmental impacts of the model is shown in Fig. (4).

4.2 The results per construction elements

The 3D Revit model includes floors, walls and the structural system (foundations, columns and beams). The results of environmental impact comparing these elements are shown in Figure. (5). This shows that the structural elements of the building dominate the high values of all environmental impact; with 3.71E+03 (kg SO_2eq), 2.09E+02 (kg Neq), 1.09E+06 (kg CO_2eq), 6.59E+04 (kg O_3eq) and 1.14E-3 (kg CFC-11eq). The Primary Energy Demand presented 1.14E+07 (MJ); of which the non- renewable energy is responsible for 1.06E+07 (MJ) and renewable energy with 7.83E+05 (MJ). The structural elements were responsible for approximately 7%-24% more than of other elements as walls and floors, with a total mass of 3.93E+06. Furthermore, These results are consistent because an assembly that uses more materials would, therefore, have more environmental impacts than other assemblies as stated by Zhang et al. (2014) who found that walls which have a huge quantity, thus they have a total GWP of approximately 30,948 kg CO_2 eq and a total fossil-fuel consumption (FFC) of about 434,571 MJ, which is the highest of the assemblies.

4.3 The results per construction materials

Then a sensitivity analysis is performed to help stakeholders decide which material has the greatest impact on the environmental performance of a building. Using the analysis results shown in Fig. (6), designers could choose to reduce the amount of these materials in the building or find substitutes that have fewer environmental impacts.
This shows that the reinforcement concrete has the highest contribution for all environmental categories (except ODP) because of the amount of concrete with 7.2*106 kg, which it represents an increase of 17.2 over steel reinforcing rod quantity. That mainly due to significant transportation, fossil fuel consumption as well as the use of energy-intensive materials during the manufacturing

process that compatible with studied of Lu et al. (2017). The structural concrete has 4.04E+03 (kg SO2eq), 3.18E+02 (kg Neq), 1.51E+06 (kg CO2eq), 9.19E+04 (kgO3eq), 1.24E+07 (MJ) for non-renewable energy demand and 9.94E+05 (MJ) for renewable energy demand. The structure concrete represents an increase of 1.35, 3.69, 2.57, 2.65, 1.59, 1.98 over steel reinforcing rod for AP, EP, GWP, SEP, NREP and RED, respectively.

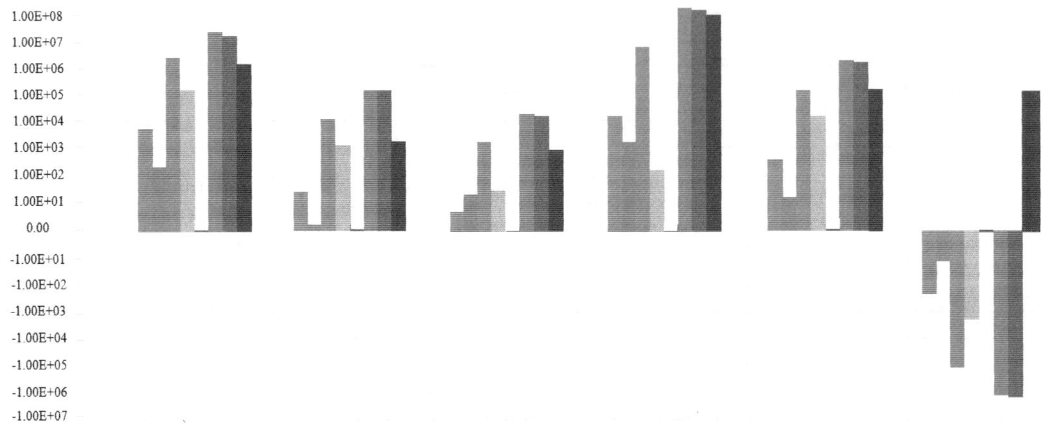

	Product	Transportation	Construction	Operational energy	End of life	Module D
AP	6.69E+03	4.68E+01	7.28E+00	1.76E+04	6.92E+02	-3.88E+02
EP	3.81E+02	3.81E+00	3.07E+01	1.84E+03	3.51E+01	-1.51E+01
GWP	2.15E+06	1.01E+04	1.54E+03	7.05E+06	1.50E+05	-2.17E+05
SEP	1.16E+05	1.55E+03	4.84E+01	2.10E+05	1.38E+04	-4.24E+03
ODP	3.64E-03	3.46E-10	7.96E-10	8.88E-06	2.75E-08	1.58E-03
PED	2.09E+07	1.47E+05	2.74E+04	1.81E+08	2.56E+06	-1.89E+06
NRED	1.97E+07	1.43E+05	1.82E+04	1.02E+08	2.40E+06	-2.01E+06
RED	1.20E+06	3.55E+03	9.21E+03	7.98E+07	1.69E+05	1.19E+05

Figure 4.Distribution of the environmental impacts of the building life cycle studied.

	Walls	Structures (foundations, columns and beams)	Floors
AP	2.21E+02	3.71E+03	3.11E+03
EP	1.41E+01	2.09E+02	1.81E+02
GWP	7.09E+04	1.09E+06	9.34E+05
SEP	4.29E+03	6.59E+04	5.64E+04
ODP	4.45E-05	1.14E-03	8.77E-04
PED	6.96E+05	1.14E+07	9.60E+06
NRED	6.47E+05	1.06E+07	8.94E+06
RED	4.92E+04	7.83E+05	6.64E+05
Total mass (kg)	2.90E+05	3.93E+06	3.48E+06

Figure 5. Distribution of the environmental impacts of different building components.

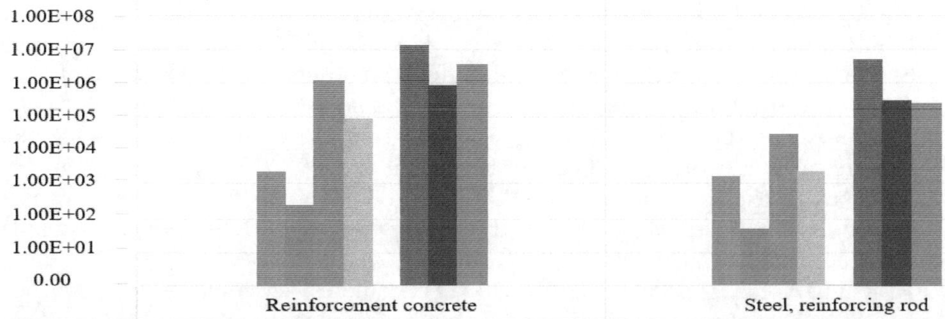

	Reinforcement concrete	Steel, reinforcing rod
AP	4.04E+03	3.00E+03
EP	3.18E+02	8.62E+01
GWP	1.51E+06	5.87E+05
SEP	9.19E+04	3.47E+04
ODP	2.71E-07	2.07E-03
NRED	1.24E+07	7.78E+06
RED	9.94E+05	5.02E+05
Total mass (kg)	7.28E+06	4.23E+05

Figure 6. Distribution of the environmental impacts of concrete structural systems materials

5. Conclusion

This study performed an LCA for a residential building in New Cairo, Egypt. This research evaluated and analyzed adverse environmental impacts through the building life cycle. The operating energy phase alone produced more than half of the total environmental impacts, with a share of approximately 60%-76% for concrete structural elements. Another significant life cycle phase was manufacturing, with a share of approximately 18 to 34% of total impacts. A breakdown of the building components indicated that the structure element (foundation, columns and beams) and floor element resulted in significant environmental impacts in the given impact categories. Moreover, according to the sensitivity analysis, the concrete was the most critical construction material; not only in terms of consumed quantities, but also in the associated environmental impact categories.

6. Reference

1 J. Randolph, G. M. Masters, J. Randolph, and G. M. Masters, "Energy Efficiency for Buildings," *Energy Sustain.*, vol. 33, no. 1, pp. 173–213, 2018.
2 L. Guan, M. Walmsely, and G. Chen, "Life Cycle Energy Analysis of Eight Residential Houses in Brisbane, Australia," *Procedia Eng.*, vol. 121, pp. 653–661, 2015.
3 L. C. M. Itard, "Embodied and operational energy use of buildings," *Lifecycle Des. Build. Syst. Mater.*, pp. 77–84, 2009.
4 The International Standards Organisation, "INTERNATIONAL STANDARD assessment — Requirements and guilelines," *Int. J. Life Cycle Assess.*, vol. 2006, no. 7, pp. 652–668, 2006.
5 W. Ismaeel, "Drawing the operating mechanisms of green building rating systems," *J. Clean. Prod.*, vol. 213, pp. 599–609, 2019.
6 M. A. Elsayed and W. Ismaeel, "Environmental assessment for major development projects: A case study 'Qattara Depression,'" *J. Clean. Prod.*, vol. 215, pp. 522–533, 2019.
7 W. S. E. Ismaeel, "Midpoint and endpoint impact categories in Green building rating systems," *J. Clean. Prod.*, vol. 182, p. , 2018.
8 J. Bare, "TRACI 2.1: User Manual," *US Environ. Prot. Agency*, pp. 1–24, 2012.
9 A. B. Robertson, F. C. F. Lam, and R. J. Cole, "A Comparative Cradle-to-Gate Life Cycle

Assessment of Mid-Rise Office Building Construction Alternatives: Laminated Timber or Reinforced Concrete," *Buildings*, vol. 2, no. 4, pp. 245–270, 2012.

10 S. Suh, S. Tomar, M. Leighton, and J. Kneifel, "Environmental Performance of Green Building Code and Certification Systems," *Environ. Sci. Technol.*, vol. 48, no. 5, pp. 2551–2560, 2014.

11 T. Ramesh, R. Prakash, and K. K. Shukla, "Life cycle energy analysis of buildings: An overview," *Energy Build.*, vol. 42, no. 10, pp. 1592–1600, 2010.

12 X. Gong, Z. Nie, Z. Wang, S. Cui, F. Gao, and T. Zuo, "Life cycle energy consumption and carbon dioxide emission of residential building designs in Beijing: A comparative study," *J. Ind. Ecol.*, vol. 16, no. 4, pp. 576–587, 2012.

13 H. R. Lu, A. El Hanandeh, B. Gilbert, and H. Bailleres, "A comparative life cycle assessment (LCA) of alternative material for Australian building construction," *MATEC Web Conf.*, vol. 120, pp. 1–9, 2017.

14 B. Xia, T. Ding, and J. Xiao, "Life cycle assessment of concrete structures with reuse and recycling strategies : A novel framework and case study," *Waste Manag.*, vol. 105, pp. 268–278, 2020.

15 Athena Sustainable Materials Institute, "A Cradle-to-Gate Life Cycle Assessment of Ready-Mixed Concrete Manufactured by NRMCA Members - Version 2.0," no. October, 2016.

16 BRE Global, "BRE Global Methodology For The Environmental Assessment Of Buildings Using EN 15978: 2011.," p. 38, 2018.

17 W. Zhang, S. Tan, Y. Lei, and S. Wang, "Life cycle assessment of a single-family residential building in Canada: A case study," *Build. Simul.*, vol. 7, no. 4, pp. 429–438, 2014.

18 K. Bawden and E. Williams, "Hybrid Life Cycle Assessment of Low, Mid and High-Rise Multi-Family Dwellings," *Challenges*, vol. 6, no. 1, pp. 98–116, 2015.

19 J. Wang, C. Yu, and W. Pan, "Life cycle energy of high-rise office buildings in Hong Kong," *Energy Build.*, vol. 167, pp. 152–164, 2018.

**Military Technical College
Kobry El-Kobbah,
Cairo, Egypt**

**13th International Conference
on Civil and Architecture
Engineering
ICCAE-13-2020**

The impact of an external obstacle daylight in deep office spaces at hot arid zone

Doha Mohamed Saeed

Teacher – Architectural Engineering Department – Faculty of Engineering and Technology, Badr University in Cairo (BUC), Egypt

Abstract. The use of effective daylighting systems is one of the main factors that inflounces energy consumption in buildings especially in deep office spaces. Therefore, the need for innovative daylighting systems with optimum overall daylighting performance - especially in term of saving energy performance - is an essential design element for such spaces. Light shelf systems are the most major potential lighting systems that can be used to provide such innovative daylighting systems.

The parametric approach was implemented where all possible combinations of the various design parameters for the main two design elements of the light shelf system (light shelf and room specifications) were computed through Diva-for-Rhino (Grasshopper). The study of optimized light shelf systems in deep office spaces has been carried out in case of an external obstacle (neighbor) and without it – according to the Egyptian building codes - in order to identify the impact of such design elements on the performance of the optimized light shelf system in the main four directions in hot arid zone climate.

Keywords: Light shelf systems – obstacle - deep office spaces - Diva-for-Rhino (Grasshopper)

1. Introduction

Light shelf is a daylight-redirecting system intended to bounce daylight to the deepest side of a room. This function suites well to all climates [1]. Light shelf was developed to create uniform indoor illuminance. However, in hot climates the unshaded clerestory above the shelf transmits high solar heat gain. In dense urban context, these advantages and disadvantages might vary regarding the context and position of the fenestration [2]. This confirms the necessity of studying the effect of the presence of a neighbor integrated with the reflectors system. employed scaled physical models and computational simulation methods to examine the daylighting performance of lightshelves under several tropical sky conditions in Subang, Malaysia, i.e., intermediate sky with direct sunlight, intermediate sky without direct sunlight and overcast sky. Under clear sky conditions, daylighting performance of light shelf depends on the dynamic movement of solar position. Franco [3] proposed tilted and automatic lightshelves to solve daylighting problems in hot tropics by adjusting the elevation of the internal shelf to the dynamic movement of solar position.Previous studies on light shelf were limited in building context. Recently, studies on urban environment have widely emerged following the awareness of the impacts on the urban climatology. Shafaghat et al. [4] established a taxonomy body of knowlege in urban climatology studies and mentioned urban street canyon, which consists of aspect ratio and street orientation, as the most important urban features. The impacts of urban height to width ratio aspect and spacing distance to length ratio on the building energy demand had been studied by Kesten et al. [5]. Additionally, canyon surfaces have been understood to be vital in determining the thermal performance of the urban canyon [4], that affected on the surrounding building energy consumption. Akbari et al. [6] found that the surrounding's surface albedo or emissivity could modify the energy balance of the buildings. High albedo materials reduce the amount of solar radiation absorbed by the building envelope. In multi-story buildings, the urban geometry and texture effects on the indoor illuminance and radiation fluxes may vary for each floor level. Therefore, it could be interesting to relate these factors to the light shelf planning in multi-story buildings in terms of building energy performance. In order to measure the effectiveness of light shelf.

2. Methadology

This study comes as a part of the completion of the study of the development of the performance of the light shelf systems. This study, based on the benefit of the outcome of what has been concluded in a previous study phase, especially the identification of design indicators and key design elements to develop the performance of light shelf systems. Based on this conclusion, The results were studied from a different perspective, which is the effect of integrating the neighbor's presence with the light shelf system in four orientations. this various cases experimental tests by using parametric design approach by using Diva for Rhino- Grasshopper program, in order to define The impact of an external obstacle (neighbor) or its absence on the performance of optimized light shelf system in deep office spaces in the main four orientation in hot arid zone climate. (Ahmed,Reham,Doha, 2017&2018) [7,8]& [9].based on CIE Standard.

2.1. Light Shelf Models

Light shelf is a parametric panel installed at a height below the clerestory and above the view window for The hypothetical office is 6x12x3.3 m, which represents a large deep space in a multi-story Building Which properties as follows. Figure (1) & Table (1)

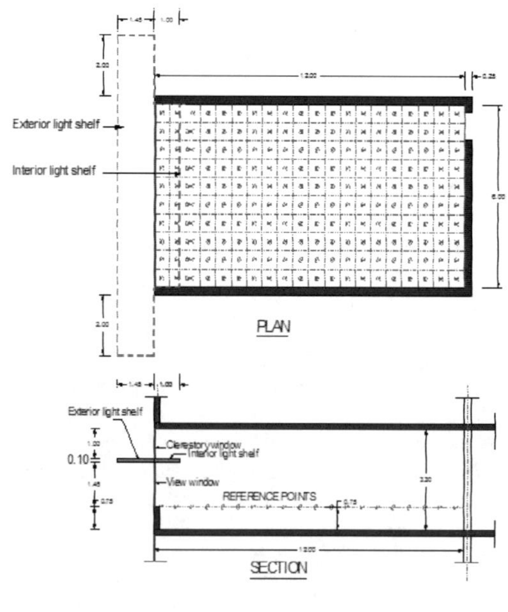

Fig. (1) Hypothetical office building room plan and section is 300 lux (daylight autonomy)

Table 1. Dimensions and properties of the tested office space

Materials and Dimensions		
Level of Floor		One level
dimensions (m)		6m x 12m x 3.30 m
Walls	Reflecting	50%
	Material	Internal-walls Off-White (Medium Colored)
Ceiling	Reflecting	80%
	Material	Ceiling in white Color
	Rotation angle	10 from half ceiling
	Curvature angle	10 from half ceiling
Floor	Reflecting	20%
	Material	generic floor
Window (1) Dimensions and Materials		
Width (m)		6 m
Sill (m)		. 75 m
Lintel (m)		2.20 m
Glazing		Double glazing, clear (VT=80%)
Clear story (2) Dimensions and Materials		
Width (m)		6m
Sill (m)		2.30 m
Lintel (m)		3.30 m
Glazing		Double glazing, clear (VT=80%)
Light Shelf Dimensions, angle, curvature and Materials		
Light shelf orientation		South, north, east and west
Light shelf position		External, Internal, External and Internal together
External light shelf depth		1.45 m
Internal light shelf depth		1 m
Internal light shelf height from finish floor level		2.20
External light shelf height from ground level		5.90
External and Internal light shelf width		6 m
External and Internal light shelf thickness		.10 m
External light shelf rotation angle		0, 15, 35
Internal light shelf rotation angle		0, -10
External light shelf curvature angle		0, 15, 35
Internal light shelf curvature angle		0, -10
Light shelf obstruction angel		49° (the neighbor distance = 18m.)
External and Internal light shelf Material		Generic ceiling 80, Silver mirror, Transparent glass and Translucent Panel
Reflecting& Material		High-Refl_80%

2.2. Light shelf obstruction angel:

The light shelf obstruction angle is measured as the degree (α) which is the angle located in between the extension line of the external light shelf's upper surface plan (a) and the line of light shelf beginning point and neighbor end point (b) fig. (2).
The obstructions angel depends mainly on the distance from the neighbor which has defined for the exploring test phase in this study, up to the Egyptian construction law in new cities (building heights = 1.5 street width), which ensure the constancy of the impact of the neighbor distance whatever changed, because it is directly proportional to the height, which mean the constancy of the obstruction angle. Science 18 meters (49^0 obstruction angel) is more commonly use in planning office building zone areas (building height = 27 m = 6 level), thus this study has define a distance of 18 meter away from the neighbor which make an obstruction angel of (49^0) for the further developed study cases in the case of neighbor Figure (2)

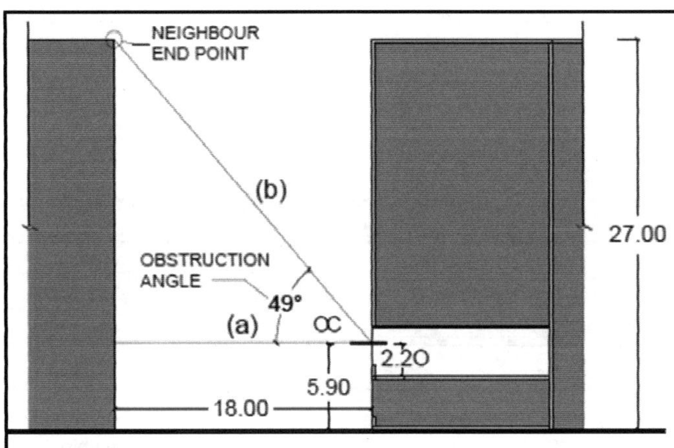

Fig. (2) This figure has define a distance of 18 meter away from the neighbour which make an obstruction angel of (49^0)

2.3. Daylighting and Energy Simulations
The calculation of energy use for lighting was based on the daylighting performance computed using Diva for Rhino- Grasshopper program based on IES/ LEED v4/ASHRAE Standard, two metrics in LEED v4 are codified for evaluating daylight autonomy design, which allow a daylight space to be evaluated for a one-year period using two different performance parameters: sufficiency of daylight luminance and the potential risk of excessive sunlight penetration [10]. These two metrics are Spatial Daylight Autonomy (sDA) and Annual Sun Exposure (ASE) metrics, which together form a clear picture of daylight performance. sDA describes how much space receives sufficient daylight, which, for office building spaces, must achieve (sDA 300 lux / 50% of the annual occupied hours) for at least 55-75% of the floor area. sDA has no upper limit on luminance levels, and, therefore, ASE is used to describe how much space receives too much direct sunlight which can cause visual discomfort (glare) or increase the cooling loads. ASE measures the percentage of floor area that receives at least 1000 lux for at least 250 occupied hours per year that must not exceed 10% of floor area [11].

3. Results and Discussions

3.1. Data Analysis Methods

6552 cases were investigated for a hypothetical deep office room for 4 orientations with 819 combinations of light shelf system parameters in two cases (with no obstructions and with obstruction angle); figure (3). Total tested cases in the four orientation= 819 x 2 x 4 = 6552 cases.

IOP Conf. Series: Materials Science and Engineering **974** (2020) 012029 doi:10.1088/1757-899X/974/1/012029

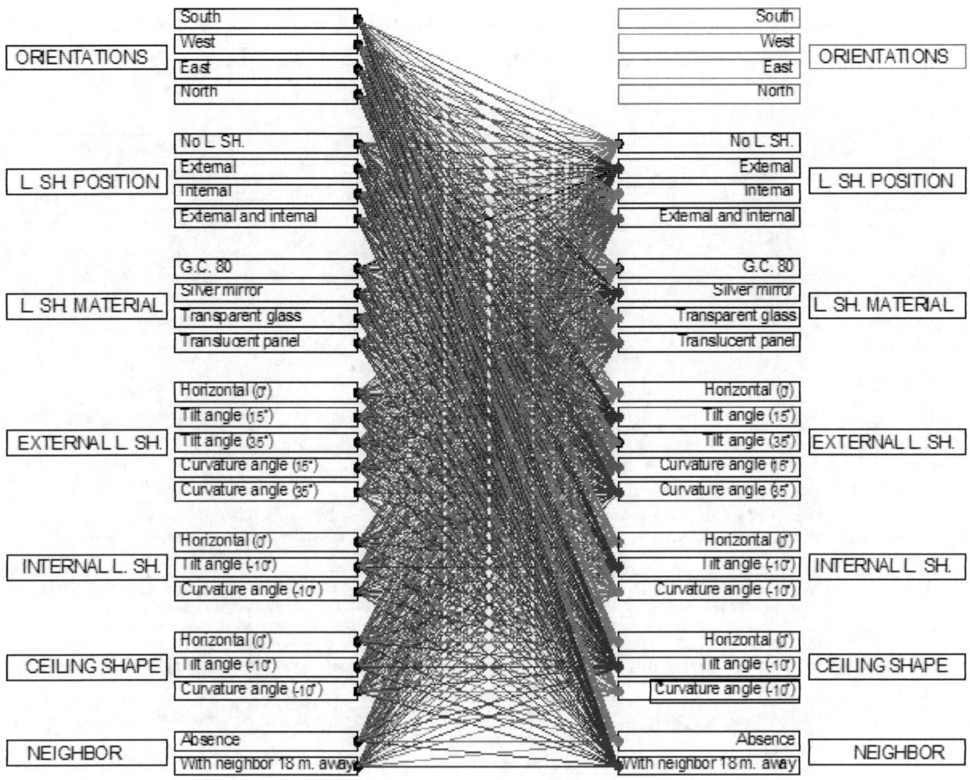

Fig. (3) Modifications parameters for the study modelling

Algorithm is inspired by natural selection to generate high-quality optimization by depending mutation, cross-reference and selection. It is based on the concept of Darwin's theory that was extended to GA in 1989. figure (4)

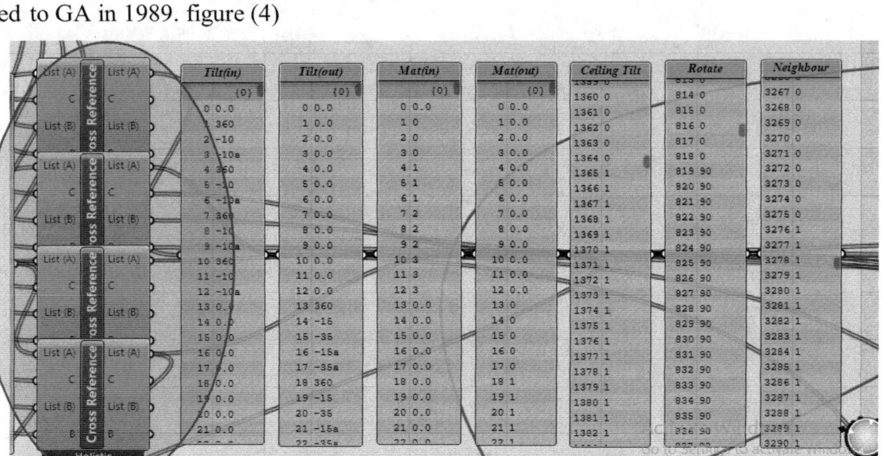

Fig. (4) Modifications parameters for the study modelling which are applied by cross-reference component in Rhino grasshopper.

After conducting the simulation experiments, the focus in analyzing the results was in four stages in the four orientations, which are as follows: Table (2)

Table 2. The results was in four stages in the four orientations

n.	Stage one		Stage two		Stage three		Stage four	
The Stages	Light shelf	neighbour	Light shelf	neighbour	Light shelf	neighbour	Light shelf	neighbour
Yes/No	No	No	Yes	No	No	Yes	Yes	Yes

3.2. Define Selected Cases and Result Analysis:

Table 3. Digital results for Daylighting and Energy consumption in four orientations

Case No.	the case situation	Orientaio n	Neighbor	Outer LS. Material	Outer LS. Angle	Inner LS. Material	Inner LS. Angle	Ceiling	sDA	ASE	Daylit	P.Daylit	Overlit	T AEC	A H L	A C L	A E E E	ALEE
0	Stage 1	South	N/A	N/A	N/A	N/A	N/A	Horizontal	80.5	27.5	44	20	36	161.1351	5.503875	149.9614	31.974	5.669833
321	stage 2	South	N/A	Generic ce	Horizonta	Silver min	Horizonta	Tilted	76.5	7	53	24	24	168.0951	5.206064	153.5096	31.974	9.379438
3276	stage 3	South	18 m	N/A	N/A	N/A	N/A	Horizontal	43	10	30	57	14	200.2854	5.187073	168.5752	31.974	26.52313
3842	stage 4	South	18 m	Silver min	35	N/A	N/A	Curved	53	5	39	47	14	189.9325	5.167231	163.8131	31.974	20.95214
2457	Stage 1	West	N/A	N/A	N/A	N/A	N/A	Horizontal	75	39.5	30	25	44	156.765	6.547382	142.4243	31.974	7.793364
2778	stage 2	West	N/A	Generic ce	Horizonta	Silver min	Horizonta	Tilted	72	18.5	45	28	27	161.0629	6.148321	144.6088	31.974	10.30584
5733	stage 3	West	18 m	N/A	N/A	N/A	N/A	Horizontal	53	12	34	47	19	164.0778	6.523888	145.9362	31.974	11.61767
6021	stage 4	West	18 m	Generic ce	35	N/A	N/A	Tilted	64.5	6	46	36	18	160.9599	6.287348	144.6418	31.974	10.03077
819	Stage 1	East	N/A	N/A	N/A	N/A	N/A	Horizontal	72	39.5	26	28	46	146.4902	5.02029	136.089	31.974	5.380912
1140	stage 2	East	N/A	Generic ce	Horizonta	Sheet Mir	Horizonta	Tilted	72	16	43	28	29	149.3177	4.777921	137.7194	31.974	6.820353
4095	stage 3	East	18 m	N/A	N/A	N/A	N/A	Horizontal	53	12	34	47	19	181.4815	4.733934	152.6352	31.974	24.11232
4383	stage 4	East	18 m	Generic ce	35	N/A	N/A	Tilted	64	6	46	36	18	175.0902	4.563516	149.853	31.974	20.67367
1638	Stage 1	North	N/A	N/A	N/A	N/A	N/A	Horizontal	68	0	49	32	19	91.60071	7.257533	74.15991	31.974	10.18328
1972	stage 2	North	N/A	Generic ce	+35 Curve	Silver min	-10	Tilted	75.5	0	56	24	19	90.01158	6.823174	73.50326	31.974	9.685148
4914	stage 3	North	18 m	N/A	N/A	N/A	N/A	Horizontal	60.5	0	42	40	19	101.8132	7.185491	79.03686	31.974	15.59084
5249	stage 4	North	18 m	Generic ce	+35 Curve	Silver min	-10	Tilted	74	0	55	26	19	90.52819	7.048315	73.81528	31.974	9.664603

3.2.1 Result analysis for Daylighting and Energy Simulations in four orientation:

Table (3) & Figure (5 and 6)

The first stage:

The basic case (without any obstruction& light shelf system) in south, west, east and north orientations achieved (80.5% , 75% , 72% and 86% sDA 300, 50%) Respectively and (27.5%, 39.5% , 39.5% and 0% ASE1000, 250h) Respectively . We find that the amount of sunlight entering is acceptable, but the amount of direct light is higher than acceptable.

The second stage:

This case (without obstruction & with light shelf system) in south orientation achieved (76.5 % , 72%, 72% and 75.5% sDA 300, 50%) Respectively and (7% , 18.5% , 16% and 0% ASE1000, 250h) Respectively . We find that the amount of sunlight entering is minimal, but the amount of direct light is minimal also the reason for existence Light shelf.

The third stage:

This case (with obstruction & without light shelf system) in south orientation achieved (43%, 53% ,53% and 60.5% sDA 300, 50%) Respectively and (10%, 12%, 12% and 0% ASE1000, 250h) Respectively. We find that the amount of sunlight entering is minimal, but the amount of direct light is minimal also the reason for existence neighbor.

The fourth stage:

This case (with obstruction & without light shelf system) in south orientation achieved (53% , 64.5% , 64% and 74% sDA 300, 50%) Respectively and (5%, 6% ,6% and 0% ASE1000, 250h) Respectively. We find that the neighbor integrated with light shelf system caused amount of sunlight entering more about the third case and the amount of direct light is minimal. Table (4,5,6 and 7)

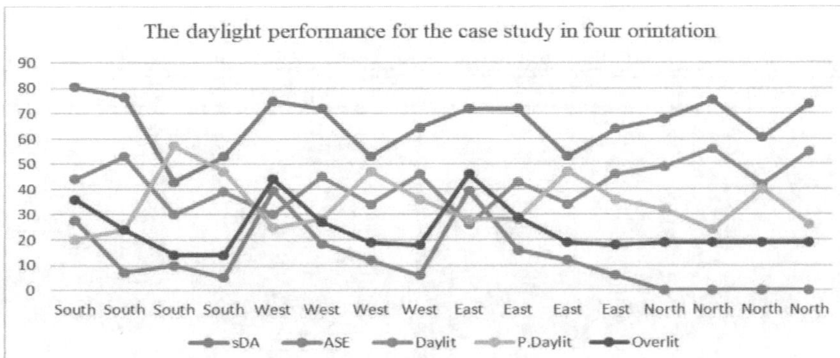

Fig. (5) Daylight performance in four orientation in four stages

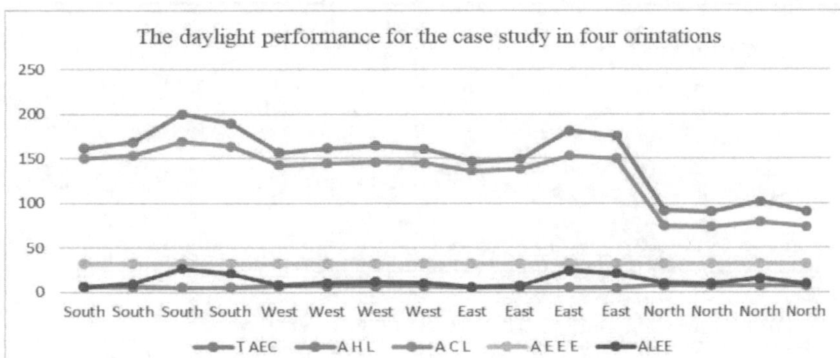

Fig. (5). Energy consumption in four orientations in four stages

Table 3. Visual results for Daylighting performance in South orientations

0000	The first stage	0321	The second stage:
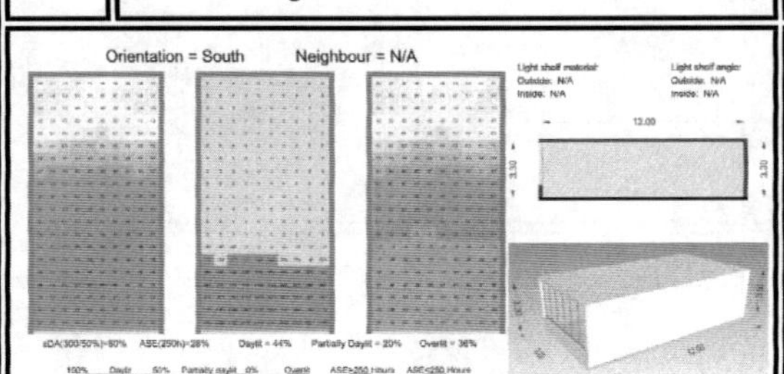		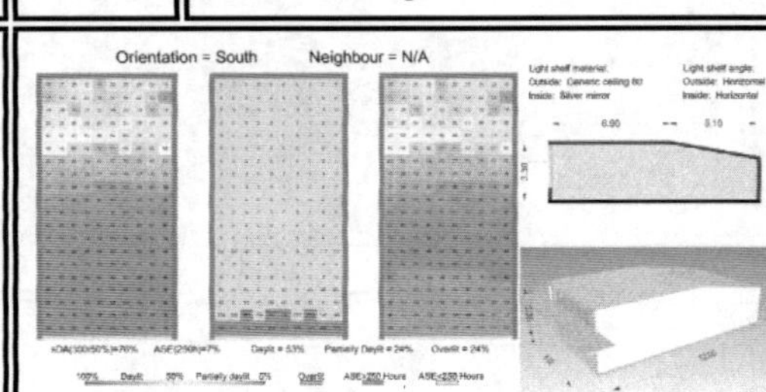	
3276	The third stage	3842	The fourth stage
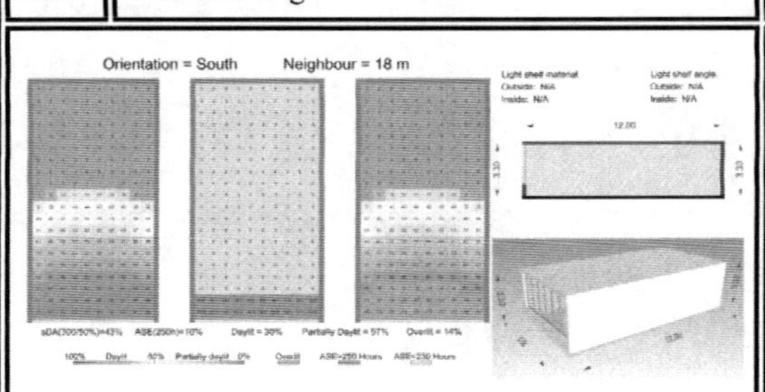		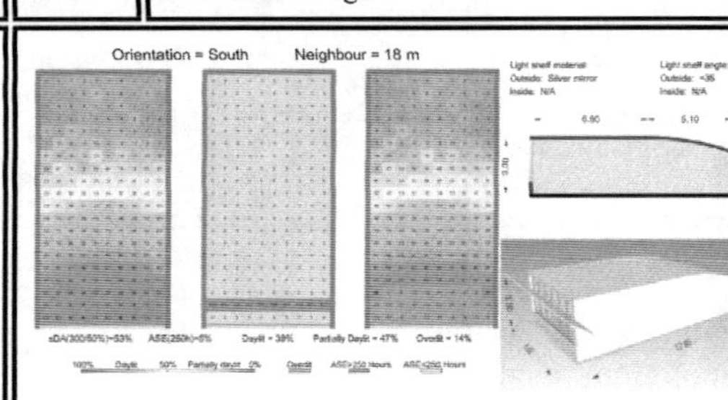	

Table 4. Visual results for Daylighting performance in West orientations

2457	The first stage	2778	The second stage:
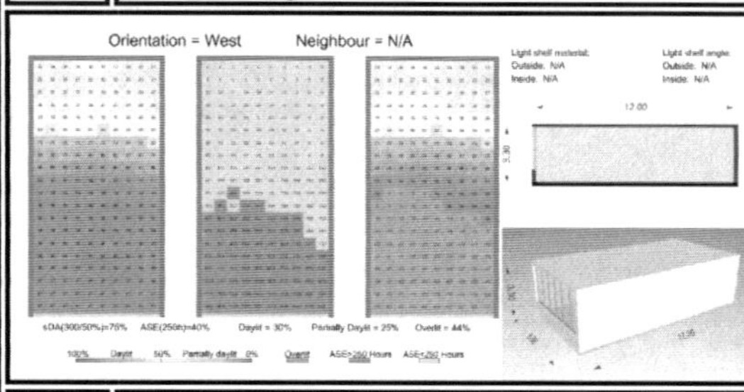		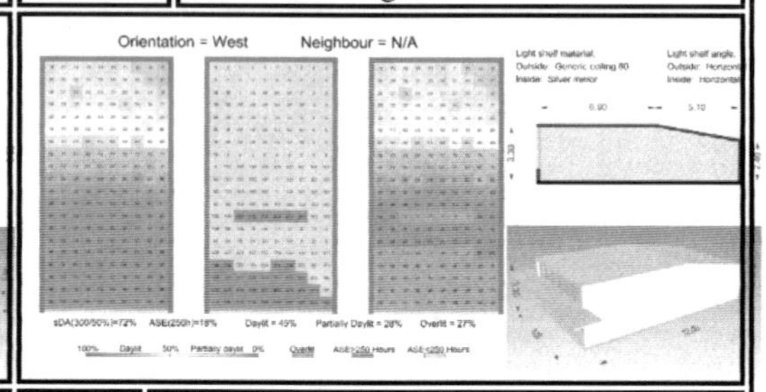	
5733	The third stage	6021	The fourth stage
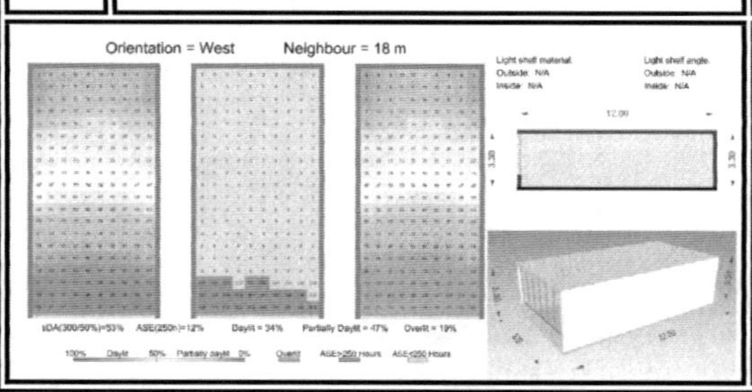		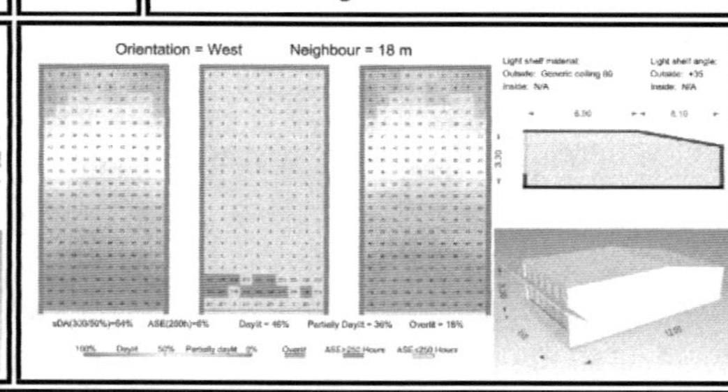	

Table 5. Visual results for Daylighting performance in East orientations

819	The first stage	1140	The second stage:
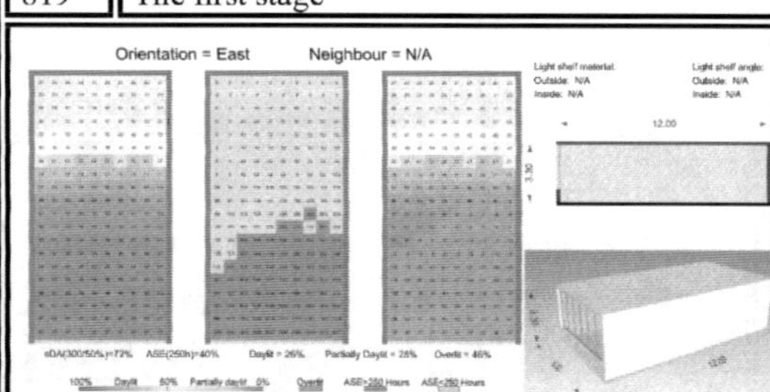		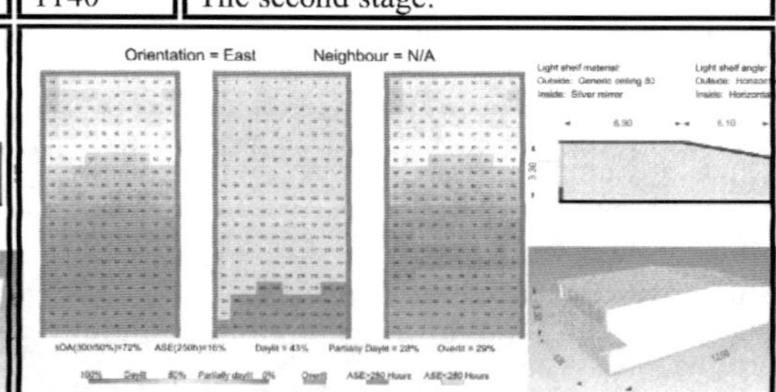	
4095	The third stage	4383	The fourth stage
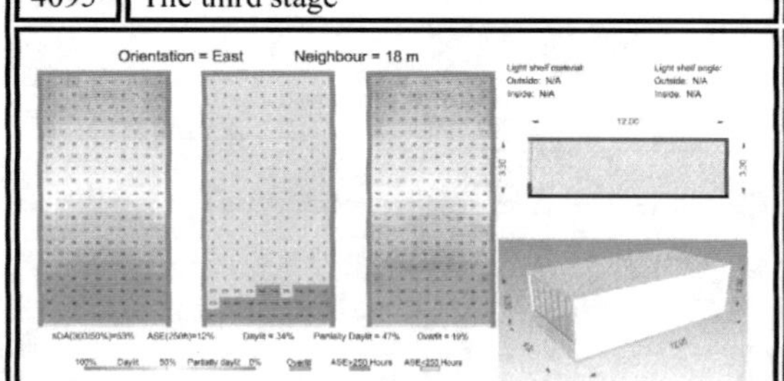		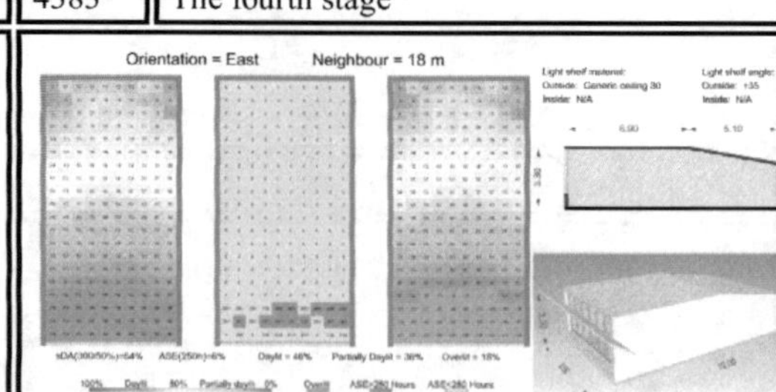	

Table 6. Visual results for Daylighting performance in North orientations

1638	The first stage	1972	The second stage:
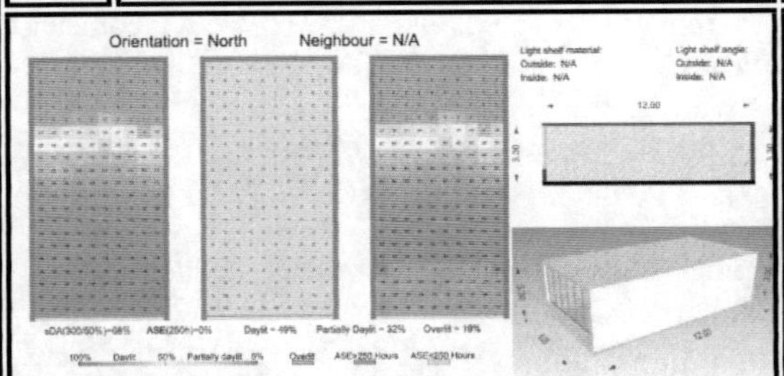		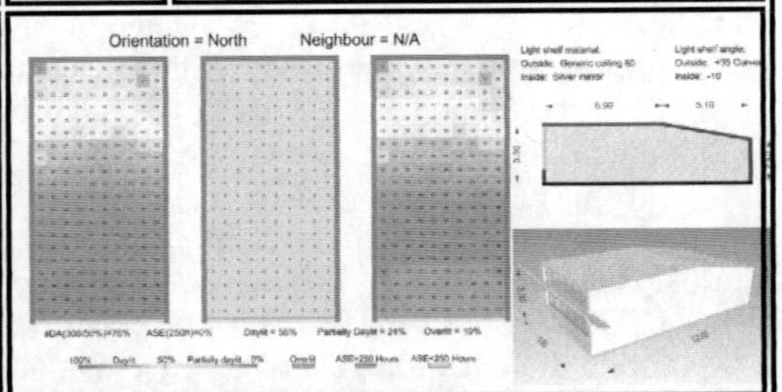	
4914	The third stage	5249	The fourth stage
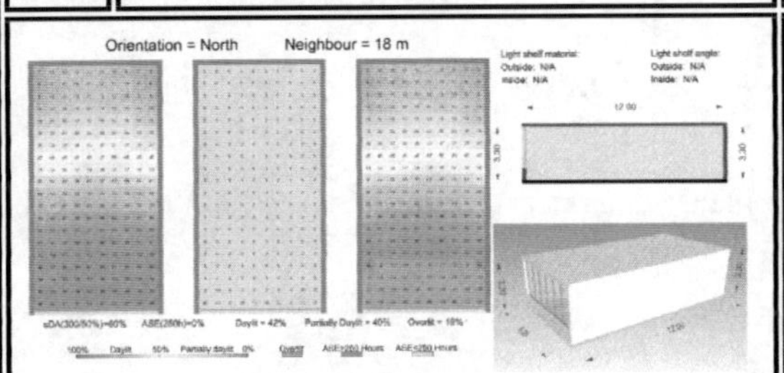		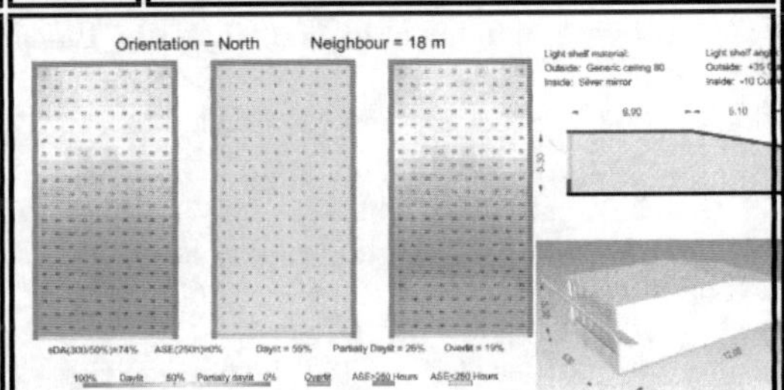	

2. Summary and conclusion

In the total study of the four stages of four orientations, we find that the best results in terms of Daylight performance and energy consumption were in the presence of the light shelf only without the presence of the neighbour (the second stage), but if the neighbour is found, it is better to integrate with the light shelf (the fourth stage)
As for the effect of the presence of the neighbour with the light shelf system on the total energy consumption, it is a result of energy consumption in the electric lighting loads that we resort to as an alternative to the lack of quality of the available daylight and the reason was not the change in cooling or heating loads.
We find that the arrangement of the positive effect of the presence of the neighbour is integrated with the light shelf system in the four directions as follows, firstly the northern direction, because it collects light from the south and sends it to the north, then the eastern and western direction, then the southern direction, where light is collected from the north, it is weaker.
In the end, the study determined the effect of the presence of the neighbour integrated with the presence of the light shelf system, but it is recommended that the design of the light shelf needs further development in the all orientations of achieving balance between the amount of daylight entering and blocked to reach the harmony of daylight and prevent the glare and thus reduce dependence on electric lighting during the day and thus reduce energy consumption with Achieving daylight efficiency and this is the optimum goal.

3. References

[1] International Energy Agency. Energy Conservation in Building Community Systems (IEA-ECBCS). Annex 29/Solar Heating and Cooling Programmes Task 21, 2010.
[2] The Effectiveness of Light Shelf in Tropical Urban Context, 2016 Floriberta Binarti, Sinta Dewi. This is an open access article licensed under the Creative Commons Attribution License (http://creativecommons.org/licenses/by/4.0), in the manner agreed with De Gruyter Open.

IOP Conf. Series: Materials Science and Engineering **974** (2020) 012029 doi:10.1088/1757-899X/974/1/012029

[3] Franco I. M. Efficacy of light shelves: passive, dynamic and automatic devices related to light and thermal behavior. Presented at Thermal Performance of Exterior Envelopes of Whole Buildings, 2007

[4] Shafaghat A., Manteghi G., Keyvanfar A., Bin Lamit H., Saito K., Ossen D.H. Street Geometry Factors Influence Urban Microclimate in Tropical Coastal Cities: A Review. Environmental and Climate Technologies 2016:17:61–75. doi:10.1515/rtuect-2016-0006

[5] Kesten D., Kereci A., Strzalka A., Eicker U. A method to quantify the energy performance in urban quarters. Indoor Air Quality, Ventilation 2012:18(1–2):100–111. doi: 10.1080/10789669.2011.583307

[6] Akbari H., Menon S., Rosenfeld A. Global cooling increasing world-wide urban albedo to offset CO2. Climate Change 2009:95(3–4). doi:10.1007/s10584-008-9515-9

[7] Ahmed f., Reham D., Doha m., Investigating of the light shelf system performance in the main four geographical direction in hot arid zone. Egyption Engenering magazen 2017.

[8] Ahmed f., Reham D., Doha m., Aparamtric study of light shelf system performance for south orientation in hot arid zone. Egyption Engenering magazen 2017.

[9] Ahmed f., Reham D., Doha m., Sophisticated Light Shelf System in South Orientation in Hot Arid Zone. Proceedings of Science and Technolgy , Print ISSN: 2357-0849,2018, http://www.ierek.com/press

[10] IES. Spatial daylight autonomy (sDA) and annual sunlight exposure (ASE). New York (NY): IES. LM-83-12. Lighting Measurement # 83, 2012.

[11] USGBC. LEED Reference Guide for Building Design and Construction v4. USGBC. Washington DC 2013.

Military Technical College
Kobry El-Kobbah,
Cairo, Egypt

13th International Conference on Civil and Architecture Engineering ICCAE-13-2020

National projects as development poles and their impact on urban development`

Asmaa Gabr[1], Kamal Khalaf[2], Gamal Helewa[3]

[1]Assistant Lecturer, Al-Azhar University,Urban Planning Department, Cairo, Egypt.
[2]Prof. Dr. Professor of Planning, Al-Azhar University,Urban Planning Department, Cairo, Egypt.
[3]Lecturer of planning, Al-Azhar University,Urban Planning Department, Cairo, Egypt.

Corresponding author:Assmaa.gabr2015@yahoo.com

The Egyptian state is witnessing a strong impetus and development boom in national projects because of its effective role in spreading development in the lagging regions and attracting the population, and in this context the research aims to shed light on national projects in Egypt and measure the impact of these projects in terms of their development returns from being development poles that help in the achievement of urban development , so as to measure the extent of the ability of projects to polarization actual Imran , by identifying factors affecting the national projects , the ability of the urban polarization, and then determine the shape of the relationship between these factors and the ability polar metric urban and exit mathematical equation from which the ability to measure Urban polarization of any existing national project .

Key words: national projects, urban development, development poles, polarization, common strategies for development.

1. Introduction: National projects consider Egypt's vision to invest its various resources and one of the most important means used to create urban development on all development sectors, starting from the local level through the regional level to the national level, which takes place in a significant development in its urban environment, especially in the scope of those projects, which is represented in its role in terms of being the poles of development Help in spread urban development. From this standpoint, the research seeks to answer the following question:

Do national projects have the impact of the yield in the achievement of urban development? How can this effect be measured?

IOP Conf. Series: Materials Science and Engineering **974** (2020) 012030 doi:10.1088/1757-899X/974/1/012030

This question can be answered by identifying the factors affecting the ability of national projects to attract and reach the relationship between them and urban polarization ability through a statistical mathematical scale based on comparative theoretical studies, and through this scale, the urban polarization ability can be measured for any existing national project or any An urban gathering in the light of the number and size of national projects , and predicting the potential population of the project to determine development and investment priorities, which in turn reflects on spreading urban development throughout the republic.

2. Methodology

The research deals with a set of the following axes: -

The first converter: clarifying the concept of national projects and their classifications. The axis second bass review of some local experiences and analysis for to identify the elements of success of national projects in urban development. The third axis is concerned with studying the developmental role of the new city of El Alamein in the axis of the northwestern coast: as an application model for a national project, and the fourth axis ends. By presenting some results to clarify the relationship between national projects and their polarizing capacity, so that their urban returns on urban development can be evaluated.

2.1. The first axis: the basic concepts and their classifications

The concept of national projects, their classifications, and concepts for achieving urban development is addressed, as well as identifying the major projects currently underway.

2.1.1. The concept of national projects

Major projects with national goals, which work to achieve some of the sectorial development goals, and that are based on addressing national and regional issues and exploiting competitive capabilities and opportunities, with a national developmental return (economic - social - service - strategic - security) , in the context of broad community participation[(1)], Under real political and administrative will. One of its characteristics is that it consists of national projects that constitute it, during a time schedule of up to twenty years [(2)]. Like the project (developing the axis of the Suez Canal - the axis of the northwestern coast).The geographical scope of these projects extends to more than one governorate, Projects that have a direct return on the national product, and that these projects should have an impact on drawing the urban map, as it includes more than one national project

2.1.2. Classification of national projects

To get to know the levels of national urban development projects, we find that they are graded under three levels related to the administrative division of Egypt (administrative economic rather than development), differ from each of them in terms of (project size, beneficiaries, investment return, geographical scope and level of development), [(1)] national projects have been classified for several Categories are as follows:

- **According to the planning level:** (national projects - regional projects - regional projects).
- **Depending on the type of project:** National projects usually appear through them, which are three patterns as follows [(1)]:

– Pattern of national projects in the form of points (Spots): The new administrative capital.
– Pattern of national projects in the form of perimeter regions (Zones): The West Bay of Suez.
– The pattern of national projects in the form of development hubs (Corridors): The axis of the Suez Canal.

- **Depending on the function of the project and activity leader :** has been classified as follows: (projects with dye agricultural project one and a half acres - projects with a dye industrial project west of the Gulf of Suez development - projects with a character of urban project new administrative capital - with dye tourists projects J of As a project for the new city of El Alamein - projects supporting development such as (axes - energy projects - food projects, ...) , which had a clear impact and a noticeable return on urban development through the establishment of a group of urban communities to serve these projects, their success and continuity .

2.1.3. Basic concepts related to urban development:

include the concepts of a group (pole Development- growth point- Growth Point- Polarization - common strategies for development) [4] [5] [6], which include more detailed meanings that integrate with them in order to reach the desired urban development[3]through the settlement of a set of activities related to technical and economic relations in a homogeneous space and extend its development impacts to the surrounding areas so that these areas grow and grow It contributes to advancing economic development and achieving national goals in light of urban development strategies.

2.1.4.The major national projects currently underway

The major projects that were proposed in the previous periods, of which part has come into effect and the rest under study through the regions in which these projects are located, have been handled, and according to the National Plan for Egypt 2050 CE, there are promising development regions figure(1) that were covered by the national plan[7] . They are as follows:

1- Developing the Suez Canal Region as a global logistics and industrial center 2- Sinai Development.3- Development of the northwestern coast and its desert back . 4- Development of North Upper Egypt governorates. 5- Development of southern Egypt, Nubia, the region of Halayeb and Shalateen. 6- New projects on the national level. 7- National, regional and local transportation projects.

According to the approved national plan 2050 2012, these previously monitored regions are the most promising development regions / areas on the Egyptian map with the ingredients and development opportunities that have the potential to lead development in Heliopolis in the next stage. [11]

IOP Conf. Series: Materials Science and Engineering **974** (2020) 012030 doi:10.1088/1757-899X/974/1/012030

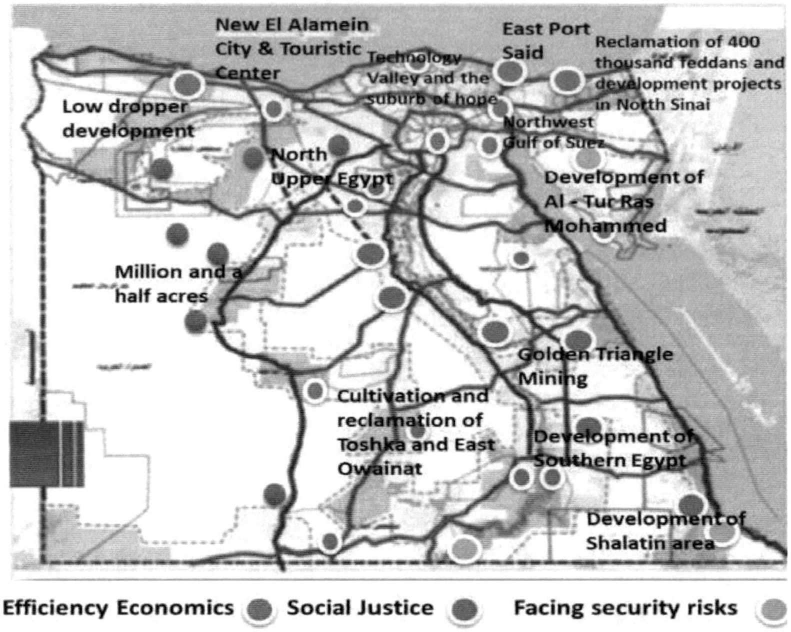

Figure1:Major national projects proposed during the first phase 2012-20120 (8)

2.2. The second axis: a review of some local experiences

National projects in Egypt represent development poles to serve as a launching pad for creating new urban nuclei that attract promising population density with a broad national trend in which national efforts come together and this is evident in the Suez Canal axis project, which is a driving project for growth in all sectors.

2.2.1. Northwest Gulf Development Project Industrial

The Gulf of Suez region was chosen to create a special economic zone to correct the path of industrial zones and advance them towards production and export.

Project location:

The North West Bay development project is considered as a development pole and the project is located 35 km south of the city of Suez on the western coast of Suez Bay near the location of AinSokhna at the intersection point between each of the Cairo / AinSokhna and Suez roads /Quick Saffron (9) .

2.2.2. The project's development potential and its impact on urban development

Where the settlement features are available in the region in terms of raw materials, labor and goods needed to establish a pivotal port and a distinctive industrial back that depends on the economic forces required to be established. Looking at the site of the project, we find that it has several components, which are considered a growth pole that attracts Imran as follows:

- The possibility of establishing a distinctive axial port at Ataqa, whether it is a commercial port or an advanced fishing port with all necessary services[11].
- The possibility of establishing an industrial zone that leads to the development and advancement of the local economy.
- Availability of an advanced infrastructure network in the area, which serves the industrial area and the port.
- The presence of natural resources and resources such as limestone, dolomite, coal, petroleum, and fish resources, which creates a huge economic base to attract and settle the work force .
- The region 's ability to absorb the population increase to the establishment of a residential city by expanding the horizontal and vertical as a result of the availability of basic infrastructure also provides the opportunity to establish urban communities based on those distinctive economic activities the land site , and thus moving the number of manpower to the area of the project as a result of the availability of jobs up to 25,000 opportunities , and is expected to reach 360,000 opportunities, Figure (2) .

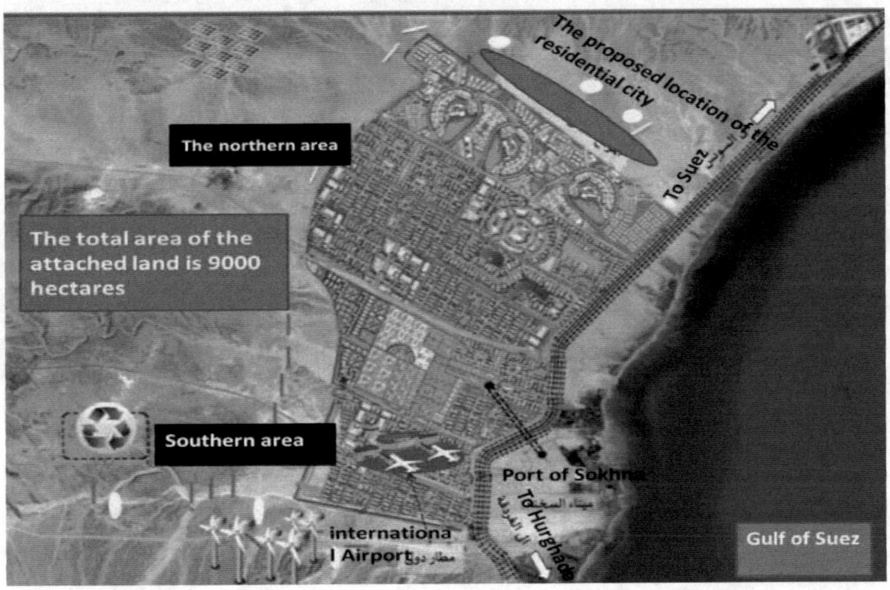

Figure 2: Proposed Master Plan for the Northwest Gulf of Suez [12]

- The availability of various economic activities in the relative importance, which provides an opportunity for the leading activity factor to appear and its impact on the rest of the activities through front and back relations.

Therefore, the role of the Suez Canal region becomes clear and considered a development pole with multiple components capable of polarizing the population and achieving urban development and disseminating it at the axis of the Suez Canal.

2.2.3. Elements for the success of national projects in urban development from previous experiences

From studying the experiments, we will draw several elements for the success of the national project according to the project type, whether it is a project in the form of axes, regions, or points as follows:

1- Contacting the international hubs. 2- The efficiency of the road network. 3- Contacting the ports. 4- The availability of the rail network. 5- Creating developmental poles. 6- Proximity to arable areas. 7- Traffic in the industrial areas. 8- Proximity to the tourist and archeological sites. 9 - Proximity to new urban communities. 10- To correspond with natural factors. 11- Avoidance of environmental risks and having a national strategy. 12- Enhancing practical aspects of research and studies. 13- Efficiency of the institutional framework for axis management.

2.3. The developmental return of national projects on urban development

2.3.1. Case Study (El Alamein City New in the framework of the North West Coast hub)

It is called the Alexandria-Matrouh axis and was chosen as a pattern of development hubs, where the northwestern coast occupies an important location on the map of Egypt, and represents the link between Egypt and the Arab Maghreb, and Egypt's hope is to absorb the population increase. Therefore,the state took care during the previous period to give a development boost to this axis, and the region as a whole which can be invested by putting it at the forefront of the tourist map of Egypt. The developmental outcome of the centers located on the axis is studied, and the axis includes the following centers [15] 7 centers: (Salloum - SidiBarani – Al Nejeila - MarsaMatrouh - AlDabaa – El Alamein – Al Hamam) . As shown in Figure (3) in order to measure the impact of the new city of El Alamein and its developmental role in achieving development on the axis of the northwestern coast as a growth pole and an important model for national projects.

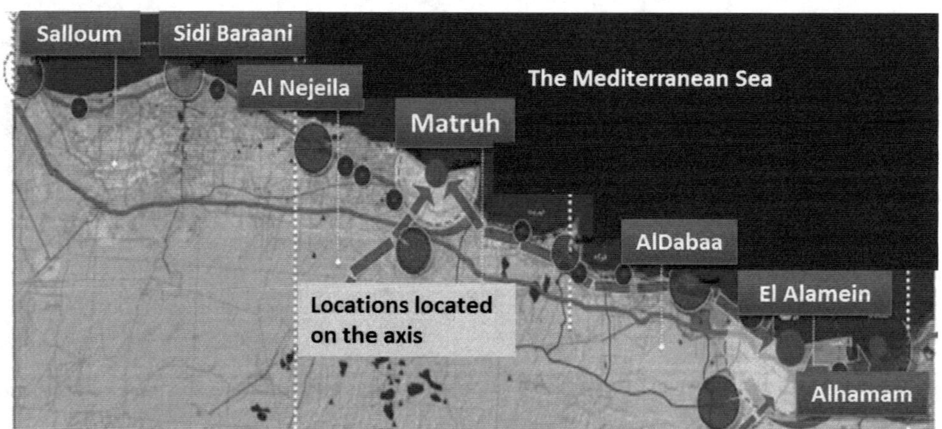

Figure3:Gatherings located on the axis [16]

2.3.2. Measuring the developmental returns in the centers located on the axis of the northwestern coast.

It is worth noting that the study and research have set several standards for the evaluation of national projects and their impact on urban development, where a set of indicators that have been measured to measure the development returns of the centers of the axis are exposed .

2.3.2.1. Standards for measuring the urban polarized capacity for national projects:

- The use of a set of independent variables due to a standard set of criteria , some of the views of planners and others to choose the researcher of a mole all aspects and variables of economic, social and urban[4] to serve as a standard for evaluation of projects to the nation in terms of being the poles of development. They are as follows: Growth potential standard [17] Capacity absorption standard [18] Regional multiplier standard [19] the extent of the region's contribution to achieving development goals.

- Based on the previous criteria and the variables it contained to formulate them in a standard way to show the extent of their impact on the urban polarizing ability, the variables for each urban gathering (7 clusters) in the Northwest Coast axis have been defined within the framework of a specific time period starting from 1996 through 2006 and 2017, and a year 2030 according to what was expected by urban development strategiesbased on the data of the Central Agency for Public Mobilization and Statistics of the Population Censuses for the years 1986-1996-2006-2017 and the expected census of 2030, the strategic plan - Matrouh Governorate - Final Report - 2015-2030 - November 2017 and those variables were classified (as independent variables) (Independent Variables(As follows:) Population numbers. Immigration rates for centers. Employment opportunities. Number of tourist rooms. Average per capita income in pounds.National projects numbers. National projects areas. Cultivated areas. Population inhabited area. Desert back area. Distance between the community and the mother governorate. Distance between the community and Alexandria. The distance between the assembly and Cairo targeted investments). Considering population numbers (dependent variable) expresses the polarization capacity of the population.

2.3.2.2. Applying development standards

By applying the development criteria according to the parameters of the standard model on these groupings in the academic axis through the program SPSS(The statistic Statistical Package For The Social Sciences) Use of linear regression battery (Linear Regression) And factor analysis Factor Analysis "" [17] and analyzed in order to obtain Athletic scale can be invoked to gain access to measure polar metric capability of the urban communities of urban. Using the mathematical expression of regression:

$$Y = a + b1X1 + b2 X2 + b3 X3 + \ldots \ldots + bnXn$$

Where Y = the value of the dependent variable that we predict (population). X1, X2, X3, Xn represent the values of the different independent variables from (1 - n, ..,bn, b1, b2, b3 ,. The regression coefficients for the independent variables are either negative or positive.a= (Constant),**was reached following**:

First: The urban polarization ability of each of the groupings located on the axis has been reached, and the form of the relationship between each of the variables and the polarizing power has been determined as follows:

- Increasing the polarizing power factor during the specified periods for the city of El Alamein, then Matrouh, followed by Dabaa. This increases the population of these communities, as shown in Figure (4) .

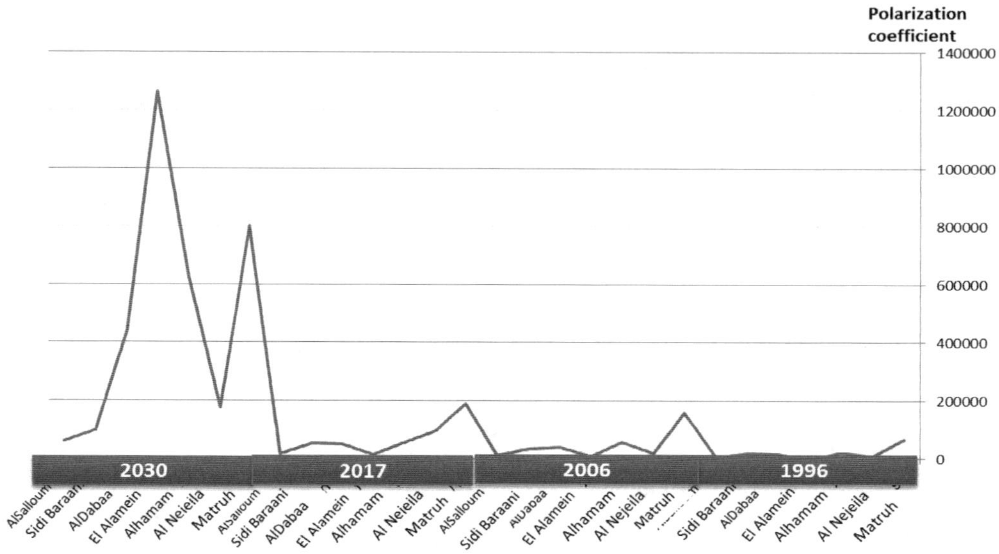

Figure 4: shows the illustrative relationship between population and polarization factor for each cluster in the years of study

Second: A statistical formula has been concluded that determines the form of the relationship between each of the specified variables according to the specified criteria and the urban polarization ability. And through it, the polarizing power can be measured

By applying this statistical experiment to urban gatherings in the academic axis, it was possible to know the standard parameters for the statistical model, which gives us the opportunity to apply it to any national project in Egypt or any urban gathering in the light of the proposed national projects, and on this the inferred equation is:

Y_{MPA} = -102680.00 + (- 4.0974 X_1 - 36.554 X_2 + 23.416 X_3+469 4X_4 - 36912 X_5 1024.1 X_6 - 488.2 X_7 - 278.44 X_8 + 2.1018 X_9 + 424.18 X_{10} -2,462.8 X_{11}+ 2589.8 X_{12} + 3671.6 X_{13})

Whereas:-

Y $_{MPA}$ the number of the population (urban polarization capacity) of the urban gathering is one thousand people.

X_1 numbers of immigrants. X_2 employment numbers. X_3preparations of tourist rooms. X_4average per capita income in pounds.X_5numbers of national projects in each grouping. X_6 areas of national projects in km^2. X_7area cultivated in $km^2$and it can be compensated by the area of the activity that leads to this gathering whose physical polarization capacity is measured. $X_8$the Populated spacein km^2. X_9area of the desert hinterland, in km^2. $X_{10}$the distance of assembly from the governorate affiliated in km^2. $X_{11}$the distance ofassembly from Alexandria, in km^2. X_{12} the distance ofassembly from Cairo, in in km^2. X_{13} pool investments.

These numbers in the equation are constants that were derived as a result of the statistical analysis of each of the variables included in the study

2.3.2.3. Apply the inferred equation to the new city of El Alamein

IOP Conf. Series: Materials Science and Engineering **974** (2020) 012030 doi:10.1088/1757-899X/974/1/012030

By applying the inferred equation to the new city of El Alamein to forecast population according to 2030 and to assess its developmental role in achieving development as a national project that represents a development pole, it is clear that:

Y_{MPA} = - 102680.00 + (- 4.0974 *120- 36.554 * 313.2 + 23.416 * 49500 + 2.469 * 30438 - 36912 * 36 + 1024.1 * 612.7- 488.2 * 49.36- 278.44 * 130 + 2.1018 * 24370 +424.18 * 175 -2462.8 * 62 +2589.8 * 236 + 3671.6 * 90) =1272336.69

We find that the prediction of the population represents from the calculation of the equation an amount of (**1272336.69**), while the population calculated on the nature is **(1265491)** , and by applying the equation to the rest of the urban gatherings of the coastal axis we find that the result has been interpreted with a very high percentage that is almost original, and from here we can pass The equation, and therefore we can start applying it to any city or urban cluster in the light of national projects in this community in Egypt to predict the number of population (measuring the urban polarization ability) as an indicator of the impact of national projects on the polarization of development in the urban map.

2.4 Results

There is a strong direct relationship between the urban polarization ability of any city or grouping and the national projects that make up this grouping. The more the number of national projects for a grouping or city increases, its greater polarization capacity and thus increases the population numbers as shown in Figure (5) .

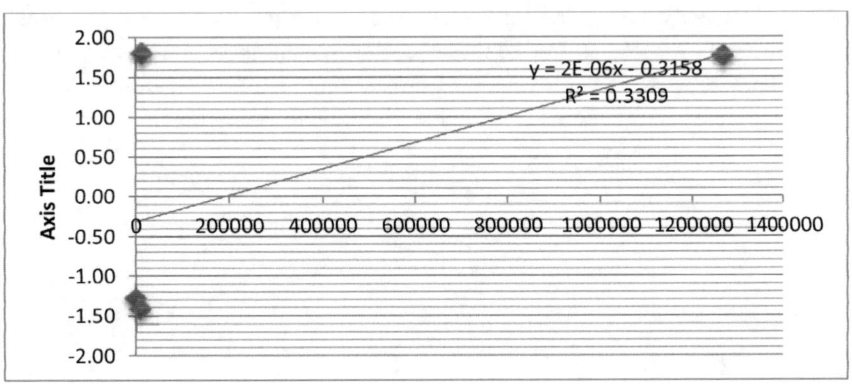

Figure 5: shows the linear regression relationship between the population and the polarization factor for the city of El Alamein during the study years

- The polarization capacity of the new city of El Alamein is suddenly increasing from 2017 to 2030, although it decreased completely in 2006 due to the increasing numbers and areas of national projects, and therefore it is considered a major development pole that attracts the population. The following figures illustrate the evolution of the population numbers of school groups, as shown in Figures (6), (7) and (8). The evolution of the numbers of the population of the urban communities during the age of Watt study

IOP Conf. Series: Materials Science and Engineering **974** (2020) 012030 doi:10.1088/1757-899X/974/1/012030

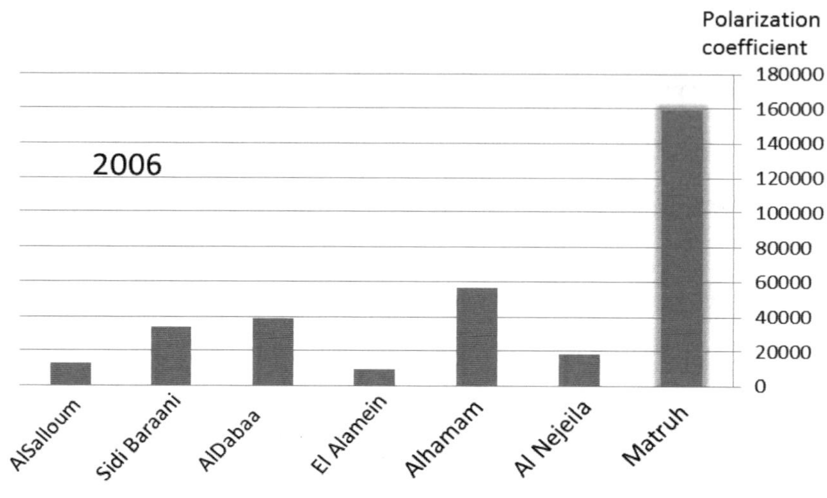

Figure 6: the evolution of population numbers for urban communities during the year 2006

- The increase in the developmental return of the axis of the northwestern coast increases with the settlement of the new city of El Alamein as a result of achieving the highest rate of polarization of the city compared to other clusters located on it according to the target year 2030. Figure (8).

- The new city of El Alamein is considered a successful national project, as the city has achieved, from the inferred application formula, the number of potential population to occur by no less than98And in this way, investment can be directed to it according to the state's investment plan to push the wheel of urban, economic and social development to the scope of national projects and achieve the desired benefit from them. The following figures (6), (7) and (8) show the evolution of population numbers for urban communities during the school years.

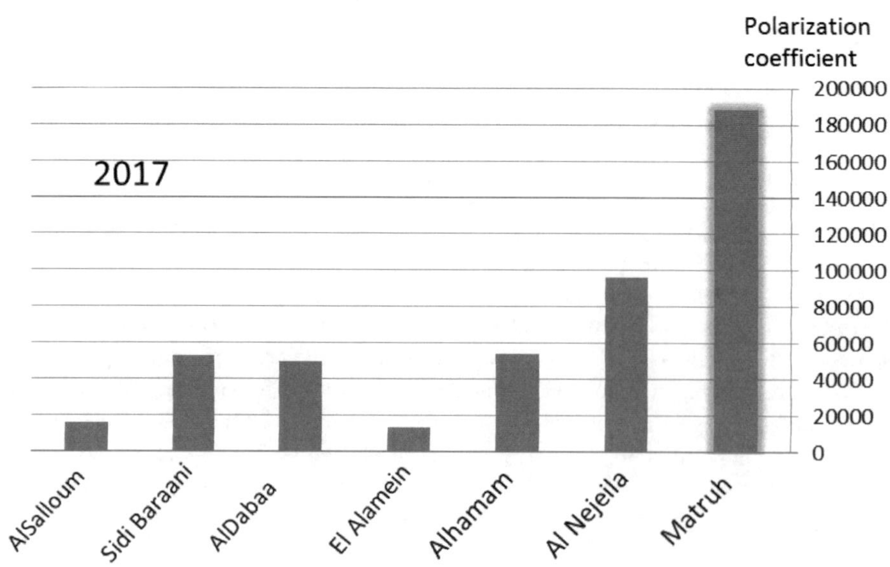

Figure 7: The evolution of population numbers for urban settlements during the year 2017

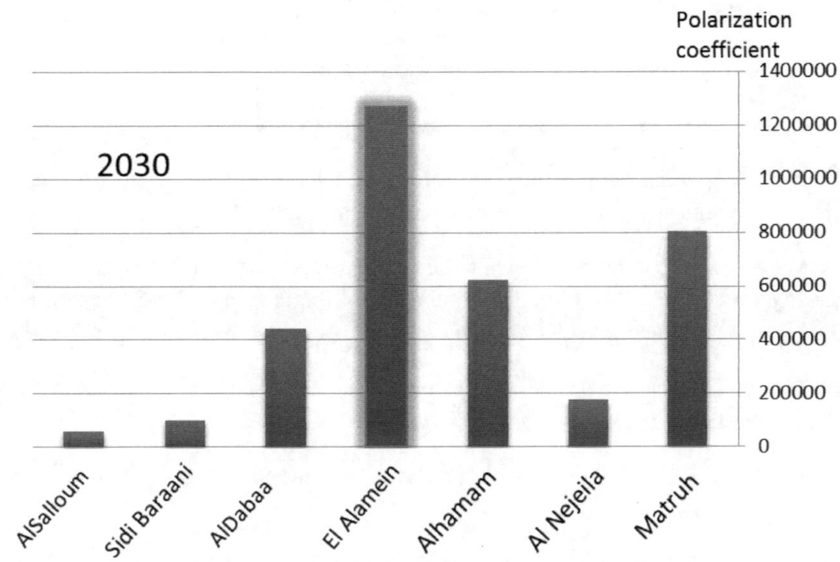

Figure 8: development of population numbers for urban communities during the year 2030

- The sizes of clusters increase in the axis of the northwestern coast by increasing the number of national projects in each gathering as a result of attracting the population, and reaching the maximum of El-Alamein, Matrouh and Dabaa.

- The need to evaluate national projects or urban agglomerations in light of the proposed projects through the inferred equation, so that the problems of incomplete project and achieve the desired urban attraction is not eliminated.
- Addressing the issues of the country's regions: through applying the equation to the various national projects in the regions of Egypt, so that the results can be evaluated and based on this, reallocating the investments, which works to achieve the regional balance of urban development on the map.

3. Conclusion

Is the national various projects levels, patterns and Their rankings need for urgent and inevitable, which is the hope of Egypt in the reconstruction of the abandoned space, and re - exploitation of its resources and potential and attracting new urban communities, and to find out the role developmental national projects must measure the ability of these projects to Polarization urban so that we can advancement of these projects and evaluate the true resolution Direct investment to it, and to measure the impact of the new city of El Alamein on development , a set of standard criteria was used, and each standard includes a set of variables that have been statistically analyzed for clusters located on the axis of the northwestern coast according to the years (1996-2006-2017-2030), and the polarizing capacity was concluded For every urban gathering and the relationship of polarizing ability to increase the population numbers for each population and come up with the statistical equation to measure the polarization capacity of any urban gathering in light of the numbers of national projects and their area in this grouping or any national project.

4. References

IOP Conf. Series: Materials Science and Engineering **974** (2020) 012030 doi:10.1088/1757-899X/974/1/012030

1. Mohamed Abou El Hassan. Developmental 3ffect of major projects on regional development - master thesis, Cairo University, 2015.

2. Institute of National Planning - giant national projects, Cairo, 2001.

3.Ahmed Allam and others.Regional Planning - Anglo Egyptian Library, 1995.

4.Marwa Mustafa. Measuring the polarization of urban development poles as an indicator of implications for development and investment - Ph.D. unpublished - Cairo University, 2003.

5.Mamdouh Mohamed Moustafa. Strategy for the Settlement of Industrial Projects in Egypt, Case Study of Upper Egypt Region - Unpublished PhD Thesis - Faculty of Engineering - Department of Urban Planning - Ain Shams University - 2224 p. and growth Centers in regional planning " paris , 1972, p. 70.

6.Nadia Anas Mohamed Kenawy.Spatial Dimension in National Development Plans in Egypt - PhD Thesis - National Planning Institute, 1995.

7.General Authority for Urban Planning - National Strategic Plan for Urban Development and Priority Development Zones, 2014.

8.General Authority for Urban Planning - National Strategic Plan for Urban Development, 2017.

9. Report - the project of economic and social planning for spatial development - Ministry of Planning, 2002.

10. The Potential of North West Gulf of Suez Project Source: Report - Council of Ministers, June 2013.

11. Suez Canal Regional Development Project,2016.

12. El-Bailey, A. The Future of Coastal Urban Development Axes in Egypt, Unpublished Doctorate Thesis - Al-Azhar University, 2011.

13. Geographic Information Systems database for the Urban Planning Authority 2017 project.

14.JefTe. G. Williamson: "Regional in Equality} and The Process of national Development, 1975, 159-170.

15.Emerson .M .JarinandF .Charles .Urban and Regional Economics Structure And Change. 1975.

16.Muhammad SarmFath al-Nur "Regional planning areas and explanatory methods" note no. 01040 National Planning Institute Cairo, 1973.

17. Richard Johnson, Dean and Chern, Abdul-MardhaAzzam - Statistical Analysis of Variables in Multiple Applications - King Saud University - QassimBranch - Saudi Arabia, 1998 AD.

**Military Technical College
Kobry El-Kobbah,
Cairo, Egypt**

**13th International Conference
on Civil and Architecture
Engineering
ICCAE-13-2020**

Architectural Design Firms' Social Role to develop Sustainable Heritage Communities in Egypt

Dalia A Elweshahy [1] And Ayman A E Othman [2]

[1] Senior Student, Architecture Engineering Department, Faculty of Engineering, The British University in Egypt, Cairo-Suez Desert Road, El-Shorouk, Cairo, Egypt. E-mail [1]: Dalia145555@bue.edu.eg
[2] Professor of construction and project management, head of architectural engineering department, faculty of engineering, The British University in Egypt, Cairo-Suez Desert Road, El-Shorouk, Cairo, Egypt. E-mail [2]: ayman.othman@bue.edu.eg

Abstract. Cultural heritage buildings are a crucial element for future generations, as they preserve the cultural identity unique to where they are located. Moreover, they create one of the most dominant and memorable architectural achievement that have significant historical and cultural values. However, heritage buildings in Egypt suffer from a surge in demolition due to the deteriorated conditions of their surroundings areas, as well as, the maintenance of only the physical structures without considering the surrounding communities. Egypt constantly encounters many challenges such as poor coordination from the government due to shortage of skilled people, poor heritage awareness of local communities and lack of finance needed for achieving sustainable development. On the other hand, Architecture Design Firms (ADFs) act as a supporting tool for the government as they have a vital role in the local community development; which further act as the cornerstone of developing the historical regions. Hence, in order for the ADFs to achieve such development, they need to activate their Corporate Social Responsibility (CSR) role which is a driver for developing the heritage community. Consequently, this research aims to develop a matrix that correlates CSR with ADFs as an approach to achieve sustainable heritage communities in Egypt. A research methodology consists of literature review to investigate the topics of CSR, Heritage community development and ADFs. Finally, developing a matrix showing the relation between the architect's social responsibility and the indicators of sustainable development of local communities.

Keywords: *Corporate Social Responsibility, Heritage Community Development, Sustainable Development, Architecture Design Firms, Heritage Buildings.*

1. Introduction

Egypt has some of the world's most important historical sites as it hosts six cultural heritage sites and one natural heritage site [1]. Yet these historical sites fall under great pressure due to a range of aspects, of which include the poor coordination of the government, (b) failure to involve different stakeholders, (c) the lack of financial contribution and mainly, (d) lack of awareness of the local community towards the significant value of these historical heritage buildings. In the meantime, the lack of the governmental role resulted in the involvement of the private sectors to support the government in sustaining heritage community development. Moreover, local community is considered one of the root aspects that surround such area and shape it, yet very little attention has been given to such communities of these historical sites. Only focusing on the conservation of their exterior aspects of such physical structures, results in a loss of value to these sites and the physical and socioeconomic deterioration of the surrounding areas [2]. Hence the heritage development of such areas have been recently brought to the attention of the Architecture Design Firms (ADFs), which target their focus mainly on the process of engaging with social and environmental causes, known as corporate social responsibility (CSR). As a result, this further leads to achieving the societal needs and meet their expectations.

2. Literature Review
2.1. Heritage Buildings
2.1.1. Definition of Heritage

Heritage is considered a significant source of values, tradition and cultural identity. Heritage helps the community to sustain their local identity. Heritage can be defined as a chosen or understood connection to a history that is greatly influenced by both its surrounding environment and by the values, beliefs and motivations of those who live in it [3]. Meanwhile, heritage is classified into both cultural and natural heritage [1]. Cultural heritage is the inheritance of a community and society's physical structures or sites and intangible qualities retained by past generations, preserved in the current and given to future generations [4]. Hence, Cultural heritage includes both tangible and intangible heritage.

2.1.2. Types of Heritage
2.1.2.1. Physical Tangible Heritage

Tangible heritage includes a) structures b) historic sites landmarks and buildings, that are of a great value and deemed worthy of potential protection shown in Figures 1 and 2. Moreover, these include artefacts that are significant to a specific community's archaeology, architecture, research, and engineering [3].

Figure (1) Amir Aqsunque Mosque [6].

Figure (2) St. George church [5].

2.1.2.2. Intangible Heritage

Intangible heritage comprises of a) traditions and images b) words and information, c) abilities d) tools, items and resources, which are connected with cultural areas and are accepted as part of the cultural heritage by communities, groups and, in some instances, individuals [4]. Furthermore, the intangible heritage includes alive cultural transmission practices between generations, and these practices are based on intergenerational communication such as story telling shown in Figure 3, Dancing shown in Figure 4, art, clothing, food and memories of the heritage community.

Figure (3) Story telling [2].

Figure (4) Dancing [5].

2.1.3. Problems impacting the Physical Heritage in Egypt

Egypt is well known of its architectural and cultural heritage [5]. Egypt is distinguished by some of the world's significant heritage areas, as it includes seven cultural historical sites and two natural historical sites [1]. Moreover, Egypt and especially historic Cairo has a distinctive architectural style and features the largest concentration of Islamic temples in the country, both in quality and quantity [6]. Yet, Egypt falls under great urban pressures which is the vandalism and deterioration of its historical sites and heritage buildings due to many factors. There are three factors that cause deterioration of the physical heritage in Egypt which are: Human-based factors, Environmental factors and governmental factors [5]. However, the human-related factor is considered the main factor due to the discrepancy between the physical heritage structure and the surrounding local community needs. Whereas, community are the main factor affecting the surrounding heritage areas therefore, further consideration is in order to achieve sustainable development of heritage communities.

2.1.1. Human-based Factors

Eight main problems are caused by the human factor affects the physical heritage [5]:

1. Lack of public concern for antiquities due to that most buildings are unused and therefore the local citizens of this area don't benefit directly from it shown in Figure 5.
2. Lack of social awareness of the importance of heritage buildings due to low education levels that cause great demolition to heritage areas as shown in figure 7.
3. Poverty and unemployment shown in Figure 6.
4. Changes in values leading to new uses of heritage resources therefore, applying renovations without previous experience, which spoil the value of the heritage style.
5. Migration to or from site and deterioration of the heritage buildings in this site due to lack of usage or improper usage.
6. Shortage of constant maintenance to the physical structures of heritage areas due to lack of awareness and shortage of financial resources which further on cause demolition to their exterior architectural style.
7. Applying renovations by using modern high tech facilities such as air conditioning which cause deterioration of the cultural identity.
8. View heritage buildings as discrete components of the surrounding environment which spoils the visual image as a whole shown in Figure 7.

Figure (5) Lack of awareness of heritage value [6].

Figure (6) Writing on Citadel of Qaitbay in Alexandria [2].

Figure (7) Garbage in heritage areas [2].

2.1.3.2. Environmental Factors

There are 4 main environmental factors that negatively affect heritage areas [5] which are:

1. Dust accumulated on ceilings and walls of heritage buildings that damages painting and decoration and cause demolition to the architectural style.
2. The presence of polluted air in the area that could cause great harm to stones and building materials, due to transportation in this area or presence of industrial areas nearby the heritage site.
3. Humidity and rain due to poor maintenance that can cause severe demolition to the structure of buildings.
4. Ground water table that negatively affect the buildings.

2.1.3.3. Governmental Factors

1. Shortage of financial resources that cause lack of maintenance due to leaving insufficient financial resources for renovations.
2. Lack of needed skills to renovate the heritage buildings.
3. Renting some of the heritage areas to private sectors, which reuse them in other activities that may not harmonize with the site and cause serious damage.

2.2. Sustainable Development
2.2.1. Definition of Sustainability

Currently, no accurate definition of the term has emerged, in spite of the great influence of the sustainability concept. However, according to Brundtland report [7], Sustainability can be described as the improvement that fulfils current needs without undermining future generations ' ability to meet their needs. Moreover, one preference defines sustainable development as the development that improves the long-term welfare of human and ecological communities and then move on to a more specific discussion of particular strategies [8]. Mainly, sustainability could be described as the philosophy that provide awareness to the community about consequences of their actions towards the environment

IOP Conf. Series: Materials Science and Engineering **974** (2020) 012031 doi:10.1088/1757-899X/974/1/012031

and encourage the society to have disciplines and to conserve the vitality of the earth's ecosystem. Thus, sustainable development comprises of three main aspects which are Social, Environmental and Economic as shown in Figure 8 [7].

Figure (8) Sustainable development [7].

2.2.2. Sustainable Development Principles

There are a set of principles conducted in order to achieve sustainable development. basic truths about sustainability [9]. These principles reflect a philosophy which shapes the theory of brickwork or which acts as a policy about the existence of sustainability. Through following the implementation of the principles mentioned below, and others that may grow with further public discussion, decision-making could better serve the security and fair distribution of resources in order to achieve equity, through prioritizing real needs before needing while keeping choices available for future generations.

1. Social Equity.

   Social equity is considered to be the most significant element to achieve sustainable development. Whereas all economic, environmental and social systems depend on a well-balanced ecosystem, in which the community contribute to be an essential part of this ecosystem. Thus, the relationship of the community and the environment must achieve the ecological integrity in order to enhance the quality of life of the society.

2. Sufficiency and Opportunity.

   Currently, human development is encouraged, this is achieved by providing opportunities and raise awareness on using the resources available efficiently. Thus a way to start applying this is by doing more with less. This requires a reduction, and recycling system.

3. Efficiency.

   One of the main aspects on sustainable development is efficiency specially in the construction industry. This is achieved by eliminating waste generation by analysing and maximizing the full life-cycle study of buildings and processes in order to address the status of natural systems where waste is not present.

4. Communication and Cooperation

   Mainly, systems that affect the need of the society have the responsibility to provide accountability. Communication can be considered as the catalyst for social improvement as public confidence could be boosted by sincere communication. Therefore, Commitment to achieve better social communication and cooperation should speed up the development process and achieve sustainability.

5. Citizen Engagement and Democracy.

   One of the main principles of sustainable development is community engagement especially in the architectural field. Whereas, most of sustainability problem could be solved by engagement of the society in the decision making process. Long - term change and sustainable development involves the need for community participation.

IOP Conf. Series: Materials Science and Engineering **974** (2020) 012031 doi:10.1088/1757-899X/974/1/012031

2.3. Heritage Local Community

2.3.1. Definition of Heritage Local Community

The term "Local community" and community participation had attracted several researcher's and architect's attentions in the recent decades. "Community" can be defined as a bunch of people engrained in a sense of place in which they have an extremely mutual and trusting relationship with one another and their surrounding environment [9]. Moreover, since the society is considered to be a self-organization, it affects its surrounding environment. Thus, while the community changes its surrounding environment in its lifetime, in reaction the environment changes the community as well. In addition, Local communities that are existing within heritage areas are considered to be cultural guardians and associated users of these areas, and they are identified as a core community [10]. Accordingly, their everyday activities and rituals are connected with their surrounding environment and local cultural heritage. Hence, they have an abundant sense of belonging to heritage and also they sustain a constant connection with local identities and values [8]. This makes them a core stakeholder of importance to heritage areas, as they have the ability to preserve cultural heritage and sustain local identities [11]. Moreover, to strengthen their commitment to protect cultural heritage, the core community's desires and views need to be considered and this is achieved by sustainable community development.

2.3.2. Heritage Community and Sustainable Development

According to the previous, the heritage community is of a great importance and need to be well considered and sustainably developed. The term of community development, mainly, it is a process involving facilitation, coordination and involvement that allows individuals to guide forms of influencing the community they desire to live in and be part of [6]. Furthermore, sustainable community development has arisen as a convincing alternative which involves planning and development, community participation, comprehensive and equitable system that leads to meaningful, substantive improvements to the community. This is achieved through job opportunities, poverty reduction, conservation of natural environment, and empowerment of community [9].

According to [9], community development according to sustainability has three aspects:

1. Environmental aspect, it means attaining the present community needs without negatively affecting the future needs by optimizing the use of natural and environmental resources.
2. Economic aspect, it is considered to be the most important precondition for the real fulfilment of human needs and raising community living standards.
3. Social aspect, includes the enhancement of the quality of life of the society by: addressing its demands, increasing social justice and improving well-being rates.

2.3.3 Indicators of Sustainable Community Development

After examining and studying many literature, four indicators of sustainable community development are concluded as shown in figure 9 [6].

Figure (9) Indicators of Sustainable Community Development [6].

1. Quality of life

It's considered as the main indicator. Currently, cultural heritage improvement may make a significant contribution to enhancing the lives of residents. In addition, enhancing the quality of life of local communities positively affect the surrounding heritage areas.

2. Security

This can be achieved by the restoration and renovation of heritage areas, while the demolition of such areas contributes to general poverty, violence and destruction of infrastructure due to lack of maintenance.

3. Participation

It is one of the main indicators which means allowing community engagement and participation in the decision making process, which gives the local community a sense of belonging and empowerment.

4. Social capital

Social capital means creating opportunities for social integration, providing workshops and training to local communities. Which helps in sustaining their local identity and pride.

5. Social Welfare

It mainly means the empowerment of the community and improving their standards of living and enhancing whole conditions of the local community.

2.4. Corporate Social Responsibility (CSR)

2.4.1. What is CSR?

Corporate social responsibility (CSR) has extensively started to incorporate with worldwide business' corporates. During the 20th century, the society's expectations from corporates has changed from appreciation of corporate's philanthropy to demanding of social justice. Hence, corporates goals started to go beyond profit maximizing and focus more on the local community and costumer's satisfaction [12]. Although CSR in the architectural field has gained many research's attention, yet little attention has been given to it in Egypt. The concept of CSR applies to what the interaction between global corporations, national governments and local society should be [13]. More precisely, CSR focuses on how the relationship between a corporate and the society should operate. Additionally, CSR could be defined as the responsibility of a corporate to act ethically, improve the quality of life of local society and contribute to the economic growth. Generally, CSR is concerned with the way corporates integrate into their values the following:

IOP Conf. Series: Materials Science and Engineering **974** (2020) 012031 doi:10.1088/1757-899X/974/1/012031

- Social responsibility.
- Environmental responsibility.
- Economic responsibility.

CSR is a management approach that positions corporates to handle uncertainties proactively and to take advantage of opportunities, particularly with regard to their reputation and the large stakeholder engagement [14]. This may include shareholders, employees, 8 clients, communities, suppliers, governments, and others influenced by the corporate. CSR of corporate should include several activities such as:

- Sustainable development.
- Corporate ethics.
- Security and health.
- Maintenance of environmental factors.
- Human rights (including the fundamental rights of workers).
- Working conditions (including safety, working hours, salaries).

2.4.2. Principles of CSR

The principles of CSR

- Sustainability.
- Accountability.
- Transparency.

2.5. Architectural Design Firms' CSR

Architects have a significant role, through their practices as experts and members of the local community as well. Architects are the main point of communication in the construction industry with customers, thus they always have the motivation, inspiration and spirit of creativity to play this role [6]. In addition, architects primarily design structures and service which acts as a catalyst towards improving the cultural development while contributing in economic growth.

Although architects have a significant contribution in improving the environment and boosting the economy. In Egypt, they neglect their social responsibility that they need to play besides the environmental and economical responsibility. Hence, ADFs need to activate their social responsibility in order to achieve sustainable development [14]. In the meantime, architect's commitment to heritage areas extends beyond the boundaries of maintaining the physical heritage to include all facets of sustainable community development [5]. Therefore, to attain sustainable development, the role of ADFs within the scope of CSR is classified under the following:

1. Social responsibility.
2. Economical responsibility.
3. Environmental responsibility.

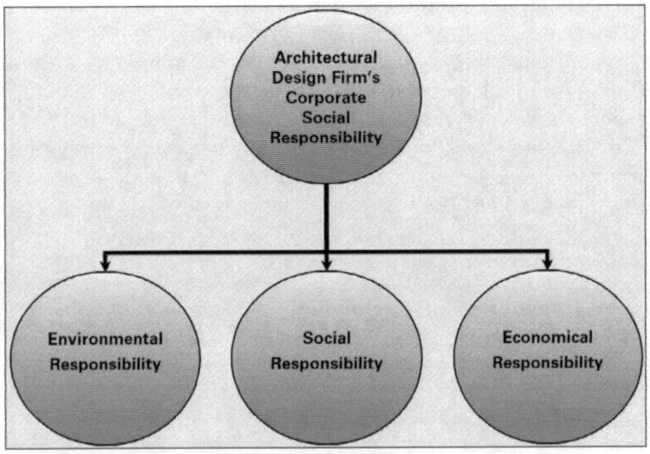

Figure (10) ADF's CSR [15].

2.5.1. Architectural Design Firms and the Society

The social aspect according to [6], includes enhancing the quality of life of the society by: meeting its desires, increasing social justice and rising well-being rates. Moreover, architect has a vital role towards the society which is: raising awareness of the important role that architects can play in improving society, guaranteeing that the final design meets the needs of community, such as preferences, traditions and cultures, involving community in the decision making of the design process and getting input from other corporates and motivating them to fulfil their social responsibility [14].

2.5.2. Architectural Design Firms and the Economy

Mainly, economical responsibility is the most significant requirement in order to fulfil society's needs and increase the community's living standards. Through the position of built environment, this contributes to economic growth in particular. Furthermore, the economic responsibility of any design firm can be seen from two different viewpoints. The first one, is to provide job opportunities and increase the percentage of gross domestic product by encouraging the growth and development of the construction industry. The second one, is increasing the profit and return on investment for the client. Hence, ADFs have a significant responsibility towards the economy such as ensuring that the resources of society are used effectively, joining any other sales from other businesses that are committed to improving their goods ' quality and minimizing the total running and maintenance costs and generate innovative ideas [14].

2.5.3. Architectural Design Firms and the Environment

The environmental responsibility of the ADFs can be defined as the contribution to optimize the use of the environmental resources in order to ensure the human development. Mainly, the architect's responsibility towards the environment is to encourage the community to save the environment and to encourage the sustainable usage of resources during the architectural design process [5].

IOP Conf. Series: Materials Science and Engineering **974** (2020) 012031 doi:10.1088/1757-899X/974/1/012031

2.5.4. ADFs' CSR and heritage community development

In order to achieve the sustainable community development in heritage areas, ADFs have to play their social role toward the community [6]. Whereas, heritage community development does not require only their social responsibility but also their economic and environmental responsibility. Therefore, this could be achieved by applying several activities of ADFs which are: Public engagement, Community empowerment, strengthening social structure, enhancement of local culture and awareness and enhancement of liveability.

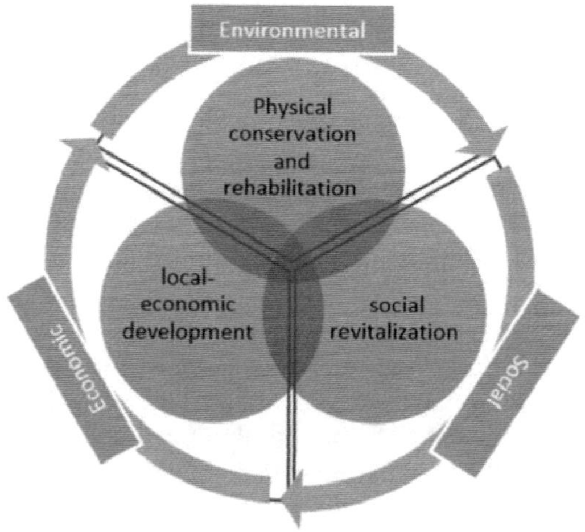

Figure (11) Architect's role in heritage community development [6].

2.5.4.1. Public Engagement

Public engagement is considered to be the most significant aspect in enhancing the quality of life of the community [6]. Whereas, it encourages social enclosure and sustainability as well as preserving the cultural identity. It is significant for the communities to have the awareness about the importance of heritage conservation. Hence, one of the main roles of architects is to raise awareness among the communities about their important role in the decision making and in heritage preservation.

2.5.4.2. Community Empowerment

Mainly, community empowerment is described as a system in which the community has full control of their existence and society and can understand their surrounding objectively. Therefore, architect should provide an encouragement for community members to demonstrate spirit of leadership in order to guide their society [5].

2.5.4.3. Strengthening Social Structure

Social structure is assisted by a participating strategy, which is highly effective for community reconstruction [10]. Thus, architects may create opportunities for volunteers in the area of cultural heritage and play an active role in heritage conservation projects.

2.5.4.4. Enhancement of Local Culture and Awareness

Mainly, local culture is considered to be local pride and social understanding to the surrounding heritage community. Therefore, these cultural values reinforce the sense of pride and gratitude of the people for their heritage. Consequently, local community should have the awareness to protect and preserve their heritage. ADFs and architects can use their skills to demonstrate the strengths and beliefs that shape society and increase public members ' awareness [10].

3. Results

Indicators of sustainable community development / ADF's CSR Role			Indicators of sustainable community development												
			Quality of life					Participation	Social capital		Social welfare			Others	
			Local history	Education	Public health	Housing	Safety	Civic engagement	Poverty	economic	Community identity	Social Change	Land use	Material conservation	Ecosystem conservation
ADF's CSR Role	Social	Engage the public in planning processes.													
		Preserve the community's open space.													
		Enhancement of Livability .													
		Seeking feedback from people.													
	Economic	Economic opportunities for the local members.													
		Provide or facilitate public services .													
		Creating innovative ideas and using sustainable materials.													
		Attract businesses, new residents, and visitors.													
	Environmental	Responding positively to different environmental effects.													
		Planning to conserve natural environment.													
		Encouraging pedestrians , cycling and public transport.													
		Promoting waste reduction.													
	Cultural	Preserve the historic and cultural heritage of a place.													
		Preserve a place's local identity and character.													
		Promote community sense of loyalty of place.													
		The quality of architectural product.													

Figure (12) Design Matrix. Developed by Author.

4. Conclusion

In conclusion, the literature reviews along within the case-study proposed have highlighted the need of ADFs to activate their CSR role. It emphasized the importance of ADFs CSR in the sustainable development of heritage communities. Heritage buildings demolition is resulted from the focus of only the conservation of the physical structures of heritage and not focusing on local communities. CSR was introduced in the 20th century an approach for ADFs to behave ethically and not only maximize the profit but also enhance the social conditions of the community. Hence, CSR is proposed approach for ADFS to achieve sustainable heritage communities' development. The paper's methodology consists of comprehensive data collection from several reliable and valid sources in order to create literature review that would help form a design matrix showing the relation between ADF's CSR and local communities needs shown in figure 12.

IOP Conf. Series: Materials Science and Engineering **974** (2020) 012031 doi:10.1088/1757-899X/974/1/012031

5. References

[1] UNESCO World Heritage Sites in Egypt. (2017, July 28). Retrieved from UNESCO:https://everything-everywhere.com/unesco-world-heritage-sites-egypt/ WCED. (1987). Sustainable Development. Berlin.

[2] Hanna, M. (2017). Egypt: Cologne Aswan Summer School on Heritage Management and Storytelling.

[3] Marmion, M. M. (2012). Understanding Heritage: Multiple Meanings and Values.

[4] Rudolff, B. (2006). 'Intangible' and 'tangible' heritage. Germany.

[5] Mostafa, S. L. (2014). Historic Cairo: Causes of Deterioration and Suggestions for Preservations.

[6] Hasan, A. E., Al-hagla, K. S., & Alana, H. a. (2018). A framework for architects' role in attaining sustainable community development in heritage areas Al-Darb AL-Ahmar, Islamic Cairo, Egypt as a case. Alexandria Engineering Journal, 11.

[7] Darwish, M. M. (2010). incorporating Sustainable Development and Environmental Ethics. Eighth LACCEI Latin American and Caribbean Conference for Engineering and Technology (LACCEI'2010), 10.

[8] Phillips, R., & Pittman, R. H. (2009). AN INTRODUCTION TO COMMUNITY DEVELOPMENT. USA and Canada: Routledge.

[9] Flint, R. W. (2013). Practice of Sustainable Community Development a Participatory Framework for Change. Napa, CA, USA: Springer New York Heidelberg Dordrecht London.

[10] Poulios, I. (2014). Discussing strategy in heritage conservation Living heritage approach as an example of strategic innovation. Cultural Heritage Management and Sustainable Development, 19.

[11] Wesemael, P. v., Roders, A. P., & Krishnamurthy, S. (2019). unity participation in cultural heritage management: A systematic literature review comparing Chinese and international practices. ELSEVIER, 9.

[12] Keinert, C. (2008). Corporate Social Responsibility as an International Strategy. Vienna, Austria.

[13] Crowther, D., & Aras, G. (2008). Corporate Social Responsibility. ventus..

[14] Hohnen, P., & Potts, J. (2007). Corporate Social Responsibility an Implementation Guide for Business. Canada.

[15] Othman, A. A. (2012). An Innovative Protocol for Improving the Ethical Behaviour of the Quantity Surveying Profession in South Africa. Research gate.

IOP Conf. Series: Materials Science and Engineering **974** (2020) 012032 doi:10.1088/1757-899X/974/1/012032

Military Technical College
Kobry El-Kobbah,
Cairo, Egypt

13th International Conference on Civil and Architecture Engineering ICCAE-13-2020

Evaluation of Slum Upgrading Interventions - Methodological Approaches

Hesham El-Maradny[a] **Mohamed M. Mahdy**[a] **Hesham El-Kady**[a]

[a]Architecture Engineering Department, Military Technical College, Cairo, Egypt

hisho2504@gmail.com helkady@mtc.edu.eg

Abstract. This work addresses slums definition according to UN-Habitat, and provides some objectives of impact evaluation of slum upgrading projects. The main objective is to clarify the importance behind the slum upgrading projects (interventions), through over viewing the slum upgrading interventions and their main characteristics then summarize methodology of evaluation studies. This paper gives an overview of counterfactual question, qualitative and quantitative methodologies discusses main strategies to control bias in Quasi-experimental approaches, and provides a summary about common past evaluation challenges and recommendations which can be applied to future work.

1. Introduction

The normal growing of the urban residents combined with the adding of rural immigration to towns formed a fast urbanization in the developing world over the last several years.Rural to urban migration is driven by the better labouropportunities and higher standards of living that are available in the cities [1]. Cities let people and companies enjoy the profits that come from the agglomeration of resources, although cities offer several economic chances and contact to better facilities and services, these benefits do not reach all city residents. Glaeser[1]classifies the three great scourges of urban life: crime, disease, and congestion.

The United Nations (UN) estimates that nearly one billion people worldwide currently live in slums (one-sixth of the planet's population). By 2030, nearly five billion people will live in urban areas, compared to 3.2 billion in 2007.Upgrading projects focus on providing basic facilities to improve poor communities. Due to expected growths in slum residents, the request for urban upgrading interventions is probable to increase.Investing resources in slum upgrading developments must perfectly be constructed on strong indication of which exact interventions are more effective. Since the several methods of upgrading interventions and the complications met in implementation, assessing their effect can be complex.

IOP Conf. Series: Materials Science and Engineering **974** (2020) 012032 doi:10.1088/1757-899X/974/1/012032

According to UN-Habitat, a slum household is a group of individuals living under the same roof, lacking one or more of the following conditions (UN-Habitat)[2]: (i) access to safe water: sufficient amount of it (20 liters/person/day), at an affordable price (less than 10 percent of the total household income), available without being subject to extreme effort (less than one hour a day of walking time); (ii) access to improved sanitation: access to an excreta disposal system, either in the form of a private toilet or a public toilet shared with a reasonable number of people; (iii) sufficient living area: fewer than three people per habitable room; (iv) structural quality/durability of dwellings: a house built on a nonhazardous location and with a permanent structure adequate enough to protect its inhabitants from the extremes of climatic conditions; and (v) security of tenure: the right to effective protection by the State against arbitrary unlawful evictions

1.1 Slum upgrading projects

Provision of basic services to improve the lives of poor communities, including both infrastructure and social interventions.

1.1.1 Infrastructure interventions:

Providing slums with physical basic facilities such as water, sanitation, waste collection, housing, access roads, footpaths, lighting, public telephones, schools, hospitals, etc.

1.1.2 Social interventions:

Focus on providing insecure tenure, health services, education services, day care, training, social protection programs, etc.

1.2 Why are slum upgrading projects (interventions) important?

The need for slum upgrading interventions grow as a result of continuous increases in slum population, which makes it an important component of the development process, that's why governments have undertaken several approaches to deal with informal settlements.

1.3 What are the main strategies of slum upgrading interventions and their basiccharacteristics?

Table 1. Slum upgrading interventions and their main characteristics

Intervention	Main characteristics
Eradication strategy	- It is an expensive method, as governments must provide homes for displaced population. - It is still practiced in some countries.
Site and services	- It was a popular approach in the 1970s - The main idea of this strategy is to provide urban lots for families that were removed from slums to build their homes step by step. - This strategy was not successful because it mostly leaves families in worse conditions than they were in the original slums.

In situ slum upgrading	- This concept was experienced by governments since the 1980s. - This strategy based on the idea that it is socially and economically more effective for residents to stay in their communities. - The aim of this intervention is the integration of poor communities with their larger urban context. - It keeps the social networks of the dwellers, while improving their living standards. - Due to the success of this strategy, it was implemented in many programs of slum upgrading projects. - the implementation of this strategy ranged from land tenure and ranging all the way to more complete versions including infrastructure projects, housing improvement, access roads and others.

By author

- Which is the most effective slum upgrading intervention for both: politicians and slum residents?
- How can officials improve slum upgrading interventions to meet the requirements of the urban poor? These questions can be replied by running suitable impact evaluation studies.

IOP Conf. Series: Materials Science and Engineering **974** (2020) 012032 doi:10.1088/1757-899X/974/1/012032

2 Methodology of evaluation studies:

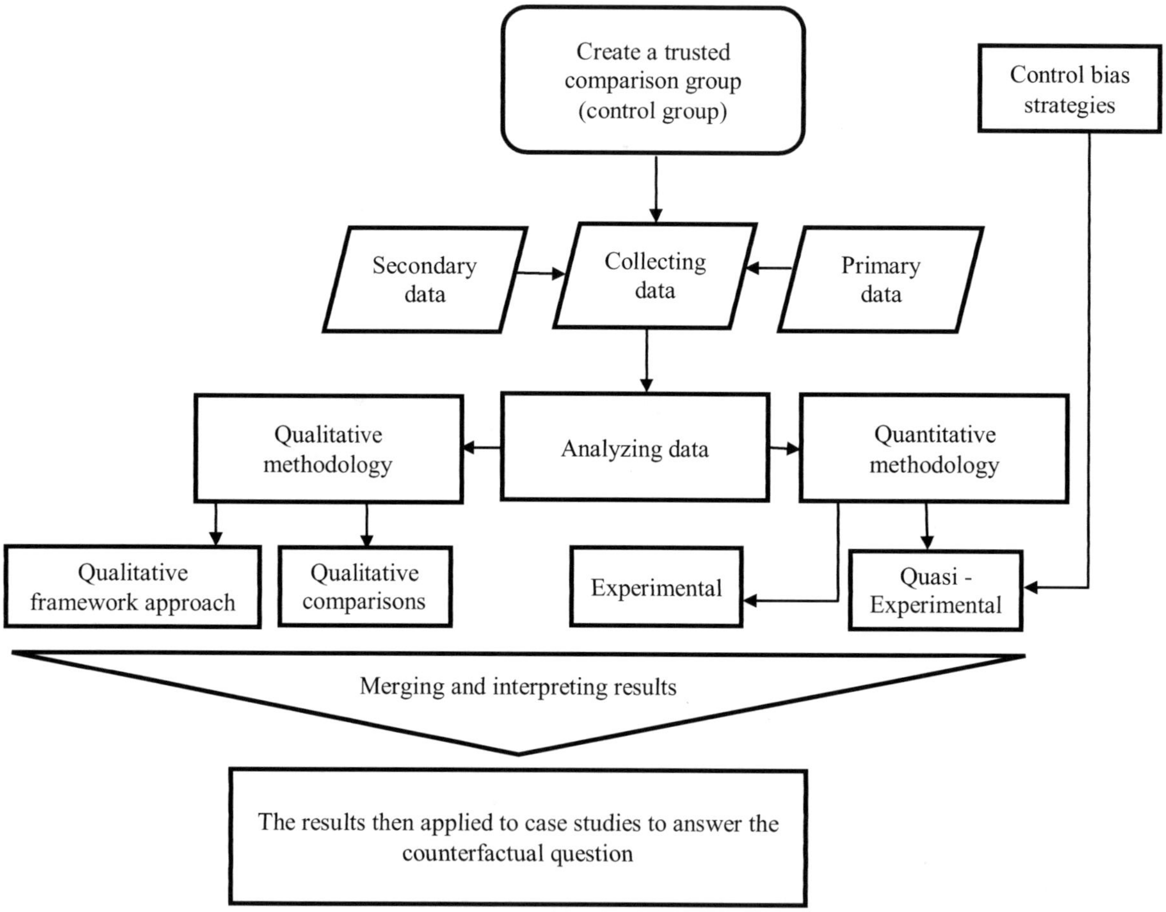

Figure 1: Methodology of evaluation studies

Although there are many slum upgrading programs, there are few studies designed to identify the causal effects of a program. Evaluating the causal effects of slum upgrading interventions on different outcomes is a complex task. Any impact evaluation tries to answer a fundamentally counterfactual question[3]:How would individuals who participated in the program have fared in the absence of the program? How would those who were not exposed to the program have fared in the presence of the program?

IOP Conf. Series: Materials Science and Engineering **974** (2020) 012032 doi:10.1088/1757-899X/974/1/012032

Answering these questions is difficult because:

- Slum household is either exposed or not exposed to the upgrading intervention program.
- Comparison for the same person by time will not, usually, give a dependable estimation of the effect the program had on him or her, as several other circumstances possibly will have differed at the same time as the program was presented.
- So, it is obvious that seeking for getting an estimation of the effect of the program on each person is not practical.
- To be able to get the average effect of the program on a group of people by making comparison between them and other similar group of individuals who were not exposed to the program is all we can hope for.
- The main goal of impact evaluation is to create a trusted comparison group, which allows us to get an estimation of the average impact on the group in question.
- So, the idea of establishing a reliable comparison group is crucial to any impact evaluation.
- *But, in reality it is mostly unlike because:* programs may be located in poorer or richer areas for example. Individuals might be chosen for participation in the program on different basis like poverty, motivation or any other basis. For these reasons, those who were not exposed to a program are often not a good comparison group for those who were.

2.1 Qualitative and quantitative methodologies:

To resolve the evaluation problem, program evaluations naturally require to be wisely planned previously to specify which group is a probable control group.Many different empirical approaches and strategies have been used in impact evaluation, involving both *qualitative* and *quantitative* methodologies.

2.1.1 *Qualitative methodologies:* due to past impact evaluations, be broken into four broad categories as shown in Fig.1.

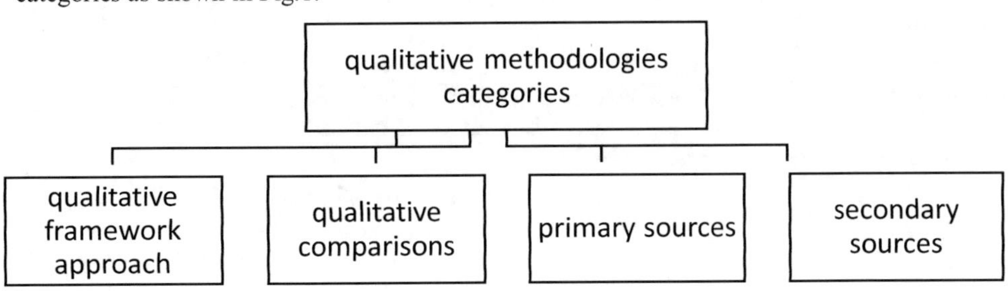

Figure2. Qualitative methodologies main categories

Table 2. Qualitative methodologies categories

<table>
<tr><th>Qualitative framework approach</th><th>Qualitative comparisons</th><th>Primary sources</th><th>Secondary sources</th></tr>
<tr><td rowspan="2">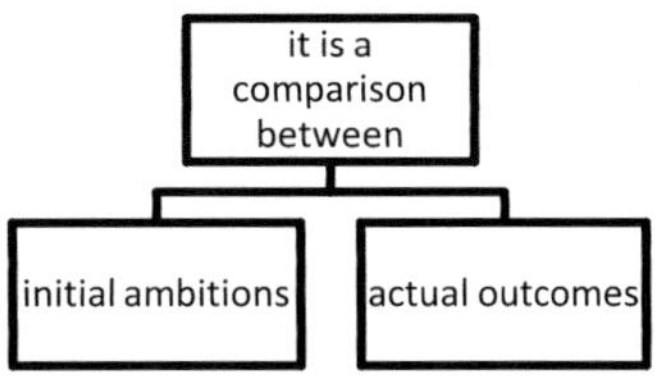
this comparison is made by:

-Defining concepts.
-Describing and mapping dynamics between actors and institutions
-Categorizing residents' attitudes and perceptions.
The resulting evaluation framework is then applied to available data and case studies[2, 4, 5].</td><td rowspan="2">Comparing case studies of two or more slum upgrading projects[2].

This comparison is made by: considering the relevant merits of different slum approaches on specific indicators of interest, such as:
- physical topography [6].
- land ownership and residential mobility [2].
- long-term sustainability[7]
-Qualitative comparisons have also been used to measure institutional roles and community cooperation, including residents' roles and relationships in different communities[4, 5].</td><td>Based on information gathered from:site visits, fieldwork, focus group, in-depth interviews with participants, technicians, officials, NGO staffs, household surveys of participants' satisfaction and perceptions, and direct observation [2, 4, 8].</td><td>Rely on:news articles, reports, government documents, and other evaluations to draw conclusions about project impact[7-13].</td></tr>
<tr><td colspan="2">Most commonly used to support results from quantitative assessment.</td></tr>
</table>

By author

2.1.2 *Quantitative approaches:*

Table 3. Quantitative methodologies categories

Experimental	Quasi-experimental
- Implemented prospectively before project implementation. - Evaluators choose one or more project components like (individuals, communities, schools or classrooms). - Evaluators measure changes overtime in treatment relative to control populations with follow-up data. - Differences between the outcomes of the groups attributed to the program.	- Implemented retrospectively after project implementation, based on available data. - They may contain large and unknown biases resulting from specification errors. - Researchers have developed alternative techniques to control for bias in retrospective evaluations as well as possible like: comparable strategy, difference-in-difference strategy (DID) and regression discontinuity design strategy.

By author

IOP Conf. Series: Materials Science and Engineering **974** (2020) 012032 doi:10.1088/1757-899X/974/1/012032

Table 4. Summary of Main Strategies to control bias in Quasi-experimental approaches

Control bias strategy	comparable strategy	difference-in-difference strategy (DID)	Regression discontinuity design strategy
methodology	- Depends on identifying all the differences between treatment and control groups. - Attempt to find a control group that is as "comparable" as possible to the treatment group along observable dimensions.	Compare the growth in the variables of interest between program and non-program regions.	Takes advantage of the fact that program rules sometimes generate discontinuities that can be used to identify the effect of the program by comparing those who made it to those who "almost made it." That is, if resources are allocated on the basis of a certain number of points, it is possible to compare those just above to those just below the threshold.
limitations	In cases where the treatment is assigned on the basis of a variable that is not observed by the researcher (demand for the service, for example), this technique can lead to misleading inferences.[3]	- Not granted. - Cannot be tested. - To achieve reasonable results, we need to collect long time series of data before starting the program. Evaluators should be sure that no other program was implemented at the same time. Bertrand, Duflo and Mullainathan[14]found that difference-indifference estimations can severely bias standard errors.	Often cannot be implemented, especially in developing countries because often the case that rules are not obligatory strictly enough to generate discontinuities that can be used for identification purposes.

2.2 Drawbacks of Quasi-experimental approaches

They may contain large and unknown biases resulting from specification errors.Glazerman et al. found that retrospective estimators often produce results dramatically different from randomized evaluations, that the estimated bias is often large, and that they were unable to identify any strategy that could consistently remove bias and still answer a well-defined question. A number of comparative studies have been conducted - some of which suggest omitted variables bias is a significant problem, others which suggest non-experimental estimators may perform well in certain contexts.

Several studies in Kenya have provided evidence that estimates from prospective randomized evaluations can often be quite different from estimated effects in a retrospective framework, suggesting that omitted variable bias is a serious concern[15, 16].

Glewwe et al. [14] study found that: it is not clear that such a difference-in-difference approach has general applicability.
Similar differences between retrospective and prospective randomized estimates arise in studies of the impact of de-worming in Kenya[14].

To address these concerns, future researchers should conduct retrospective evaluations before the results of randomized evaluations are released or conduct blind retrospective evaluations without knowledge of the results of randomized evaluations or other retrospective studies.

Project evaluation with non-randomized approaches must be undertaken with extreme attention because they are less transparent and more subject to deviation of estimation than projects evaluated with randomized approaches. Moreover, the differences between respectable and non-respectable non-randomized evaluations are hard to be found out, especially to strategy makers, because of all the cautions that must be taken with concern with the outcomes. These cautions may never be given to strategy makers, and even if the cautions are given, they might be unnoticed: so, strategy makers are completely misled[17].

It is recommend that while non-randomized approaches will remain to be required; there had better to be an obligation to run randomized evaluations where possible.

3 Common evaluation challenges

Past evaluations have met difficulties in the following parts of impact assessment:

3.1 Complementary programs:

Importance of complementary programs	A serious factor of the profits of slum interventions is the extent to which complementary services and programs reply to public progress. For example, the value of road construction relies deeply on the degree to which bus service station continue to arrive the districts. Here, the appropriate result measure in a program evaluation would relay directly on complementary services taken into the district as an effect of upgrading. Another example of importance of complementary programs is micro-credit programs, as, owners may not be in a situation to respond to increases in tenure security through housing investment if they are credit constrained[18, 19] . In these examples, the value of upgrading might depend on complimentary private or public interventions as, housing materials banks or micro-credit programs.
Difficulties of assessment	Good evaluation must detect not only the outcome of the program but also the exact substances of failure or success[20].
Recommendations	To do so, impact evaluations need to take care of all possibly vital complimentary facilities that would increase the value of the intervention. This is important both for evaluating benefits (as in the case of roads), and for identifying columns to project success.

	Such analysis can generally be done prospectively, based on careful consideration of estimated results[21]. However, assessment of these channels must then be merged into information collection efforts and participants' subjective assessment of project success.

3.2 Sustainability:

Importance of sustainability	Pre-project cost-benefit analyses depend on long-time cost recovery or welfare benefits.
Difficulties of assessment	When evaluations occur shortly after those projects are finished, it is very problematic to debate long-term effects.
Recommendations	- Evaluators should follow interventions at regular intervals at least five years after project completion. - To minimize costs of evaluation, evaluators should follow project participants one year and five years after project completion. - A good method to evaluate sustainability by making an analysis of project participation and implementation through variations in political rules and financial shockwaves, which can offer vital data about the probability those interventions, will be sustainable in the future[9].

3.3 Administrative costs:

Importance of administrative costs	Upgrading projects was unsuccessful several times in the past because unexpected charges of running the program. For example, a large materials loan program in Zimbabwe met administrative problems and costs, the loans had been kept small to avoid troubling families, but the results were normally inadequate to finance construction[22].
Difficulties of assessment	Many assessments did not succeed to track administrative problems and costs in follow-up studies and monitoring efforts, depending on pre-intervention estimates of management costs to assess the net effect of the program.
Recommendations	To avoid this sort of assessment mistake, and to detect difficulties with implementation that could be avoided in the future (as was the situation in Zimbabwe), evaluators should keep an eye on project implementation charges over time by talking to project managers and administrators through the project sequence.

3.4 Quality control:

Importance of quality control	When local experts are unavailable, construction quality can be assessed indirectly by asking households or community groups to report the amount and types of inputs used in construction, although this does not avoid the potential problem of imperfect recall and intentional misreporting in the case that corrupt project administrators skim project funds by skimping on materials.
Difficulties of assessment	Past assessments have confirmed that it is hard and expensive to control infrastructure quality. This confuses analyses of upgrading completion rates, as

	well as changes in housing values and attributes and investment motivations.
Recommendations	In the past, evaluators have often trusted self-reported information from households and project managers. More correct information could be gathered by engineers and architects to collect complete data on the quality of buildings, infrastructure and public investments.

3.5 Local participation:

Importance of local participation	This is an important characteristic of project design that has not been carefully inspected; it would be highly helpful for many poverty interventions to examine these hypotheses by setting a wisely planned experimental study.
Difficulties of assessment	It is unclear from past studies the extent to which successful project implementation relies on local participation.
Recommendations	Because interventions are most effective when ran by the municipal authority and implemented at the community level through a wide-ranging set of agencies including community-based organizations, NGO's, and UN agencies such as UNICEF and Habitat Evaluators can potentially shed light on that by designing evaluations that randomize the level of community participation and integration.

4 Conclusions

All programs should be subject to a "process evaluation" – that is, resource flows should be monitored, and there should be a very basic attempt to gather information on what did and did not go well with the program, in an attempt to learn lessons which can be applied to future work.

When we express slum upgrading evaluations in particular, more attention should be given to the wide range of results that can be monitored, particularly as there are many vital questions which past studies of slum upgrading projects have not been able to answer (as sustainability).

Subject's specific to urban situations, public goods, and implementation must be taken into consideration when planning evaluations of slum upgrading projects.

5 References

[1] E. L. Glaeser, "Distort property values and urban growth," Harvard University2011.
[2] I. Imparato and J. Ruster, "Slum upgrading and participation: Lessons from Latin America," World Bank, Washington, D.C2003.
[3] *impact evaluation for slum upgrading interventions*. harvard: harvard university, 2008.

[4] Skip Krige, Michael Schur, and G. Sippel, "The identification of towns in the Free State for an urban upgrading and development programme: a proposed method for consideration," *Development Southern Africa* vol. 15, pp. 361-377, 1998.

[5] K. Coit, "Housing policy and slum upgrading in Ho Chi Minh City," *Habitat International* vol. 22, pp. 273-280, 1998.

[6] O. Dundar, "Models of urban transformation: Informal housing in Ankara," *Cities,* vol. 18, pp. 391-401, 2001.

[7] H. Werlin, "The Slum Upgrading Myth," *Urban Studies,* vol. 36, pp. 1523-1534, 1999.

[8] G. D. Verma, "Indore's Habitat Improvement Project: Success or failure?," *Habitat International* vol. 24, pp. 91-117, 2000.

[9] J. Lu, "Beijing's Old and Dilapidated Housing Renewal," *Cities,* vol. 14, pp. 59-69, 1997.

[10] Abramson and D. Benjamin, "Marketization and Institutions in Chinese Inner-city Redevelopment: a Commentary of Lu Junhua's Beijing's Old and Dilapidated Housing Renewal," *Cities,* vol. 14, pp. 71-75, 1997.

[11] V. Mukhija, "Upgrading Housing Settlements in Developing Countries: The Impact of Existing Physical Conditions," *Cities,* vol. 18, pp. 213-222, 2001.

[12] V. Mukhija, "Enabling Slum Redevelopment in Mumbai: Policy Paradox in Practice," *Housing Studies* vol. 16, pp. 791-806, 2001.

[13] L.W.Yun and K Yusof, "Services for urban poor families in Kuala Lumpur, Malaysia: A case study," *Child Welfare,* vol. 70, pp. 293-292, 1991.

[14] Mudit Kapoor, Somik V.Lall, Matthias K.A.Lundberg, and Z. Shalizi., "Location and Welfare in Cities: Impact of Policy Interventions on the Urban Poor," Washington, D.C2004.

[15] J. W. Schoorl, J. Van der Linden, and K. S. Yap, "Between Basti Dwellers and Bureaucrats: Lessons in Squatter Settlement Upgrading in Karachi," 1983.

[16] OED, "Kenya - Development of Housing, Water Supply and Sanitation in Nairobi," World Bank, Washington, D.C1996.

[17] Michael Bamberger, Bish Sanyal, and N. Valverde, "Evaluation of Sites and Services Projects: The Experience from Lusaka, Zambia," Washington, DC1982.

[18] Douglas H Keare and S. Parris, "Evaluation of Shelter Programs for the Urban Poor: Principal Findings," World Bank, Washington, D.C. 1982.

[19] Douglas H. Keare and S. Parris, "Evaluation of Shelter Programs for the Urban Poor: Principal Findings," Washington, D.C1982.

[20] Abdul Quader Miah and K. E. Weber, "Feasible slum upgrading for Dhaka," *Habitat International* vol. 14, pp. 145-160, 1990.

[21] A. Kumar, "Delivery and management of basic services to the urban poor: the role of the urban basic services, Delhi," *Community Development Journal,* vol. 26, pp. 50–60, 1991.

[22] H. U. Bijlani, "Strategies for urban shelter: The improvement of slums, squatter settlements and sites-and-services," *Habitat International,* vol. 12, pp. 45-53, 1988.

Military Technical College
Kobry El-Kobbah
Cairo, Egypt

13th International Conference on Civil and Architecture Engineering ICCAE-13-2020

Managing of city transition toward sustainable Eco urbanism (Proposed paradigm).

Mohamad F . Fayad[1], Maged Elmahdy[2], Sherif Sabri [3], A. Alrahim Kasem [2]

[1] Doctoral Researcher, Urban planning Department, Engineering Faculty of Al-Azhar University, Cairo, Egypt.
[2] Professor, Urban planning Department, Engineering Faculty of Al-Azhar University, Cairo, Egypt
[3] Professor Emeritus, Urban planning Department, Engineering Faculty of Al-Azhar University, Cairo, Egypt.
Corresponding author: engmohamad.farag@gmail.com

Abstract: The research clarifies transition strategy towards sustainable Eco urbanism, and what spinoff the reliable Universal Criteria, in addition to concept the managing of sustainability transition in cities.

The aim of the research is to present and elaborate a proposed model for managing the sustainable transformation of Egyptian cities, so that they are suitable to deal with and apply to most existing and new urban communities, with a view to transforming them into sustainable environmental cities and communities, by working to develop a future vision for the sustainable image that the city will reach, dependent on Scientific foundations and an understanding of the problems these regions suffer from, to explain sustainability strategies in light of the general data for these areas and the current and future needs of the population and the available capabilities.

The proposed paradigm consists of ten consecutive stages of operations and is related to each other, where the subsequent stage is based on the data of the previous stage, and the success of the previous stage has a positive impact on the success of the subsequent stage and vice versa.

Keywords: transition, transformation, Eco urbanism, sustainable city,

1. Introduction

The transformations of cities are among the urban, social and economic phenomena that followed the first industrial revolution, and therefore the transformation is considered as one of the main features of the twenty-first century, and in Egypt the absence of the institutional dimension and the absence of a regulatory framework for urban transitions taking place in Egyptian cities led to a loss of the ability to manage and plan the transformation and control processes by directing them towards sustainability To achieve the requirements of sustainable eco urbanism and renewal towards sustainability.
Many countries of the world have started setting principles and standards for environmental city planning as an inevitable result of sustainable transformation, and thus the principles of sustainable eco urbanism came with ideas capable of overcoming the negatives of cities and working on compatibility between the environment and Urbanism in addition to improving environmental performance through a set of principles including (sustainable transport, water conservation, energy conservation, innovation and renewal, emissions reduction, pollution reduction, waste management, social and cultural progress, economic development).

2. Sustainable Eco urbanism

Many planners consider that environmental urbanism is the urbanism that is based on achieving the principles of living while taking into account the environmental impacts [1].while the planner Deng believes that environmental urbanism emphasizes the importance of the ecological and environmental aspects of urbanism, and is based on the complex interactive effects and relationships between different economic activities in those urban and rural societies, their ecosystem, and their local environment [2], then the urban environmental expansion aims to societal development to sustain urban and rural development, to efficiently consume resources, and to minimize environmental impacts in the context of a dynamic and interactive process.
Fiona Harvey has argued in his Green Vision that for Urbanism to be sustainable, it must have the following elements: [3]. (See Figure 1)
•An urban ecosystem which is ecologically sound
•Reduction of ecological footprint, waste management and its utilization

•Achieving environmental sustainability through reduced greenhouse gas emissions, and Maintaining high air and water quality standards.
•Actors whose drive the eco-Urbanism must be part of sustainable development management

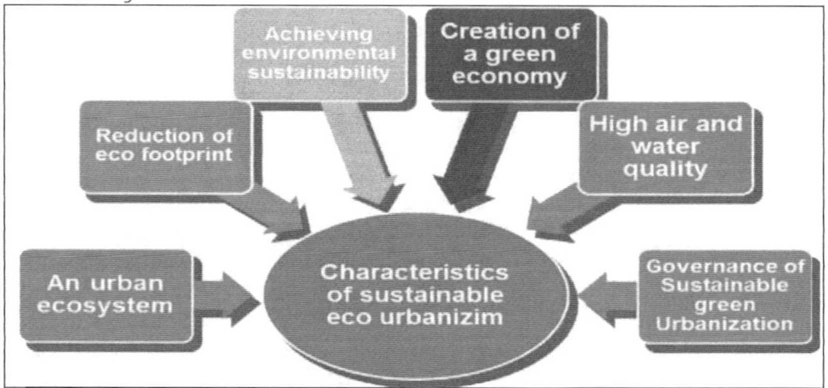

Figure 1. Characteristics of sustainable eco urbanism. [3]

2.1. Basics of the Sustainable Eco Urbanism

The principles of eco urbanism depend on the triple-zero systems. These are zero petroleum product vitality utilize, zero waste and zero discharges particularly went for low-to-no-carbon outflows. [4] (See Figure 2).
1) Climate and Context: Based on climatic condition prior to selected city, every sustainable design project needs to maintain a complexity within eco-system or neighborhood layout.
2) Renewable Energy for Zero CO2 Emissions: Transform city districts into local power stations of renewable energy sources.
3) Zero Waste City: Cities should adopt zero-waste urban planning in line with the manufacturing of metals, glass, plastics into new products.
4) Water: Cities can be used as a water catchment area by educating the inhabitants in water efficiency, promoting rainwater collection.
5) Landscape, Gardens and Biodiversity: Introduce inner-city gardens, urban agriculture and green roofs to maximize the resilience of the eco-system.
6) Sustainable Transport and Good Public Space: Compact and Poly-Centric Cities: An integration of non-motorized transport, such as, cycling or walking and bi-cycle or pedestrian-friendly environment
7) Local and Sustainable Materials with Less Embodied Energy: City construction by using regional, local materials with less embodied energy and applying pre-fabricated modular systems

8) Density and Retrofitting of Existing Districts: The city is with retrofitted districts, urban infill, and densification/intensification strategies for existing neighborhoods.
9) Green Buildings and Districts, Using Passive Design Principles: The city, here, applies deep green building design strategies and offers solar access for all new buildings.
10) Livability, Healthy Communities and Mixed-Use Programs: The prime concern of the city is for affordable housing, mixed-use programs and a healthy community.
11) Local Food and Short Supply Chains: High food security and urban agriculture by introducing 'eat local' and 'slow food' initiatives.
12) Cultural Heritage, Identity and Sense of Place: A sustainable city with high air quality.
13) Urban Governance, Leadership and Best Practices: The city applies best practice for good urban governance through combined management and approaches.
14) Education, Research and Knowledge: The city includes technical training and up-skilling, research, exchange of experiences and knowledge dissemination for all in sustainable urban development.
15) Strategies for cities in developing countries: Particular sustainability strategies are needed for cities in developing countries, such as train local people to empower communities, creating new jobs.

Figure 2. Basics of sustainable eco urbanism. [4]

3. Governance of urban sustainability transitions

The process of change during a transition is highly non-linear within which slow change is followed by rapid change when things reinforce each other which is again followed by further change in the new equilibrium [5].

The nature and speed of change differ in each of the transition stages [6]: predevelopment phase, take-off phase, acceleration phase and stabilization phase. Figures 3 show the Theory of sustainability transitions.

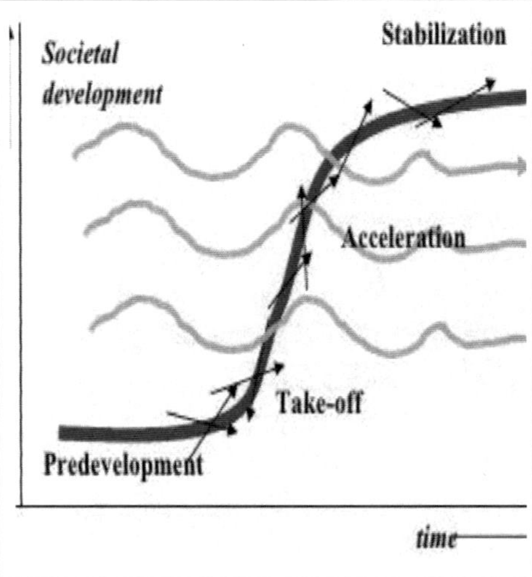

Figure 3. Theory of sustainability transitions. [6]

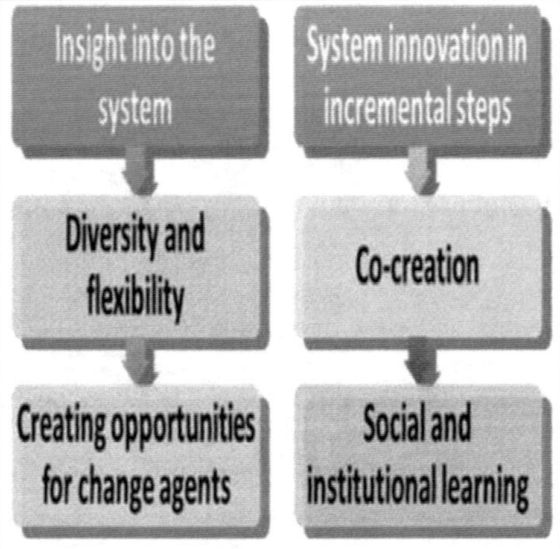

Figure 4. Principles of transformation management. [7]

3.1 Transformation management

Next to the aim of analyzing and understanding transitions, part of the field of transition studies is about how actors (can) influence transitions - their governance.

One of these governance approaches is Transition management, which is based on insights from complex systems, governance and sociological theories.

Transition management aims at influencing the direction and pace of societal change dynamics in the context of contributing to sustainability.

System innovation in incremental steps: Apart from system improvements and optimizations, aim for system innovation. This includes building

IOP Conf. Series: Materials Science and Engineering **974** (2020) 012033 doi:10.1088/1757-899X/974/1/012033

analytical capacity for transitions this includes building networking capacity for transitions.
The transition arena is a temporary setting that provides an informal and well-structured space to a small group of change agents from diverse backgrounds (businesses, government, and citizens).
The group engages in a series of meetings, jointly elaborates a transition challenge, drafts a long-term vision, and develops transition pathways to realize this vision.
Figures 4 clarifies the principles of transformation management which consists of these steps :(insight into the system - system innovation - diversity and flexibility -co-creation -creating opportunities- institutional learning). [7]

3.2 The experiment of applying Transformation management in Carnisse, Rotterdam, Netherlands

Transformation management was applied in Carnisse in the context of decreasing public service provisions, closing of public facilities, high unemployment and crumbling social cohesion to facilitate the self-organization of inhabitants to address persistent sustainability problems. While urban neighborhoods mark opportunities for self-organization – through for example emotional connectedness – the political playing field can provide support, or restrain it. It provided a new perspective for the development of that community, increased social capital, supported reflexivity on roles and own behaviors and revealed opportunities for self-organization. [8]
Therefore, the overall process design was partly done in a participatory and interactive meeting in November 2011 with five inhabitants and professionals in Carnisse who were motivated to think along about the future of the neighborhood. During this meeting, the participants and researchers defined the overall aim of the transformation management process as "supporting and stimulating inhabitants to shape and take ownership of the future of their neighborhood and formulate desired (government) activities" [9], [23]. The main focus however was on discussing the methodology developed as part of the In Context project and its contextualization for Carnisse. [23]
The former was open to invited participants only and focused on developing a shared problem description, futures images and possible pathways. The latter was open to everyone and focused on experimenting, mainly the re-opening of a closed-down community center. [10]
Figures 5 and 6 show the adapted process design of Transition management and adapted overview of community arena process in Carnisse.

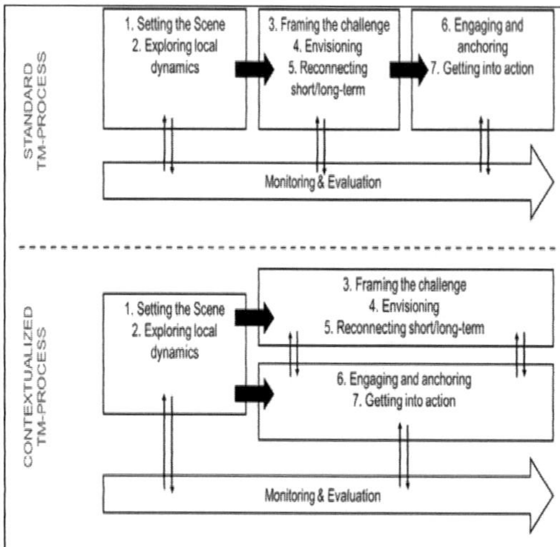

Figure 5. Adapted process design of Transition management In Carnisse. [8]

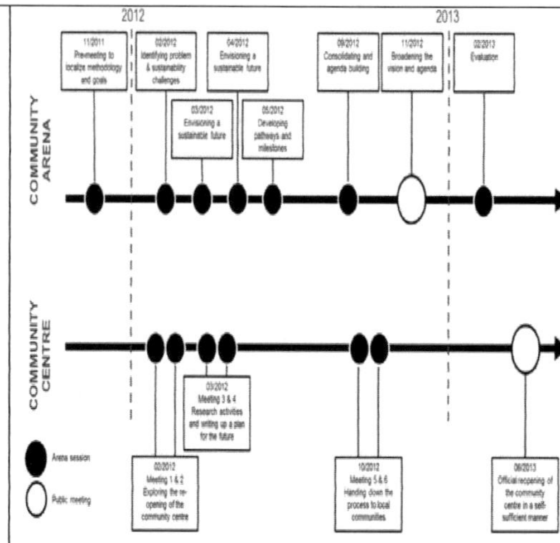

Figure 6. Adapted overview of community arena process in Carnisse. [9]

4. Egyptian cities and sustainability transitions challenges

The population of Egypt is currently estimated at about (100) million people, and the population growth rate of the population is (2.65%), and about (45%) of the population live in urban areas, where the rate of urbanization is (2.2%) and Egypt is classified among (16-20) A country experiencing the highest population growth in the world, and Egypt suffers from a delay in the application of the idea of sustainability at the urban and regional levels, as it suffers from many development problems and is subject to depletion of its natural resources and wealth without planning for future needs. In addition to that, cities in Egypt moved away from the local environmental trends that preceded it, which led to the creation of urbanization that is not suitable for the environment and does not meet the needs of its users.

The rapid change in Egyptian lifestyles has had a negative impact on Egyptian urbanization, and one of the manifestations of this is his exposure to many threats that have a negative impact on the standard of living of the population. Among these threats are the lack of drinking water, sanitation and energy facilities in addition to the spread of slums, and Egypt has witnessed during the last decades an increase It is steady in urban projects, through expansion in the construction of new cities and communities, and others, in order to provide urban services and bridge the accumulated deficit in the residential sectors. [11]

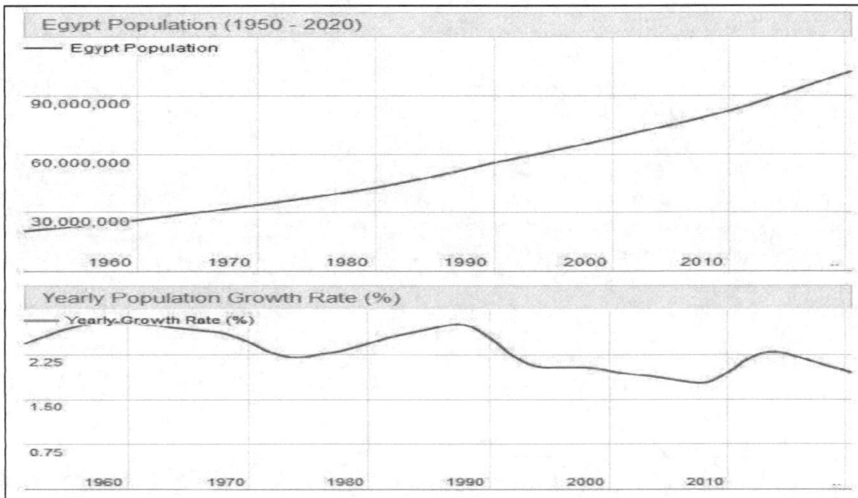

Figure 7: Egypt population (1950 - 2020) and yearly population growth rate [11].

4.1 Sustainability transitions challenges for Egyptian cities

The main challenges facing Egyptian cities transitions into sustainable eco. Urbanism can be grouped into four groups. [12] :(See Figure 8).

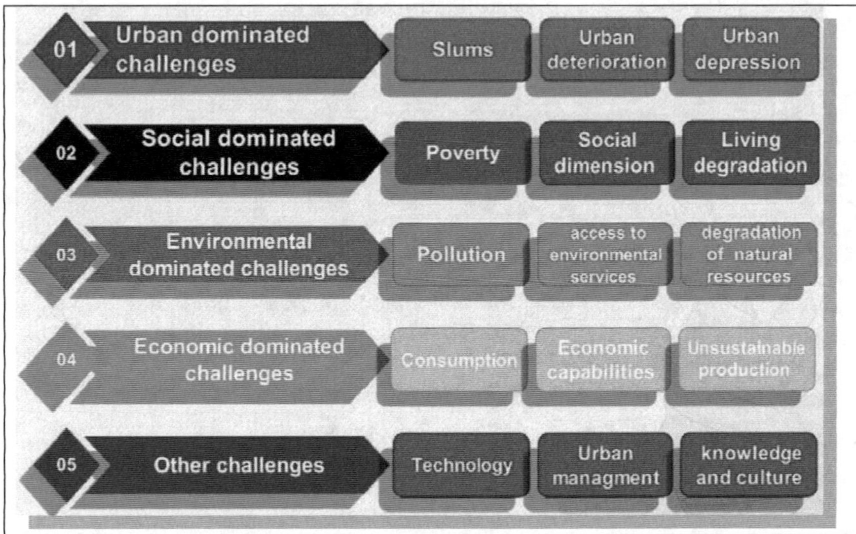

Figure 8: The sustainability transitions challenges of Egyptian cities [12].

<u>A) Urban dominated challenges:</u> It means risks and issues that threaten

the built environment such as:
- The spread of dangerous areas, slums, and encroachment on agricultural lands
- Urban deterioration of the old areas.
- Urban sprawl and unorganized expansion in suburban areas.
- Urban depression represented by the lack of cities in the distinct urban jobs and urban attractions.
- Poor state of infrastructure networks such as networks (roads, streets, water, sewage, etc.)

B) Social dominated challenges:
- Poverty, social conflicts, lack of social justice and increased crime rates population mobility and rural-to-urban migration that overwhelms the ability of cities to provide basic services.
- The absence of a social dimension in the development of cities.
- The deterioration of the standard of living of the population and lack of educational and health quality.

C) Environmental dominated challenges:
- Pollution from industrial production, urban waste and emissions with climate change.
- Lack of access to environmental services and infrastructure (water, waste collection, transportation)
- The degradation of various natural resources.

D) Economic dominated challenges:
- Excessive consumption of resources and energy (the high use of those renewable resources that can only be renewed within limited limits. Limits, and high use of non-renewable resources).
- Development needs constrain economic capabilities and local resources.
- Unsustainable production and consumption patterns.

E) Other challenges:
- Technology issues, this group has the following challenges (using outdated and inappropriate technologies - resettling technology polluting the environment in developing countries - difficulties in transferring modern and clean technology)
- Management and political issues, this group focuses on the following main issues (ineffective institutional and organizational frameworks - not possible legal and legislative frameworks - administrative centralization)
- Awareness issues, knowledge and culture, this group focuses on three prominent challenges: (challenges of lack of awareness and values - challenges of vision and change - challenging the lack of decision-making tools)

5. The proposed paradigm for city transition toward sustainable Eco urbanism.

Managing urban sustainability transitions requires integration with strategic urban planning processes in order to integrate diverse sources of knowledge and perspectives, while linking the links between local urban challenges and broader global, national and regional developments.

In this context, the researcher presents a proposed framework as a new way to deal with the ongoing problems and risks facing cities transitions, the basic argument is that these problems are strongly rooted in the structure of our community systems themselves, which means that marginal changes (i.e. improving procedures or temporary solutions) cannot be effective It will lead to suboptimal results. Therefore, managing sustainable transformation calls for fundamental change that is transformation, in order to address the root causes of persistent problems rather than their symptoms.

5.1 Framework of the proposed paradigm

Governance of urban sustainability transitions seeks to influence the interactions between these activities of different scales by creating soft structural processes, involving different types of actors at each level: (See Figure 9)

1. Strategic level activities

This group represents a vision of the desired future, the focus is on long-term aspirations and goals towards sustainable urban environments, the future vision is discussed collectively and imagined; for example: vision and formulation of long-term goals, including setting collective goals and setting standards.

2. Tactical level activities

The focus is on developing strategies, networks, and coalitions that draw attention to sustainability goals while gaining community support to achieve medium-term goals, and overcoming socio-institutional barriers to innovation / change, so these activities seek to target changes in existing structures.

3. Operational level activities

Governance activities at the operational level focus on short-term and concrete activities - they are part of the continuous flow of daily decisions and procedures, with an emphasis on experiences and application of work programs that can form new structures and procedures that can ultimately lead to system innovations

4. Reflexive level activities

The reflexive level relates, by its nature, to the identification of activities and thus their adaptation at various different levels. It includes activities related to monitoring and evaluating experiences, agenda, vision, policies and processes of change.

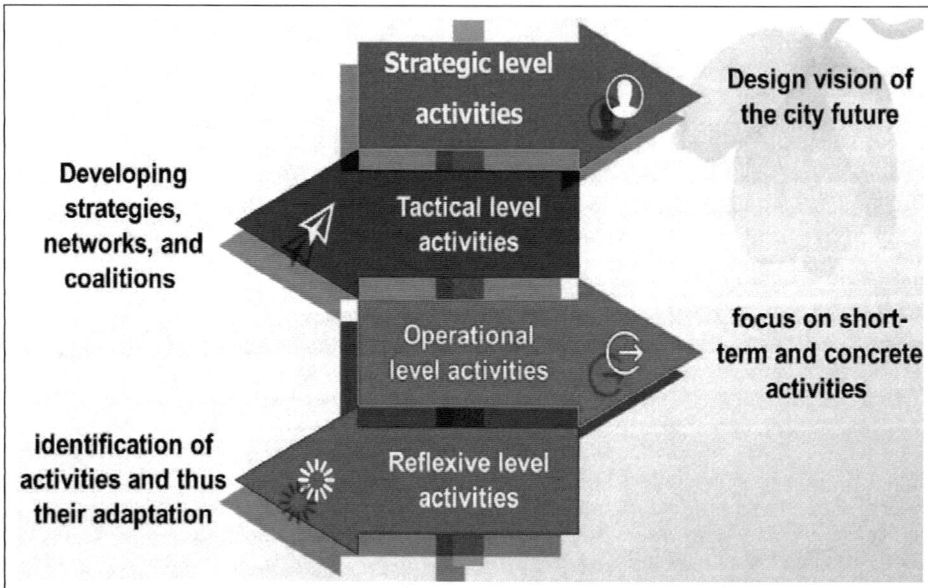

Figure 9: The levels of proposed paradigm [Researcher].

5.2 Process stages of the proposed paradigm

Governance The sustainability transitions management wheel", as shown below, is a simple periodic roadmap explaining the paths and tools available for managing transitions of city structures to more sustainable practices, with a comprehensive vision focused on the local situation, and it is expected that multiple courses will be completed, with a repetition of the set of activities Mentioned above again and again. The completion of each session is referred to here as one "cycle" of the wheel, and the first session of the wheel may be directed to overcoming urban, social or environmental risks, and the next cycle to confront the risks and other issues that impede sustainability

1) First stage. Preparation and identification of stakeholders

The main activities at this stage include identifying the organizational and coordination team (the transition team) and designing the process. The main results for this phase include the formation of the transformation team, process plan and support network .Scene identification is the important first stage of a transformation management process: it is the work that precedes a participatory process that develops in the following stages. As in any project, this is a critical stage, as it can largely determine the remainder of the process. Depending on how the transformation team is formed, the process tailored to a particular situation, and the support received from stakeholders (such as city management), the sustainable transformation management process can take root or remain marginal.

2) Second stage. Diagnose the current situation of the city
The main activities at this stage include studies of the current state of the city (or system), analysis of representatives, identification and selection of prominent candidates to participate in this process. Key findings include analyzing the status quo and coming up with an insight into the key issues and tensions to focus on, identifying and categorizing actors and shaping the scene, once the transformation team is formed and equipped with space and resources, it is ready to start exploring local dynamics. In urban areas, this could mean a city or neighborhood. There are three main aspects to this stage of the transformation management process: analyzing the current situation and actors and bringing together a group of active individuals ("pioneers" or "agents of change") in the transformation arena.
3) Third stage. Analyzing problems and issues
The main activity of this stage is the general analysis of the city or urban area with the structuring of participatory problems and then an analysis of the problems, risks and issues facing the city's shift towards sustainability. Key outcomes for this phase include perceptions of common problems and topics of change. The first issue on the agenda is the formulation of the "problems" that the group and process will work on. Usually a common understanding of the challenges of transformation is initiated from the discussion of an analysis of the current conditions presented to the city as inputs to the process by the transformation team and based on research and interviews, this analysis will be reconfigured until the group reaches a common perception of problem-solving
4) Forth stage. Preparation of the city future vision
The main activity of this stage is building a participatory vision, the main result is a shared vision, the group begins to prepare a vision for sustainability - a sustainable contextual understanding of the city transformation, through which the basic principles of long-term development are clarified, and after intense discussions in which views on the transformation are exchanged For a sustainable city, the group accepts a number of guiding principles of sustainability - which can be based on the transformation challenges set. These principles will form the basis of the future vision of the group. Elements that will be used to create these visions include the following (exchanging views on the future - formulating guidelines for sustainability - creating an integrated vision)
5) Fifth stage. Building city strategic plans
The main activities for this phase include participatory backcasting, defining transformation pathways and formulating the transformation agenda and its specific actions. The main results of this stage include analysis of backward prediction and transformation paths, as well as agenda measures and formation of possible experiences. The challenge for this stage is to reconnect this long-term vision in the short term through more ideas on how to get there, including the method of backcasting, and then the tracks are merged into a transitional agenda, as the second result of this stage. It must be endorsed by all

IOP Conf. Series: Materials Science and Engineering **974** (2020) 012033 doi:10.1088/1757-899X/974/1/012033

participants in the field of transformation, because it aims to provide a common approach.

6) Sixth stage. Implementation of the city transition strategy

The main activity of this stage is the actual short-term operational practice and concrete activities. Instead of projects,

The management of transformation phases at this level focuses on creating a set of related transformation experiences that complement and strengthen each other, and contribute to the goal of ability to be scalable, potentially scalable, and important and measurable. The main activity at this stage involves the implementation of experiences, policies and transformation projects. Key outcomes for this phase include learning and implementing the transformations agenda.

7) Seventh stage. Involve additional parties and develop a marketing campaign

We should be aware of the benefits of engaging other parties (groups from civil society, academics, commercial and economic institutions, and the media), and who and at what stage of the process should be invited to join them. They can participate in multiple ways, such as participation in meetings, conferences and initiatives.

The main activities for this phase include disseminating the vision and paths, the transformations agenda, and expanding the network. Key outcomes for this phase include broader public awareness and expanded participation in the transformational agenda, the network of agents of change, spin-offs and new groups.

8) Eighth stage. Monitoring and documenting works and data

Process documents capture and track what happens during the change process and how it happened. Good process documents enable stakeholders to articulate the reason for the changes, analyze them, and organize and publish the results. Meeting minutes, photographs, audios, videos and other ways to capture information.

With regard to capacity building, training courses and workshops that enable middle management / clients / community, etc., will encourage change of non-sustainable thinking and ensure engagement with new sustainable technologies and solutions.

9) Ninth stage. Evaluation, learning and feedback

The transformation management aims to facilitate the process of research and community learning where continuous actions are taken by a group of actors as a starting point for building new collaborative transformation networks. As such, the Transformation Department opens up a method for introducing and experimenting with alternative social relationships, such as between local governments and citizens, or between citizens and companies.

Each stage of the sustainable transformation management cycle must be evaluated to see the strengths and weaknesses of the various activities and to measure the degree of progress of transformation practices.

10) Tenth stage. Modification and prepare for the second round of the transition path

The cycle of managing sustainability transitions is almost complete at this stage, but the final journey towards the desired future has not yet ended, and the first round may be directed to overcoming the risks and issues prevailing in the city (urban - social - environmental) then the next round of transformation begins to overcome the other risks in the city (As previously explained), then the gaps will be identified and the vision adjusted if necessary. New insights may be needed, and new actors may need to be found willing to become pioneers of change in the future.

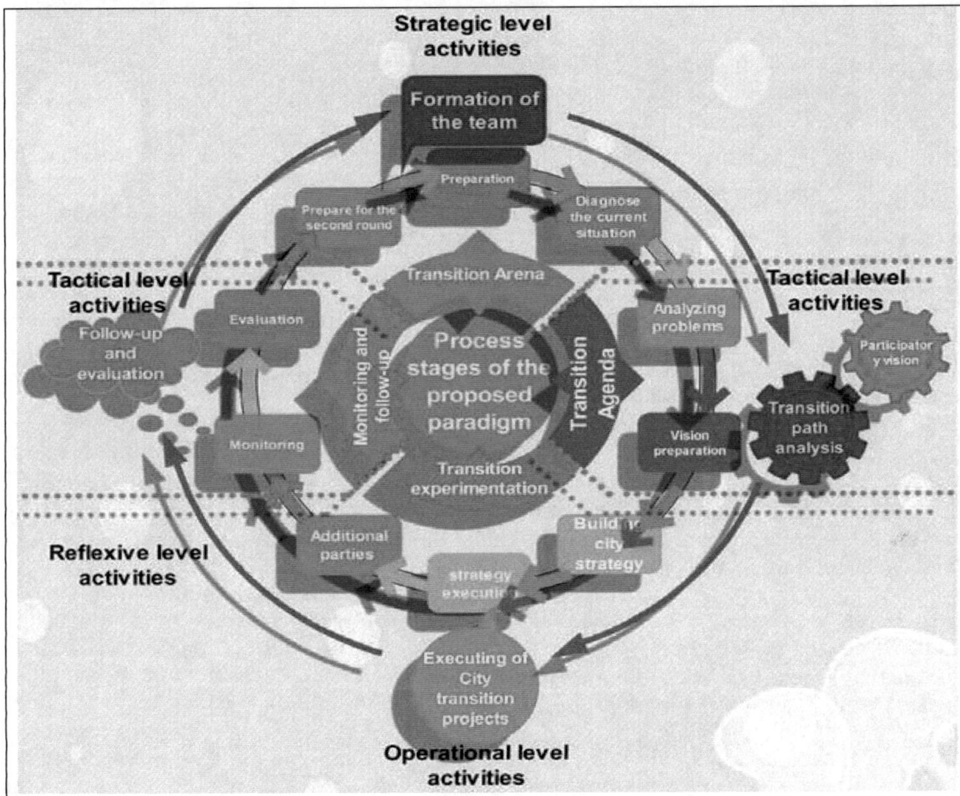

Figure 10: The cycle and stages of the proposed paradigm [Researcher].

6. Conclusion

1) Managing city transitions can be applied as an approach to transforming or moving the city and society, from the current state to the desired sustainable state. It is an organizational process that aims to assist actors, stakeholders, involved in the transformation of social technology systems with a view to managing balances between the human condition and its built and natural environment.

2) In the strict sense of the term, no city has yet reached high levels

IOP Conf. Series: Materials Science and Engineering **974** (2020) 012033 doi:10.1088/1757-899X/974/1/012033

of sustainability. Cities have played critical roles in the direction of rapid global transformation since the mid-1980s, especially those in middle- and high-income countries. Cities will follow traditional scenarios, which assume that current trends play non-stop significantly in the evolution of institutions, ecosystems, and human values. So that comprehensive and coordinated government actions are taken to achieve sustainability.

3) The Department of Sustainable Transformation seeks to influence the interactions between these activities of different domains by establishing facilitated structural processes, involving different types of actors at each level, which requires a range of diverse skills and competencies. They are all important and therefore the levels do not represent any hierarchy

4) The proposed framework for sustainability transition managing links a number of tools to each level of governance. The periodic nature of the model implies that activities at the strategic level are followed by tactical and operational tools and the course is closed with reflex tools. However, the cycle must be understood as iterative; activities can be initiated at every level of governance, and therefore at the operational level rather than the strategic level, for example, and may operate in parallel. Thus, activities and tools interact with each other.

5) The proposed "sustainable urban transformation management wheel" is a simple periodic roadmap that outlines the pathways and tools available for managing transitions of city structures to more sustainable practices, with a comprehensive vision focused on the local situation, and multiple courses are expected to be completed, with the set of activities repeatedly repeated Repeatedly. The completion of each session is referred to here as one "cycle" of the wheel, and the first session of the wheel may be directed to overcoming urban, social or environmental risks, and the next cycle to confront the risks and other issues that impede sustainability.

6) The researcher applied a holistic approach to define the processes of urban transformation through taking into account the urban, social, environmental and economic dimensions, and accordingly four groups of prevailing risks and issues (or threats or obstacles) were identified, which confront the urban transitions towards sustainability which are (urban issues - social issues - Environmental Issues - Economic Issues - Technology Issues - Institutional Frameworks

7) Sustainability challenges and issues may differ in many countries of the world. In developing countries, the economic recession is the most important constraint to sustainable development. In developed countries, cities suffer as a result of high overcapacity or mass consumption (of resources and energy), and comprehensive disposal that also depends on production. Huge, which increases the amount of inputs and outputs, usually far exceeding the environmental capacity of the city, and due to the exorbitant results or possible reduction of the quality of life, governments or individuals have local incentives to improve and introduce sustainable patterns of production and consumption

8) Requirements for preparing and implementing the management of Egyptian cities transformation into sustainable cities :include the flowing:

- Building an institutional framework to manage urban sustainability transitions at the national level
- Prepare a strategy for existing and new cities
- Provide the necessary financial resources for implementation
- Information integration and capacity support
- Institutional support and capacity building
- Support capabilities to implement multilateral environmental agreements technical and material support.

IOP Conf. Series: Materials Science and Engineering **974** (2020) 012033 doi:10.1088/1757-899X/974/1/012033

7. Conclusion

1) Jing yuan Li, China's Eco-city Construction, Springer, 2015.
2) Wang, Z.; Deng, X.; Wang, P.; Chen, J., Ecological intercorrelation in urban-rural development: An eco-city of China, 2016.
3) Fiona Harvey, Green vision, the search for the ideal eco-city, 2011.
4) Steffen Lehmann, what is Eco urbanism? Holistic Principles to Transform Cities for Sustainability. 2010.
5) Grin, J.; Rotmans, J.; Schot, J.W. Transitions to Sustainable Development: New Directions in the Study of Long Term Transformative Change; Routledge: New York, NY, USA, 2010.
6) ˙Idil Gaziulusoy, Elif Erdo˘gan Öztekin, Design for Sustainability Transitions: Attitudes and Future Directions Attitudes and Future Directions, Sustainable Design Research Group, Department of Design, School of Arts, Design and Architecture,Aalto University, 2015, Espoo, Finland
7) Julia Wittmayer, Chris Roorda, Frank van Steenbergen, Governing Urban Sustainability Transitions – Inspiring examples, Guidance manual - collaborative evaluation version, Dutch Research Institute for Transitions, Erasmus University Rotterdam. 2014.
8) Wittmayer JM, Mock M, van Steenbergen F, Baasch S, Omann I, Schäpke N (, Taking stock – three years of addressing societal challenges on community level through action research Pilot specific synthesis report.2013.
9) Wittmayer JM, Roorda C, van Steenbergen F, Governing urban sustainability transitions – inspiring examples. DRIFT, Rotterdam, 2014.
10) Schäpke N, Omann I, Wittmayer JM, van Steenbergen F, Mock M, Linking transitions to sustainability: a study of the societal effects of transition management, 2017.
11) The current population of Egypt, [Online], Available online https://www.worldometers.info/world-population/egypt-population/ (accessed on 12/1/2020.).
12) Mohamad Farrag Fayad, An analytical study of land uses for Egyptian city core, MSc. in urban planning engineering, Faculty of Engineering, Al Azhar University, 2014.
13) Sara Hughes, Eric K. Chu, Susan G. Mason,Innovations in Multi-Level Governance, Climate Change in Cities, ,2018 Springer
14) Suzanne Maas, Karen Fortuin, Niki FrantzeskakiChris Roorda, Starting Up Transition Management, A Closer View on the Systems Analysis and How It Initiated ,Co-creating Sustainable Urban Futures Springer , 2018, pp 159-185.
15) Matthias Ruth, Shobhakar Dhakal, Creating Low Carbon Cities, 2016. Springer
16) Thomas Sauer, Susanne Elsen, Cristina Garzillo , Social Innovation for Europe's Urban Sustainability Cities in Transition ,Routledge, 2016.
17) Adrien Krauz, Transition Management in Montreuil: Towards Perspectives of Hybridisation Between, 'Top-Down' and 'Bottom-Up' Transitions, Governance of Urban Sustainability Transitions, 2017, pp 133-150.
18) Allison Bridges, The role of institutions in sustainable urban governance, natural resources forum 40, 2016.
19) Steffen Lehmann, Low Carbon Cities, Transforming Urban Systems, Earthscan series on sustainable design, Routledge, 2014.
20) Albiez, Marius, Banse, Gerhard, Lindeman, Kenyon C, Designing Sustainable Urban Futures, Concepts and Practices from Different Countries, KIT Scientific Publishing, 2016.
21) Jeroen van der Heijden, Governance for Urban Sustainability and Resilience, Responding to Climate Change and Relevance of the Built Environment, Edward Elgar Publishing, 2014.
22) Braden Ryan Kay, Developing and Testing Transition Strategies for Urban Sustainability, Case Studies in Transition Research in Phoenix, Arizona, PHD Dissertation Presented in Partial Fulfillment, Arizona State University, 2012.
23) Wittmayer, J., et al., 2011b. Pilot projects getting started. Year 1 status report. Deliverable 4.2. In Context, 2011.

Military Technical College
Kobry El-Kobbah,
Cairo, Egypt

13th International Conference on Civil and Architecture Engineering ICCAE-13-2020

Static and Dynamic Performance of Two Stage Concrete

Mohamed A.E.M. Ali

Department of Civil Engineering, Military Technical College, Cairo, Egypt

Abstract. Concrete is arguably the most important construction material in the world. In the last decades, it has been reported that infrastructures faced extreme static and dynamic loads which caused structural catastrophic disasters [1]. The increase usage of conventional concrete in engineering industries led to high consumption of mineral resources which caused several environmental damages related to reduction in the rocket layers [2], besides, the economical damages due to the high demand of natural resources materials which rises the cost of construction [3]. Therefore, the utilization of recycled waste rubber (steel wires) in concrete can be a sustainable solution for waste construction developments [4]. Recently, two stage concrete (TSC) (non-traditional concrete) was introduced. Placing of coarse aggregate followed by injecting the cementitious materials among the aggregate layers in the formwork are the basic steps of producing TSC. Evaluating the static and dynamic performance of TSC were the objectives of this study. Cement, aggregates, water, recycled steel wires of rubber tires were utilized in manufacturing TSC specimens. The compressive characteristic and impact resistance of TSC test specimens were evaluated. Results showed that steel wires improved the dynamic behaviour of TSC, while a slight to no effect was observed on its compressive characteristics. This study demonstrates the potential to engineer sustainable TSC mixtures.

Keywords: Two-stage concrete; impact; sustainability

1. Introduction

The dynamic behaviour of concrete is considered one of the most important priorities in infrastructure construction [5]. Due to its brittleness, concrete can be damaged by the impact of sudden heavy loads [6]. Many research work have demonstrated the significant enhancements in the dynamic performance of metallic fibre reinforced concretes [7]. The incorporation of metallic fibres in concrete production also enhances various properties such as; fatigue, toughness, and energy absorption capability [8]. This behaviour is attributed to fibres capability to arrest crack initiation through concrete elements [7, 9]. Furthermore, incorporation of waste rubbers in concrete production have provided very promising characteristics to concrete under static and dynamic loading [10]. Many research work have demonstrated that the rubberized concrete performed efficiently under static and dynamic loading by partial replacement of sand and cement with rubber by up to 20% by volume fraction [11].

Recently, two stage concrete (TSC) was introduced. The TSC can be made by direct injecting of cementitious grout to fill the voids among the aggregate layers in the formwork. A sustainable rubberized TSC can be produced by adding waste truck tire rubber as a partial replacement of sand or aggregate. This can be considered as one of the solutions to reduce time and mineral resources consumption of construction. On the other hand, the ease of gravitational placement of TSC in formworks differs from pumping process in traditional concrete. This can reduce the probability of honeycombing and segregation also [12].

Although, various research work had been conducted in concrete industries especially on the static behaviour of TSC [12, 13]. However, there is lack of information on the impact resistance of TSC. Therefore, the impact resistance of TSC mixtures incorporating randomly dispersed recycled steel wires as fibres was investigated in this study. The study aims to produce a sustainable cheap concrete mixtures with impact resistance, to present novel alternative to conventional concrete.

2. Experiments

2.1 Proportions of Utilized Materials

The TSC mixtures was produced using ordinary portland cement (OPC) Type I with a specific gravity of 371 m^2/kg, in accordance with ASTM C150 standards [14]. Also, traditional sand was utilized. Furthermore, the TSC mixtures was prepared using 20-45 mm recycled steel wires, with 0.2 mm diameter, and different volume fractions of 0%, 0.5%, 1%, and 1.5%. To control the workability, a water reducing admixture (WRA) as per ASTM C494 specifications [15] was added. Table 1 shows the different TSC mixture proportions.

Table 1. Proportions of Utilized Materials.

ID	OPC	Fine sand	w/c	WRA	Recycled steel wires (%V_f)
TSC0	1.00	1.00	0.45	0.0004	0.00
TSC0.5	1.00	1.00	0.45	0.0004	0.50
TSC1	1.00	1.00	0.45	0.0004	1.00
TSC1.5	1.00	1.00	0.45	0.0004	1.50

2.2 Mixture Preparation, Casting and Curing

A cylindrical mold of 150 mm x 300 mm was prepared from each TSC mixture. First, the aggregates were placed at the bottom of the cylinders followed by the distribution of the recycled steel wires, and then, another layer of aggregates as shown in Fig. 1.

Fig. (1) TSC placing technique

A Cement Mixer was used to dry mix cement and fine sand. Thereafter, a homogeneous mixture was achieved by adding the mixing water and WRA gradually to the dry mixture for three minutes. Finally, the cementitious mixture was injected into the formworks. The formworks were removed after 24 hours then left in clean water up to testing age of 28 days.

2.3 Test Procedures

2.3.1 Compressive strength

Th compressive strength was determined using a 2000 kN compression machine. Three cylindrical speceimens of 150 mm x 300 mm were placed and tested at 28 d following ASTM C39 guidelines [16].

2.3.2 Tensile strength

To evaluate the tensile properties of the different TSC mixtures, three cylindrical specimens of 150 mm x 300 mm were tested at 28 d following ASTM C496 standards [17]. The splitting tensile strength was obtained using Eq. 1:

$$T = \frac{2*P}{\pi*l*d} \qquad \textbf{(Eq. 1)}$$

Where *T* and *P* are the tensile strength and applied load in MPa and Newton, respectively, *l* and *d* are the length and diameter of tested cylinders in mm.

2.3.3 Impact test

Furthermore, impact test was applied on the different TSC specimens at 28 d following the American Concrete Institute (ACI) Committee 544 [18] guidelines as displayed in **Fig. 2**. The impact load was applied on the different TSC specimen in which each hit was able to produce 20.167 J. This is due to the free fall of a 45 N steel weight falls from 457 mm height. The number of drops induced up to visibility of crack propagation (N1), and fracture (N2) were recorded, respectively. The impact energy was evaluated using Eq. 2 as per ASTM D5628 guidelines [19]:

$$I = N_i . h . w . f \qquad \textbf{(Eq. 2)}$$

Where the impact energy in Joule is represented by *I*, the number of hits, the falling height in mm, and the steel hammer mass in kg are represented by *Ni*, *h*, and *w*, respectively, and *f* is a 9.806 x 10^{-3} constant.

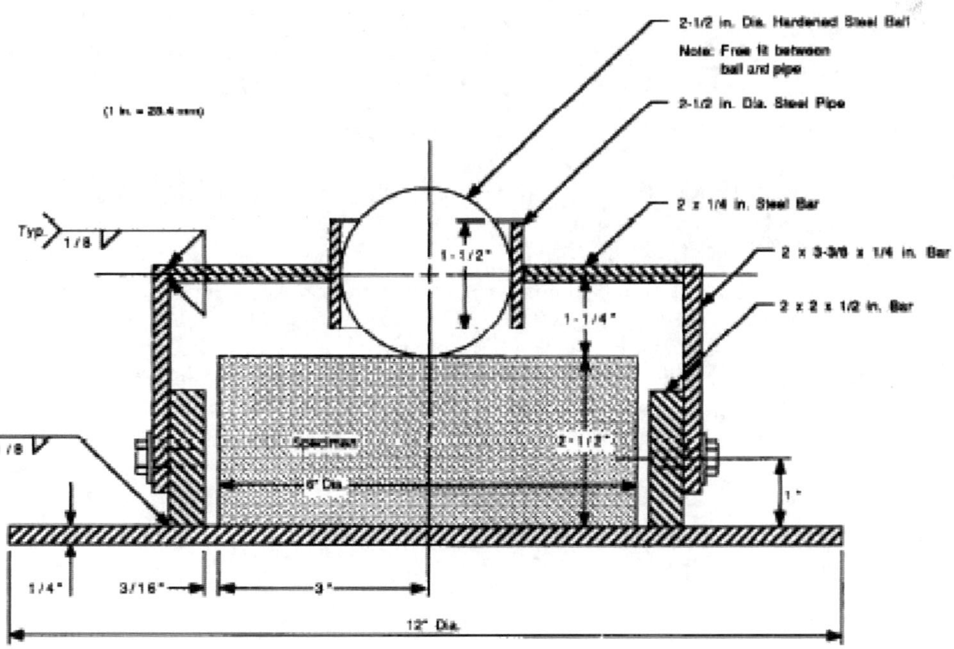

Fig. (2) Test setup of drop weight impact test [18]

IOP Conf. Series: Materials Science and Engineering **974** (2020) 012034 doi:10.1088/1757-899X/974/1/012034

3. Results and discussion

3.1 Compressive strength

The compressive strength data was measured at 28 days for the different TSC mixtures. As shown in Fig. 3, the compressive strength of the different TSC specimens varies from 30 to 34 MPa. As expected, the compressive strength generally decreased with increasing fibers content. Incorporating steel fibers in TSC mixtures induced a slight decrease in the compressive strength of TSC. For example, a steel wires addition of 0.5%, 1%, and 1.5% by volume fraction in TSC slightly reduced the compressive behavior by 5.1%, 6%, and 6.4%, respectively, compared to TSC fibreless mixture. This is due to increased porosity accompanied to steel wires addition, which in turn resulted in decreased compressive behavior.

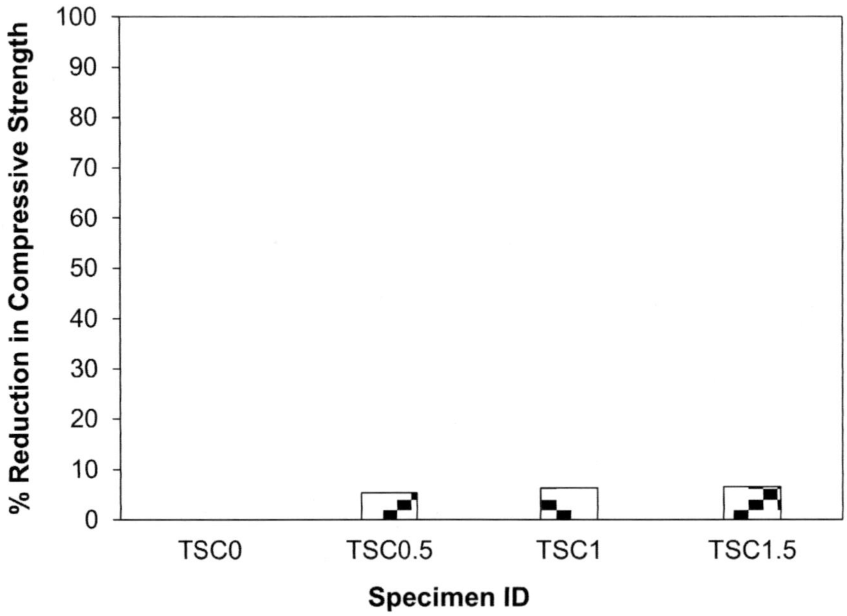

Fig. (3) Compressive behaviour of steel wired TSC specimens

3.2 Splitting tensile strength

Figure 4 shows the variation in splitting tensile strength of TSC specimens at testing age of 28 days. The achieved tensile strength was 3.8 to 6 MPa, depending on fiber content. The tensile behavior of the TSC mixtures was improved owing to incorporation of the steel wires in TSC. For example, the tensile strength of TSC0.5, TSC1, and TSC1.5 specimens increased by about 44.7%, 50.8%, and 60.5%, respectively, relative to the fibreless TSC mixture as shown in Fig. 4. This can be attributed to improved fiber-matrix interfacial bond which improved the overall tensile characteristics.

IOP Conf. Series: Materials Science and Engineering **974** (2020) 012034 doi:10.1088/1757-899X/974/1/012034

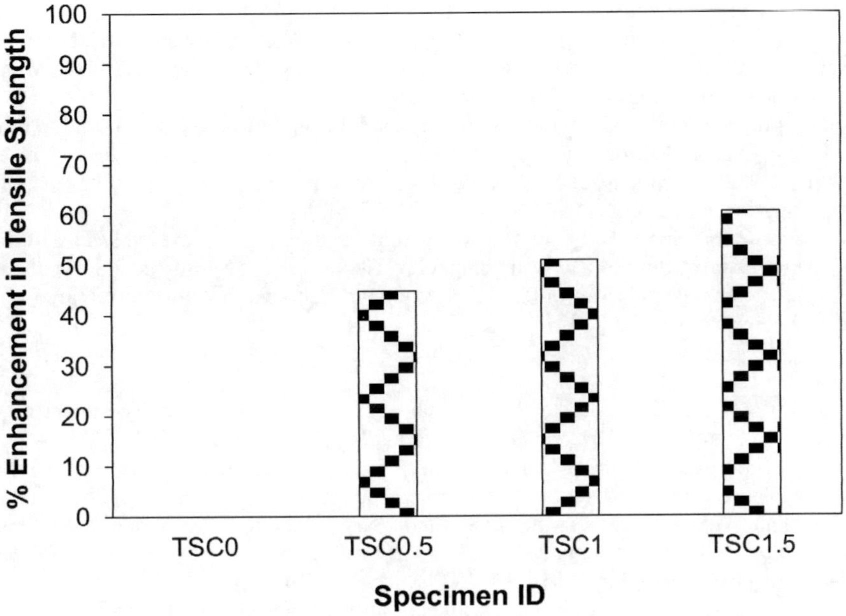

Fig. (4) Splitting tensile characteristics of steel wired TSC specimens

3.3 Impact resistance

The impact resistance of the various TSC mixtures were tested and evaluated under the drop weight impact test [18]. Figure 5 illustrates the impact resistance of TSC mixtures up to first crack and failure. Due to its brittle nature, the fibreless TSC specimen failed after only one drop and split into multiple fragments. Countering that, including recycled steel wires in TSC production significantly increased the sustained impacts by TSC cylinders. The dynamic behavior of recycled steel wired TSC reinforced specimens was approximately 40 times that of fibreless TSC specimens. For example, steel wires addition by 0.5%, 1%, and 1.5% increased the first crack and fracture impact resistance of TSC specimen by up to 3, 4, and 5, and 22, 25, and 40 times that of fibreless TSC specimens, respectively. This is ascribed to the crack arresting capability of steel wires.

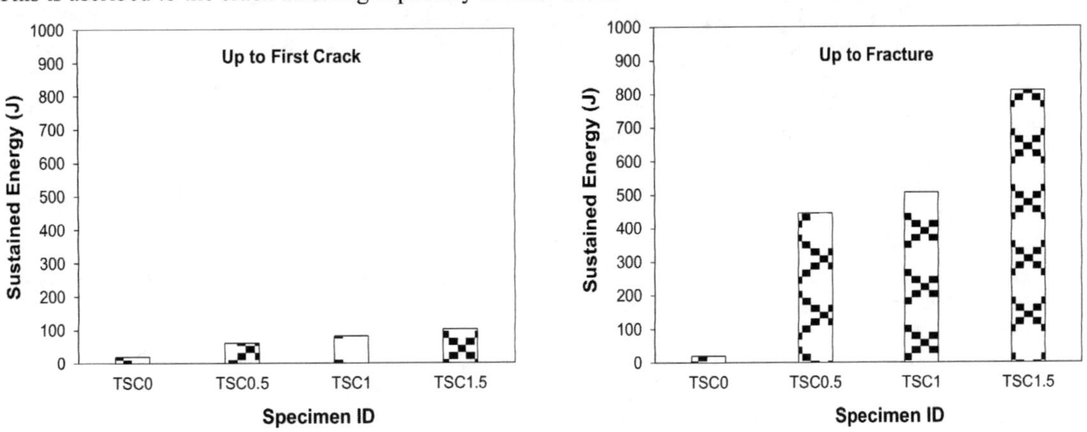

Fig. (5) Dynamic response of recycled steel wired TSC specimens

4. Summary and conclusion

In this study, an experimental investigation was conducted to evaluate the static and dynamic behaviour of two stage recycled steel wired concrete. The main findings are summarized as follows:

- A general reduction in the compressive strength of TSC specimens was observed owing to steel wires addition relative to fibreless TSC mixture.
- The tensile properties of TSC mixtures significantly enhanced owing to incorporation of recycled steel wires in TSC production.
- Among all tested specimens, incorporating 1.5% recycled steel wires in TSC manufacturing acquired superior tensile properties.
- The dynamic performance of TSC mixtures was improved owing to recycled steel wires addition. It was approximately 40 times that of the impact resistance of non-reinforced TSC specimens.
- The static and dynamic behavior of TSC concrete paving the way for further research to produce sustainable cementitious materials.

5. References

[1] S.D. Adhikary, B. Li, K. Fujikake, "State-of-the-art review on low-velocity impact response of reinforced concrete beams." *Magazine of Concrete Research*, 2016, 68, 701.

[2] T.S. Vadivel, R. Thenmozhi, M. Doddurani, "Experimental behavior of waste tire rubber aggregate concrete under impact loading." *Transactions of Civil Engineering*, 2014, 38, 251.

[3] S.R. Abdullah, W.R.W. Zainal-Abidin, S. Shahidan, "Strength of concrete containing rubber particle as partial cement replacement." *3rd Int. Conference on Civil and Environmental Engineering for Sustainability,* 2016, 47, 1.

[4] S. Yehia, K. Helal, A. Abusharkh, H. Istaitiyeh, "Strength and durability evaluation of recycled aggregate concrete." *International Journal of Concrete Structures and Materials,* 2015, 9, 219.

[5] Murali G., Santhi A. S., and Ganesh G. M. "Impact resistance and Strength Reliability of Fiber-reinforced Concrete in Bending under Drop Weight Impact Load." *International Journal of Technology*, 2014, 5, 120.

[6] Murnal P. B. and Chatorikar R.N. "Impact resistance of steel fiber reinforced concrete." *International Journal of Research in Engineering and Technology,* 2015, 4, 246.

[7] Ali M.A.E.M., Nehdi M.L., and Soliman A.M. "Exploring behavior of novel hybrid-fiber reinforced engineered cementitious composite under impact loading." *Materials and Design*, 2017, 117, 149.

[8] Ali M.A.E.M. and Nehdi M.L. "Innovative self-healing hybrid fiber reinforced engineered cementitious composite." *Construction and Building Materials*, 2017, 150, 702.

[9] Rao B.K., Ravindra V., and Rajagopal A. "Experimental investigation on impact strength of steel fiber reinforced normal and self-compacting concrete." *International journal of Civil Engineering and Architecture*, 2012, 2, 505.

[10] Khalil E., Abd-Elmohsen M., and Anwar A.M. "Impact resistance of rubberized Self-compacting concrete." *Water Science,* 2015, 29, 53.

[11] Al-Tayeb M.M., Abu-Bakar B.H., Ismail H., and Akil H.M. "Impact resistance of concrete with partial replacements of sand and cement by waste rubber." *Polymer-Plastics Technology and Engineering*, 2012, 51, 1236.

[12] Nehdi M.L., Najjar M.F., Soliman A.M., and Azabi T.M. "Novel steel fibre-reinforced preplaced aggregate concrete with superior mechanical performance." *Cement and Concrete Composites*, 2017, 82, 251.

[13] Liu F., Chen G., Li L., and Guo Y. "Study of impact performance of rubber reinforced concrete." *Construction and building materials*, 2012, 36, 616.

[14] ASTM Standard, Standard specification for portland cement, *American Society for Testing and Materials,* ASTM Standard C150/C150M, ASTM International, West Conshohocken, USA, 2015.

[15] ASTM Standard, Standard Specification for Chemical Admixtures for Concrete, *American Society for Testing and Materials,* ASTM Standard C494/C494M, ASTM International, West Conshohocken, USA, 2015a.

[16] ASTM Standard, Standard Test Method for Compressive Strength of Cylindrical Concrete Specimens, *American Society for Testing and Materials,* ASTM Standard C39/C39M, ASTM International, West Conshohocken, USA, 2015a.
[17] ASTM Standard, Standard Test Method for Splitting Tensile Strength of Cylindrical Concrete Specimens, *American Society for Testing and Materials,* ASTM Standard C496/C496M, ASTM International, West Conshohocken, USA, 2011.
[18] ACI Standard, Measurement of properties of fiber reinforced concrete, *American Concrete Institute,* ACI Standard 544.2R-89, Farmington Hills, MI 48331-3439 USA, Reapproved 2009.
[19] ASTM Standard, Standard test method for impact resistance of flat, rigid plastic specimens by means of a falling dart (tup or falling mass), *American Society for Testing and Materials,* ASTM Standard D5628, ASTM International, West Conshohocken, USA, 2015.

Military Technical College

Kobry El-Kobbah,

Cairo, Egypt

13th International Conference on Civil and Architecture Engineering

ICCAE-13-2020

Thermal Conductivity for Foamed Paste made of Cement and Cement kiln dust

Rafik K. Abdel Wahab[1,2]

1 - A.Prof.Civil Engineering Department, Faculty of Engineering,Qena Branch- Al-Azhar University - Egypt
2- Vice Dean of High Institute for Engineering and Technology In Obour City –Qalyobia -Egypt
Dr.rkhairy@gmail.com

Abstract

As with most large manufacturing industries, by-product materials are generated. Cement kiln dust is a by - product material of cement manufacturing process. The aim of this paper is to investigate the properties of foamed paste made of CKD in addition of different percentages of cement. Compressive strength and thermal conductivity were determined. Sixty three cubes in three groups with different densities, amount of cement, CKD, water and foaming agent were cast. The results were encourage to use the Foamed(Cement kiln dust – cement) Paste (FCKDC) as an insulation materials. The results indicates' a comparable value for thermal conductivity which vary from (**0.074- 0.3)** W/mk.

Keywords :Cement ; Cement kiln dust (CKD) ;Thermal conductivity ; Compressive strength ; Foamed Paste

1- Introduction

Cement manufacturing is a critically important industry in Egypt and throughout the world. In2006, U.S. cement plants produced 99.8 million metric tons of cement[20] Worldwide production accounted for about 2.5 billion metric tons. As with most large manufacturing industries, by-product materials are generated. Cement kiln dust (CKD) is a significant by-product material of the cement manufacturing process. Cement kiln dust production is estimated at 15-20% of cement production , estimated at 420-560 million tons annually throughout the world[15] .

This industrial waste is characterized by high softness (20 - 100 microns) and high proportions of chlorides, sulfates, alkalis and lime is a source of high risk in terms of health and environmental aspects. These industrial by-product and waste materials must be managed responsibly to ensure a clean and safe environment. Over the past several years dramatic advances have been achieved in the management and use of cement kiln dust, thus reducing its dependency on landfill disposal.[5].

It was found that due to its high alkaline content , large quantities of CKD could not be returned to the kiln [2] ,but the higher alkalinity and finer particles in addition to their (sometimes)

cementitious properties ,make these materials usable for several applications such as waste solidification[13],replacement of Portland cement in concrete block manufacturing and ready mix concrete [16] and use as agricultural soil amendments[19] .The presence of free lime (CaO) ,the high alkali content and the high fineness of CKDs also make them potentially valuable materials for stabilizing soils [6]. R.Siddique[17] concluded that Cement Kiln Dust and wood ash could be successfully used as a partial replacement of cement in making controlled low –strength materials (CLSM). El – Sayed et al.[7] have investigated the effect of CKD on the compressive strength of cement paste and on the corrosion behavior of embedded reinforcement The Study reported that up to 5% substitution of CKD by weight of cement had no adverse effect either on cement paste strength nor the reinforcement passivity. Batis et al.[4] Where it was found that when CKD and blast furnace slag are in proper ratio in ordinary Portland cement, the compressive strength and corrosion resistance of the mix increase. Salem et al.[18] have studied the hydration of cement pastes containing granulated slag and CKD made with and without silica fume. It was reported in their study that the hydraulic reactivities of granulated slag and silica fume as activated by raw CKD are relatively high as compared with those activated by washed CKD. Maslehuddin et al[12]suggested to limit the amount of CKD in concrete to 5% since the chloride permeability and electrical resistivity data indicate that the chances of reinforcement corrosion would increase in case of 10%and 15% CKD. Wang et al [11] mentioned that The test results show that replacing 5 percent of cement by weight with kiln dust produces a mix exhibiting slightly greater shrinkage and creep and greater setting time than plain concrete but does not adversely affect other properties of mortar or concrete.

One means of obtaining lightweight concrete is to introduce gas bubbles into the plastic mix of mortar (cement and sand) in order to produce a material with cellular structure.[3]. The strength of lightweight concrete may be reduced by high air contents. At normal air contents (4 to 6 percent),the reduction is small if slumps are 5 in. or less and cement contents are used as recommended.[1]. Foamed concrete is a versatile building material with a simple production method that is relatively inexpensive compared to autoclave aerated concrete .[10]

Foam concrete is created by uniform distribution of air bubbles throughout the mass of concrete . For very low densities 300-500 kg/m^3 made with cement and foam only such densities is used in roof and floor as insulation against heat and sound ,and is applied on rigid floors(because it is not a structural material). It is used interspaces filling between brickworks ,insulation in hollow blocks and any other filling situation where high insulation properties are required.

As sustainability involves that the needs of present generation are met without wasting , pollution , harming or destroying the environment .[14] .

It was found very useful to use waste material as CKD instead of cement to decrease the bad impact on the environment and converting these wastes into useful products for sustainability .

This study was conducted to recycle cement kiln dust with small amount of cement Not exceed 30% to produce a very good and cheap insulation materials.

2- Research significant

The importance of research lies in two things, the first is to take advantage of the cement kiln dust that pollutes the environment and the second is to obtain an economical material that is cheap and used for thermal insulation, which in turn leads to a reduction in the energy consumption needed to make buildings well air conditioned .

3- Materials

3-1 Cement

IOP Conf. Series: Materials Science and Engineering **974** (2020) 012035 doi:10.1088/1757-899X/974/1/012035

Ordinary Portland Cement From Qena Factory of cement had been used in this study and the chemical composites are shown in Table(1). The compressive strength test for cement had been done and the results was
(182 -274)kg/cm^2 for 3days and 7 days respectively.

3-2 Cement Kiln Dust (CKD).

CKD is a fine , highly alkaline powder from the same factory and Table(1) shows the chemical composites of CKD and cement used in this study.

Table(1)Chemical composites for Cement and CKD

Component	Portland cement (%)	Cement Kiln Dust (%)
SiO_2	20.6	15.8
AL_2O_3	4.5	3.6
Fe_2O_3	3.6	2.8
CaO	62.5	63.8
MgO	2.6	1.9
SO_3	2.7	1.7
K_2O	0.5	3.0
Na_2O	0.2	0.3
CL^-	0.01	1.1

3-3 Euco –Foaming- Agent

This is the name of the foaming material which is a product from Swiss-Kem company.
Euco-Foaming-Agent is a chloride free foaming agent to produce light weight concrete or mortar by entraining a controlled amount of air bubbles to the mix. It is high thermal and acoustic insulator. The dosage differ according to the density needed .The dosage used in this study is mentioned in Table (2).

4- Methods

4-1 Mix Proportioning

As mentioned in Table (2), three groups with different densities, amount of cement, CKD, water and foaming agent were tabulated. Number of cubes to be tested in compression test and thermal conductivity .

Table (2) Mix proportioning by weight

Group No.	Mix No.	Target Density (Kg /m^3)	Cement (Kg)	CKD (Kg)	Water (Lit.)	Foamed agent	No. of cubes for compressive strength	No. of cubes for thermal conductivity
G1	G11	500-550	6	24	24	10ml	6	1

IOP Conf. Series: Materials Science and Engineering **974** (2020) 012035 doi:10.1088/1757-899X/974/1/012035

	G12		9	21	22	10ml	6	1
	G13		12	18	20	10ml	6	1
G2	G21	600-650	6	24	24	8ml	6	1
	G22		9	21	22	8ml	6	1
	G23		12	18	20	8ml	6	1
G3	G31	700-750	6	24	24	6ml	6	1
	G32		9	21	22	6ml	6	1
	G33		12	18	20	6ml	6	1

4-2 Sample preparation

Sixty three cubes with size 100X100X100 mm had been cast with nine mix proportioning as mentioned in Table (2) . The cubes were divided to three main groups according to their density. Groups (G1),(G2)and (G3) represented mix with target density range from (500 -550),(600-650)and(700-750)kg/m^3. A batch of 30kg of cement and CKD were prepared with different proportioning ratios .Mixes (G11),(G21)and (G31) had the same cement to CKD ratio which is 80% CKD and 20% cement and the foamed agent was 10 ml. Mixes (G12),(G22)and (G32) had the same cement to CKD ratio which is 70% CKD and 30% cement and the foamed agent was 8 mL. Mixes (G13),(G23)and (G31) had the same cement to CKD ratio which is 60% CKD and 40% cement and the foamed agent was 6ml.The produced paste was casted in the cubic mold without compaction and trawled .After 24 hours the specimens demolded and subjected to fresh air with average temperature of (25° C) ,average relative humidity 60% and subjected to curing twice a day by sprinkling water till the date of testing.

4-3 Test procedures

4-3-1 Compressive strength test.

Compression test was conducted by crushing six cubes, three after 7 days and the rest after 28 days All specimens had been tested under compression static load by using 2000kn compression testing machine with rate (0.6N/mm^2) until failure .According to the Egyptian code C203.[8]

4-3-2 Thermal conductivity test

By using (KD2 Pro) picture (1) instrument nine specimens represented nine mixes had been tested to measure the thermal conductivity of each mix. The test was conducted after 28 days.

The test was conducted according to the standard specification ASTM C-518.The temperature was 24°C and R.H.55%.

IOP Conf. Series: Materials Science and Engineering **974** (2020) 012035 doi:10.1088/1757-899X/974/1/012035

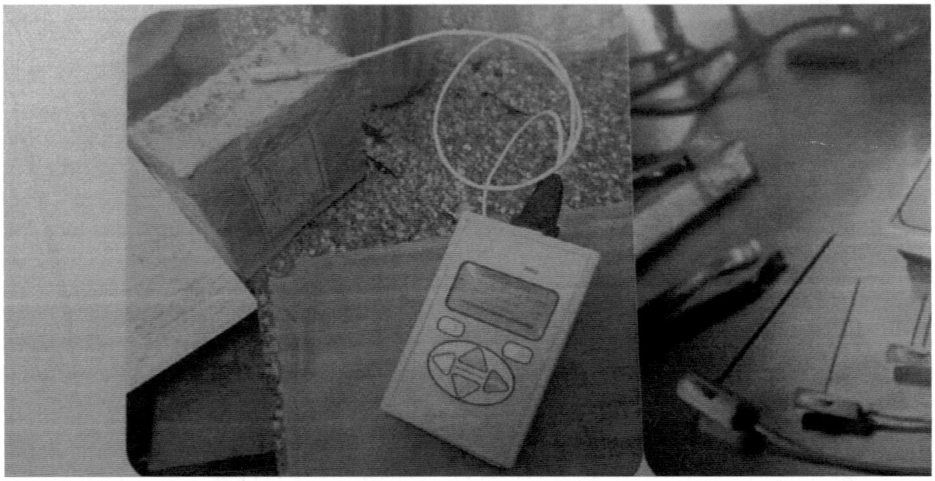

Picture (1) KD2 Pro Instrument

5- Results and discussion

5-1 Results

Results had been tabulated in Table (3) as follows :

Table (3) shows the results of compressive strength tests after 7 &28 days and determination of thermal conductivity for all groups and different mixes .

Thermal resistance **(R)** had been calculated from the relationship[9]

$$R= L/K \quad \text{.....(1)}$$

Where

R is thermal resistance

L is thickness of insulation material

K is the thermal conductivity of the insulation material

The equation (1) indicates that the lesser the thermal conductivity the higher thermal resistance . A relationship between density and compressive strength had been blotted in Figures (1) and (2).

Table(3) Compressive strength ,thermal conductivity Vs, Denisty and cement content

IOP Conf. Series: Materials Science and Engineering **974** (2020) 012035 doi:10.1088/1757-899X/974/1/012035

No. of Group	G1			G2			G3		
Density	500-550 kg/m^3			600-650 kg/m^3			700-750 kg/m^3		
Cement Percent by weight	**20%**	**30%**	**40%**	**20%**	**30%**	**40%**	**20%**	**30%**	**40%**
CKD substitution Percent by weight	**80%**	**70%**	**60%**	**80%**	**70%**	**60%**	**80%**	**70%**	**60%**
strength after 7 days (kg/cm^2)	**2**	**2.2**	**4.08**	**2.7**	**2.9**	**4.3**	**5.7**	**5.99**	**7.67**
Strength after 28 days (kg/cm^2)	**3.4**	**4.3**	**5.87**	**4**	**5.3**	**6.22**	**7.2**	**7.89**	**9.23**
Thermal Conductivity (W/m.K) after 28 days	**.074**	**.092**	**.28**	**.083**	**.096**	**.29**	**.087**	**.1**	**.3**
Heat insulation for 5 cm thickness(R)	**68**	**54**	**18**	**60**	**52**	**17**	**57**	**50**	**17**
Heat insulation for10cm thickness (R)	**135**	**109**	**36**	**120**	**104**	**34**	**115**	**100**	**33**
Heat insulation for15cm thickness(R)	**203**	**163**	**54**	**180**	**156**	**52**	**172**	**150**	**50**

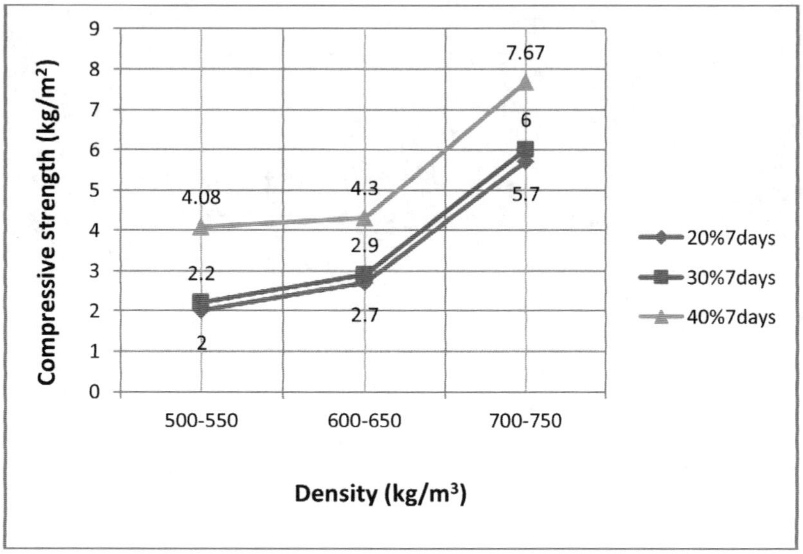

Fig(1) Compressive strength after 7 days Vs density and cement content

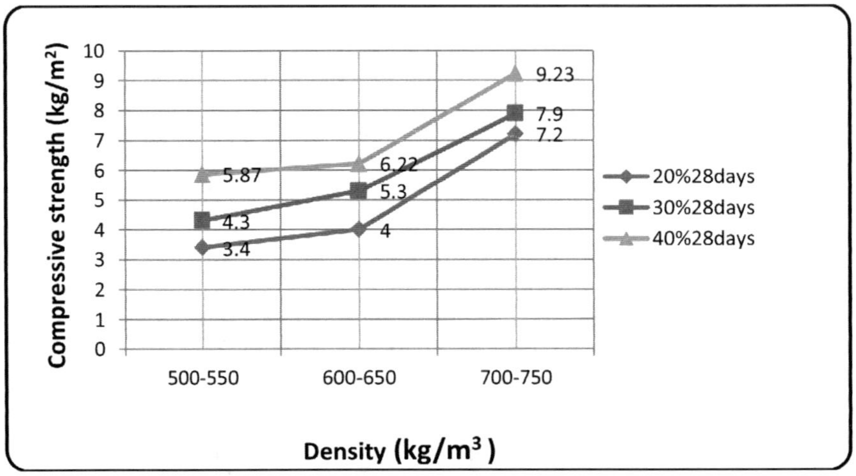

Fig(2) Compressive strength after 28 days vs Density and cement content

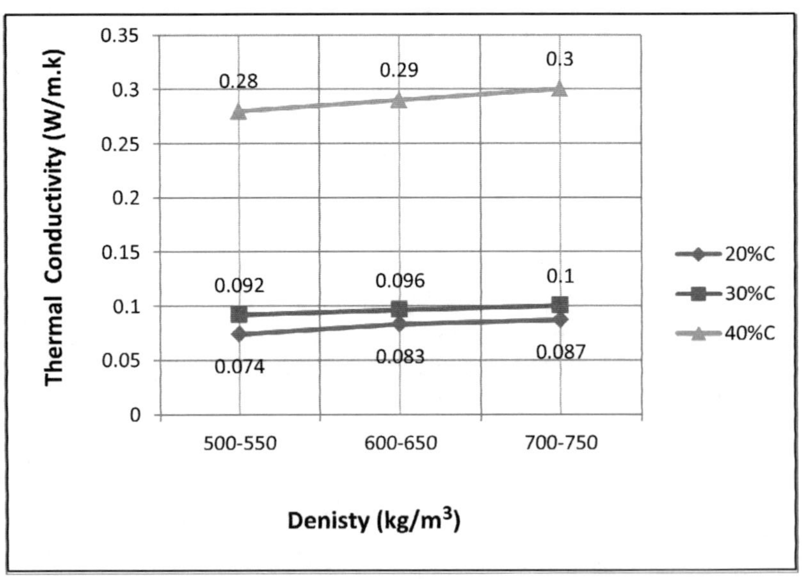

Fig(3) Thermal conductivity vs Density and cement content

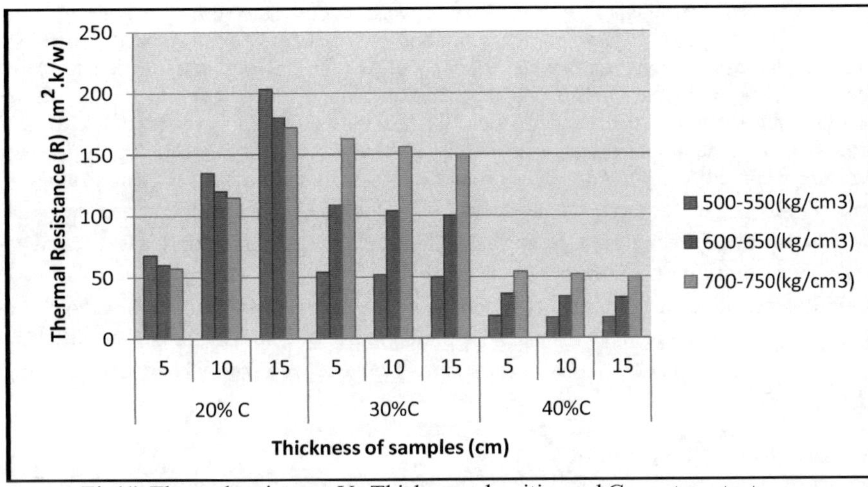

Fig(4) Thermal resistance Vs Thickness ,densities and Cement content

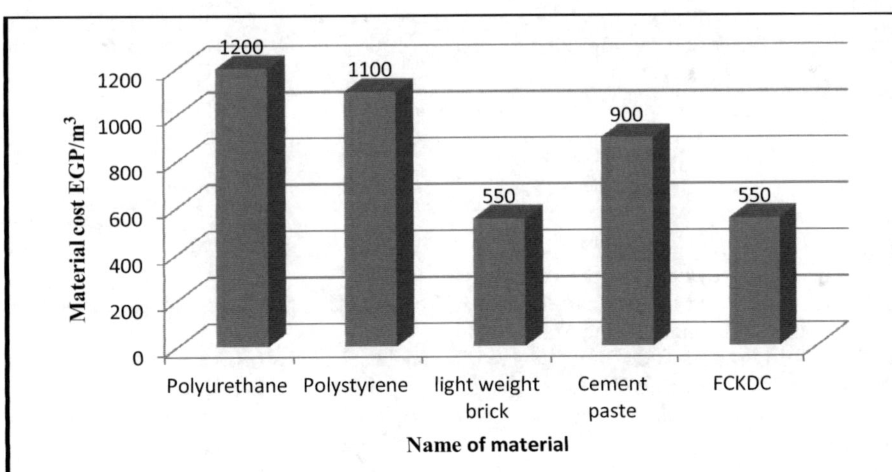

Fig(5)Comparison of cost for different insulation materials Vs FCKDC

5-2 Discussions

The results as mentioned in Table (3) showed that the compressive strength for all specimen vary from 2 kg/cm^2 to 9 kg/cm^2 which is very low compared with the normal compressive strength for cement paste . But this compressive strength is accepted for the foamed paste from CKD and cement when it is compared with the strength of other insulation materials such as Expanded polystyrene which has compressive strength 1.73kg/cm^2 , polyurethane (1-2 kg/cm^2) and foamed concrete which has in very low density a compressive strength about 5 kg/cm^2 as mentioned in the Egyptian specifications for thermal insulation works[9] . Also, the minimum compression strength

IOP Conf. Series: Materials Science and Engineering **974** (2020) 012035 doi:10.1088/1757-899X/974/1/012035

which is 2 kg/cm^2 is enough to carry a person weigh more than 100 kg as this material is not structural material but it is required for insulation only and not to be crushed under the foot of workers during application on ceiling .When applied this material in sandwich brick walls the property of compression strength is not important at all. As per Fig (1) and Fig(2) it is obviously observed that the compressive strength for (FCKDC) paste directly proportional with density .As regardless of the type of foamed paste or mortar ,the strength increased when density increased . The results recorded a very good values of thermal conductivity which is varying from (0.074-0.3) W/mk . The low value for thermal conductivity (K) which means very high thermal resistance refer to the low density . The change of percentage of cement and CKD in mixes leads to change the thermal conductivity for the same density as observed from Fig.(3) . for example at density equal to (500-550)kg/m^3 the value of thermal conductivity for cement content(20% ,30%and 40%) were (0.074,0.92and 0.28) respectively . This change was probably related to the increase of CKD fineness than cement since the larger interface area would act as a thermal barrier decreasing the amount of heat transfer.[20]

6- Cost analysis for FCKDC paste

The cost of cubic meter of FCKDC is 550EGP, compared this value with pure cement paste which cost approximately 900EGP according to Fig(5) it could be safely save about 60%

7- Conclusion

- Foamed paste of Cement and CKD had a reasonable value in compressive strength as an insulation material.
- Foamed paste of cement and CKD had an excellent thermal resistance.
- Increasing the amount of CKD in Cement CKD paste leads to increase thermal resistance .
- The optimum value of cement percentage in Cement CKD paste from author point of view is 30% this is because the low thermal conductivity and the reasonable compressive strength.

References

[1] ACI manual of concrete practice committee 211.2-98

[2] Al-Harthy AS, Taha R, Al-Maamary Effect of cement Kiln dust (CKD) on mortar and concrete mixtures Construction and Building Materials 17(2003)353 -360

[3] NEVILLEA. M., J.J. BROOKS “Concrete Technology” ISBN 0 582 24913 9

[4] Batis G, Rakanta E, Sideri E, Chaniotakis E, Papageorgiou A. Advantage of simultaneous use of cement kiln dust and blast furnace slag, In proceeding of the international conference of the challenges of concrete construction. Dundee, (UK): University of Dundee;2002.

[5] Beneficial Uses Of Cement Kiln Dustby: Ieee- Ias Cement Industry Committee Wayne S. Adaska, P.E., Director, Public Works, Portland Cement Association Donald H. Taubert, Director, Promotion & Technical Service, Capitol Cement Presented at 2008 IEEE/PCA 50th Cement Industry Technical Conf., Miami, FL, May 19-22, 2008.

[6] Bhatty JI. Alternative uses of cement kiln dust,Skokie, Illinois USA: RP327 portland Cement Association 1995.

[7] ElSayed, H,A., Gaber, N.A., Hanafi, S., Mohran M.A Re-Utilization of by-pass kiln dustin cement manufacture International Conference on blended Cement in Construction, Sheffield, UK, Septemper 1991.

[8] Egyptian code for design of concrete structures C203.

IOP Conf. Series: Materials Science and Engineering **974** (2020) 012035 doi:10.1088/1757-899X/974/1/012035

[9] Egyptian specifications for insulation works 176/1998
[10] Foamed concrete leaflet(http://www.theconcretinstitute.org.za/wp-content/uploads/2013/10/Foamed-concret.pdf) the concrete Institute ,Midrand,2013[CrossRef.]
[11] M.L. Wang, V. Ramakrishnan, Evaluation of blended cement and mortar and concrete made from type III Cement and kiln dust, construction and Building Materials 4(2)(1990) 78-97.
[12] Maslehuddin M., O.S.B. Al-Amoudi, M.K. Rahman, M.R. Ali, M.S. Barry Properties of cement kiln dust concrete Construction and building materials 23 (2009)2357 -2361.
[13] Mckay M., J. Emery, Stabilization/solidification of contaminated soils and sludgesusing cementitious systems, proceeding of the Cement Industry Solutions to Waste Management, Calgary, Alberta, Canada Oct.7-9 1992, pp.135- 151.
[14] Mohamed M.El-Attar,Dina M.Sadek,Amir M.Salah,"Recycling of high volumes of cement Kiln dust in bricks industry"Journal of cleaner Production 143(2017)506-515.
[15]Rafik K.Abdel Wahab "Effect of adding Silica Fume to the Cement kiln Dust(CKD) Journal of American Science2013;9(12),pp.274-281
[16] Bhatty M.S.Y, Properties of blended cement made with Portland cement, cement kiln dust, fly ash and slag, proceeding of the 8th International Congress on the chemistry of cement, Rio de Janeiro Brazil, Sept.22-28 1986, pp. 118-127.
[17] Siddique R. "Utilization of waste materials and by – products in producing controlled low – strength materials" Resources, Conservation and Recycling 54(2009)1-8.
[18] Salem Thm, Ragi Sh M. Electrical Conductivity of granulated slag cement kiln dust-silica fume pastes at different porosities Cement and Concrete research 2001;31:781-7.
[19] Sulapha Peethamparan, Jan Olek, Janet Lovell Influence of chemical and physical characteristics of cement kiln dusts (CKDs) on their hydration behavior and potential suitability for soil stabilization Cement and Construction Research 38(2008) 803-815.
[20]Xiao Zhao,Siong-Kang Lim,Cher-Siang Tan,Bo Li,Tung-Chai Ling,Runqiu,and Qingyuan Wang "Properties of Foamed Mortar Prepared With Granulated Blast-Furnace Slag" Materials (Basel).2015 Feb;8(2):462-473.

Military Technical College
Kobry El-Kobbah,
Cairo, Egypt

13th International Conference on Civil and Architecture Engineering ICCAE-13-2020

A Review on Investigating the experimental process for partial replacement of cement with sugarcane bagasse in the construction industry

Nadin Dawoud[a], Amany Micheal[b], Rania Rushdy Moussa[c]

[a, b & c]The British university in Egypt (BUE), Suez Desert Road El Sherouk City, 11837, Cairo Egypt

Nadin150210@bue.edu.eg

Abstract. In the last few decades there has been speedily increasing in the agriculture and industrial wastes. This causes many environmental issues and raises the potential to contaminate the natural resources of living such as water, air and soil. Recently, the amount of organic waste produced daily has been rising, while it is poorly managed. It is either burned or disposed improperly, which effect negatively the environment and public health. On the other hand, during the cement production process many wastes, and pollutants are generated which have major negative impacts on the environment. Cement is considered as a substantial constituent of building materials in the construction industry. Many research's intended to explore the potential of utilizing organic waste products in the construction industry by partially replacing cement with organic wastes such as sugarcane bagasse to create an eco-friendly brick with appropriate structure properties. Previous research's used different treatment techniques to treat the organic waste and prepare it for construction industry. It was noticed that the treatment method used in previous research's affected the structural properties of the new bricks with organic waste partially replacing. This research intends to study and analyse the process and techniques used by other researcher's experimental work to treat and replace cement with sugarcane bagasse. This research will present the best procedures for partially replacing cement with sugarcane bagasse in cement bricks without compensating the structural properties of regular cement bricks. This study will analyse and compare previous researcher's experimental work to obtain the best experimental program and to utilize sugarcane bagasse in brick industry.

Keywords: *Sugarcane bagasse; Cement; Eco-friendly block; Partial replacement; Thermal conductivity; Strength*

1. Introduction

Agriculture and industrial wastes generation are becoming a dangerous and significant problem in the world (Xu, Ji, Gao, Yang, & Wu, 2018). This problem is becoming more significant by time due to growth in population rate. The accumulation of wastes causes many environmental issues and raises the potential to contaminate natural resources of living such as water, air and soil. So, the appropriate disposal of these wastes and by products is a serious burden of every country (Moussa, 2019). The safe disposal of such industrial wastes is very expensive. Furthermore, there is a lack of disposal sites which can appropriately handle such wastes without causing harmful effects on the environment.The

understanding of waste management process requires the understanding of wastes and how can they be managed. Therefore, researchers from all over the world are focusing on ways to utilize these wastes, to reduce their harmful effects (Singh, Saxena, Bharti,& Singh,2017).
Lately, burning the organic waste or the inappropriate disposal methods for these organic wastes became a huge international problem as shown in Figs. 1 and 2. Today, Egypt recycling only 20% of its organic wastes (Elfeki & tkadlec, 2014; Guirguis & Moussa, 2019). The wastes affect negatively the environment by creating visual pollution, the black cloud, increasing the insect population in the environment as well as it affects the human's health. Sugarcane is one of the most promising organic residues around the world and in Egypt. It produces cane trash and bagasse as wastes materials. Bagasse is a fibrous material left after cane milling. Most bagasse is burned and combusted to generate energy and electricity; however, this contribute to the emission of carbon dioxide and to the formation of black cloud. As the government and people became aware of recycling, researchers started to explore methods of incorporating bagasse in the construction industry of manufacturing bricks by replacing some of its components (Mangi, et al., 2017). The brick industry is one of the places where the beneficial use of agricultural waste could have goodfuture.''

Figure 1: sugarcane bagasse burning (afrik21, 2019)

Figure 2: Sugarcane bagasse disposal, (afrik21, 2019)

At present, brick industry is cursed with its environment pollution from cement production as shown in Fig. 3. Cement is considered one of the most important industries worldwide as it is the backbone for the development of any country. It is the world's most consumed construction material because of its excellent mechanical and durability properties. According to (Stajanča M., 2012) for every one human being in the world, 1 ton of cement is being produced every year and as a result to its significant, its production impacts need to be understood. According to the Industrial Development Authority, Egypt alone has produced 83.5 metric ton (M/T) of cement in 2016 and it is also named as the 14th country in the list of the cement producing countries as cement Industry estimates to be 5.9% of the Egyptian economy (2019). Cement production is one of the major energy consuming industries in the construction system and generates harmful emission, odours and noise thus harming the environment and its inhabitants. As demonstrated in Figs. 4 and 5, some of the emissions produced from the process are dust, carbon dioxide, nitrogen oxides and Sulphur dioxide (Stajanča M.,2012).

IOP Conf. Series: Materials Science and Engineering **974** (2020) 012036 doi:10.1088/1757-899X/974/1/012036

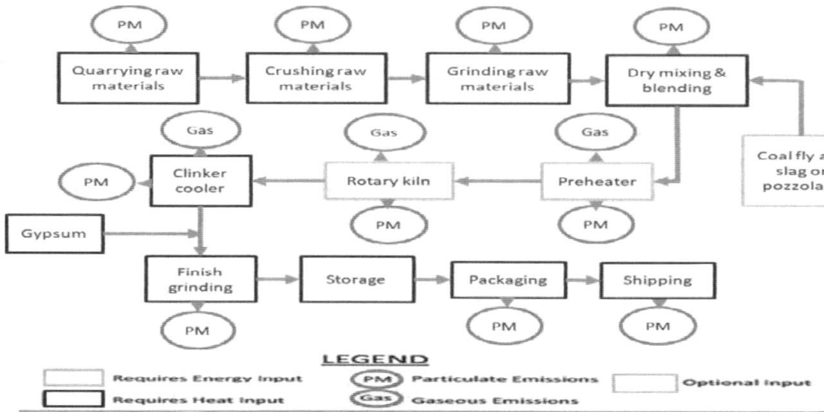

Figure 3. Cement production process and emissions, (Eštoková, 2012)

Figure 4. Air pollution in cement production industry, (Eštoková, 2012)

Figure 5. Dust emissions in cement production industry, (Eštoková, 2012)

In recent years, various studies have been conducted to produce bricks having an appropriate strength and better mechanical properties with low environmental impacts with the integration of agricultural waste. Many researchers started testing the process of partially replacing cement by sugarcane bagasse to reduce waste generated from cement production industries and utilize the usage of bagasse in various areas, this is briefly explained in Fig.6.

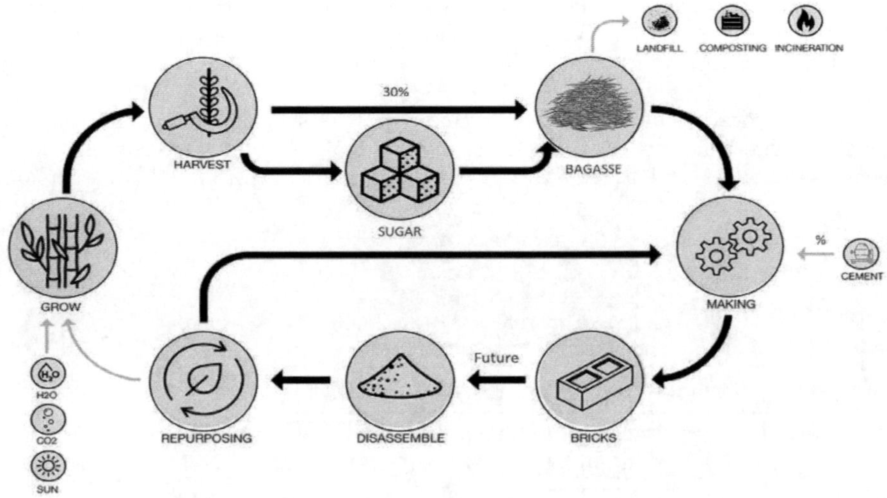

Figure 6. Utilizing sugarcane bagasse in brick industry, (developed by author, 2019)

This study aims to investigate the production of a block by using an organic waste Material; sugarcane bagasse fibres as a partial replacement of cement to improve its physio-mechanical properties.
Achieving the aim of this study, there are some of integrated objectives that must be applied:

- Utilizing the usage of organic waste materials such as sugarcane bagasse in the construction industry
- Analysing previous researchers' experimental work and compare theirresults
- Obtain the best experimental program for cement replacement by sugarcanebagasse

The presented research suggest that the sugarcane bagasse fibres can be used as a cement replacement to produce a new model of eco-friendly bricks with lower thermal conductivity of the brick and appropriate strength as shown in Fig. 7.

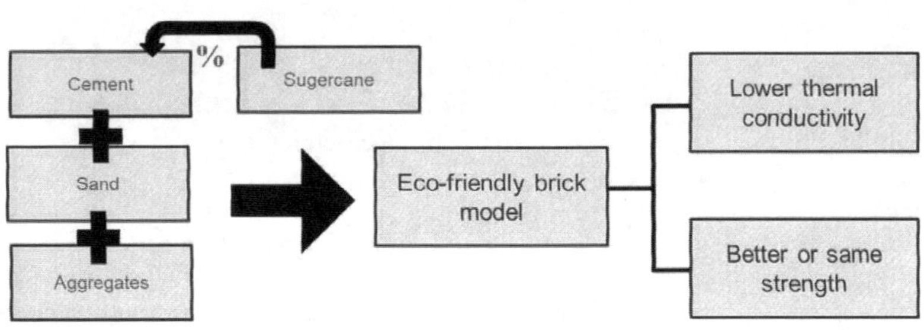

Figure 7. Research hypothesis

The results of this research are developed by two Theoretical qualitative methods (Fig. 8):

- Literature review descriptivemethod
- Analytical comparative previous experimental workmethod

The first theoretical method section will be covered through the detailed background information in the literature review. The literature review will be composed of detailed information about organic waste, its environmental impact and its use in the construction industry. Then the sugarcane bagasse component, properties and its negative impacts on the environment will be stated. In addition, the cement production and its impacts will be explained with the properties of the ordinary Cementous brick.

The second theoretical method will be covered through analyzing and comparing previous researchers' experimental work that was done on this topic. Several case studies with several experimental alternatives will be compared to obtain the best experimental program for future research.

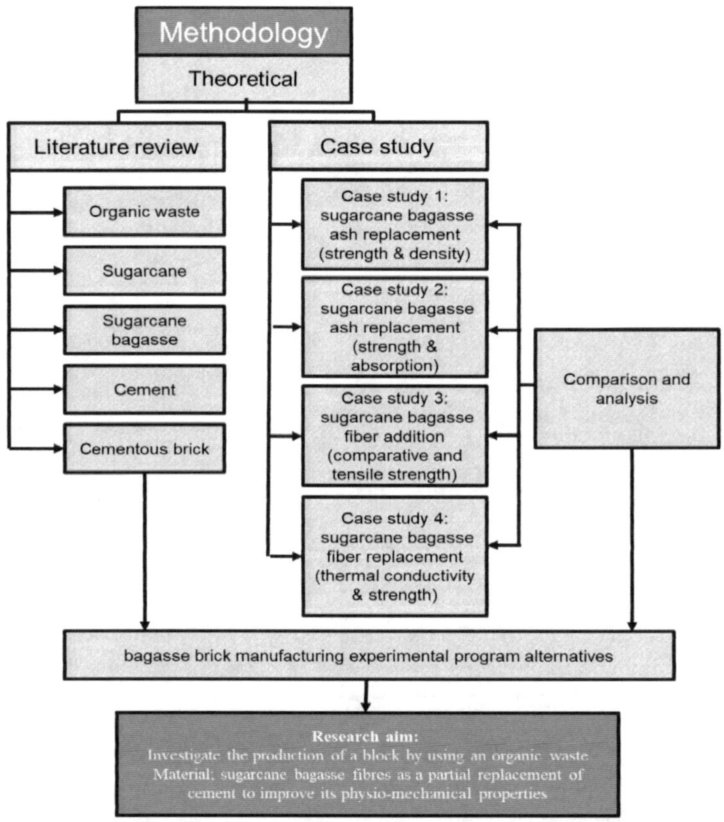

Figure 8. The research outline

2. Organic waste

In this part, the sugarcane bagasse is introduced in detail with description and analysis of its chemical, physical and mechanical properties. In addition, its negative impacts to the environment and people's health are presented to highlight its recycling and management necessity. This will assist in understanding of its importance in the utilization in other industries especially the construction industry.

2.1. Sugarcane bagasse inEgypt

Around the world, 38 billion metric tons of organic wastes approximately are produced annually. This dramatic increase is the result of numerous factors such as human actions, consumption rate, and population explosion (Kiyasudeen, et al., 2016). According to Elfeki & tkadlec 56% of the total Egyptian municipal solid wastes of 12.88 × 106 tons are organic wastes which is considered the main source of solid wastes in Egypt as shown in Fig. 9 and 83.5% is being dumped and poorly disposed.

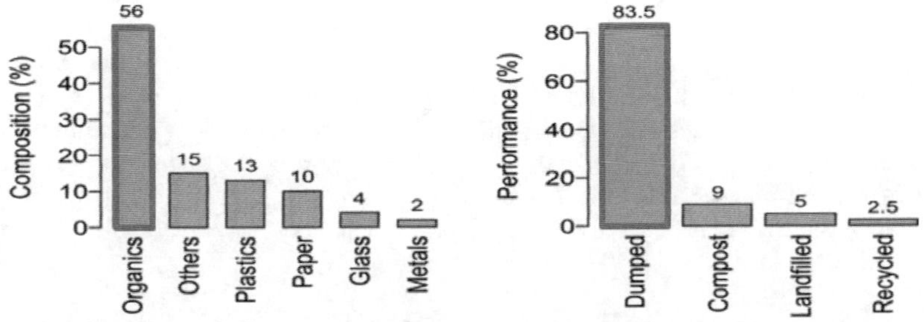

Figure 9. Composition (left) and performance (right) of MSW in Egypt as reported in 2010 (Elfeki & tkadlec, 2014)

Despite the existence of 66 composting plants in Egypt, the recycled organic waste still does not exceed the 20%, therefore managing organic waste is crucial to avoid serious environmental and public health problems (2014). By 2025, it is predicted that the annual organic waste generation will exceed 30 million tons annually. Poor disposal and combustion of organic waste have always been the result due to the poor management of these wastes which lead to numerous environmental problems. In addition, improper disposal of organic wastes in dump sites and waterways contaminated air and water supplies, this hindered the Egyptians natural resources, heritage, and population welfare.

Organic wastes have been utilized in many sectors in the construction industry. In 10 are the six main fields of applications identified are (Arub, 2017): Interior partitions and finishes such as flat boards, furniture, acoustic absorption, thermal insulation, carpets, moquette and envelope systems.

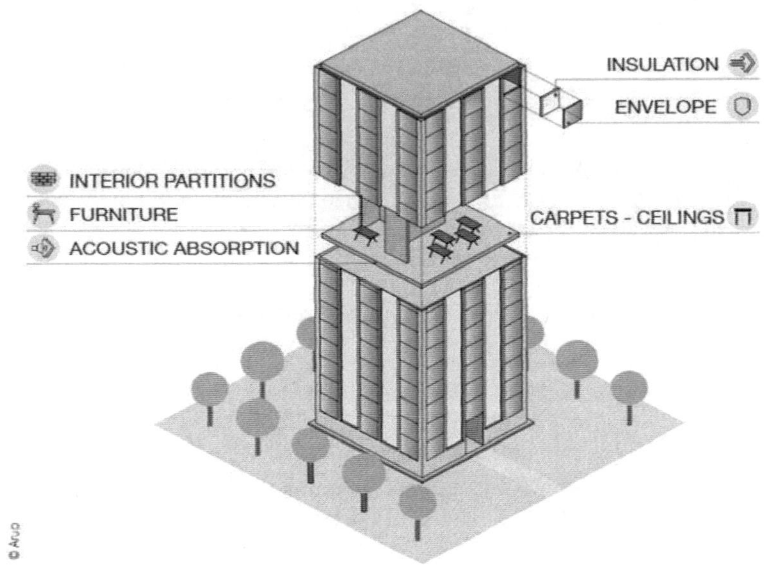

Figure 10. Fields of application for products made from organic waste (Arub, 2017)

IOP Conf. Series: Materials Science and Engineering **974** (2020) 012036 doi:10.1088/1757-899X/974/1/012036

2.2. Types of agriculture waste inEgypt

Agricultural wastes are the highest type of organic waste source in Egypt, Egyptian crops produce numerous amounts of residues every year, Table 1 lists the types of major agricultural crop residues in Egypt and its generation amounts annually. The table illustrates the sugarcane residue production is more than 16.8 million tons annually. Therefore, sugarcane bagasse recycling is a promising field in Egyptian neighbourhoods as it provides a great opportunity for creating a sustainable building material.

Table 1. Types of agricultural waste and percentage of it in Egypt. (Megeed, 2012)

Crop residues	Amount (million tons)
Cotton stalks	1.6
Rice straw	3.6
Maize residues	4.5
Sugar cane residues	**16.8**
Wheat straw	6.7
Barely straw	0.2
Sugar beet residues	0.32
Trees trimming residues	1.7
Vegetable residues	0.71
Banana residues	1.7
Beans straw	0.35
Lentil straw	0.012
Pea straw	0.042
Public garden residues	1.14
Sorghum residues	1.2
Sesame straw	0.56
Date palm residues	0.66
potato	0.317
tomato	1.11

2.3. Sugarcane inEgypt

Sugar is one of the main substrates of human diet. India, Brazil, Thailand, Australia and China are the world's five top sugar producing countries. Sugar production industry in Egypt started from the year 710 AD (Hassan and Nasr, 2008). Distribution percentages are shown in Fig. 11. Cane plantations are concentrated in the area of Upper Egypt in Menia, Sohag, Qena, Luxor and Aswan. Annually, about 16 million tons of cane are cultivated in Upper Egypt (Hamada, 2011). Among Arab countries, Egypt is considered the largest producer of sugarcane followed by Sudan, at 7.5 million tons annually (ESCWA, 2009). During the sugar production process in the mill, several by-products and residues are generated. These are (Fig.12):

- 30% Bagasse, which is the fibrous remaining material produced from sugarcane chopping and milling.
- 3.5% Filter mud/cake resulting from cane juicefiltration.
- 0.4% Furnace ash, in case the bagasse is combustion for steam and electricitygeneration.

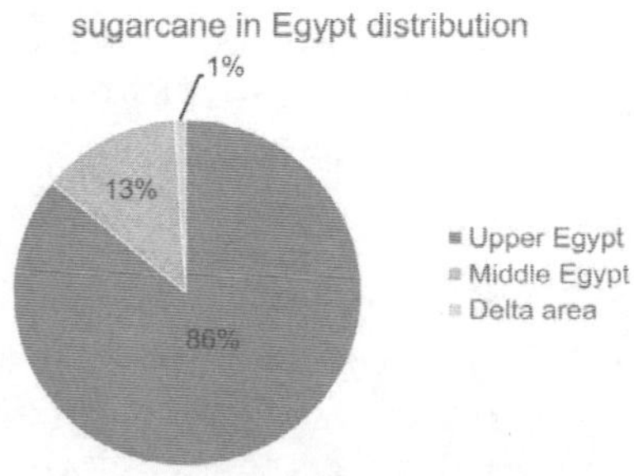

Figure 11. Percentage of sugarcane in Egypt, (Megeed, 2012)

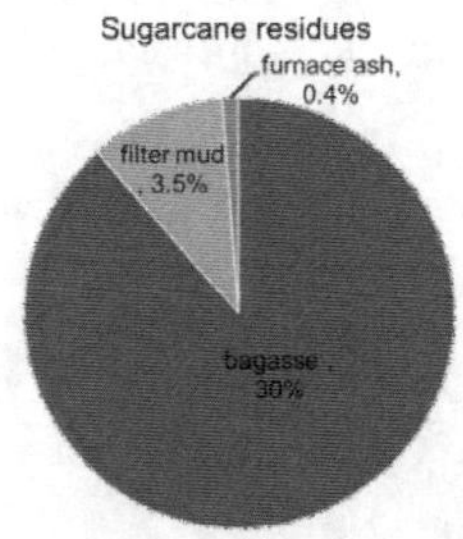

Figure 12. Percentage of sugarcane residues, (developed by author, 2019)

3. Sugarcane bagasse(SCB)

In this part, the sugarcane bagasse is introduced in detail with description and analysis of its chemical, physical and mechanical properties. In addition, its negative impacts to the environment and people's health are presented to highlight its recycling and management necessity. This will assist in understanding of its importance in the utilization in other industries especially the construction industry.

3.1. Sugarcane bagasse inEgypt

Sucrose extraction for sugar and ethanol production is the main reason sugarcane has been grown. The sugarcane culture accountable for huge bagasse generation quantities, its generation of bagasse is 140 kg bagasse for every ton of sugarcane (Faria, 2011). Nowadays, the by-product; sugarcane bagasse is valued by sugar-alcohol sector producers as it is the main raw material for bioenergy and biofuel production, this grants the use of bagasse an economic and environmental importance for the producing countries (Mulay, Vesmawala, Patil, & Gholap, 2017). Till this day, the use of bagasse for energy generation in industrial ovens is continuous, however its industries have expanded as its value is remarkably growing g due to the increase in recycling awareness and it is nowadays used to produce building materials, packaging materials, and disposable tableware. In addition, the paper industry started to introduce sugarcane bagasse fibres as a replacement of wood fibres for the production of napkins, toilet paper and cardboards (Sales & Lima, 2010).

Bagasse a heterogeneous in size and particles shape with respect to its three dominant constituents, the polymers: cellulose, hemicellulose, and lignin. Their chemical ingredients and compositionare listed in Table 2:

Table 2. Composition of bagasse (Rabelo, 2015)

Component	Composition (%)
Glucose	19.5
Xylose	10.5
Arabinose	1.5
Galactose	0.55
Lignin	9.91
Organo soluble	2.7
Reducing sugars	1.85
Uronic acids	1.91
Ash	1.6
Cellulose	50
Total hexoses	20.04

Table 3 presents a list of the physical properties of bagasse. The fibres with the highest aspect ratio have the highest tensile properties and provide a high surface area which is useful for reinforcement purposes.

Table 3. Physical properties of bagasse (Gupta, 2015)

Property	Value
Diameter (μm)	10-34
Length (mm)	0.8-2.8
Aspect ratio (1/d)	76
Moisture content (%)	49

Bagasse exhibits excellent mechanical properties such as high strength, flawless finish, easy manufacture, and extremely low thermal conductivity. In table 4, the mechanical properties of bagasse fibres are presented

Table 4. Mechanical properties of bagasse (Gupta, 2015)

Property	Value
Tensile strength (Mpa)	180-290
Young's Modulus (Gpa)	15-19
Failure strain (%)	1-5
Density (kg/m^3)	880-270

3.2. *Negative impact of SCBdisposal*

The dramatic residue generation during sugarcane harvest mismanagement, insufficient tools and data assessment for using waste management alternatives regardless economic criteria and lack of cheap sustainable waste management alternatives to reach the optimum environmental balance results in numerous negative impacts. Sugarcane bagasse is poor disposing and burning during the harvest season cause various many detrimental impacts for example air quality degradation and harmful combustion products emission such as, carbon monoxide (CO) and volatile organic carbons (VOC). This negatively affect neighbouring communities' health and contribute to black cloud formation. This contamination is not limited to only the surrounding environment, but also the production of fly ash damages the soil microbial diversity. In addition, water stream polluted by fertilizers and waste from sugarcane harvesting are subjected to reduction of water-oxygen content which negatively impacts the aquatic animals (Kiyasudeen et al., 2016). Air and water pollution occur mainly from agricultural disposal and burning. Various harmful emissions are produced due to bagasse combustion this causes numerous health issues to the community such as lung problems and breathing issues. Moreover, fertilizers' chemicals and pesticides leaking into ground drinking water contributes to many health-related problems such as blue baby syndrome which causes death in infants (Hussein & Sawan, 2010).

3.3. *Current uses ofSCB*

Sugarcane bagasse as shown in Fig. 13 is currently used in various industries. Some of its uses in Egypt are steam and electricity generation as it is a free, secure and reliable fuel generated originally as a waste product, fibreboard production which positive effect on deforestation as it avoids the use of virgin wood and pulp and paperproduction.

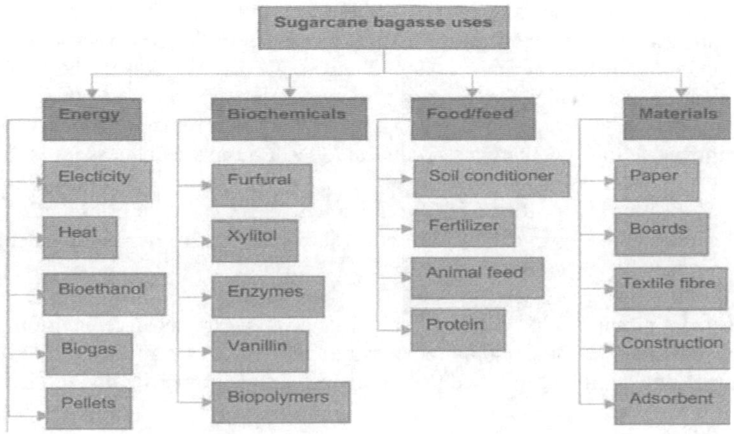

Figure 13. Sugarcane bagasse uses (WADE, 2004)

3.4. The use of SCB in the constructionindustry

Recent studies are made to use organic fibres in the construction industry in brick industry as shown in the table below, insulation boards, fibrous building panels and cement boards. Organic fibres are produced from various solid wastes for example bamboo, coconut, date palm, oil palm, sugar palm, sugarcane, and vegetable wastes. These fibres have inert chemical properties than either steel or glass fibres in addition to their lower cost and natural form. Table 5 demonstrate the different types and sources of solid wastes worldwide and their recycling and utilization potentials for construction materials.

Table 5. Different types and sources of solid wastes and their recycling and utilization potentials for construction materials (Pappu et al., 2007)

Type of solidwastes	Source details	Recycling and utilization potential
Agro-waste (organic)	Baggage, rice and wheat straw and husk, sawmill waste, ground nutshell, jute, sisal, cotton stalk, vegetable residues	Cement boards, particle boards, insulation boards, wall panels, roof sheets, binder, fibrous building panels, bricks, acid-proof cement, coir fibre, reinforced composites, polymer composites
Industrial waste (inorganic)	Coal combustion residues, steel slag, bauxite red mud, construction debris	Bricks, blocks, tiles, cement, paint, fine and coarse aggregates, concrete, wood substitute products, ceramic products
Mining/mineral waste	Coal washeries waste; mining waste tailing from iron, copper, zinc, gold and aluminium industries	Bricks, fine and coarse lightweight aggregates, tiles
Non-hazardous waste	Waste gypsum, lime sludge, limestone waste, broken glass and ceramics, marble processingresidues, kiln dust	Blocks, bricks, cement clinker, hydraulic binder, fibrous gypsum boards, gypsum plaster,super-sulphated cement
Hazardous waste	Contaminated blasting materials, galvanizing waste, metallurgical residues, sludge from wastewater and waste treatment plants, tannerywaste	Boards, bricks, cement, ceramics, tiles

The principle causes of environmental concerns and burdens of the continuous growth of agriculture and industrial waste generation, therefore using organic wastes in the construction industry such as brick production is highly beneficial (Xu, Ji, Gao, Yang, & Wu, 2018). Researchers from around the world started focusing and searching for ways to utilize these wastes, to reduce their harmful effects. Many studies nowadays try to approach this issue by partially replacing cement with bagasse. According to a study, one ton of sugarcane produces 280 kilo grams of bagasse waste which is also considered as an economical issue also as an environmental issue due to poor disposal and handling (Xu, Ji, Gao, Yang, & Wu, 2018). Recycling the sugarcane bagasse in the brick industry to partially replace cement is considered as inexpensive and will reduce both environmental and economic problems.

4. Cement

In this part, the cement production process will be introduced, and its negative environmental impacts will be evaluated. This will help understand the cement's properties and application in brick industry. In addition, Cements brick will be explained to help understand its properties and usage.

4.1. Cement productionindustry

Cement is a fine, soft, powdery-type substance, used mainly for binding sand and aggregates together for concrete mixing. Cement acts as a hydraulic binder; hardens during water addition, and it is the key ingredient in concrete and mortar to be used for building durable structures. According to the European cement association (2016) The cement-making process can be divided into two basic steps (Fig.14):

- The main constituent of cement is clinker, is manufactured in a kiln with gas at temperature of 2000 °C, this heats raw materials such as limestone and other materials for example clay to 1450°C. in this process, limestone is transformed into calcium oxide (lime) then reacts with the other components to form new minerals. This semi-molten material is then rapidly cooled to a temperature of 100 -200°C.
- Cement grey powder is produced by grounding clinker with gypsum and othermaterials

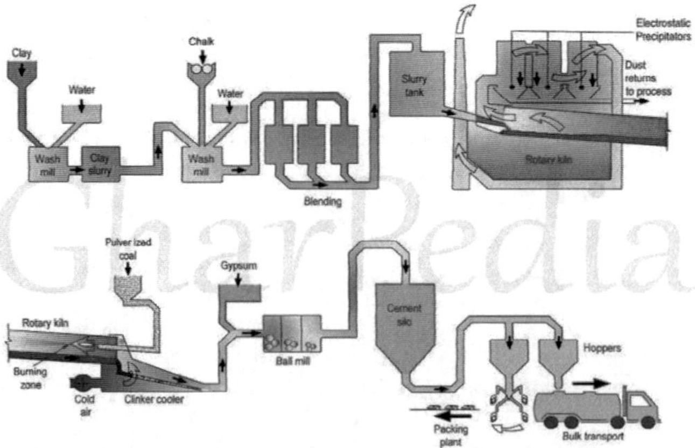

Figure 14. cement manufacturing process (gharpedia , 2018)

4.2. Impact on theenvironment

4.2.1. Energyconsumption:

Energy consumption is the biggest environmental concern with cement and concrete production. Manufacturing process of cement is considered to be the most energy intensive industries around the world. Cement production, direct fuel use for mining and transporting raw materials consumes around 1,758 kWh per cement ton (Parga, Rocco, Christoforo, & Panzera, 2012). Fig. 15 present the percentage of energy demand during cement production process.

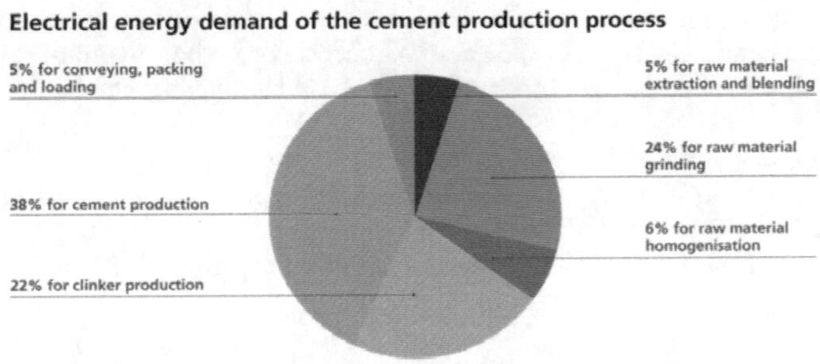

Figure 15. Energy demand of the cement production process (cembureau, 2018)

4.2.2. Air emissions:

Carbon dioxide emissions are produced from two different sources during cement production. The largest source is fossil fuels combustion for rotary kiln operation which is approximately 3/4 t of CO_2 per ton of cement followed by limestone calcining into lime chemical process in the cement kiln. Cement industry is accountable for about of 5% of global CO_2 emissions (Torrie, 2017). Fig. 16shows emissions percentages of CO_2 comes from cement bricks production in Egypt in 2006. Cement and concrete production generate considerable quantities of air-pollutant emissions besides CO_2 such as dust and other air pollution emissions. The dust particulate emits around 360 pounds per ton of cement produced as a result of raw material handling, cement clinker grinding and packaging or loading finished cement. Air pollutants emitted during cement manufacturing are sulphur dioxide (SO_2) and nitrous oxidesemissions.

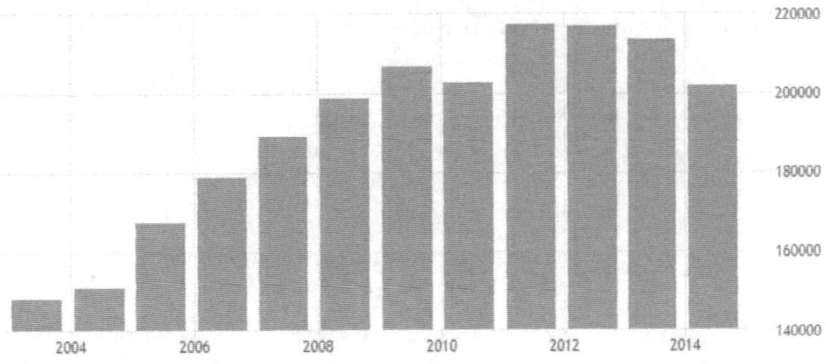

Figure 16. Percentage of CO_2 comes from cement bricks in Egypt in 2006 (Torrie, 2017)

4.2.3. Waterpollution:

Water pollution is another environmental issue with cement and concrete production. At the batch plant, water used for equipment cleaning is sometimes discharged into setting ponds where the solids settle out (Hussein & Sawan, 2010).

IOP Conf. Series: Materials Science and Engineering **974** (2020) 012036 doi:10.1088/1757-899X/974/1/012036

4.3. Cementsbrick

Cement is used by architects and structural engineers to design structures with high structural capacity, fire resistance, water resistance and insulating and acoustical advantages. In many cases, this minimum-maintenance material provides economical advantage to building codes and clients'needs.

4.3.1. Structuralproperties:

Cement blocks vary by type however; its primary structural property is compressive strength. The Building Code Requirements for Structural Concrete include the structural property standards for cement bricks at 7-day and 28-day (Dlamini, 2014). According to the Egyptian standard specification 1292, for load bearing brick to be 5.6 N/m² and for non-load bearing brick to be 2 N/m².

4.3.2. Thermalconductivity:

The insulating properties of various concrete blocks vary depending on manufacturer and on block density. As the density of the block decreases by volume producing low weight block, the lower the thermal conductivity of the block, which enhance the insulation properties of the block (Babor, Plian, & Judele, 2010)

4.3.3. Cost:

Cement blocks prices in Egypt are expensive in which in 2018 for every

- 1000 brick from Cement hollow 40 × 20 × 12 cm cost 3729 Egyptianpound
- 1000 brick from Cement hollow 40 x 20 x 20 cm cost 4746 Egyptianpounds
- 1000 brick from Cement solid 25 x 12.5 x 6 cm cost 630 Egyptianpounds

4.3.4. Waterresistance:

Permeability and porosity vary by unit type, but generally cement blocks water absorbents. The mixture of course and fine aggregates creates a better waterproofed cement block finish. The block's permeability is affected by the amount of cement used in the manufacturing process. As the mixture gets rich in cement, the less the block's permeability (Torrie, 2017).

4.3.5. Aestheticproperties:

Most cement blocks come in greyish colour. Nowadays, manufactures produce blocks with variety of colours, textures and finishes for different purposes. This eases the architect's job for utilizing this type of block in various uses, design and to meet the ever-evolving constructionmarket.

5. Case studies

In this part, several case studies that experimented sugarcane bagasse as a partial cement replacement will be analysed to obtain different experimental alternative for the different experiment's components. This helps to obtain the best suitable experimental program to be adapted in this research.

5.1. Experimental Study on Partial Replacement of Cement by Sugarcane Bagasse Ash (Kumari, 2015)

A. Location: India
B. **Date:** July2015
C. **Type of model experimented:** concrete cylindrical specimens with dimensions of 150 mm x 300 mm
D. **Cement type:** Portland Pozzolana cement with trade name "LAFARAGE DURAGUARD" cement referring to IS code 1489-1-1991(part 1). Table 6 shows itsproperties:

IOP Conf. Series: Materials Science and Engineering **974** (2020) 012036 doi:10.1088/1757-899X/974/1/012036

Table 6. Portland Pozzolana cement physical properties (Kumari, 2015)

Tests	Results
Fineness (%)	6%
Soundness (mm)	5 mm
Specific gravity	3.057
Normal consistency (%)	34%
Initial setting time (min.)	106 min.
Final setting time(min.)	310n.

E. **Bagasse treatment method:** Combustion; The bagasse was heated at 700 ºC for one hour until it became dry and black ash as shown in Fig.17.

Figure 17. Combustion treatment process (Kumari, 2015)

F. **Percentage of replacement:** 5%, 10%, 15% and20%
G. **Water-cement ratio:** 0.44, 0.45,0.46
H. **Factors tested:** Compressive strength and density of concrete
I. **Methodology:**theconcretemixwasexperimentedatwater-cementratioof0.44,0.45and 0.46. Several samples with varied sugarcane bagasse ash percentages of 5, 10, 15 and 20%. Samples then undertake the proposed tests to obtain results (Fig.18).

Figure 18. Concrete blocks after curing (Kumari, 2015)

J. **Results and findings:** the following graphs show the strength and density results obtained through this experimentalprogram

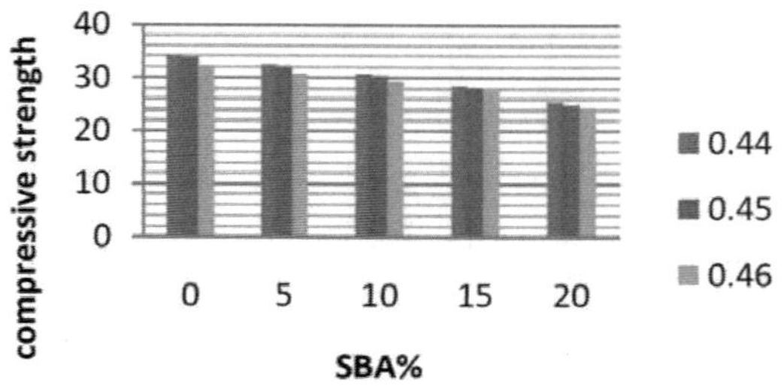

Figure 19. Compressive strength of different SBA percentages and different W-C ratio (Kumari, 2015)

Fig.19 display the compressive strength comparison between concrete and SBA samples at different W-C ratio of 0.44, 0.45 and 0.46. The graph shows that compressive strength decreases with increase in w/c ratio.

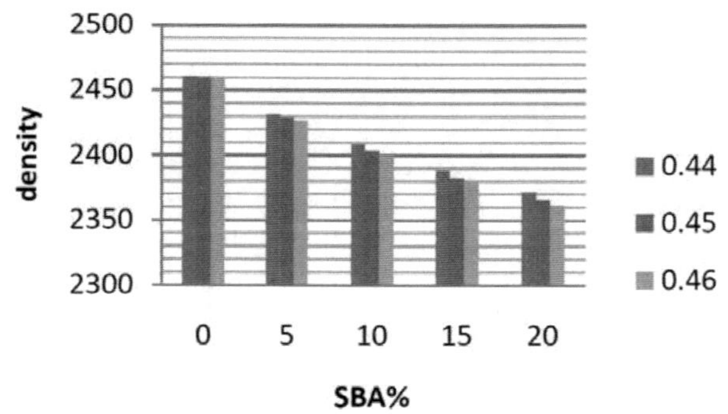

Figure 20. Density at different SBA percentages and different W-C ratio (Kumari, 2015)

Fig. 20 shows the density comparison between concrete and SBA samples at different W-C ratio of 0.44, 0.45 and 0.46. The graph shows decrease in density with the increase in w/c ratio.

K. **Conclusion:**

- Concrete containing SBA showed a decrease in workability as SBA increases due to the higher water absorption of bagasseash.
- The concrete's compressive strength decreases with the increase of SBApercentage
- Density and compressive strength of concrete showed slight reduction as water to cement ratioincreased
- Sugarcane bagasse ash concrete is more economical and have lowerdensity.
- The burning of bagasse altered the mechanical, chemical and physical properties of the bagasse thus reducing itsproperties.

5.2. Utilization of sugarcane bagasse ash in concrete as partial replacement of cement (Mangi,2017)

A. **Location:**Pakistan
B. **Date:**2017
C. **Type of model experimented:** concrete cylindrical specimens with dimensions of 150 mm x 300 mm with 2 different mixes M20 (1:1.5:3) and M15(1:2:4).

D. **Cement type:** The ordinary Portland cement (ASTM C150 Type-I). Table 7 shows its properties

Table 7. Ordinary Portland cement physical properties (Mangi, 2017)

Tests	Results
Fineness (kg/m²)	330 kg/m²
Soundness (mm)	2.5 mm
Specific gravity	3.12
Normal consistency (%)	29%
Initial setting time (min.)	65 min.
Final setting time(min.)	275 min.

E. **Bagasse treatment method:** Combustion; The bagasse was heated and burned is a close drumfornon-controllableburninguntilfullyturningintoblackishgreyashasshowninFig.

Figure 21. Bagasse ash after being burned in a close drum (Mangi, 2017)

21.

F. **Percentage of replacement:** 5%, and10%
G. **Water-cement ratio:**0.4
H. **Factors tested:** Compressivestrength
I. **Methodology**: the 2 concrete mixes are prepared and poured then cured for the period of 7, 14 and 28 days. Cement is then partially replaced by bagasse ash by 5%, and 10% and cured for the same period then all specimen was tested for compressive strength after the completion of the curing period. The steps are shown in Fig.22.

Figure 22. steps of concrete specimen from curing to testing and its final shape (Mangi, 2017)

J. **Results and findings:** the following graphs show the compressive results obtained from the 2 mixes M20 and M15 through this experimentalprogram

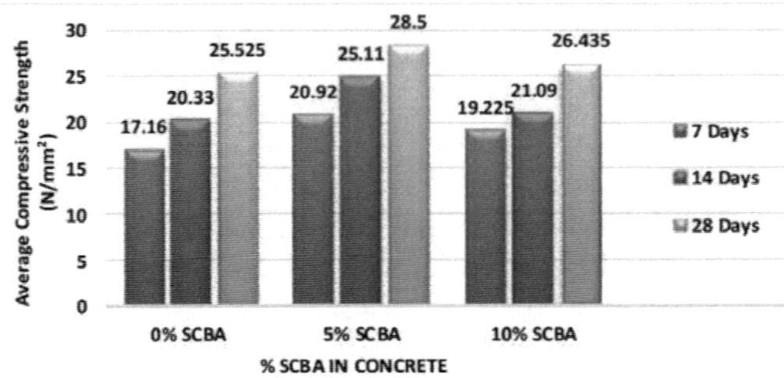

Figure 23. Compressive strength of concrete M20 mix with different SCBA percentages and different curing periods (Mangi,2017)

The compressive strength in M20 mix results are presented in Fig. 23. It is observed that at 5% replacement the strength is at the highest strength of 28.5 N/mm² then it decreases to 26.4 N/mm² at 10% replacement of SCBA. It is 12% higher than the normal concrete sample.

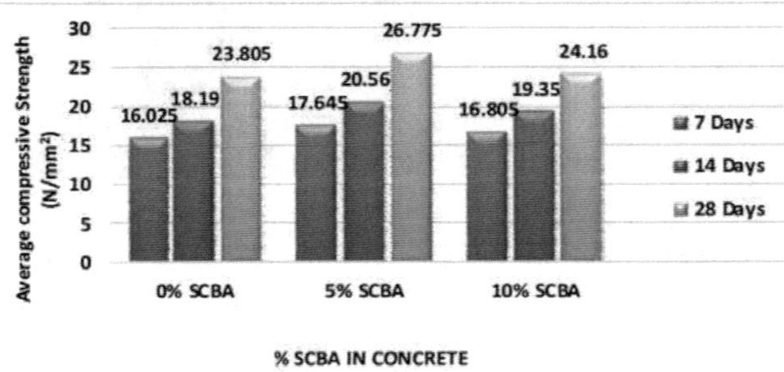

Figure 24. Compressive strength of concrete M15 mix with different SCBA percentages at different curing periods (Mangi, 2017)

Fig. 24 shows the concrete sample of M15. The compressive strength showed at 5% replacement the strength was recorded the highest at 26.7 N/mm² which results of an increase of about 11.5% compared to the normal concrete strength.

K. **Conclusion:**

- Concrete with SCBA shows higher compressive strength compared to normal concrete and the optimum replacement is5%.
- Usage of SCBA reduces cement and reduce wastes by itsproduction
- Cement replacement decreases the workability ofconcrete.

5.3. Utilization of sugarcane bagasse ash in concrete as partial replacement of cement (Mangi,2017)

A. **Location:**Malaysia
B. **Date:**2017
C. **Type of model experimented**: cube with size of 100 mm x 100 mm x 100mm
D. **Cement type**: normal concrete and light weightconcrete

IOP Conf. Series: Materials Science and Engineering **974** (2020) 012036 doi:10.1088/1757-899X/974/1/012036

E. **Bagasse treatment method:** Sodium Hydroxide NaOH treatment; after drying bagasse completely under the sun for 7 days it is cut into small pieces 5cm until 10cm, then it is immersed in 50% dilute NaOH for 3 days. This is done to remove the impurities and withstand longer to use in concrete. After this, it is completely dried under the sun as presented in Fig.25.

Figure 25. Sodium Hydroxide NaOH bagasse treatment method steps (Khalid, 2017)

F. **Percentage of addition**: 0.5%, 1.0% and1.5%
G. **Water-cement ratio:**0.4
H. **Factors tested:** Compressive strength and tensilestrength
I. **Methodology:** the concrete curing in water at room temperature 7 and 28 days in order to harden to achieve maximum strength. For each sample of normal concrete and light weight concrete, 32 samples were prepared then compressive and tensile tests wereconducted.
J. **Results and findings**: the following graphs show the compressive results obtained from both of normal concrete and light weight concretesamples

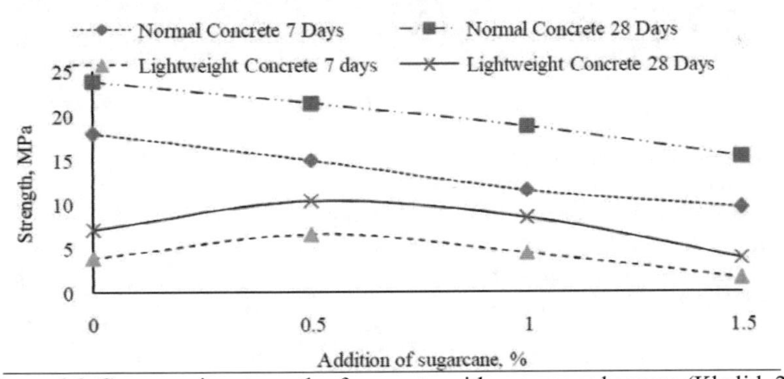

Figure 26. Compressive strength of concrete with sugarcane bagasse (Khalid, 2017)

The compressive strength results are presented in Fig. 26. The sample with 0.5% sugarcane bagasse had the highest strength of 10.3 MPa, while the sample with 1.5% addition of bagasse showed the lowest value of 3.8 MPa.

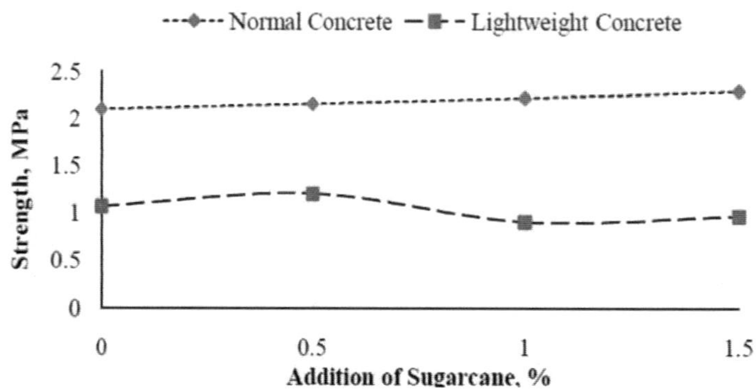

Figure 27. tensile strength of concrete after 28 days (Khalid, 2017)

Fig. 27 present the tensile strength of normal concrete and light weight concrete samples. In normal concrete, the sample with 1.5% sugarcane represented the highest tensile strength for 28 days curing of 2.29 MPa. While in light weight concrete, the addition of 0.5% bagasse showed the maximum tensile strength of 1.21 MPa in relation with 1% and 1.5% bagasse addition. The addition of bagasse results in lowering the tensile strength.

K. Conclusion:

- The optimum bagasse addition for the higher compressive strength is 0.5% addition for both normal and light weightconcrete.
- In normal concrete, the tensile strength increase as bagasse addition increases and the optimum percentage is 1.5%, while in light weight concrete the tensile strength decreases as the bagasse percentage increases and the optimum addition is0.5%.

5.4. Utilization of bagasse waste-based materials as improvement for thermal insulation of cement brick (Aminudin,2017))

A. **Location:**Malaysia
B. **Date:**2017
C. **Type of model experimented:** samples are with two different dimensions of 50 mm x 50 mm x 50 mm and 215mm x 102.5mm x65mm
D. **Cement type:** The ordinary Portland cement (ASTM C150 Type-I). Table 8 present its chemicalcompounds

Table 8. physical properties of ordinary Portland cement (Aminudin, 2017)

Tests	Results
Fineness (kg/m²)	330 kg/m²
Soundness (mm)	2.5 mm
Specific gravity	3.12
Normal consistency (%)	29%
Initial setting time (min.)	65 min.
Final setting time(min.)	275 min.

E. **Bagasse treatment method:** hot water treatment; as displayed in Fig. 28, the bagasse is soaked in hot water for 4-6 days to remove excess sugar and impurities then it is washed thoroughly and placed dry in the sun. after drying the bagasse is cut into smallpieces.

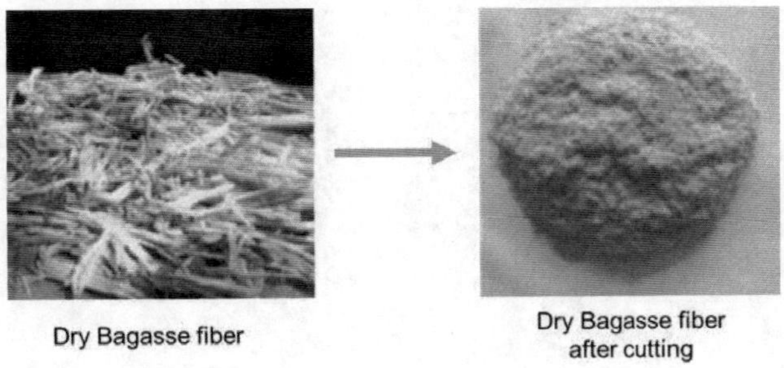

Figure 28. Hot water bagasse treatment method after cut into small pieces (Aminudin, 2017)

F. **Percentage of replacement:** 2%, 4%, 6%, 8% and10%
G. **Water-cement ratio:**0.45
H. **Factors tested:** Compressive strength and thermalconductivity
I. **Methodology:** the samples presented used cement to sand ratio of 1:3 and cement-water ratio of 0.45 for cement brick production. The bagasse was introduced by 2%, 4%, 6%, 8% and 10%. A control sample with 0% bagasse was introduced for comparison. This is to ensure the optimum results based on the most favourable. 50mm × 50mm × 50mm samples were produced for compressive strength test, while 102.5mm × 215mm × 65mm samples tested thermalconductivity.
J. **Results and findings:** the following graphs show the compressive results and the thermal conductivity- compressive strength correlation obtained from the 2 dimensions samples of 50mm × 50mm × 50mm and 102.5mm × 215mm × 65mmbricks

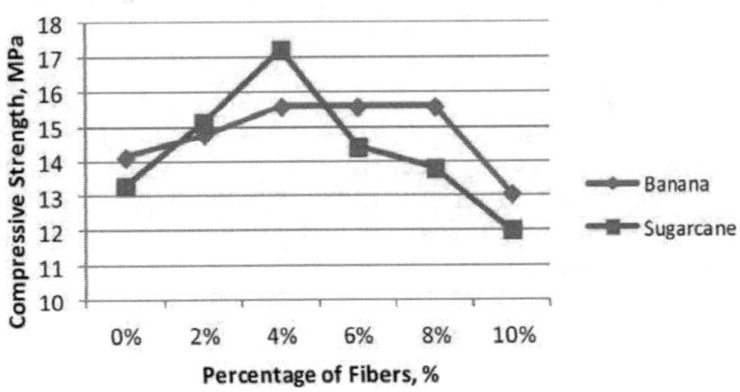

Figure 29. Compressive strength against percentage in comparison with banana trunk fiber and sugarcane bagasse (Aminudin, 2017)

The compressive strength of 50mm × 50mm × 50mm brick results are presented in Fig. 29. More than 4% replacement results in a decrease of strength of the brick. This happens due to poor workability and poor boding between bagasse and mortar.

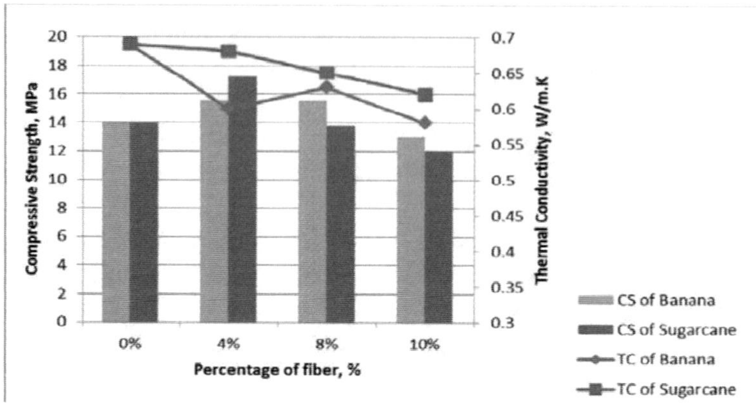

Figure 30. Compressive strength and thermal conductivity relationship of sugarcane bagasse and banana trunk fiber (Aminudin, 2017)

Fig. 30 shows that the 102.5mm × 215mm × 65mm cement brick sample compressive strength and thermal conductivity correlation. The thermal conductivity decreases as the percentage of sugarcane bagasse increases. At 0% bagasse the strength was lower, but the thermal conductivity was higher compared to 4% and 8%. Although 10% replacement shows the best result for reduction of thermal conductivity, the compressive strength decreases lower than the required strength, so it is not favourable for permeantstructure.

K. Conclusion:

- Sugarcane bagasse has a potential of partially replacing cement in cement brick industry.
- Sugarcane bagasse can lower the thermal conductivity of the brick if the strength was maintained
- 4%- 8% preplacement is the optimum result before the strength drops below the standardstrength

4.6. Case studies conclusion

After analysing the previous four case studies, Table 9 summarizes all the results. Analysing Table 9, it was deducted that cement replacement with sugarcane bagasse is not limited to any specimen dimensions and most case studies used ordinary Portland cement. For bagasse treatment 3 methods were used combustion, NaOH treatment and hot water treatment, both combustion and NaoH treatment alter the properties of the bagasse, whereas the hot water treatment does not affect the bagasse's properties and it is a cheap method. Preferably the replacement should begin at 4% and does not exceed 10% to maintain standard strength with W-C ratio of 0.4 and 0.45 since as W-C ratio increase strength decrease.

Table 9. Comparison between case studies (developed by author, 2019)

Case study name	Experimental Study on Partial Replacement of Cement by Sugarcane Bagasse Ash (Kumari, 2015)	Utilization of bagasse materials as improvement for thermal insulation of cement brick (Aminudin,2017)	Properties of sugarcane fibre on the strength of the normal and lightweight concrete (Khalid, 2017)	Utilization of bagasse materials as improvement for thermal insulation of cement brick (Aminudin,2017)
Location	India	Pakistan	Malaysia	Malaysia

Year	2015	2017	2017	2017
Model type	150 mm x 300 mm concrete cylindrical specimens	150 mm x 300 mm concrete cylindrical specimens 2 different mixes M20 (1:1.5:3) and M15(1:2:4).	cube with size of 100 mm x 100 mm x 100 mm	50 mm x 50 mm x 50 mm for compressive strength 215mm x 102.5mm x 65mm for thermal conductivity
Cement	Portland Pozzolana cement	Ordinary Portland cement	normal concrete and light weight concrete	Ordinary Portland cement
Bagasse treatment	Combustion	Combustion	Sodium Hydroxide NaOH	Hot water
Hazardous waste	5%, 10%, 15% and 20% replacement	5%, and 10% replacement	0.5%, 1.0% and 1.5% addition	2%, 4%, 6%, 8% and 10% replacement
Water-cement ratio	0.44, 0.45, 0.46	0.4	0.4	0.45
Factors tested	Compressive strength and density ofconcrete	Compressive strength	Compressive strength and tensile strength	Compressive strength and thermal conductivity
Curing period	7 and 28 days	7, 14 and 28 days	7 and 28 days	28 days
Results	compressive strength decreases with increase in w/cratio. decrease in density with the increase in w/c ratio	M20 mix: at 5% replacement the strength is at the highest strength then it decreases slightly at 10%replacement. M15 mix: at 5% replacement the strength was the highest and decreased slightly at 10%.	Compressive strength: 0.5% sugarcane bagasse had the highest strength Tensile strength in normal concrete: the 1.5% sugarcane sample had the highest tensile strength Tensile strength in light weight: 0.5% bagasse had the highest tensile strength	Compressive strength: More than 4% replacement decreases the bricks' strength Thermal conductivity: The thermal conductivity decreases as the percentage of sugarcane bagasse increases. At 10% replacement thermal conductivity is the lowest
Conclusion	Workability of concrete decreases with the increase in bagassepercentage As the W-C ratio increases the strength and density decrease slightly	Concrete with SCBA shows higher compressive strength compared to normal concrete the optimum replacement is5%. Cement replacement decreases the	The addition of bagasse results in lowering the tensile strength. The optimum bagasse addition for the higher compressive strength is 0.5% additionfor both normal and	Brick strength decreases after 4% replacement due to poor workability and poor boding between bagasse andmortar. 4%- 8% preplacement is the optimum result before the strength

	Optimum strength and density are at 10% bagasse replacement	workability of concrete.	light weight concrete.	drops below the standardstrength
Comments	Low density concrete isn't used for all building types Bagasse combustion alters its mechanical, chemical and physicalproperties	Usage of SCBA reduces cement and reduce wastes by its production Bagasse combustion alters its mechanical, chemical and physicalproperties	Using bagasse fibre NaOH treatment alters chemical properties but does not affect the environment Handling NaOH can be risky Expensive treatment process and takes time	Brick with low thermal conductivity and low strength is not favourable for permeant structure Hot water treatment is favourable as it does not alter the bagasseproperties Hot water treatment is easy and cheapbut takestime

6. Conclusion

As a result, to the negative environmental impacts from cement production industry, the government with the aid of scientist, engineers and researchers are searching for alternatives to decrease cement production and usage in the construction industry. The huge production of agricultural waste provided these parties with an alternative to utilize them in the construction industry. sugarcane bagasse is highly produced in Egypt; therefore, it is a good candidate for cement replacement in the brick industry. As a conclusion, sugarcane bagasse is a great candidate to be used as cement replacement to reduce the environmental impacts of cement production industry and poor disposal and burning of bagasse and to develop cheap alternative for cement brick despite its strength and durability. Sustainability will be achieved through reusing sugarcane bagasse to produce useful products like cement bricks. These proposed utilizing techniques are simple, available, cheap and can be by small-scale workers. This approach has a significant impact on achieving environmental sustainability factors, socially and economically. This type of bricks can be used for temporary and non-bearing structures. The results of previous experiments have been encouraging as in most of them strength increases and thermal conductivity and density decreases. Adding fibre to concrete bricks has reduced the density of cement bricks, which can be attributed to low fibre density this lowers the total structure weight. Although past certain percentage results drop, more research and development in this sector may alter these results in the future. As a conclusion, the use of sugarcane made cement bricks a great future in the field of construction in Egypt because it is able to achieve the three factors environmentally, socially and economically. As a result, from the previous analysis of the literature review and case studies, the following SWOT analysis (Fig. 31) shows the strengths, weakness, opportunities, and threats of the brick with cement replacement, then experimental program alternatives are shown in Table10.

IOP Conf. Series: Materials Science and Engineering **974** (2020) 012036 doi:10.1088/1757-899X/974/1/012036

Table 10. Experimental program alternatives matrix (developed by author,2019)

CEMENT	Cement type	Ordinary Portland cement
		Portland Pozzolana Cement
		Rapid Hardening Cement
		Quick Setting Cement
		Ordinary Portland cement
		Sulphates resisting cement
		High Alumina Cement
		White and coloured Cement
		Air Entraining Cement
	Cement: Sand: Aggregates	M15- 1:2:4
		M20- 1:1.5:3
	Cement: Water ratio	1: 0.4
		1: 0.45
		1: 46
		1: 0.5
Sugarcane bagasse	Bagasse treatment method	Combustion
		Sodium Hydroxide NaOH
		Hot water (preferably)
	Bagasse piece dimension	3 < D < 5 mm (best)
		D < 1 cm (preferable)
		1 < D cm
	Bagasse status	Fibre
		Ash
	Bagasse replacement percentage (%)	R < 5%
		5 < R < 10% (preferable)
		10 < R%
Mould	Mould type	Cube
		Cuboid
		Cylindrical
		Hexagon
		Interlocking
		Customized
	Mould material	Iron
		Stainless steel
		Wood
		Plastic
	Mould dimension (mm)	150 mm x 300 mm cylindrical
		100 mm x 100 mm x 100 mm
		50 mm x 50 mm x 50 mm
		215mm x 102.5mm x 65mm
		250mm x 125mm x 60mm
		Concrete mix
	Mould numbering/ pouring (unit)	Diesel oil
		Rubber lubricant
		6
		9
		12
		N > 12

Aggregate size (mm)	9.5 < A mm (not in bricks)
	4.75 < A < 9.5 mm (preferable)
	A < 4.75 mm
Curing period	7 days
	14 days
	28 days (preferable)
Tests	Compressive strength (must)
	Tensile strength
	Density
	Absorption
	Slump test
	Thermal conductivity
	Durability

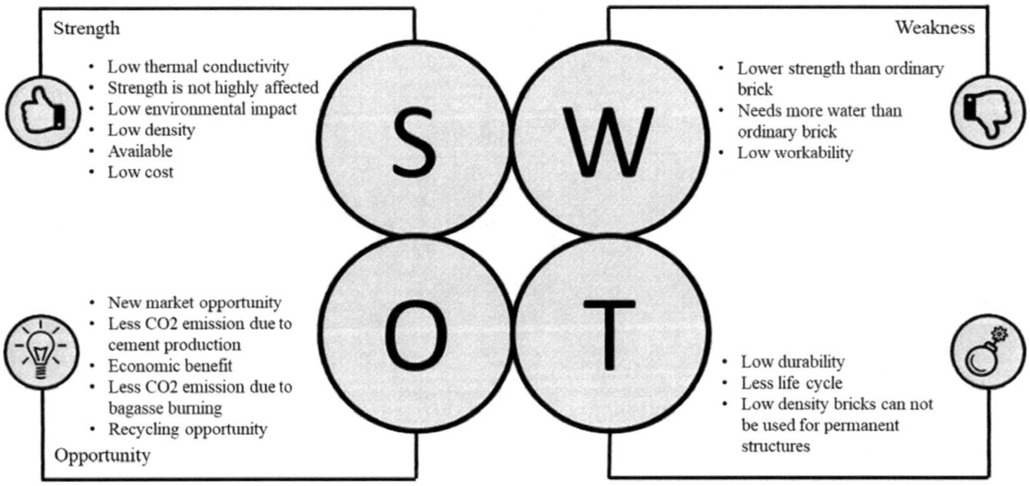

Figure 31. SWOT analysis of brick with bagasse replacement (developed by author, 2019)

References

[1] Aminudin, E., Khalid, N. H. A., Azman, N. A., Bakri, K., Din, M. F. M., Zakaria, R., & Zainuddin, N. A. (2017). Utilization of Bagasse Waste Based Materials as Improvement for Thermal Insulation of Cement Brick. *MATEC Web of Conferences,* 103, 1–8. Retrieved from: https://doi.org/10.1051/matecconf/201710301019 [Last Accesses 16 December2019].

[2] Babor, D., Plian, D., & Judele, L. (2009). Environmental impact of concrete. *Buletinul Institutului Politehnic Din Iaşi; Universitatea Tehnică, Gheorghe Asachi*. Retrieved from: https://www.researchgate.net/publication/40497595_Environmental_Impact_of_Concrete [Last Accesses 21 October2019].

[3] Balaji, A., Karthikeyan, B., & Sundar Raj, C. (2015). Bagasse fiber – The future bio composite material: A review. *International Journal of Chemtech Research,* 7(1),223–233.

[4] Cara, G., Ilardi, S., Perkins, C., & Acharya, D. (2017). The Urban Bio-Loop. Retrieved from: https://www.arup.com/publications/research/section/the-urban-bio-loop?utm_source=LinkedIn&utm_medium=Social&utm_campaign=UrbanBioLoop&utm_content=C0000039[Last Accesses 12 December2019].

[5] Chauhan M.K., Chaudhary V.S, Samar S.K., 2011. Life cycle assessment of sugar]]]] industry: A review. *Renewable and Sustainable Energy Reviews*, 15(7):3445-3453.

[6] Dlamini, M. N. (2014). Clay and concrete brick. *CSIR BE* 1–6. Retrieved from:https://researchspace.csir.co.za/dspace/bitstream/handle/10204/7575/Dlamini_2014.pdf?sequence=1&isAllowed=y[Last Accesses 25 December 2019].

[7] Elfeki, M., & Tkadlec, E. (2015). Treatment of municipal organic solid waste in Egypt. *J. Mater. Environ. Sci.* 6, (3); 756-764. Retrieved from:https://www.jmaterenvironsci.com/Document/vol6/vol6_N3/88-JMES-1184-2014-Elfeki.pdf[Last Accesses 24 October2019].

[8] Faria, K.C.P., Gurgel, R.F. & Holanda, J.N.F., (2012). Recycling of sugarcane bagasse ash waste in the production of clay bricks. *Journal of Environmental Management*, 101:7-12.

[9] Gupta, G., & Bhatnagar, R. (2015). A Review on Composition and Properties of Bagasse Fibers. *International Journal of Scientific & Engineering Research,* 6(5),143–148.

[10] Guirguis, M.N. and Moussa, R.R., 2019. Investigation on utilizing garbage as a resource for a sustainable neighbourhood; Case study of a neighbourhood in New Cairo, (November, 28 -30). (BSCairo2019) Simulation for a Sustainable Built Environment, Cairo_ Egypt. *IOP Conf. Series: Earth and Environmental Science* 397- 012018. Retrieved from: doi:10.1088/1755-1315/397/1/012018 [Last Accesses 24 October2019].

[11] Hassan, S.F. & Nasr, M.I., (2008). Sugar Industry in Egypt. *Sugar Tech*, 10(3):204-209.

[12] Hamada, Y.M., (2011). Water Resources Reallocation in Upper and Middle Egypt. *EWRA European Water, EW Publications*, 33:33-44.

[13] Hussein, S.D.A, Sawan, O.M., (2010). The Utilization of Agricultural Waste as One of the Environmental Issues in Egypt (A Case Study). Journal of Applied Sciences Research, 6(8): 1116-1124. Retrieved from:http://www.aensiweb.com/old/jasr/jasr/2010/1116-1124.pdf[Last Accesses 25 December2019].

[14] Jorge Parga Silva, A., Antonio Rocco Lahr, F., Luis Christoforo, A., & Hallak Panzera, T. (2012). Properties of Sugar Cane Bagasse to Use in OSB. *International Journal of Materials Engineering,* 2(4), 50–56. Retrieved from:https://doi.org/10.5923/j.ijme.20120204.04[Last Accesses 20 December2019].

[15] Kiyasudeen, K., Ibrahim, M. H., Quaik, S., & Ismail, S. A. (2016). Prospects of Organic Waste Management and the Significance of Earthworms. *Applied Environmental science and Engineering for a sustainable future. Springer International Publishing Switzerland* 2016. Retrieved from: https://doi.org/10.1007/978-3-319-24708-3 [Last Accesses 24 October2019].

[16] Mangi, S.A., Jamaluddin, N., Ibrahim, M.H.W., Abdullah, A.H., Awal, A.S.M.A, Sohu, S. & Ali, N. (2017). Utilization of sugarcane bagasse ash in concrete as partial replacement of cement. *IOP Conference Series: Materials Science and Engineering,* 271; pp. 1-8. Retrieved from: https://doi.org/10.1088/1757-899X/271/1/012001 [Last Accesses 24 October2019]

[17] Martinez-Hernandez, E., Amezcua-Allieri, M. A., Sadhukhan, J., & Anell, J. A. (2018). SugarcaneBagasseValorizationStrategiesforBioethanolandEnergyProduction.*Sugarcane-Technology and Research*. Retrieved from: https://doi.org/10.5772/intechopen.72237

[18] Moussa, R.R., 2019. The Reasons For Not Implementing Green Pyramid Rating System In Egyptian Buildings.*Ain Shams Engineering Journal,* 10(4);917-927.

[19] Nakhla, D. A.H., (2014). ACHIEVING ENVIRONMENTAL SUSTAINABILITY OF SUGARCANE INDUSTRY IN EGYPT: AN APPLICATION OF LIFE CYCLE ASSESSMENT. Doctor of Philosophy thesis in Engineering, *The American university in Cairo.* Retrieved from:https://pdfs.semanticscholar.org/05b7/621677f60befa406336df9f556dc54b22be4.pdf[Last Accesses 28 December 2019]

[20] Rabelo, S.C., Costa, A.C.D., & Rossel, C.E.V. (2015). Chapter 17 - Industrial Waste Recovery.*SugarcaneAgricultural Production, Bioenergy and Ethanol*, *Academic press* 365-381. Retrieved from: https://www.sciencedirect.com/science/article/pii/B9780128022399000177?dgcid=scitechconnect_foodsci_starlink#![Last Accesses 28 December 2019]

[21] Singh, D. (2016). STUDY ON THE PERFORMANCE OF CONCRETE USING SUGARCANE BAGASSE ASH AND COAL. Master thesis: *Department of Civil Engineering.* Punjab Agricultural University. Retrieved from:

https://pdfs.semanticscholar.org/93e7/e4f70e0a67dd7e8ca50349743a4028238a9e.pdf[Last Accesses 3 November2019].

[22] Singh, J., Saxena, R., Bharti, V., & Singh, A. (2017). The Importance of Waste Management to Environmental Sanitation: A Review. *Advances in Bioresearch,* 9[2]; 202-207. Retrieved from: https://www.researchgate.net/publication/329864740_The_Importance_of_Waste_Management_to_Environmental_Sanitation_A_Review [Last Accesses 20 October2019].

[23] Stajanča, M. E. and Eštoková, A. (2012). Environmental impacts of cement production. *Technical University of Košice, Civil Engineering Faculty, Institute of Architectural Engineering*. p. 296-302. Retrieved from: http://ena.lp.edu.ua:8080/bitstream/ntb/16692/1/55-Stajanca-296-302.pdf [Last Accesses 25 October2019].

[24] Xu, Q., Ji, T., Gao, S. J., Yang, Z., & Wu, N. (2018). Characteristics and applications of sugar cane bagasse ash waste in cementitious materials. Materials, 12(1), 1–19. https://doi.org/10.3390/ma12010039

[25] Wang, M. et al., 2008. Life cycle energy analysis and greenhouse gas emission implications of Brazilian sugarcane production. International Sugar Journal, 110(1317):527-545.

[26] Economic and Social Commission for Western Asia (ESCWA), 2009. Increasing the competitiveness of small and medium-sized enterprises through the use of environmentally sound technologies: assessing the potential for the development of second-generation biofuels in the ESCWA region, United Nations, NewYork.

[27] World Alliance for Decentralized Energy (WADE), 2004. Bagasse Cogeneration- Global Review and Potential, Scotland, UnitedKingdom.

[28] Torrie, C. (2017). Properties of Concrete Blocks. Retrieved fromhunker.

Military Technical College
Kobry El-Kobbah,
Cairo, Egypt

13th International Conference
on Civil and Architecture
Engineering
ICCAE-13-2020

Structural Behavior of Light Weight Ferrocement Walls

Prof.Yousry B. I. Shaheen[1], Dr.Mohsen Mousa[2,] and Eng.Eman Gamal[3]

Department of Civil Engineering, Faculty of Engineering, Menoufia University, EGYPT.
3Civil Engineer and Postgraduate Fellow

Abstract

This paper presents the experimental results of research to develop ferrocement sandwich panels for use as wall bearing units. The proposed panels are lighter in weight relative to the conventional reinforced concrete panels The sandwich panels consisted of two thin ferrocement layers reinforced with one or two layers of closely spaced galvanized welded wire mesh and expanded steel mesh. The core of the panel was made of light weight brick. Steel wires were used to tie the steel meshes of the two skin layers together and to act as shear connectors to transfer shear between the two ferrocement skin layers. These steel wires were embedded in the mortar joints of the brick. The thickness of the ferrocement skin layer was 10mm when single layer of wire mesh was used and 15mm when two layers of mesh reinforcement were used. Experimental investigation was conducted on the proposed panels. A total of 10 sandwich panels Having the dimensions of 600mm width and 700mm height were tested under compressive loadings until failure. The deformation characteristics and cracking behavior were measured and observed for each wall panel. The results showed that high ultimate and serviceability loads, crack resistance control, high ductility, and good energy absorption properties could be achieved by using the proposed panels. This could be of great construction advantages for both developed and developing counties.

Keywords: Ferrocement; Sandwich Panels; Deformation characteristics; Serviceability load; Ductility
and Energy absorption properties; Cracking pattern; Ultimate loadings.

1. INTRODUCTION

Ferrocement is a construction material that proved to yield superior properties in terms of crack control, impact resistance, and toughness, largely due to the close spacing and uniform dispersion of reinforcement within the material. Many investigators have reported the physical and mechanical properties of this material and numerous test data are available to define its performance criteria for design and construction [1], [2]. Recently, ferrocement has received attention as a potential building material in developing countries, especially for roofing of housing construction [3]. The behavior and strength of wall elements under axial loading were studied by Joshi [4]. Swamy and Spanos [5] presented a proposed method of fabrication of ferrocement slabs by grouping the reinforcement meshes at top and bottom and tying them with skeletal steel to form a rigid cage. The proposed method proved to be effective in obtaining high ultimate strength and better deformation characteristics. Fahmy et al [6] presented the experimental results of first phase of the current research [6]. In that phase, twelve ferrocement sandwich panels and hollow core panels were developed and evaluated for applications in walls, slabs and roofing systems. The sandwich panels were manufactured from two outer ferrocement skins separated by a core of light -weight materials. Two types of core material were investigated, namely; light weight brick and foam concrete. For the hollow cored panels, the core was made of the same mortar matrix as the skin ferrocement layer with three circular holes running along the length of the specimen. The experimental results showed that the behavior of the light brick specimens was superior to that of the foam concrete core specimens. The results also showed that high ultimate and serviceability loads crack resistance control, high ductility, and good energy absorption properties were achieved by using the proposed panels.

Ferrocement is proposed is being used for thin radial gates with high strength, crack resistance, high ductility and energy absorption suitable for irrigation works to replace the commonly used steel gales. This paper presents the results of analytical stud y of the strength of a proposed ferrocement radial gate. The cement-sand matrix of the ferrocement composite was designed to achieve high compressive, tensile and flexural strength for the ferrocement radial gates. This was achieved by adding silica fume as replacement percentage of the cement content. The results of an experimental investigation of the effect of silica fume on the strength of the matrix are also included. Replacement of 15% of cement with silica fume proved to be the optimum percentage for water-binder ratio of 0.28. The suitable gate thicknesses for different water heights and waterway widths are also presented in this payer. E.H. Fahmy, Y.B. Shaheen and W.M. El Dessouki, 1995, [8].

Tensile Behavior of Thin Ferrocement Plates presents comprehensive test data on the tensile behavior of ferrocement plates, 12.5mm thick. The main variables investigated were mesh geometry, specific surface, volume fraction, mesh yield strength and skeletal bars. The specimens were specially designed to ensure

IOP Conf. Series: Materials Science and Engineering **974** (2020) 012037 doi:10.1088/1757-899X/974/1/012037

failure in the gage length. The matrix was proportioned for high strength, high workability and high durability with low water to binder ratio, and contained 50% cement replacement with fly ash, Cracking and deformation were monitored throughout the loading range. The results showed that the composite properties of elastic modulus and ultimate strength could be very satisfactorily predicted by a single unique relationship, there was, however, a good correlation between the composite properties of ultimate tensile strength and ultimate flexural strength. The results show that by suitable design of the matrix and the reinforcement, high strength ferrocement sheets with high crack resistance can be developed for a variety of structural applications, Swamy R N and Y B I Shaheen, 1990, [9].

Sandwich wall bearing units presents the experimental results of the second phase of the previously published research to develop ferrocement sandwich wall bearing units. The sandwich panels consisted of two thin ferrocement layers reinforced with one or two layers of closely spaced welded wire mesh. The core of the panel was made of light weight brick. Z-shaped steel bars were used as shear connectors to transfer shear between the two ferrocement skin layers. The shear connectors were tied to the mesh reinforcement of the ferrocement skin layer and were embedded in the mortar joints of the brick. The thickness of the ferrocement skin layer was 10mm when single and double layers of wire mesh was used. The core material was 100 mm thick. Experimental investigation was conducted on a total of 10 sandwich panels. The experimental results showed that the panel thickness, type of steel mesh and the number of reinforcing steel meshes have major impact on the behavior of the panels. The results also showed that all tested panels had high ultimate and serviceability loads, crack resistance control, high ductility, and good energy absorption properties.

2. EXPERIMENTAL PROGRAM

The experimental program was designed to investigate the effect of strengthening light weight wall with ferrocement of wall bearing panels. These parameters are: thickness of the ferrocement layer, type of reinforcing steel mesh, and the number of steel reinforcement meshes in each ferrocement layer were tested under axial compression loadings. Steel reinforcement provided number of the steel layers at each face, and the thickness of the two thin skin layers, as shown in Table 1. Each designation comprised of 5 typical specimens. The width and the height of all specimens were 600mm and 700mm respectively. Different types of steel meshes were used to provide the steel reinforcement in the two thin skin layers at each face of the specimens. These types included welded wire meshes (WWM), expanded steel meshes (ESM). Figure 1 shows welded, expanded meshes, Poly propylene fibers e300, and tensile steel of meshes according to, (Swamy and Shaheen, 1990).

.

2.1 Light Weight Brick Core Panels

For this type of panels, a 100mm thick wall was built of this light brick to provide the core material between the two thin skin layers. The outer two skin layers were tied together by 3mm diameter wires to act as shear connectors to transfer the shear between these two layers. These steel wires were embedded inside the mortar of the core wall and were tied to the steel wire mesh of each ferrocement layer. Figure 2 shows ferrocement wall sandwich wall panels and the steel mesh before casting the mortar of the ferrocement skin layers. While Figs. 3 and 4 show reinforcement details of light weight walls.

2.2 Mix Design and Material Properties

The sand-cement mortar of ferrocement consisted of sand, ordinary Portland cement, and silica fume with sand/cement ratio of 2.0. 12% of the cement was replaced by silica fume. The water cement/silica fume ratio used was 0.30. Super plasticizer with ratio of 2.0% by weight of cement/silica fume was used to improve workability. The strength of the ferrocement mortar was

determined by testing 50x50x50mm cubes at 1, 7, and 28 days. The average compressive strength of the ferrocement mortar after 28 days, $(f_{cu})_{ferr}$, was found to be 38 MPa.

Commercially produced light weight brick of dimensions 100x200x600mm was used as the core material for the sandwich panel specimens. Variety of the published technical data of this type of brick shows that it has strength of 2-3 MPa, dry unit weight of 400-500 kg/m^3, porosity of 22-30%, and thermal conductivity (K) of 0.27-0.34 W/m^{o}c. It is worth mentioning that standard compression tests were performed on three units of the used light weight brick and the average compressive strength was found to be 3 MPa.

Different types of steel meshes were used to provide the reinforcement inside the two thin skin layers of each specimen. In the present work, two types of steel meshed were used including: high tensile steel welded wire mesh with 12.5x12.5m opening and average diameter of 0.7 mm, and mild steel type expanded steel mesh. Tensile test was performed on samples of the three steel types. The results showed that the proof stress and the tensile strength of the high tensile steel were 400 MPa and 600 MPa respectively while the proof stress and tensile strength of expanded steel mesh were 250 and 300 MPa respectively. Mild steel bars had yield stress of 238 MPa and tensile strength of 361 MPa. Figs. 3 and 4 show reinforcement details of light weight walls.

2.3 Test Setup

Compression test was performed on each of the test specimens. The compression test setup is shown in Figure 5. Vertical strain was measured on front face at mid-height of the specimen at one location, 200mm gauge length. on the left and right of the centerline (mid-width) of the specimen. Demec points and mechanical strain gauge extensometer were used to record the strain at these locations. Two dial gauges were located in the horizontal front and the other in the horizontal side for measuring displacement for each tested wall panel.

3. Experimental Results

Table 2 summarizes the experimental results of the axial compressive loading of the panels tested. The results includes first crack load, ultimate and serviceability loads, compression stresses for all the tested light walls, ductility ratio and energy absorption properties for all the tested walls. Fig. 6 shows comparison first crack loads of all the tested light walls while Fig.7 emphasizes comparison of serviceability load of tested walls according to CP110. Fig. 8 shows comparison of ultimate loads of all tested light walls. Fig. 9 shows comparison of calculated ductility ratio of all tested light walls. Fig. 10 presents comparison of calculated energy absorption of all tested light weight walls.

Fig. 11 shows load horizontal displacement and vertical compressive strains for tested light weight walls, W1-W3. Fig. 12 shows load horizontal displacement and vertical compressive strains for tested light weight walls, W4-W6. Fig. 13 shows load horizontal displacement and vertical compressive strains for tested light weight walls, W7-W9. Fig. 14 shows load horizontal displacement and vertical compressive strains for tested light weight wall W10.

It is interesting to note that Wall W5 shows higher first crack load reached 120 Kn. and Wall W10 reached higher serviceability load, higher ultimate load and higher energy absorption, while Wall W8 achieved higher ductility ratio

- *Ductility ratio*

The ductility ratio was calculated as the mid height displacement at the ultimate load to that of the first cracking loads. Walls reinforced with expanded metal mesh

IOP Conf. Series: Materials Science and Engineering **974** (2020) 012037 doi:10.1088/1757-899X/974/1/012037

and welded steel mesh was given higher ductility ratio compared with wall W1. Fig. 8 shows ductility ratios for all tested light weight walls.

- *Energy absorption*

The energy absorption was obtained by calculating the area under the load displacement curve for each wall. Walls reinforced with expanded steel mesh were achieved higher energy absorption than control wall W1. Fig. 9 emphasizes energy absorption for all tested slabs. Higher ductility and energy absorption properties are very useful for dynamic applications.

- *Serviceability load*

The l serviceability load was calculated using the equation according to CP110:

Se_{rvice} load = ($Exc_{remental}$ load - 1.4 D.L)/1.6. Where experimental load is equal to the maximum load reached.

4. Cracking Behavior

Fig. 15 shows cracking patterns of all the tested light weight panels under compression loadings. It is interesting to note that the role of the specific surface area of reinforcing materials and yield stress of the reinforcement in controlling the crack width, the load at crack width, the smaller the opening of mesh used; the smaller; the opening of crack width; the better the result. The obtained results show conclusively that mesh reinforcement is far superior in controlling crack width. It is excellent to notice that is there is no spalling of concrete mortar cover at failure that is predominant.

5. CONCLUSIONS

Based on the results and observations of the experimental investigation presented in this paper, the following conclusions could be drawn:

1. Ferrocement panels show good potential for use in several construction applications as reflected by their relatively high ultimate and serviceability loads, crack resistance control, high ductility, and remarkable energy absorption which are very useful for dynamic applications.

2. Irrespective of sandwich panels are lighter in weight when compared to conventional reinforced concrete construction of the same thickness.

3. For light weight panels subjected to axial compressive loadings the same exhibited higher ultimate loads irrespective of the type of steel mesh employed compared with the conventional reinforced panel.

4. For the same core material, ultimate load, ductility ratio, energy absorption, and deflection were all increased when increasing the volume fraction of reinforcing materials.

5. The overall results demonstrate good performance for the ferrocement panels which can be of true construction merits in both developed and developing countries. Therefore, further research work needs to be conducted as well as through economic feasibility needs to be performed.

6. sound recommendations for applications.

IOP Conf. Series: Materials Science and Engineering **974** (2020) 012037 doi:10.1088/1757-899X/974/1/012037

7-It is interesting to note that Wall W5 shows higher first crack load reached 120 Kn. and Wall W10 reached higher serviceability load, higher ultimate load and higher energy absorption, while Wall W8 achieved higher ductility ratio

6. REFERENCES

[1] Fahmy, E., and Shaheen, Y. "Laminated ferrocement for strengthening and repairing of reinforced concrete beams". Proceedings of the Annual Conference of the Canadian Society for Civil Engineering, 475-483., 1994.

[2] Fahmy, E.H., Shaheen, Y.B.I, and Korany, Y.S. "Repairing Reinforced Concrete Columns Using Ferrocement Laminates", Journal of ferrocement, 29, No.2, 115-124, 1999.

[3] National Academy of Sciences. "Ferrocement: Applications in Developing Countries. A Report of an Adhoc Panel of the Advisory Committee on Technological Innovation Board on Science and Technology for International Development Office of the Foreign Secretary", Washington, D.C., 1973.

[4] Joshi, A.D. "Strength and Behavior of Ferrocement Load Bearing Wall Elements". thesis presented to the Indian Institute of Science in partial fulfillment of the requirements for the degree of Master of Engineering, Bangalore, India. 1974.

[5] Swamy, R.N. and Spanos, "Deflection and cracking Behavior of Ferrocement with Grouped Reinforcement and Fiber Reinforced Matrix". ACI Journal 79-91, 1988.

[6] Ezzat H. Fahmy, Yousry B. Shaheen, Mohamed N. Abou Zeid "Development of Ferrocement Panels for Floor and Wall Construction". Proceedings of the Annual Conference of the Canadian Society for Civil Engineering, St1-ST10, 2004..

[7] American Concrete Institute. 1999. Building Code Requirements for Structural Concrete (ACI 318-99) And Commentary (ACI 318R-99) reported by ACI Committee 318

[8] E.H. Fahmy, Y.B. Shaheen and W.M. El Dessouki, "Application of Ferrocement Construction of Radial. Gates", the Journal of Ferrocement, Vol. 25, No. 2, Thailand, April 1995.

[9] Swamy R N and Y B I Shaheen". Tensile Behaviour of thin Ferrocement plates ". A C I, SP-124, September 1990, PP 357-387.

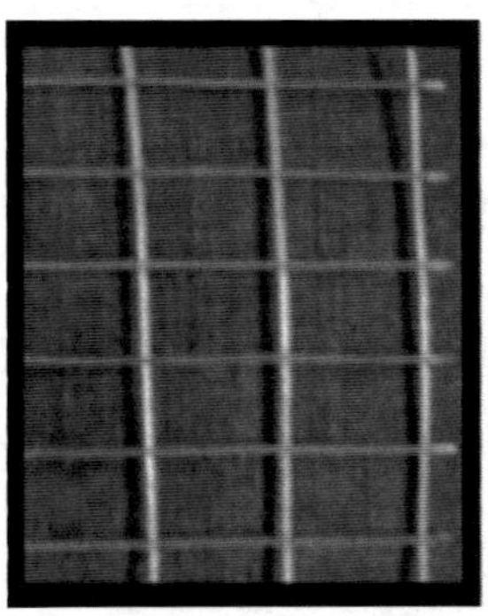

Welded steel mesh

Expanded steel mesh

Tensile steel meshes, (Swamy and Shaheen, 1990)

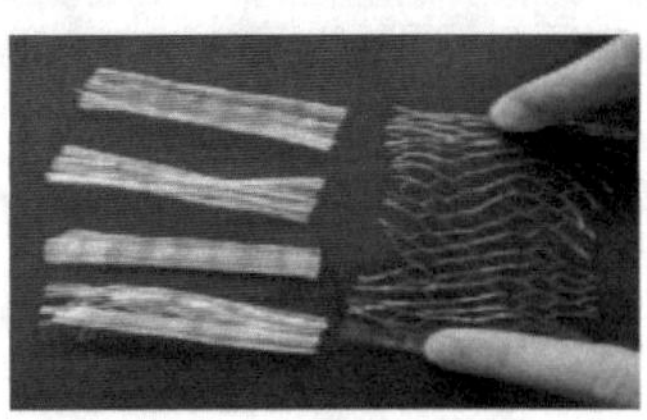

IOP Conf. Series: Materials Science and Engineering **974** (2020) 012037 doi:10.1088/1757-899X/974/1/012037

Fig.1 Welded, expanded meshes, Poly propylene fibers e300, and Tensile steel meshes

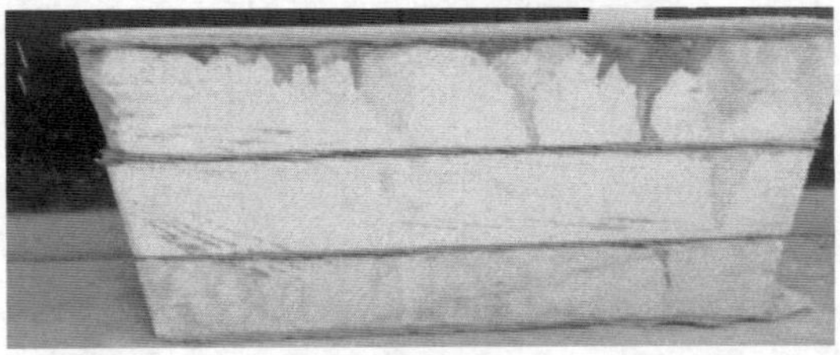

Wall 1

Wall 2

Wall 3

Wall 4

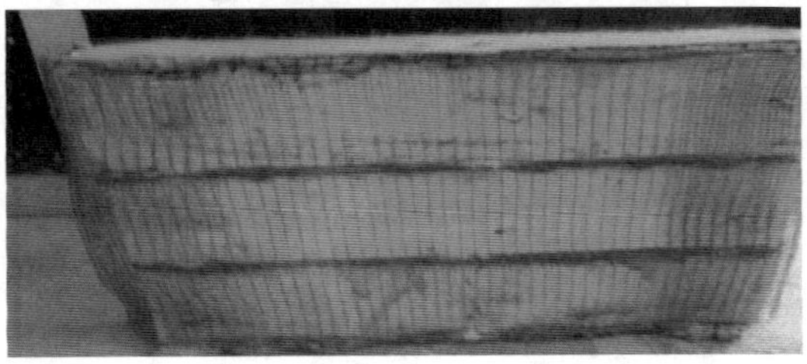

Wall 5

Wall 6

IOP Conf. Series: Materials Science and Engineering **974** (2020) 012037 doi:10.1088/1757-899X/974/1/012037

Wall 7

Wall 8

Wall 9

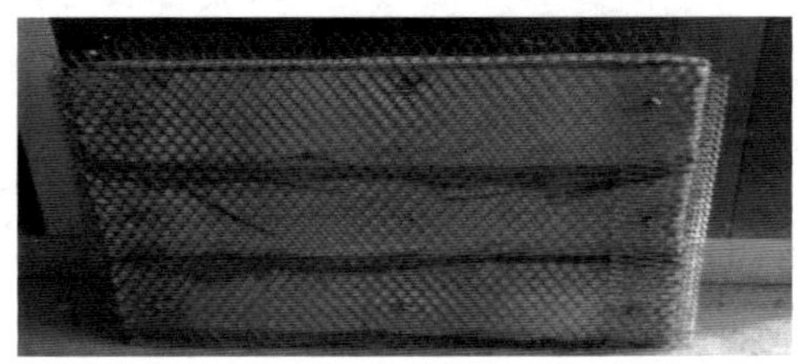

Wall 10

Figure 2: Ferrocement wall sandwich panels.

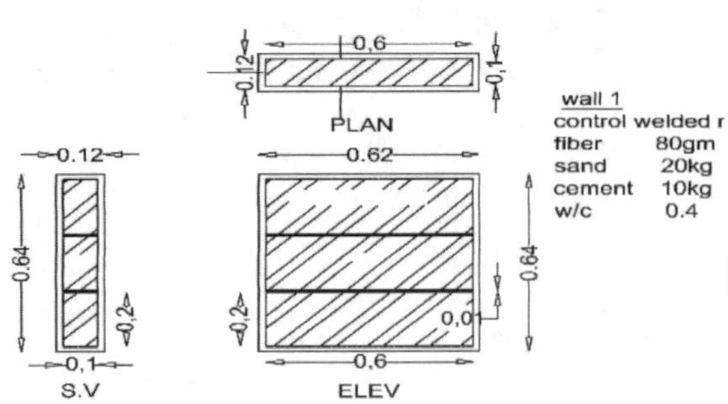

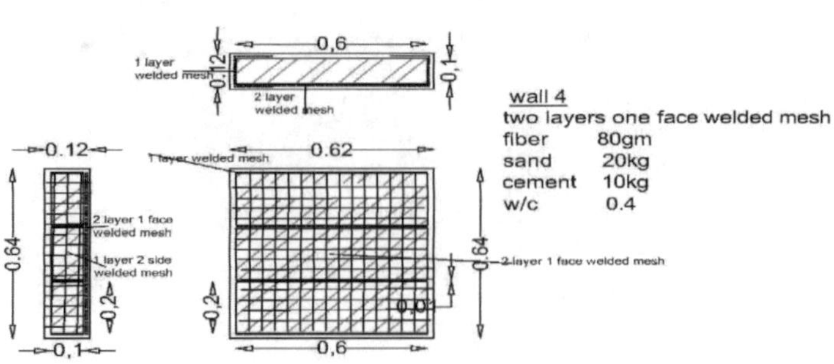

IOP Conf. Series: Materials Science and Engineering **974** (2020) 012037 doi:10.1088/1757-899X/974/1/012037

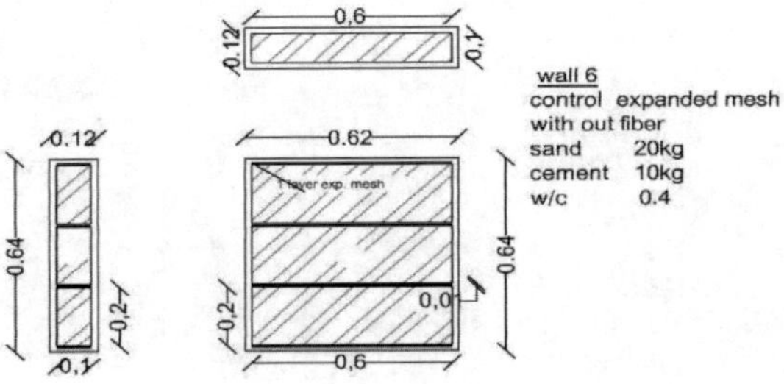

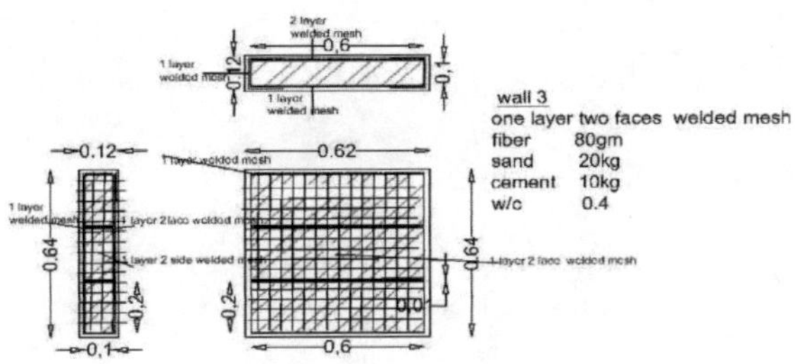

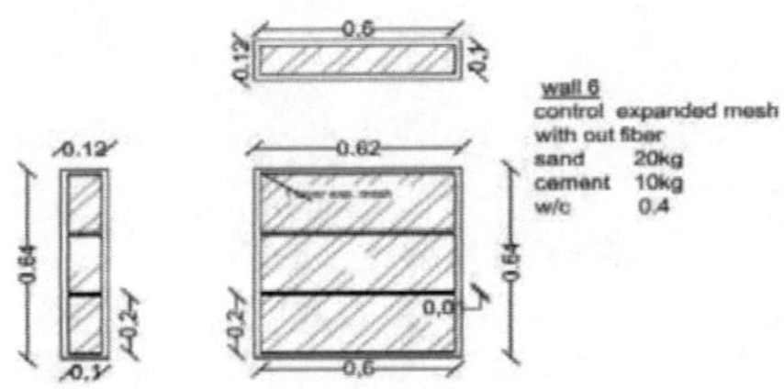

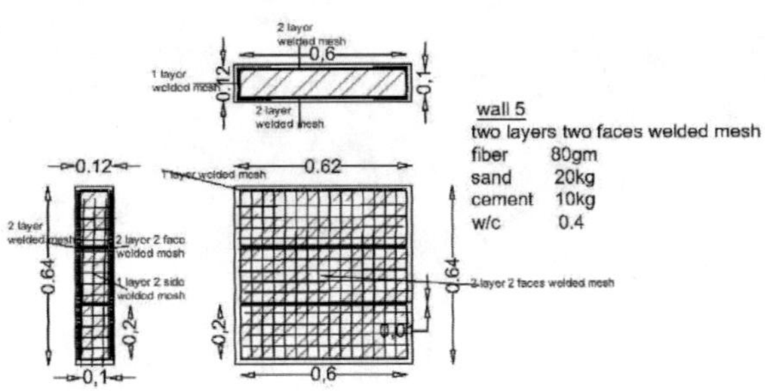

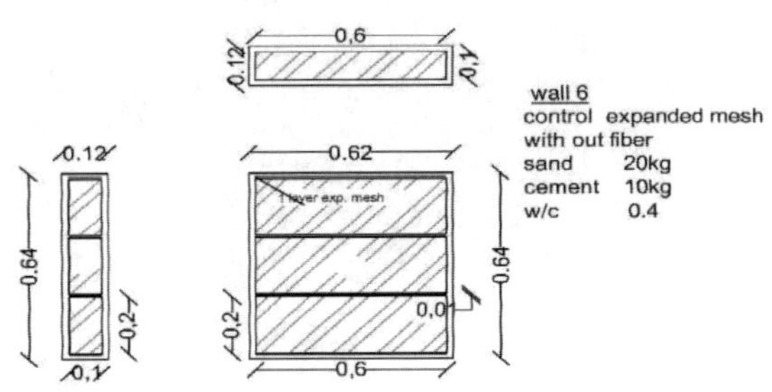

IOP Conf. Series: Materials Science and Engineering **974** (2020) 012037 doi:10.1088/1757-899X/974/1/012037

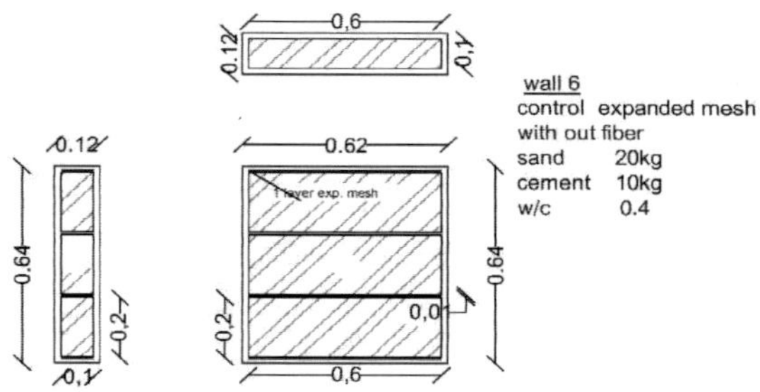

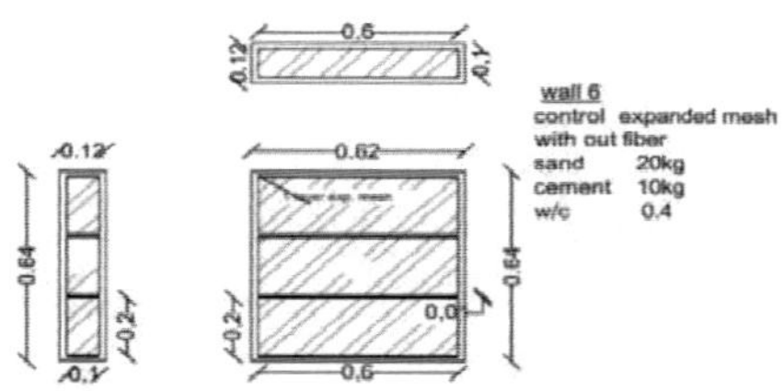

Fig. 3 Reinforcement details of the light weight walls W1-W6.

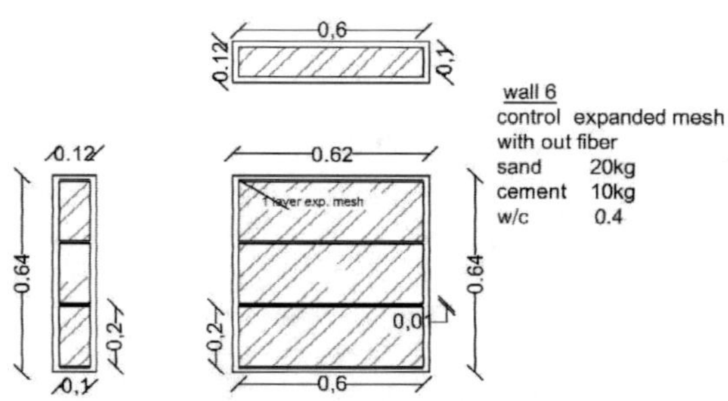

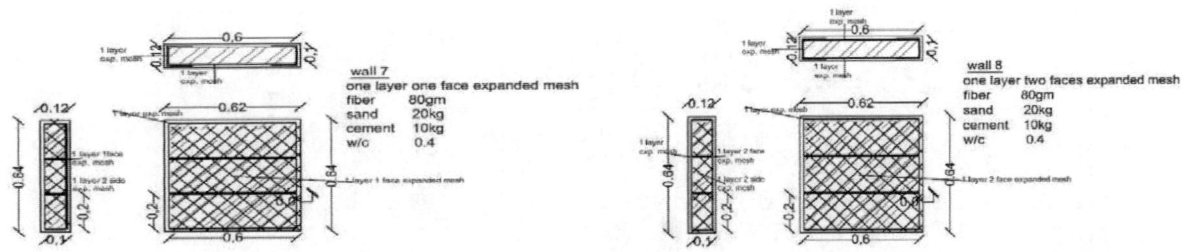

Fig. 4 Reinforcement details of the light weight walls W7-W10

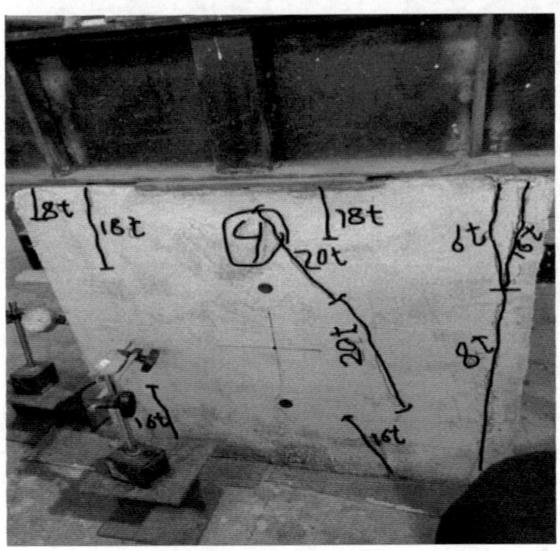

Fig.5 Test set up

IOP Conf. Series: Materials Science and Engineering **974** (2020) 012037 doi:10.1088/1757-899X/974/1/012037

Table 1: Designations of ferrocement lightweight walls.

Designations	**No. of Layers of Reinf. at each face**	**Type of Reinforcing Steel**	Vol. fraction%	**Total thickness (mm)**
Wall 1	Without layers	Welded wire mesh (WWM)	**0.509**	120
Wall 2	One layer one face +**U** WWM each face	WWM	**1.019**	120
Wall 3	One layer two faces+**U** WWM each face	WWM	**1.019**	120
Wall 4	Two layers one face+**U** WWM each face	WWM	**1.528**	120
Wall 5	Two layers two faces+**U** WWM each face	WWM	**2.667**	120
Wall 6	Without layers	Expanded Steel Mesh (ESM)	**1.273**	120
Wall 7	One layer one face+**U** E S M each face	(ESM)	**2.548**	120
Wall 8	One layer two faces+**U** E S M each face	(ESM)	**3.822**	120
Wall 9	Two layers one face+**U** E S M each face	(ESM)	**3.822**	120
Wall 10	Two layers two faces+**U** E S M each face	**(ESM)**	**3.822**	**120**

IOP Conf. Series: Materials Science and Engineering **974** (2020) 012037 doi:10.1088/1757-899X/974/1/012037

Table 2 First crack, ultimate, serviceability loads, stresses, ductility ratio and energy absorption properties for all tested Walls.

Beam Design .	Vol. fraction%	F.C.L KN	Ultimate .load, KN	stresses, N/mm^2	Service. Load, KN	Def.at F.C.L, mm	Max. defl., mm	Duct. ratio	Energy Absorp. . KN.mm
Wall 1	0.509	40	100	1.39	29.375	0.01	0.045	4.5	2.65
Wall 2	1.019	60	140	1.95	54.375	0.1	0.18	1.8	10.6
Wall 3	1.019	100	180	2.5	79.375	0.035	0.07	2	6.5
Wall 4	1.528	110	200	2.78	91.875	0.05	0.12	2.4	13.7
Wall 5	2.667	120	240	3.33	116.875	0.12	0.8	6.67	145.2
Wall 6	1.273	60	120	1.67	41.875	0.02	0.07	3.5	5.2
Wall 7	2.548	60	180	2.5	79.375	0.02	0.12	6	13.5
Wall 8	3.822	100	220	3.45	104.375	0.025	0.176	7.04	25.12
Wall 9	3.822	100	240	3.33	116.875	0.15	0.64	4.27	95.2
Wall10	3.822	120	280	3.89	141.875	0.35	1.18	3.37	185.45

IOP Conf. Series: Materials Science and Engineering **974** (2020) 012037 doi:10.1088/1757-899X/974/1/012037

Fig. 6 First Crack of tested Walls.

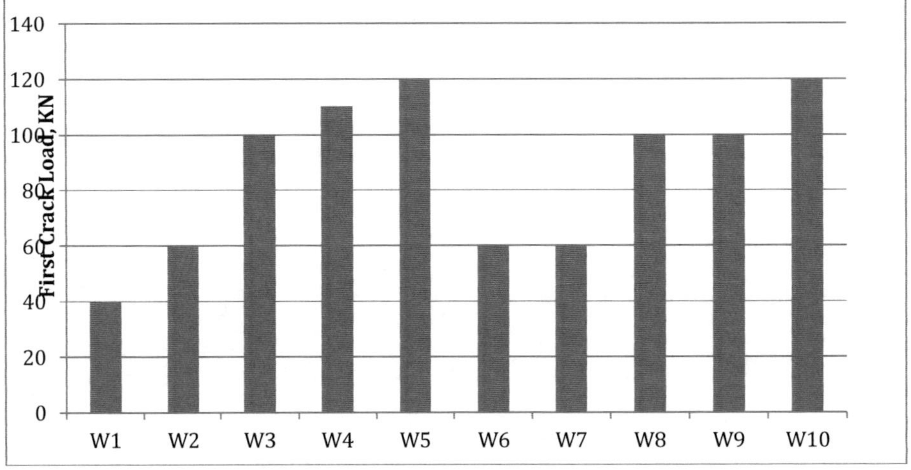

Fig. 7 Serviceability Load of Tested Walls.

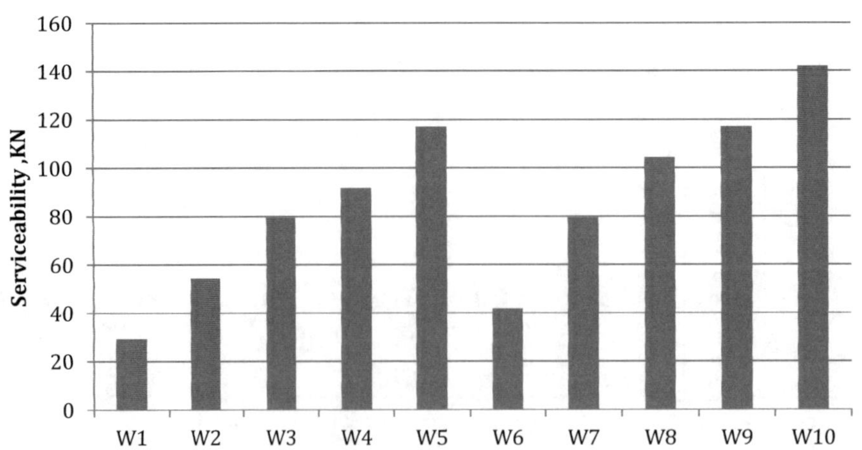

Fig. 8 Ultimate Load of Tested Walls.

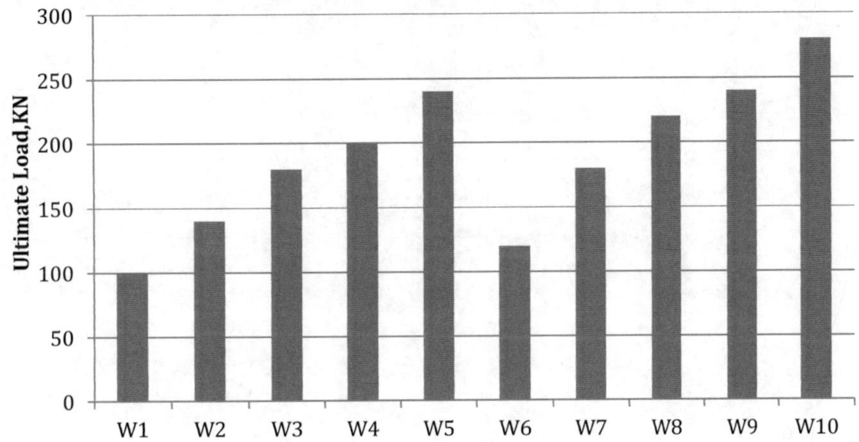

Fig. 9 Ductility Ratio of Tested Walls.

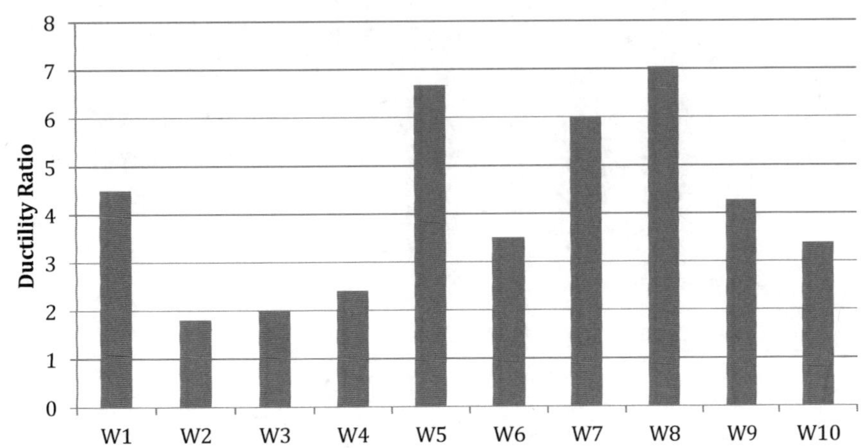

IOP Conf. Series: Materials Science and Engineering **974** (2020) 012037 doi:10.1088/1757-899X/974/1/012037

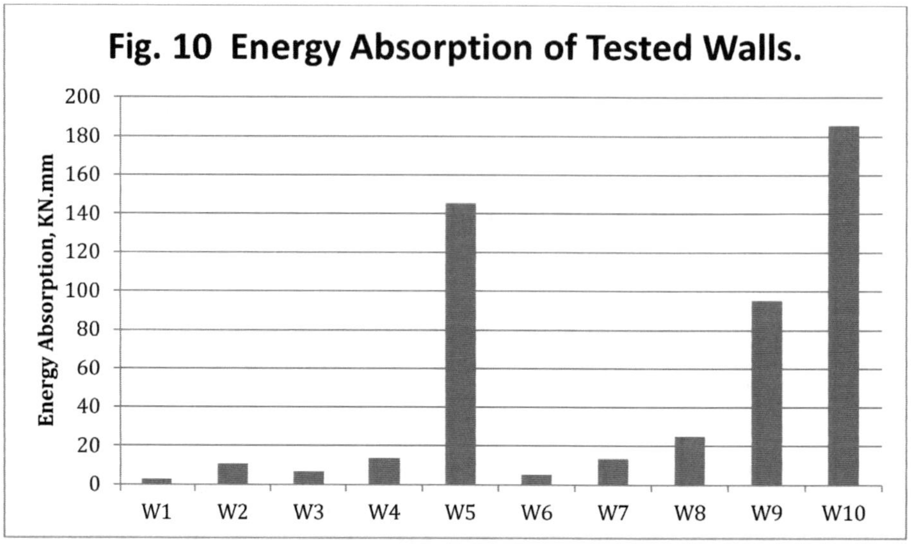

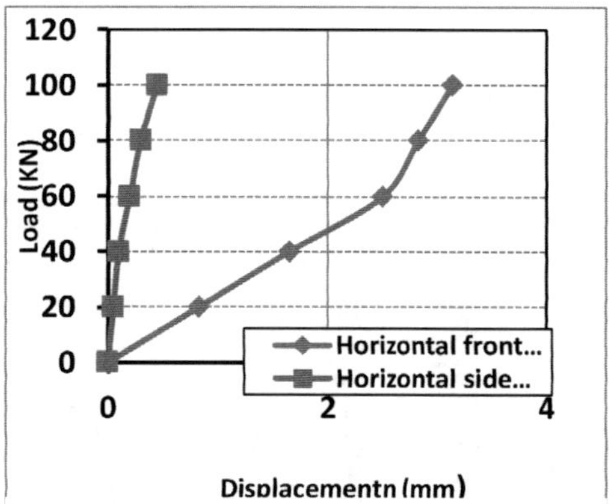

Load horizontal displacement of wall W1

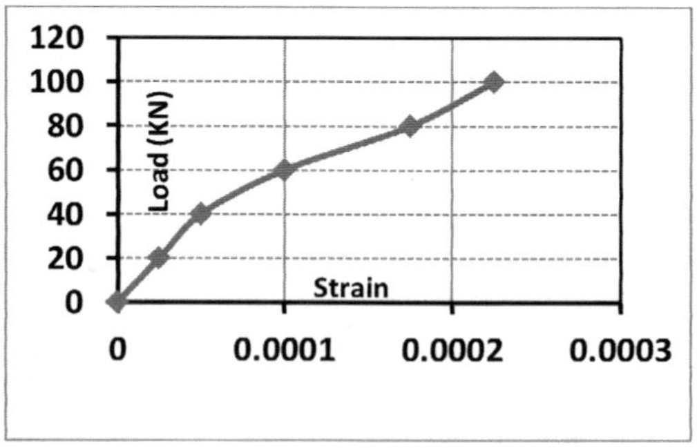

Load vertical strain of wall W1

Load horizontal displacements and vertical compressive strains for tested light weight walls W1

IOP Conf. Series: Materials Science and Engineering **974** (2020) 012037 doi:10.1088/1757-899X/974/1/012037

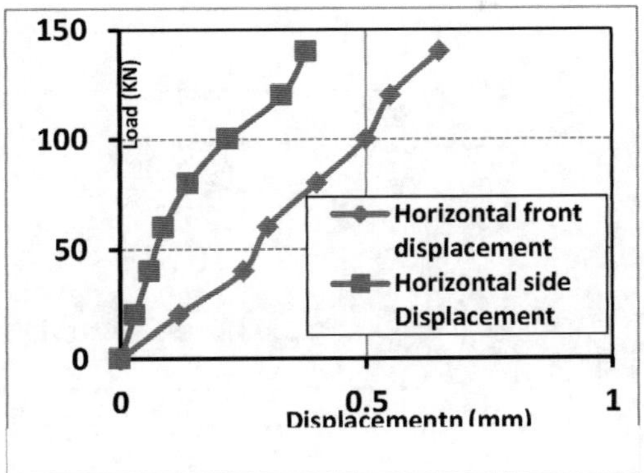

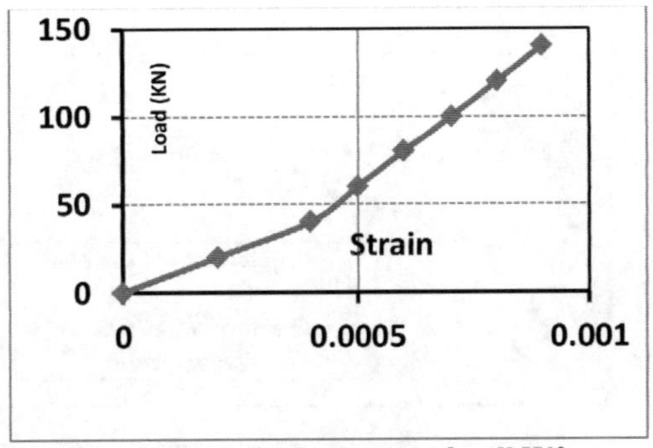

Load horizontal displacement of wall W2
Load vertical strain of wall W2

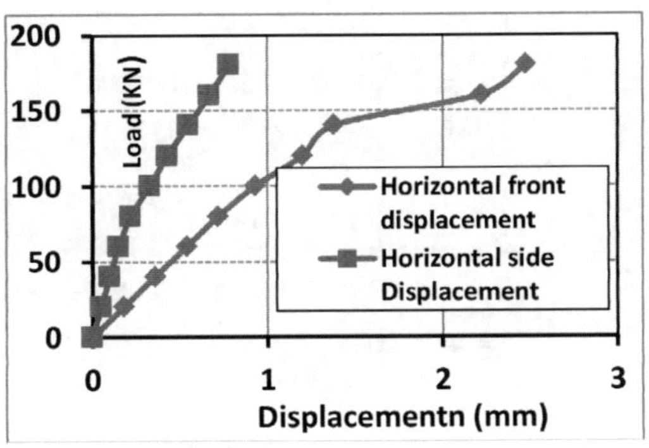

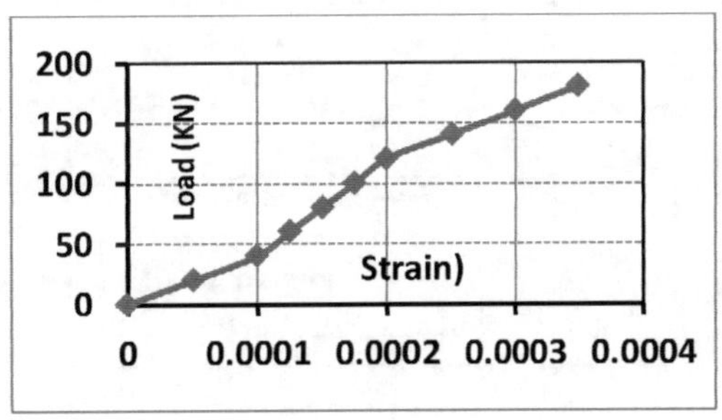

Load horizontal displacement of wall W3
Load vertical strain of wall W3

Fig.11 Load horizontal displacements and vertical compressive strains for tested light weight walls W1-W3.

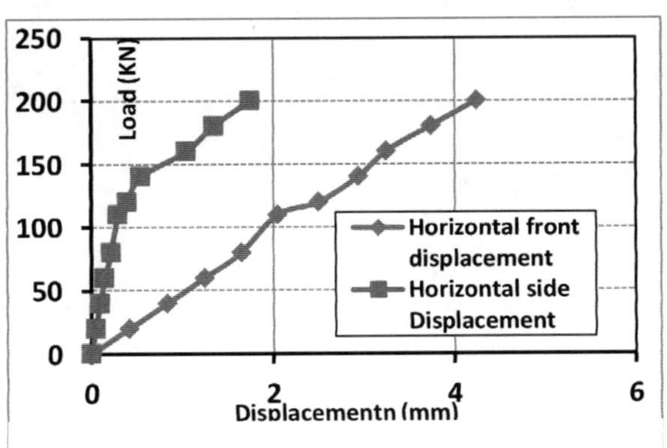

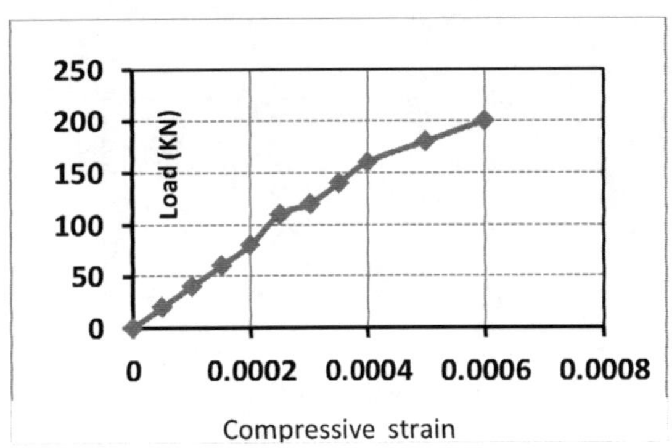

Load horizontal displacement of wall W4

Load vertical strain of wall W4

Load horizontal displacements and vertical compressive strains for tested light weight walls W4

IOP Conf. Series: Materials Science and Engineering **974** (2020) 012037 doi:10.1088/1757-899X/974/1/012037

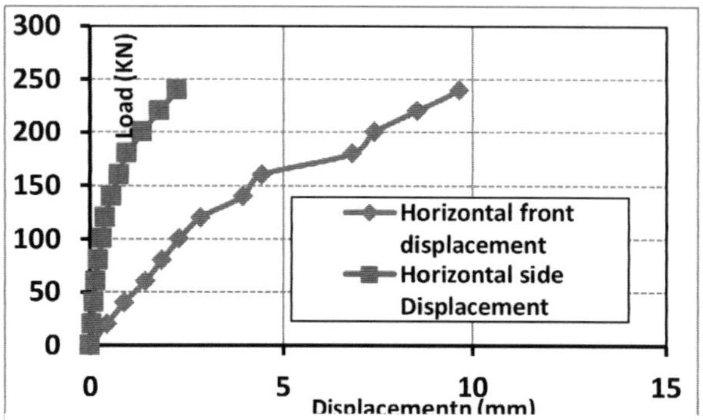

Load horizontal displacement of wall (W5).

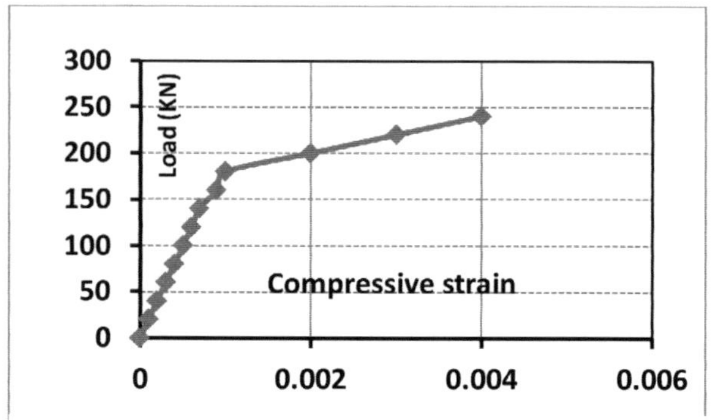

Load vertical strain of wall (W5).

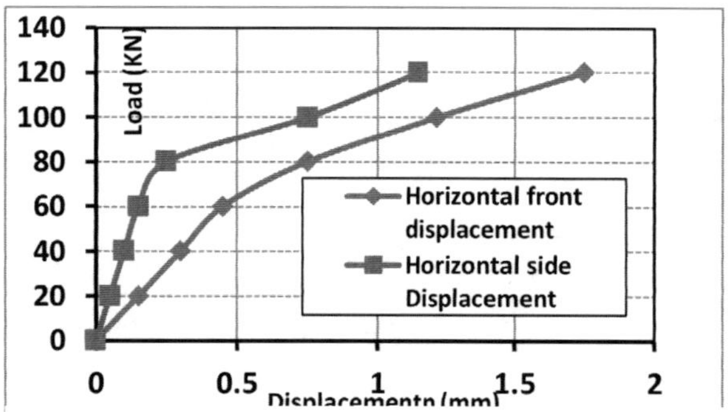

Load horizontal displacement of wall (W6).

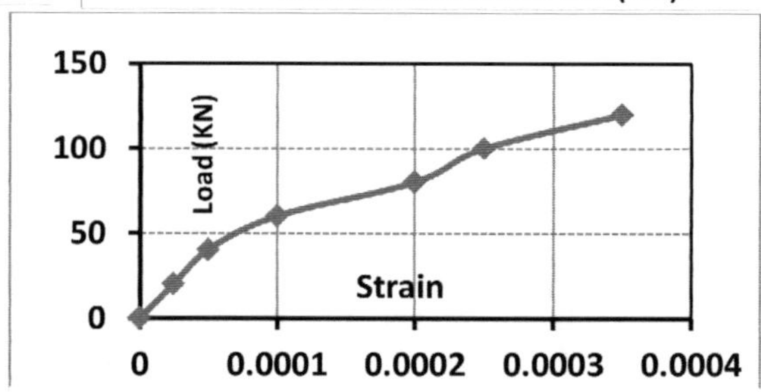

Load vertical strain of wall (W6).

Fig.12 . Load horizontal displacements and vertical compressive strains for tested light weight walls W4-W6.

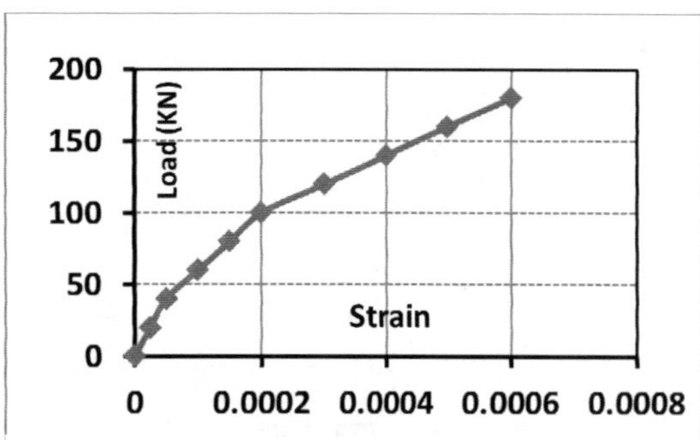

Load vertical strain of wall (W7).

Load horizontal displacements and vertical compressive strains for tested light weight walls W7

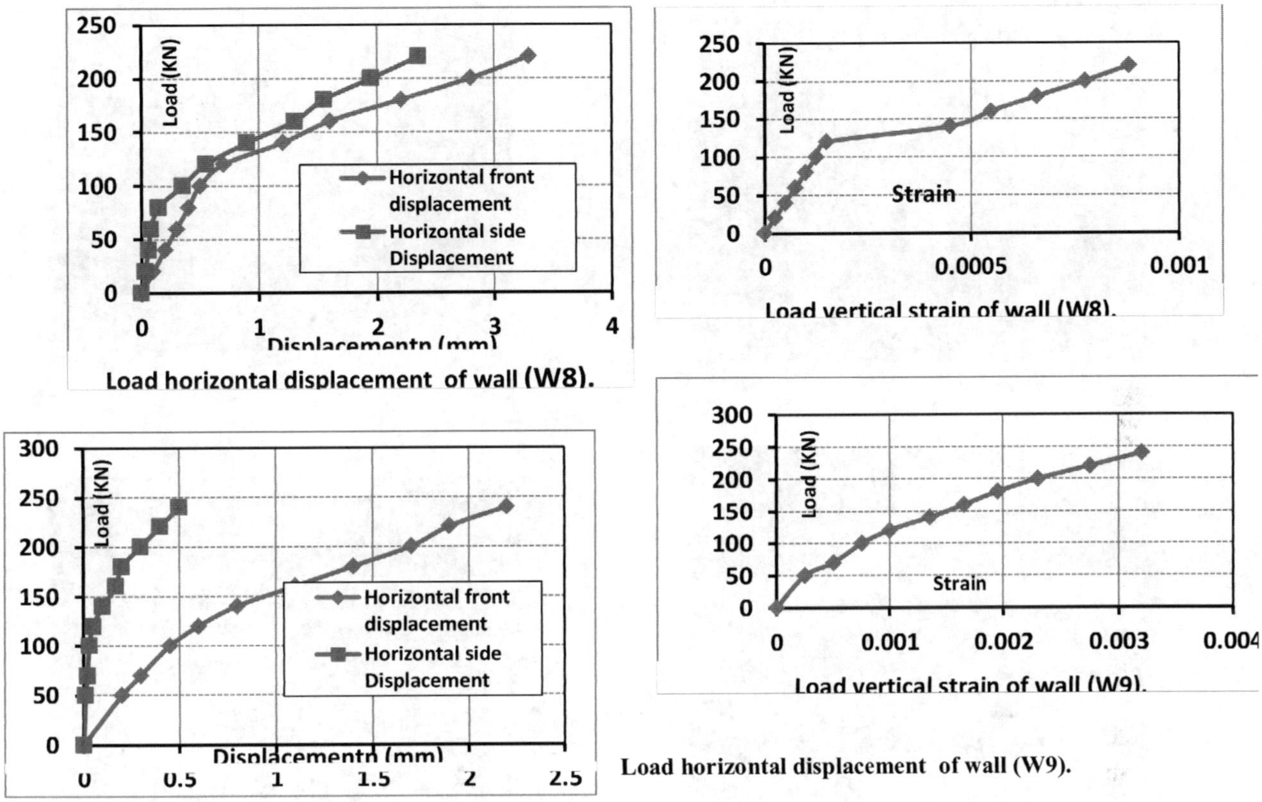

Fig. 13 Load horizontal displacements and vertical compressive strains for tested light weight walls W7-W9.

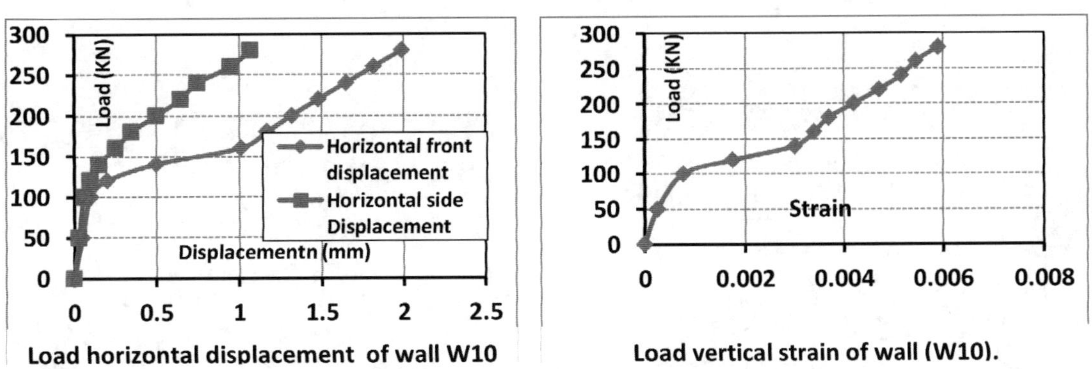

Fig. 14 Load horizontal displacements and vertical compressive strains for tested light weight wall W10.

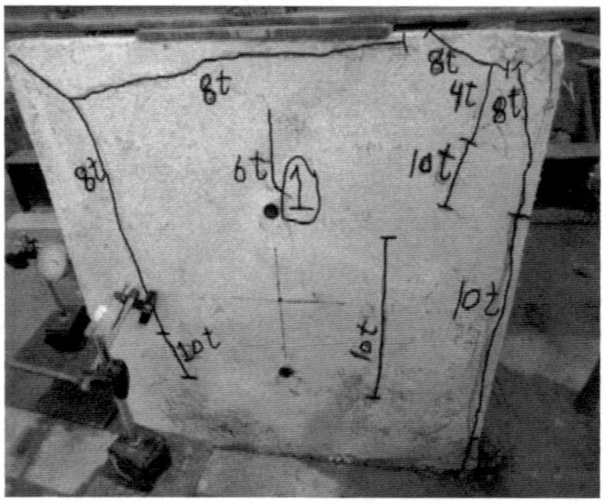

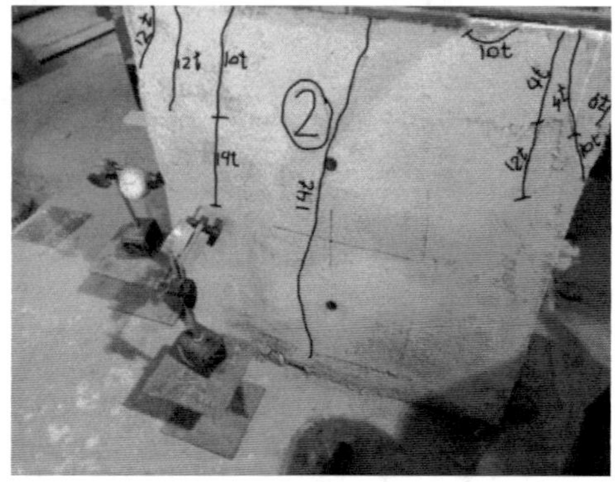

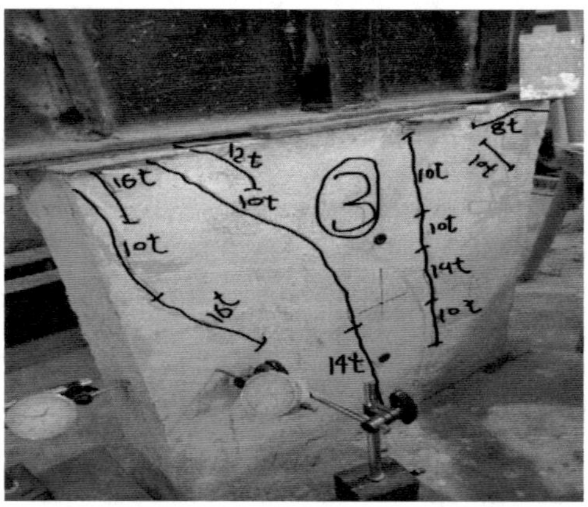

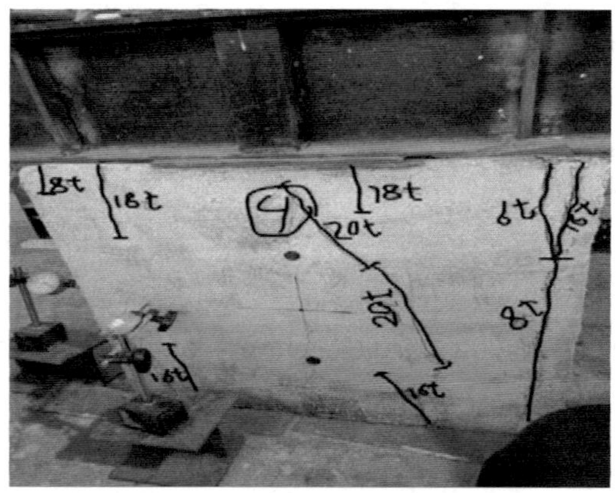

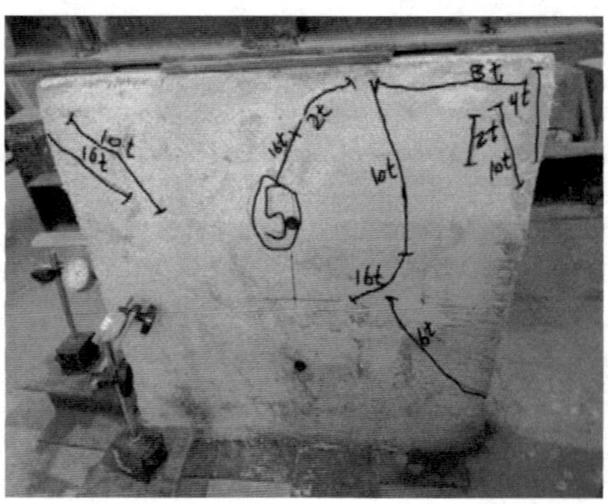

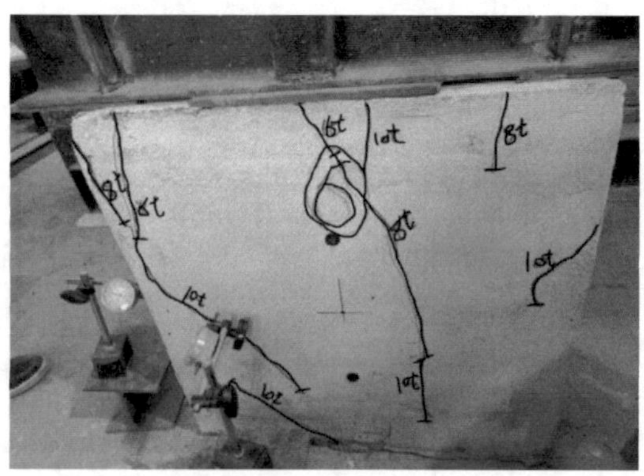

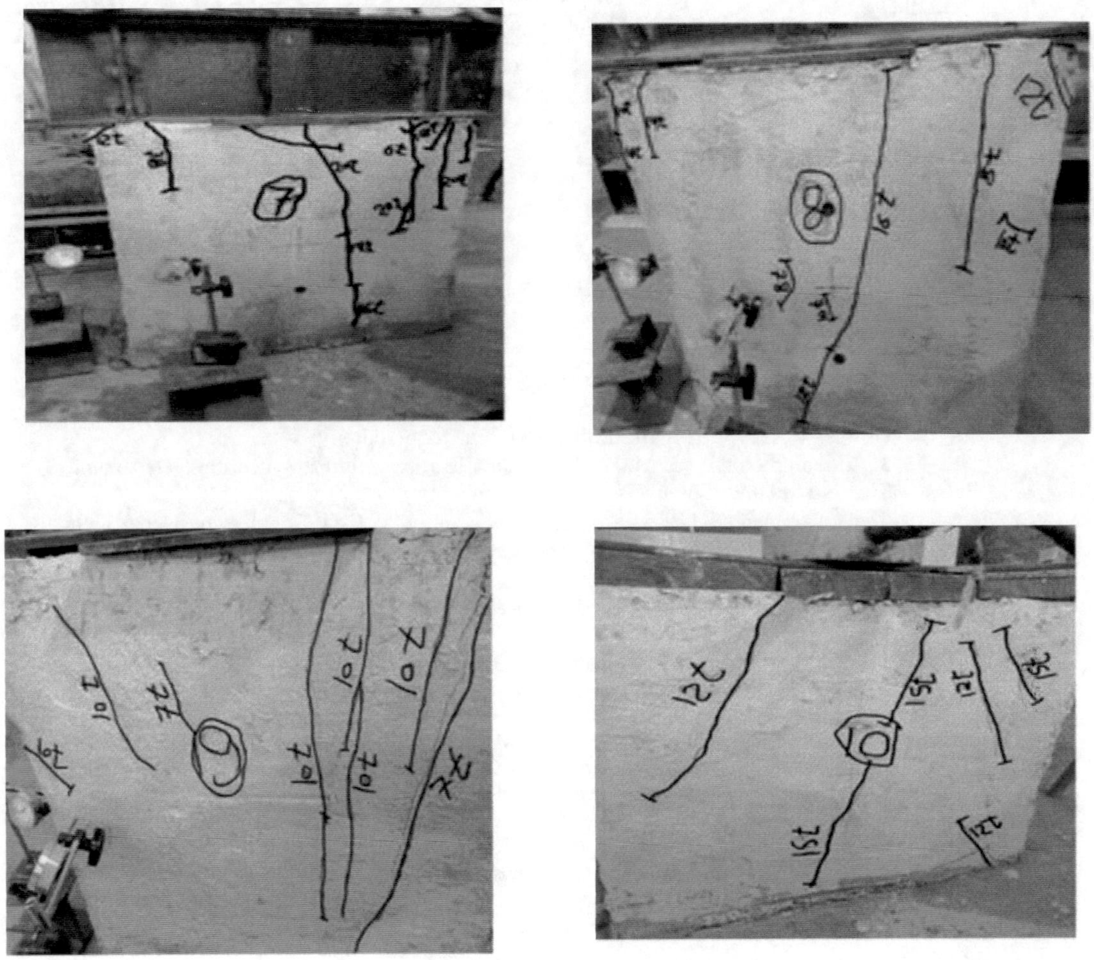

Fig. 15 Cracking patterns of all the tested light weight panels under compressive.

Page | 1Military Technical College Kobry El-Kobbah, Cairo, Egypt

13th International Conference on Civil and Architecture Engineering ICCAE-13-2020

Effect of crumbed rubber Treatment method on the Flexural Behavior of High Strength Rubberized Concrete Slabs

A A Bashandy [1], F M Eid[2],*, M A Mazyad [3] and M A Arab [4],*

1 Alaa Ali Bashandy, Associated prof., Civil Engineering Department, Minoufiya University, Egypt (E-mail: eng_alb@yahoo.com)
2 Fatma M. Eid, Associated prof., Civil Engineering Department, Minoufiya University, Egypt (E-mail: dr.fatmaelzahraa@gmail.com)
3 Mohammed Ahmed Mazyad, Assistant lecturer, Civil Engineering Department, Thebes academy, Egypt (E-mail: mohammedmazayd79@gmail.com)
4 Mohammed Abd El-Salam Arab, Assistant prof., Civil Engineering Department, Beni-Suef University, Egypt (Corresponding Author; E-mail: emarab@eng.bsu.edu.eg)
*Correspondence: dr.fatmaelzahraa@gmail.com , fatma_elzahraa2002@yahoo.com

The most important research trends that the world is currently interested in are those that have a positive impact on the environment. The burning of the waste tyres is one of the most used means to get rid of these tyres, which causes a lot of gases to the air. The use of crumbed tyre rubber in concrete is considered one of the methods with a positive environmental impact. Using crumbed rubber as a partial replacement material of concrete aggregate is found to possess many engineering applications and holds promise within the future. Replacing 5% of the sand volume by crumbed tyre rubber was used. Flexural behaviour of treated and untreated rubberized concrete slabs was investigated. Experimental program shows that the behaviour of the high strength rubberized concrete slabs can be improved by the used treatment methods. Beside that the coarser of the rubber particles, the more negative effect on the behaviour of the high strength rubberized concrete slabs.

1. Introduction

Across the world, one billion tyres wasted every year and 0.5 billion more are expected to be discarded annually by 2030. Waste tyres need a bigger space for storing than other waste due to their large volume and fixed shape, this creates a serious problem[1,2]. To control the undesirable effects of disposal tyres, such as fire hazard and environmental problems.Many researchers have many trials in recycling and reusing end-of-life tyres by partial replacement of the concrete aggregates with crumb or scrap tyre rubber[3-5].

Siddique and Naik (2004)[4], El-Sherbini et al., (2010)[6], Nor et al., (2015)[7], More et al., (2015) [8], Selvakumar, et al., (2015) [1], Sofi, A., (2016)[9], Pardeshi et al., (2017)[10], Hassanli, et al., (2017)[3], and Al-Nasra et al., (2013)[11]researches shown that there was a significant increase in crumbed rubber concrete's density than normal concrete and that normal strength concrete containing crumbed tyre rubber exhibited a ductile failure.

Gerges et al., (2018)[12] described that softness of the surface of the rubber particles reduced the concrete strength because the rubber particles behaves like voids within the cement and aggregate matrix. Initiation of cracks quickly around the rubber accelerates the failure while loading.
More et al., (2015) [8]investigated the concrete tensile strength for 0, 3, 6, 9 and 12% fine aggregate replacement percentage with crumbed rubber. Their results showed that 3% replacement ratio of sand decreased split tensile strength of concrete by about 30% and the weak bond between cement and rubber particles was the reason of this reduction.
Al-Tayeb, et al., (2012)[13]replaced the volume of sand with 5 %, 10 % and 20% waste crumb rubber. Their study concluded that the impact and bending loads increased with the increase in the percentage of sand replacement by crumb rubber. Beside that the compressive and tensile stresses of concrete by the partial replacement of sand by crumb rubber, reduce by 5–20 %, and 11–17 % respectively.
Sallam, et al., (2008)[14]reported that 10% of fine rubber in the concrete mix was suitable for having an acceptable decrease in compressive strength. In addition, the presence of crumb rubber of small size in concrete increased the resistance of concrete to crack initiation under impact load.
Hassanli, et al., (2017) [3]studied the behavior of rubberized concrete at the structural application level and introduced a numerical prediction of the behavior of rubberized concrete beams and columns.It could be observed that the deflection capacity of the rubberized concrete beam was increasedby 7.7% to 27.9% compared tothat has no rubber content. Beside that the reduction in flexural strength due to rubber content is not as much as the reduction in compressive strength.Recommendationto use rubberized concrete in flexural members with ductile response was introduced.
Sivakumar, et al., (2013) [1]studied the replacement of coarse aggregate in reinforced concrete beams. According to their study the cracks have been occurred slightly near to the support of the beam, also recycled coarse aggregate replacement up to 30%in concrete has shown good improvement in flexuralstrength.
The present work focused on studying the effect of using different treating methods for crumbed rubber using different chemicals and techniques on the flexural behavior of treated and untreated high strength rubberized concrete slabs.

2. Materials and Methods

2.1. Materials

Cement: The used cement was ordinary Portland cement CEM I N 52.5 confirming to (E.S.S. 4756-1/2009)[15].

Coarse aggregate: The coarse aggregate was a natural dolomite with a maximum nominal size of 12.5mm. The coarse aggregate has a specific gravity 2.65.

Fine aggregate: The fine aggregate was natural siliceous sand with a specific gravity 2.65.

Crumb tyre rubber: Two grades of rubber were used. The first grade was the finer and mixes contain this type has a symbol started with F. The second grade was the coarser and mixes contain this type has a symbol started with C. Grade F has a grain size less than 0.6mm and larger than 0.15mm. Grade C has a grain size less than 2.36 mm and larger than 0.6mm.

Silica fume: The density and fineness of the used silica fume were 2210 kg/m^3 and 23.52m^2/gm respectively confirming to ASTM C 1240[16].

Super-plasticizer: Sika Viscocrete 3425 was used according to ASTM-C-494, types G and F[17].

Sodium hydroxide (NaOH): Sodium hydroxide is a strong caustic alkali (known as caustic soda). It is highly soluble in water liberating intense heat. It is an odorless white crystalline solid.

Potassium permanganate ($KMnO_4$): Potassium permanganate is a strong oxidizing agent. It is purplish crystalline solid. It is soluble in water giving an intensely purple or pink solution.

Sodium bisulfite ($NaHSO_3$): Sodium bisulfite is a compound with a chemical formula $NaHSO_3$.This salt of bisulfate is prepared by bubbling sulfur dioxide in a solution of sodium carbonate in water.

IOP Conf. Series: Materials Science and Engineering **974** (2020) 012038 doi:10.1088/1757-899X/974/1/012038

Steel Reinforcement: Mild steel of 8 mm diameter was used as the main reinforcement for the concrete slabs.The yield and ultimate strength of the steel bars are 2950 Kg/cm^2 and 3990 Kg/cm^2 respectively.

2.2. Mix Proportions

A high strength concrete mix was used in this research. As indicated from the literature review, 5% rubber replacement as a percentage of fine aggregates volume was chosen. Mix proportions used in this test program for control mix concrete are summarized in Table (1).

Table 1. Concrete mix proportions in kg/m^3.

Cement	Silica fume	Water	Super-Plasticizer	Coarse Aggregate	Fine Aggregate
550	27.5	184.8	4.91	850	670

2.3. Slabs Dimensions and Reinforcement

The tested slabs were with dimensions of 1100×450×50 mm. 11 reinforced high strength concrete slabs were cast. The slab reinforced with 4 steel bars of 8mm diameter were in the longitudinal direction and 7 steel bars of 8mm diameter were in the short direction.

2.4. Testing Procedures and Equipment

Impact test: Impact test was carried out on 100x100x500mm prisms. A 1.150 kg metallic ball was dropped from 1.12 m height at the center of the prisms and this for all prisms. For each prism, the number of blows was recorded for the failure crack then impact energy was calculated. Impact test procedure and apparatus was set according to (CSTB/ MELTM (1989) and ISO 7892 (1988))[18].

IE (Impact energy) (J) = No. of blow at the Final crack × M ×H× g

Where M: the weight of ball= 1.15 Kg. N: No. of blow at the final crack

H: is the drop height of the body (1.12 m) g: is the acceleration of gravity (9.81 m/s^2).

Slab Flexural Test: Three points load flexural test wereexecuted on the reinforced high strength concrete slabs to investigate the load deflection behavior and study the different factors affecting this behavior. Slabs were tested on a steel frame with two support. The load was applied by a hydraulic cylinder acting with 150 ton capacity and 150 mm maximum Stroke which was connected to a hydraulic electric oil pump for feeding the hydraulic jack. The applied load was measured by a load cell of 225 ton capacity. A strain gauge for measuring displacements up to at least 100 mm was placed under each slab at the center of its span to measure deflections were used for deflection measurements within the region of pure bending. Figure (1) shows the machinery used in the slabs testing.

Figure 1. Three Point Load Flexural Test.

2.5. Rubber Treatment Methods

IOP Conf. Series: Materials Science and Engineering **974** (2020) 012038 doi:10.1088/1757-899X/974/1/012038

2.5.1. Using NaOH alone:

In this method, 20%NaOH solution was used. Crumb rubber was soaked in the solution for 24 h after that the rubber was rinsed in distilled water to remove excess NaOH. Sample with this treatment method abbreviated by N.

2.5.2. Using NaOH, $KMnO_4$, and $NaHSO_3$:

In this method, two samples were used. The first sample treated by soaking in 20% NaOH concentration for 24h and then treated crumbed rubber was soaked in the solution of 10% $KMnO_4$ for 24h then treated by 5% $NaHSO_3$ for 24h. The second sample was soaked in 5% NaOH and then soaked in5% $KMnO_4$ by stirring for 2h with heating and treated by 5% $NaHSO_3$ stirring for 0.5h with heating respectively.The two samples rinsed in distilled water. The first sample abbreviated by NKS24 and the second sample abbreviated by NKS.

2.5.3. Subjected the treated and untreated rubber to 200°C temperature:

This technique aimed to investigate the effect of high temperature on rubber surface modification. In this method, untreated crumbed rubber was placed in tightly closed containers and heated at 200°C for 0.5h. Samples with this treatment were abbreviated as CF and FF for grade C and F respectively.

2.6. Mechanical Properties of the Tested Concrete Mixes

Table 2 shows the mechanical properties results of the treated and un treated rubberized concrete mixs.

Table 2. Mechanical Properties of the used mixes.

	Compressive Strength Mpa	Tensile Strength Mpa	Flexural Strength Mpa	IE J
Control	73	8.4	10.95	2095
C	63	7.2	9.92	3900
CN	74	8.53	11.4	3421
CNKS	78.5	8.75	10.96	3320
CNKS24	79.5	8.99	11.54	3446
CF	77	8.5	11.25	1994
F	71	8.16	11.36	2287
FN	63.5	7.1	10.4	2338
FNKS24	65.5	7.6	9.2	2148
FNKS	57.5	6.2	8.1	1971
FF	80	9.1	11.7	2146

3. Slab Results and Discussion

3.1. Load deflection curve

The comparison of load deflection curves for tested slabs containing treated crumb rubber and control slab compared in figure 2 and figure 3. The reasons for the difference between the curves will be addressed in the following items.

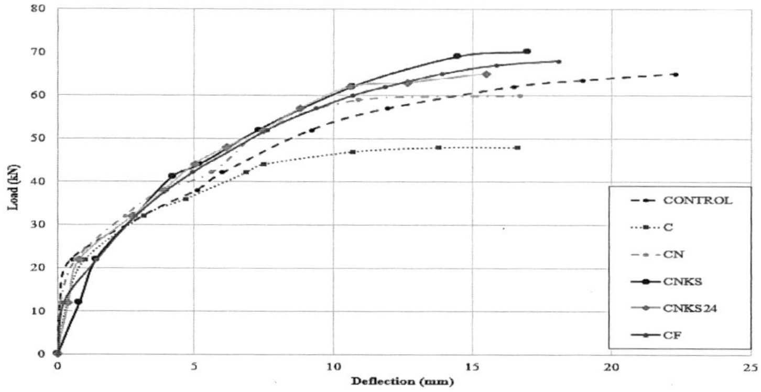

Figure 2. Load deflection curves for slabs containing grade C rubber.

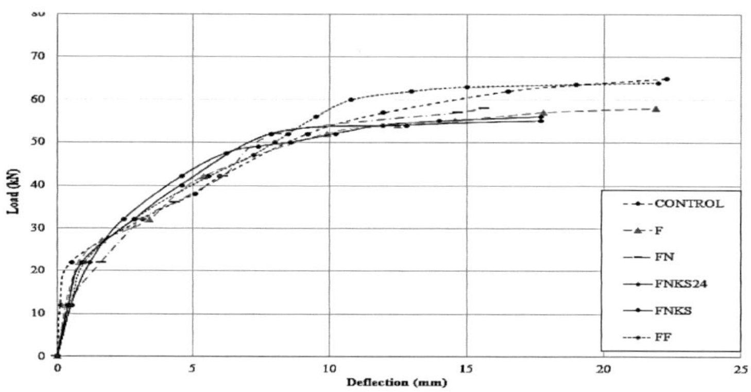

Figure 3. Load deflection curves for slabs containing grade F rubber.

3.2. Absorbed energy

The ductility is a structural property that is governed by the fracture and depends on the structure element's size.It could be calculated by considering the area below the load-deflection curve.The more increase in absorbed energy values refers to more slab ductile behavior.

Figure 4 shows the absorbed energy values for the tested slabs. As shown in Figure 4it could be noticed that the control slab was more ductile than others and the CNKS gave the best result which decreased by 1.81% compared with the control mix.

3.3. Ultimate and cracking load

Figure 5 and Figure 6 show the ultimate and cracking load for the tested slabs respectively. It is clear that the results are close in the initial cracking load, and the difference was in the ultimate load. Beside that using treatment methods improved the ultimate load results compared to the control slab. The ultimate load decreased by 26.15% and7.69% for used C and CN. While, the ultimate load increased by 7.69%, 10.76% and 4.61 % for CNKS, CNKS24 and CF, respectively. However for grade F the ultimate load decreased by 10.76%, 10.76%, 13.84 %, 15.38% and 1.5% for mixes F, FN,

IOP Conf. Series: Materials Science and Engineering **974** (2020) 012038 doi:10.1088/1757-899X/974/1/012038

FNKS24, FNKS and FF respectively. The pervious results is a good indication of the effectiveness of the used chemical treatment methods. Chemical modification methods modify the surface of rubber to introduce strong polarity groups such as polar carbonyl(C=O), hydroxyl (O–H) and sulfonate (SO_3) groups on the rubber surface. Bonded groups made a hydrogen bond and ionic bond with large numbers. The bond between the cement matrix and rubber greatly improved which leads to improving the mechanical properties of the rubberized concrete. The results of CF and FF slabs in ultimate loading is an indication for the improvement happened to the rubber by Subjecting the untreated rubber to 200°C temperature.

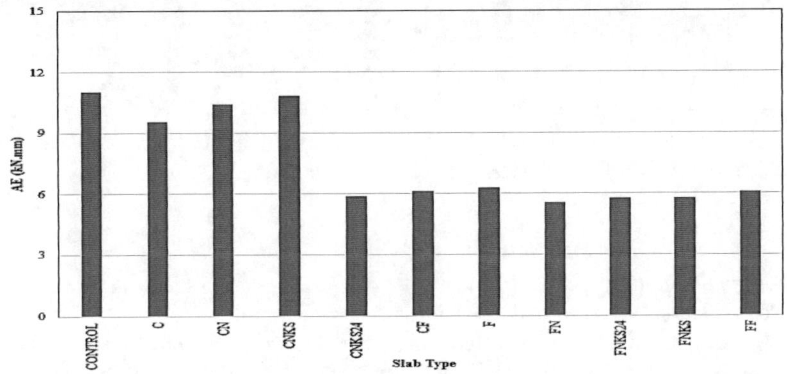

Figure 4. Absorbed energy for high strength rubberized concrete slabs.

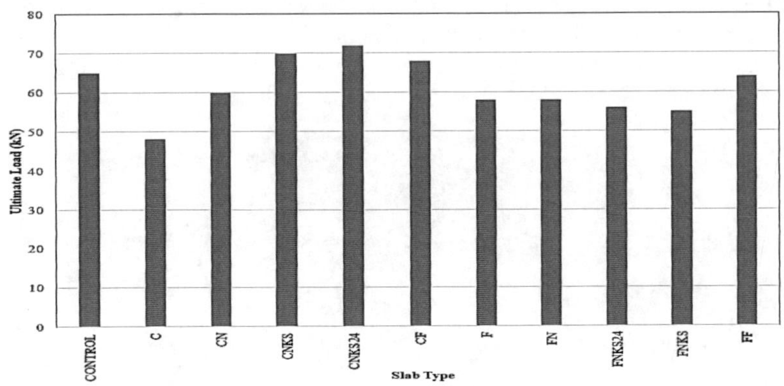

Figure 5. Ultimate load for high strength rubberized concrete slabs.

IOP Conf. Series: Materials Science and Engineering **974** (2020) 012038 doi:10.1088/1757-899X/974/1/012038

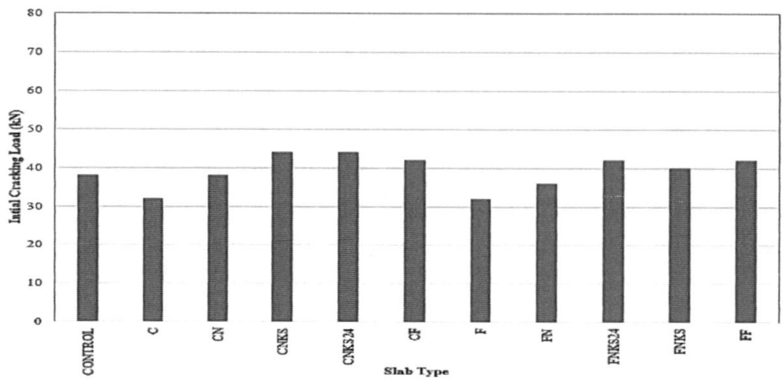

Figure 6. Initial cracking load for high strength rubberized concrete.

3.4. Cracks pattern and failure

The cracks' patterns were recorded at each load increment; all the tested slabs were failed shear. Figure 7 shows examples of 9 types of tested slabs after failure. For control slab, the flexural cracks were initiated at the bottom of the slab but stopped at a certain distance before penetrating compression zone and with load increasing; shear crack appeared at both sides of the slabs to form with the load increasing the same shape and direction of the inclined compression strut till sudden failure. All slabs were initiated by flexure cracks and propagated vertically upward, but did not meet the inclined compression strut. As the applied load increased, diagonal shear cracks appeared through the inclined compression strut. With increasing load, other shear cracks develop to the lower area near the support and the slab reached sudden failure through the inclined compression strut.

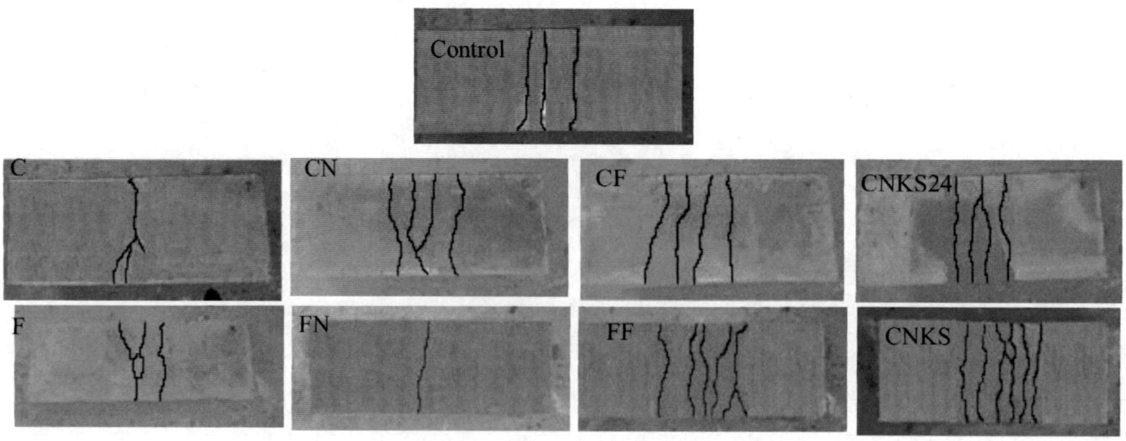

Figure 7. Crack pattern of tested slabs.

4. Conclusions

IOP Conf. Series: Materials Science and Engineering **974** (2020) 012038 doi:10.1088/1757-899X/974/1/012038

Based on the results obtained from the current research, the following conclusions could be obtained:

1- Partiallyreplacingof fine aggregate with untreated crumbed rubber has a negative effect on the behavior of high strength concrete slabs.
2- The coarserof the rubber particles, the more negative effect on the behavior of the high strength rubberized concrete slabs.
3- The behavior of the high strength rubberized concrete slabs can be improved by the used treatment methods.
4- For the same grade of crumbed rubber, the used treatment method has a slightly effect on the initial cracking load of the high strength concrete slabs and a significant effect on the ultimate load.
5- The ultimate load decreased by 26.15% and 1.5% when used grade C and grade F on the concrete mix respectively.
6- CNKS24 gave the best results on the ultimate and cracking loads than other treatment methods.

5. References

1. Selvakumar, S.; Venkatakrishnaiah, R. Strength properties of concrete using crumb rubber with partial replacement of fine aggregate. *International Journal of Innovative Research in science engineering and technology* **2015**, 1171-1175.
2. Kumar, G.N.; Sandeep, V.; Sudharani, C. Using tyres wastes as aggregates in concrete to form rubcrete-mix for engineering applications. *International Journal of Research in Engineering and technology* **2014**, *3*, 500-509.
3. Hassanli, R.; Youssf, O.; Mills, J.E. Experimental investigations of reinforced rubberized concrete structural members. *Journal of Building Engineering* **2017**, *10*, 149-165.
4. Siddique, R.; Naik, T.R. Properties of concrete containing scrap-tire rubber–an overview. *Waste management* **2004**, *24*, 563-569.
5. Boudaoud, Z.; Beddar, M. Effects of recycled tires rubber aggregates on the characteristics of cement concrete. **2012**.
6. El-Sherbini, Y.; Abdel-Gawad, A.; Shalaby, A.; El-Gammal, A. Compressive strength of concrete utilizing waste tire rubber. *Journal of Emerging Trends in Engineering and Applied Sciences* **2010**, *1*, 96-99.
7. Nor, N.S.M.; Ishak, D.M.Y.B. Effect of various sizes of waste tyre Rubber shred on concrete strength. *University Tekndogi Malaysia*. **2012.**
8. More, T.R.; Jadhao, P.D.; Dumne, S. STRENGTH APPRAISAL OF CONCRETE CONTAINING WASTE TYRE CRUMB RUBBER. **2015**.
9. Sofi, A. Sustainable Concrete with Waste Tyre Rubber—An Overview. *J. Chem. Pharm. Sci* **2016**, *9*, 848-853.
10. Pardeshi, N.R.; Singh, D.P.; Patil, S.R.; Gorde, P.J.; Janrao, P.P. Performance and evaluation of rubber as concrete material. *International Research Journal of Engineering and Technology (IRJET)* **2017**.
11. Al-Nasra, M.; Torbica, Z. Concrete Made for Energy Conservation Using Recycled Rubber Aggregates. *International Journal of Engineering Science Invention* **2013**, *2*, 10-16.
12. Gerges, N.N.; Issa, C.A.; Fawaz, S.A. Rubber concrete: Mechanical and dynamical properties. *Case Studies in Construction Materials* **2018**, *9*, e00184.
13. Al-Tayeb, M.; Bakar, B.A.; Akil, H.; Ismail, H. Performance of rubberized and hybrid rubberized concrete structures under static and impact load conditions. *Experimental Mechanics* **2013**, *53*, 377-384.
14. Sallam, H.; Sherbini, A.; Seleem, M.; Balaha, M. Impact resistance of rubberized concrete. *Engineering Research Journal* **2008**, *31*, 265-271.

15. Quality, E.O.f.S. E.S.S. 4756-1/2009: Egyptian Standard Specification for Ordinary Portland Cement. Cairo, Egypt, 2009.
16. (ASTM), A.S.f.T.a.M. ASTM C1240-05: Standard Specification for Silica Fume Used in Cementitious Mixtures. ASTM International, West Conshohocken, PA, USA, 2005.
17. ASTM, C. Standard specification for chemical admixtures for concrete. *Annual book of ASTM standards* **2013**.
18. Y.B.I. Shaheen, N.M. Soliman, D.E.M. Kandil, Influence of reinforced ferrocement concrete plates under impact load, Int. J. Curr. Eng. Technol. 3 (**2013**) 1528–1540.

AUTHOR INDEX

Institute of Physics
Dirac House, Temple Back
Bristol BS1 6BE UK

ISSN: 1757-8981
ISBN 978-1-7138-3296-6